Air Pollution Control and Design

Pen-Chi Chiang · Xiang Gao

Air Pollution Control and Design

Springer

Pen-Chi Chiang
Distinguished Visiting Professor at The Graduate Program of Post-Plastics
Department of Environmental Engineering
University of Seoul
Seoul, Korea

Xiang Gao
College of Energy Engineering
Zhejiang University
Hangzhou, Zhejiang, China

ISBN 978-981-13-7487-6 ISBN 978-981-13-7488-3 (eBook)
https://doi.org/10.1007/978-981-13-7488-3

This Springer imprint is published by the registered company Springer Nature Singapore Pte Ltd.
The registered company address is: 152 Beach Road, #21-01/04 Gateway East, Singapore 189721, Singapore

Contents

Chapter 1
Introduction

This book focused specifically on the environmental issues related to air pollution control and design. The book is intended to introduce the concepts of air pollution control and design to implement the main task-forces, namely (1) Fundamentals of Air Pollution Control, including SOx Control, NOx Control, VOCs Control, PM2.5 Control, and CO_2 Control; (2) Fundamentals of Energy Utilization, including Thermodynamics, Heat Transfer, Advanced Combustion, Renewable Energy, and Energy Materials; (3) Gaseous Control and Design, including Adsorption, Absorption, Combustion, and Stack; and (4) Particulate Control and Design, including Cyclone, Wet Scrubber, Baghouse Filter, and Electrostatic Precipitator.

In addition, the book features regulations and policy instruments pervasively adopted worldwide and several unanimous case studies. We believe this book will attract a wide range of readers such as researchers, students, engineers, policymakers as well as entrepreneurs since it provides readers with emerging challenges, new concepts, innovative methodologies, and resolving strategies. Under this vision, it is expected to allow engineers and policymakers to understand new techniques, thereby implementing sustainable energy policy, introducing green technology, and utilizing appropriate abatement techniques. The insights from the case studies are also considered illustrative and inspiring. Based on the contents in this book, the development of technology and infrastructure can reduce air pollution emissions, promote resource/energy utilization efficiency, and lessen environmental degradation, thereby resulting in "improving human well-being and social equity, while significantly reducing environmental risks and ecological scarcities" toward sustainable development goal (SDG). More detailed discussions on the book regarding (1) Challenges in Air Pollution Control, (2) Green Chemistry Principles: Roadmap to Air Pollution Control and Design, (3) Air and Energy Nexus, and (4) Structure and Contents of this book will be stated as follows.

P.-C. Chiang and X. Gao, *Air Pollution Control and Design*,
https://doi.org/10.1007/978-981-13-7488-3_1

1.1 Challenges in Air Pollution Control

With the rapid development of modern society, energy and environmental issues are two major global challenges which attract much attention. Air pollution, which is mainly from energy utilization processes, is a major public concern worldwide. It is noted that increased temperatures due to climate change have the potential to increase the formation of photochemical smog. Climate change threatens human health including mental health, and access to clean air, safe drinking water, nutritious food, and shelter. US EPA developed climate change adaptation strategies to reduce the environmental and public health impacts, especially for ambient air quality management by implementing energy efficiency pollution prevention and source reduction programs. Thus, Air Division will need to adapt to this reality and will focus on reducing air quality impacts of climate change through efforts to reduce NOx and other smog and PM2.5-forming pollutants, leveraging additional public and private resources for zero and low-emission technology deployment.

Newer scientific studies have shown that some pollutants would harm public health and welfare even at very low levels. In addition, the other big challenge of air pollution is attributed to global climate change and GHG emissions. As a result, to mitigate air pollution emissions as well as huge anthropogenic CO_2 emissions, different strategies and tools from various aspects have been proposed and implemented. Action plans and practical technologies have been executed to pursue scientific solutions for overcoming the challenges of air pollution control.

The extent to which the engineering profession responds to this challenge depends largely on the limits imposed by three principal considerations: (1) legal limitations imposed for of public health and welfare, (2) social limitations imposed by the pollution source, and (3) economic limitations imposed by marketplace constraints.

The control strategy for air pollution often focuses on five alternatives including the reduction of pollutant emissions, elimination of the operation, modification of the operation, relocation of the operation, and application of appropriate control technology.

There are 3E factors, i.e., Environmental, Engineering, and Economic to be considered prior to the selection of the best available control technology. For the environmental evaluation, the equipment location, available space, ambient conditions, availability of adequate utilities and ancillary system facilities, air pollution regulations, aesthetic considerations and secondary pollution, i.e., wastewater solid waste, and noise levels should be considered. For the engineering aspect, the contaminant characteristics including physical and chemical properties, concentration, particulate shape and size distribution, chemical reactivity, corrosivity, abrasiveness, and toxicity should be investigated. For the economic consideration, the capital cost (equipment, installation, engineering, etc.), operation cost (utilities, maintenance, etc.), and expected equipment lifetime and salvage value should be evaluated.

Great concerns about serious damages caused by air pollutant emissions and rapid GHG accumulation in the atmosphere are raised. Another challenge to the ambient air pollution control may be the issue of cost allocation in the case of GHG emission reductions. For instance, the management of a variety of CO_2 emission sources within a single industrial plant and the selection of appropriate process and technology for CO_2 sequestration should be the major concern to lower the industrial capture costs.

1.2 Green Chemistry Principles: Roadmap to Air Pollution Control and Design

The green chemistry principles (GCPs) are a set of twelve principles including (1) Waste Prevention, (2) Atom Economy, (3) Less Hazardous Chemical Synthesis, (4) Designing Safer chemicals, (5) Safer Solvent and Auxiliaries, (6) Design for Energy Efficiency, (7) Use of Renewable Feedstocks, (8) Reduce Derivatives, (9) Catalysis, (10) Design for Degradation, (11) Real-Time Pollution Prevention, and (12) Safer Chemistry for Accident prevention which aim to maximize the efficiency on chemical process and production and minimize the negative effect on the environment and human health which should be considered as a roadmap to air pollution control and design.

In order to be simple and easy to implement, the 12 principles can be classified into three categories as, i.e., (1) Pollution and Accident Prevention, (2) Energy and Resource Sustainability, and 3) Safety and Security Assurance. The pollution and accident prevention category consists of GCPs 1, 3, 4, and 5 because those principles are a focus on the elimination of potential pollution and accident in advance. While, for the energy and resource sustainability category, GCPs 6 and 7 can be fitted in this category as these principles specialize the importance of energy and raw material sustainability. Finally, for the safety and security assurance category, GCPs 8, 9, 10, and 11 are chosen in this category as they relate to the technology and method to solve the pollution which already has been produced.

For instance, by implementing those GCPs in the VOC control technology, the corresponding green VOC control technology can be classified as cleaner production (CP), recovery technology, and destruction technology. The CP technology represents source prevention, the innovative technology, or processes that are used to prevent the formation of VOC in advance rather than treat VOC after generation. For resources utilization, the recovery technology, especially the adsorption process, for VOC removal is focused. The emission reduction aspect will focus on the destruction of VOC treatment process including plasma and catalytic oxidation processes. After implementing these technologies, VOC can be maintained by good engineering practice.

1.3 Air and Energy Nexus

Under the influence of the concept of energy saving and environmental protection, it is necessary to optimize and reform the thermal system of the power plant, improve the energy saving and emission reduction capacity of the thermal system of the power plant, and adopt the strategy of energy saving and emission reduction, so as to effectively improve operational efficiency. Combustion science is now evolving from a traditional empirical science into a systematic one involving thermodynamics, hydrodynamics, chemical dynamics, heat and mass transfer, physics, and a comprehensive theoretical system based on mathematics. Practical systems and standards for heat exchanger design make great significance for enhancing heat transfer, and energy conservation resulted in reducing air emissions.

With the above scientific knowledge, reduction of air pollution (as well as GHG) emissions from both industries and coal-fired power plants should be a portfolio option. Therefore, the establishment of an integrated air pollution control should provide a method of simultaneously addressing the problems of energy efficiency, pollutant emissions, and abatement costs. As we discussed in this book, for instance, the emissions of SO_x and NO_x from energy industries will react in the atmosphere to form PM (including $PM_{2.5}$).

In other words, innovative technologies and tools to enhancing SOx, NOx, and PM emissions reduction for coal-fired power plants can be developed as follows: (1) anti-poison, wide-temperature window and low SO_2/SO_3 conversion SCR catalysts; (2) for NOx and SO_2 control, pH looping technology (pH value adjustment, dual-loop) and enhanced absorber for SOx control; (3) reducing particle resistivity, increasing migration velocity, avoiding gas sneakage/re-entrainment, and enlarging particle size demonstrated for PM2.5 control via wet high-temperature ESPs.

An integrated way to multiple air pollutants control as green solutions is an attractive approach to achieving direct and indirect reduction of air pollutants in industries. However, barriers from the regulatory, institutional, financial, and technological aspects can be encountered while implementing the multiple air pollutants control. These barriers are hard to be distinctly separated; e.g., green policies and regulations often act on more than one barrier simultaneously. The institutional and financial barriers also can habitually be closely related. Portfolio solutions can be provided along with the multiple air pollutants control.

1.4 Structure and Contents of This Book

This book is dedicated to a wide-angle review of air pollution control and design in 20 chapters. The book is intended to introduce the concepts of air pollution control and design to implement the main task-forces including (1) gaseous pollutant control such as VOC, SO_x, NO_x, and CO_2; (2) particulate matter control; (3) removal mechanism; and (4) equipment design criteria. Under this vision, it is expected to allow policymakers to understand new techniques, thereby implementing sustainable development policy, introducing green technology, attracting the interest of the public, and utilizing appropriate evaluation tools.

The topics covered in this book are meant not only to introduce the basic concepts of air pollution control and design, but also to address the fundamentals of energy utilization regarding good engineering practice and policy instruments. More importantly, we present several innovative, technologies and integrated methodologies concerning gaseous and particulate matter control and design. To facilitate technology integration and meet the needs for comprehensive education on sustainable development, the book covers a wide range of sectors of major concern in the principles, applications, and assessment for air pollution control and design and thermodynamics, heat transfer, advanced combustion, renewable energy, and energy materials for energy utilization.

In part I, a broad review on air pollution control is provided, including SO_x Control (Chap. 2), NO_x Control (Chap. 3), VOCs Control (Chap. 4), $PM_{2.5}$ Control (Chap. 5), and CO_2 Control (Chap. 6). In part II, the fundamentals of energy utilization are systematically presented, in terms of Sustainable Energy (Chap. 7), Thermal Dynamics and Heat Transfer (Chap. 8), Advanced Combustion (Chap. 9), Energy Materials (Chap. 10), and Integrated Energy Systems (Chap. 11). In part III, the gaseous control and design are illustrated, including Adsorption (Chap. 12), Absorption (Chap. 13), Combustion (Chap. 14), and Stack (Chap. 15). In part IV, the particulate control and design are discussed, including Cyclone (Chap. 16), Wet Scrubber (Chap. 17), Baghouse Filter (Chap. 18), and Electrostatic Precipitator (Chap. 19). Finally, the prospective and perspective on the strategies toward zero waste for sustainability are provided in Chap. 20.

Based on the principles of air pollution control and design, the development of technology and infrastructure can reduce air pollution emissions, promote resource/energy utilization efficiency, and lessen environmental degradation, thereby resulting in "improving human well-being and social equity, while significantly reducing environmental risks and ecological scarcities" toward sustainable development goal (SDG). We believe this book will attract a wide range of readers such as researchers, students, policymakers as well as entrepreneurs since it provides readers with emerging challenges, new concepts, innovative methodologies, and resolving strategies. The insights from the case studies are also considered illustrative and inspiring.

Chapter 2
SOx Control

Nomenclature

A	Outer surface area of absorbent
B	Adsorption equilibrium constant
c_A	Concentration of component A
c_i	Concentration at the interface
c^*	Liquid concentration in equilibrium with gas pressure
D	Diffusion coefficient
H	Henry's constant
K	Overall coefficient
k_g	Gas film absorption coefficient
k_L	Liquid film absorption coefficient
K_g	Overall coefficient of gas-phase absorption
K_L	Overall coefficient of gas–liquid absorption
N_A	Flux of component A
p	Gas partial pressure
p_i	Partial pressure at the interface
p^*	Partial gas pressure in equilibrium with liquid concentration
p_0	Saturated vapor pressure
r_A	Adsorption rate
V	Adsorption volume
V_m	Volume of adsorbate at full monolayer coverage
x	Molar ratio of the gas in the liquid phase
Y	Concentration of adsorbate
Y^*	Equilibrium concentration

P.-C. Chiang and X. Gao, *Air Pollution Control and Design*,
https://doi.org/10.1007/978-981-13-7488-3_2

2.1 Introduction

2.1.1 Importance and Significance

Sulfur oxides (SOx) contains two kinds of highly reactive gases, SO_2 and SO_3, which have a lot of impact on human and environment. Coal combustion is a major source of SOx emission. Other sources of SOx emissions include petroleum refining, oil and gas production, sulfur and sulfuric acid manufacturing, ore smelting, waste incineration, and petroleum coke calcining.

Gaseous SOx generated from coal combustion are primarily sulfur dioxide (SO_2), with a much lower quantity of sulfur trioxide (SO_3). SO_2 is a colorless but toxic gas. It can easily react with other substances to form harmful compounds, like sulfuric acid, sulfurous acid, and sulfate particles. When SO_2 is breathed in, it will damage human health; irritate the nose, throat, and airways, and cause coughing, wheezing, shortness of breath, or a tight feeling around the chest. Controlled studies suggest that exposure to SO_2 within as short as 10 min could lead to a proportion experience changes in pulmonary function and respiratory symptoms. WHO recommended that with a SO_2 concentration of 500 $\mu g/m^2$, it should not be exceeded over averaging periods of 10 min duration [1]. For long-term exposures, there is a basis for the new 24-h guideline for SO_2 downwards using a prudent precautionary approach to 20 $\mu g/m^2$. In addition, the emission of SO_2 will also result in acid rain as shown in Fig. 2.1.

Sulfur trioxide (SO_3), also known as H_2SO_4 or sulfuric acid mist, is one of the least discussed air pollutants in coal-fired power plant emissions [2–5]. The public attaches great importance to atmospheric pollutants such as nitrogen oxides (NOx), sulfur dioxide (SO_2), and particulate matter (PM) [6], but often ignores SO_3. Although the

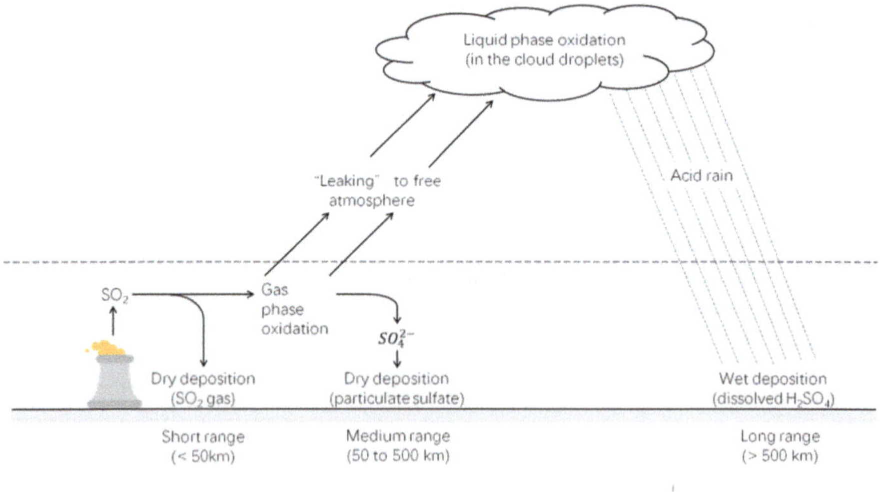

Fig. 2.1 The impact of SO_2 on the environment

public has discovered that SO_3 is emitted during the production of sulfuric acid at an early age, the sulfur mist is rarely emitted in other fields, and it is not emitted on a large scale globally compared to the air pollutions discussed frequently. In recent years, the selective catalytic reduction (SCR) technology [7] has been widely used and has also exacerbated SO_3 emissions, which makes the public pay more attention to SO_3 emissions. The adverse effects of SO_3 emissions on the performance of coal-fired power plants will become more pronounced with the increasing SO_3 concentration.

A small amount of SO_3 is emitted into the atmosphere, possibly causing serious environmental pollution. The SO_3 in the flue gas reacts with water to form sulfuric acid. When the temperature of the flue gas is reduced to a certain extent, the H_2SO_4 mist in the flue gas will condense to form a submicron aerosol (acid mist). Once the sulfuric acid mist is formed, it will react with some other materials to form condensable substances such as ammonia and sodium salts, which will increase the total particulate matter emissions [8, 9]. Under certain atmospheric conditions, when the SO_3 concentration is as low as 6–10 ppm (dry volume), visible plumes will form. There is also evidence that SO_3 emissions may cause negative effects such as eye burning, sore throat, and headache [5, 10, 11]. Some evidence of impaired mucociliary clearance and modest changes in lung function has been identified [12]. In sensitive populations [13], exposure to H_2SO_4 may lead to more serious problems in combination with other pollutants [14, 15] compared with exposure to H_2SO_4 alone. Therefore, it is of great urgency and importance to effectively control SO_3 emission, although no consensus on the required emission standards for SO_3 exists at present.

Since the 1990s, there have been many serious air pollution disasters around the world, such as the 1948 Donora smog and the Great Smog of London. There are many reasons for these disasters, and the excessive emission of sulfur oxides is one of the most important causes.

Fig. 2.2 The Great Smog of 1952 in London

Known as the Great Smog in 1952, the London Great Smog, as shown in Fig. 2.2, was another serious disaster of air pollution that came to London in early December 1952. Due to the cold weather, Londoners used a lot of coal in order to keep themselves warm. However, after World War II, domestic coal tended to be low-grade coal with high sulfur contents, such as lignite coal, while high-grade coal, such as anthracite coal, was often used for export. Such a situation increased SO_2 emissions. What's more, there were also several coal-fired power plants next to London, such as Bankside, Fulham, Greenwich, Battersea, and so on. These all increase the emissions of pollutants. On the basis of the UK Meteorological Administration, the pollutants emitted during the Smog were as follows: 1,000 tons of particle matters, 140 tons of hydrogen chloride, 14 tons of fluorides, and 370 tons of SO_2. These SO_2 may be eventually converted to 800 biennium of sulfuric acid. Due to the large amount of pollutants emissions, combined with the effects of the anticyclone and windless conditions, the pollutants produced by coal combustion could not be spread in time. This led to the occurrence of the Great Smog of London. In four days, the incident caused about 4,000 deaths. Thousands of people died in the next few months. It is the most serious air pollution incident in British history, which gave rise to the amendment of a number of regulations, such as the 1956 Clean Air Act.

2.1.2 Regulations and Standards

The first federal legislation on air pollution in the United States was the US Air Pollution Control Act in 1955. In 1963, the Clean Air Act was issued, which was the first federal legislation in the United States to control air pollution. This act achieved its goal by establishing a federal program within the united states public health services and committed to researching technologies that can monitor and reduce air pollution. As one of the earliest and most influential modern environmental laws in the United States, it is also one of the most comprehensive air quality laws in the world. The U.S. Congress established the basic framework of the Clean Air Act in 1970, and major revisions were made in 1977 and 13 years later. The 1977 amendment to the Clean Air Act required compliance with national ambient air quality standards (NAAQS) to prevent significant deterioration of air quality (PSD) and required additional conditions in areas that did not meet these standards. Procedures for controlling acid deposition (acid rain) and stationary source operating permits were added to the Clean Air Act 1990. The goal was to reduce sulfur dioxide emissions, a provision that proposed a cap-and-trade program compared with earlier versions of the Clean Air Act. The plan gave power companies more flexibility and feasibility.

Amendments added Title IV-A of the Clean Air Act in 1990. The problem of acid rain was discussed in Title IV-A. Nitrogen oxides (NO_x) and sulfur dioxide (SO_2) emitted from fossil fuel power plants and other industrial sources are the main causes of acid rain. The 1990 amendment proposed more other technologies for industrial pollution control, including using low-sulfur coal and adding equipment to reduce harmful pollutants. In some extreme cases, the plant must be shut down in order to prevent the atmosphere from receiving hazardous chemical contamination.

Title IV-A consisted primarily of two steps to reduce SO_2 emissions. More than 100 power generation facilities were set up in the first phase with a total power of more than 100 megawatts, with the aim of reducing emissions of 3.5 million tons of sulfur dioxide before the first half of 1995. More than 75 MW of facilities were set in the second phase, with the goal of meeting the sulfur dioxide emissions target by January 2000.

The Clean Air Act requires US EPA to establish national ambient air quality standards (NAAQS). Six common air pollutants were within the national air quality standards, including particulate matter, carbon monoxide, sulfur oxides, photochemical oxidants (ozone), nitrogen oxides, and lead. Protecting human health is the primary standard and requires adequate safety boundaries, especially for sensitive populations such as children, the elderly, and respiratory patients. Secondary standards cover a wide range of issues, including traffic safety, public welfare, property safety, economic value, personal comfort, and well-being which are not adversely affected by any known or expected contaminants.

Stringent standards have been established in order to reduce SO_2 emissions in the world. The present SO_2 emission limits for coal-fired power plants of some countries are shown in Fig. 2.3. It is observed that the SO_2 emission limits in coal-fired power plants in China, including both the special limit and general limit, are the most stringent, followed by that in the US, Australia, the EU, Japan, and other countries. In order to control the emission of SO_2 effectively, advanced technologies are required to be developed and installed for combustion processes to meet the increasingly stringent emission standards.

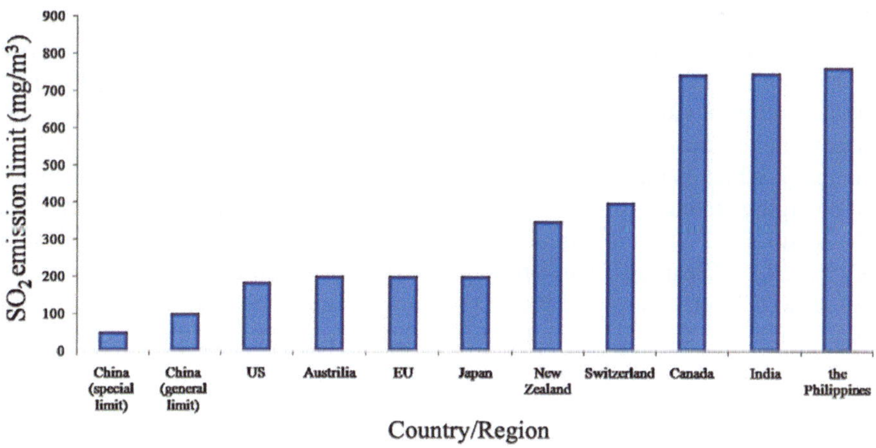

Fig. 2.3 SO_2 emission limits for coal-fired power plants in different countries

2.1.3 Technology Development

Stringent standards have been established in order to reduce SO_2 emissions. In order to control the emission of SO_2 effectively, advanced technologies are required to develop for combustion processes to meet the increasingly stringent emission standards. In the 1930s, wet limestone–gypsum FGD technology, which is widely used nowadays, was first industrialized in Battersea, Bankside, and Fulham power plants in the UK. In the 1970s, seawater FGD technology was first industrialized in oil-fired boilers in Porsgrunn, Norway, and wet ammonia FGD technology was first developed by Krupp Koppers in Germany. In the 1980s, circulating fluidized-bed FGD technology was first developed by Lurgi in Germany, and limestone injection into the furnace and activation of calcium oxide FGD technology were first developed by Tempella and IVO in Finland. In the 1990s, biological FGD technology was first developed by HTS E&E and Paques in the Netherlands, and yellow phosphorus desulfurization and denitration technology were first developed by Lawrence Berkeley Laboratory in the USA. In the 2000s, low-temperature oxidation system was industrialized by BOC Gases and Cannon Technology Inc. Synergistic removal of multiple pollutants had become a research hotspot in the 2010s.

2.1.4 Contents of Chapters

In this chapter, we mainly focus on the SO_x formation mechanisms, control principles, and control technologies. The formation mechanisms including SO_2 formation and transformation of SO_2 to SO_3 will be discussed. In consideration of the SO_x control principles, the fundamentals of gas diffusion, absorption, and adsorption will be introduced. For SO_2 control technologies, we will focus on three types of flue gas desulfurization (FGD) technologies. Limestone injection into furnace and activation of calcium oxide (LIFAC)is a typical dry FGD process suitable for water-deficient areas. Wet limestone–gypsum FGD has the highest SO_2 removal efficiency, which is the most widely applied technology in coal-fired power plants. An emerging technology, circulating fluidized-bed FGD (CFB-FGD) as well as an advanced CFB-FGD, and Circulating Suspension and Multi-stage Humidification Dry Flue Gas Cleaning Technology (CSMHD-FGC) will also be introduced. We will as well briefly introduce the SO_3 control technologies. At last, comparisons between the SO_2 and SO_3 control technologies in consideration of the technical and economic aspects are provided.

2.2 SOx Formation

2.2.1 *Oxidation of Sulfur*

The sulfur in the fuel is generally classified into organic sulfur and inorganic sulfur, depending on its form of existence. Organic sulfur comes from the combination of sulfur and the organic structure of the fuel, and inorganic sulfur refers to sulfur in the form of inorganic substances. In addition, there is a small amount of elemental sulfur present in the elemental state in some coals and oils. The presence of sulfur in three different types of fuels, gases, and solids has different characteristics. In gaseous fuels, most of the sulfur species are inorganic sulfur species where hydrogen sulfide (H_2S) dominates. A small fraction of organic sulfur also exists, including carbon disulfide (CS_2), thioalcohol (CH_3SH), thioether (CH_2SCH_3), thiophene (C_4H_4S), carbonyl sulfide (COS), etc. In liquid fuels, especially heavy oils, the sulfur mainly includes hydrogen sulfide, elemental sulfur, and various organic sulfur. The organic sulfur mainly exists in the functional groups, including thiophene, thioalcohol, and thioether. In solid fuels such as coal, the sulfur exists as pyrite (FeS_2), mineral sulfates, and elemental sulfur, and in organic compounds and mercaptans, pyrite is the dominant inorganic sulfur species. Organic sulfur forms are mainly sulfides, thiophene, and thiols. The composition and distribution of sulfur in coal are shown in Fig. 2.4.

The conversion of sulfur into SO_2 is phased in the combustion process. In the former stage, volatilization and ignition lead to the decomposition of some unstable organic sulfur, and the decomposition process will move forward due to the increase in temperature. In the latter stage, the organic sulfur and inorganic sulfur decompose into SO_2.

1. Conversion of organic sulfur

The composition of organic sulfur in the fuel is complicated, and the bond energy of each component is different. Therefore, the pyrolysis temperatures of various organic sulfur are also different. It is generally believed that when the coal is heated to 400 °C, the organic sulfur begins to decompose, which is slightly different for various coals.

The main intermediate for the thermal cracking of organic sulfur under a reducing atmosphere is H_2S. Studies show that 90% (volume fraction) of sulfur-containing components in coal pyrolysis gas at 1000 °C is H_2S, and the rest is CS_2 and a small amount of COS, mercaptans, and thiophene compounds. Moreover, the H_2S content in coal volatiles is basically proportional to the sulfur content in coal. Most of the H_2S is further burned during the actual combustion process to produce SO_2 and water.

2. Conversion of inorganic sulfur

Inorganic sulfur in the fuel mainly includes sulfate sulfur, pyrite, and elemental sulfur. The main component of inorganic sulfur which thermally decomposed at a high temperature is pyrite (FeS_2). The pyrite in the coal, in an oxidizing atmosphere, will react with O_2 to form SO_2, as shown in Eq. (2.1):

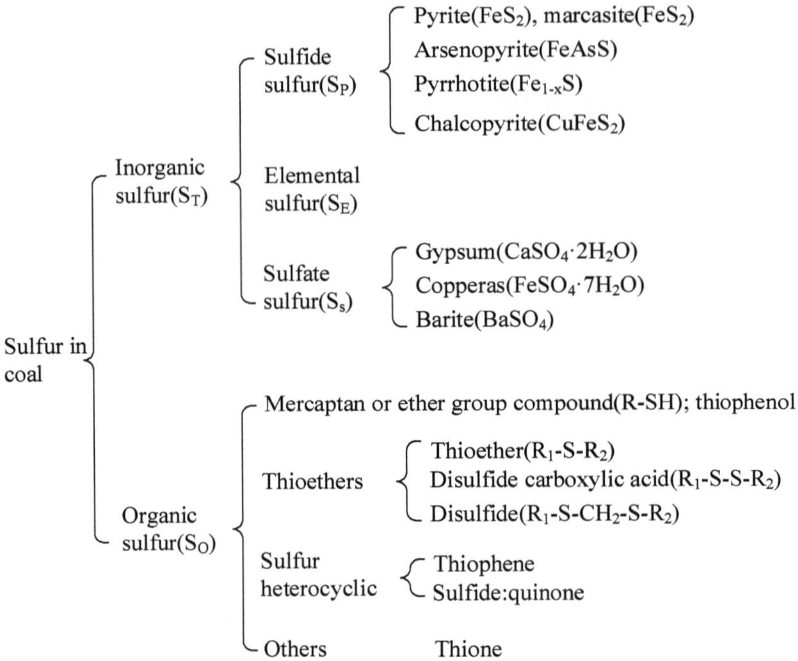

Fig. 2.4 S species existing in the coal

$$4FeS_2 + 11O_2 \rightarrow 2Fe_2O_3 + 8SO_2 \tag{2.1}$$

In reducing atmosphere, the transformation process of FeS_2 follows Eqs. (2.2)–(2.7), as shown in Fig. 2.5:

$$FeS_2 \rightarrow FeS + \frac{1}{2}S_2 \tag{2.2}$$

$$FeS_2 + H_2 \rightarrow FeS + H_2S \tag{2.3}$$

Fig. 2.5 Transformation of FeS_2 species in reducing atmosphere

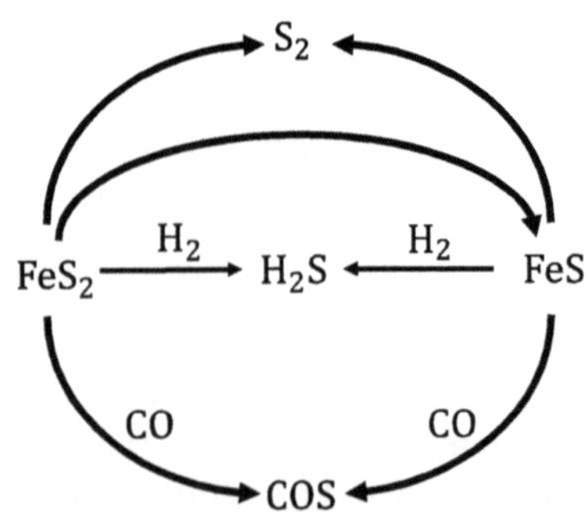

$$FeS_2 + CO \rightarrow FeS + COS \tag{2.4}$$

$$FeS \rightarrow Fe + \frac{1}{2}S_2 \tag{2.5}$$

$$FeS + H_2 \rightarrow Fe + H_2S \tag{2.6}$$

$$FeS + CO \rightarrow Fe + COS \tag{2.7}$$

For the transformation of organic sulfur, SO_2 is formed in the oxidizing atmosphere while in the reducing atmosphere, the product is H_2S and/or COS. It is generally believed that the sulfur in the coal undergoes the following process to gradually form SO_2, as shown in Eq. (2.8):

$$S \rightarrow H_2S \rightarrow HS \rightarrow SO \rightarrow SO_2 \tag{2.8}$$

In addition, another reaction pathway is the oxidation of COS, the intermediate product, to form SO_2, as shown in Eqs. (2.9)–(2.11):

$$H_2S + CO \rightarrow COS + H_2 \tag{2.9}$$

$$H_2S + CO_2 \rightarrow COS + H_2O \tag{2.10}$$

$$COS + \frac{3}{2}O_2 \rightarrow CO_2 + SO_2 \tag{2.11}$$

The reactions mentioned above are the overall reaction pathways. In fact, there are many intermediate reaction pathways.

2.2.2 *SO_2 Conversion to SO_3*

The sulfur in the coal is mostly oxidized to SO_2 during combustion in the furnace, and only a small part of sulfur is oxidized to SO_3. If the SCR system is used to control NOx emissions, a small portion of SO_2 will be further oxidized by the SCR catalyst to form SO_3 as it enters the SCR system. The formation mechanism of SO_3 is elaborated in this section. Based on different working conditions and operating parameters, the SO_3 concentration can be estimated. The control of SO_3 is also based on the SO_3 formation mechanism, and the optimization of operating parameters can effectively reduce the SO_3 formation in the boiler and SCR system.

1. SO_3 formation in the boiler

In an oxygen-enriched environment, a small fraction (ca. 0.5%–2.0%) of SO_2 will be oxidized into SO_3, as shown in Eq. (2.12):

$$SO_2 + \frac{1}{2}O_2 \rightleftharpoons SO_3 \tag{2.12}$$

However, it should be noted that in actual combustion conditions, SO_2 does not directly react with molecule oxygen. SO_3 is formed primarily by Eq. (2.13):

$$SO_2 + O + M \rightarrow SO_3 + M \tag{2.13}$$

where M is the intermediate that can absorb energy.

At high temperatures, SO_3 can also be consumed, as shown in Eqs. (2.14)–(2.16):

$$SO_3 + O \rightarrow SO_2 + O_2 \tag{2.14}$$

$$SO_3 + H \rightarrow SO_2 + OH \tag{2.15}$$

$$SO_3 + M \rightarrow SO_2 + O + M \tag{2.16}$$

2. SO_3 formation in the SCR system

Besides the combustion process, SO_3 can also be formed via the catalytic effect by the catalyst for NO_x removal, which has the maximum contribution to the plant's SO_3 generation. The values for the conversion rate in an SCR unit range from 0.25% to 1.25%. The reaction rate depends on the catalyst compositions and the operating conditions, which can be expressed as follows:

$$\alpha = \frac{K}{AV} \tag{2.17}$$

where K is the constant of proportionality, a function of catalyst material, design, and flue gas properties,

$$K = f\,(\text{catalyst material, catalyst design, flue gas properties}). \tag{2.18}$$

AV is the ratio of the flue gas volumetric flow rate to the geometric catalyst surface area. This variable implies that the conversion rate is proportional to catalyst volume (and hence, geometric surface area) and gas residence time in the catalyst.

In commercial SCR systems for stationary source, the active component of the catalyst is mainly V_2O_5, which can catalyze SO_2 to form SO_3 in the presence of O_2, as shown in Eqs. (2.19)–(2.21):

$$V_2O_5 + SO_2 \rightarrow V_2O_4 + SO_3 \tag{2.19}$$

$$V_2O_4 + 2SO_2 + O_2 \rightarrow 2VOSO_4 \tag{2.20}$$

$$2VOSO_4 \rightarrow V_2O_5 + SO_3 + SO_2 \tag{2.21}$$

In addition, other metal oxides, like iron oxide, can also play a role as the catalyst to convert SO_2 to SO_3.

Since a fraction of water vapor is contained in the flue gas, it will react with SO_3 to form a sulfuric acid vapor, as shown in Eq. (2.22):

$$SO_3 + H_2O \rightarrow H_2SO_4 \tag{2.22}$$

The formation of H_2SO_4 is highly dependent on the temperature. When the temperature decreases, the transform of SO_3 to H_2SO_4 increases. H_2SO_4 will cause corrosion to the downstream facilities. At ambient temperatures, H_2SO_4 vapor condenses rapidly to form a fine sulfuric acid aerosol. It always appears near the stack as a blue fume.

2.3 Control Principles

There are plenty of flue gas desulfurization techniques at present. Based on the amount of water consumption, they can be classified as wet flue gas desulfurization and dry flue gas desulfurization techniques.

For wet FGD techniques, the SO_2 is absorbed in a wet state by solution or slurry containing absorbent. There are lots of choices for absorbent, for example, limestone, seawater, ammonia, sodium hydroxide, etc. Besides diffusion of SO_2, the absorption of SO_2 is one of the key processes of wet desulfurization, which largely determines the desulfurization performance of the system.

For dry FGD techniques, flue gases containing sulfur dioxide contact the alkaline sorbent, producing a dry waste, which is easier to dispose than the waste of wet FGD processes. The sorbent, which is either hydrated lime or limestone, can be used in the form of aqueous slurry or as dry powder. The key factors of these technologies are the diffusion and adsorption of SOx. And both physical and chemical reactions exist in the adsorption of SOx in flue gas by the sorbent.

2.3.1 Process Chemistry

2.3.1.1 Calcium-Based Sorbents

Calcium-based sorbent mainly refers to three kinds of sorbents, limestone ($CaCO_3$) and hydrated lime ($Ca(OH)_2$).

In the wet FGD process, SOx will react with sorbent in the liquid phase. As a result, the reaction can be described as follows:

$$CaCO_3 + 1/2\,H_2O + SO_2 \rightarrow CaSO_3 \cdot 1/2H_2O + CO_2 \quad (2.23)$$

$$Ca(OH)_2 + H_2O + SO_2 \rightarrow CaSO_3 + 2H_2O \quad (2.24)$$

$$CaSO_3 + 1/2O_2 + 2H_2O \rightarrow CaSO_4 \cdot 2H_2O \quad (2.25)$$

In the dry/semi-dry FGD process, the reactions are influenced by reaction conditions. At optimum temperatures between 980 and 1230 °C in the boiler, calcination of the limestone and dehydration of the hydrated lime occur to form quicklime (CaO), and then reacts with SO_2/SO_3 and oxygen in the flue gas, which can be described as follows:

$$CaCO_3 \rightarrow CaO + CO_2 \quad (2.26)$$

$$Ca(OH)_2 \rightarrow CaO + H_2O \quad (2.27)$$

$$CaO + SO_2 + 1/2O_2 \rightarrow CaSO_4 \quad (2.28)$$

$$CaO + SO_3 \rightarrow CaSO_4 \quad (2.29)$$

However, in the spray dry absorption process, the calcium hydroxide is injected as slurry. The heat in the flue gas evaporates the water from the slurry, by which the flue gas is cooled meanwhile. The calcium hydroxide then reacts directly with SO_2 and SO_3 as the temperature is too low for its dehydration:

$$Ca(OH)_2 + SO_2 \rightarrow CaSO_3 \cdot 1/2H_2O + 1/2H_2O \quad (2.30)$$

$$Ca(OH)_2 + SO_2 + H_2O + 1/2O_2 \rightarrow CaSO_4 \cdot 2H_2O \quad (2.31)$$

$$Ca(OH)_2 + SO_3 + H_2O \rightarrow CaSO_4 \cdot 2H_2O \quad (2.32)$$

2.3.1.2 Sodium-Based Sorbents

In the wet FGD process, solid sodium carbonate (Na_2CO_3), critic sodium (NaOH), and sodium sulfite ($NaSO_3$) are the dominant sodium-based sorbents. SOx will react with these sorbents in the liquid phase. As a result, the reaction can be described as follows:

$$2NaOH + SO_2 \rightarrow Na_2SO_3 + H_2O \quad (2.33)$$

$$2Na_2CO_3 + SO_2 + H_2O \rightarrow 2NaHCO_3 + Na_2SO_3 \quad (2.34)$$

$$2NaHCO_3 + SO_2 \rightarrow Na_2SO_3 + H_2O + 2CO_2 \quad (2.35)$$

$$Na_2SO_3 + SO_2 + H_2O \rightarrow 2NaHSO_3 \quad (2.36)$$

The produced sodium hydrogen sulfite can react with alkali to regenerate the sorbent, which can be expressed as follows:

$$Ca(OH)_2 + 2NaHSO_3 \rightarrow CaSO_3 + Na_2SO_3 + 2H_2O \quad (2.37)$$

$$Ca(OH)_2 + Na_2SO_3 \rightarrow CaSO_3 + 2NaOH \quad (2.38)$$

In the dry/semi-dry process, solid sodium carbonate (Na_2CO_3), sodium bicarbonate ($NaHCO_3$), and trona are the dominant sodium-based sorbents used. Sodium bicarbonate decomposed to sodium carbonate when injected into hot flue gas (above ~125 °C), and then fast reactions between sodium carbonate and Sox occur, which can be described as follows:

$$2NaHCO_3 \rightarrow Na_2CO_3 + CO_2 + H_2O \quad (2.39)$$

$$2(Na_2CO_3NaHCO_32H_2O) \rightarrow 3Na_2CO_3 + CO_2 + 5H_2O \quad (2.40)$$

$$Na_2CO_3 + SO_2 \rightarrow Na_2SO_3 + CO_2 \quad (2.41)$$

$$Na_2CO_3 + SO_2 + 1/2O_2 \rightarrow Na_2SO_4 + CO_2 \quad (2.42)$$

$$Na_2CO_3 + SO_3 \rightarrow Na_2SO_4 + CO_2 \quad (2.43)$$

Sodium bisulfite ($NaHSO_3$) and sodium sulfite (Na_2SO_3) are injected as solutions to remove SO_3 according to the following chemical reactions:

$$NaHSO_3 + SO_3 \rightarrow NaHSO_4 + SO_2 \tag{2.44}$$

$$Na_2SO_3 + 2SO_3 + H_2O \rightarrow 2NaHSO_4 + SO_2 \tag{2.45}$$

$$2NaHSO_3 + SO_3 \rightarrow Na_2SO_4 + 2SO_2 + H_2O \tag{2.46}$$

$$Na_2SO_3 + SO_3 \rightarrow Na_2SO_4 + SO_2 \tag{2.47}$$

2.3.1.3 Ammonia-Based Sorbents

In the wet FGD process, the ammonia is introduced into the absorption tower and brought into contact with the flue gas containing sulfur dioxide for absorption. The chemical reaction is as follows:

$$2NH_3 + SO_2 + H_2O \rightarrow (NH_4)_2SO_3 \tag{2.48}$$

$$NH_3 + SO_2 + H_2O \rightarrow NH_4HSO_3 \tag{2.49}$$

As the absorption proceeds, the NH_4HSO_3 in the circulating absorbent liquid gradually increases, the absorption capacity decreases, and ammonia needs to be added to convert some NH_4HSO_3 into $(NH_4)_2SO_3$. The reaction is as follows:

$$(NH_4)_2SO_3 + SO_2 + H_2O \rightarrow 2NH_4HSO_3 \tag{2.50}$$

$$(NH_4)_2SO_3 + SO_2 + H_2O \rightarrow 2NH_4HSO_3 \tag{2.51}$$

$$NH_3 + NH_4HSO_3 \rightarrow (NH_4)_2SO_3 \tag{2.52}$$

2.3.1.4 Magnesium-Based Sorbents

Currently, the most commonly used magnesium-based sorbent is magnesium oxide. Magnesium oxide is mainly used as a sorbent for the wet FGD process, which can be described as follows:

$$MgO + H_2O \rightarrow Mg(OH)_2 \tag{2.53}$$

$$Mg(OH)_2 + SO_2 \rightarrow MgSO_3 + H_2O \tag{2.54}$$

$$MgSO_3 + SO_2 + H_2O \rightarrow Mg(HSO_3)_2 \quad (2.55)$$

$$Mg(HSO_3)_2 + Mg(OH)_2 \rightarrow 2MgSO_3 + 2H_2O \quad (2.56)$$

In addition, the magnesium dihydroxide slurry or magnesium oxide particle is injected as solutions to remove SO_3 according to the following chemical reactions:

$$Mg(OH)_2 + SO_3 \rightarrow MgSO_4 + H_2O \quad (2.57)$$

$$MgO + SO_3 \rightarrow MgSO_4 \quad (2.58)$$

2.3.2 *Mass Transfer*

In the FGD process, complex reactions are involved. During these reactions, the reactants and products are transferred among the gas–liquid–solid three phases. The gas–liquid mass transfer process is also called the absorption process, and the gas–solid mass transfer process is also called the adsorption process.

2.3.2.1 Absorption

In the wet desulfurization process, acid gas such as SO_2 in the flue gas enters the liquid phase through the gas–liquid interface and reacts with the alkaline ions in the liquid phase to achieve the purpose of desulfurization. Figure 2.6 shows the mass transfer process in a typical limestone–gypsum wet desulfurization process.

The gas absorption process is a complicated process. A variety of absorption models have been proposed. Herein, we are introducing the two-film theory which is simple and easy to understand. As shown in Fig. 2.6, there is a stable phase interface between the gas–liquid two-phase fluids, and a very thin stagnant film layer exists on both sides of the interface. The acid gas such as SO_2 in the flue gas enters the liquid phase through these two films and then is removed.

The basic views of the two-film theory are as follows:

(1) There is a stable interface between the gas–liquid two-phase fluids, and a very thin stagnant film layer exists on both sides of the interface. The solute enters the liquid phase from the gas phase through the two membrane layers via molecular diffusion.
(2) At the interface, the gas phase and the liquid phase reach the phase equilibrium, and the mass transfer resistance at the interface is 0.

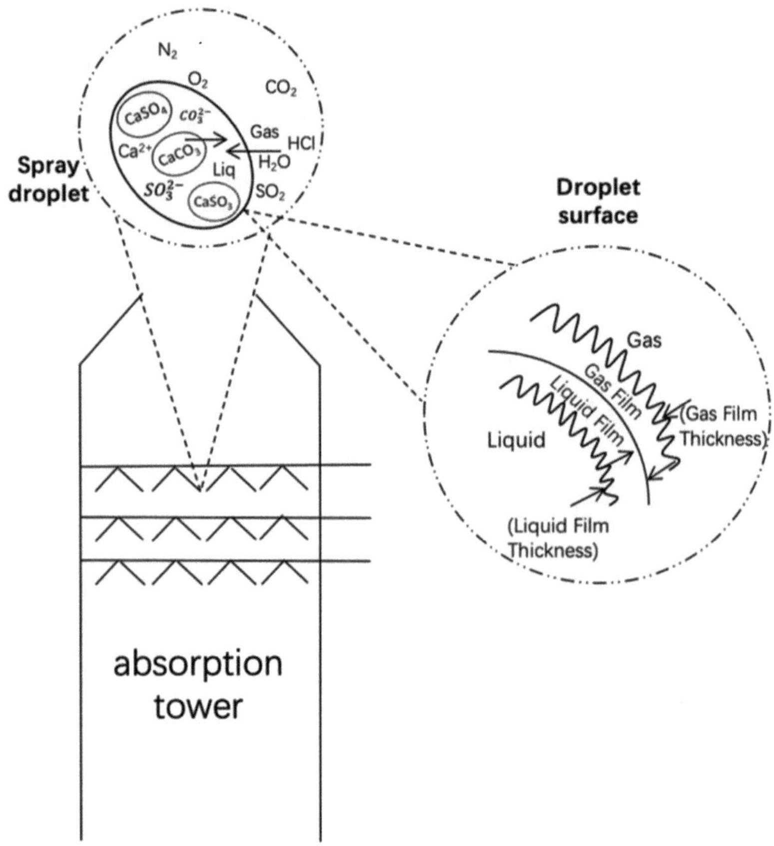

Fig. 2.6 The mass transfer process in a typical limestone–gypsum wet desulfurization process

(3) Other than the gas film and the liquid film layers, since the turbulent flow is sufficient, the concentration of the solute is considered to be uniformly equal, that is, all the concentration gradients are concentrated in the two stagnant film layers.

The theoretical model is shown in Fig. 2.7.

As we can see from Fig. 2.7, there is no concentration gradient in gas or liquid bulk. The concentration gradient exists only in the two films. The pressure drops from p to p_i when the absorbate goes through the gas film. The concentration decreases from c_i to c when the absorbate passes through the liquid film. Therefore, the resistance during the whole absorption process is only present in two laminar films.

It can be seen from the two-film theory that the absorption process can be regarded as a molecular diffusion process of the absorbate passing through the gas film and liquid film. Therefore, the absorption rate equals the molecular diffusion rate of the substance A passing through the two films. The absorption rate of the substance A through the gas film can be expressed as in Eq. (2.59):

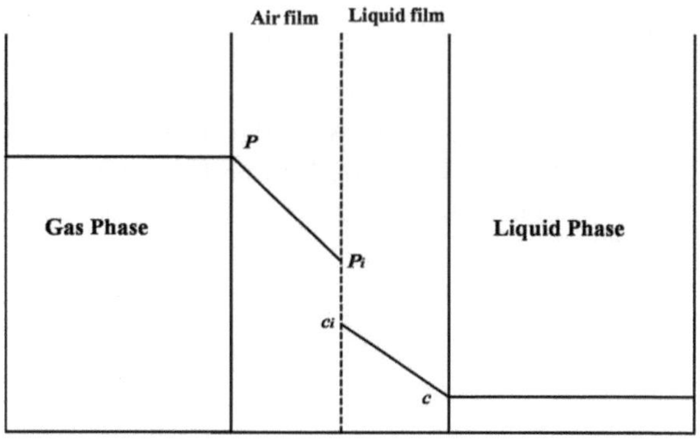

Fig. 2.7 Reaction and mass transfer model between gas–liquid–solid phases

$$N_A = k_g(p - p_i) \tag{2.59}$$

where

N_A = the absorption rate, kmol/(m^2·s);
p, p_i = the partial pressure of A at the gas phase and the interface, Pa;
k_g = the gas film absorption coefficient, kmol/(m^2·s·Pa).

The absorption rate of A through the liquid film is as in Eq. (2.60):

$$N_A = k_L(c_i - c) \tag{2.60}$$

where

c_i, c = the concentration of A at the interface and liquid phase, kmol/m^3;
k_L = the liquid film absorption coefficient, m/s.

From Eqs. (2.59) and (2.60), we can find that $1/k_g$ and $1/k_L$ are the mass transfer resistances of the substance A through the gas film and liquid film, respectively.

In a steady absorption process, the absorption rate of the substance A through the gas film is equal to that through the liquid film. Therefore, we can obtain Eq. (2.61):

$$N_A = k_g(P - P_i) = k_L(c_i - c) \tag{2.61}$$

However, it should be noted that the k_g, k_L, c_i, and p_i values are difficult to measure, and hence, the overall rate equation is used in practical applications to calculate the absorption rate, as shown in Eq. (2.62):

$$N_A = K_g(p - p^*) = K_L(c^* - c) \tag{2.62}$$

where

K_g = overall coefficient of gas-phase absorption, kmol/(m^2·s·Pa);
K_L = overall coefficient of liquid-phase absorption, m/s;

p^* = partial gas pressure in equilibrium with liquid concentration c, Pa;
c^* = liquid concentration in equilibrium with gas pressure p, kmol/m^3.

In Eq. (2.62), (p-p$*$) and (c$*$-c) can be regarded as the driving forces of the process, and $1/K_g$ and $1/K_L$ are the overall resistance of the absorption process.

When the gas–liquid equilibrium follows Henry's law, the relationship between the absorption coefficient and the total absorption coefficient is shown in Eqs. (2.63) and (2.64):

$$\frac{1}{K_g} = \frac{1}{k_g} + \frac{H}{k_L} \tag{2.63}$$

$$\frac{1}{K_L} = \frac{1}{k_L} + \frac{1}{Hk_g} \tag{2.64}$$

The above two equations illustrate that the overall resistance of the absorption process is the sum of the resistances of the gas film and liquid film.

In the desulfurization tower, the mass transfer between gas and liquid is usually expressed by the number of mass transfer units, according to the two-film theory, which can be expressed as in Eq. (2.65):

$$\mathrm{NTU} = \ln\left(\mathrm{SO}_{2,\mathrm{in}}/\mathrm{SO}_{2,\mathrm{out}}\right) = \mathrm{KgAPV}/\mathrm{G} \tag{2.65}$$

Where:

NTU is the number of mass transfer units;

G is the molar rate of the gas, mol/s;

SO_2, in is the volume concentration of SO_2 at the inlet of the desulfurization tower;

SO_2, out is the volume concentration of SO_2 at the outlet of the desulfurization tower;

Kg is the overall coefficient of gas phase absorption, mol/(m^2·s·atm);

A is the unit volume interface area, m^2/m^3;

P is the absolute pressure of the desulfurization tower, atm;

V is the volume of the desulfurization tower, m^3.

Thus, the desulfurization rate can be expressed as:

$$\eta = 1 - \exp(\mathrm{NTU}) \tag{2.66}$$

2.3.2.2 Adsorption

In the dry desulfurization process, the gas reacts directly with the solid particles, and a product layer is formed on the periphery of the particles, and the unreacted absorbent is wrapped in the product layer. At this point, the absorbent needs to pass through the product layer by diffusion to further react with the absorbent. The following figure shows the mass transfer process in a typical LIFAC process in the furnace (Fig. 2.8).

Figure 2.9 is a schematic diagram of a model of a nuclear reduction reaction. R_P and R_C represent the particle and unreacted core radius, respectively, m; C_{AO}, $C_{AP,}$ and C_{AC} represent the concentration of the reaction gas at the gas phase, the particle surface, and the particle reaction interface, respectively, mol/m^3.

As the moving velocity of the reaction interface is much smaller than the gas diffusion velocity, the pseudo-steady-state method was used to establish the differential equation.

$$\frac{d}{dR}\left(D_{eff}\frac{dC_A}{dR}\right) + \frac{2}{R}D_{eff}\frac{dC_A}{dR} = 0 \quad (2.67)$$

where

D_{eff} is the diffusion coefficient of the gas in the product layer, m^2/s;
C_A is the concentration of A in the product layer, mol/m^3;
R is the reaction radius, m.

Fig. 2.8 The mass transfer process in a typical LIFAC process

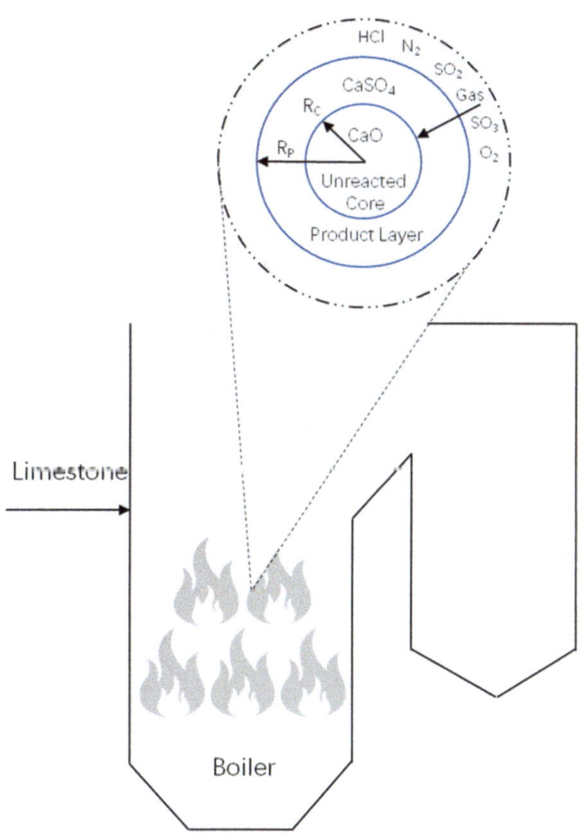

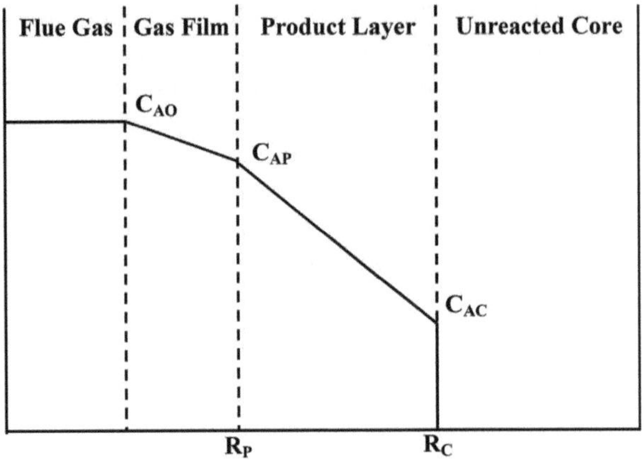

Fig. 2.9 Reaction and mass transfer model between gas–liquid–solid phases

The gas diffusion from the main gas phase to the reaction interface can be expressed as follows:

$$v_1 = 4\pi R_p^2 h_D (C_{AO} - C_{AP}) \quad (2.68)$$

$$v_2 = \frac{4\pi D_{eff}}{\frac{1}{R_C} - \frac{1}{R_P}} (C_{AP} - C_{AC}) \quad (2.69)$$

$$v_3 = 4\pi R_C^2 k_s C_{AC} \quad (2.70)$$

where

v_1 is the diffusion rate of A in the gas film, mol/s;
v_2 is the diffusion rate of A in the product layer, mol/s;
v_3 is the reaction rate of A, mol/s;
h_D is the gas film diffusion coefficient, m/s;
k_s is the chemical reaction rate constant, m/s.

According to pseudo-steady-state assumption,

$$v_1 = v_2 = v_3 \quad (2.71)$$

According to the formula (2.35)–(2.39), the differential equation of the reaction is obtained as

$$\frac{dR_C}{dt} = -\frac{C_A}{\rho_P \left[\frac{1}{h_D} \frac{R_C^2}{R_P^2} + \frac{1}{D_{eff}} \left(R_C - \frac{R_C^2}{R_P} \right) + \frac{1}{k_s} \right]} \quad (2.72)$$

where

t is time, s;

ρ_P is the molar density of sorbent, mol/m^3.

Boundary condition: $t = 0$, $R_C = R_P$

The conversion rate is defined as follows:

$$x = 1 - \left(\frac{R_C}{R_P}\right)^3 \tag{2.73}$$

As a result, the relationship between the conversion rate and time can be described as follows:

$$t = \frac{\rho_P R_P}{3h_D C_{AO}} x + \frac{\rho_P R_P^2}{6D_{eff} C_{AO}}\left[3 - 3(1-x)^{\frac{2}{3}} - 2x\right] + \frac{\rho_P R_P}{k_s C_{AO}}\left[1 - (1-x)^{\frac{1}{3}}\right]. \tag{2.74}$$

Thus, the conversion rate of the absorbent can be determined from the residence time, and then the desulfurization efficiency can be determined according to the addition ratio of the absorbent.

In the semi-dry desulfurization process, the droplets of the absorbent slurry are dried in the flue gas to form product particles, thus having the above two mass transfer processes.

2.4 Control Technologies

The flue gas desulfurization (FGD) technologies can be divided into furnace desulfurization, dry/semi-dry FGD, and wet FGD technologies, as shown in Fig. 2.10.

2.4.1 Wet FGD

2.4.1.1 Process

Several types of materials can be used as absorbents in the wet FGD process to remove SO_2 from flue gas, including limestone ($CaCO_3$), calcium hydroxide ($Ca(OH)_2$), magnesium hydroxide ($Mg(OH)_2$), ammonia, seawater, and so on. The flue gas with high sulfur content enters the absorption tower, reacting with the alkaline absorbents, and thus the SO_2 is absorbed and removed. Consequently, the clean gas is discharged from the tower, and the waste liquid is subjected to other treatments.

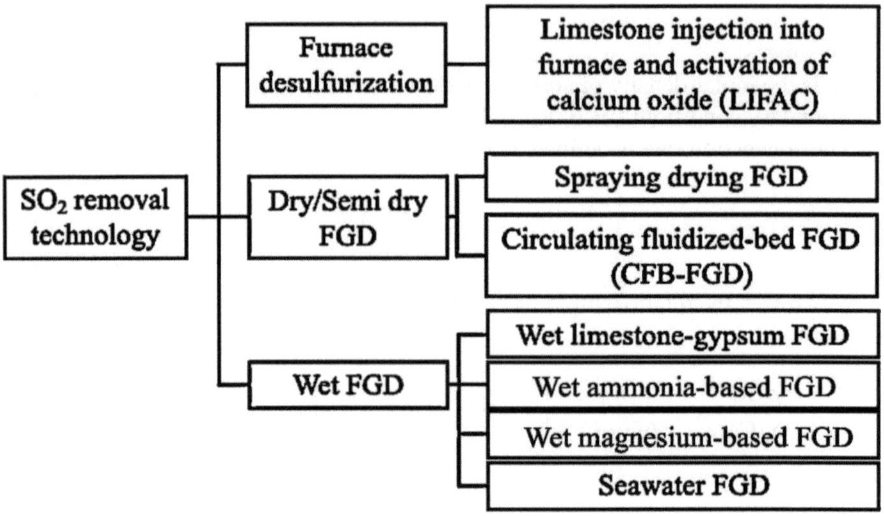

Fig. 2.10 Summary of SO_2 control technologies

1. Wet limestone–gypsum FGD process

The wet limestone–gypsum FGD technology is the most common commercial process for SO_2 removal from flue gas due to the abundant sources of absorbents, low cost, and high SO_2 removal efficiency.

The SO_2 absorption process is very complicated. Typically four key steps are mainly involved: (1) SO_2 absorption. In the wet FGD process, SO_2 is first absorbed into the water and exists in ion forms, i.e., HSO_3^- and SO_3^{2-}. (2) Limestone dissolution. The solid limestone is dissolved and exists as Ca^{2+}, CO_3^{2-}, and HCO_3^-. (3) Oxidation. In this process, the sulfite and bisulfite are oxidized by oxygen in the air to form SO_4^{2-}. (4) Crystallization. In this process, calcium cations and sulfate ions react to form gypsum, which can be recycled for further use.

Limestone is used as the absorbent in this process. The activity of limestone is a significant index to evaluate its absorbent quality. It affects not only the selection of an appropriate absorbent in system design, but also the optimal operation parameters, which can reduce the investment and operating costs. The main factors influencing the activity of limestone are physical properties ($CaCO_3$ content, particle size, geological age) and the operating environment (slurry pH, ion contained in the slurry, CO_2 partial pressure, temperature, stirring rate, etc.). In the limestone selection, $CaCO_3$ content, particle size, and grindability are the major factors that should be taken into consideration.

(1) The main active ingredient in limestone is $CaCO_3$, the content of which has an important influence on the activity of the absorbent. Generally, the $CaCO_3$ content is higher than 90%.

(2) The limestone activity increases with the smaller particle size of limestone, larger specific surface area, and more liquid–solid contact. In consideration of the impact of particle size on dissolution and grinding energy consumption, the general requirement for limestone powder is 250–325 mesh.
(3) The $MgCO_3$ component contained in limestone directly affects the desulfurization effect and the wastewater discharge volume of the system. Generally, $MgCO_3$ should be less than 4%.

In addition, it should be noted that the content of SiO_2 in limestone is also an important factor due to the possibility of equipment wear by high SiO_2 content.

In the wet limestone–gypsum FGD technology, the absorption of SO_2 is relatively fast and the removal efficiency is generally higher than 95%. In addition, the absorbent resource is abundant on Earth and the product gypsum can be recycled and utilized. Therefore, this technology is most widely applied in coal-fired power plants for SO_2 removal all over the world.

The schematic flow sheet of this process is shown in Fig. 2.11. Generally, the absorbent, i.e., the limestone slurry, is mixed with the flue gas in the absorber. SO_2 in the flue gas reacts with $CaCO_3$ in the slurry and O_2 to form gypsum. The flue gas after desulfurization then passes through the demister and reheater and is emitted through the stack.

The core part of the wet desulfurization process is the absorption tower. Harmful gases in the flue gas are absorbed in contact with absorbents in the absorption tower. Different reaction towers are designed with different absorption zones, usually including the following: the slurry sprayed from the top, the grid arranged in the tower, the tray arranged in the tower, the jet bubbling, and the bottom liquid jet sprayed.

Droplet, liquid film, or liquid column are formed in different towers. Three gas–liquid contact types are usually concluded: countercurrent flow, co-current flow, and cross flow. There are usually five types of absorption tower: spray tower, packed tower, bubble tower, double contact flow tower, and High-Gravity Rotating Packed

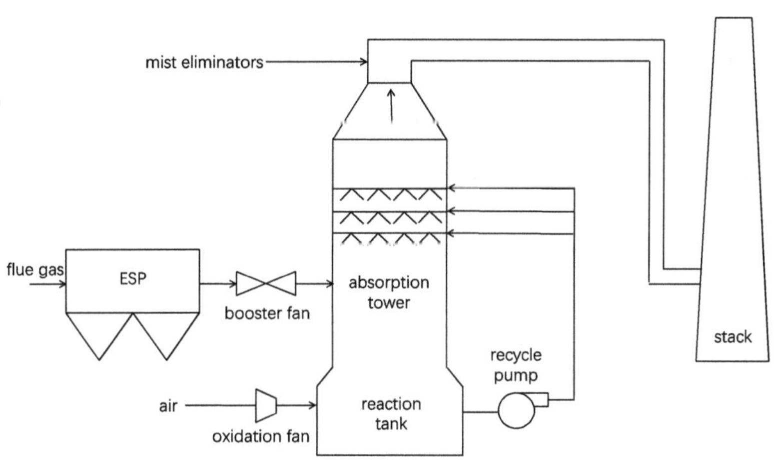

Fig. 2.11 Schematic flow diagram of the wet limestone–gypsum FGD process

Bed. High SO_2 removal efficiency can be achieved in these absorption towers. Compared to others, the volume and land usage of RPB are much less. Detailed characteristics are listed in Table 2.1.

2. Dual alkaline FGD

The wet limestone–gypsum FGD process is, however, easy to scale. To overcome this drawback and to further improve the desulfurization efficiency, a dual-alkali flue gas desulfurization process has been developed. In this process, SO_2 is first absorbed by aqueous solutions of alkali metal salts such as sodium salts (mixture of NaOH and Na_2CO_3 solutions in a certain proportion). The spent absorbent solution then enters another reactor and is regenerated with lime or limestone, during which the final product is precipitated in the form of calcium sulfite and gypsum. The absorbent after regeneration returned to the absorber for reuse. As a result, the actual consumption is lime or limestone which is inexpensive (Fig. 2.12).

3. Wet ammonia-based FGD process

Wet ammonia-based FGD process takes alkaline ammonia as absorbent to neutralize SO_2. The by-product ammonium sulfate is an important fertilizer raw material and has a high utilization value. Since ammonia has higher alkaline than limestone, the desulfurization system can achieve more than 98% desulfurization efficiency at a small liquid-to-gas ratio. Meanwhile, the circulating slurry volume is small and the system energy consumption is low. It is widely used in the sulfur recovery of the acid-making industry.

The pH of the absorbent determines the direction of the reaction. When $pH < 2$, the absorbed SO_2 is mainly present in the form of H_2SO_3. When the pH rises to about 4, it is mainly HSO_3^-. When the pH is <7.5, it is substantially in the form of SO_3^{2-}. It can also be seen from the above reaction formula that $(NH_4)_2SO_3$ is mainly absorbent for SO_2. $(NH_4)_2SO_3$ has a good absorption capacity for SO_2, and NH_4HSO_3 is formed after absorption. Ammonia water is added to adjust the pH value to convert NH_4HSO_3 to $(NH_4)_2SO_3$, and the absorption capacity of the absorption liquid for SO_2 is maintained (Fig. 2.13).

Wet ammonia FGD has the advantages of high desulfurization efficiency, high added value of desulfurization by-products, and low system energy consumption. However, the equipment corrosion is obvious.

4. Wet Magnesia-based FGD process

Wet magnesia-based FGD is also a common method in flue gas desulfurization systems. After pretreatment, the flue gas enters the absorber and reacts with the absorbent $Mg(OH)_2$ and $MgCO_3$ in the column.

In production, a small amount of $MgSO_3$ is oxidized to $MgSO_4$. When $MgSO_3$ and $MgSO_4$ settle down, they are in the hydrated crystalline state, and their crystals are large and, thus, easy to separate. After separation, they are sent to the dryer to make dry $MgSO_3$/$MgSO_4$ for transportation to the regeneration section. In the regeneration section, $MgSO_3$ is decomposed at a high temperature of 1500 °F during forging, and carbon is used as a reducing agent to react with $MgSO_4$.

Table 2.1 Summary of the absorption towers

Tower types	Spray tower	Packed tower	Bubble tower	Liquid column tower	Dual-pH	High-Gravity Rotating Packed Bed
Scheme	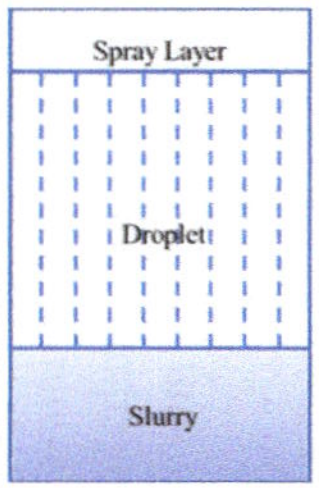	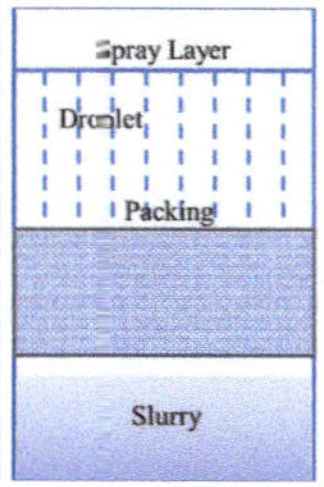	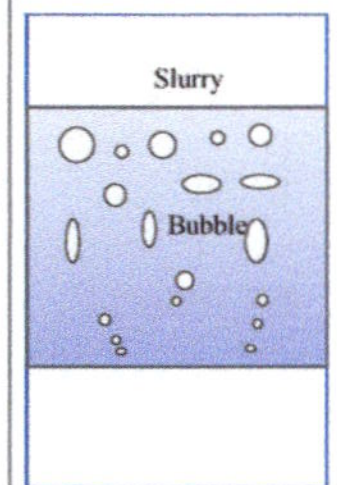	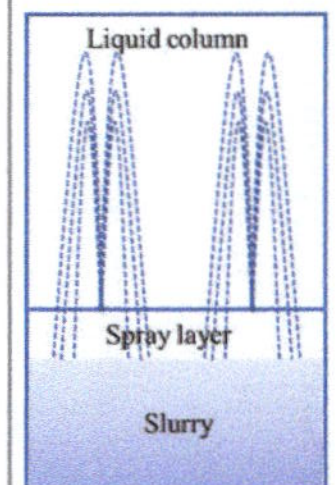	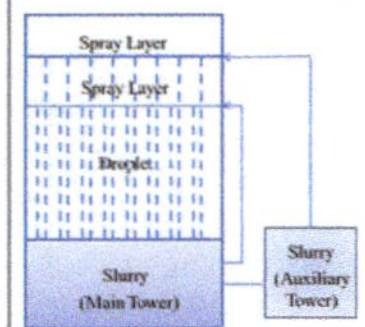	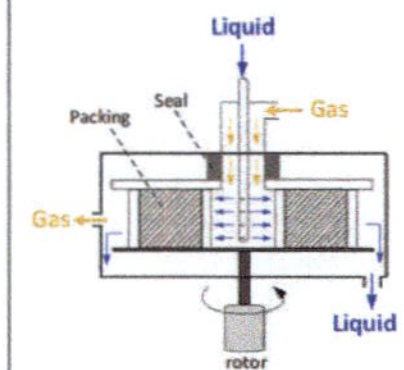 [18]
Features	Smallest liquid–gas ratio Large liquid–gas contact area Simple structure Small pressure loss Nozzle blockage High particle size	Easy to block High requirements on absorbent particle size and smoke dust content	No circulation pump No nozzle Large gas–liquid contact area Large system resistance Large device volume	Large aperture nozzle Not easy to block Low requirements on the particle size and the dust content of the flue gas Stable in operation	Simple structure Stable in operation Easy to transform High efficiency	High mass transfer Low liquid hold-up High micromixing efficiency High hydraulic and flooding capacity Small reactor size Better self-cleaning of the rotor

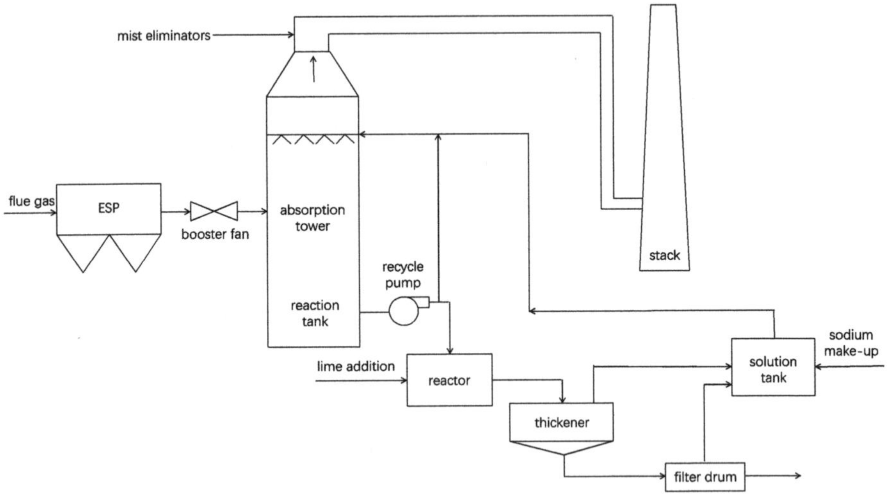

Fig. 2.12 Schematic flow diagram of the wet limestone–gypsum FGD process

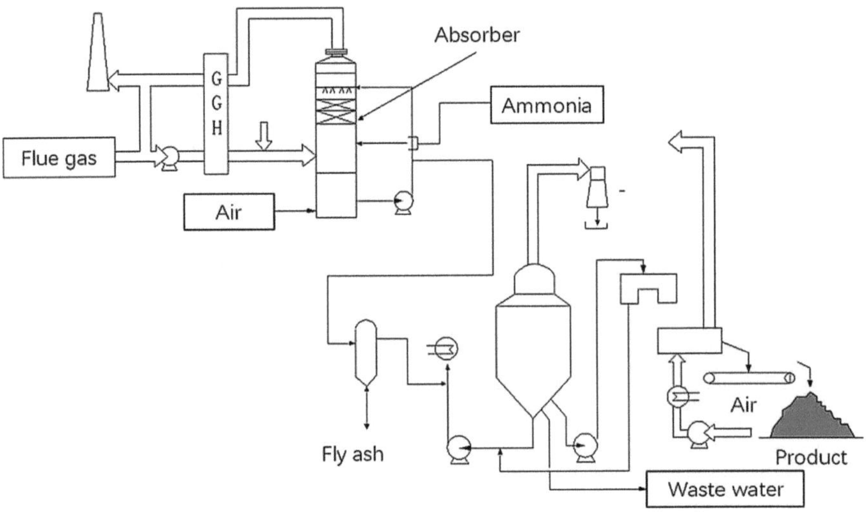

Fig. 2.13 Schematic flow diagram of the wet ammonia-based FGD process

The SO_2 gas from the calciner is sent to the sulfur or acid production after dedusting. The regenerated MgO, together with the newly added MgO, ripens to magnesium hydroxide by adding water and recycled to the absorption tower (Fig. 2.14).

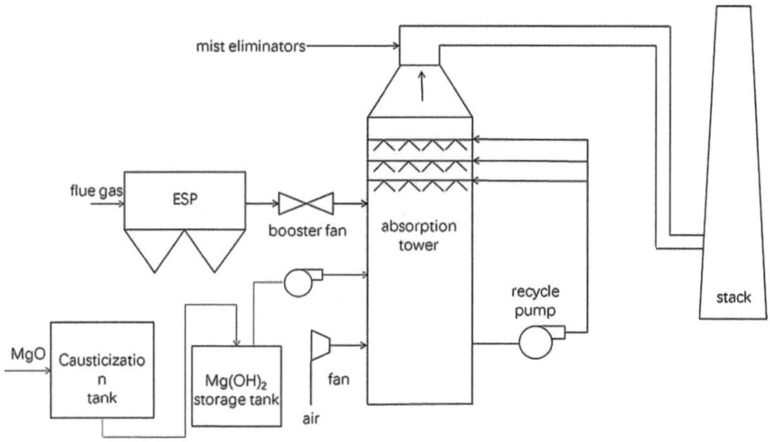

Fig. 2.14 Schematic flow diagram of the wet magnesia-based FGD process

The wet magnesia-based FGD is more complicated and costly, but it is viable. This is mainly due to the high desulfurization efficiency of the method (generally above 90%). And both $MgSO_3$ and $MgSO_4$ have a large solubility. Therefore, there is no scaling problem (commonly appearing in lime/limestone systems) for the wet magnesia-based FGD process. The final product can save absorbent by the regeneration method, and it also saves the trouble of waste disposal.

5. Seawater FGD process

Natural seawater contains a large amount of soluble salts, the main components of which are chlorides and sulfates, and also contains certain soluble carbonates. Seawater is usually alkaline, with a natural alkalinity of about 1.2–2.5 mmol/L. This gives seawater a natural acid–base buffering capacity and the ability to absorb SO_2. Some desulfurization companies have utilized this characteristic of seawater to develop seawater FGD and successfully absorb SO_2 with seawater to achieve the purpose of flue gas purification.

The main process and reaction principle of seawater FGD is that after the flue gas is discharged from the boiler and dedusted by the dust collector, it is sent to the hot side of the gas–gas heat exchanger by the FGD system booster fan to cool down. Then, it is sent to the absorption tower. In the absorption tower, the flue gas is washed by seawater (Fig. 2.15).

The seawater after washing the flue gas in the absorption tower is acidic and contains more sulfite, which cannot be directly discharged into the sea. The wastewater discharged from the absorption tower flows into the seawater treatment plant by gravity, mixes with the seawater from the cooling circulation system, and then a large amount of air is blown into seawater with a blower to oxidize the sulfite to sulfate and drive out the carbon dioxide in the seawater. After mixing and treatment, the pH value and COD of the seawater reach the discharge standard and are discharged into

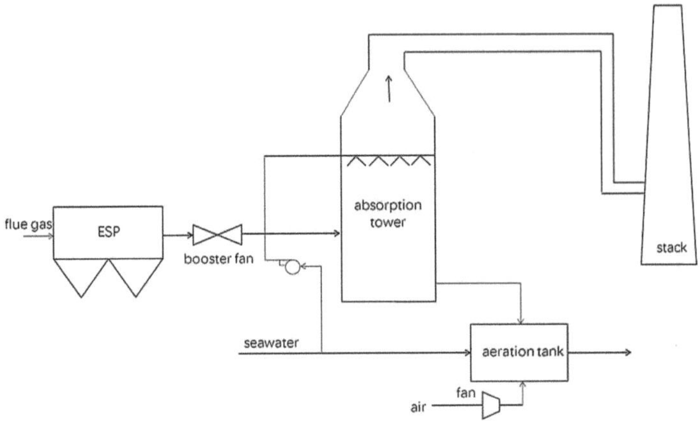

Fig. 2.15 Schematic flow diagram of the seawater FGD process

the sea. The purified flue gas is heated by GGH and discharged into the atmosphere through the chimney.

Seawater FGD is attractive for power plants located by the sea, but it requires a large amount of electricity used by the plant and may have an impact on marine life after extensive use of seawater desulfurization technology.

2.4.2 *Semi-dry FGD*

2.4.2.1 Spray Dry Absorption (SDA)

Lime is generally used as an absorbent in the spray drying flue gas desulfurization process. Lime is digested into lime slurry, which is then atomized into fine pulp drops in a spray dry absorption tower. The flue gas has sufficient residence time in the absorber to make SO_2 and other acid gases such as SO_3 and HCl react with calcium hydroxide, forming calcium sulfite/calcium sulfate and calcium chloride. Because the moisture is evaporated completely in the spray drying tower, the wastewater treatment system is not needed in this process. It is worth noting that the removal rate of SO_3 and HCl in the spray drying flue gas desulfurization process is 95%, which is much higher than that of SO_3 and HCl in the wet desulfurization process. First, the lime is digested and hot water is added to make the slurries.

The slurry is filtered from the pump to the atomization device in the absorber, in which the slurry is atomized into fine droplets of absorbent and mixed with flue gas. The reaction with SO_2 and SO_3 in flue gas produces calcium sulfite ($CaSO_3$) and calcium sulfate ($CaSO_4$). At the same time, the moisture brought in by the absorbent is evaporated and dried quickly, which, thus, forms the powder reaction by-product (mostly calcium sulfite). The flue gas temperature decreases accordingly, but it is still

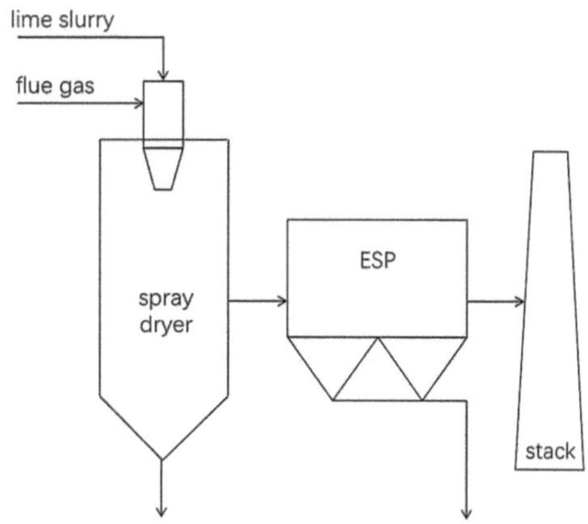

Fig. 2.16 Schematic flow diagram of the spray dry absorption process

higher than the acid dew point, so it can be discharged directly. And no wastewater is produced in the process.

A part of the product of the desulfurization reaction and the unused desulfurizing agent are taken out of the absorption tower with the flue gas in the form of dry particulate matter and are collected into the dust collector. At the same time, due to the effect of gravity, a part of the larger solid product is deposited at the bottom of the absorption tower, which still contains unreacted calcium hydroxide. In order to improve the utilization rate of the desulfurization absorbent, this part of the solid particles and the partially desulfurized ash and dust mixture collected by the precipitator are ground together again and then recycled to the pulping system. Therefore, the desulfurization absorbent used in the desulfurization process is actually a mixture of slaked lime-based, a small amount of fly ash, and a by-product of the recycling reaction (Fig. 2.16).

The advantage of the method is that the system is simple, the operation is reliable, the investment is low, the energy consumption is small, the desulfurization product is a dry powder, and no wastewater is discharged. The weakness is that it is necessary to use lime instead of limestone as the absorbent, the inner wall of the drying absorption tower is easy to scale, the atomizing device is easy to wear or block, and the desulfurization ash is difficult to use.

2.4.2.2 Circulating Fluidized-Bed FGD (CFB-FGD)

CFB-FGD is an emerging dry/semi-dry desulfurization technology. Under the basis of circulating fluidized bed, the absorbents can be recycled several times, which helps to prolong the contact time of the absorbent with the flue gas, and thereby, the SO_2 removal efficiency is obviously enhanced. The schematic flow sheet of this process is shown in Fig. 2.17. The flue gas enters the CFB from the bottom and then mixes and reacts with the finely pulverized absorbent and spraying water. Due to the recycling

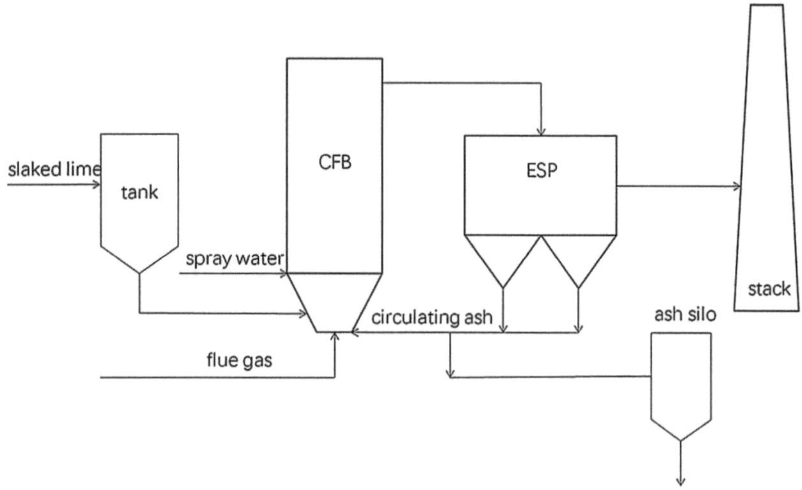

Fig. 2.17 Schematic flow diagram of the CFB-FGD process

of absorbent, a large number of particulate matters are formed, the concentration of which is as high as 800–1,000 g/Nm^3. The flue gas with these particulates comes out from the top of the CFB, followed by entering into the particulate removal facility. Most of the captured particles are then fed again into the CFB reactor to make full use of the absorbent.

The absorbent used is mainly $Ca(OH)_2$, which can react with SO_2 and SO_3. In addition, other acidic gases like HCl and HF can also be removed.

2.4.2.3 Circulating Suspension and Multi-stage Humidification Dry Flue Gas Cleaning Technology (CSMHD-FGC)

To further increase the SO_x removal efficiency and to achieve the simultaneous removal of multi-pollutants, a modified CFB-FGD technology, i.e., circulating suspension and multi-stage humidification dry flue gas cleaning technology (CSMHD-FGC), has been developed by Zhejiang University. It is adaptable to various boiler loadings and different types of coals and can remove SO_2, HCl, HF, heavy metals, and dioxins simultaneously.

The schematic flow sheet of the CSMHD-FGC technology is shown in Fig. 2.18. Compared with the conventional CFB-FGD technique, the CSMHD-FGC uses multi-stage humidification, which facilities the reaction of absorbent and SO_2 by properly wetting the absorbent at the final reaction step. By designing new absorbent, multi-pollutants including SO_2, HCl, HF, heavy metals, and dioxins can be absorbed simultaneously. By tuning the ash/water ratio combined with multi-stage humidification, acidic pollutants can be effectively removed and dust accumulation in the CFB can

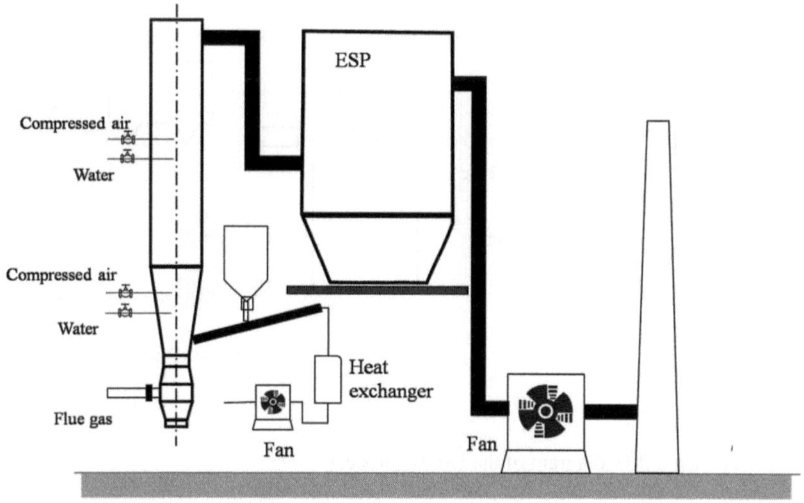

Fig. 2.18 Schematic flow diagram of the CSMHD-FGC technology

be avoided. Therefore, it is more reliable. Results showed that when the ratio of Ca/S was around 1.2–1.5, the SO_2 removal efficiency can be higher than 92%.

2.4.3 *Dry FGD*

2.4.3.1 Limestone Injection into Furnace and Activation of Calcium Oxide (LIFAC)

The LIFAC technology is a typical dry FGD process. The schematic flow sheet of this process is shown in Fig. 2.19. Two processes are mainly involved in this technology: 1 injection of pulverized limestone to the high-temperature boiler and humidifying in the reactor.

The finely pulverized limestone is injected into the boiler where the temperature is around 900–1150 °C for the decomposition of limestone to form more reactive lime, which readily reacts with SO_2 and a small amount of SO_3. In this process, the SO_x removal efficiency is about 25–35% and the investment is about 10% of that of the total system.

The flue gas, which contains the unreacted lime and SO_2, exits the boiler and passes through the air preheater, followed by entering the LIFAC activation reactor. In the reactor, water is sprayed to humidify the gas, and meanwhile, hydrated lime is formed, which rapidly reacts with SO_2. In this process, the investment accounts for about 85% of that of the total system.

When the flue gas leaves the reactor, it enters the electrostatic precipitator (ESP) where the spent sorbent and fly ash are removed. A fraction of the solids is recycled

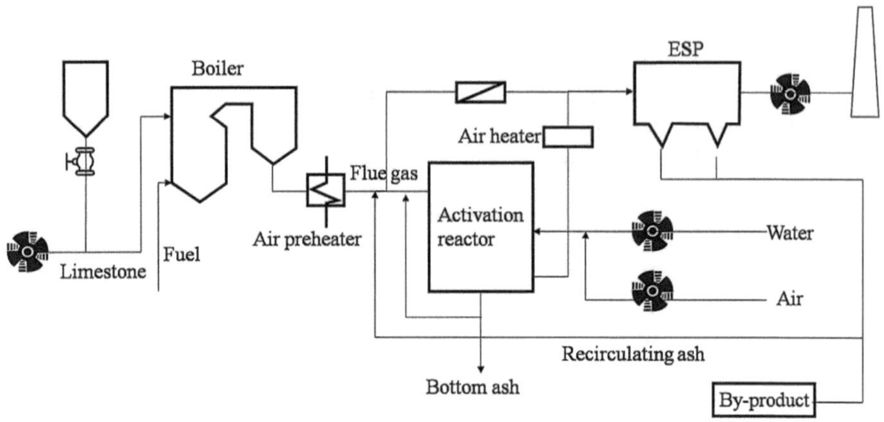

Fig. 2.19 Schematic flow diagram of the LIFAC process

to the activation reactor and the rest is sent to a landfill for disposal. In the LIFAC process, the overall SOx removal efficiency is higher than 70%.

2.4.3.2 Active Carbon Method

The activated carbon adsorption method is a method for removing SO_2 in flue gas by utilizing the adsorption performance of activated carbon. It is generally believed that the adsorption of SO_2 by activated carbon includes physical adsorption and chemical adsorption. When there is no vapor and O_2 in the flue gas, physical adsorption occurs mainly, and the amount of adsorption is small. When the flue gas contains enough vapor and O_2, a series of chemical reactions occur with physical adsorption proceeding. Firstly, physical adsorption occurs on the surface of the activated carbon, and then the SO_2 adsorbed on the surface of the activated carbon is catalytically oxidized to sulfuric acid in the presence of vapor and O_2, so that the adsorption amount of SO_2 is increased. The total reaction of chemisorption is as follows:

$$2SO_2 + 2H_2O + O_2 \rightarrow 2H_2SO_4 \tag{2.75}$$

After the activated carbon adsorbs SO_2, the sulfuric acid is in the micropores of the activated carbon. The adsorption capacity of the activated carbon is decreased, so the activated carbon needs to be regenerated. Regeneration methods include the following: (1) Washing regeneration: washing out the sulfuric acid in the activated carbon micropores with water. (2) Heating regeneration: heating the loaded activated carbon to promote the reaction of activated carbon with sulfuric acid in the micropore

$$2H_2SO_4 + C \rightarrow 2SO_2 + 2H_2O + CO_2 \tag{2.76}$$

The active carbon method has many advantages: (1) low desulfurizer consumption, reusable desulfurizer, saving raw materials, reducing operating costs; (2) desulfurization products can be recycled; (3) simple and easy to operate; and (4) no secondary pollution. However, the active carbon method also has disadvantages: (1) Large consumption of activated carbon, (2) frequent regeneration, (3) high price, and (4) short life. Therefore, the promotion of the active carbon method is limited.

2.4.3.3 Alkaline Sorbent Injection

SO_3 can be removed by injecting an alkaline adsorbent. Wet sorbent injection and dry sorbent injection are the two main methods. In the wet sorbent injection, the wet sorbent solution is injected through a two-fluid atomizing nozzle and mixed with hot flue gas. When the moisture in the solution evaporates by absorbing heat, the dried small particles form. These particles are calcined into a porous structure due to the evolution of water vapor and other gases. The dried adsorbent particles react with SO_3 or H_2SO_4 in the flue gas to form sulfate. The reaction product and the unreacted adsorbent can be removed by arranging particle control devices such as an ESP or a fabric filter downstream. In the dry sorbent injection, solid sorbent particulates are directly injected into the flue gas. Table 2.2 lists the existing data of alkaline adsorbents for SO_3 removal.

2.5 Comparison of SOx Control Technologies

2.5.1 4E Evaluation

To address the techno-commercial constraints with SO_2 control in industry, a host of alternative technologies is being evaluated. 4E evaluation, which included environmental, energy, engineering, and economic aspects, is consequently proposed to access different technologies. Analysis of SO_2 control technologies based on the 4E Evaluation is further summarized in Table 2.3.

The removal efficiency of Wet FGD technologies is more than 90%, while those of the dry and semi-dry technologies are lower than the Wet FGD method. Furnace desulfurization shows the poorest SO_2 removal performance. Limestone–gypsum FGD technology needs larger space compared to other control technologies.

The electricity consumption of limestone–gypsum FGD technology is highest among the SO_2 control technologies. Water consumption of magnesium oxide FGD technology is 60t/h. For limestone–gypsum FGD technology, ammonia FGD technology, and circulating fluidized bed, the water demand is about 40t/h. While the water consumption of seawater FGD technology is 10t/h, spray dryer absorber and furnace desulfurization need low water consumption. The water consumption of seawater desulfurization technology is 10t/h, and the water consumption of spray dryer absorption furnace desulfurization is lower.

Table 2.2 Summary of sorbent injection results

Sorbent	Main reactions	Medium Phase	Scale	Injection position	Injection rate (molar ratio)	Respective removal efficiency	Total removal efficiency	References
$Ca(OH)_2$	$Ca(OH)_2 + SO_3 \rightarrow CaSO_4 + H_2O$	Solid	Field	Pre SCR		75%		[17]
			Field	Pre APH		70%		[18]
			Field	Pre ESP	Sorbent:SO_3 4.2:1		83%	[19]
			Field	Post APH	Sorbent:SO_3 3.24:1		98%	[19]
			Field	Pre ESP	Sorbent:SO_3 3:1		90%	[20]
			Field	Pre APH		49.3%		[21]
			Pilot	Pre ESP	Sorbent:SO_3 7:1	90%		[4]
			Field	Pre baghouse	Sorbent:SO_3 1.71:1		95%	[22]
CaO	$CaO + SO_3 \rightarrow CaSO_4$	Solid	Laboratory	Fixed bed	Sorbent:SO_3 8:1		95%	[23]
$CaCO_3$	$CaCO_3 + SO_3 \rightarrow CaSO_4 + CO_2$	Solid	Laboratory	Fixed bed			95%	[24]
			Pilot	Pre ESP	Sorbent:SO_3 7:1	53%		[4]
$Mg(OH)_2$	$Mg(OH)_2 + SO_3 \rightarrow MgSO_4 + H_2O$	Slurry	Pilot	Pre APH	Sorbent:SO_3 4:1	90%		[25]
MgO	$MgO + SO_3 \rightarrow MgSO_4$	Solid	Pilot	Pre ESP	Sorbent:SO_3 8:1	63%		[4]

(continued)

Table 2.2 (continued)

Sorbent	Main reactions	Medium Phase	Scale	Injection position	Injection rate (molar ratio)	Respective removal efficiency	Total removal efficiency	References
Trona	$2(Na_2CO_3.NaHCO_3.2H_2O)(s) \rightarrow 3Na_2CO_3 + 5H_2O + CO_2$ $Na_2CO_3 + SO_3 \rightarrow Na_2SO_4 + CO_2$ $Na_2SO_4 + SO_3 + H_2O \rightarrow 2NaHSO_4$	Solid	Field	Pre FGD			90%	[26]
			Field	Pre ESP		80%		[26]
			Field	Pre ESP	Sorbent:SO_3 3.5:1	85%		[27]
SBS	$SO_3 + 2NaHSO_3 \rightarrow Na_2SO_4 + 2SO_2 + H_2O$ $SO_3 + NaHSO_3 \rightarrow NaHSO_4 + SO_2$ $SO_3 + Na_2SO_3 \rightarrow Na_2SO_4 + SO_2$ $2SO_3 + Na_2SO_3 + H_2O \rightarrow 2NaHSO_4 + SO_2$ $SO_2 + Na_2SO_3 + H_2O \rightarrow 2NaHSO_3$	Solution	Field	Pre APH	Na:SO_3 1.5:1		90%	[28]
			Field	Pre APH or pre ESP	Na:SO_3 1.0:1		90%	[29]
			Field		Na:SO_3 1.5:1		99%	[30]
$NaHCO_3$	$2NaHCO_3 \rightarrow Na_2CO_3 + H_2O + CO_2\uparrow$ $Na_2CO_3 + SO_3 \rightarrow Na_2SO_4 + CO_2$	Solid	Pilot	Pre ESP	Sorbent:SO_3 10:1	70%		[4]

Table 2.3 4E evaluation of SO_2 control technologies

4E evaluation	Classification	Wet FGD				Semi-dry FGD		Dry FGD
		Limestone -FGD	Ammonia FGD	Magnesium oxide FGD	Seawater FGD	CFB-FGD	SDA	LIFAC
Environmental	Removal efficiency	>95%	>95%	>90%	>95%	85%	85%	60%–80%
	Land usage	2,700-3,000	1000–1500	low	low	1,000-1,200	1500–1800	900–1100
Energy	Electricity consumption (kWh/h)	1.5%–2%	1.2%–1.6%	0.8%–1%	1%–1.5%		0.9%–1.4%	0.5%–1%
	Water consumption (t/h)	45	40	60	10	40	low	low
Engineering	Process complexity	Complicated	Complicated	Complicated	Simple	Simple	Average	Simple
	Ration of Ca/S	~1.1	/	/	/	~2.3	~1.5	~2.0
Economic	Capital cost	13%–18%	15%–20%	10%–15%	10%–15%	5%–10%	6%–14%	5%–10%
	Operating cost ($/t SO_2)	145–205	110–150	115–160	110–150	100–145	110–150	85–130

The process of limestone–gypsum FGD technology, ammonia FGD and magnesium oxide FGD technology is complicated, while the process of other technologies is simpler. Dry or semi-dry method and furnace desulfurization require a higher ratio of Ca/S than limestone–gypsum FGD technology.

Economic costs consist of capital costs and operating costs. The capital cost of wet FGD technologies is higher than that of dry/semi-dry FGD technology and furnace desulfurization. Except for limestone–gypsum desulfurization technology, the operating cost of other technologies is basically the same.

2.5.2 *Technology Selection*

There are lots of technologies that can achieve SO_2 emission standards for coal-fired power plants. However, for different desulfurization processes, due to the various types of absorbent type, the arrangement of the absorbent in the desulfurization tower, and the method of transportation, etc., the application scope of different desulfurization processes is also different, as shown in Table 2.3.

A variety of wet desulfurization processes (traditional empty towers, composite towers, pH partitioning) based on limestone–gypsum FGD are suitable for coal-fired power plants of various coal types, with the desulfurization efficiency of 95.0%–99.7%. Since the desulfurization slurry used in different processes has different mass transfer absorption modes in the tower, the indicators such as desulfurization efficiency, energy consumption, and operational stability are different. According to different SO_2 inlet concentrations of flue gas, the suitable process should be considered comprehensively.

Circulating fluidized-bed FGD technology mainly uses quicklime powder or quicklime slurry as the absorbent, and the desulfurization efficiency is generally between 93 and 98%. For medium and low-sulfur coal with a SO_2 concentration below 3000 mg/m^3 in flue gas, the SO_2 emission concentration can meet the requirement of 100 mg/m^3. It is suitable for SO_2 pollution control of coal-fired boilers with 300 MW class and below, and it has been demonstrated in 600 MW coal-fired units. For circulating fluidized-bed boilers in water-deficient areas, it is more suitable to install desulfurization outside the furnace on the basis of desulfurization in the furnace.

The absorbent of ammonia FGD technology mainly uses ammonia water or liquid ammonia, with the desulfurization efficiency of 95.0% to 99.7%, and the desulfurization system resistance is less than 1800 Pa. The ammonia FGD technology has strong adaptability to coal types and load changes, and it is suitable for coal-fired power plants with stable ammonia sources nearby, insensitive to the surrounding environment of power plants, and unit capacity of 300 MW or less.

The seawater FGD technology utilizes the natural alkalinity of seawater to achieve SO_2 absorption, with a system desulfurization efficiency of 95 to 99%. For coastal power plants with an inlet SO_2 concentration of less than 2000 mg/m^3 and good seawater diffusion conditions, and when the environmental functional zoning requirements of offshore waters are met, seawater FGD can be selected.

2.5.3 *Challenges and Perspectives*

1. Challenges
(1) Improve SO_2 and SO_3 removal efficiency

China introduced the most stringent air quality standards in the world in 2020, known as ultra-low emission standards. The ultra-low emission standards require that the SO_2 concentration in the flue gas should be less than 35 mg/Nm^3, which imposes higher air emission requirements on the flue gas desulfurization system. Developing a high-efficiency desulfurization technology and further improving the efficiency of the flue gas desulfurization technology will become important challenges.

(2) Reduce the cost of the desulfurization system

The cost of the desulfurization system includes operating costs and fixed costs, where operating costs include material costs and energy costs, and fixed costs include investment costs and labor costs. Considering the benefits of the factory, it is unrealistic to remove the pollutants without cost. Therefore, reducing the cost of the desulfurization system will become an important challenge.

(3) Take multi-device linkage and synergistic removal of multiple pollutants into consideration

There are multiple devices in the entire flue gas treatment system. Each device removes its own target pollutants and also affects the generation or removal of other pollutants. At the same time, each pollutant is also affected by the multiple synergistic control effects of the device. Therefore, how to use it to further optimize the overall operating costs while achieving the discharge of a variety of pollutants to meet the standard is a challenge.

(4) Improve operational stability and reliability

As a peaking unit, coal-fired units have a large number of variable loading conditions. How to ensure the reliability of desulfurization system operations and the stability of export pollutant concentration is an important challenge when the inlet flue gas conditions fluctuate drastically.

(5) Improve the quality of by-products

The by-products of the desulfurization system include gypsum, ammonium sulfite, etc., which can be used as the main raw materials for building materials and fertilizers. Therefore, it is an important challenge to improve the quality of desulfurization by-products in order to facilitate secondary utilization.

(6) Improve the regeneration efficiency of renewable sorbents

The absorbent used in the partial desulfurization method can be regenerated, such as double alkali desulfurization and activated carbon desulfurization. In the regeneration process of the absorbent, there is always a loss of efficiency. Once the regeneration efficiency of the absorbent is improved, it can not only reduce the total pollutants removal costs, but also improve the removal efficiency of pollutants, which is one of the important challenges in the development of desulfurization technology.

2. Perspectives

(1) Strengthen mass transfer and reaction in desulfurization system

Since the desulfurization process involves complex mass transfer, heat transfer, and reaction processes, the factors affecting efficiency are also numerous. In order to better regulate the desulfurization process, it is necessary to understand the mechanism of the desulfurization process and to determine the influence of various factors on the desulfurization efficiency. At the same time, it is necessary to explore methods to enhance mass transfer and reaction, speed up the reaction, and thus improve the efficiency of desulfurization.

(2) Product regulation of desulfurization system

At present, most of the by-products of the desulfurization systems are of low quality. For example, most of the gypsum produced by desulfurization of limestone–gypsum will be discarded because it is too low of quality due to excessive impurities. Most of the abandoned gypsum will be landfilled or piled up, which wastes land resources and does not conform to the concept of a circular economy. Therefore, the product regulation of desulfurization by-products will be a major development direction in the future.

(3) Establish a circular economy industrial park

Based on the concept of a circular economy, by-products of desulfurization systems can be used as a raw material for many other industrial processes, such as the production process of building materials and fertilizers. The establishment of a circular economy industrial park can make better use of these by-products and improve the overall resource utilization efficiency of the park.

(4) Advanced control and intelligent global optimization

At present, the control of most desulfurization equipment is based on the traditional PID control and the manual control. With the gradual strictness of environmental protection standards, the desulfurization system is gradually difficult to adapt to the rapid changes of the load, and the concentrations of the flue gas pollutants at the exit often exceed the standard. At the same time, due to the strong coupling and non-linearity of the whole pollutant removal system, it is difficult to determine the optimal operating parameters while meeting the pollutant discharge standards by manual means.

Exercises

1. Define the following terms and explain why they should be considered for air pollution control design: the gas film absorption coefficient (k_g), the liquid film absorption coefficient (k_L), and the Henry constant (H).
2. Briefly describe the SO_3 formation process and SO_3 control technologies.
3. Describe the reaction process in wet limestone flue gas desulfurization (WLFGD).
4. Analyze SO_2 control processes and compare their advantages and disadvantages from the aspects of technology, economy, and application, respectively.
5. Alkaline sorbent sodium carbonate (Na_2CO_3) is used to remove SO_3. The chemical equation for the reaction is as follows:

$$Na_2CO_3 + SO_3 \rightarrow Na_2SO_4 + CO_2$$

The mass of Na_2CO_3 is 106 g measured at the beginning. After 20 min, the mass change of the solid is 9 g. Calculate the SO_3 absorption rate (Unit: $mol \cdot mol^{-1} \cdot min^{-1}$).

References

1. Organization TWH (2005) WHO Air quality guidelines for particulate matter, ozone, nitrogen dioxide and sulfur dioxide global update 2005
2. EPA.GOV (1998) Emergency planning & community right-to-know act section 313, guidance for reporting sulfuric acid. http://www.epa.gov/tri/guidedocs/pdf/1998/sulfuric.pdf
3. Farthing WE, Walsh PM, Gooch JP, McCain JD, Hinton WS, Heaphy RF (2004) Identification of (and Responses to) Potential Effects of SCR and Wet Scrubbers on Submicron Particulate Emissions and Plume Characteristics. Research Triangle Park, NC: US Environmental Protection Agency, Office of Research and Development National Risk Management Research Laboratory, Air Pollution Prevention and Control Division
4. Blythe G, Dombrowski K (2004) SO_3 Mitigation guide update. EPRI, Palo Alto, CA
5. EP-C ECN, Numbers WA, Walsh PM, McCain JD, Cushing KM (2006) Evaluation and mitigation of visible acidic aerosol plumes from coal fired power boilers
6. EPA.GOV (1995) Final guideline document: control of sulfuric acid mist emissions from existing sulfuric acid production units. http://www.epa.gov/ttn/naaqs/ozone/ctgact/197709_sox-epa450_2-77-019sulfuricacidmist.pdf
7. Adams B, Senior C (2006) Curbing the blue plume: SO_2 formation and mitigation. Power 150(4):39–41
8. Moretti AL, Advanced emissions control technologies for coal-fired power plants
9. Solutions BE (2012) Comprehensive acid gas management. http://breenes.com/wp-content/uploads/2014/05/Comprehensive_Condensable_Management_Program-RevA.pdf
10. Kikuchi R (2001) Environmental management of sulfur trioxide emission: impact of SO_3 on human health. Environ Manag 27(6):837–844
11. Hawthorne M (2001) Plant spewing acid clouds. Columbus Dispatch, August 3:2001
12. Leikauf G, Yeates DB, Wales KA, Spektor D, Albert RE, Lippmann M (1981) Effects of sulfuric acid aerosol on respiratory mechanics and mucociliary particle clearance in healthy nonsmoking adults. Amer Industr Hygiene Assoc J 42(4):273–282

13. Folinsbee LJ (1989) Human health effects of exposure to airborne acid. Environ Health Perspect 79:195
14. Schlesinger RB, Zelikoff JT, Chen LC, Kinney PL (1992) Assessment of toxicologic interactions resulting from acute inhalation exposure to sulfuric acid and ozone mixtures. Toxicology Appl Pharmacol 115 (2):183–190
15. Jakab GJ, Clarke RW, Hemenway DR, Longphre MV, Kleeberger SR, Frank R (1996) Inhalation of acid coated carbon black particles impairs alveolar macrophage phagocytosis. Toxicol Lett 88(1):243–248
16. Pan SY, Wang PC, Chen Q, Jiang WJ, Chu YH, Chiang PC (2017) Development of high-gravity technology for removing particulate and gaseous pollutant emissions: Principles and applications. J Clean Prod 149:540–556. https://doi.org/10.1016/j.jclepro.2017.02.108
17. Lockert CA (2010) Advances in hot side SO_3 mitigation technology
18. Lockert CA (2010) Pre-air heater control of SO_3 and related condensables
19. Curran L (2011) Using dry sorbent injection to meet the utility MACT
20. Nolan PS (2006) Developments in sorbent injection technology for sulfuric acid mist emissions control
21. Evans MAP, Gresh MJA, Birmingham MW, Benson SA, McCollor DP (2006) Benchmarking SO_3 removal–the cleanstack® process
22. Wolf DE, Seaba JP (1994) Opacity reduction using dry hydrated lime injection. Air & waste 44 (7):908–912
23. Wang Z, Huan Q, Qi C, Zhang L, Cui L, Xu X, Ma C (2012) Study on the removal of coal smoke SO3 with CaO. Energy Procedia 14:1911–1917
24. Chen P, Wang Z, Chang J, Ma C (2011) Experimental study of the reactivity of ca-based matters with SO_3. In: Power and energy engineering conference (APPEEC), 2011 Asia-Pacific, IEEE, pp 1–4
25. Benson LB (2006) Use of magnesium hydroxide for reduction of plume visibility in coal-fired power plants. In: Measurement. Citeseer
26. Kong Y, Davidson H (2010) Dry sorbent injection of sodium sorbents for SO_2, HCl and mercury mitigation. In: 18th annual north american waste-to-energy conference, 2010. American Society of Mechanical Engineers, pp 317–320
27. Maziuk J (2005) Successful mitigation of SO_3 by employing dry sorbent injection of trona upstream of the ESP. In: Mega symposium
28. EPRI, Alto P (2002) Results of full-scale testing of sodium bisulfite injection for flue gas sulfuric acid control
29. Wilhelm JH (2004) SBS injection fights off SO_3. Power engineering international 12 (12):28–30
30. Corporation U (2008) Ultra-high SO_3 removal: SBS injection™ as an alternative to WESP

Chapter 3
NO_x Control

Nomenclature

A	Pre-exponential factor
A_c	The catalyst typical floor area, m/h
ASR	The molar ratio of NH_3/NO_x
A_v	The face velocity, m/h
c	The concentration of the specific substance, mol/cm^3
c_B	The concentration, mol/L
C_f	The unit conversion factor
E_a	The activation energy, J/mol
F	The flow rate, mol/s
G	Gibbs free energy
k	The reaction rate constant
k'	The backward reaction constant
K	the equilibrium constant
m_{cat}	The mass of the catalyst, g
n_B	The number of moles, mol
Q_f	The amount of flow gas, m^3/h
r	The reaction rate, mol/L·s
$\Delta_r H_m^\circ$	Standard molar enthalpy
R	The universal gas constant, J/(mol K)
S	The entropy of a system, J/K
S_m°	Standard molar entropy
t	The time, s
T	Reaction temperature, K
v_s	The surface velocity, m/s
V	The volume, L
X	The conversion, %
α	The reaction orders of NO
β	The reaction orders of NH_3

P.-C. Chiang and X. Gao, *Air Pollution Control and Design*,
https://doi.org/10.1007/978-981-13-7488-3_3

γ	The reaction orders of O_2
ν_B	The stoichiometric number
ξ	The reaction extend, mol
Ω	The number of microscopic configurations, ≥ 1
η	The designed deNO_x efficiency

3.1 Introduction

3.1.1 Importance and Significance

Nitrogen oxides (NO_x) refer to various species of oxides of nitrogen [1]. Due to the various valence states of nitrogen, there are several types of NO_x, including nitrous oxide (N_2O), nitric oxide (NO), nitrogen trioxide (N_2O_3), nitrogen dioxide (NO_2), and nitrogen pentoxide (N_2O_5). NO_x can result in photochemical smog and nitric acid rain, as shown in Fig. 3.1. From the point of view of air pollutants, two oxides of nitrogen are elements of concern with regard to environmental quality and public health: NO and NO_2 [2]. In many applications relating to high-temperature heating, the NO_x emitted from the exhaust stack is mainly in the form of nitric oxide (NO). NO is a kind of colorless gas and can contribute to the formation of ozone. It also exists as the predominant NO_x formed during combustion processes. NO will be oxidized to NO_2 in the atmosphere. NO is harmful to human health and may result in inflammation of the eyes and throat, headache, nausea, tightness of the chest, and gradual loss of strength. Exposure to NO for a long period may also lead to difficulty in breathing, severe coughing, and cyanosis.

NO_2 is a kind of reddish-brown gas, which can result in opacity in flue gas plumes from stacks [2]. NO_2 is not only an important primary air pollutant, but also it

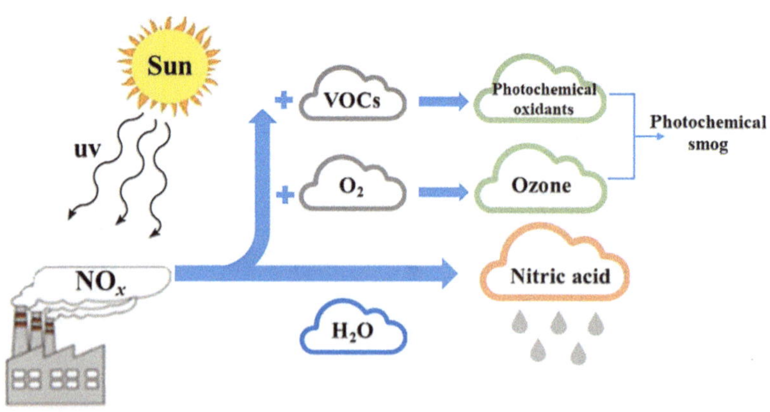

Fig. 3.1 Sketch diagram of various processes which determine the fate of NOx in the atmosphere

reacts in the atmosphere. NO_2 can cause many severe diseases such as delayed chemical pneumonitis and pulmonary edema. The NO_2 vapor is a strong irritant to the pulmonary tract. Severe symptoms may be developed, including lassitude, cyanosis, irregular respiration, breathing difficulty, and even death due to pulmonary edema. Intermittent low-level NO_2 exposure can cause the debility of the liver, kidney, spleen, red blood cells, and immune system. Additionally, the formation of acid rain can also be via the reaction of NO_2 with water in the atmosphere. Long-term or repeated exposure to NO_2 can lead to a permanent decrease in pulmonary function [3]. Finally, NO_x and volatile organic compounds (VOCs) can also react photochemically to produce photochemical smog, including ozone, NO_x, peroxybenzoyl nitrate (PBN), peroxyacetyl nitrate (PAN), and other trace oxidizing agents [2].

3.1.2 *Regulations and Standards*

With increasing concerns about adverse public health impacts of nitrogen oxides, governments around the world began to promulgate more and more strict emission standards and regulations, as shown in Fig. 3.2. As mentioned in Sect. 2.1.2. of Chap. 2, the United States Clean Air Act (CAA) requires the Environmental Protection Agency (EPA) to establish national ambient air quality standards (NAAQS) for nitrogen oxides and five other pollutants (O_3, PM, CO, $SO_{x,}$, and Pb) in consideration of their harmful impacts to public health and the environment. In 1971, under the US 36 FR 8186 regulation, the primary and secondary standards of NO_2 were set at 53 ppb (or 100 $\mu g/m^3$) as the annual average. In 2010, under 75 FR 6474

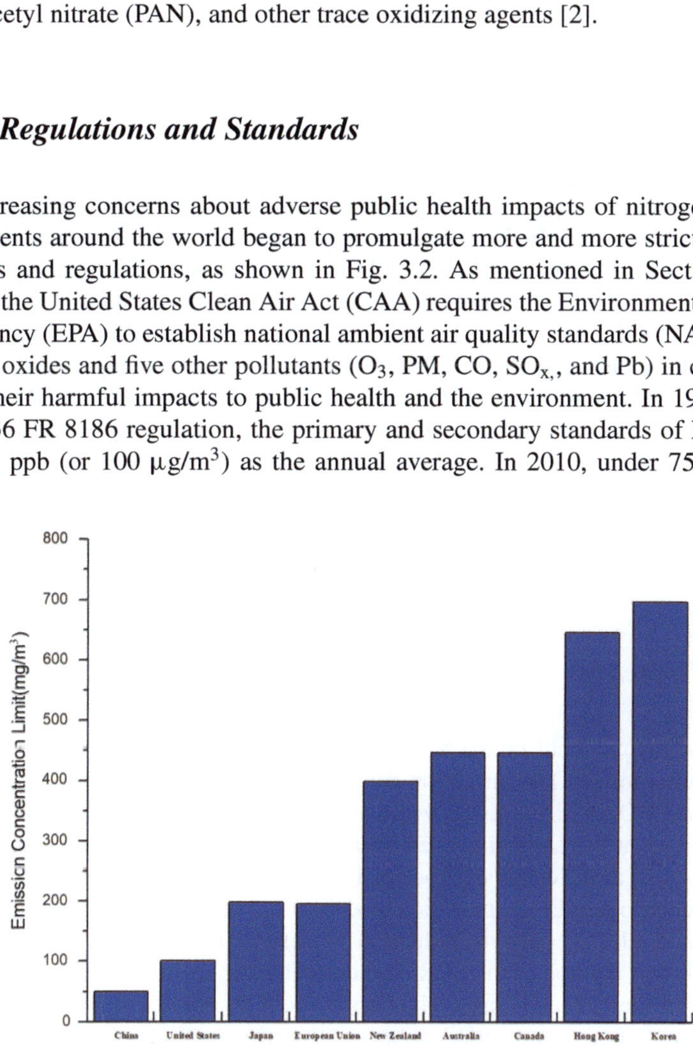

Fig. 3.2 NO_x emission standards in some countries and regions

regulation, the primary standard was revised to 100 ppb. However, the secondary standard remained at 53 ppb as an annual average. In 2001, the European Union revised the emission limits of pollutants and issued a large combustion plant directive (2001/80/EC) which limits the emission concentration of nitrogen oxides at 200 mg/Nm^3. In 2011, the Ministry of Environmental Protection of China issued emission standards for air pollutants for thermal power plants and proposed the emission concentrations of nitrogen oxides not to exceed 100 mg/Nm^3. At the same time, the Chinese government required the use of ultra-low emission technology to reduce the emission concentration of nitrogen oxides to be less than 50 mg/Nm^3. These strict emission standards of nitrogen oxides imply that the development of nitrogen oxides control technology is imminent.

3.1.3 Technology Development

In many combustion processes, high temperatures and high air concentrations are required for operation. In the combustion chamber, the products of combustion may have long residence times. These combustion conditions have advantages of high thermal efficiencies and product throughput rates. However, these conditions also contribute to NO_x formation. In order to reduce NO_x emissions, some stringent standards have been established. Different NO_x control technologies will be used in order to meet the new emission standards. In the 1970s, SCR and SNCR technology were first developed in Japan and the U.S., respectively. Over the next few decades, researchers have carried out a series of studies for technological innovation. In the 1990s, low-NO_x burners rapidly developed and industrialized. In order to meet the increasingly stringent emission standard, low-NOx burners and SCR technology are being routinely combined to reduce NO_x emission.

3.1.4 Objectives of Chapter

In this chapter, we focus on the NO_x formation mechanisms, control principles, and control technologies. The formation mechanisms including thermal NO_x, fuel NO_x, and prompt NO_x will be discussed in detail. For the NO_x control principles, we are predominantly introducing the fundamentals of thermodynamics and reaction kinetics as well as the catalysis process. Three types of the most widely applied NO_x control technologies, i.e., combustion modification, selective non-catalytic reduction (SNCR), and selective catalytic reduction (SCR) will be discussed. We will also illustrate the design method of the SCR reactor, which is the most important component of an SCR system. Finally, an evaluation of the NO_x control technologies from the point of view of environmental, energy, engineering, and economic (4E) aspects is provided.

3.2 NO_x Formation

As a kind of air pollutants, NO_x represents mainly nitric oxide (NO) and nitrogen dioxide (NO_2). In the flue gas emitted from fossil fuel combustion, 90–95% of the total NO_x is NO, while NO_2 accounts for about 5–10% due to the kinetic limitation of NO oxidation to NO_2 [4]. During the combustion process, NO_x is generated from three mechanisms [2] (Fig. 3.3).

(1) Thermal NO_x. Thermal NO_x comes from the thermal dissociation and the following reaction of N_2 with O_2 in the air;
(2) Fuel NO_x. Fuel NO_x arises from the evolution and reaction of fuel-bound nitrogen compounds with oxygen;
(3) Prompt NO_x. Prompt NO_x involves the formation of HCN intermediate, which is subsequently oxidized to NO.

For natural gas and most distillate oils, the formed NO_x is the most thermal NO_x since they do not have chemically bound fuel nitrogen. For residual oils and coals, NO_x is formed by all three mechanisms since both have fuel-bound nitrogen. The prompt NO_x formation becomes significant in fuel-rich flames [4]. These mechanisms are described below in more detail.

3.2.1 Thermal NO_x

Thermal NO_x is generated via the reaction of N_2 with O_2 at high temperatures. It is given by Eqs. (3.1, 3.2, and 3.3). The forward and backward reaction constants shown in the above equations are listed in Table 3.1 [5].

The rate-determining step for thermal NO_x formation is Eq. (3.1). The formation of NO_x strongly depends on the combustion temperature. The reaction rate and temperature exhibit an exponential relationship. Additionally, the formation of thermal NO_x is closely related to the residual time in the high-temperature zones.

The kinetic constants of the reactions associated with Eqs. (3.1) and (3.2) in Table 3.1 could be expressed as in Eq. (3.4):

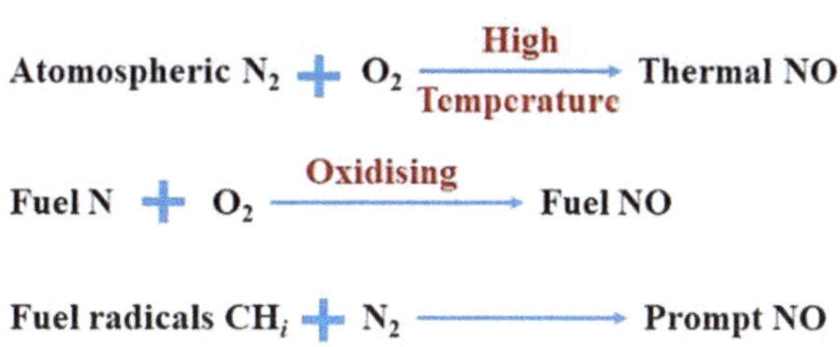

Fig. 3.3 NOx formation in the combustion process

Table 3.1 Reaction rate constants and temperature regions [5]

Reaction equations	Reaction rate constants/m^3 (mol s)$^{-1}$		Temperature regions/K
$N_2 + O \rightleftharpoons NO + N$ (3.1)	k_1 (forward)	$7.6 \times 10^{13}.\exp[-3800/T]$	2000–5000
	k_1' (backward)	1.6×10^{13}	300–5000
$O_2 + N \rightleftharpoons NO + O$ (3.2)	k_2 (forward)	6.4×10^9 T $\exp[-3150/T]$	300–3000
	k_2' (backward)	1.5×10^9 T $\exp[-19500/T]$	1000–3000
$N + OH \rightleftharpoons NO + H$ (3.3)	k_3 (forward)	1.0×10^{14}	300–2500
	k_3' (backward)	2.0×10^{14} $\exp[-23650/T]$	2200–4500

$$\frac{dc_{NO}}{dt} = k_1 c_{N_2} c_O - k_1' c_{NO} c_N + k_2 c_N c_{O_2} - k_2' c_{NO} c_O \tag{3.4}$$

where

k_1 and k_2 = the forward reaction constants;
k_1' and k_2' = the backward reaction constant;
c = the concentration of the specific substance, mol/cm^3;
t = the time, s.

In the reactions, the concentration of N atoms, the intermediate products, is 10^{-5}–10^{-8} times lower than that of NO. Thus, it could be assumed that the concentration of N atoms remains almost unchanged in the entire reaction. Therefore, Eq. (3.4) could be expressed as Eq. (3.5):

$$\frac{dc_N}{dt} = k_1 c_{N_2} c_O - k_1' c_{NO} c_N - k_2 c_N c_{O_2} + k_2' c_{NO} c_O = 0 \tag{3.5}$$

Therefore, the concentration of N_2 can be calculated as in Eq. (3.6):

$$c_{N_2} = \frac{k_1 c_{N_2} c_O + k_2' c_{NO} c_O}{k_1' c_{NO} + k_2 c_{O_2}} \tag{3.6}$$

When Eq. (3.6) is introduced to Eq. (3.4), we can obtain Eq. (3.7):

$$\frac{dc_{NO}}{dt} = \frac{2k_1 k_2 c_O c_{O_2} c_{N_2} - k_1' k_2' c_{NO}^2 c_{NO}}{k_2 c_{O_2} + k_1' c_{NO}} \tag{3.7}$$

As in real flue gas, the concentration of O_2 is much higher than that of NO, i.e., $c_{NO} \ll c_{O_2}$ and k_2 and k_1' are in the same order; therefore, $k_1' c_{NO} \ll k_2 c_{O_2}$. Equation (3.7) could be simplified as Eq. (3.8):

$$\frac{dc_{NO}}{dt} = 2k_1 c_{N_2} c_O \tag{3.8}$$

Provided that the dissociation of O_2 exists in an equilibrium state, as shown in Eqs. (3.9) and (3.10):

$$O_2 \rightleftharpoons O + O \tag{3.9}$$

$$c_O = K \cdot c_{O_2}^{1/2} \tag{3.10}$$

where
K = the equilibrium constant.

When Eq. (3.10) is introduced to Eq. (3.8), Eq. (3.11) is obtained as follows:

$$\frac{dc_{NO}}{dt} = 2Kk_1 c_{N_2} c_{O_2}^{1/2} \tag{3.11}$$

And through experiments, Eq. (3.12) could be obtained

$$k = 2Kk_1 = 3 \times 10^{14} \times \exp\left(-\frac{54200}{RT}\right) \tag{3.12}$$

where
k = the constant;
R = the universal gas constant, 8.314 J/(mol K).

The relationship between the formation rate of thermal NO_x and temperature could be further simplified as in Eq. (3.13):

$$\frac{dc_{NO}}{dt} \propto \exp(0.009RT) \tag{3.13}$$

3.2.2 *Fuel NO_x*

Fuel NO_x is formed through the oxidation of nitrogen compounds, which generally exist in the fuel. All the coal, fossil oil, and natural gas contain a certain amount of nitrogen compounds. Taking coal as an example, the nitrogen content in coal varies typically between 0.5 and 2.0% [6]. Volatile N compounds (HCN, NH_i) formed during the combustion process, together with char-N, react with radicals (O, OH, and O_2) to form fuel NO_x (mainly NO). The as-formed NO then reacts with volatile HCN, NH_i to produce N_2, as illustrated in Fig. 3.4. Combustion temperatures, excess air coefficient,

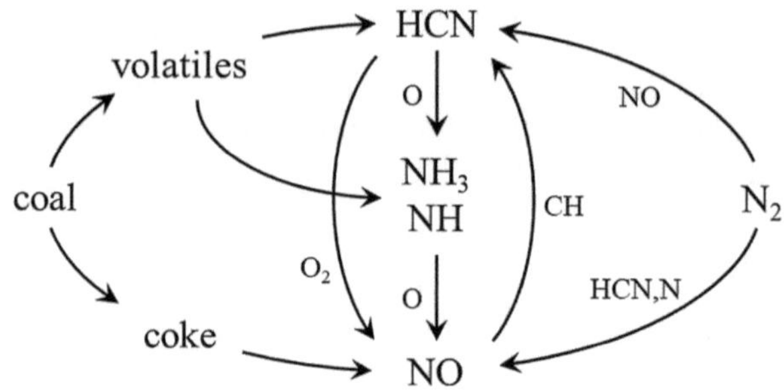

Fig. 3.4 Reaction pathway of fuel N compound to NO [7]

types of coal, particle sizes, and fuel–air conditions can influence the distribution of radicals in the combustion chamber, which determines fuel NO_x formation. The volatile N accounts for most of fuel NO_x formation, while char-N for a few.

The formation process of fuel NO is quite complicated, which is related to several radicals, including OH, O, H, NH, NH_2, NCO, CH_i, and so on. Volatile HCH can react with NH_3 according to Eqs. (3.14)–(3.18):

$$2HCN + 2.5O_2 \rightarrow 2NO + 2CO + H_2O \tag{3.14}$$

$$HCN + H_2O \rightarrow CO + NH_3 \tag{3.15}$$

$$2HCN + 3NO \rightarrow 2.5N_2 + 2CO + H_2O \tag{3.16}$$

$$3NO + NH_3 \rightarrow 3H_2O + 2.5N_2 \tag{3.17}$$

$$2.5O_2 + NH_3 \rightarrow 3H_2O + 2NO \tag{3.18}$$

NO is reduced to N_2 over the surface of coke, as shown in Eq. (3.19):

$$C + H_2O \rightarrow 0.5N_2 + CO \tag{3.19}$$

CH_i compounds can also react with NO, as shown in Eq. (3.20):

$$CH_i + 2NO \rightarrow CO + N_2 + OH + \cdots \tag{3.20}$$

During the coal combustion process, the oxidation reaction between volatile HCN and NH_3 is more complicated than the reaction equations mentioned above, which are related to several intermediates and NO reduction reaction.

As mentioned above, the formation and destruction of fuel NO_x is a complicated oxidation–reduction reaction process. There are more than 250 reaction routines and equations. During the combustion process, the destruction of NO_x would occur because of the variations in the local atmosphere. Additionally, intermediates would sometimes interact with each other. The quantity of NO_x coming from fuel N compound is independent of the types of N-containing organic matters, while it is closely related to volatiles and excess air coefficient instead of temperature, as shown in Figs. 3.5, 3.6 and 3.7 [5].

Given that the excess air coefficient is higher than 1, an increase in the volatile content would lead to an increase in the formation amount of NO_x. When the excess

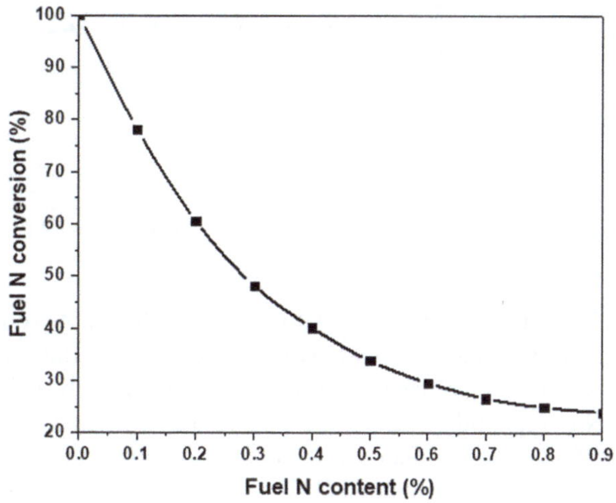

Fig. 3.5 Relationship between conversion efficiency of the fuel N compounds and N content of the fuel [5]

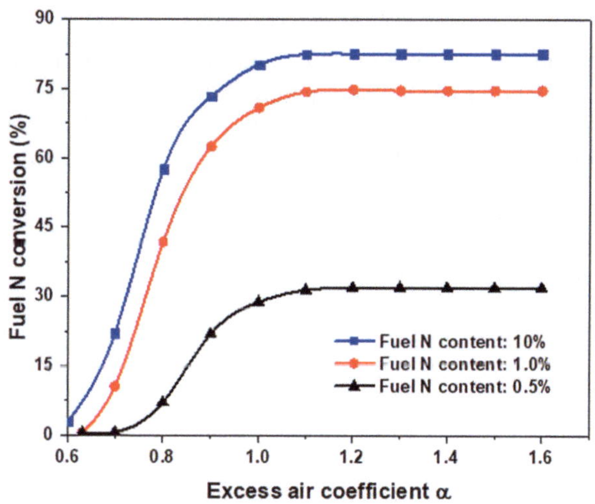

Fig. 3.6 Relationship between conversion efficiency of the fuel N compounds and the excess air coefficient α [5]

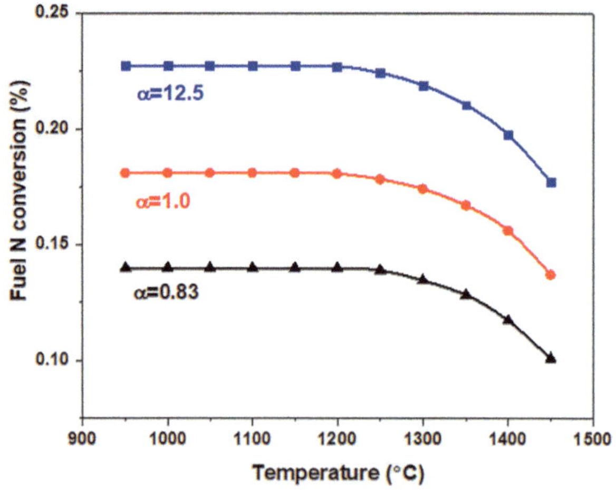

Fig. 3.7 Relationship between the conversion efficiency of the fuel N compounds and the combustion temperature [5]

air coefficient is lower than 1, a higher volatile content would not promote the formation of NO_x, because the rapid combustion of volatiles would reduce the local oxygen amount, which would have a negative effect on the formation of NO_x. When the content of fuel N compounds is fixed, the conversion reaches a stable value. With an excess air coefficient higher than 1.1, the conversion dramatically decreases with the excess air coefficient lower than 1 [5].

3.2.3 *Prompt NO_x*

In the formation process of prompt NO_x, intermediate HCN is produced at first, followed by its oxidation to NO. The reaction routine is illustrated in Fig. 3.8.

The reaction process of prompt NO_x could be simplified as the following two steps, as shown in Eqs. (3.21 and 3.22):

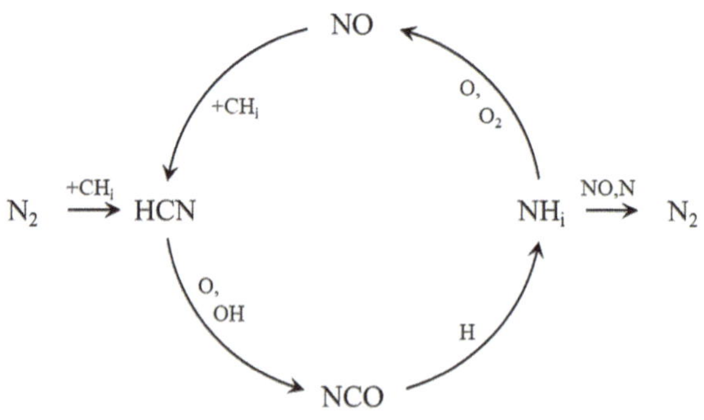

Fig. 3.8 Formation mechanism of prompt NOx [4]

$$CH + N_2 \rightarrow HCN + N \tag{3.21}$$

$$CH_2 + N_2 \rightarrow HCN + NH \tag{3.22}$$

The formation of prompt NO_x could be expressed using Eqs. (3.21) and (3.22) through the flame diagnosis experiments. The concentration of fast NO_x is in proportion to the concentration of CH. Equation (3.21) is regarded as the key step in the formation process of prompt NO_x. The active group CH comes from the following reaction process shown in Eqs. (3.23, 3.24 and 3.25):

$$CH_3 + X \rightarrow CH_2 + HX \tag{3.23}$$

$$CH_2 + X \rightarrow CH + HX \tag{3.24}$$

$$CH + X \rightarrow C + HX \tag{3.25}$$

where

X = H or OH.

The activation energies of the reaction (3.21) and (3.22) are quite low, suggesting that the reaction rates are rather high. The as-formed HCN reacts with O and OH groups to form NO, as shown in Eqs. (3.26, 3.27, 3.28, 3.29, 3.30, 3.31 and 3.32):

$$HCN + OH \rightarrow CN + H_2O \tag{3.26}$$

$$CN + O_2 \rightarrow CO + NO \tag{3.27}$$

$$CN + O \rightarrow CO + N \tag{3.28}$$

$$HN + OH \rightarrow N + H_2O \tag{3.29}$$

$$NH + O \rightarrow NO + H \tag{3.30}$$

$$N + OH \rightarrow NO + H \tag{3.31}$$

$$N + O_2 \rightarrow NO + H \tag{3.32}$$

In the combustion processes at lower temperatures, prompt NO_x is considered as an important NO_x formation mechanism. However, in most industrial combustion processes, due to high temperatures, it is much less important compared with the thermal NO_x formation mechanism [1].

3.3 Control Principles

3.3.1 Thermodynamics

The laws of thermodynamics help us to understand why some reactions can occur while others cannot. It can also tell us whether a reaction is absorbing heat (endothermic reaction) or releasing heat (exothermic reaction). When 1 mol product is produced for a chemical reaction process under a standard state and temperature T, the value of enthalpy change is defined as the standard molar enthalpy of reaction, denoted as $\Delta_r H_m^\circ(T)$, kJ/mol. Therefore, considering a chemical reaction Eq. (3.33) at temperature T, the $\Delta_r H_m^\circ(T)$ can be calculated using Eq. (3.34).

$$\sum_{i=1}^{n} x_i A_i \rightarrow \sum_{j=1}^{m} y_j B_j \tag{3.33}$$

$$\Delta_r H_m^\circ(T) = \sum_{j=1}^{m} y_j \Delta_f H_m^\circ(B_j, T) - \sum_{i=1}^{n} x_i \Delta_f H_m^\circ(A_i, T) \tag{3.34}$$

For a chemical system, entropy is a measure of the randomness or disorder of this system. It is defined as Eq. (3.35):

$$S = k \ln \Omega \tag{3.35}$$

where

S = the entropy of a system, J/K;
k = the Boltzmann constant, 1.38×10^{-23} J/K;
Ω = the number of microscopic configurations, ≥ 1.

Under a standard state at a specific temperature T, the entropy of 1 mol of substance is denoted as the standard molar entropy, denoted as S_m° (substance, T), J/mol·K. At 298.15 K, the S_m° values of the substances relating to NO_x reduction at 298.15 K are summarized in Table 3.2.

Similarly, with $\Delta_r H_m^\circ$, under a standard state and temperature T, the value of entropy changes when the reaction extend of 1 mol is defined as the standard molar entropy of reaction, denoted as $\Delta_r S_m^\circ(T)$, kJ/mol·K. Therefore, considering reaction (3.33), the $\Delta_r S_m^\circ(T)$ can be calculated as in Eq. (3.36):

$$\Delta_r S_m^\circ(T) = \sum_{j=1}^{m} y_j S_m^\circ(B_j, T) - \sum_{i=1}^{n} x_i S_m^\circ(A_i, T) \tag{3.36}$$

In thermodynamics, the Gibbs free energy is defined by Eq. (3.37). The Gibbs free energy can be used to decide the direction of a chemical reaction and to determine

Table 3.2 The standard molar enthalpies of formation, standard molar entropies, and standard molar Gibbs function for the formation of several gases at 298.15 K

Substance	$\Delta_f H_m^\circ / kJ\ mol^{-1}$	$S_m^\circ / J\ mol^{-1}\ K^{-1}$	$\Delta_f G_m^\circ / kJ\ mol^{-1}$
NO	91.3	210.8	87.6
NO_2	33.2	240.1	51.3
NH_3	−45.9	192.8	−16.4
H_2	0	130.7	0
CH_4	−74.6	186.3	−50.5
CO	−110.5	197.7	−137.2
N_2	0	191.6	0
H_2O (g)	−241.8	188.8	−228.6
H_2O (l)	−285.8	70.0	−237.1
CO_2	−393.5	213.8	−394.4
O_2	0	205.2	0

the equilibrium concentrations or phases of chemical species in equilibrium.

$$G = H - TS \tag{3.37}$$

When, at constant temperature and reversible conditions, the change of Gibbs free energy between two states is described using Eq. (3.38):

$$\Delta_r G = \Delta_r H - T\Delta_r S \tag{3.38}$$

$\Delta_r G$ can be used to determine whether a reversible reaction (at constant temperature and pressure) can take place spontaneously. If $\Delta_r G < 0$, the reaction is a spontaneous process. If $\Delta_r G = 0$, the reaction reaches equilibrium. If $\Delta_r G > 0$, the reaction cannot take place spontaneously although its reverse reaction can.

Similarly, with $\Delta_f H_m^\circ$, under a standard state and temperature T, the Gibbs function for the formation of 1 mol of substance is defined as the standard molar Gibbs function for formation, denoted as $\Delta_f G_m^\circ$ (substance, T), kJ/mol. The $\Delta_f G_m^\circ$ values of the substances relating to NO_x reduction at 298.15 K are summarized in Table 3.2.

For example, for a NO reduction reaction using CO as a reducing agent $CO + NO \rightarrow \frac{1}{2}N_2 + CO_2$, the $\Delta_r H_m^\circ$ at 298.15 K is calculated to be $\frac{1}{2} \times 0 + (-393.5) - (-110.5) - 91.3 = -374.3$kJ/mol. We know, therefore, that this reaction is an exothermic reaction.

The $\Delta_r S_m^\circ$ at 298.15 K is calculated to be $\frac{1}{2} \times 191.6 + 213.8 - 197.7 - 210.8 = -98.9$kJ/mol $\cdot K$. And, the $\Delta_r G_m^\circ$ at 298.15 K is calculated to be $-374.3 - 298.15 \times \left(-98.9 \times 10^{-3}\right) = -344.81$kJ/mol < 0. Therefore, we know that this reaction at 298.15 K is a spontaneous process.

As illustrated above, temperature plays a decisive role in the thermal NO_x formation rate. When the temperature is below 1800 K, the amount of thermal NO_x produced is quite low. When the temperature is higher than 1800 K, the reaction rate increased by 6 –7 times as the temperature increases by 100 K.

The concentration of O_2 also plays an important role in the formation of thermal NO_x. As shown in Fig. 3.10, the increase in the concentration of O_2 initially leads to the increased formation amount of thermal NO_x. Further increasing the O_2 concentration, the formation rate of thermal NO_x decreases due to the decrease in the combustion temperature [8].

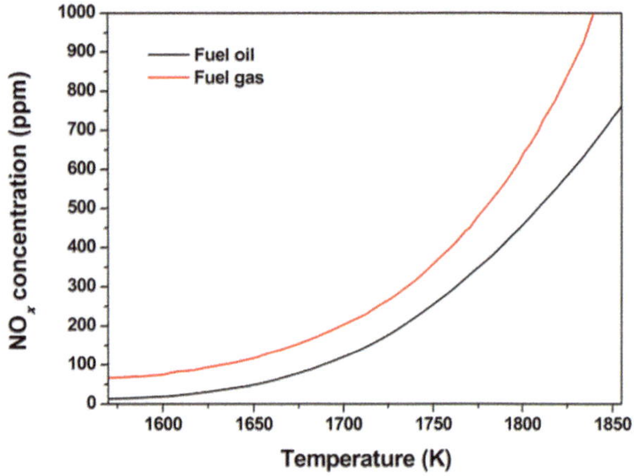

Fig. 3.9 Relationship between the formation of thermal NOx and temperature [7]. (Conditions: reaction time = 5 s)

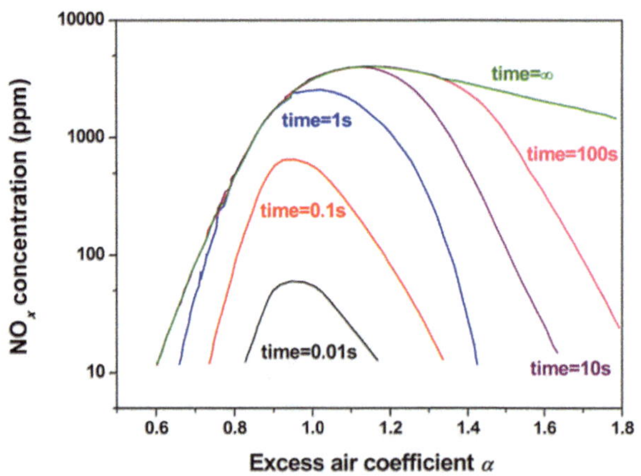

Fig. 3.10 Relationship between the formation of thermal NOx and excess air coefficient [8]

3.3.2 Reaction Kinetics

It is noteworthy that thermodynamics tells whether a chemical reaction can take place spontaneously at a standard state. However, it cannot tell the rate of a reaction. Taking the reaction $CO + NO \rightarrow \frac{1}{2}N_2 + CO_2$ as an example, we know it can take place spontaneously and it is an exothermic reaction. However, in fact, the reaction rate is rather slow. A catalyst should be used to accelerate the reaction. Reaction kinetics can be introduced to solve this problem.

For a chemical reaction process, as shown in Eq. (3.39), the reaction extend is defined as in Eq. (3.40):

$$0 = \sum_B \nu_B B \tag{3.39}$$

$$\xi = \frac{dn_B}{\nu_B} \tag{3.40}$$

where
ν_B = the stoichiometric number;
ξ = the reaction extend, mol;
n_B = the number of moles, mol;

Hence, the reaction rate per unit volume could be defined as in Eq. (3.41):

$$r = \frac{1}{V}\frac{d\xi}{dt} = \frac{dc_B}{dt} \tag{3.41}$$

where
r = the reaction rate, mol/L·s;
V = the volume, L;
t = the reaction time, s;
c_B = the concentration, mol/L.

From Eq. (3.41), we can see that the reaction rate keeps changing while the reaction is in process. For the reactant, the concentration is decreasing. While for the product, the concentration increases. Here, we will mainly focus on the reaction rate of the reactant. For an elemental reaction Eq. (3.42), the reaction rate can be expressed as Eq. (3.43). This is called the law of mass action.

$$pP + qQ \rightarrow products \tag{3.42}$$

$$r = kc_P^p c_Q^q \tag{3.43}$$

where

k = the reaction rate constant;

p, q = the reaction order of the substance P and Q, respectively.

If the reaction is not an elemental reaction, the reaction order will not be the stoichiometric number and should be determined by experiments.

Now, we consider a typical reaction for NO_x abatement, i.e., the selective catalytic reduction (SCR), as shown in Eq. (3.44):

$$4NH_3 + 4NO + O_2 \rightarrow 4N_2 + 6H_2O \tag{3.44}$$

The reaction rate r_{NO} for such an SCR process can be supposed to be dependent on the concentrations of the reactants (NO, NH_3, and O_2). Accordingly, the reaction rate can be expressed as Eq. (3.45):

$$r_{NO} = k_{NO} c_{NO}^{\alpha} c_{NH_3}^{\beta} c_{O_2}^{\gamma} \tag{3.45}$$

where

r_{NO} = the reaction rate of SCR reaction, mol/L·s;

k_{NO} = the reaction rate constant of SCR reaction;

c = the concentration of the chemical (NO, NH_3, O_2), mol/cm^3;

α, β, γ = the reaction orders of NO, NH_3, and O_2.

Several studies were performed to measure the reaction orders. Different data are obtained over different catalysts and under different reaction conditions. However, in many practical conditions, the reaction order with respect to NO, NH_3, and O_2 could be 1, 0, 0, respectively, which are suitable for most of the actual conditions. Therefore, the kinetics of an SCR reaction can be modeled using simple power-law kinetics, as shown in Eq. (3.46) [9]:

$$r_{NO} = k_{NO} c_{NO} \tag{3.46}$$

Generally, a reaction rate constant could be used to compare the reactivity at the same temperature. For the reaction over different catalysts, it is meaningful to use the reaction rate constant to compare the reactivity, solely if the reaction rate constant equation has the same form. As discussed above, the NH_3-SCR reaction could be regarded as a first-order reaction. Therefore, the reaction rate constant could be calculated using Eq. (3.47):

$$k_{NO} = -\left[\frac{F_{NO}}{m_{cat} c_{NO}}\right] \ln(1 - X) \tag{3.47}$$

where

k_{NO} = the reaction rate constant of SCR reaction, L/(g.s);

F_{NO} = the NO flow rate, mol/s;
m_{cat} = the mass of the catalyst, g;
c_{NO} = the mole concentration of NO in the flue gas, mol/L;
X = the NO conversion, %.

In a chemical reaction process, the reactant molecules should initially get activated by obtaining a minimum definite amount of energy. This energy is the activation energy. Its relationship with the reaction rate constant and the reaction temperature is known as the Arrhenius equation, as shown in Eq. (3.48):

$$k = A \cdot \exp\left(-\frac{E_a}{RT}\right) \tag{3.48}$$

where
k = the reaction rate constant;
A = pre-exponential factor;
E_a = the activation energy, J/mol;
R = molar gas constant, 8.314 J/ (mol·K);
T = reaction temperature, K.

Based on Eq. (3.48), the reaction rate constant at a certain temperature is determined by the pre-exponential factor A and the activation energy E_a. The primary condition of the synergetic interaction between two molecules is the collision. Although the frequency of the collision between the two molecules is very high, only a few of these collisions are effective. E_a reflects the energy demand for the occurrence of effective collision of the two molecules. For the non-elementary reactions, E_a has no physical meaning. It is a particular combination of the activation energies of each elementary reaction in the overall reaction. In general, one reaction proceeding over a catalyst has high activation energy, indicating this catalyst has low reactivity and vice versa. The activation energy of an overall reaction is always used to compare the reactivity.

3.3.3 Catalytic Reaction

The SCR reaction is a catalysis process, where gaseous reactants react at the catalyst surface. A catalysis process mainly includes the following steps, as shown in Fig. 3.11:

(1) Reactants diffuse from the gas main phase to the outer surface of catalysts via gas film (external diffusion);
(2) Reactants diffuse to the inner surface through the micropores of catalysts (internal diffusion);
(3) Reactants adsorb onto the catalyst inner surface (adsorption);
(4) Chemical reactions take place on the catalyst inner surface (surface reaction);

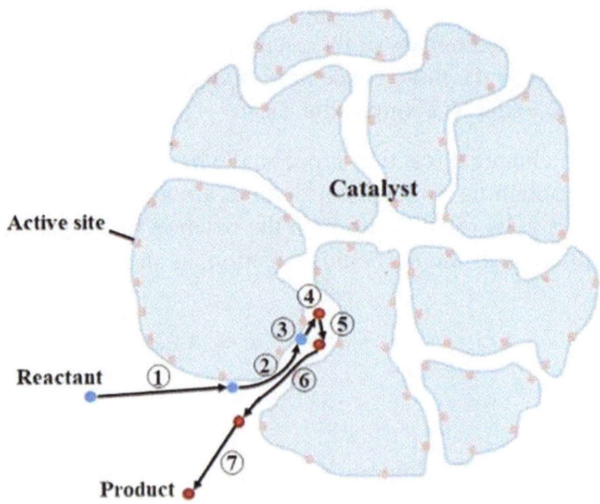

Fig. 3.11 Catalysis reaction process: (1) external diffusion; (2) internal diffusion; (3) adsorption; (4) surface reaction; (5) desorption; (6) internal diffusion; (7) external diffusion

(5) Products desorb from the catalyst inner surface (desorption);
(6) Products diffuse to the outer surface through the micropores of catalysts (internal diffusion);
(7) Products diffuse from the catalyst outer surface to the gas main phase (external diffusion).

Among the previously mentioned seven steps, step 1 and step 7 are the external diffusion process, which is determined by the gas flow conditions. Step 2 and step 6 are the internal diffusion process, which is affected by the micropore structures of the catalysts. Step 3, step 4, and step 5 are the surface chemical reaction kinetics process, which is influenced by chemical reaction, catalyst activity, temperature, gas pressure, etc. Different catalysis reactions might be controlled by certain steps: (1) external diffusion control, (2) internal diffusion control, and (3) chemical kinetics control.

External diffusion resistance mainly comes from the gas film, while internal diffusion resistance is due to the sizes of the pore structures. The resistance in the chemical reaction process is related to the concentrations of certain gases and the reaction rate. The control process would, therefore, be varied through changing the catalysis reaction conditions. For external diffusion, the flow rate could be increased. For internal diffusion, catalyst particle size could be decreased. For chemical kinetics control, the activity of the catalyst could be enhanced. Since the chemical reaction in the SCR process could be complete in a short time, diffusion is the main control factor, which is affected by the uniform mixing of gas and NH_3, uniform distribution of the flue gases on the catalyst surface, and the physicochemical properties of the catalysts.

Conversion is the apparent indicator to express the reactivity. Thus, conversion is used to compare the apparent activity of the catalysts, which is defined as Eq. (3.49):

$$\chi_A = \frac{\text{reacted amount of substance A}}{\text{initial amount of substance A}} \tag{3.49}$$

The conversion, which is used to evaluate the SCR catalytic activity, could be expressed as Eq. (3.50):

$$NO\ conversion\ (\%) = \left(1 - \frac{NO_{x\ (out)}}{NO_{x\ (in)}}\right) \times 100\% \tag{3.50}$$

When using this parameter, the contact time of reactant and catalyst should be clarified. The concept of the gas hourly space velocity should, therefore, be introduced. In the flowing systems, the flow rate of the reactant divided by the catalyst volume is named as volume space velocity, the unit of which is s^{-1} or h^{-1}.

Turnover frequency (TOF) is also used to give an intrinsic insight into the catalyst activity. TOF is related to the number of reactant molecules converted per active site per second [10, 11]. This value is the ratio of the reaction rate of a given catalysis reaction system and the exact number of active centers participating in the reaction. Turnover frequency reflects the intrinsic activity of catalysts, the measurements of which need a clear insight into the structure and concentration of the catalyst active center.

3.4 Control Technologies

3.4.1 Combustion Modification

Low excess air operation

Excess airflow for combustion has been correlated to the concentration of NO_x generated. As a result, one basic method to reduce NO_x emissions is to reduce the excess air level to the burners. In reducing fuel NO_x formation, low excess air operation is effective, but it is, however, limited in reducing the emissions of thermal NO_x.

Figure 3.12 illustrates the relationship between the formation amount of NO_x and oxygen content in the flue gas. Normally, this level is set to some constraints including flame length, flame stability, or CO formation [12]. It should be noted that low excess air technology will increase the number of pollutants including carbon monoxide, hydrocarbons, and carbon black. The content of fuel in the fly ash may also increase so that the combustion efficiency drops [5]. In real operation, the adjustments of fuel injector positions, air registers, and overfire air dampers can help to reduce the minimum excess air level with adequate air/fuel distribution [4].

Most power plants are operating at the excess air coefficient of 1.17–1.20 (O_2 concentration: 3.5–4.0%). Currently, the concentration of CO is about 30 to 40 ppm. Provided that the O_2 content decreases to 3%, the concentration of CO dramatically increases. That not only causes the fuel uncomplete utilization, but additionally

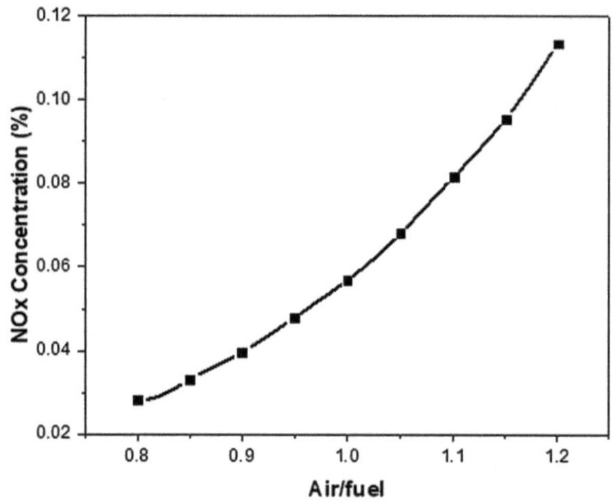

Fig. 3.12 Relationship between the formation of NOx and the ratio of air/fuel [4]

causes the slag and corrosion in the furnace. Therefore, an O_2 concentration higher than 3% or a CO concentration lower than 2×10^{-4} could be regarded as the guideline for choosing the minimum excess air coefficient [5]. Combustion air tuning can reduce an NO_x emission of 0 to 25% [2], depending on the prior operating conditions.

Flue gas recirculation

In this method, a fraction of the flue gas is separated from the main flow, commonly after the PM control system. The circulated flue gas is mixed with combustion air in the burner, primary air, or other staging air in the furnace [6]. The flue gas acts to dilute, so flue gas recirculation reduces NO_x emission [12]. Additionally, the reduced partial pressure of oxygen also limits the NO_x formation. The recirculation rate of flue gas is the main parameter affecting the NO_x formation. The ratio of the recirculation flue gas volume to the boiler flue gas volume without recirculation is called the flue gas recirculation rate. The recirculation rate is normally limited to 15–25% [12] considering the stability of flame. Flue gas recirculation technology can be used alone or in conjunction with other low-NO_x combustion technologies. The scheme of the flue gas recirculation is shown in Fig. 3.13.

Figure 3.14 shows the effect of gas recirculation on the concentration of NO_x emissions for systems with or without air preheating [1]. It can be found that with increasing cycles, NO_x emissions are dramatically reduced. It also should be noted that preheating the inlet air obviously increases NO_x emission concentrations, because the flame temperatures increase as well. While air preheating can increase fuel efficiency as well as flame temperatures, which is beneficial for high-temperature applications like glass melting.

Staged fuel combustion

This technology is also called reburning, which is designed to reduce NO_x without generating CO [4]. It involves the process of passing the burner zone through a

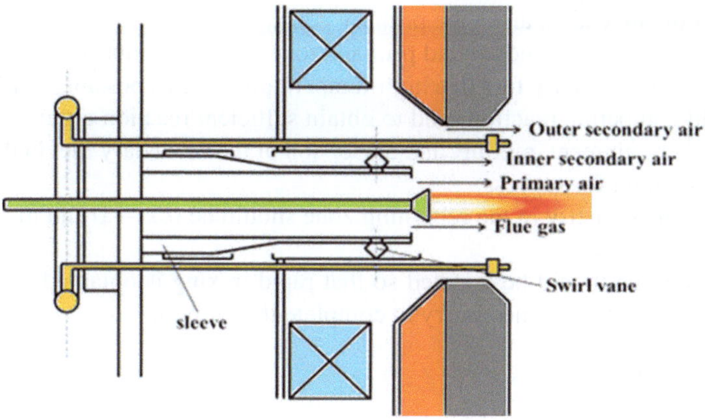

Fig. 3.13 Flue gas recirculating low-NOx combustor

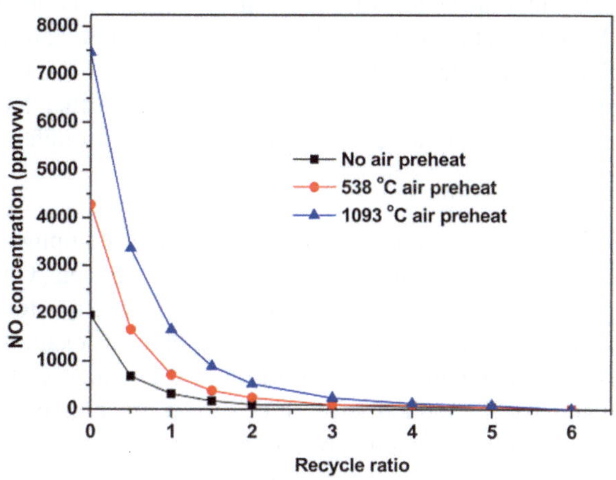

Fig. 3.14 Relationship between NO concentration and the recycle ratio for the stoichiometric combustion of methane with or without air preheating with the recycle gases at 1093 °C [1]

secondary flame or fuel-rich combustion. Combustion can be divided into three zones. In the primary zone, the main fuel, i.e. coal, is combusted in an oxidizing or a slightly reducing atmosphere [6]. In the secondary combustion zone, additional hydrocarbon fuels from outside of the primary flame zone are added. In this zone, NO_x is reduced via reaction with the hydrocarbon radicals [2]. This second zone is called the "reburning zone" with the secondary fuel as the "reburning fuel" [6]. The unburned hydrocarbons generated in this zone are fully combusted in the third stage by the introduction of additional air.

Reburning can be implemented by means of re-distributing the air and fuel via the existing burner pattern or installing additional fuel and air ports above the burner pattern [2]. It is important that the operational conditions are optimized in order to succeed in reburning [6]. The methods to optimize conditions for reburning are summarized as follows [6]:

(1) Minimizing excess air in the primary zone;
(2) Separating the secondary and primary zones so that combustion is complete;
(3) Injecting secondary fuel in a high-temperature zone as possible (>1200 °C) to avoid competing reactions and to obtain sufficient reaction time;
(4) Good entrainment, mixing, and dispersion of the secondary fuel in the flue gas is essential l for success;
(5) The stoichiometry in the reburning zone should be 0.7 –1.0 and the residence time 0.4–1.5 s;
(6) Burnout air should be injected so that rapid mixing is obtained. A residence time of 0.7–0.9 s is necessary to complete the burnout.

Staged air combustion

Fuel is injected into the primary zone and mixed with the primary air. This zone is rich in fuel and partial combustion occurs. NO_x is minimized in this zone as the nitrogen is reduced and converted into reducing agents, which are then oxidized to elemental nitrogen. Additional air is introduced outside of the primary flame where the temperature is lower, keeping the thermodynamic equilibrium NO_x concentration at a low level [2]. Generally, 10–20% of the total combustion air is used for staging, while 20~30% is investigated for further NO_x reduction. In furnaces, the most commonly used air staging method is through over fire air [6].

Low-NO_x burners

A burner is an important part of the boiler equipment and ensures the stable ignition, combustion, and burnout process of the fuels. On the other hand, from the NO_x generation mechanisms, fuel NO_x, which accounts for a large part of NO_x, is formed in the dual ignition process. Taking coal as the example, the special design of burner structure, changing the proportion of the wind coal burner and using other mechanisms for the reduction of NO_x, can as much as possible reduce the concentration of oxygen and the temperature in the fire area, thereby achieving the goal of utmost inhibiting the formation of NO_x. The low-NO_x burner is the one that works most effectively. The low-NO_x combustor can not only ensure the ignition and combustion of coal powder, but also can effectively inhibit the generation of NO_x. Since the low-NO_x burner can suppress the production of NO_x during the ignition process of coal powder, it can achieve lower NO_x emissions, so the low-NO_x combustors have been widely used and developed.

Low-NO_x burners are designed to either stage the air or stage the fuel, as shown in Fig. 3.15 a and b. The principles are similar to staged air or staged fuel in a furnace, as mentioned above. Low-NO_x burners have also been developed for coal-fired applications, where NO_x concentrations are obviously higher than those in liquid and gaseous fuel applications. Through using low-NO_x burners, a NO_x emission level of about 90–140 ppm at 3% O_2 can be achieved [2].

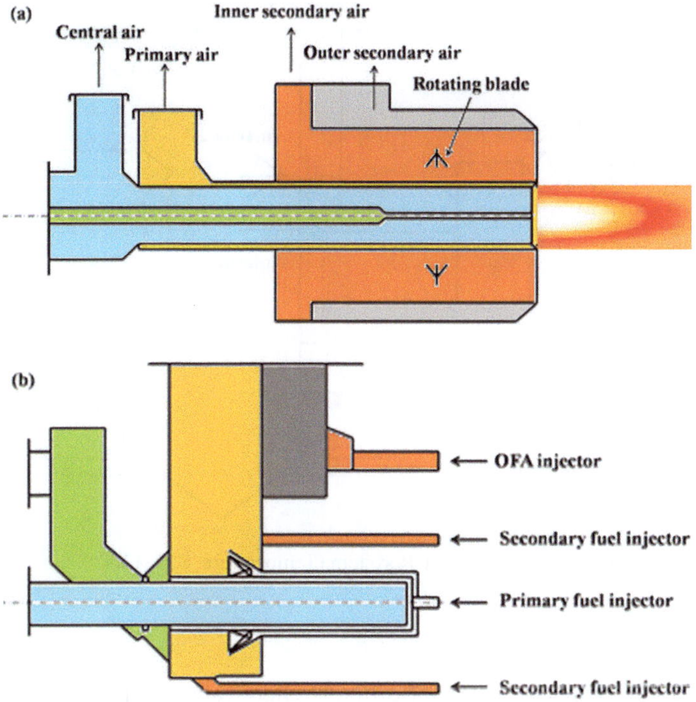

Fig. 3.15 Low-NOx burners. **(a)** staged air low-NOx burner; **(b)** staged fuel low-NOx burner

3.4.2 SNCR Technology

SNCR is an attractive technology for the abatement of NO_x where no catalyst is needed. The reaction process is the same as the SCR reaction, as shown in Eq. (3.44). In a pulverized coal combustion boiler, SNCR accounts for about 30–50% NO_x reduction, while in the fluidized-bed boiler, SNCR accounts for about 60–70% [2]. When urea is used as the reducing agent, NH_3 is generated through the decomposition of urea, as shown in Eqs. (3.51) and (3.52):

$$NH_2 - CO - NH_2 \rightarrow NH_3 + HCNO \tag{3.51}$$

$$HNCO + H_2O \rightarrow NH_3 + CO_2 \tag{3.52}$$

The schematic diagram of SNCR system is shown in Fig. 3.16. In designing SNCR systems, the selection of the reductant temperature, the ratio of NH_3/NO_x, and the use of additives are the three main factors. Among them, the most important factor is the selection of the reductant injection point which is closely related to the reaction temperature.

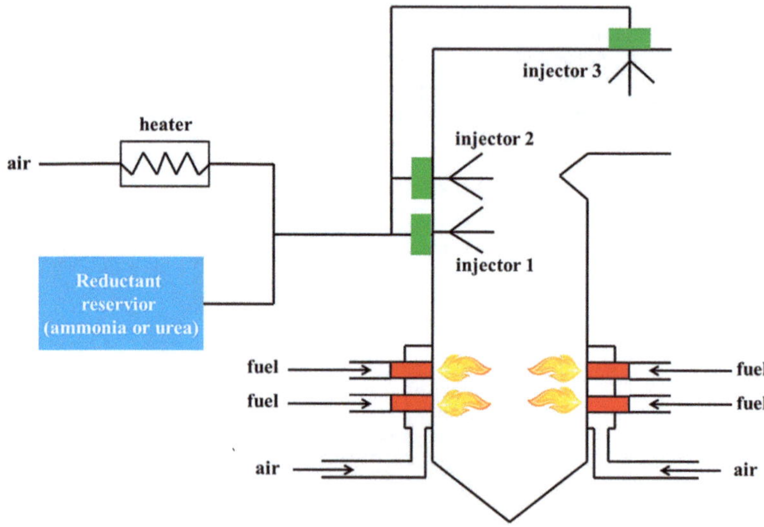

Fig. 3.16 Schematic diagram of an SNCR system (Temperature: 850–1100 °C)

Effect of temperature

The temperature window varies for different chemicals, and for the most common chemicals, it is 850–1100 °C. Ammonia can be formed below the temperature window. At higher temperatures above the temperature window, more NO_x can actually be produced [6]. The effect of temperature on NO_x conversion over an SNCR system is illustrated in Fig. 3.17 [2]. The NO_x conversion reaches the highest value at 1000–1025 °C. At higher temperatures, the reagent, NH_3, can actually oxidized

Fig. 3.17 Relationship between the NOx conversion and the reaction temperature

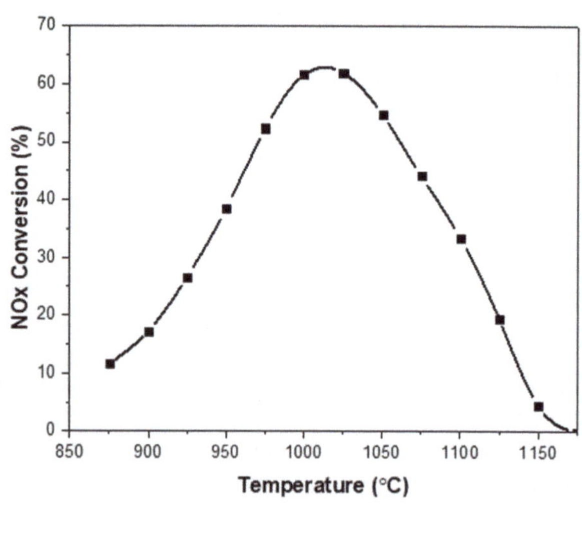

to NO [1]. When the temperature is below 1000 °C, more ammonia slip occurs. An optimal reaction temperature exists to reach the highest NO_x conversion using ammonia as the reductant.

The key issue is finding an injection point or points where the temperature is right under all operating conditions for the chemicals used. The chemicals then have to be mixed well with the flue gases to reach the maximum possible NO_x reaction without producing unacceptable concentrations of ammonia as a by-product [6]. The second complicating factor in affecting the SNCR performance is the availability of residence time at the appropriate temperature. The available residence time may also vary with loading conditions due to the boiler configuration and the flue gas flow [2]. It has been confirmed that the maximum deNO_x efficiency could be obtained with the residual time longer than 1 s. In fact, the residual time of NH_3 and urea between 0.3 and 0.4 s could lead to a superior deNO_x efficiency [5].

Effect of the ratio of NH_3/NO_x

The stoichiometric ratio of NH_3/NO_x means the mole number of reductant NH_3 or urea required in the removal of 1 mol NO_x. The used amount of reductant could be calculated from the amount of reductant injected into the system and the removal amount of NO_x. The stoichiometric ratio of NH_3/NO_x for the SNCR systems is much higher than that for the SCR systems, because a high temperature would cause the decomposition of the reductant, thereby leading effective utilization to be decreased.

Provided that urea is utilized as the reductant, a relatively large amount of N_2O would be produced in the SNCR reactions; given an increase in the NO_x conversion, the N_2O formation amount increases, suggesting that part of the removed NO_x is transformed to N_2O. Therefore, the consumption amount of reductant increases, and the stoichiometric ratio increases. Figure 3.18a, b illustrates the relationship between the deNOx efficiency/ammonia slip and urea flow rate. The stoichiometric ratio of NH_3/NO_x greatly affects the deNO_x efficiency.

A high stoichiometric ratio of NH_3/NO_x could result in a high deNOx efficiency. However, the concentrations of NH_3 and N_2O in the flue gas increase with the

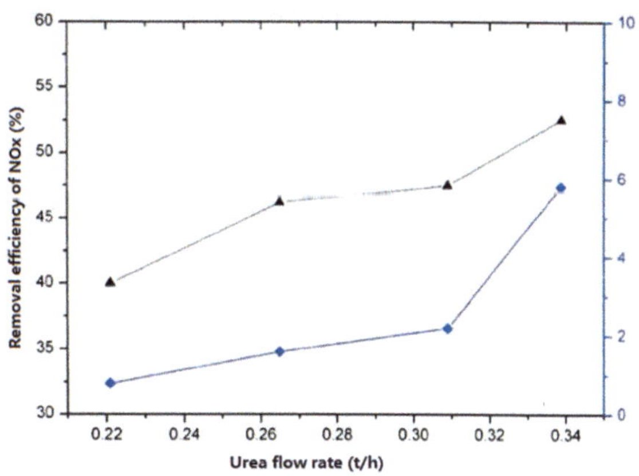

Fig. 3.18 Relationship between **a** the NO conversion efficiency versus urea flow rate (black line) and **b** ammonia slip vs urea flow rate (blue line)

increase in the stoichiometric ratio of NH_3/NO_x. Generally, the optimal value of the stoichiometric ratio of NH_3/NO_x is between 1.0 and 1.6.

Effect of additives

Additives can be used in order to reduce the reaction temperature and decrease ammonia slip. Using appropriate additives, the operational temperature window may be tuned to the region of 800–1200 °C. The chemical used includes hydrogen (H_2), hydrogen peroxide (H_2O_2), and ethane (C_2H_6) [6].

3.4.3 SCR Technology

SCR reaction

In an SCR process, a reducing agent (generally NH_3 or urea) is injected into an exhaust stream. The reaction processes are shown in Eqs. (3.44), (3.51), and (3.52). A catalyst is generally used to facilitate the reaction. The optimum temperature is usually between 300 and 400 °C. Given an increase in the temperature of flue gas, some side reactions will occur, as shown in Eqs. (3.53), (3.54), (3.55), (3.56), (3.57) and (3.58):

$$4NH_3 + 4NO + 3O_2 \rightarrow 4N_2O + 6H_2O \tag{3.53}$$

$$4NH_3 + 4NO_2 + O_2 \rightarrow 4N_2O + 6H_2O \tag{3.54}$$

$$6NH_3 + 8NO_2 \rightarrow 7N_2O + 9H_2O \tag{3.55}$$

$$4NH_3 + 5O_2 \rightarrow 4NO + 6H_2O \tag{3.56}$$

$$4NH_3 + 3O_2 \rightarrow 2N_2 + 6H_2O \tag{3.57}$$

$$2NH_3 + 2O_2 \rightarrow N_2O + 3H_2O \tag{3.58}$$

The mechanisms of the SCR reaction over commercial vanadia-based catalysts have been extensively investigated [13]. Different intermediates and reaction routes were proposed. In general, three key steps are involved: (1) NH_3 adsorption, (2) reaction of adsorbed NH_3 with gaseous NO or adsorbed NO species, and (3) regeneration of the active site by O_2. A typical reaction mechanism is shown in Fig. 3.19 [14, 15].

SCR denitrification catalyst

The catalyst is the core of SCR technology, and its cost accounts for about 50% of the total engineering cost of SCR [16]. The performance of the catalyst directly

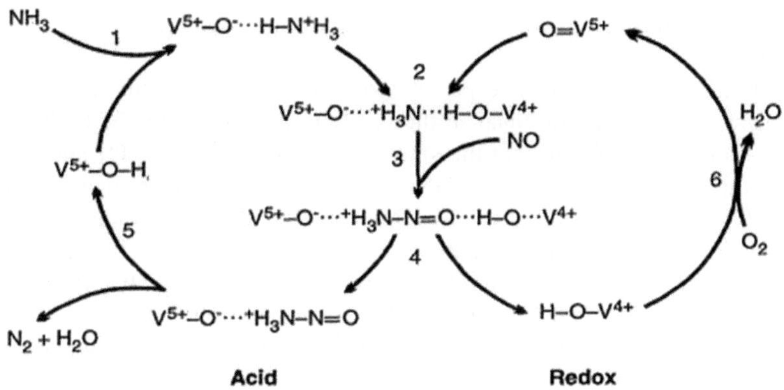

Fig. 3.19 Mechanism of the SCR reaction [14, 15]

determines the NO_x removal efficiency. According to the classification of active components, SCR denitrification catalysts can be classified into noble metal catalysts, metal oxide catalysts, and molecular sieve catalysts.

1. Noble metal catalyst

Noble metal catalysts have been found to have good catalytic activity as developed early in the 1970s. Its main active components include Pt, Pd, Rt, and other precious metals, and the main carrier is alumina and other integrated ceramics. Due to the strong oxidation capacity of noble metal catalysts for NH_3 [17], the reducing agent used for SCR is mainly hydrocarbons (HC) [18]. Noble metal catalysts are expensive and prone to be poisoned by SO_2 and other components of flue gas; therefore, with the development of metal oxide catalysts, they were quickly replaced.

2. Metal oxide catalysts

The researchers found that the metal oxides showed good denitrification activity under oxygen-rich conditions with NH_3 as a reducing agent. Bosch and Janssen [19] summarized in detail the NH_3-SCR activity of many metal oxides, including V_2O_5, Fe_2O_3, CuO, Cr_2O_3, Co_3O_4, NiO_2, CeO_2, and the results show that vanadium oxide has the best activity and selectivity. Bauerle [20] found, when V_2O_5 was supported on a TiO_2 or Al_2O_3 support, the catalyst exhibited better activity than pure V_2O_5.

The performance of the carrier has an obvious influence on the activity of the catalyst. Shikada et al. [21, 22] systematically studied the supported oxide catalysts and found that for vanadium oxide, the carrier properties are: TiO_2-SiO_2 > Al_2O_3 > SiO_2. In addition to better activity, TiO_2 also has good resistance to SO_2. It is reported that the catalyst containing Al_2O_3 and Fe_2O_3 easily reacts with SO_2 in the flue gas to produce sulfate, which in turn causes blockage of the catalyst pores and decreases the specific surface area; however, TiO_2 has good resistance to SO_2. V_2O_5/TiO_2 catalyst is not affected by SO_2 and in a wide temperature range (200–600°C) shows good activity and selectivity [23]. Different forms of TiO_2 support also have an effect on the catalytic performance, and the TiO_2 anatase form is found to be more active than the spinel type [24].

In addition to improving the performance of the support, the addition of the co-catalyst component can enhance the catalytic activity of the V_2O_5/TiO_2 catalyst. Morikawa et al. found that the simultaneous addition of V_2O_5 and WO_3 can promote the activity of the catalyst. Chen et al. [25] studied the role and mechanism of V_2O_5/TiO_2 catalyst addition of WO_3 in detail and found that WO_3 has the following three functions: (1) increasing the activity and broadening the SCR reaction temperature window, (2) significantly improving the resistance of the catalyst to alkali metal and arsenic poisoning, and (3) reducing the oxidation of NH_3. In addition to WO_3, other metal oxides have also been found to improve the denitrification performance of V_2O_5/TiO_2 catalysts. Amiridis et al. [26] systematically studied the effects of added components such as MoO_3, WO_3, Nb_2O_5, GeO_2, Fe_2O_3, CeO_2, and MnO_2. It was found that MoO_3 and WO_3 have the best promoting effect, while ZnO poisons the catalyst, resulting in decreased activity. V_2O_5/TiO_2 to which WO_3 or MoO_3 is added has become a major component of a widely used commercial catalyst.

Although vanadium-based catalysts have good catalytic performances, they have exposed some problems in practical applications, including the toxicity of vanadium itself, the difficulty in the disposal of spent catalysts, and the high oxidation rate of SO_2. Therefore, researchers all over the world have been working on a new type of SCR catalyst with no toxicity, high efficiency, and low SO_2 oxidation rate. Among them, a catalyst containing an oxide component such as Fe [27], Mn [28], Cu [29], and Ce [30] exhibits good activity. Also among them, the cerium-containing oxide catalyst is one of the powerful substitutes for vanadium-based catalysts because of its good denitrification performance and lack of biotoxicity.

3. Molecular sieve catalyst

Metal ion-exchanged zeolite, which has a molecular sieve as a skeleton, has good catalytic activity, a wide temperature window, and is non-toxic and easy to dispose of after use. Therefore, it has attracted extensive attention from researchers all over the world. The molecular sieve structure with NH_3-SCR activity mainly includes ZSM-5, FAU, BEA, MOR, USY, SSZ-13, etc. The main exchanged metal ions include Cu, Co, Fe, Ce, Pt, Ga, Ag, In, and so on. Iwamoto et al. [31] first discovered that Cu^{2+}-exchanged ZSM-5 had catalytic activity for the decomposition of NO. Later, the researchers found that the catalyst exhibited better SCR performance than the commercial V-W-Ti catalyst after the addition of the reducing agent NH_3 [32]. Over the past 20 years, researchers have conducted extensive research on the properties, influencing factors, reaction mechanisms, and active components of ZSM-5 molecular sieves. However, ZSM-5 has an unavoidable problem. Its hydrothermal stability is poor. At high temperatures (above 600°C) and in the presence of water vapor, Cu^{2+} ions will detach and migrate from the active sites, and the molecular sieve skeleton structure will collapse causing a decline in specific surface area and disappearance of active sites [33]. This is very unfavorable for practical application. In recent years, researchers have proposed molecular sieve structures with higher hydrothermal stability, better activity, and selectivity than ZSM-5. Among them,

Chabazite (SSZ-13 [34], SSZ-39 [35], etc.) has received extensive attention. It has smaller pores than ZSM-5 molecular sieves and can dope more Al, thus having more active sites and better hydrothermal stability. The reaction mechanism and active components of the chabazite structure molecular sieve for NH_3-SCR have not been ascertained, and researchers all over the world are conducting significant amounts of research on this problem.

SCR system arrangement

A SCR system consists primarily of an injection grid that mixes ammonia with a flue gas stream and a reactor which contains a fixed bed for the catalyst. The key component in the SCR systems is the SCR reactor. The position of the reactor in the process effluent of a thermal power plant strongly influences the selection of the catalyst composition [36].

1. High-dust location

As shown in Fig. 3.20, for high-dust/coal-fired boiler applications, the catalyst is placed between the economizer and air preheater (with temperatures 320–400 °C) before any flue gas particulate or scrubbing operations. As a result, the flue gas containing all the sulfur oxides and the fly ash from combustion directly passes through the catalyst. This can cause catalyst deactivation, leading to a decrease in NO_x reduction efficiency [6]. So the catalyst must have large channels or holes to avoid plugging and excessive backpressure [36].

2. Low-dust location

As shown in Fig. 3.21, a low-dust location means that the catalyst is placed after

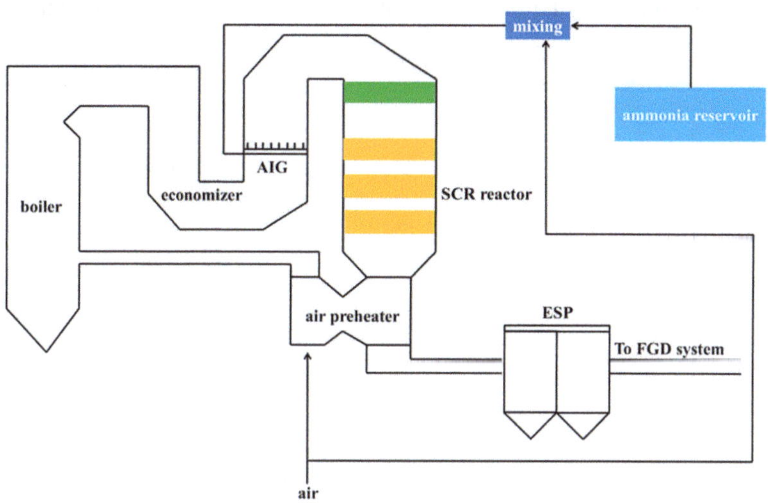

Fig. 3.20 Flowchart of SCR reactor with high-temperature, high-dust installation

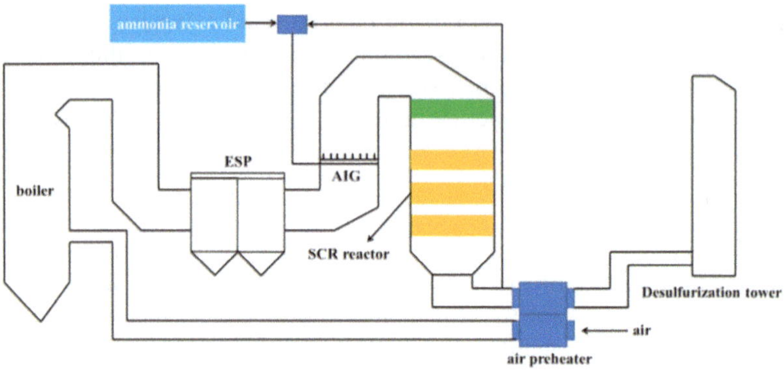

Fig. 3.21 Flowchart of SCR reactor with high-temperature, low-dust installation

the electrostatic precipitator (ESP) and before the air preheater, which could provide the catalyst with a low-dust environment, where the catalysts would have a longer lifetime. Moreover, dust-cleaning apparatus is no longer needed, which dramatically decreases the cost. However, the temperature of the flue gas after passing through the electrostatic precipitator decreases. So, the size of the bypass in the economizers would increase to make sure the temperature of the flue gas is within the operation temperature window of the catalysts. A major issue in the low-dust system SCR is that the catalyst might be covered by fine dust particles, which are small enough to pass through the electrostatic filter.

3. Tail end location

As shown in Fig. 3.22, the catalyst is placed at the end of the pollutant control

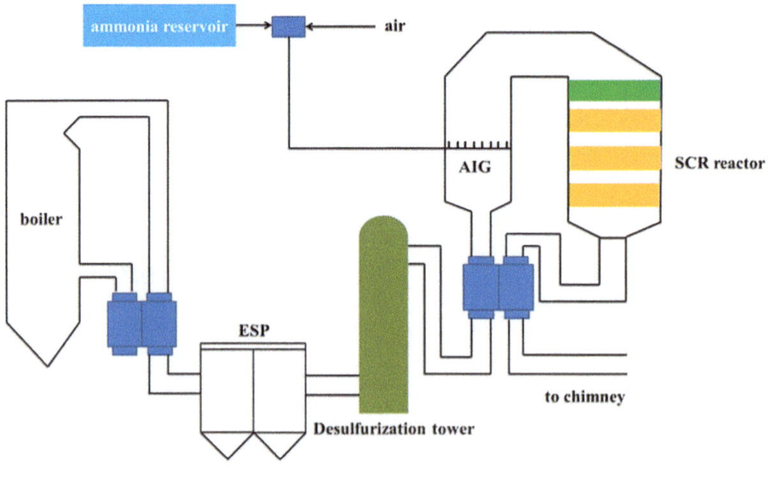

Fig. 3.22 Flowchart of SCR reactor with low-temperature, low-dust installation

devices. The flue gas has passed through an electrostatic precipitator and a flue gas desulfurizer. As a result, the flue gas contains only small amounts of sulfur oxides and particulates when passing through the catalyst. However, the temperature of the as-obtained flue gas (50–60 °C) is far lower than the required temperature of the catalysts. Consequently, the flue gases are normally heated first by utilizing residual heat from the cleaned gases after the catalyst in a rotating heat exchanger. Extra heat is also needed, by direct or indirect heating, which increases the consumption of energy and operation costs.

Comparison of SCR unit locations

There are many advantages and disadvantages to be considered in a comparison of different positions for the catalyst. The optimum is site-specific with the main considerations being whether the installation is new or retrofitted, availability of space and time for retrofitting, fuel and combustion type, and operating philosophy among others [6]. The comparative factors are listed in Table 3.3.

Design of SCR reactor

SCR reactor is related to the place where SCR reactions occur. The generalized SCR deNO_x reactor contains the connected inlet and outlet channel systems. It is generally defined that the SCR reactor starts from the economizers and ends at the inlet of the air preheaters. The design of the SCR reactor is critical in determining the deNO_x

Table 3.3 Comparison of the three SCR installation locations

Items	High-dust location	Low-dust location	Tail end location
Location	Between economizer and air preheater	Between high-temperature electrostatic precipitator and air preheater	At the end of pollutant abatement facilities
DeNO_x efficiency	75–95%	75–95%	75–95%
Flue gas temperature	320–420 °C	320–420 °C	Reheated to 300 °C
Ash concentration	High	Relatively low	Very low
Operation condition	Bad	Medium	Good
Catalyst pitch	Large	Little	Very little
Catalyst volume	Large	Little	Very little
Accessory equipment modification	Induced draft fan, air preheater	Induced draft fan, air preheater	Air preheater
Modification space requirements	High	Very high	High
Initial investment	Low	High	Very high
Operation cost	Low	High	Very high
Catalyst replacement cycle	Short	Long	Very long

performance. Many factors, including the installation of catalyst module and flue gas flow field, should be taken into consideration.

The sectional dimensions of SCR reactors can be determined by the amount of flow gas and surface velocity. Surface velocity means the velocity of flue gas when passing through the catalyst surfaces, which can be defined as Eq. (3.59):

$$A_v = \frac{Q_f}{V_c \times \beta} \tag{3.59}$$

where
A_v = the face velocity, m/h;
Q_f = the amount of flue gas, m^3/h;
V_c = the volume of the catalyst, m^3;
β = the catalyst surface area, m^2/m^3.

Catalyst sectional dimensions can be calculated based on Eq. (3.60).

$$A_c = \frac{Q_f}{C_f \times v_s} \tag{3.60}$$

where
A_c = the catalyst typical floor area, m/h;
Q_f = the amount of flow gas, m^3/h;
C_f = the unit conversion factor;
v_s = the surface velocity, m/s.

Considering the gap between the catalyst modules and the seal between the reactor surfaces, the sectional dimensions of the SCR reactor are larger than those of the catalyst by around 15%. Thus, the sectional dimensions of reactors A_R could be calculated by Eq. (3.61):

$$A_R = 1.15A_C \tag{3.61}$$

In the actual project, the typical dimensions of the catalyst module are around 1 × 2 m. The sectional dimensions of SCR reactors could be determined based on the times of those belonging to the catalyst modules. Additionally, there is limited space available for the installation of SCR reactors for the deNO_x reconstruction projects. Considering the installation of catalyst modules in a matrix way, the reactor could be designed in a rectangular square shape. The designed volume of the catalyst could be calculated by Eq. (3.62):

$$V_c = \frac{-(Q_f \ln[1 - \frac{\eta}{ASR}])}{K_c \times \beta} \tag{3.62}$$

where

η = the designed $deNO_x$ efficiency;

ASR = the molar ratio of NH_3/NO_x;

K_c = the catalyst activity constant.

The layer number of catalysts could be calculated according to the catalyst volume V_c, sectional area A_c, and height h (Eq. (3.63). The catalyst layer number should be adjusted to the nearest integer.

$$n = \frac{V_c}{h \times A_c} \tag{3.63}$$

In most of the actual projects, at least two layers of catalysts would be installed in the initial installation process. And one extra layer of catalyst would also be necessary. Therefore, the total layers of catalysts could be calculated using Eq. (3.64):

$$n_t = n + n' \tag{3.64}$$

where

n_t = the total layers of catalysts;

n = the layers of the initially installed catalysts;

n' = the layers of the standby catalysts.

The façade dimensions of SCR reactors can be determined by the layers of catalysts, the installation space of catalysts, and the height of rectifying layer, the value of which could be calculated using Eq. (3.65):

$$H = n_t \times (c_1 + h) + c_2 \tag{3.65}$$

where

H = the height of the reactor, m;

c_1 = the installation height of the catalysts, m;

c_2 = the height and installation height of the rectifying layers, m.

In that equation, c_1 and c_2 are the empirical data, which are related to the catalyst installation manners and soot blower forms. For easy installation, the height of the catalyst layer is no smaller than 3.5 m.

Additionally, there exist many variable cross-sections and turning sections in the SCR channel systems. Without any interference, the velocity distribution would be uneven, which would lower the $deNO_x$ efficiency and increase the NH_3 slip. Moreover, the large angle of incidence would erode and wear the catalysts; a high system resistance and serious ash deposition would occur. Therefore, the flow field should be optimized to obtain a relatively uniform velocity, NH_3/NO_x, temperature distribution, and lower angle of incidence and system resistance. The deposition of ash would be also impeded. As a result, high performance of catalysts would be observed, which lays a solid foundation for the secure and economic operation of SCR systems.

Catalyst poisoning and disposal

The lifetime of catalysts is limited by different types of degradation, causing losses in catalytic activity and reaction efficiency [6]. A substance that causes the catalyst to lose its catalytic action is called poisoning. In general, poisons that selectively react with the active sites of the catalysts are known as chemical intoxication, while poisons that non-selectively cover or block the catalyst surfaces or pore structures are known as physical intoxication. In the case of SCR catalysts-used coal-fired power stations, the factors that result in the inactivation of its use are both chemical poisoning and physical poisoning. The main effect of physical poisoning is to cover the active sites and block the pore structures of the catalysts, which adversely affects the catalyst activity, as shown in Fig. 3.23. Chemical poisoning means that the poisons cause a decrease in the catalyst active sites, as shown in Fig. 3.24. Chemical poisoning is mainly divided into alkali metal and alkaline earth metal poisoning; SO_2, HCl, and

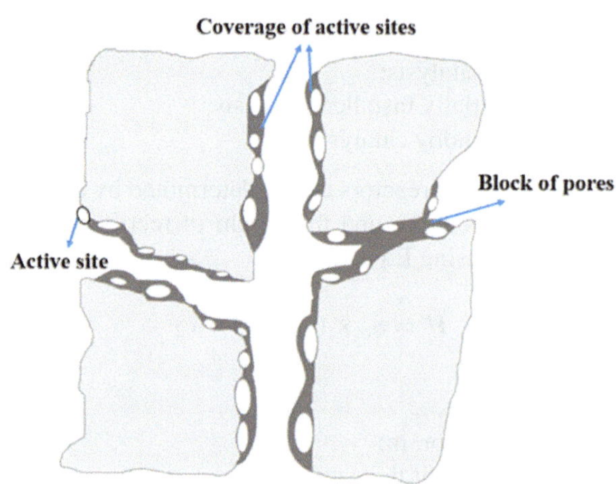

Fig. 3.23 Coverage of active sites and block of pores of the catalyst

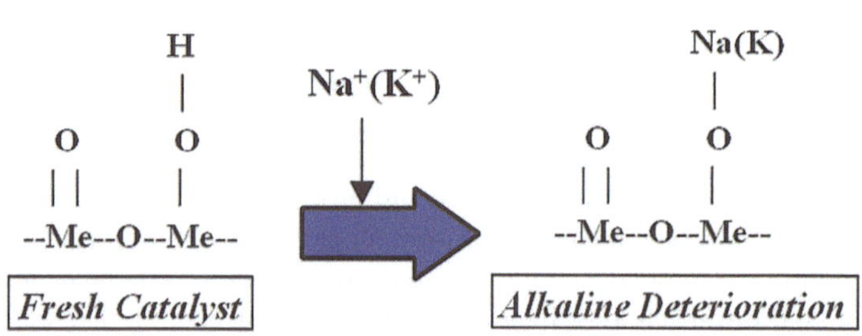

Fig. 3.24 Deactivation mechanism of SCR catalyst by alkaline metal

other acidic gas poisonings; phosphorus poisoning; and arsenic poisoning. Usually, the catalyst is considered to have been deactivated when the relative activity of the catalysts is lower than half of that belongs to fresh catalysts. The so-called deactivation is the loss of the chemical reaction capacity of the catalyst and the physical surface used in the reaction. This makes it possible to regenerate the catalyst. The generation of SCR catalyst means that material structure remains almost unchanged and catalyst activity could be recovered through the external physical and chemical methods. Provided that largely irreversible changes occur in the poisoned catalysts after the reactions, such as wear and tear (the mechanical strength decreases), sintering (permanent loss of active centers), and the regenerated catalysts lacking the required intensity and activity for applications, catalyst resource treatment is needed. In short, in the aspect of processing and disposal of waste catalyst, regeneration is encouraged as a priority. The catalyst regeneration, utilization, and disposal abilities should be improved upon. Non-renewable and unavailable waste catalysts should be kept in the corresponding hazardous waste management unit, such as a hazardous waste landfill.

3.5 Comparison of NO_x Control Technologies

3.5.1 4E Evaluation of NO_x Control Technologies

In the above sections, we discussed the principles and technologies for NO_x abatement. We have summarized the characteristics and costs of these technologies, considering environmental, energy, engineering, and economic aspects, representing the average values for coal-fired power plant, as shown in Table 3.4. In evaluating the use of energy by the NOx control technologies, the combustion modification may cause heat loss in the coal-fired power plant, but there is less heat loss found in the SNCR and SCR systems. It can be observed that the combustion modification technologies have relatively low NO_x reduction efficiency and low costs. Among them, low excess air operation is a burner optimization strategy which is also part of a normal furnace operation. No significant capital costs are required for a low excess air operation. SCR technology has the advantages of (1) low reaction temperature, (2) high NO_x removal efficiency, and (3) small NH_3 slip compared with SNCR. However, the use of a catalyst will also result in pressure drop and SO_2 conversion to SO_3.

Table 3.4 4E evaluation of NOx control Technologies

4E Evaluation	Classification		Combustion modification					Flue gas denitrification	
			Low excess air operation	Flue gas recirculation	Staged fuel combustion	Staged air combustion	Low-NOx burners (LNB)	SNCR	SCR
Environmental	NOx removal efficiency		5–15%	40–50%	40–50%	20–30%	40–60%	20–40%	80-90%
	NH_3 slip (ppm)		–	–	–	–	–	5–10	1–3
	SO_2/SO_3 conversion		–	–	–	–	–	–	0.5–1%
	Space needed		Small	Small	Small	Small	Small	Small	Large
Energy	Heat loss		0.25–1%					<0.5%	<0.3%
Engineering	NH_3/NO_X Molar ratio		–	–	–	–	–	0.8–2.5	0.6–0.95
	System complexity		Low	Low	High	Low	Low	Low	High
	Range of unit		–	–	–	–	>300 MW	<300 MW	>300 MW
	Energy loss		–	–	–	–	–	–	400– 600
Economic	Capital cost	$/ton	N/A	300–500	500–800	N/A	300–500	500–1,000	1,000–10,000
		$/kW	N/A	5–10	10–12	N/A	5–10	10–20	20–200
	Operating cost		Low	Low	Low	Low	Low	Low	High

Table 3.5 NOx ultra-low emission technology

Furnace type	Inlet concentration (mg/m^3)	Removal efficiency (%)	Technology Options	Technical evaluation	Catalyst layers
Pulverized coal-fired boilers	<200	80	Low-NOx burner + SCR	Fair + Excellent	2 + 1
	200–350	80–86			3 + 1
	350–550	86–91			
Circulating fluidized-bed boilers		60–80	Low-NOx burner + SNCR	Fair + Good	If necessary

3.5.2 Technology Selection

Combustion modification is a preferential technique to control the NO_x pollution for boilers. It reduces the emission concentration of NO_x as much as possible on the premise of ensuring boiler efficiency and safety.

For pulverized coal-fired boilers, NO_x emission concentration at the boiler outlet is ensured to be less than 550 mg/m^3 through burner modifications and combustion optimization. SCR technology is adopted after the burners to further reduce NO_x emission. By rationally selecting catalyst layers, accurately injecting ammonia, and evenly distributing the flow field, high NO_x removal efficiency can be achieved.

For circulating fluidized-bed boilers, the emission concentration of NO_x is ensured to be less than 200 mg/m^3 through combustion modification. The use of an SNCR system can efficiently remove NO_x. One layer of SCR catalyst after the boiler will be used when necessary.

For W-type flame boilers burning anthracite, both low-nitrogen combustion and SCR denitrification should be used. It remains challenging to achieve ultra-low emission of nitrogen oxides.

The NO_x ultra-low emission routes of various furnaces are shown in Table 3.5. The application of the NO_x control technologies is summarized in Fig. 3.25.

Figure 3.26 shows the integration of NO_x control technologies in a coal-fired power plant, including the combustion modification and SNCR in the boiler and SCR unit after the unit. In the process of combustion modification, the NO_x reduction potential over low-NO_x burners and staged air combustion should be fully exploited. Concurrently, measures should be taken in the design procedure to prevent the problems such as a reduction of boiler efficiency, corrosion, and coking.

3.5.3 Challenges and Perspectives

For other industrial boilers, the outlet flue gas temperatures are always below 200 °C. Even for a coal-fired power plant, when the boiler is working at low loadings, the

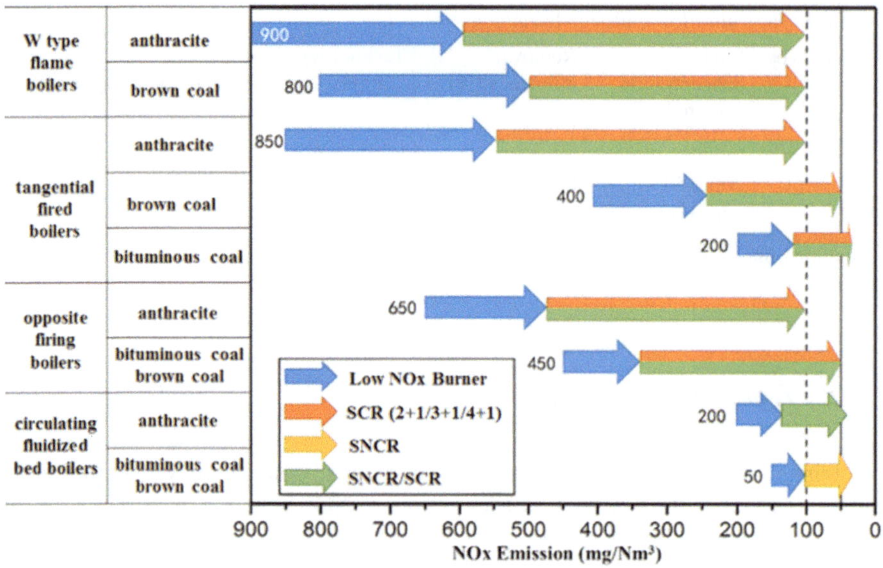

Fig. 3.25 Feasible technology of NOx control in a thermal power plant

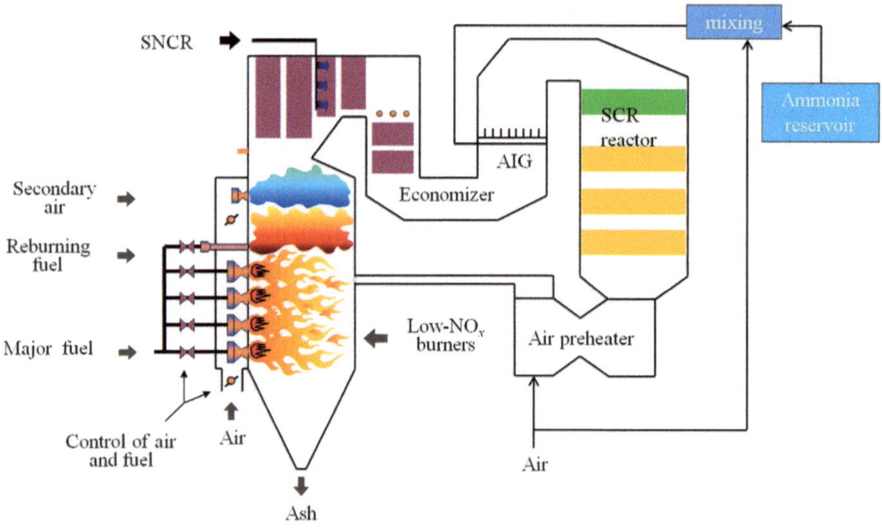

Fig. 3.26 Integration of NOx control technologies

outlet gas temperature is decreasing. At low temperatures, the SCR catalyst cannot function. This is due to two reasons. (1) The catalytic activity of commercial V_2O_5–WO_3 / TiO_2 (denoted as V-W-Ti) catalyst at low temperatures (<350 °C) is relatively low [30, 37]; (2) Ammonia bisulfate (ABS) will form at temperatures lower than 360 °C and deposit on the catalyst surface [38–40]. Therefore, it is important to develop a low-temperature SCR catalyst to further meet the stringent NO_x emission reduction standard. After combinational control of the active centers and enhancing the decomposition of ammonia bisulfate, the activity at a low temperature increased significantly. In the Wenzhou power plant, a series of cerium–niobium-based low-temperature catalysts have been used under 275 °C. Further improvement on the low-temperature activity and ABS decomposition rate in the SCR process is imperative in the future.

Additionally, except for the SCR, SNCR, and combustion modification for NO_x abatement, one novel technology is to develop the control device for the simultaneous removal of multi-pollutants (including SO_2, NO_x, other acid gases, and CO_2). The related technology has been tested in DOE's National Energy Technology Laboratory, which uses ammonia solution to remove acid gases in a wet scrubber. CO_2 can also be removed using ammonia solution in a packed bed, in which CO_2 removal efficiency can be higher than 90%.

Exercises

1. Define the following terms and explain why they should be considered for air pollution control and design: thermal NOx, fuel NOx, and prompt NOx.
2. Briefly describe the principle and process of NOx emission control.
3. Briefly describe the major factors affecting the SNCR systems.
4. Analyze NOx control processes and compare their advantages and disadvantages from the aspects of technology, economy, and application, respectively.
5. When the feed gas flow rate is 200 mL/min, the concentration of NO in the flue gas is 1000 ppm, the mass of the catalyst is 100 mg, and the conversion of NO is 45% at 200 ºC. Calculate the reaction rate constant for the SCR reaction.

References

1. Charles E, Baukal J (2004) Industrial combustion pollution and control. Marcel Dekker, Inc
2. Karl B, Schnelle J (2002) Air pollution control technology handbook. CRC Press
3. Wang LK (2005) Advanced air and noise pollution control. HumanaPress
4. Campbell LM (1991) Sourcebook: NOx control technology data. US EPA
5. Yang Y (2007) Nitrogen oxides emission reduction technology and flue gas denitration engineering. Metallurgical Industry Press
6. Sloss LL (1992) Nitrogen oxides control technology fact book. Noyes Data Corp, New Jersey
7. Su Y (2005) Coal-fired nitrogen oxides emission control technology. Chemical Industry Press
8. Zhong Q (2007) Coal-fired flue gas desulfurization and denitration technology and engineering cases. Chemical Industry Press
9. Busca G et al (1998) Chemical and mechanistic aspects of the selective catalytic reduction of NOx by ammonia over oxide catalysts: a review. Appl Catal B 18(1):1–36

10. Boudart M (1995) Turnover rates in heterogeneous catalysis. Chem Rev 95(3):661–666
11. Kozuch S, Martin JML (2012) Turning over definitions in catalytic cycles. ACS Catal 2(12):2787–2794
12. Pfafflin JR (2006) Encyclopedia of environmental science and engineering. CRC Press
13. Busca G et al (1998) Chemical and mechanistic aspects of the selective catalytic reduction of NOx by ammonia over oxide catalysts: a review. Appl Catal B-Environ 18(1–2):1–36
14. Inomata M et al (1982) Activities of V2o5/Tio2 and V2o5/Al2o3 catalysts for the reaction of No and NH3 in the presence of O2. Ind Eng Chem Prod Res Dev 21(3):424–428
15. Inomata M, Miyamoto A, Murakami Y (1980) Mechanism of the reaction of NO and NH3 on vanadium-oxide catalyst in the presence of oxygen under the dilute gas condition. J Catal 62(1):140–148
16. Beeckman JW, Hegedus LL (1991) Design of monolith catalysts for power plant nitrogen oxide (No.+-.) emission control. Indus Eng Chem Res 30(5): 969–978
17. Forzatti P (2001) Present status and perspectives in de-NOx SCR catalysis. Appl Catal A-Gen 222(1–2):221–236
18. Nikolopoulos AA et al (1999) Selective catalytic reduction of NO by propene in excess oxygen on Pt- and Rh-supported alumina catalysts. Catal Today 54(4):439–450
19. Bosch H, Janssen F (1988) Formation and control of nitrogen oxides. Catal Today 2(4):369–379
20. Bauerle GL, Wu SC, Nobe K (1978) Parametric and durability studies of NOX reduction with NH3 on V2O5 catalysts. Ind Eng Chem Prod Res Dev 17(2):117–122
21. Shikada T et al (1983) Reduction of nitric-oxide with ammonia on silica-supported vanadium-oxide catalysts. J Chem Technol Biotechnol A-Chem Technol 33(8):446–454
22. Shikada T et al (1981) Reduction of nitric-oxide with ammonia on vanadium-oxide catalysts supported on homogeneously precipitated silica-titania. Ind Eng Chem Prod Res Dev 20(1):91–95
23. Nakajima F, Hamada I (1996) The state-of-the-art technology of NOx control. Catal Today 29(1–4):109–115
24. Pearson IM et al (1983) Chemical mixed catalysts. Ind Eng Chem Prod Res Dev 22(2):381–382
25. Chen JP, Yang RT (1992) Role of WO3 in mixed V2O5-WO3/TIO2 catalysts for selective catalytic reduction of nitric-oxide with ammonia. Appl Catal A-Gen 80(1):135–148
26. Amiridis MD, Duevel RV, Wachs IE (1999) The effect of metal oxide additives on the activity of V2O5/TiO2 catalysts for the selective catalytic reduction of nitric oxide by ammonia. Appl Catal B-Environ 20(2):111–122
27. Liu F, He H, Zhang C (2008) Novel iron titanate catalyst for the selective catalytic reduction of NO with NH3 in the medium temperature range. Chem Commun 17:2043–2045
28. Tang XL et al (2007) Low temperature selective catalytic reduction of NOx with NH3 over amorphous MnOx catalysts prepared by three methods. Catal Commun 8(3):329–334
29. Si ZC et al (2010) Structure, acidity and activity of CuOx/WOx-ZrO2 catalyst for selective catalytic reduction of NO by NH3. J Catal 271(1):43–51
30. Qu R et al (2013) Relationship between structure and performance of a novel cerium-niobium binary oxide catalyst for selective catalytic reduction of NO with NH3. Appl Catal B 142–143:290–297
31. Iwamoto M et al (1986) Copper(II) ion-exchanged ZSM-5 zeolites as highly-active catalysts for direct and continuous decomposition of nitrogen monoxide. J Chem Soc Chem Commun 16:1272–1273
32. Sullivan JA et al (1995) Conditions in which Cu-ZSM-5 outperforms supported vanadia catalysts in SCR of NOxby NH3. Appl Catal B 7(1):137–151
33. Park J-H et al (2006) Hydrothermal stability of CuZSM5 catalyst in reducing NO by NH3 for the urea selective catalytic reduction process. J Catal 240(1):47–57
34. Kwak JH et al (2010) Excellent activity and selectivity of Cu-SSZ-13 in the selective catalytic reduction of NOx with NH3. J Catal 275(2):187–190
35. Moliner M et al (2012) Cu–SSZ-39, an active and hydrothermally stable catalyst for the selective catalytic reduction of NOx. Chem Commun 48(66):8264–8266

36. Fahmy YM (2007) Air pollution control technologies compendium. United National Industrial Development Organization
37. Li J et al (2011) Low-temperature selective catalytic reduction of NO(x) with NH(3) over metal oxide and zeolite catalysts-a review. Catal Today 175(1):147–156
38. Wang X et al (2018) Promotion of NH4HSO4 decomposition in NO/NO2 contained atmosphere at low temperature over V2O5-WO3/TiO2 catalyst for NO reduction. Appl Catal A 559:112–121
39. Song L et al (2016) Promotion of ceria for decomposition of ammonia bisulfate over V2O5-MoO3/TiO2 catalyst for selective catalytic reduction. Chem Eng J 303:275–281
40. Kobayashi M, Kuma R, Morita A (2006) Low temperature selective catalytic reduction of NO by NH_3 over V2O5 supported on TiO2–SiO2–MoO3. Catal Lett 112(1):37–44

Chapter 4
Volatile Organic Compounds (VOCs) Control

Nomenclature

A	Heat transfer area
B, F, D	Molar velocity
c_i	Concentration
c_m	Explosive limit of mixture
c_w, c_p, c'_p	Heat capacity
E	Activation energy
E_i	Henry's coefficient
f	Liquefied fraction
g	Gravity acceleration
G', L'	Mass flow velocity
ΔH	Latent heat
H_i	Solubility coefficient
k, m	Empirical parameter
k_1, k_2	Rate constant of kinetic model
K	Heat transfer coefficient
n	Reaction order
P	Pressure
q_e, q	Adsorption capacity of activated carbon
Q	Quantity of heat
r	Reaction rate
R	Universal gas constant
Δt_m	Log mean temperature difference
T	Reaction temperature
x_i, y_i, z_i	Molar fraction
η	Removal efficiency
μ	Liquid viscosity
ρ_g, ρ_L	Density
Φ	Packing factor

P.-C. Chiang and X. Gao, *Air Pollution Control and Design*,
https://doi.org/10.1007/978-981-13-7488-3_4

4.1 Introduction

4.1.1 Importance and Significance

VOCs, which is short for Volatile Organic Compounds, are organic chemicals with low boiling point and high vapor pressure at room temperature. They are numerous and ubiquitous, making them very complicated to define. The most widely accepted definition of VOCs is the organic compound with boiling point ranging from (50–100 °C) to (240–260 °C) and saturation vapor pressures greater than 102 kPa at 25 °C [1], proposed by the World Health Organization (WHO). Because of their tremendous hazard to human health, air environment, industrial safety, and atmospheric condition, it is important to limit the emission of VOCs [2].

The toxicity of VOCs to human health is mainly in the nervous system, respiratory system, blood system, and skin system, as shown in Fig. 4.1. Exposure to high levels of part VOCs within a short period of time can cause headache, drowsiness, nausea, dizziness, light-headedness, and irritation to the eye and respiratory organs. Long-term exposure to some VOCs may cause cancer and ruin the nervous system of human beings [3].

The emission of odorous VOCs is more harmful, including dimethyl sulfide, styrene, trimethylamine, ethyl acrylate, etc. They not only bring discomfort to people's sense of smell, but can also cause diverse indirect health effects such as nausea and vomiting and reactions of hypersensitivity for people who have been in the odorous polluted environment for a long time [4].

Some flammable VOCs (gasoline, benzene, alcohol, etc.) also have some hidden dangers to industrial production safety. VOCs combustion is a strong exothermic reaction, which increases the gas temperature. When the concentration range of

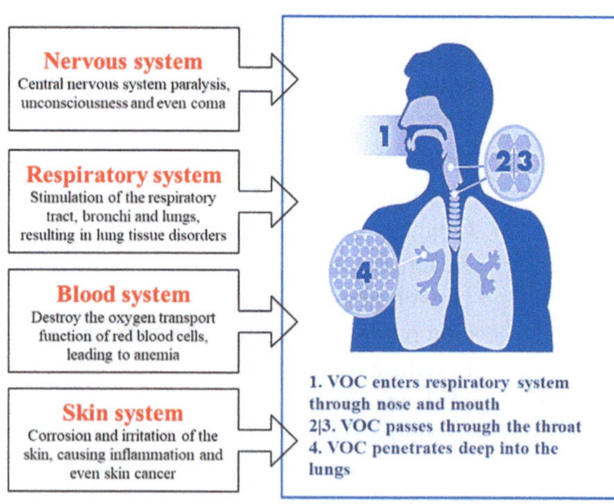

Fig. 4.1 The toxicity of VOCs to human health [3]

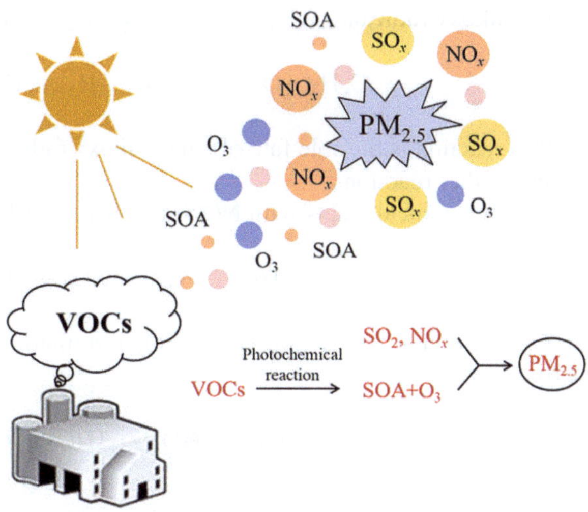

Fig. 4.2 Sketch of the various processes which determine the fate of VOCs in the atmosphere [7]

VOCs is between the upper explosion limit and the lower explosion limit, the mixed gas will explode when it encounters a fire source [5].

Once emitted into the atmosphere, VOCs could be the precursors for the formation of aerosol and ozone, as shown in Fig. 4.2. It has been discovered that the photochemical smog should be caused by the oxidation of VOCs and nitrogen oxides (NO_x), indicating that VOCs are also the precursors of secondary pollution [6].

In the early 1950s, it was confirmed that the VOCs emission was one of the key roles in causing photochemical smog, which was believed to be the main cause of "Los Angeles smog". When the VOCs and nitrogen oxides (NO_x) are mixed together under ultraviolet (UV) light, complex reactions happened and converted them into products indicated as photochemical pollutants [8]. The atmosphere could be seen as a photochemical reactor. The reagents of VOCs are provided by both natural and anthropogenic emissions, and the energy is supplied by sunlight.

The type of photochemical reaction was mainly radical, and the chain finally leads to the emission of ozone, aldehydes, peroxyacetyl nitrate, hydrogen peroxide, organic/inorganic acids, and fine particles. The reaction of radical (OH) with many hydrocarbons (C_xH_y) leads to alkyl peroxy radicals ($C_xH_{y-1}O_2$), as shown in Eqs. (4.1) and (4.2)

$$C_xH_y + OH \rightarrow C_xH_{y-1} + H_2O \tag{4.1}$$

$$C_xH_{y-1} + O_2 \rightarrow C_xH_{y-1}O_2 \tag{4.2}$$

The alkyl peroxy radicals ($C_xH_{y-1}O_2$) reacts with NO, as indicated in Eq. (4.3)

$$C_xH_{y-1}O_2 + NO \rightarrow C_xH_{y-1}O + NO_2 \rightarrow C_xH_{y-1}ONO_2 \tag{4.3}$$

The alkoxy radicals ($C_xH_{y-1}O$) react with O_2, as shown in Eq. (4.4)

$$C_xH_{y-1}O + O_2 \rightarrow C_{x-1}H_{y-3}CHO + HO_2 \tag{4.4}$$

This reaction is the sole fate of small alkoxy radicals. Larger alkoxy radicals can undergo other reactions;

The HO_2 radical reacts with NO to regenerate OH, as shown in Eq. (4.5):

$$HO_2 + NO \rightarrow NO_2 + OH \tag{4.5}$$

The main HO_x –HO_x and HO_x –NO_x termination reactions are shown in Eqs. (4.6)–(4.8):

$$HO_2 + HO_2 \rightarrow H_2O_2 + O_2 \tag{4.6}$$

$$C_xH_{y-1}O_2 + HO_2 \rightarrow C_xH_{y-1}OOH + O_2 \tag{4.7}$$

$$OH + NO_2 \rightarrow HNO_3 \tag{4.8}$$

4.1.2 Control Regulations

The control of VOCs emission began from 20th century in developed countries located in Europe and America. After more than 20 years of development, the standards have become increasingly stringent. In 1990, the United States promulgated the Clean Air Act Amendments (CAA), which provided 189 kinds of air pollutants, most of which were VOCs. According to different types of emission sources, it was divided into five categories: process exhaust, equipment leakage, wastewater volatilization, storage tank, and loading operation. This measure is targeted and effectively controls the discharge of pollutants from polluting enterprises or facilities. The U.S. subsequently issued Clean Air Interstate Rule (CAIR) and National Emission Standard for Harmful Air Pollutants (NESHAPs) in 2005 and 2006, requiring to minimize air pollution in the next ten years by reducing the release of atmospheric pollutants during trans-state transmission.

The VOCs emission control of EU is issued in the form of directives, including the industry directive (integrated pollution prevention and control command 96/62/EC) and the general directive (using the organic solvents and paint directives 1999/13/EC). In 2004, 2004/42/EC directive was issued which refers to the limitation of VOCs emissions from organic solvents in decorative paints, varnishes and vehicle refinishing products. Germany established a classification of pollutant emission control standards, implementing classification and grading control according to the health hazards of toxic pollutants (carcinogenic, sensory irritation) or other environmental hazards (ozone formation potential, greenhouse effect). Technical Instructions on Air Quality Control (TA Luft) in 2002 was promulgated by Germany. The fractionation of gaseous organic pollutants was divided into two grades (176 for I and 10 for II) according to their toxicity, and the emission limits were 20 and 100 mg/m^3, respectively.

Regulation of VOCs in Japan can be traced back to 1972 with Odor Control Law stipulating more than 10 kinds of odor VOCs emission standards. This law continuously improved to 53 species of VOCs by 1996 and further increased to 149 species by 2002. Air Pollution Prevention Law in 2006 regulated the control on six key VOCs emission sources (chemical manufacturing, industrial cleaning, coating, adhesive, printing, and VOCs material storage) and listed benzene, trichloroethylene, and four vinyl chloride as priority pollutants. In 2006, Thailand launched a research project on environmental emission standards for VOCs and promulgated National Ambient Air Quality Standard in 2007, which contains nine VOCs emission standards.

In China, VOCs emission reduction and control are relatively lagging behind, but with the rapid development of industrialization, the corresponding policies and regulations are gradually put into effect. In 2002, National Standard for Indoor Air Environmental Quality was released, which proposed requirements of indoor air benzene, toluene, xylene, formaldehyde, TVOCs, and other pollutants in the prescribed range. In May 2010, the general office of the State Council issued the Notice of Environmental Protection Departments on Promoting the Joint Prevention and Control of Air Pollution and Improve Regional Air Quality Guidelines, which was officially made from the national level to carry out pollution prevention and control work of VOCs for the first time. Some of VOCs control laws and regulations are listed in Fig. 4.3.

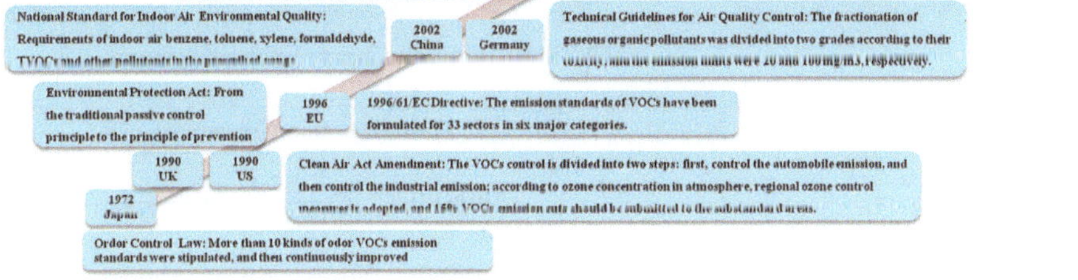

Fig. 4.3 VOCs control laws and regulations in some countries and regions

4.1.3 Technology Development

Reductions in VOC emitted by stationary sources could be achieved via two approaches: clean production and end-of-pipe technology [9]. Clean production includes resource management (improving the control of processes to minimize emissions), product reformulation (the reduction of organic solvents from products such as coatings), and process modification (reducing emissions from a process by optimization or otherwise making modifications to the equipment used). In the industrial production processes of petroleum, chemical industry, printing and dyeing, coating, and glue which involve VOC pollution, the use of closed integrated production technology should be encouraged. For leakage-prone equipment, regular inspection and repair should be carried out to prevent running, leaking, dripping and leakage [7].

End-of-pipe technology of VOCs is the treatment of emissions from processes by means of specialized technologies which either destroy or recover the emission. Traditional end-of-pipe abatement systems include recovery technologies of adsorption, absorption, and condensation, together with destruction technologies of thermal combustion and catalytic oxidation. Emerging technologies developed in recent years include biological treatment, plasma-assisted oxidation, and photocatalytic oxidation. With the more and more stringent emission standards and policies to control VOCs, a single VOC control technology is difficult to meet the emission requirements. The development of main VOCs control technologies is summarized in Fig. 4.4.

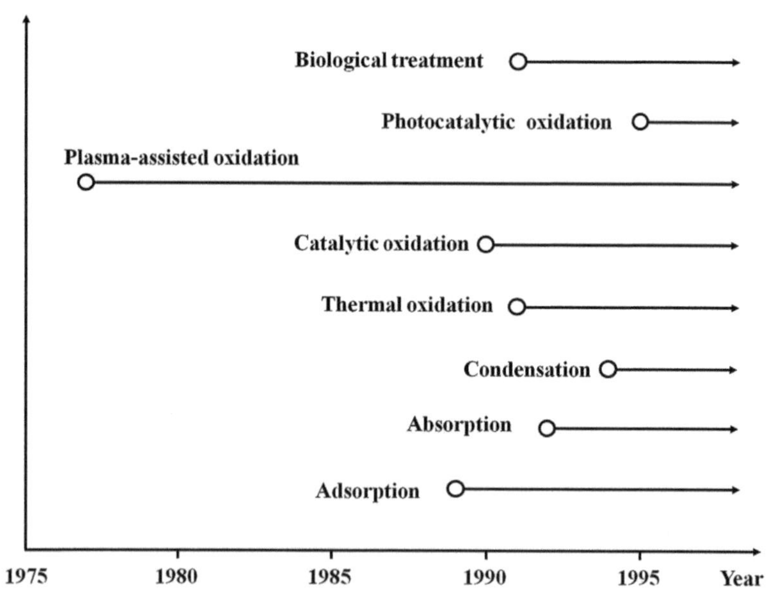

Fig. 4.4 Development of main VOCs control technologies

The combination of different VOCs technologies has been more widely used today. For example, a dust catcher could be added before VOCs control device to remove the dust from waste gas; the adsorption concentration and regenerative catalytic oxidation combined technology could be chosen in the treatment of organic waste gas of large volume, low concentration, and no recovery value [10]; the adsorption enrichment technology and condensation technique could be chosen in the recovery of valuable organic waste gas [11]; condensation and adsorption technology, adsorption concentration and condensation recovery, or combustion technology could be chosen in the treatment of high-concentration organic waste gas [12]; biological treatment and photo-catalytic or low-temperature plasma technology could be chosen in the treatment of malodorous gases [13].

4.1.4 Objectives of Chapter

Overall, it is of great value and significance to study the emission and control technologies of VOCs. In the following sections, we will discuss the formation of VOCs, the principles of control technologies, the process routes of VOCs control, and the evaluation of different VOCs control technologies. The objectives of this chapter are mainly in the following three aspects:

1. Formation: The emission sources of VOCs are discussed, especially for the industrial source. VOCs emission inventory are further listed, which refers to the effect of various human activity on pollutant emissions. In addition, the discharged VOCs are divided into several groups according to chemical composition, and the characteristics of VOCs such as boiling point and solubility are also summarized.
2. Control: The principles and processing of numerous control technologies by either recovery or destruction of VOCs are discussed. For VOCs with recovery value, adsorption, condensation, and absorption techniques are commonly used. If the VOC stream contains other substances or concerning disposal, the appropriate techniques should be destruction like oxidation and biological treatment. Calculation formulas and design parameters in VOCs control technologies are introduced, which can be used to guide the design work.
3. Evaluation: Evaluation of VOCs control technologies in aspects of engineering, environmental, energy, and economic (4E) is proposed, and the selection criteria for processing engineering are discussed. The application of different technologies under different operating conditions is also listed. Finally, challenges and perspectives of VOCs control development are also proposed according to the current control situation.

4.2 Formation

4.2.1 Emission Sources of VOCs

VOC emissions are mainly from organic solvents, the oil and chemical industries, combustion, motor vehicles, and nature. Other minor sources of VOCs include the metals industries, food and drink manufacture, waste disposal, and straw and stubble burning. Generally speaking, VOCs of chimney and escape emissions from industry are rich, with a variety of components and characteristics. Understanding the properties of VOCs from industrial emissions is helpful to diagnose the influence of emission sources on the atmosphere and to formulate effective control methods. VOCs emission inventory refers to the effect of various human activity on pollutant emissions, including product output, raw material consumption, population, etc. It is a key factor affecting the accuracy of emission inventory. Official statistics for some activity data are listed in Table 4.1, some of which are from Industry Association statistics.

4.2.2 Classification and Characteristics

Due to the complexity of emission sources and the variety of organic compounds, the classification of VOCs is needed. Three distinct groups of volatile organic compounds are defined according to WHO, which differ from the definition based on the boiling point of each chemical, including Very Volatile Organic Compounds of the boiling point below 50 °C (VVOCs), Volatile Organic Compounds with the boiling point between 50 and 260 °C (VOCs), and Semi-Volatile Organic Compounds with the boiling point between 260 and 400 °C (SVOCs). For instance, formaldehyde and acetaldehyde belong to the VVOC grouping based on boiling point, while benzene, toluene, and xylenes fit within the VOC grouping more accurately.

In consideration of chemical groups, VOCs can be further classified into alkanes, alkenes, aromatics, halides, esters, aldehydes, ketones, alcohols, and others. Alkane molecule consists of carbon atoms and multiple hydrogens [15]. Typical examples are methane, ethane, and propane. In chemistry, vinyl is the functional group with the formula $-CH=CH_2$. Typical examples are ethylene, propylene, and 1-butene. Aromaticity is used to describe a conjugated system or molecule with resonance bonds in a ring, including benzene, toluene, and ethylbenzene. Aldehyde is a specific class of organic compound with the functional group of $-$ CHO, i.e., a carbon atom connected to an oxygen atom with a double bond and a single bond with a hydrogen atom and an R group, which is any generic alkyl or side chain. Typical examples are formaldehyde, acetaldehyde, and ethyl two aldehyde. Ketone is an organic compound with the RC(=O)R' structure, in which R and R' can be various carbon-containing substituents, such as acetone, 2-butanone, and 2- amyl ketone. Alcohol is a kind of organic compound with a functional group of –OH bounded to a

Table 4.1 The links, emission sources, and activity data of VOCs emission inventory in the industry [14]

Links	Emission sources	Activity data
Production of VOCs	Gas exploration of crude oil/natural	Exploration of crude oil/natural gas
	Petroleum refining	Tank/Transport/Leakage loss
		Volatile refining wastewater
	Basic chemical raw materials manufacturing	Ethylene
		Methyl alcohol
		Benzene
Storage and transport	Storage of crude oil/gasoline	Production/Import/Export
	Storage of solvent	Production/Import/Export
	Manufacture of adhesive	Water-based/Other adhesive
	Manufacture of food and drink	Vegetable oil/Sugar refining
		Fermentation alcohol/Wine
	Manufacture of commodity	Manufacture of commodity
Use of VOCs-containing products	Textile dyeing	Textile dyeing
	Synthetic leather	PU size
	Shoemaking industry	Shoe adhesive
	Furniture manufacturing	Wood paintings
	Architectural ornament	Coating for interior walls/other building
		Wood paintings
		Assembling adhesive
	Dry cleaning of clothing	Tetrachloroethylene
	Electronic component manufacturing	Diode/Transistor
		Printed circuit board
		Copper clad laminate
	Coke production	Coke
	Paper production	Pulp/Paper products
	Garbage pollution treatment	Sanitary landfill/Composting
	Thermal power/Heat supply	Coal/Fuel Oil/Liquefied petroleum gas
	Industrial consumption	Coal/Fuel Oil/Coal gas/Liquefied petroleum gas

saturated carbon atom [16], consisting of methanol, ethanol, and propyl alcohol. Ester is the compound derived from an acid in which the –OH group is substituted by the –O– group. Halogenated hydrocarbons are a group of chemical compounds derived from alkanes containing one or more halogens. It becomes possible by far to identify and classify over 300 kinds of VOCs, including benzene, toluene, formaldehyde, and so forth [17].

The physicochemical properties of VOCs also have an impact on the selection of control technology. Generally speaking, organic compounds in the same category exhibit similar properties. Because of the importance of boiling point in definition, VOCs in the same category are sorted by boiling point. The characteristics of commonly used VOCs are listed in Table 4.2.

4.3 Control Principles

4.3.1 Adsorption

Adsorption is an effective method to remove VOCs at low concentrations in the waste gas, which can meet the requirement of a stringent outlet concentration. VOCs of molecular weights from 40 to 500 and boiling points from 40 to 260 °C are suitable for this technique. Adsorption is a phenomenon that occurred at the surface. When gas molecules pass through solid particles, they will be captured selectively by attractive forces. Compared with chemical bonds, those forces are weaker and less specific. Gas molecules migrating from the gas phase to the adsorbent surface release energy, i.e., the "heat of adsorption", which is typically equal to or exceeding the condensation heat. The adsorption capacity for gases over a solid usually increases with the increase of the concentration of the gas, diffusivity, polarity, molecular weight, and boiling point.

The VOC concentration in the outlet of the adsorption process at different times is measured to obtain the breakthrough curves. At the beginning of adsorption, the VOCs are completely adsorbed, so the outlet concentration is zero. However, with the progress of the adsorption, the adsorbent is saturated and the bed can no longer retain all VOCs. Therefore, the outlet concentration of VOCs increases. In the end, the maximum adsorption capacity of the adsorbent is reached and VOCs are no longer retained. So the outlet concentration of VOCs is equal to the inlet concentration [18]. Figure 4.5 exhibits the shape of the breakthrough curve in VOCs adsorption from the gas phase. C_0 refers to the inlet concentration, C refers to the outlet concentration, C_b refers to the outlet concentration at time t_b, and MTZ is short for the mass transfer zone. The concentration distribution of VOCs at different sections of the adsorbent bed is shown in Fig. 4.6.

In the adsorption process, the substance that needs to be removed or collected is called adsorbate, while the agent that adsorbs adsorbate is called adsorbent. The adsorbents which are widely applied in the industrial sector contain activated carbon, zeolites, activated alumina, silica gel, and other clays. The carbonaceous adsorbents are widely used in the VOCs adsorption, especially for activated carbon, owing to their large specific surface area, consequently the high adsorption capacity. Generally, the adsorption capacity of activated carbon is affected by the temperature and pressure

Table 4.2 Physicochemical characteristics of common VOCs

Type	Name	Chemical formula	Molecular structure	Boiling point (°C)	Water solubility	Organic solvents solubility	Molecular weight	Combustion heat (kJ/mol)
Alkanes	Ethane	C_2H_6		−88.6	Insoluble	Soluble	30	−1158.3
	Propane	C_3H_8	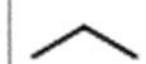	−42.1	Slightly soluble	Soluble	44	−2217.8
	Isobutane	C_4H_{10}	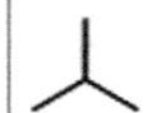	−11.8	Slightly soluble	Soluble	58	−2871.1
	N-butane	C_4H_{10}		−0.5	Soluble	Soluble	58	−2637.8
	Methylbutane	C_5H_{12}	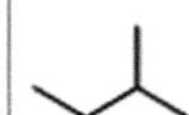	27.8	Slightly soluble	Soluble	72	−3504.1
	N-pentane	C_5H_{12}		36.1	Slightly soluble	Soluble	72	−3245.0
	N-hexane	C_6H_{14}		68.9	Insoluble	Soluble	86	−4159.2
	Cyclohexane	C_6H_{12}	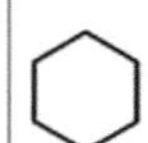	80.7	Insoluble	Soluble	84	−3919.6
	Heptane	C_7H_{16}		98.5	Insoluble	Soluble	100	−4806.6
	Octane	C_8H_{18}		125.6	Insoluble	Soluble	114	−5445.3

(continued)

Table 4.2 (continued)

Type	Name	Chemical formula	Molecular structure	Boiling point (°C)	Water solubility	Organic solvents solubility	Molecular weight	Combustion heat (kJ/mol)
Alkenes	Ethylene	C_2H_4		−103.9	Insoluble	Slightly soluble	28	−1323.8
	Propylene	C_3H_6	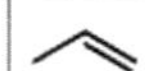	−47.4	Insoluble	Soluble	42	−1927.3
	1-butene	C_4H_8	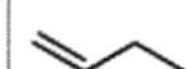	−6.4	Insoluble	Soluble	56	−2719.1
	2- Butene	C_4H_8		1.0	Insoluble	Soluble	56	−2536.0
	1-hexene	C_6H_{12}		63.4	Insoluble	Soluble	84	−3981.9
	Cyclohexene	C_6H_{10}	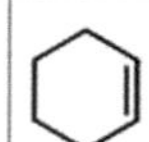	83.0	Insoluble	Soluble	82	−3751.7
Aromatics	Benzene	C_6H_6	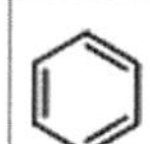	80.1	Insoluble	Soluble	78	−3264.4
	Fluobenzene	C_6H_5F	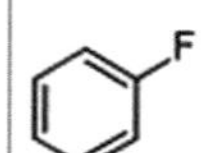	85.2	Insoluble	Soluble	96	−3123.3
	Toluene	C_7H_8	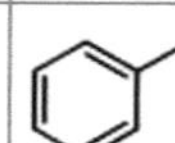	110.6	Insoluble	Soluble	92	−3910.3

(continued)

Table 4.2 (continued)

Type	Name	Chemical formula	Molecular structure	Boiling point (°C)	Water solubility	Organic solvents solubility	Molecular weight	Combustion heat (kJ/mol)
	Chlorobenzene	C_6H_5Cl	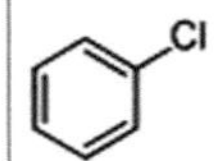	132.2	Insoluble	Soluble	112.5	−3100
	Ethylbenzene	C_8H_{10}	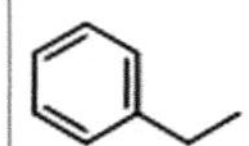	136.2	Insoluble	Soluble	106	−4390.1
	p-Xylene	C_8H_{10}	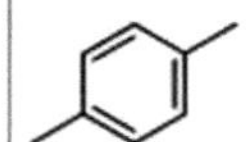	138.3	Insoluble	Soluble	106	−4559.8
	Meta xylene	C_8H_{10}	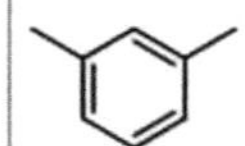	139.1	Insoluble	Soluble	106	−4830.7
	O-xylene	C_8H_{10}	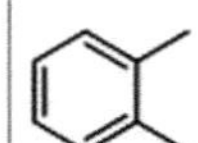	144.4	Insoluble	Soluble	106	−4845.3
	Styrene	C_8H_8	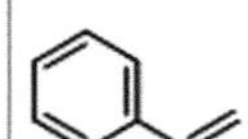	145.2	Insoluble	Soluble	104	−4376.9
	Bromopheryl	C_6H_5Br	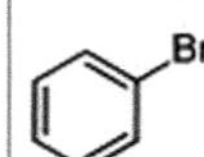	156.2	Insoluble	Soluble	157	−3124.6

(continued)

Table 4.2 (continued)

Type	Name	Chemical formula	Molecular structure	Boiling point (°C)	Water solubility	Organic solvents solubility	Molecular weight	Combustion heat (kJ/mol)
	1,3,5-Trimethylbenzene	C_9H_{12}	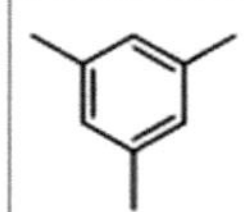	164.7	Insoluble	Soluble	120	−4979.8
	Benzaldehyde	C_7H_6O	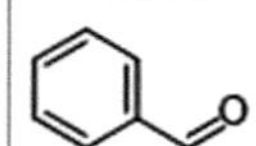	178.1	Slightly soluble	Soluble	106	−3575
	Phenol	C_6H_6O	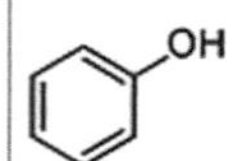	181.9	Slightly soluble	Soluble	94	−3050.6
	SBR	$C_{10}H_{14}$	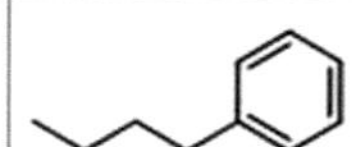	183.3	Insoluble	Soluble	134	−5872.5
	Naphthalene	$C_{10}H_8$	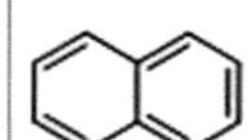	217.9	Insoluble	Soluble	128	−4983
	Biphenyl	$C_{12}H_{10}$	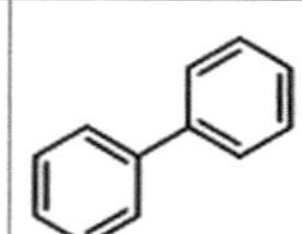	255.2	Insoluble	Soluble	154	−6252.4
Aldehydes	Formaldehyde	CH_2O	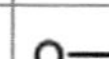	−19.5	Soluble	Soluble	30	−570.77

(continued)

Table 4.2 (continued)

Type	Name	Chemical formula	Molecular structure	Boiling point (°C)	Water solubility	Organic solvents solubility	Molecular weight	Combustion heat (kJ/mol)
	Acetaldehyde	C_2H_4O	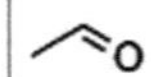	20.8	Soluble	Soluble	44	−1166.37
	Propaldehyde	C_3H_6O		48	Soluble	Soluble	58	−1822.7
	Ethyl two aldehyde	$C_2H_2O_2$	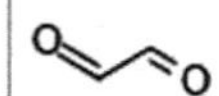	50.5	Soluble	Soluble	58	−860.88
	Acrolein	C_3H_4O	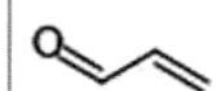	52.5	Soluble	Soluble	56	−1625.74
	Butyraldehyde	C_4H_8O	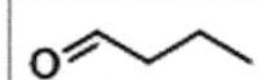	74.8	Slightly soluble	Soluble	72	−2479.34
Ketones	Acetone	C_3H_6O	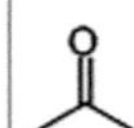	56.5	Soluble	Soluble	58	−1788.7
	2-butanone	C_4H_8O	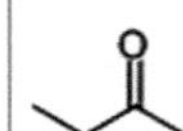	79.6	Soluble	Soluble	72	−2261.7
	2- amyl ketone	$C_5H_{10}O$	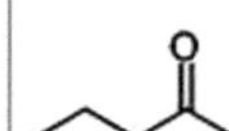	101.7	Slightly soluble	Soluble	86	−3099.4
	3- amyl ketone	$C_5H_{10}O$	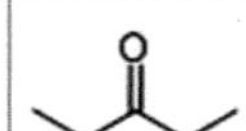	102.0	Slightly soluble	Soluble	86	−3100.2

(continued)

Table 4.2 (continued)

Type	Name	Chemical formula	Molecular structure	Boiling point (°C)	Water solubility	Organic solvents solubility	Molecular weight	Combustion heat (kJ/mol)
	Cyclohexanone	$C_6H_{10}O$	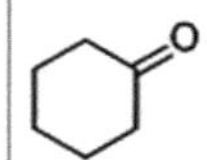	155.6	Slightly soluble	Soluble	98	−3521.3
	Isophorone	$C_9H_{14}O$	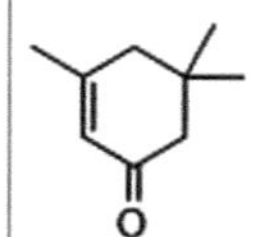	215.2	Slightly soluble	Soluble	138	−5272
Alcohols	Methyl mercaptan	CH_4S	—SH	6	Insoluble	Soluble	48	−1235
	Methanol	CH_4O	—OH	64.7	Soluble	Soluble	32	−723
	Ethanol	C_2H_6O	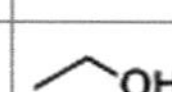	78	Soluble	Soluble	46	−1365.5
	Isopropyl alcohol	C_3H_8O	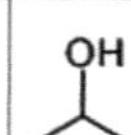	82.5	Soluble	Soluble	60	−1995.5
	Allyl alcohol	C_3H_6O	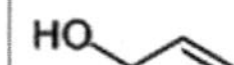	96.9	Soluble	Soluble	58	−1853.9
	Propyl alcohol	C_3H_8O		97.1	Soluble	Soluble	60	−2021.3

(continued)

Table 4.2 (continued)

Type	Name	Chemical formula	Molecular structure	Boiling point (°C)	Water solubility	Organic solvents solubility	Molecular weight	Combustion heat (kJ/mol)
	Isobutyl alcohol	$C_4H_{10}O$	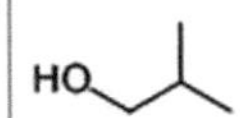	107.9	Soluble	Soluble	74	−2667.7
	N-butanol	$C_4H_{10}O$	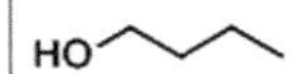	117.6	Slightly soluble	Soluble	74	−2673.2
	Isoamyl alcohol	$C_5H_{12}O$	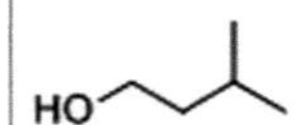	132.5	Slightly soluble	Soluble	88	−3323
	Amyl alcohol	$C_5H_{12}O$		137.3	Slightly soluble	Soluble	88	−3316.2
	Cyclopentaol	$C_5H_{10}O$	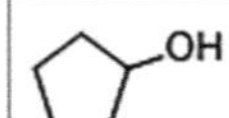	140.8	Slightly soluble	Soluble	86	−3154.2
	Hexyl alcohol	$C_6H_{14}O$		156.5	Slightly soluble	Soluble	102	−4046.2
	Cyclohexanol	$C_6H_{12}O$	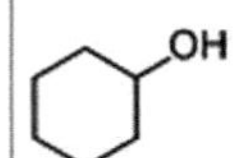	161.1	Soluble	Soluble	100	−3726.7
	1,2- propylene glycol	$C_3H_8O_2$	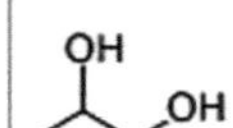	187.3	Soluble	Soluble	76	−1827.5
	Glycol	$C_2H_6O_2$	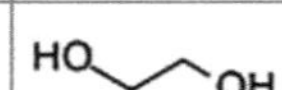	197.3	Soluble	Soluble	62	−1186.1

(continued)

Table 4.2 (continued)

Type	Name	Chemical formula	Molecular structure	Boiling point (°C)	Water solubility	Organic solvents solubility	Molecular weight	Combustion heat (kJ/mol)
	1,2,3- glycerin	$C_3H_8O_3$	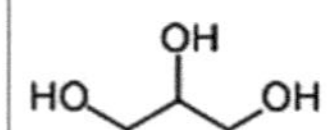	290.9	Soluble	Soluble	92	−1656.4
Esters	Methyl formate	$C_2H_4O_2$	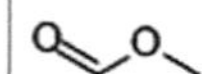	31.5	Soluble	Soluble	60	−973
	Ethyl formate	$C_3H_6O_2$	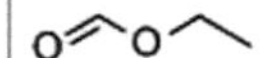	54.3	Slightly soluble	Soluble	74	−1630
	Methyl acetate	$C_3H_6O_2$	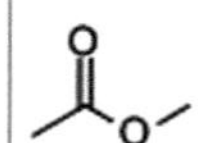	56.8	Slightly soluble	Soluble	74	−1592.7
	Ethyl acetate	$C_4H_8O_2$	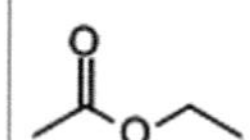	77.2	Slightly soluble	Soluble	88	−2072
	Methyl acrylate	$C_4H_6O_2$	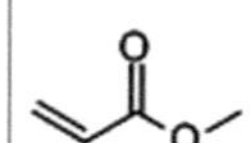	80.5	Slightly soluble	Soluble	86	−2102
	Isopropyl acetate	$C_5H_{10}O_2$	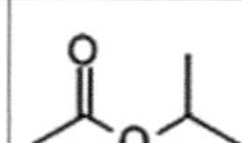	88.4	Slightly soluble	Soluble	102	−2236.6
	Ethyl propionate	$C_5H_{10}O_2$	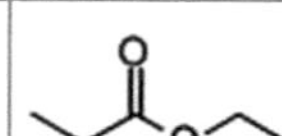	99.1	Insoluble	Soluble	102	−2892.7

(continued)

Table 4.2 (continued)

Type	Name	Chemical formula	Molecular structure	Boiling point (°C)	Water solubility	Organic solvents solubility	Molecular weight	Combustion heat (kJ/mol)
	Methyl methacrylate	$C_5H_8O_2$	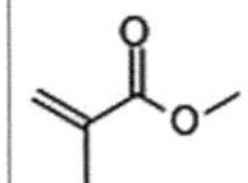	100.5	Slightly soluble	Soluble	100	−2642.9
	Butyric acid ethyl ester	$C_6H_{12}O_2$	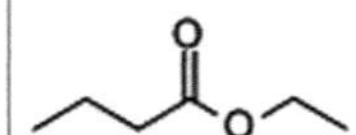	121.3	Insoluble	Soluble	116	−3558
	Acetic acid butyl ester	$C_6H_{12}O_2$	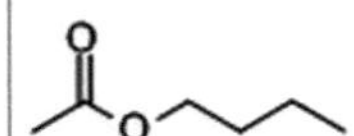	126.1	Slightly soluble	Soluble	116	−3463.5
	Isoamyl acetate	$C_7H_{14}O_2$	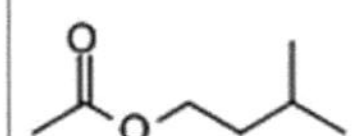	142.5	Slightly soluble	Soluble	130	−4191.8
	Butyl acrylate	$C_7H_{12}O_2$	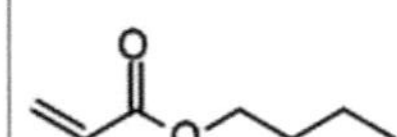	145.7	Insoluble	Soluble	128	−4073.2
	Ethyl lactate	$C_5H_{10}O_3$	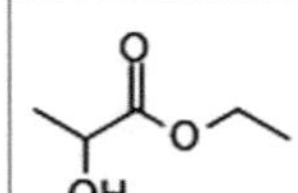	154	Soluble	Soluble	118	−3190.1

(continued)

Table 4.2 (continued)

Type	Name	Chemical formula	Molecular structure	Boiling point (°C)	Water solubility	Organic solvents solubility	Molecular weight	Combustion heat (kJ/mol)
	Phosphoric acid three ethyl ester	$C_6H_{15}O_4P$	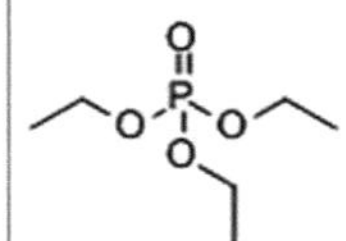	215	Soluble	Soluble	182	−4117.3
Halides	Monochloromethane	CH_3Cl		−23.7	Slightly soluble	Soluble	50	−620.27
	Ethyl chloride	C_2H_5Cl		12.3	Slightly soluble	Soluble	65	−1323.8
	Dichloromethane	CH_2Cl_2	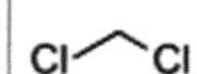	39.8	Slightly soluble	Soluble	85	−604.9
	Chloroform	$CHCl_3$	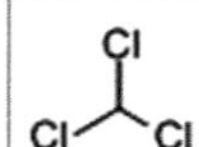	61.3	Insoluble	Soluble	119	−402.23
	Carbon tetrachloride	CCl_4	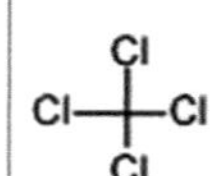	76.8	Soluble	Soluble	154	−258.24
	1,2-dichloroethane	$C_2H_4Cl_2$		83.5	Slightly soluble	Soluble	99	−1243.9
	Trichloroethylene	C_2HCl_3	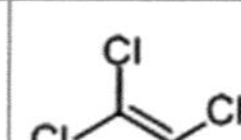	87.1	Insoluble	Soluble	131	−961.4

(continued)

Table 4.2 (continued)

Type	Name	Chemical formula	Molecular structure	Boiling point (°C)	Water solubility	Organic solvents solubility	Molecular weight	Combustion heat (kJ/mol)
	Methylene bromide	CH_2Br_2	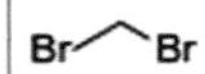	96	Slightly soluble	Soluble	174	–
	Four vinyl chloride	C_2Cl_4	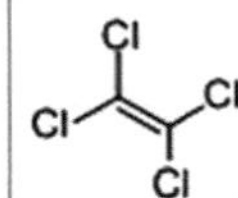	121.2	Insoluble	Soluble	166	−679.3
	1,3- two chloropropane	$C_3H_6Cl_2$		122	Slightly soluble	Soluble	113	−1751.8
	1,2-dibromoethane	$C_2H_4Br_2$		131.4	Slightly soluble	Soluble	188	−1217.9
	Three methyl bromide	$CHBr_3$	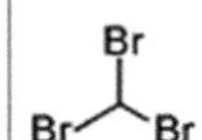	149.5	Slightly soluble	Soluble	253	–
	Four methyl bromide	CBr_4	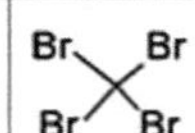	189.5	Insoluble	Soluble	332	
Others	Ethylene oxide	C_2H_4O	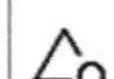	10.8	Soluble	Soluble	44	−306.1
	Isopropyl amine	C_3H_9N	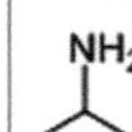	33.5	Soluble	Soluble	59	−2345.5

(continued)

Table 4.2 (continued)

Type	Name	Chemical formula	Molecular structure	Boiling point (°C)	Water solubility	Organic solvents solubility	Molecular weight	Combustion heat (kJ/mol)
	Ether	$C_4H_{10}O$	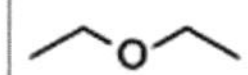	34.6	Slightly soluble	Soluble	74	−2748.4
	Methyl sulfide	C_2H_6S	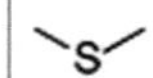	37.3	Insoluble	Soluble	62	−1907.7
	Diethylamine	$C_4H_{11}N$	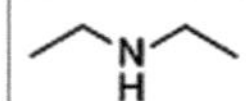	55.5	Soluble	Soluble	73	−3001.7
	Tetrahydrofuran	C_4H_8O	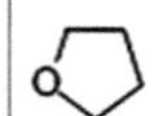	65.4	Soluble	Soluble	72	−2515.2
	Acrylonitrile	C_3H_3N	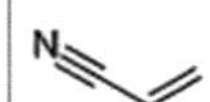	77.3	Slightly soluble	Soluble	53	−1761.5
	Triethylamine	$C_6H_{15}N$	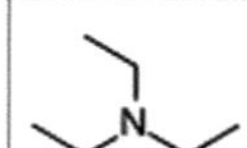	89.5	Slightly soluble	Soluble	101	−4334.6
	3-picoline	C_6H_7N	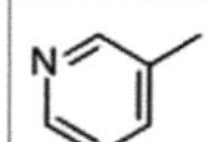	143.5	Soluble	Soluble	93	−3420
	N,N-Dimethylformamide	C_3H_7NO	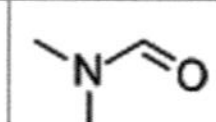	153	Soluble	Soluble	73	−1921

(continued)

Table 4.2 (continued)

Type	Name	Chemical formula	Molecular structure	Boiling point (°C)	Water solubility	Organic solvents solubility	Molecular weight	Combustion heat (kJ/mol)
	Anisole	C_7H_8O	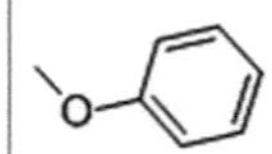	153.8	Insoluble	Soluble	108	−3788.6
	Ethanolamine	C_2H_7NO	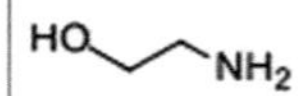	170.5	Soluble	Soluble	61	−924.99
	Terephthalonitrile	$C_8H_4N_2$	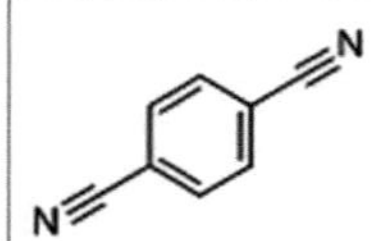	289.6	Insoluble	Soluble	128	–

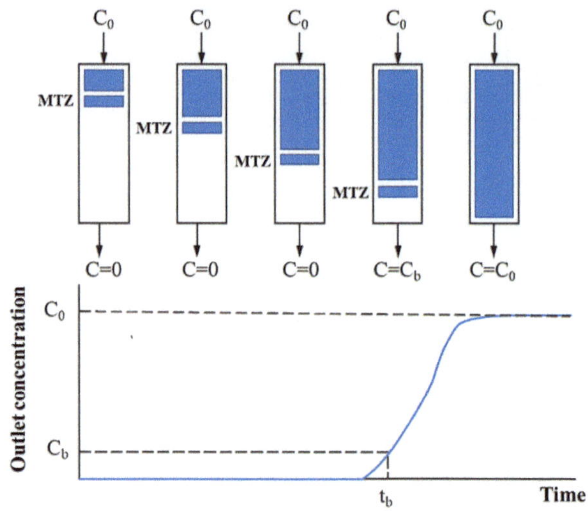

Fig. 4.5 Breakthrough curve of VOC in gas-phase adsorption

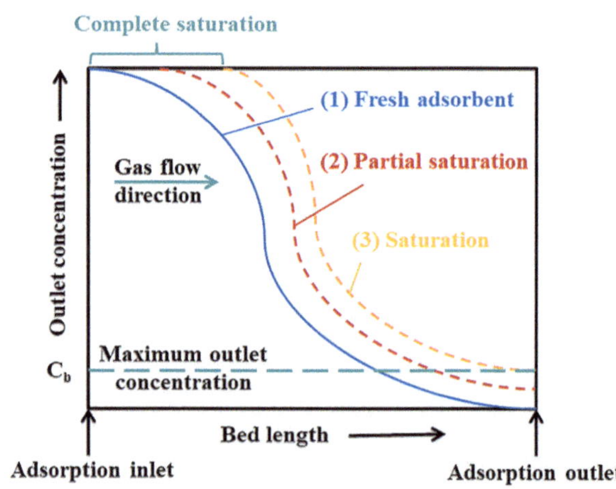

Fig. 4.6 Concentration distribution of VOCs at different sections of the adsorbent bed

[19], the adsorbed species, along with the characteristics of the activated carbon, like particle sizes and pore structures.

Various models have been proposed to study the kinetics of the VOCs adsorption process. Pseudo-first-order rate equation is shown in Eq. (4.9):

$$\frac{dq}{dt} = k_1(q_e - q) \tag{4.9}$$

Another model for the analysis of the VOCs adsorption is pseudo-second-order. The flow rate for this model is expressed as in Eq. (4.10):

$$\frac{dq}{dt} = k_2(q_e - q)^2 \tag{4.10}$$

where

q_e = the equilibrium adsorption capacity (VOCs/adsorbent);
q = the adsorption amount at any time (VOCs/adsorbent);
k_1, k_2 = the rate constant of first-order and second-order adsorption.

An adsorption isotherm of a specific adsorbent-VOCs combination at a specific temperature can be developed. Equation (4.11) is usually used in industrial design [20].

$$q_e = kP^m \tag{4.11}$$

where

q_e = the equilibrium adsorption capacity (VOCs/adsorbent);
P = the VOCs pressure, Pa;
k, m = the empirical parameters.

The equilibrium adsorption capacity is the maximum amount of VOCs which can be held in an adsorbent at a given temperature and pressure. While there are two continuous adsorption beds operating in actual adsorption systems, the adsorbent can never reach equilibrium. Actually, when the outlet concentration of VOCs reaches a limitation, the adsorption is shut down. The limitation in the adsorption system may only be 50% or less of the equilibrium concentration.

4.3.2 *Absorption*

Absorption is the selective transfer of one or more species from a gas mixture to a suitable liquid solvent. The absorption system can be designed to handle recyclable VOCs with high solubility in solvent and outlet concentrations from 500 to 5,000 ppm or higher. The driving force for absorption is the difference between the actual gas-phase concentration of the species and the gas-phase concentration which is in equilibrium with the liquid-phase concentration. Conventional absorption equipment includes packed tower, spray tower, and plate column, which can provide the necessary gas–liquid contact area for mass transfer to take place. In absorption, the recovery efficiency is highly dependent on the solvent and the equipment design. For a given system, absorption efficiency can be increased by the following: (1) increasing the gas–liquid contact area in the equipment, (2) chilling the absorbent liquid, pressurizing the gas stream, and (3) increasing the flow rate of the absorbent, while increasing the cost or complexities of the process.

It is now widely known that no matter how liquid and gas contact, there exists a layer of liquid on the gas side of the interface, which is virtually free from convective

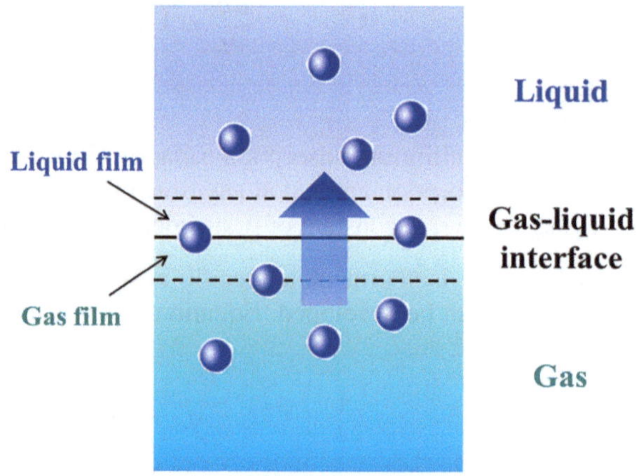

Fig. 4.7 Gas–liquid absorption mechanism (blue spheres are solute molecules)

mixing. This phenomenon is often represented by the assumption that there are fixed membranes of gases and liquids on both sides of the interface. The mechanism of gas–liquid absorption is shown in Fig. 4.7.

The diffusion of the gas film is carried out at a rate proportional to the difference of solute concentration in the gas inside and outside the gas film. The diffusion is controlled by the difference between the solute concentration in the gas and the solute concentration in the liquid. Since the surface films are very thin, the amount of solute contained in them is usually negligible when comparing to the amount diffusing.

The potential or driving force leading to diffusion is expressed by the concentration difference of any of these films. The only resistance that diffusion molecules encounter is because they collide more frequently with interfering molecules of gas or liquid, and diffusion is taking place [21].

Gases are soluble in liquids, which are related to the properties of solvents and are affected by temperature and pressure. The Henry law can be used to describe the phase equilibrium between gas and liquid for physical absorption as Eqs. (4.12) and (4.13). At a given temperature, when the total pressure is low, the solute in dilute solution is proportional to the pressure in the gas phase

$$p_i^* = E_i x_i \tag{4.12}$$

$$c_i = H_i p_i^* \tag{4.13}$$

where

p_i^* = the partial pressure of a species in the gas phase under equilibrium conditions, Pa;
x_i = the molar mixing ratio in the aqueous phase;
E_i = Henry's coefficient, Pa;
H_i = solubility coefficient, kmol/m^3 Pa;
c_i = the concentration of a species in the aqueous phase, kmol/m^3.

Most of the VOCs absorption systems are gas–liquid reactors, which generally require a large gas–liquid contact area, high gas–liquid turbulence flow, and low-pressure loss of the equipment. The design index for the filled tower includes liquid-to-gas ratio, packed bed height, and tower diameter. The liquid–gas ratio is an important design parameter, which affects the gas–liquid equilibrium under operating conditions and the economy of the absorption process. Packed bed height is calculated by the number of mass transfer units and the height of mass transfer units. For an industrial absorption tower, the height of the mass transfer unit is usually in the range of 1.5–1.8 m. Tower height mainly depends on the volume flow rate of gas and the gas velocity of the empty column. The gas flow rate is generally controlled within 75% of the flooding velocity. Flooding velocity can be calculated by semi-empirical equation or found in correlation diagram, as shown in Figure 4.8 [7],

4.3.3 Condensation

A condensation system is usually used on gas streams that contain only VOC compounds, which condenses the organic compounds in gas stream on cold surfaces. It is efficient for recyclable VOCs of boiling points above 40 °C at VOCs concentrations above 5000 ppm. The cold surfaces could be generated by different cooling methods: (1) cooling water flowing through the indirect heat exchanger, (2) using a refrigerant to create very cold coils, (3) spraying cold liquid to a chamber with the gas stream, and (4) injecting cryogenic gases into the gas stream. Therefore, the VOC concentration is reduced to the level equivalent to the corresponding vapor pressures at the operation temperature. Generally, condensation systems are used in conditions with high concentration and low gas flow rate sources [22]. The mechanism of the VOCs condensation is shown in Fig. 4.9.

Through chilling and/or pressurization of the VOC control system, over-saturation is achieved. For VOCs with a low boiling point, extensive cooling is required in which the operating costs obviously increased. For condensation systems, it is more common to exceed the 25% lower explosive limit (LEL). Some systems even start operation above the upper explosive limit (UEL). In that situation, it is dangerous since the concentration probably falls through the explosive range during the condensation process. Therefore, it is suggested to use an inert gas-coated container or unattended process shell to avoid exposure to hazardous VOCs [23], though causing

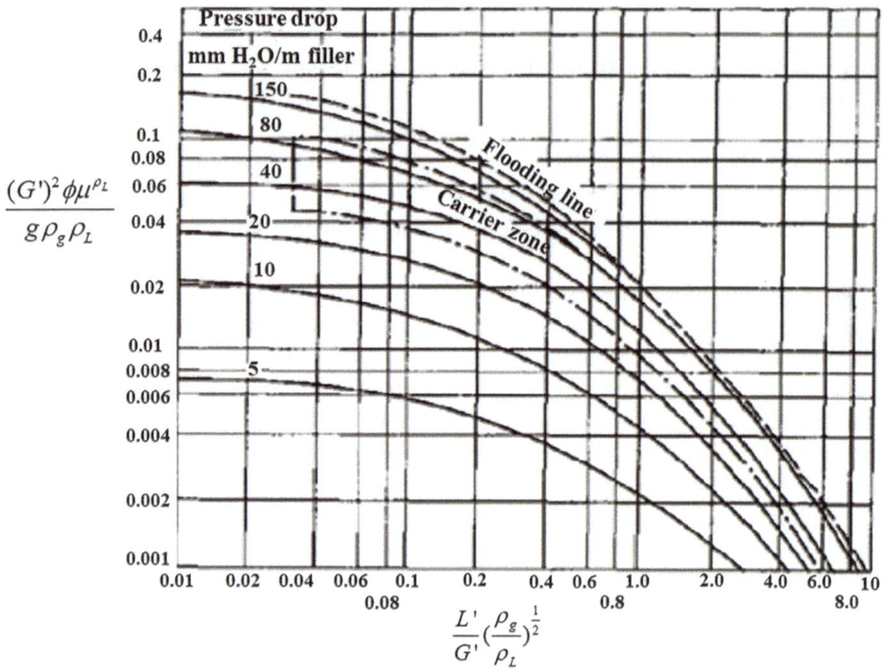

Fig. 4.8 Correlation diagram of flooding velocity and pressure drop in the packed tower. Where G', L' = mass flow velocity of gas and liquid, kg/(m^2·s); ρ_g, ρ_L = density of gas and liquid, kg/m^3; Φ = packing factor, 1/m; μ = liquid viscosity, cP; g = gravity acceleration, m/s^2

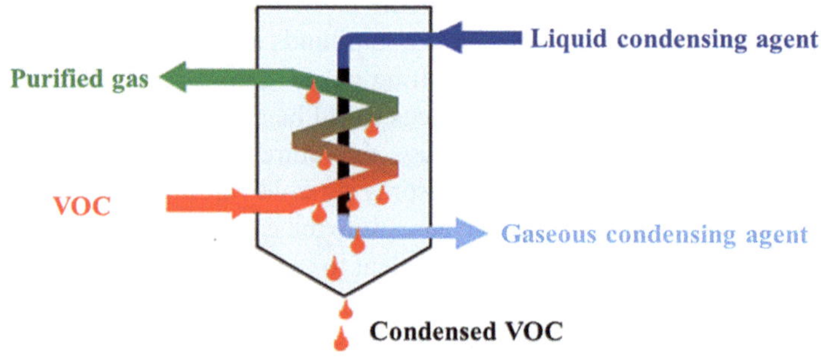

Fig. 4.9 Condensation mechanism of VOCs

additional costs for operation. Polymerization materials should also be avoided to use in the condensation system in order to prevent fouling the heat-exchanger surface. As the recovery method for VOCs control, condensation is suitable for mono solvent systems. The liquid product is produced during a condensation process, which requires the dehydration process for removing the condensed water and the separation process for separating the species. The VOCs after recovery can be handled in the following ways: (1) shipped off-site for disposal, (2) used as washing solvents during equipment cleanup, and (3) combusted as an alternative fuel.

The partial pressure of VOCs, together with the initial concentration and the condensing temperature, play important roles in the removal efficiency. Equations (4.14) and (4.15) can be obtained as follows:

$$p_p = m_0 p_1 \tag{4.14}$$

$$m_0 = \frac{m_i(1-\eta)}{1-\eta m_i} \tag{4.15}$$

where

p_p = VOC partial pressure in the outlet gas, Pa;
p_1 = system pressure, Pa;
m_0 = VOC mole fraction in outlet gas;
m_i = VOC mole fraction in inlet gas;
η = removal efficiency, %.

The condensation system is shown in Fig. 4.10. The molar fraction of component i in inlet gas is z_i. The composition of condensed and uncondensed VOCs is x_i and y_i, respectively. Liquefied fraction f can be calculated by Eq. (4.16):

$$f = \frac{B}{F} \tag{4.16}$$

The material balance equation in the condenser is shown in Eq. (4.17):

$$F = B + D \tag{4.17}$$

where

B = Molar velocity of condensate discharge, kmol/h;
F = Molar velocity of inlet VOCs, kmol/h;

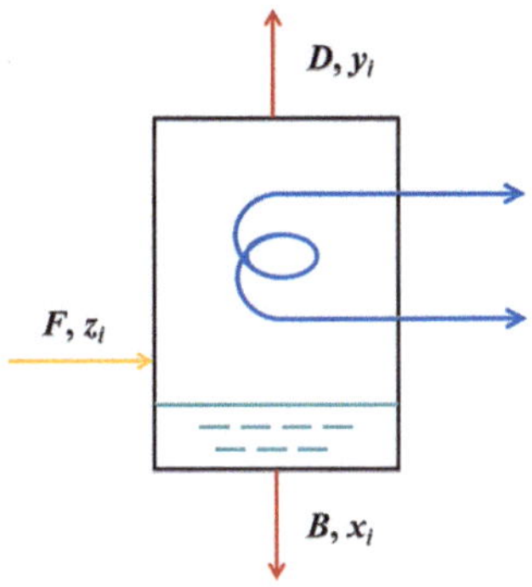

Fig. 4.10 Material balance calculation of condensation system

D = Molar velocity of outlet VOCs, kmol/h.

Material balance for component i is calculated by Eq. (4.18):

$$F \cdot z_i = (1 - f)F \cdot y_i + f \cdot F \cdot x_i \tag{4.18}$$

Other balance equations include Eqs. (4.19–4.21):

$$y_i = m_i \cdot x_i \tag{4.19}$$

$$\sum_{i=1}^{n} x_i = 1 \tag{4.20}$$

$$\sum_{i=1}^{n} y_i = 1 \tag{4.21}$$

When water is used as the cooling medium, the total heat exchange Q in the process of contact condensation can be described as Eq. (4.22):

$$Q = q_{m.w} \cdot c_w \cdot (t_2 - t_1) = q \cdot \Delta H + q_m \cdot c_p \cdot (t_2 - t_1) + q_{m,g} \cdot c_p' \cdot (t_2 - t_1) \tag{4.22}$$

where

$q_{m,w}$, q_m, $q_{m,g}$ = cooling water consumption, kg/h;
c_w, c_p, c'_p = heat capacity of water, condensate, exhaust, kJ/(kg·°C);
t_1, t_2 = temperature of cooling water inlet and outlet, °C;
ΔH = the condensation latent heat of VOCs, kJ/kg.

The heat calculation of the surface condenser is the same as that of the common heat exchanger. The heat exchanger is calculated according to the heat transfer theory, and the total heat exchange Q can be calculated by Eq. (4.23):

$$Q = K \cdot A \cdot \Delta t_m \tag{4.23}$$

where

K = heat transfer coefficient, kJ/(m^2·h·°C);
A = heat transfer area,m^2;
Δt_m = log mean temperature difference, °C.

4.3.4 Oxidation

Control technologies by oxidation of VOCs include thermal combustion, catalytic oxidation, plasma-assisted oxidation, and photo-catalytic oxidation. It is the most suitable treatment technology for unrecyclable VOCs which are toxic, harmful, and are in high concentration. When the concentration of the stream is between 15 and 20% of the lower explosive limit (LEL), then both thermal combustion and catalytic oxidation are the better technologies. For the control of gas between 20 and 25% LEL, the thermal combustion is the preferred choice.

When various combustible VOCs are mixed with air, the approximate formula for calculation of LEL is shown in Eq. (4.24):

$$c_m = \frac{100}{\frac{a}{c_1} + \frac{b}{c_2} + \cdots + \frac{m}{c_i}} \tag{4.24}$$

where

c_m = explosive limit of mixture;
c_i = explosive limit of component i;
a, b, m = percentage content of component.

During the oxidation process, the combustible VOCs react with an oxidant or auxiliary oxidation agent at a certain temperature. The final products of complete oxidation of VOCs are mainly CO_2 and H_2O. The energy change during the reaction is shown in Fig. 4.11. The amount of reaction heat released in combustion can be expressed by ordinary thermochemical equations, taking hydrocarbons as an example, and the specific formula is expressed as Eqs. (4.25) and (4.26):

$$C_xH_y + \left(x + \frac{y}{4}\right)O_2 \rightarrow xCO_2 + \frac{y}{2}H_2O + Q \tag{4.25}$$

$$r = -\frac{dc_{VOCs}}{dt} = kc_{VOCs}^n \tag{4.26}$$

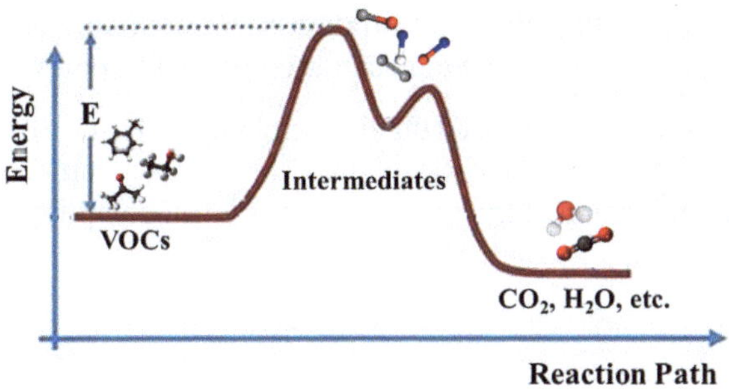

Fig. 4.11 Energy change during the oxidation reaction of VOCs

where

Q = the amount of heat released by the reaction, kJ;
r = the reaction rate, mol/L·s;
k = the reaction rate constant;
c_{VOC} = the concentration of the VOC, mol/L;
n = the reaction order of the VOC.

For most oxidation reactions, the relationship between the kinetic rate constant k and temperature T is expressed by the Arrhenius equation:

$$k = A\exp\left(-\frac{E}{RT}\right) \tag{4.27}$$

where
A = pre-exponential factor;
E = activation energy, kJ/mol;
R = universal gas constant, J/(mol·K);
T = reaction temperature, K.

Although this method cannot recycle useful organic compounds, the energy can be recovered since a large amount of heat is released during oxidation. The total energy released as heat when VOCs undergo complete oxidation with oxygen under standard conditions is called the combustion heat (kJ/mol). The combustion heat of common VOCs at 25 °C and 101 kPa is shown in Table 4.2.

The oxidation system should control the amount of VOCs released to the environment. The evaluating indicator of the VOCs oxidation system can be described by the removal efficiency (%), as defined in Eq. (4.28):

$$\eta = \frac{c_{inlet} - c_{outlet}}{c_{inlet}} \times 100 \tag{4.28}$$

where
η = the removal efficiency of VOCs, %;
c_{inlet} = the inlet concentration of VOCs, ppmv or mg/g;
c_{outlet} = the outlet concentration of VOCs, ppmv or mg/g.

It should be noticed that incomplete combustion of VOCs may result in the gather of other by-product. This situation requires comprehensive chemical analysis of the incoming and outgoing airflow to confirm compliance.

4.4 Control Technologies

Several end-of-pipe technologies including physical, chemical, and biological treatments can achieve the removal of VOCs by either recovery or destruction [1]. If the exhaust stream contains only one single kind of relatively concentrated VOC and if the cost of recovery is less than that of purchasing a new VOC, the VOC stream is of recovery value. Adsorption, condensation, and absorption are typical recovery techniques. VOCs with molecular weights from 40 to 500 and boiling points from 40 to 260 °C are suitable for adsorption. The absorption system is designed to control recyclable VOCs with high solubility in solvent and outlet concentrations from 500 to 5,000 ppm or higher. While condensation is an efficient technology for recyclable VOCs of boiling points above 40 °C at VOCs concentrations above 5000 ppm.

If the VOC stream contains other substances, such as toxic compounds, then the appropriate techniques should be destruction. Oxidation and bio-filtration are typical techniques in this case. An organizational diagram of various VOC control techniques is shown in Fig. 4.12. Because of the different characteristics and application scope of different technologies, there is no absolute advantage of one technology in practical application. The nature of industrial VOCs that need to be treated in practical engineering is also diverse, resulting in the co-existence of a variety of technologies. Processing engineering does not necessarily choose the most popular mainstream technology, but chooses the most appropriate technology according to the processing requirements and different technical characteristics.

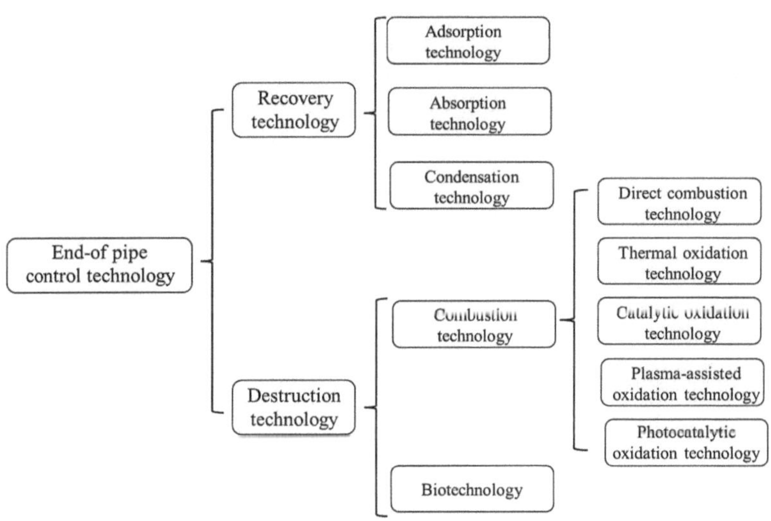

Fig. 4.12 Classification of end-of-pipe VOCs control techniques

4.4.1 Adsorption

As discussed before, adsorption is commonly used in VOCs emission control. It could be divided into two types including physisorption and chemisorption according to the interaction between VOCs and adsorbents [2]. The VOC removal efficiency by adsorption usually exceeds 95%. For solvent recovery and pre-concentrator-type system, the removal efficiency is often in the range from 98 to 99%. Figure 4.13 shows the diagram of an adsorption system used for the collection and recovery of VOCs.

The inlet VOC gas is usually at high temperature and should be cooled to room temperature before adsorption to increase the adsorption capacity [24]. When the VOCs gas passes through the adsorption bed, they are adsorbed onto the adsorbent surface, such as activated carbon and zeolite. Desorption of VOCs is always taken by the water stream and/or nitrogen to remove the adsorbed organics by sweeping and heating. VOCs are then recovered by the stream or nitrogen flow from the desorption cycle. After desorption, the adsorbent on the bed is back to service with another bed being isolated and desorbed [22]. A comparison of various adsorbents for VOCs adsorption is shown in Table 4.3.

A case-to-case basis should be adopted when considering the suitability of an adsorption system for specific situation. Generally, adsorption systems are suitable for treating VOCs with molecular weights ranging from 40 to 150 [2]. When the molecular weight of VOCs is low, they usually do not sufficiently adsorb. While for high molecular weight VOCs, they strongly adsorb, making them difficult to be removed from the adsorbents. Adsorption systems are suitable for VOC stream of concentrations from 10 to 10,000 ppm, in which the upper concentration limit is in case of the possible hazards of explosion when the total concentration of VOCs is higher than 25% of the lower explosive limit (LEL) [22]. In addition, the adsorption

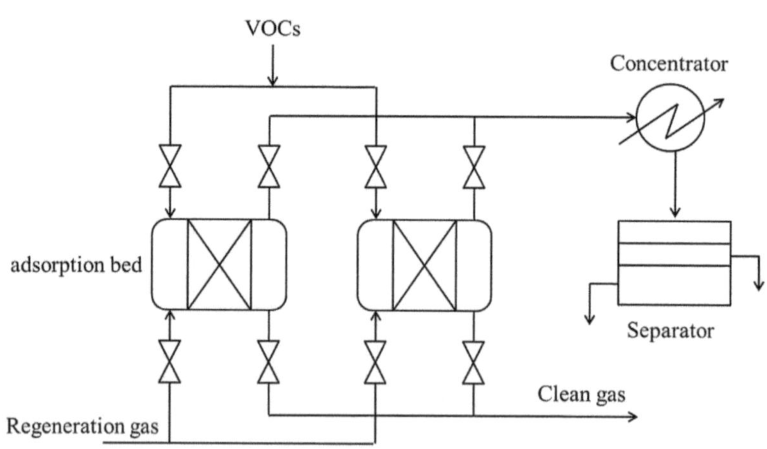

Fig. 4.13 Adsorption system for VOCs recovery

Table 4.3 Comparison of various adsorbents for VOCs adsorption [25–27]

Adsorbent	Activated carbon	Molecular sieve	Silica gel
Material source	Coal, shell, wood, etc	Silicon and aluminum compound	Sodium silicate and vitriol
Operating temperature	<50 °C	< 100 °C	< 50 °C
Regeneration	Thermal, electrochemistry, wet oxidation	Thermal regeneration	Thermal regeneration
Advantages	Large specific surface area, high removal efficiency	Stability, high adaptability to operation condition	Low price, stable chemical properties
Disadvantages	Combustible, difficult to regeneration	Narrow in pore size, unsuitable for macromolecules	Low efficiency at high humidity gas

system is not appropriate when the pending gas stream contains particulate matter (PM) and/or the moisture concentration of gas stream is relatively high. The reason is due to the competitive adsorption in pore spaces between the PM and moisture with gaseous pollutants.

4.4.2 Absorption

Absorption is different from adsorption, in which the VOC component condenses on the cold surface of the solid phase. In Absorption, the VOCs-containing gas contacts liquid solvent directly and soluble VOCs could transfer to the liquid absorbent. The diagram of an absorption system is shown in Fig. 4.14.

When gas flow containing VOCs enters from the bottom of the tower, it is absorbed by the adsorbent through countercurrent contact, and the gas after purification is discharged from the top of the tower. An adsorbent with VOCs passes through the heat exchanger, and desorption occurred when the temperature is higher than the absorption temperature or pressure is lower than the desorption pressure. The absorbents after desorption are condensed into the absorber tower. The desorbed VOCs pass through the condenser and are further recycled by a gas–liquid separator.

The absorption system is easy to operate and has high removal efficiency. This method is suitable for the treatment of most VOCs and is widely used in industry. While most of the absorbents are toxic organic solvents. Consequently, secondary pollution would be generated in the subsequent absorption process. Hence, it is important to select the solvent carefully. The following factors are general to be considered in the choice of absorbents: (1) high solubility of gas; (2) low viscosity; (3) low saturated vapor pressure and low volatility; (4) low melting point, high boiling point, and stable chemical properties; and (5) non-toxic, low corrosion to equipment. The common absorbents are mineral oi and oil based organic solvent

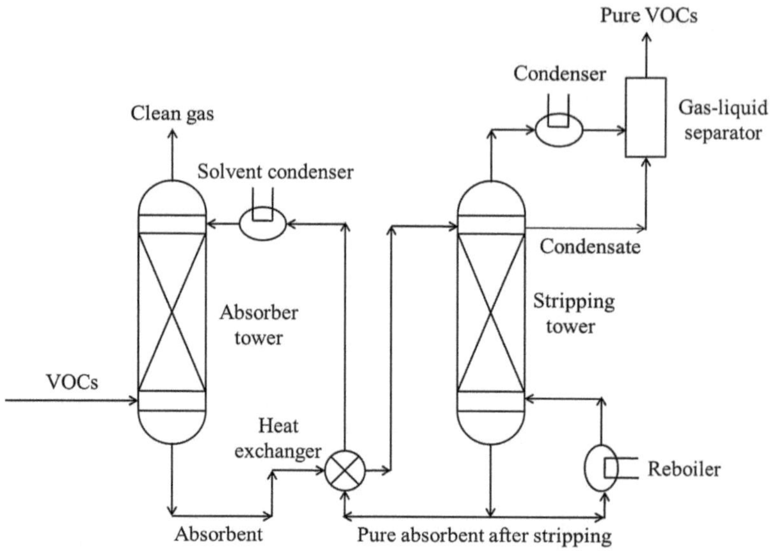

Fig. 4.14 Absorption system of VOCs recovery

with high boiling point (water-oil, water surfactant additives). The main absorbents of different VOCs are shown in Table 4.4.

However, the applicability of absorption is limited by the suitable solvent for VOCs in the waste gas. In addition, a single solvent may not be efficient for all VOCs present in the stream. Consequently, the final recovery efficiency is limited by not only the absorption capacity but also by the subsequent regeneration of VOCs. The absorption system also requires various auxiliary units like chiller, heat exchanger, vacuum units, condenser, and several pumps.

Table 4.4 Main absorbents of different VOCs

VOCs	Absorbent	Characteristic
Mineral oil	Light diesel oil, engine oil, etc	High absorption capacity, complex components, volatility, and high price
Water mixed solvent	Water–oil, water–water–oil, waste oil, water and surfactant, water–acid, water and alkali, etc	The absorption effect is slightly lower, the price is low, the volatilization loss is less, and there is no secondary pollution
Organic solvent of high boiling point	Phthalic acid esters, silicone oil, phenols, etc	The utility model has the advantages of high absorption efficiency, great difficulty, uneven liquid distribution, large equipment pressure difference, and high price

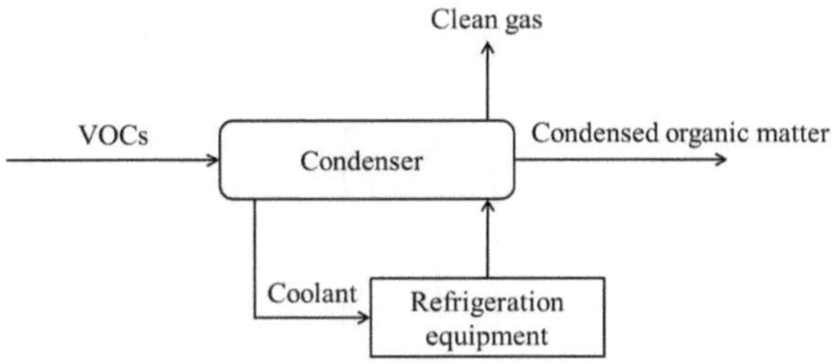

Fig. 4.15 Condensation system of VOCs recovery

4.4.3 Condensation

Cryogenic condensation can achieve great economic benefit by the VOCs components enrichment from the low-concentration gas stream of chemical processing facilities under certain conditions. Figure 4.15 shows the diagram of a condensation system used for the removal of VOCs, which is the process worked by cooling down the temperature of the gas stream.

Generally, a refrigerator is involved in the condensation system to cooling the exhaust gas until the temperature is below the dew point of VOCs. Then, the gaseous VOCs condense on the cold surface of the condenser and are separated with the gas stream [28]. The condensate is similar to the moisture on the inside of a window in winter. The temperature of the window is below the dew point temperature, making water vapor in the gas stream condense on the cold surface. If the gas stream contains a single VOC component, the single VOC component will condense and be separated at a constant temperature. If the gas stream contains other non-condensable gases or VOC components with various volatilities, the condensation and separation temperature will fluctuate in a certain range. The decrease of outlet temperature can convert more VOCs from the vapor phase to the liquid phase [29].

VOC condensation is dependent on the following parameters: (1) mass flow rate and molecular weight, (2) operating temperature, (3) operating pressure, (4) VOC vapor pressure, (5) characteristics of other condensable components, and (6) pressure drop across the heat exchanger. Generally, the reclamation decreases with (1) higher non-condensable flow rate, (2) lower operating pressure/ higher pressure drop across the condenser, (3) higher vapor pressure of a VOC, and (4) higher concentration of other VOCs. With low temperature and/or low vapor pressure, the reclamation increases [29].

4.4.4 Oxidation

1. Thermal combustion

Thermal combustion of regenerative thermal oxidizer (RTO) is one of the most widely used VOCs emission abatement technologies in industrial emissions. The system

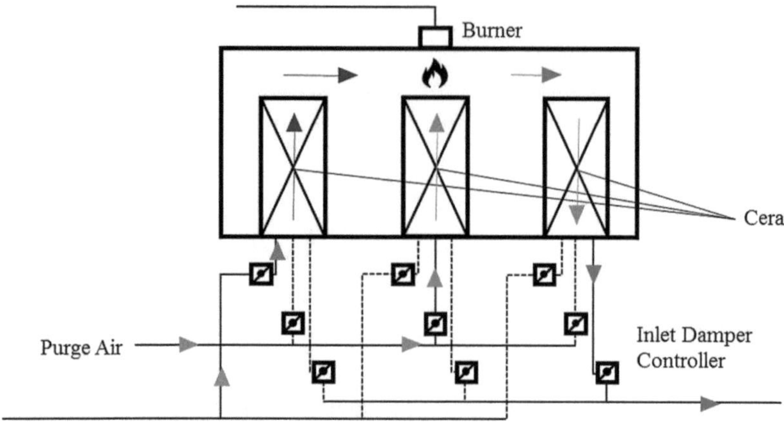

Fig. 4.16 Diagram of a thermal combustion system

achieves VOCs destruction through thermal combustion at a very high temperature in a bed of ceramic material. A thermal combustion system can convert various VOCs into carbon dioxide and water vapor by different combinations of oxidation temperature, residence time, and oxygen concentration. Sometimes, the system may need additional auxiliary heating due to the low VOCs concentration in the exhaust gas. The captured heat is used to preheat waste gas to reduce the operating cost.

A typical thermal combustion system is shown in Fig. 4.16. The production of gas streams is usually via the processes which need ventilation, including printing, paint booths, and paper mills. The features of this class of exhaust gas are low VOC concentrations and high gas flow rates. The operation of the system basically involves: (1) a hot gas stream being passed over a heated sink; (2) the heat being recovered by passing a cold gas stream through that sink in an alternate cycle.

The thermal combustion operation process contains four basic steps:

(1) Heating: exhaust gas containing VOCs and HAPs is introduced into a hot ceramic material in one tower, making the gas heated up to ~ 700 °C.
(2) Incineration: the introduced gas flow is heated by natural gas combustor and the temperature rises from 700 to ~ 800 °C, then VOCs in the introduced gas are oxidized to form CO_2 and water vapor.
(3) Recovery: the superheated gas flows through the second tower containing the same ceramic material, followed by recovering most of the heat in the superheated gas.
(4) Purging: in the three-tower design only, a small amount of air is purged into the combustion chamber in order to prevent any exhausts from escaping in the third tower.

When the first tower cools and the second tower is heating up, the flow direction through the towers reverses periodically so that the heat can be recovered from the burner.

Table 4.5 Main parameter types of thermal combustion incinerator

Parameter types	Value	Unit	Value	Unit
Air inlet air volume	59,500	Nm^3/h	35, 000	scfm
Exhaust gas inlet temperature	21	°C	70	°F
Solvent inlet air volume	0	g/m^3	0	g/ft^3
Combustion temperature	816	°C	1,500	°F
Heat recovery efficiency	96	%	96	%

Fig. 4.17 Diagram of a catalytic oxidation system

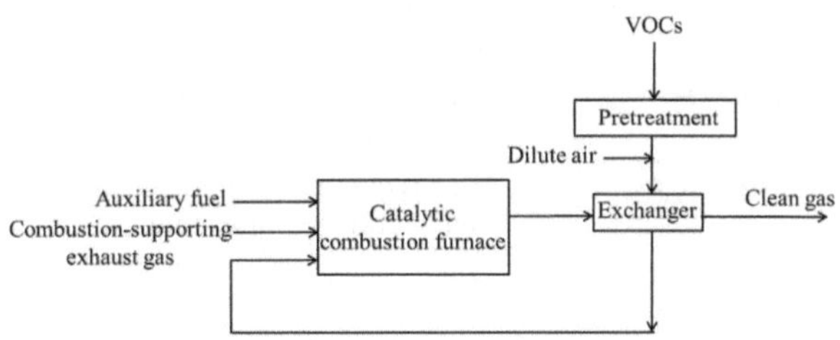

The design parameters of a typical incinerator are shown in Table 4.5.

2. Catalytic oxidation

In a regenerative catalytic oxidizer (RCO) system, VOCs are finally oxidized similar to thermal combustion. While the catalytic oxidation system operates at a lower temperature, usually in the range of 250–550 °C, whereby catalysts are used to save energy. The inlet VOC is heated in the heat exchanger, along with passing through a catalyst-coated honeycomb or monolithic support. Catalysts can be designed to treat VOC concentrations in the range of 100–2,000 ppm. Catalytic oxidation is suitable for VOCs of low concentration. Therefore, the application of catalyst in VOCs oxidation can handle the variable flow rates and VOCs contents [2]. Figure 4.17 shows a diagram of the typical catalytic oxidation system.

Though catalytic oxidation system can achieve great abatement efficiency, this technology is not as popular as thermal combustion in industrial sector. The high cost of catalyst replacement is the main barrier. In addition, catalytic oxidation can produce secondary combustion wastes. Up to now, catalytic oxidation for gaseous pollutants control is limited to the VOCs which only contain C, H, and O elements. While gases containing S, Cl, and other elements, which could result in the deactivation of the catalysts, are not suitable to be treated using catalytic oxidation systems. Most catalysts are composed of mono- or multi-metal oxides as the active component, which are always supported on a carrier with good mechanical strength, e.g., alumina. The development of poison-resistant catalysts mainly focuses on chlorinated VOCs (CVOCs), since these compounds are widely applied as degreasers and solvents. Chromium oxide, copper oxide, and manganese oxide are often used as the active component for the CVOC oxidation [30]. Platinum-based catalysts are found to be efficient for the oxidation of VOCs with S element. However, Pt-based

Table 4.6 Comparison of thermal combustion and catalytic oxidation technology [33]

Technology	Thermal combustion	Catalytic oxidation
Removal efficiency	>98%	>90%
Equipment investment	Middle	High
Running cost	High	Low
Maintenance cost	Middle	High
Area	Middle	Middle
Safety	High	High
Utility condition	Compressed air, electricity, gas, or fuel	Compressed air, electricity, gas, or fuel
Advantages	High removal efficiency, low energy consumption	Low ignition temperature, recovery waste heat
Disadvantages	High requirements for insulation at high working temperature	The catalyst cost is high and the removal effect is not stable

catalysts are easy to be deactivated by chlorine [31]. Compounds containing arsenic, lead, and phosphorous are considered to cause deactivation for most of the oxidation catalysts, though the concentration is always low. Hence, the deactivation rate is also low, along with a longer life span and lower catalyst replacement costs [32].

The comparison in engineering application of thermal combustion and catalytic oxidation technology is shown in Table 4.6.

3. Plasma-assisted oxidation

Plasma is the fourth state of matter, which is different from solid, liquid, or gas. The ionization makes the plasma affected by electromagnetic fields, which also obviously influence the behaviors of the plasma. Plasma can only be artificially created. For example, when being heated, gas will become plasma until the atoms lose all the electrons. As a result, only highly charged nuclei and free electrons are left.

Corona discharge plasma technology is regarded as a promising new technology in VOCs control. Corona discharge is produced by applying negative pulse voltage to the wire. The advantage of a plasma-assisted oxidation system on the DC corona is the possibility to acquire high-energy electrons even using small power through applying the voltage as fast-rising pulses. In fact, the electron energy is dependent on two issues: (1) average time of collision between electrons and other molecules; (2) electric field strength.

Gaseous VOCs can be destructed via oxidation to harmless small molecule products (mainly CO_2 and H_2O). The most efficient method to oxidize organics is to use species with high oxidizing potential, such as ·H, ·O, ·OH, O_3, and H_2O_2. Figure 4.18 shows the reactor of the plasma treatment system for VOCs removal.

4. Photo-catalytic oxidation

Photo-catalytic oxidation (PCO) is cost-effective and has recently attracted considerable attention for air pollution abatement. Removal of VOCs could be achieved

Fig. 4.18 Reactor of VOCs purification by plasma

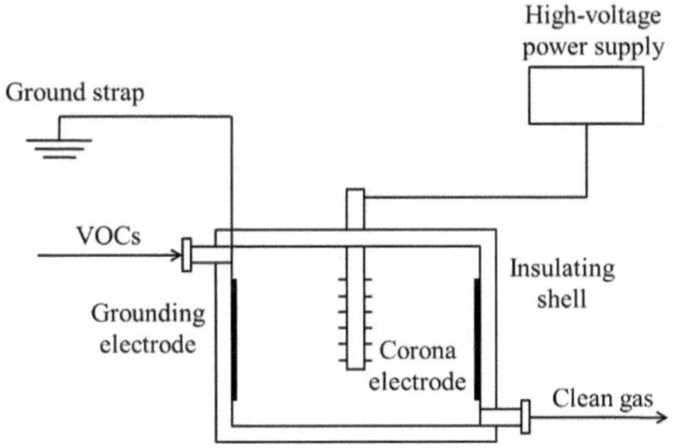

by using PCO, which is a promising and cost-effective approach. In a PCO process, VOCs can be oxidized to harmless carbon dioxide and water. Furthermore, in the low operating temperature and pressure, inexpensive semiconducting catalysts in the PCO system make it suitable for the oxidation of various pollutants including both organic and inorganic compounds.

The key to the PCO process is the photo-catalysts. The development of active catalyst systems is crucial. TiO_2 has the advantages of strong oxidizing power, high photo-catalytic activity, chemical inertness, and physical stability, making it being considered as the best semiconductor photo-catalyst available at the present. The overwhelming majority of the literature concerning photo-catalysis at present is mainly focusing on identifying the properties, applications, and theories of using titanium dioxide as a semiconductor.

The removal of VOCs on the TiO_2-based photo-catalyst is illustrated in Fig. 4.19. A typical photo-catalytic process is very complicated, with many factors affecting photo-catalytic efficiency. The PCO reaction process can produce various intermediates with different oxidation states. These organics can occupy the active sites on the photo-catalyst surface temporarily or permanently, resulting in suppression or even

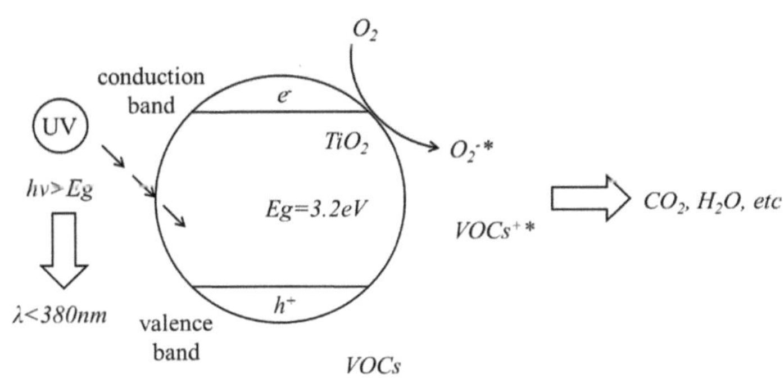

Fig. 4.19 Photo-catalytic degradation of VOCs by TiO_2

termination of the reaction. The high removal efficiency of these intermediates is required because these components are often more toxic than the initial VOCs. The emission standard or ambient concentration value is especially critical if these components can be discharged into the indoor environment. Advanced approaches for re-activation and regeneration of photo-catalysts are also very important for extending their lifetime so as to allow long-term VOC degradation.

4.4.5 Biological Treatment

The biological treatment has proven to be an effective and cheap technology for the removal of industrial VOCs. In this technique, microorganisms under aerobic conditions have the ability to convert VOCs to H_2O, CO_2, and biomass [2]. Aqueous liquid or wet media inoculated with microorganisms can absorb and consume the dissolved and/or adsorbed organic compounds through microbial decomposition. Then the VOCs abatement can be achieved. An irrigated packed bed is used to host the microorganisms. A pre-saturator is used to increase the relative humidity (>95%) of VOCs-containing gas stream and is often placed in front of the biological system [34].

Biological treatment is gaining increasing attention in recent years. Compared with other waste gas purification methods, this technique has some important advantages: (1) mild operating conditions, (2) generally not transferring the pollution to another environmental compartment. Many other purification methods cannot achieve the latter one. In addition, biological treatment is very effective for low-concentration (i.e., at few ppm levels) odors or toxic waste gases [2].

There are three main industrial techniques, as shown in Table 4.7 and Fig. 4.20. (1) Bio-trickling filtration. When the waste gas is humidified by the tank, the microorganisms are responsible for purification when the gas passes through the medium filter; (2) Bio-filtration. After the exhaust gas is firstly wetted by a circulating–spraying liquid storage tank, the exhaust gas is introduced to the top of the packing medium, and then the purified gas is discharged from the top of the biological tank; (3) Biological washing method. VOC enters the filter bed from the bottom of the biological washing tank through the circulation liquid and contacts with the microorganism. After the gas is purified, the gas is discharged the upper part of the washing tank, and the circulating liquid directly enters the regeneration circulating pool to be recycled.

Table 4.7 Comparison of three biological treatments [35]

Technology	Application	Operating characteristics	Comments
Bio-trickling filtration	VOCs with high gas content, low concentration, high organic loading, and acid degradation process	The utility model has the advantages of large processing capacity and easy operation, but the operation requirement is high, and the utility model is not suitable for the VOCs with high concentration and large gas fluctuation	Bacteria tend to flow away with the mobile phase
Bio-filtration	VOCs with high gas content and low concentration	The utility model has the advantages of large processing capacity, convenient operation, simple process, low energy consumption, low running cost, and high degradation rate of mixed VOCs	The bacteria reproduce and metabolize rapidly and do not lose with the mobile phase
Biological washing	VOCs with low gas content, high concentration, and low metabolic rate	The system pressure is large, and the bacteria easily lose with the continuous phase	For refractory gas, absorption equipment with the long contact time, such as bubble column and porous plate tower, is adopted

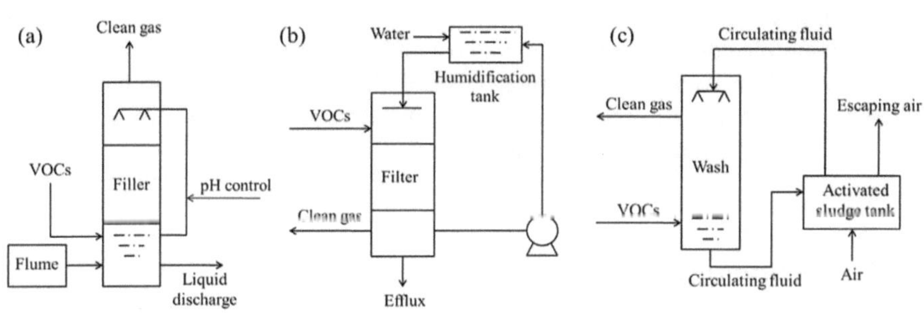

Fig. 4.20 Biological treatment of VOCs: (**a**) Bio-trickling filtration; (**b**) Bio-filtration; (**c**) Biological washing

4.5 Technology Comparison

4.5.1 4E Evaluation

To address these technical and commercial constraints of VOC control in industry, many alternative technologies are being evaluated. For this purpose, the 4E Evaluation (environmental, energy, engineering, and economic aspects) analysis should be used to evaluate different VOC control technologies. Analysis of the VOCs control technologies based on the 4E Evaluation is further summarized in Table 4.8 [2, 36–40].

The removal efficiencies of adsorption, thermal combustion, catalytic oxidation, and photo-catalytic oxidation can reach as high as 99%, while those of absorption and plasma-assisted oxidation are slightly lower. Little secondary pollutants are discharged during the process of catalytic oxidation, plasma-assisted oxidation, and photo-catalytic oxidation. For other technologies, it is necessary to consider how to eliminate secondary pollutants.

The operation temperatures of thermal combustion and catalytic oxidation are 700 °C and 300–500 °C, respectively. While the operation temperatures of other technologies are all lower than 150 °C. Therefore, the energy consumptions of thermal combustion and catalytic oxidation are higher, due to the high temperature required during the processing. The heat recovery of thermal combustion can reach 70–85%, which is 60–75% for catalytic oxidation.

Adsorption, plasma-assisted oxidation, photo-catalytic oxidation, and biological treatment are suitable for exhaust gas with low concentration and high flow rate. Absorption is used for VOCs of high concentration and high flow rate. Condensation, thermal combustion, and catalytic oxidation are selectable for high concentration and low flow rate conditions.

The economic cost consists of capital cost and operating cost, which are different for various technologies. The capital cost of thermal combustion and catalytic oxidation is higher than other technologies. While operating costs of adsorption, absorption, and thermal combustion are higher.

4.5.2 Technology Selection

A holistic assessment should be made before selecting a particular VOC control technology. Basically, some issues should be taken into consideration for the technology selection [2, 41, 42].

The characteristics of VOCs emission source include compositions, temperatures, variabilities of loading, and the presence of other air pollutants such as PM and NO_x. The selection of VOCs control technique should be based on the source characteristics rather than a particular source type. These characteristics determine the variability of loading, and the mean loading determines the applicability of processes. It is difficult

Table 4.8 4E evaluation of VOC control technologies

4E evaluation	Classification	Recoverable method			Destruction method				
		Adsorption	Absorption	Condensation	Thermal combustion	Catalytic oxidation	Plasma-assisted oxidation	Photo-catalytic oxidation	Biological treatment
Environmental	Removal efficiency	90–99%	90–95%	70–85%	95–99%	90–99%	90–95%	90–99%	60–95%
	Secondary pollutants	Discard adsorbent	Wastewater	Condensing agent	Combustion by-product	Little	Little	Little	Biological membrane
Energy	Operating temperature (°C)	<50	<45	<150	700	300–500	<80	<45	<45
	Energy consumption	Low	Low	Low	High	High	Low	Low	Low
	Heat recovery	—	—	—	70–85%	60–75%	—	—	—
Engineering	Flow rates (Nm^3/h)	<60,000	2,000–100,000	100–10,000	1000–30,000	1000–20,000	1000–50,000	1000–80,000	<100,000
	VOCs concentration (ppm)	<1000	500–10,000	5000–10,000	20–25%LEL	1000–5000	<500	<1000	<500
Economic	Capital cost	Low	Low	Low	High	High	Low	Low	Low
	Operating cost	High	High	Low	High	Low	Low	Low	Low

to find a single control technology because of the complex characteristics of VOCs. The limitations upon the selection process enhanced with the increase of variability of the controlled substances. However, the existence of other pollutants may bring problems to the control system.

The recycle potential and maintenance of organic compounds is another important factor in VOCs control. It's also important to consider the cost and removal efficiency of VOCs control technology. Recycling could reduce the overall costs to a certain extend. If recycling is desirable and achievable, it is not necessary to use destruction devices. The maintenance facility level should also be considered. It is not necessary to use continual monitoring in the control system if current maintenance is not sophisticated.

It's also necessary to take into consideration the cost and removal efficiency of VOCs control technology. Annual operating costs and direct capital costs should be estimated. Application of different control techniques has different removal efficiency, which should be considered when selecting the proper VOCs control technique. For example, bio-filtration approaches have the advantages of relatively low costs and excellent operational stability, which are used widely nowadays.

The conventional control methods of VOC are smaller, lighter, and efficient than their predecessors, and the basic technologies of VOC control are still in use today [43]. Adsorption is a simple and economical method for VOC recovery, and VOCs can be recovered with suitable adsorbent and treating conditions [44]. Cryogenic condensation can achieve economic benefit by VOCs components enrichment from the low-concentration gas stream of chemical processing facilities under certain conditions [45]. Thermal combustion and condensation are selectable only for VOCs of medium to high concentrations [46]. The application scope of the VOCs control technologies is summarized in Fig. 4.21 and Table 4.9.

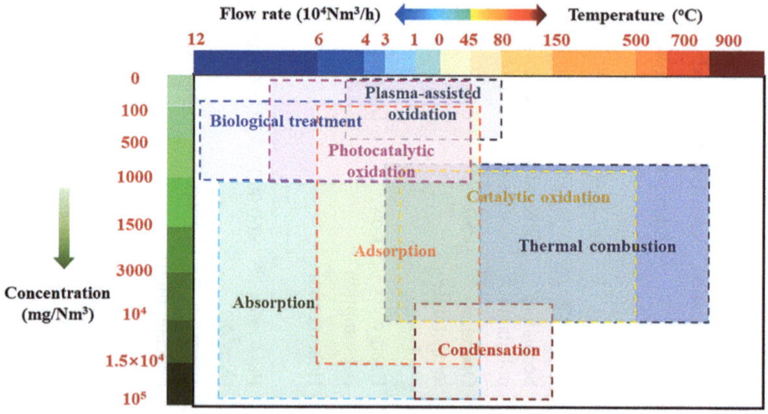

Fig. 4.21 Application scope of different VOCs control technologies

Table 4.9 Evaluation of VOCs control technologies under different operating conditions

Operating condition		Recoverable method			Destruction method				
		Adsorption	Absorption	Condensation	Thermal combustion	Catalytic oxidation	Plasma-assisted oxidation	Photo-catalytic oxidation	Biological treatment
Temperature	High	Poor	Poor	Fair	Excellent	Excellent	Good	Poor	Poor
	Low	Excellent	Excellent	Excellent	Poor	Poor	Excellent	Excellent	Excellent
Flow rate	Large	Excellent	Fair	Poor	Poor	Poor	Fair	Good	Excellent
	Small	Excellent	Excellent	Excellent	Excellent	Excellent	Excellent	Excellent	Excellent
VOCs concentration	High	Excellent	Excellent	Excellent	Good	Good	Poor	Poor	Poor
	Low	Fair	Good	Poor	Fair	Fair	Excellent	Excellent	Excellent
VOCs composition	Single	Excellent	Excellent	Excellent	Excellent	Excellent	Excellent	Excellent	Excellent
	Complex	Excellent	Excellent	Excellent	Fair	Fair	Good	Good	Fair

4.5.3 Challenges and Perspectives

At present, VOCs control technologies are developing toward energy saving and high efficiency. Adsorption, condensation, and catalytic oxidation are most suitable for industrial development. At the same time, the biological method has great potential for development in order to avoid secondary pollution [47].

The VOCs control technologies include two categories: source control and end-of-pipe treatment. In terms of source control, the most important thing is to adjust or replace the raw material formulation so as to reduce the use of highly toxic VOC. Second, it is necessary to improve the process, such as improving the automation and sealing of equipment components to reduce VOC dissipation. For the end-of-pipe treatment, the policies and regulations of VOCs control are not yet perfect. Although many countries have issued control standards, most of the policies only have a certain guiding role at the macrolevel, thereby lacking the operability of implementation. As there are more than 200 kinds of VOCs discharged by a large number of industries under complex emission conditions, the detection and monitoring of VOCs have become a key problem restricting the governance of VOCs at present. The ways to collect and centralize VOCs are also very important for emission sources which are characterized by intermittent production processes and scattered emission points. At the same time, due to the lack of evaluation guidance of prevention and control technology, enterprises often chose technology with low cost. The low-price bidding technology may not be suitable for VOCs management of the enterprise, resulting in waste of resources, and cannot achieve the effective treatment of VOCs [48].

In view of the above analysis, some suggestions are put forward as follows:

1. Policies and standards are the basis of VOCs control. It is suggested that VOCs emission standards of key industries should be formulated quickly, and the system should be gradually improved. Accelerate the research and formulation of supporting policies for VOCs emission management and implement VOCs reduction substitution for key industries. The research on VOCs emission coefficient, verification accounting methods, and pollution charges policies in key industries should be strengthened. Enterprises are encouraged to choose third-party environmental service companies to participate in VOCs pollution prevention and control, which provide environmental protection services such as treatment plan, emission declaration, and monitoring.
2. Unified VOCs monitoring methods and technical standards. More targeted and efficient monitoring methods and technical standards of VOCs emissions control need to be promoted. Develop technical methods for centralized collection and supervision of VOCs. Strengthen the supervision and control of VOC emission without organization, do a good job of monitoring key VOCs emission industries, and establish a list of pollution sources [49].
3. The difficulties in VOCs control are mainly due to the complex properties of VOCs and the various discharge condition. For the occasion of multi-point source, low VOCs emission rate, and intermittent emission, a new VOCs governance model of "decentralized collection and centralized treatment" is proposed

in order to reduce the cost of pollution control for such enterprises or single enterprises in industrial parks. VOCs disposal centers can be built in enterprises or industrial parks. Treatment and purification service charges according to the amount of air discharged by enterprises. This could reduce the one-time investment and later operation costs of each enterprise and is conducive to centralized supervision and control of the total amount of pollution emissions [50].

4. The control technology should be more energy conservation and high efficiency, such as developing VOCs catalyst at low temperature, encouraging the recovery and utilization of VOCs. Development of new materials suitable for VOCs catalysis at low temperature or even room temperature possesses significant advantages of energy saving and safety in applications. It's also necessary to consider whether the resources can be recycled in the governance of VOCs and try to achieve resource recovery at high removal efficiency [51].
5. As one of the most promising technologies for VOC abatement, catalytic oxidation should be paid more attention as it is more effective and economical than conventional thermal oxidation technology [52]. In this regard, future efforts should focus on developing highly efficient catalysts with highly dispersed active phases, especially single-atom catalysts [53] to fully utilize the active metal species and reduce the cost, and on designing universally applicable and stable catalysts with strong resistance to poisons in the gas stream.
6. It is also important to develop coupling technologies, like regenerative catalytic combustion, ozone catalytic oxidation, and non-thermal plasma-assisted catalytic oxidation to promote the VOC removal efficiency in a more efficient and cost-effective way.

Exercises

1. Briefly describe the basic principles regarding the following VOCs control methods: adsorption, absorption, condensation, and oxidation.
2. Describe the processes of VOCs recovery by absorption and condensation method.
3. When ethylene is fully combusted, the amount of required air is 125% of the theoretical base, and the standard combustion heat of ethylene is 1411 kJ/mol. The thermochemical equation for the reaction is as follows:

$$C_2H_4 + 3O_2 \rightarrow 2CO_2 + 2H_2O + Q$$

   How many moles of air are required to burn 10 mol of ethylene? How much heat is released under the standard condition?
4. Analyze VOCs control processes and compare their advantages and disadvantages in technology, economy, and application, respectively
5. The activated carbon was employed to treat 20,000 m^3/h of gas containing 3000 ppm of benzene, operated under 25 °C and 101 kPa, with the recovery efficiency of 99.5%. The adsorption capacity of activated carbon is 0.18 kg/kg. The adsorption time and the regeneration time are 4 and 3 h, respectively. Standby time is 1 h. Determine the amount of active carbon with a density of 580 kg/m^3.

References

1. Aydin Berenjian NC, Malmiri HJ (2012) volatile organic compounds removal methods: a review. Am J Biochem Biotechnol 8(4):220–229
2. Faisal I, Khan AKG (2000) Removal of volatile organic compounds from polluted air. J Loss Prev Process Ind 13:527–545
3. Nikhil Sharma AKA, Eastwood P, Gupta T, Singh AP (2018) Air pollution and control. Springer, Singapore
4. Colon J, Alvarez C, Vinot M, Lafuente FJ, Ponsa S, Sanchez A, Gabriel D (2017) Characterization of odorous compounds and odor load in indoor air of modern complex MBT facilities. Chem Eng J 313:1311–1319
5. USEPA (2002) EPA Air Pollution Control Cost Manual. USA
6. Burn J, Henk J, Bloemen T (1993) Chemistry and analysis of volatile organic compounds in the environment. Springer-Science&Business Media, New Delhi
7. Hao J, Ma G, Wang S (2010) Air pollution control engineering, 3 edn. Higher Edution Press
8. Poschl U (2005) Atmospheric aerosols: composition, transformation, climate and health effects. Angewandte Chem Int Ed 44(46):7520–7540. https://doi.org/10.1002/anie.200501122
9. Khan FI, Ghoshal AK (2000) Removal of volatile organic compounds from polluted air. J Loss Prev Process Ind 13(6):527–545. https://doi.org/10.1016/s0950-4230(00)00007-3
10. Meeyoo V, Lee JH, Trimm DL, Cant NW (1998) Hydrogen sulphide emission control by combined adsorption and catalytic combustion. Catal Today 44(1–4):67–72. https://doi.org/10.1016/s0920-5861(98)00174-6
11. Stenberg U, Westerholm R, Alsberg T (1985) Enrichment of gaseous compounds from diluted gasoline exhausts–a comparison between adsorption and cryogenic condensation. Environ Int 11(2–4):119–124. https://doi.org/10.1016/0160-4120(85)90004-2
12. Dwivedi P, Gaur V, Sharma A, Verma N (2004) Comparative study of removal of volatile organic compounds by cryogenic condensation and adsorption by activated carbon fiber. Sep Purif Technol 39(1–2):23–37. https://doi.org/10.1016/j.seppur.2003.12.016
13. YueDong M, ShaoFeng Z, XinYang X (2006) Advances in applied low-temperature plasma technology. Physics 35(2):140–146
14. Wu XC, Huang WW, Zhang YX, Zheng CH, Jiang X, Gao X, Cen KF (2015) Characteristics and uncertainty of industrial VOCs emissions in China. Aerosol Air Qual Res 15(3):1045–1058. https://doi.org/10.4209/aaqr.2014.10.0236
15. Madler L, Stark WJ, Pratsinis SE (2002) Flame-made ceria nanoparticles. J Mater Res 17(6):1356–1362
16. IUPAC Compendium of Chemical Terminology, 2nd ed. (the "Gold Book") (1997) Online corrected version: (2006–) "esters"
17. Hewitt CN (1999) Reactive hydrocarbons in the atmosphere. Academic Press, California
18. Valenciano R, Aylon E, Izquierdo MT (2015) A critical short review of equilibrium and kinetic adsorption models for VOCs breakthrough curves modelling. Adsorpt Sci Technol 33(10):851–869. https://doi.org/10.1260/0263-6174.33.10.851
19. Repelewicz M, Choma J (2008) Influence of chemical modification on porous structure and surface properties of activated carbons. Pol J Chem 82(1–2):3–9
20. SaE C, Harold M (eds) (1984) Handbook of air pollution control technology. John Wiley & Sons, New York
21. Lewis WK, McAdams WH (1928) Computation methods in countercurrent absorption systems. Ind Eng Chem 20(1):253–257. https://doi.org/10.1021/Ie50219a013
22. Fahmy YM, Fornasiero P, Zinoviev S, Miertus S (2007) Air pollition control technologies. International Centre for Science and High Technology, Trieste
23. Messerle L, Mallis LM, Hatch PJ (1989) An inexpensive sample holder for storage and introduction of air-sensitive organometallic compounds into a mass-spectrometer with inert-atmosphere blanketing. J Chem Educ 66(7):618–618. https://doi.org/10.1021/ed066p618
24. Zhang X, Gao B, Creamer AE, Cao C, Li Y (2017) Adsorption of VOCs onto engineered carbon materials: a review. J Hazard Mater 338:102–123

25. Zhao X, Ma Q, Lu G (1998) VOC removal: comparison of MCM-41 with hydrophobic zeolites and activated carbon. Energy Fuels 12(6):1051–1054
26. Yan B, Zhang H (2017) Studies on the adsorption properity of VOC by palygorskite, silica gel and activated carbon. Build Energy Environ
27. Yaakob Z, Kamarudin SK, Kamaruzaman I, Ibrahim A (2008) Adsorption equilibria of propane on activated carbon and molecular sieves. In: Wseas International conference on system science and simulation in engineering. pp 372–377
28. Lina YC, Chang FT, Bai H, Pei BS (2005) Control of VOCs emissions by condenser pre-treatmentin a semiconductor fab. J Hazard Mater 120:9–14
29. Lines JR, Smith AE (2000) Condensers control and reclaim VOCs: technological advances increase recovery of costly product. Environ Health Saf 6
30. Wang XY, Ran L, Dai Y, Lu YJ, Dai QG (2014) Removal of Cl adsorbed on Mn-Ce-La solid solution catalysts during CVOC combustion. J Colloid Interface Sci 426:324–332. https://doi.org/10.1016/j.jcis.2013.10.007
31. Rasmussen SB, Kustov A, Due-Hansen J, Siret B, Tabaries F, Fehrmann R (2006) Characterization and regeneration of Pt-catalysts deactivated in municipal waste flue gas. Appl Catal B 69(1):10–16
32. Agency USEP (2000) VOC destruction controls. vol EPA/452/B-02–001
33. Vaart DRVD, Vatvuk WM, Wehe AH (1991) Thermal and catalytic incinerators for the control of VOCs. J Air Waste Manag Assoc 41(1):92–98
34. Kaebnick EG (1989) Catching our breath: next steps for reducing urban ozone. Government Printing Office, Washington, U.S
35. Kleinheinz GT, Wright PC (2009) Biological odor and VOC control process. Humana Press
36. Wang S, Zhang L, Long C, Li A (2014) Enhanced adsorption and desorption of VOCs vapor on novel micro-mesoporous polymeric adsorbents. J Colloid Interface Sci 428:185–190
37. Biard PF, Couvert A, Giraudet S (2017) Volatile organic compounds absorption in packed column: theoretical assessment of water, DEHA and PDMS 50 as absorbents. J Ind Eng Chem
38. Asian Journal of Atmospheric Environment (2017). vol 11
39. Wang X, Chen J, Cheng T, Zhang R, Wang X (2014) Particle number concentration, size distribution and chemical composition during haze and photochemical smog episodes in Shanghai. J Environ Sci 26(9):1894–1902
40. Bai Y, Huang ZH, Kang F (2013) Synthesis of reduced graphene oxide/phenolic resin-based carbon composite ultrafine fibers and their adsorption performance for volatile organic compounds and water. J Mater Chem A 1(33):9536–9543
41. Rene ER, Mohammad BT, Veiga MC, Kennes C (2012) Biodegradation of BTEX in a fungal biofilter: Influence of operational parameters, effect of shockloads and substrate stratification. Biores Technol 166:204–213
42. Saini VK, Pires J (2017) Development of metal organic framework-199 immobilized zeolite foam for adsorption of common indoor VOCs. J Environ Sci 55 (Suppl. C):321–330. doi:https://doi.org/10.1016/j.jes.2016.09.017
43. Gupta AK, Modi BA (2018) Selection of sustainable technology for VOC abatement in an industry: an integrated AHP–QFD approach. J Inst Eng 3:1–14
44. Spivey JJ (2010) Recovery of volatile organics from small industrial sources. Environ Prog Sustainable Energy 7(1):31–40
45. Moretti EC (2002) Reduce VOC and HAP emissions. Chem Eng Prog 98(6):30–40
46. Calinescu I, Bulearca A, Ighigeanu D, Martin D, Matei C, Trifan A (2009) Hybrid technology with microwaves, electron beams and catalysts for VOCs removals. J Microwave Power Electromagn Energy A Publ Int Microwave Power Inst 43(3):4–11
47. Kiared K, Bibeau L, Brezezinski R, Viel G, Heitz M (2010) Biological elimination of VOCs in biofilter. Environ Prog Sustainable Energy 15(3):148–152
48. Long Guilin CL (2017) Thinking on the treatment of volatile organic compounds. Pollut Control Technol 30(2):77–79
49. Yue L (2018) Discussion on VOC emission and control measures. Environ Pollut Cont 9:65–66

50. Deng Rui ZF, Jiang T, Yaxiong H, Yinghao C (2014) Emission and control measures and suggestions of VOC. Mater Rev 28(24):378 384
51. Jinhu H (2018) Causes of VOCs and their control measures. Ind Prod 44(1):154–155
52. He C, Cheng J, Zhang X, Douthwaite M, Pattisson S, Hao Z (2019) Recent advances in the catalytic oxidation of volatile organic compounds: a review based on pollutant sorts and sources. Chem Rev 119(7):4471–4568. https://doi.org/10.1021/acs.chemrev.8b00408
53. Qiao B, Wang A, Yang X, Allard LF, Jiang Z, Cui Y, Liu J, Li J, Zhang T (2011) Single-atom catalysis of CO oxidation using Pt(1)/FeO(x). Nat Chem 3(8):634–641. https://doi.org/10.1038/nchem.1095

Chapter 5
PM Control

Nomenclature

C_C	The Cunningham slip correction factor
C_D	Drag coefficient
C_i	Ion thermal velocity, m/s
dp	Particle diameter, m
D	Particle diffusion coefficient
D_0	Characteristic length, m
e	Electronic charge, C
E	Electric field strength, V/m
f_c	Maximum capillary force, N
F_{ad}	Pull-off force for the bumpy particle, N
F_B	Buoyancy, N
F_c	Centrifugal force, N
F_{drag}	Drag force, N
F_D	Fluid resistance, N
F_{ed}	Dielectrophoretic force, N
F_{ei}	Image force
F_{ep}	Polarization force, N
F_{elec}	Electrostatic force, N
F_E	Electrostatic force, N
F_{ne}	Normal force of the particles under static conditions, N
F_{p-p}	VdW force between unit areas of the two opposing parallel plates, N
$F^R(t)$	Rapidly fluctuating force, N
F_s	Tangential force due to relative motion, N
F_{s-p}	VdW force between spherical particles and plates, N
F_{s-s}	VdW force between two spherical particles, N
G	Gravity, N
k	Boltzmann's constant
Kn	Continuum regime

P.-C. Chiang and X. Gao, *Air Pollution Control and Design*,
https://doi.org/10.1007/978-981-13-7488-3_5

k_n	Apparent elasticity coefficient
M_t	Hydrodynamic torque, N · m
n	Number of particles
N	Ion density, $1/m^3$
p_g	Air density at normal conditions, kg/m^3
Pe	Pikelet's number
q	Particle charge, C
q^*	Charge constant, C
q_s	Saturation charge, C
R	Dimensionless number, indicates the intercepting efficiency
Re_D	Reynolds number of the relative collector flow of the gas
St	Stokes' number
t	Charging time, s
T	Absolute temperature, K
μ	Dynamic viscosity of gas, $N{\cdot}s/m^2$
u_0	Velocity of the undisturbed upstream airflow relative to the collector, m/s
u_s	Deposition rate of the particles, m/s
Vp	Particle velocity with respect to gas, m/s
z	Distance between particles, m
Z	Ion mobility, $m^2/(V{\cdot}s)$
α, β, γ	Experimentally determined coefficients
ε_0	Vacuum dielectric constant
η	Capture efficiency caused by interception
μ	Dynamic viscosity of gas, $N{\cdot}s/m^2$
u_s	Deposition rate of the particles, m/s
ξ	Stokes' drag law coefficient
ρ	Gas density, kg/m^3
ρ_p	Particle density, kg/m^3
σ_n	Normal overlap
τ	Time constant of field charging, s
$\tau*$	Time constant, s

5.1 Introduction

5.1.1 Importance and Significance

Airborne particulate matter is ubiquitous in our environment. This matter is produced by natural processes and human activities such as windblown dust, forest fire smoke, automobile exhaust and power plant smoke, etc. However, particulate matter pollution is causing serious problems around the world. Fine particulate matter may be suspended in the air and diffused with atmospheric motion, causing permanent health

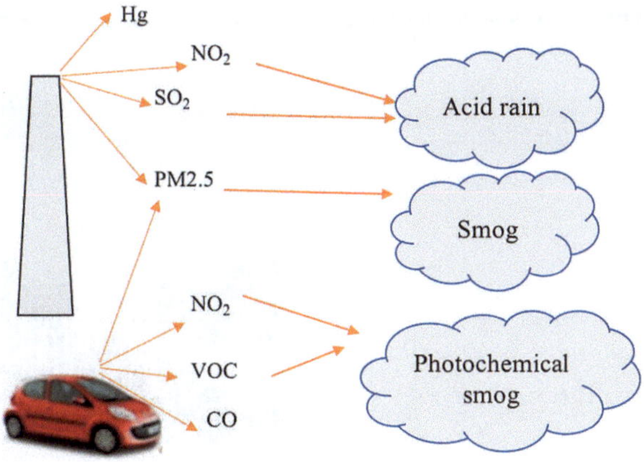

Fig. 5.1 Sketch of the various processes which determine the form of PM in the atmosphere

hazards and environmental damage [1, 2] (Fig. 5.1). For particles smaller than 1 μm in size, these particles can permeate into the lungs and affect gas exchange, which is similar to the behavior of gas molecules. The particles permeate into the lungs and penetrate further into the bloodstream, then translocate into the cell tissue and/or circulatory system. This process can seriously harm human health and cause significant health hazards, such as asthma attacks, nonfatal heart attacks, and even cancer [3]. Airborne particulate matter, which is potentially dangerous to human health and the environment, should be eliminated. For these reasons, the clean coal technologies, aiming at achieving more efficient and prudent emission control, are now becoming compelling and remarkable [4].

In the past, there have been many air pollution accidents in history. The Donovan smog in Pennsylvania in 1948 was an historic environmental accident that killed 20 people and adversely affected over 6,000 people with respiratory illnesses. Sixty years later, the New York Times described the incident as "one of the most serious air pollution disasters in American history". The direct impact of the accident lasted for more than a decade, and the death rate of the Donora military and civilians was significantly higher than that of other nearby communities. Additionally, the smog in London has also caused serious environmental pollution problems and has a huge negative impact on London. It can reduce visibility and even enter the indoor areas directly, causing serious damage. Recent studies have shown that the death toll from the smog is far more than the current estimate of approximately 12,000 (Fig. 5.2).

5.1.2 Regulations and Standards

In order to lower the ambient air pollution, many countries have issued strict regulations on pollution emissions from power plants and industrial production processes, which lead to new demands for the performance of pollution control equipment. The USA, China, and Western European countries were the first to introduce emission

Fig. 5.2 The Donovan smog in Pennsylvania in 1948

Table 5.1 PM emission standards for coal-fired power plants in the EU, China, and the US

	EU	China	US
Present standards	50 mg/m^3 (50~100 MW) 30 mg/m^3 (>100 MW)	30 mg/m^3 (General region) 20 mg/m^3 (Key region)	12 mg/m^3
Upcoming standards	10 mg/m^3 (Newly built, >300 MW)	10 or 5 mg/m^3	–

limit values for particulate matter. The most stringent particulate emission limitations are required by the new regulations to be 10–50 mg/m^3. Table 5.1 shows the PM emission standards for coal-fired power plants within the EU, China, and the US.

5.1.3 Technology Development

Dust removal devices generally separate the particles from gas or separate particles on collectors, via mechanical forces such as inertial forces and gravity. Dry and wet dust collectors are two pieces of equipment commonly used for dust removal [5]. The dry control devices include cyclones, filters (fabric, ceramic, electret), and electrostatic precipitators (ESPs). The other group is the wet control devices, including centrifugal, inertial, electrostatic scrubber or Venturi tube, foam precipitator, and irrigation ESPs. Fibrous or porous filter materials are also used for removing the particles. ESPs remove the particles from the gas by electrical energy. Over the past few decades, substantial progress has been made in new electrical technologies for gas cleaning toward the goal of improving the removal efficiency of small particles [6].

5.1.4 Contents of Chapters

In this chapter, we will be concentrating on the PM formation, control principles, and control technologies. The process of particulate matter formation includes coal combustion, ash vaporization, condensation, and particle growth. For fly ash with a particle size larger than 1 μm, the formation of particles is explained by coal particle breakup and the coalescence of molten grains of mineral matter during combustion [7, 8]. The particle with a size less than 1 μm is explained by the vaporization homogeneous condensation mechanism. There are many factors affecting the transportation of particles, including flow field characteristics, particle diameter, flow field composition, and external field. In industry, common particle collection technologies include cyclones, baghouse, wet scrubbers, electrostatic precipitators, and other technologies.

5.2 Formation

5.2.1 Emission Sources of PM

The main sources of PM emissions are the petroleum and chemical industries, combustion sources, motor vehicles, construction, and dusty weather, as summarized in Table 5.2. Different sources of emissions produce particles of various sizes that will affect the environment and health.

Figure 5.3 illustrates the typical size range of various types of aerosols, which are classified according to formation mechanisms [9]. The size of aerosol particles ranges from 0.001 μm molecular clusters to visible droplets, and the particle size of dust particles may be as large as 100 μm. The sources of airborne particulate matter can be divided into primary sources and secondary sources. Primary particles refer to those particles that are emitted directly in the form of particles from power stations, motor vehicles, and other sources. Secondary aerosols refer to particles formed in the atmosphere due to vapor condensation or chemically reactive substances [10].

Coal combustion for power generation is a major source of particulate air pollution. The size distribution of particles emitted from a power plant is very wide, usually ranging from tens of nanometers to hundreds of microns in diameter. Exploring the formation mechanisms and properties of coal combustion aerosols is essential for both the development of particulate control technologies and the assessment of their impact on human health (Fig. 5.3).

In short, most of PM10 was generated by four formation mechanisms: (1) char fragmentation, [11–13]; (2) inclusion of mineral coalescence, [14–17]; (3) exclusion of mineral fragmentation, [18]; and (4) vaporization and subsequent condensation of the volatile metallic elements [19–23]. The first three mechanisms mentioned above are mainly responsible for the formation of the majority of ultrafine ash particles. Submicron ash particles are the main reason for the final formation mechanism.

Table 5.2 Emission sources of PM

	Sources
Human-made source	Power plants
	Power plants (Coal)
	Industrial combustion
	Industrial combustion (Coal)
	Other industrial processes
	Cement
	Steel
	Residential fossil fuel combustion
	Residential biofuel combustion
	On-road transportation
	Off-road transportation
	Biomass open burning
Natural source	Soil
	Forest fire
	Volcano
	Photochemical reaction (plant life)
	Gas to Particle conversion

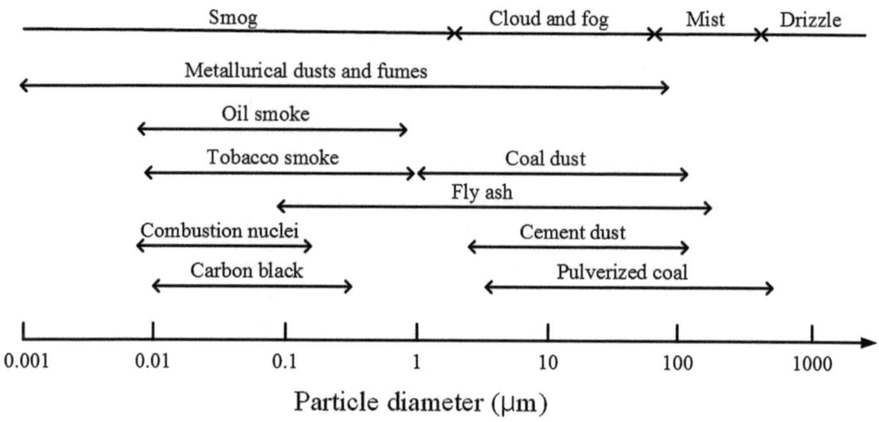

Fig. 5.3 Particle size range from different sources [9]

5.2.2 *Properties of Particles*

1. Chemical composition of particles

The main components of particles from a coal-fired power plant are Al, Si, Ca, Mg, Fe, Na, K, and Ti, and depending on the source of the coal, S and P may also be present. Most of the trace elements remain in the large ash, and in addition to mercury and halide elements (F, Cl, and Br), they are very easily released in the gas. At the same time, polycyclic aromatic hydrocarbons (PAHs), trace elements, and particulate elemental carbon (PEC) may also be present.

Zhejiang University [24] has carried out a significant amount of research on the characteristics of fly ash in coal-fired power plants. The difference in coal and combustion modes will lead to different chemical compositions of fly ash. The chemical composition (e.g., Si, Al, Mg, Li, Fe, K, Na, and Ca) of fly ash from coal-fired power plants was analyzed. The main components of the fly ash were SiO_2, Al_2O_3, Fe_2O_3, CaO, MgO, K_2O, and Na_2O. The fly ash from a coal-fired power plant is mainly SiO_2 and Al_2O_3. The content of SiO_2 is 35% ~ 55%, the content of Al_2O_3 is 10% ~ 45%, and the content of Fe_2O_3 is mainly between 2% ~ 10%. There are also several coal-fired power plants with Fe_2O_3 content exceeding 10%, that is related to the burning of coal, and the CaO content is between 3 and 15%.

2. Particle size distribution

Particle size distribution is one of the most important parameters used to characterize particle behavior. Most of the fly ash particles are usually larger than 1 μm in size and are mainly concentrated in the diameter range of 3 to 50 μm. The ash residue model can help to explain this phenomenon. In contrast, submicron ash always accounts for less than 2% of the total mass of fly ash [25]. Figure 5.4 shows the particle size distribution of the fly ash sample. Samples from glass kilns, waste incineration plants, coal gasifiers, and biomass power plants have median sizes in the range of 10–40 μm and are similar to samples from coal-fired power plants (Fig. 5.4).

3. Morphology of particles

In addition to particle size and chemical composition, the morphology of particles is also a significant parameter of particle morphology that should be given attention.

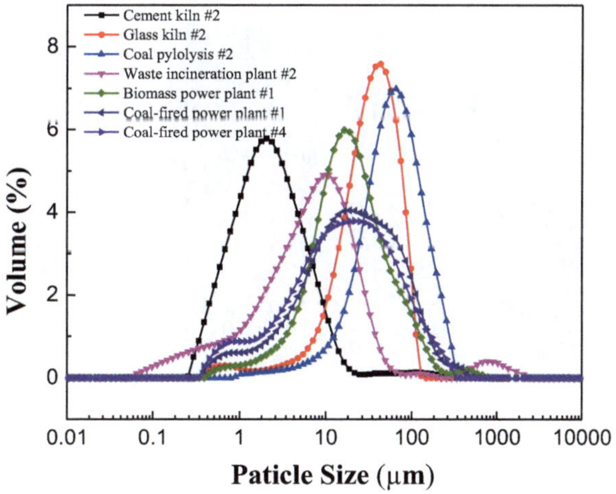

Fig. 5.4 Particle size distribution from different industries [24]

Different particle generation methods result in different particle appearances. Temperature is a key factor controlling the morphology of residual ash particles which are formed through solid-to-particle processes. The temperature reached the melting point of the fly ash in high-temperature gasification, and the residual ash particles were mainly spherical with a smooth surface, indicating that the particles had completely melted [13, 26, 27]. The morphologies of the particles from the coal-fired power plant are shown in Fig. 5.5 [28].

In contrast, there are no significant morphological differences in vaporized ash particles produced in different combustion systems. The vaporized ash particles generated from a coal-fired power plant can be single spheres (Fig. 5.6a) [28]. However, most of them exist as chain-like or cluster aggregates consisting of primary aerosols 10–50 nm in diameter (Fig. 5.6b) [29].

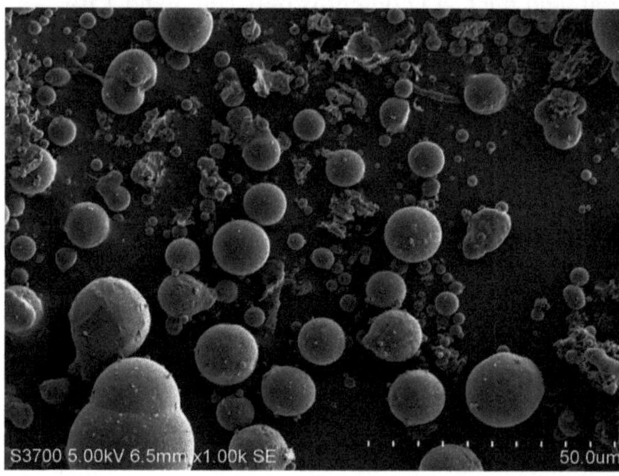

Fig. 5.5 Morphology of coarse modal particles [28]

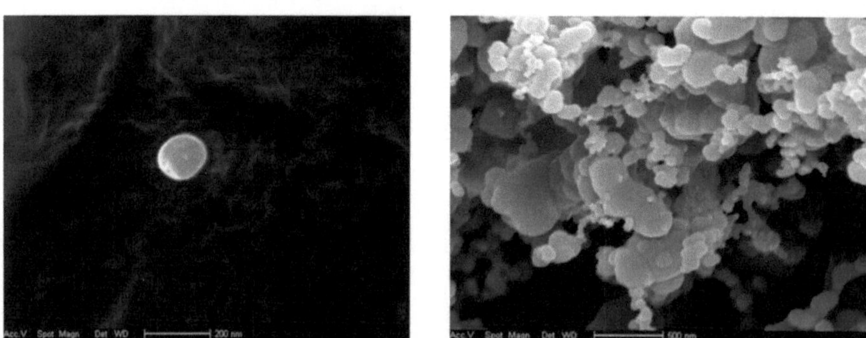

(a) **(b)**

Fig. 5.6 Morphology of vaporized ash particles [28, 29]

5.3 Control Principles

After being generated by coal combustion, particles are carried by the flue gas and travel along the ducts. They pass through all the dust removal facilities and get collected, and the remaining particles are emitted to the atmosphere. The transport of particles is affected by the flow field characteristics, particle diameter, flow composition, and external field. Understanding particle transport is of fundamental interest in designing and optimizing all particle collection devices [30].

5.3.1 Dust Removal Mechanisms

1. Centrifugal settling

The use of centrifugal force to remove particles is also one of the main dust removal mechanisms for inertial collision and interception. It belongs to unsteady motion.

Under the action of centrifugal force, the particles will produce a centrifugal radial motion (perpendicular to the tangential direction). The particles are also subjected to resistance during the movement. When the centrifugal force and fluid resistance are balanced, the particles reach a tipping speed of centrifugal sedimentation.

2. Inertial impact

It is generally believed that the particles in the gas stream move with the gas flow. However, if there are obstacles (droplets or fibers) that move in a static or slow motion, the airflow will flow around and some particles will settle to the surface. Inertial collision and interception are important mechanisms for dust removal by particle inertial migration.

(1) Inertial collision

The distribution of airflow velocity around the trap varies with the Reynolds number of the relative collector flow of the gas:

$$\mathrm{Re_D} = \frac{\mathrm{u_0 \rho D_0}}{\mu} \tag{5.1}$$

where u_0 is the velocity of the undisturbed upstream airflow relative to the collector, ρ is the gas density, D_0 is the size of the collective, and μ is the dynamic viscosity.

(2) Intercept

The particles are directly intercepted on the trapping group, which usually happens within a distance of d/2 from the trapping group. The interception efficiency is directly intercepted by the dimensionless number R to indicate the intercepting efficiency.

$$R = \frac{d_0}{D_0} \tag{5.2}$$

For particles with larger inertia, the interception effect improves the capture efficiency can be expressed as:

Cylindrical collector:

$$\eta = R \tag{5.3}$$

Spherical collector:

$$\eta = 2R + R^2 \tag{5.4}$$

For particles with less inertia:

For the potential flow around the cylinder,

$$\eta = 1 + R - \frac{1}{1 + R}(R < 0.1) \tag{5.5}$$

For the potential flow around the sphere,

$$\eta = (1 + R)^2 - \frac{1}{1 + R}(R < 0.1) \tag{5.6}$$

For viscous flow around a cylinder,

$$\eta \approx \frac{R^2}{2.002 - \ln Re_D}(R < 0.07, Re_0 < 0.5) \tag{5.7}$$

For viscous flow around a sphere,

$$\eta \approx \frac{3R^2}{2}(R < 0.01) \tag{5.8}$$

3. Diffusive settling

Small particles are subject to random collisions of gas molecules, causing them to move irregularly like gas molecules, and the particles will diffuse from a higher concentration region to a lower concentration region. The process of particle diffusion is similar to the diffusion of gas molecules and can be described by differential equations of the same form:

$$\frac{\partial n}{\partial t} = D(\frac{\partial^2 n}{\partial x^2} + \frac{\partial^2 n}{\partial y^2} + \frac{\partial^2 n}{\partial y^2}) \tag{5.9}$$

where n is the number of particles, t is time, and D is the particle diffusion coefficient.

The diffusion coefficient of the particles depends on the type and temperature of the gas, and the value is several orders of magnitude smaller than the gas diffusion coefficient.

The rate of diffusion subsidence depends primarily on the mass transfer Peclet number and the Reynolds number of the collector. The Pikelet number is the ratio of the inertial force and the particle migration produced by Brownian motion. The larger the value, the less important the diffusion and sedimentation are.

$$\mathrm{Pe} = \frac{u_0 D_0}{D} \tag{5.10}$$

For viscous flow around a cylinder,

$$\eta = \frac{1.71\mathrm{Pe}^{-2/3}}{(2 - \mathrm{lnRe_D})^{1/3}} \tag{5.11}$$

For viscous flow around a sphere,

$$\eta = \frac{8}{\mathrm{Pe}} + 2.23\mathrm{Re_D}^{1/8}\mathrm{Pe}^{5/8} \tag{5.12}$$

4. Electrostatic settling

Charged particles can be removed under the action of an electrical field. The mechanism of particle charging is that the electrons and ions are generated by the electrode discharge and then reattached to the particles. The corona is the formation of ions and the corona can be either negative or positive. In general, a gas itself owns a small number of free electrons as well as the same number of positive ions, because air is electrically neutral. This situation is used to generate a corona. When the gas is present between two electrodes, the free electrons move toward the positive electrode while the positive ions move toward the negative electrode resulting in the corona current [31].

"Field charging" was the first model proposed for particle charging. In the model, the charged particle is simply replaced by a suspended conductive sphere in the electric field, assuming that the charged ions (gaseous ions or electrons) are transported along the field lines to the particle surface. As the charging process of particles continues, the particles carry more electric charges, and field lines are gradually repelled from the surface.

Under the assumptions above, the electrostatic principle can be used to calculate the amounts of ions that are deposited on the particle surface. By integrating the current around the particle, the total charging rate of the particle in an external electric field E is determined by Eq. (5.13):

$$\frac{dq}{dt} = 3\pi d_p^2 \mathrm{ENeZ}(1 - \frac{q}{q_s})^2 \tag{5.13}$$

where q is the particle charge, t is the charging time, d_p is the particle diameter, E is the electric field strength, N is the ion density, e is the electronic charge, and Z is the ion mobility.

After integrating with the starting condition t = 0, q = 0,

$$q(t) = q_s \frac{1}{1 + \frac{\tau}{t}} \tag{5.14}$$

where

q_s = saturation charge, C, which equals to $\frac{3\varepsilon_s}{\varepsilon_s+2}\pi\varepsilon_0 d_p^2 E$;

$\tau = 4\varepsilon_0/N_0eZ$, which is the time constant of field charging.

The formula applies to conductive and dielectric particles with different specific dielectric constants.

For particles with a size larger than 0.5 μm, the "field charging" mechanism is usually considered to be predominant, but as the particle size decreases, its charging efficiency becomes lower. In the case of low electric field intensity or small size particles, the Brownian motion of charged ions (gaseous ions or electrons) cannot be ignored. At the same time, a "diffusion charging" theory was proposed to better describe this charging mechanism for small size particles. Based on the continuum transport equation [31] for the concentration of an ion around a single spherical particle, the ion current near the particle is determined by Eqs. (5.15)–(5.17).

$$q_d(t) = q^* \ln(1 + \frac{t}{\tau^*}) \tag{5.15}$$

$$q^* = \frac{2\pi\varepsilon_0 d_p KT}{e} \tag{5.16}$$

$$\tau^* = \frac{8\varepsilon_0 kT}{d_p C_i N_i e^2} \tag{5.17}$$

where q^* is the charge constant, ε_0 is the vacuum dielectric constant, k is the Boltzmann constant, T is the temperature, τ^* is the time constant, C_i is the ion thermal velocity, and N_i is the ion density.

Due to their weaker weight, submicron particles can be removed by applying electrostatic force. In ESP, electrostatic forces are used to remove aerosol particles. In aerosol particle sampling and measuring devices, electrostatic forces are used to collect or redirect all aerosol particles or aerosol particles within a specific size range.

The total charge of particles generally equals n times the elementary unit of charge (electron) e, so the electrostatic force F_{elec}, in an electric field of intensity E, can be determined by Eq. (5.18).

$$F_{elec} = neE \tag{5.18}$$

5. Gravity settling

When a single spherical particle in a stationary fluid settles under the action of the gravitational field, the forces acting on it are gravity, buoyancy, and fluid resistance. The balance of the three forces is as follows:

$$F_D = G - F_B = \frac{\pi d_p^3}{6}(\rho_p - \rho)g \tag{5.29}$$

5.3.2 Fundamental Forces in Particle Migration and Collection

The particle collection equipment essentially operated on the principle that the flue gas containing the particles separates the particles from the flue gas by the presence of external forces or intercepting obstacles. When subjected to external forces in a direction different from that of the airflow, the particles may gain an extra velocity component. To design a flue gas cleaning device for removing particles based on external forces, it is necessary to have the capacity to calculate the trajectory of a particle in the presence of an external force.

1. Centrifugal force

Centrifugal force is an inertial force that keeps particles away from the center of rotation:

$$F_c = m\frac{V^2}{r} \tag{5.20}$$

2. Dragging force

The captions below should be renumbered accordingly. When an object moves relative to a fluid (gas or liquid), the object is subject to the resistance of the fluid. This resistance is called fluid drag, as long as there is a difference in speed (size and direction) between the object and the fluid, subject to fluid drag. The drag force can be calculated using the Navier–Stokes equations for a particle moving in the air. Stokes obtained the following expression Eq. (5.21),

$$F_{drag} = 3\pi\mu V d_p \tag{5.21}$$

where

μ = the dynamic viscosity, $kg{\cdot}m^{-1}s^{-1}$
V = the particle velocity with respect to gas, m/s;
d_p = the particle diameter, m.

The continuum regime (Kn < < 1) drag force F_{drag} is derived as the equation above. While in the slip regime (Kn > > 1), the particle may pass through the gap between the molecules before colliding with other molecules or objects. This effect causes an increase in the speed of the particle and a decrease in the resistance compared to that predicted by Stokes' law. In consideration of this difference, slip correction factor C_c is introduced to accommodate for it, which is expressed as in Eq. (5.22):

$$C_C = 1 + Kn(\alpha + \beta e^{-\frac{\gamma}{Kn}}) \quad (5.22)$$

For solid particles, $\alpha = 1.142, \beta = 0.558, \gamma = 0.999$. ($\alpha, \beta$, and γ are experimentally determined coefficients) [32].

After introducing the slip correction factor, the modified drag force on the particle in the slip regime can be described as in Eq. (5.23), which is applicable for particles in all diameter range:

$$F_{drag} = \frac{3\pi \mu V d_p}{C_C} \quad (5.23)$$

3. Van der Waals Forces

The Van der Waals forces are generated when dipoles in the surface regions of two interacting bodies respond to electromagnetic radiation propagating between the surfaces of the objects. These forces are significant at separation distances of up to 40 nm, depending on the properties of the medium between the objects. There are two ways to describe these forces, Hamaker's pairwise additive approach and the Lifshitz continuum method.

$$F_{s-p} = \frac{-AR}{6d^2} \quad (5.24)$$

where A is Hamaker constant, R is the radius of particle, d is the gap distance between two spherical particles. Similarly, the expression for the Van der Waals forces for other regular geometries can be derived using Hamaker's pairwise additivity. The Van der Waals force between unit areas of the two opposing parallel plates is derived as follows:

$$F_{p-p} = \frac{-A}{6\pi d^3} \quad (5.25)$$

The Van der Waals force between two spherical particles of radii R_1 and R_2 is given as

$$F_{s-s} = \frac{-A}{6d^2}\left(\frac{R_1 R_2}{R_1 + R_2}\right) \quad (5.26)$$

4. Electrostatic Forces

The electrostatic force is one of the main forces of particles. When analyzing the movement and migration of particles, the electrostatic force needs to be considered. It always affects the flow and deposition of particles.

$$F_E = QE + \frac{Q^2}{16\pi\varepsilon_0 z^2} + \frac{QEd^3}{16z^3} + \frac{3\pi\varepsilon_0}{128z^4} \tag{5.27}$$

The first part is the Coulomb force, which is due to the extra charge on the plates or particles that produces a graded Coulomb attraction.

The second part is the image force, which occurs when there is a negatively charged (-q) particle in the space of z outside the dust collecting plate. The charge will then induce a positive charge on the surface of the plate. Charge (+q), the two positive and negative charges, will attract each other. The attraction can be calculated by the concept of a so-called image charge, that is, the Coulomb force generated by this induction is the mirror force.

$$F_{ei} = \frac{Q^2}{16\pi\varepsilon_0 z^2} \tag{5.28}$$

The third part is the Dielectrophoretic Force, which occurs when a dielectric particle is placed in a non-uniform electric field. The dipole moment is generated due to polarization, causing the particles to move due to uneven force at both ends of the particle. If the electric field is a uniform electric field, the force at both ends of the particle is equal and opposite in direction. Therefore, the resultant force of the particle is zero, and the original motion state remains unchanged. Here, you can simplify this force with the following formula:

$$F_{ed} = \frac{QEd^3}{16z^3} \tag{5.39}$$

The fourth part is the polarization force, which can be expressed as follows:

$$F_{ep} = \frac{3\pi\varepsilon_0}{128z^4} \tag{5.30}$$

5. Capillary Forces

Liquid bridges are widely used in nature and industry, and the liquid bridge force plays an important role. "Liquid bridge force" is now generally considered to be the external surface tension of a liquid or the external performance of the capillary force. For coarse particles, the formula for the capillary force is given by

$$F_c = 4\pi\sigma\beta N_c \tag{5.31}$$

where σ is the surface tension of water (0.0735 N/m, at room temperature), β is the contact angle, N_c is the perimeter.

If the liquid meniscus is considered to be formed around the contact surface, the flattening effect of the protrusion is ignored. If the liquid forms a film around the granules and immerses the bulges, the capillary force of the spherical particles of diameter d is as follows:

$$f_c = 2\pi\sigma d \tag{5.32}$$

6. Gravitational force

The gravitational force, G, is proportional to particle mass, mp, and gravitational acceleration, g, as expressed in Eq. (5.34):

$$G = mg = \rho vg = \frac{\pi}{6}\rho d_p^3 g \tag{5.33}$$

7. Other forces

The normal force of the particles under static conditions is

$$F_{ne} = -k_n\sigma_n \tag{5.34}$$

where

$k_n = 4Ea(t)/3$ is the apparent elasticity coefficient, and σ_n is the Normal overlap. The tangential force due to relative motion is

$$F_s = -(k_T\int_{t_0}^{t} v_s(\xi)d\xi + \eta_T v_s)\cdot t_s \tag{5.35}$$

where

$k_T = 8Ga(t)$ is the tangential elastic coefficient;
η_T= the tangential damping coefficient;
v_s= the tangential relative speed;
t_s= the tangential unit vector.

8. Summary

The adhesion of particles is mainly caused by Van der Waals force, electrostatic force, capillary force, and drag force. This section gives the expression formula of the main force noted above and clarifies the physical meaning of the parameters in the formula, as shown in Table 5.3. Force, friction, and other forces occur, but the above four forces are the main reasons. Under different conditions, the relative magnitudes of several forces will change, resulting in a change in the amount of force required to remove the particles.

Table 5.3 Summary of the force of particles in the dust collector

Classification		Calculation formula	Action object	Particle size	Remarks
Centrifugal force		$F = m\frac{V^2}{r}$	Single particle	Big particles	m= Particle quality (kg),V = Particle velocity (m/s^2), r = Particle trajectory radius (m)
Dragging force		$F_{drag} = \frac{3\pi\eta(V-V_P)d_p}{C_C}$ $C_C = 1 + \frac{2.54\lambda}{d_p} + \frac{0.8\lambda}{d_p}e^{\frac{-0.55d_p}{\lambda}}$ $\mu = 0.9\mu_0\left(\frac{T}{273}\right)^{1.5}\frac{273+T_S}{T+T_S}$	Single particle and fluid	0.1 ~ 0.5 μm	V and V_P= Gas and particle velocity (m/s), d_p= Particle diameter (m), μ = Dynamic viscosity coefficient (N·s/m^2),C_C = the Cunningham correction coefficient (dimensionless), λ= Mean free path of gas molecule (m)
Van der Waals force		$F_v = \frac{-A}{12z^2}\frac{d_{p1}}{\frac{d_{p1}}{d_{p2}}+1}$	Particles and collecting plate	< 0.1 μm	A= The Hamaker constant (J), for spherical particle and dust collecting plates, A =2.4 × 10^{-20} J, d_{p1} and d_{p2}= Diameter of two particles (m), z = Distance between two particles (m)
Electrostatic force	Coulomb force	$F_{ec} = \frac{kq_1q_2}{z^2}$ or $F_{ec} = qE$	Particles of opposite charge	0.5 ~ 10 μm	k= Coulomb constant (9.0 × 10^9 N·m^2/C^2), q_1 and q_2= Charge of the two particles (C)

(continued)

Table 5.3 (continued)

Classification		Calculation formula	Action object	Particle size	Remarks
	Image force	$F_{ei} = \frac{q^2}{16\pi\varepsilon_0 z^2}$	Charged particle and collection plate		q= Particles charge (C), ε_0= Vacuum dielectric constant (8.85 $\times$ 10^{-12} F/m)
	Dielectrophoretic force	$F_{ed} = \frac{qEd_p^3}{16z^2}$	Single particle		E= Electric field strength (V/m)
	Polarization force	$F_{eq} = \frac{3\pi\varepsilon_0 E^2 d_p^6}{128z^4}$	Single particle		/
Capillary Forces		$F_1 = 2\pi Cd_p\cos\theta[1-(1+\frac{2V}{\pi z^2 d_p})^{-\frac{1}{2}}$	Particles wrapped by a liquid film	/	γ= Liquid surface tension, N/m, k = Electrostatic constant, 9.0 $\times$ 10^9 N $\times$ m^2/C^2, θ= Contact angle, °, V = Liquid bridge volume, m^3
Gravitational force		$G = mg = \rho vg = \frac{\pi}{6}\rho d_p^3 g$	Single particle	Big particles	ρ= Particle density (kg/m^3), d_p= Particle diameter (m), g=Gravity acceleration (m/s^2)

5.3.3 Particle Transport Properties

1. Particle motion under centrifugal settling

The deposition rate of the particles under centrifugal force in the Stokes region is

$$u_s = \frac{d_p^2\rho_p}{18\mu}\cdot\frac{u_1^2}{R} \tag{5.36}$$

where $\frac{u_1^2}{R}$ is the centrifugal acceleration.

The equilibrium migration velocity of the particles in the smooth region is

$$u_s = \frac{d_p^2\rho_p}{18\mu}\cdot\frac{u_1^2}{R}C_D \tag{5.37}$$

Inertial transport

Under inertial collision, the trajectory of the particles depends on the mass of the particles, the resistance of the airflow, the shape and size of the trap, and the flow rate. The Stokes number is used to describe the particle motion characteristics, which is defined as the ratio of the stopping distance of the particles x to the size of the trapping group D:

$$St = \frac{xC}{D} = \frac{d_p^2 \rho_p u_0 C}{18\mu D} \tag{5.38}$$

2. Brownian motion

The brownian motion refers to the seemingly random motion of a small particle observed when it is suspended in a fluid. Among the earliest written accounts of this motion is that of Roman philosopher Lucretius, who described the random, tumbling motion of dust particles in the air as proof of the existence of atoms. Botanist Robert Brown provided a detailed account of the random motion of pollen grains when viewed under the microscope in 1827 while studying the plant life of the South Seas. Einstein and Smoluchowski provided among the first mathematical theories for Brownian motion, the former using the measured mean value of the second moment of particle displacement $< \times 2 >$ as a means of deducing Avogadro's number. A stochastic differential equation governing the velocity v of a small particle in the presence of Brownian motion, with no mean fluid velocity field, was proposed in the following form:

$$m\frac{dv}{dt} = -\xi v + F^R(t) \tag{5.39}$$

where $\xi = 3\pi\mu d$ for the Stokes drag law and $F^R(t)$ is a rapidly fluctuating force that represents the force imposed on the particle from the impact of molecules of the surrounding fluid. Instead of resolving each individual molecular impact, the forcing term $F^R(t)$ in the Langevin equation is chosen to be a random variable with the Gaussian probability distribution having zero mean and variance $2\xi k_b T$, where k_B is Boltzmann's constant (the ratio of the ideal gas constant to Avogadro's number) and T is the absolute temperature. For this choice of random forcing function, it follows that the autocorrelation of the random forcing is given by

$$< F^R(t) F^R\left(t'\right) >= 2\xi k_B T \delta(t - t') \tag{5.40}$$

where $\delta(\cdot)$ is the Dirac delta. Over long time periods, it can be shown from the system of equations presented here that the mean value of the second moment of particle displacement increases linearly with time or

$$< x^2 >= 2k_BTt/\xi \tag{5.41}$$

The Brownian dynamics (BD) was introduced as a mesoscale computational method by Ermak and McCammon as a simplification of the more general Langevin dynamics (LD) method. In LD, it is typical to add a particle interaction term to the Langevin equation, giving

$$m\frac{dv}{dt} = -\xi v + F^R(t) + \sum_j F^C(j) \tag{5.42}$$

The particle interaction force $F^C(j)$ is determined by using a conservative potential, such as the Lennard–Jones potential, which is the potential sum over particles nearby to the given particle. The Langevin dynamics is often applicable for applications where the computational elements represent small colloidal particles, sections of a long-chain molecule, or large-size atoms. The small solvent atoms are not explicitly resolved in this method, but instead, the effect of these solvent atoms on the computational particle is modeled by the sum of the Stokes drag and the random force terms. This simplification yields a considerable speed-up for computations with LD compared with those with MD, for which every atom of both the solvent and the particles is resolved.

The Brownian dynamics (BD) invokes the further simplification that the particle inertia is negligible. In the absence of particle collisions, BD thus assumes that the Brownian motion of each particle is balanced by the hydrodynamic drag. The use of the fluid continuum approximation, together with neglect of particle inertia, enables BD to use significantly larger time frames than that of MD.

3. Electrical migration

The particle motion can be changed when the particles are affected by electrostatic force. Electrostatic forces play an important role in particle removal and measurement. The electrostatic force depends on the particle charges and the electric field strength, which may be zero or may reach the largest force discussed here. If a particle is subjected to the electrostatic force and the field and drag forces are equal, it will reach a terminal velocity, V_{elec}, as shown in Eq. (5.44).

$$V_{elec} = \frac{neEC_c}{3\pi\eta d_p} \tag{5.43}$$

The electronic charge e is 1.602 × 10^{-19} coulombs (C). This terminal velocity can also be expressed by the particle mobility B, as shown in Eq. (5.45):

$$V_{elec} = neEB \tag{5.44}$$

4. Particle motion under gravity settling

For the particles in the Stokes region, substitute the resistance calculation formula to obtain the gravity settling end velocity of the particles:

$$u_s = \frac{d_p^2(\rho_p - \rho)g}{18u} \tag{5.45}$$

When the fluid medium is a gas, the effect of buoyancy can be neglected, and the settling speed can be simplified as follows:

$$u_s = \frac{d_p^2\rho_p g}{18u} = \tau g \tag{5.46}$$

For the fine particles of the Cantingham sliding area, the correction is

$$u_s = \frac{d_p^2\rho_p g}{18u}C = \tau g C \tag{5.47}$$

For large particle size spherical particles, the correction is

$$u_s = [\frac{4d_p(\rho_p - \rho)g}{3C_D\rho}]^2 \tag{5.48}$$

For turbulent transition zone,

$$u_s = \frac{0.153d_p^{1.14}(\rho_p - \rho)^{0.714}g^{0.714}}{\mu^{0.428}\rho^{0.286}} \tag{5.49}$$

For Newton, the correction is

$$u_s = 1.74[\frac{d_p(\rho_p - \rho)g}{\rho}]^{0.5} \tag{5.50}$$

5. Particle migration on the dust collecting plate

The three raised points are considered to be the contact points of the rough particles with the smooth planar substrate. The particles can leave one of the contact points and roll by joining the centers of the other two contact points. Assume that the direction of fluid drag is perpendicular to the axis. Rolling separation is determined by the torque balance generated by external forces and drag.

$$M_t + F_t\frac{d}{2} \geq (F_{ad} + F_e + F_c)(0.58\eta_B\beta) \tag{5.51}$$

And the hydrodynamic torque Mt acting on the particle is given by

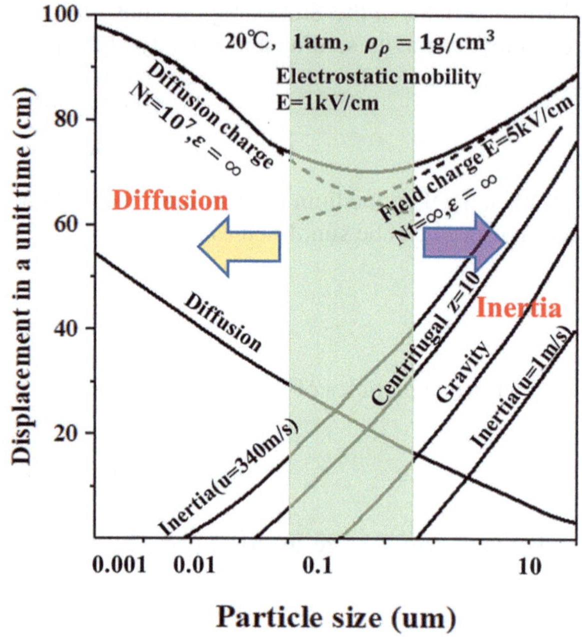

Fig. 5.7 Displacement of a particle in a unit time by an external force

$$M_t = \frac{2\pi \mu f_m d^2 V}{C_c} \tag{5.52}$$

where F_{ad} is the pull-off force for the bumpy particle with three bumps in contact with the substrate, $f_m = 0.943993$, and C_c is the Cunningham slip correction factor. Here, it is assumed that the adhesion force at the moment of incipient rolling is equal to the pull-off force. Due to symmetry, the lines of the action of adhesion force, capillary force, and the electrical forces are located at the centroid of the triangle which connects the centers of the hemispherical contact bumps.

The migration of particles was affected by gravity, electrostatic force, drag force, etc. Figure 5.7 summarizes the effect of various external forces on particle migration.

5.4 Control Technologies

In general, the removal of particulate matter in flue gas after combustion is realized by a dust removal device. By means of gravity, inertia force, electrostatic force, and other external forces, the dust removal devices enable the particles to obtain a velocity that is different from the direction of airflow, so as to separate them from the airflow. Particle control technology can be divided into four categories based on the difference of particle capture mechanism, namely mechanical collector, wet dust collector, electrostatic precipitator, and filter dust collector.

5.4.1 Cyclone

Cyclone separators are particle control devices that use rotation to generate the centrifugal force to collect particle matter from gas. The structure is shown in Fig. 5.8. The gas flow follows a curved trajectory forced by the cyclone separator. The particles are concurrently moved by the inertia out of the fluid flow direction to the outside wall, where they are collected [33]. In a cyclone separator, the particles gradually disengage from the spiral flow toward the outside wall. As the gas stream flows through the entire device, it will experience several complete spirals. To design a cyclone separator, a significant question is as follows: with known gas flow velocity and outer and inner diameters, what is the length of the body of the cyclone separator to be designed in order to ensure the desired removal efficiency of the given particles? Since the airflow speed and the number of turns performed by the airflow can determine the length of the cyclone, the number of turns required to achieve the specified removal efficiency is the primary computational focus.

There are many designs for cyclone separators, and the main difference is that the geometry of the cyclone separators makes the rotational movement impart differently from the airflow. Traditional cyclone separators can be divided into the following categories: reverse-flow cyclones (with an axial inlet and a tangential inlet), straight-flow cyclones, and impeller collectors.

In a common reverse-flow tangential-inlet cyclone, the polluted gas enters from the cyclone top, given a spinning motion due to the tangential entry. Under the action

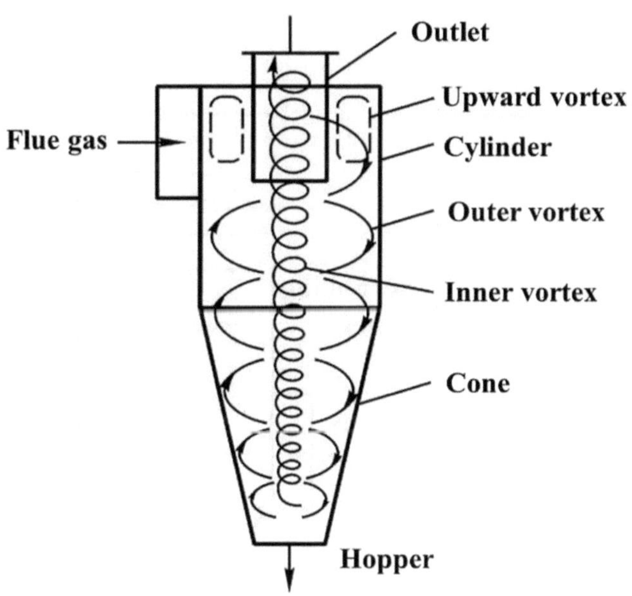

Fig. 5.8 The structure of a cyclone separator

of centrifugal force, the particles are forced to the wall, followed by falling down the wall due to gravity. The gas flow reverses at the bottom of the cyclone, forming an inner core, which leaves the cyclone from the top of the unit. In a reverse-flow cyclone with an axial inlet, the inlet gas is introduced downward along the cyclone axis and the centrifugal motion is transmitted through the permanent blades at the top.

5.4.2 *Wet Scrubbers*

Wet scrubbers use water washing to collect particles from the gas flow, as shown in Fig. 5.9. A spray scrubber is a device that breaks a liquid stream into droplets, ranging from approximately 0.1–1.0 mm in diameter, and then sends them to the dusty flue gas. The array of moving droplets becomes a set of targets for particle collection in the gas stream. Removal efficiency is calculated by considering the efficiency of an individual spherical collector and then adding the number of droplets of gas flow. The capture of particles occurs by collision with the droplets and being directly intercepted by the droplets. Therefore, the relative movement between the droplets and particles is a significant factor in the removal efficiency. Simultaneously, diffusion is also significant for smaller particles. There is a common spray scrubber that uses a pre-formed spray, and the liquid is sprayed into the gas stream through the atomizing nozzle to form droplets.

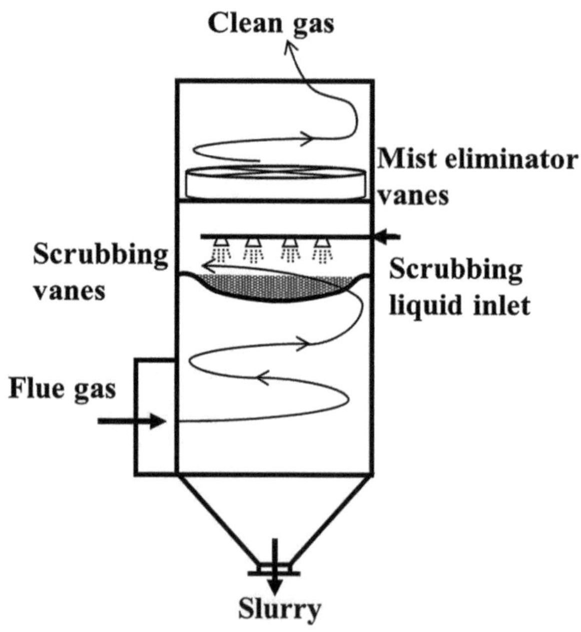

Fig. 5.9 The structure of wet scrubber

In wet scrubbers, the removal efficiency of small particles is very high. The environmental humidity in the scrubber is significant, and the small particles are easy to grow in size due to the condensation of water, so it is easier to remove them. It is important to avoid re-entrainment of particles because the particles are trapped in droplets or in a liquid layer. A scrubber can also remove soluble gas contaminants simultaneously. The used scrubbing liquid will contain particles that must be disposed, which is different from the dry methods of gas cleaning.

The removal efficiency of a wet scrubber may be correlated with the energy loss of the device; the higher the power consumed by the wet scrubber, per unit volume of gas treated, the higher the removal efficiency. Almost all of the energy is introduced into the gas. Energy loss can be measured using the pressure drop of the gas passing through the unit. The main advantage of the wet collectors is the various collector types that allow the selection of a unit appropriate to the particle collection problem. Its disadvantages are that the high-pressure drops must be maintained (and, therefore, require a lot of energy consumption), and further large amounts of scrubbing liquid also have to be processed.

5.4.3 Baghouse

Fabric media filtration is one of the oldest, simplest, and most effective methods for collecting solid particles from gas, as shown in Fig. 5.10. The fabric filter can effectively remove the particles as small as 0.1 μm and has a certain removal efficiency for the particles with a size as small as 0.01 μm. The industrial fabric filter, which is the simplest form, is made of the felted fabric or woven fabric, through which gases containing dust can be forcedly removed. A combination of various factors gives the fabric filters the ability to remove particles. When woven fabrics are applied, a dust cake will form, which predominantly acts as a sieving mechanism. When felted fabrics are used, there is little or no dust cake, so the main mechanism of particle removal is a combination of impingement, inertial forces, etc. The mechanisms are basically the same as the mechanism of particle removal on wet scrubbers, where the removal media is liquid droplets rather than solid fibers. Table 5.4 summarizes the filter materials and their characteristics.

As the collection of particles increases, the pressure drop of the flue gas flow through the fabric filtering media increases. Due to fan limitations, the filters have to be cleaned regularly. The cleaning method is generally to remove dust from the fabric by gravity and/or mechanical means. The fabric filters are usually flat or tubular.

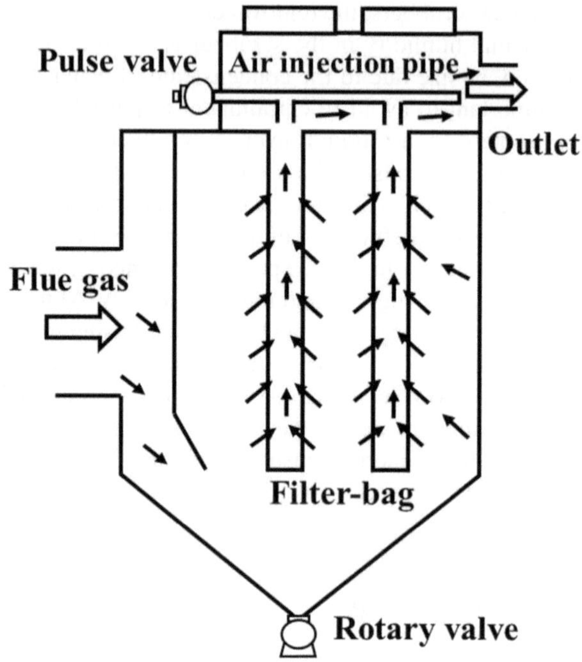

Fig. 5.10 The structure of a baghouse separator

Table 5.4 Filter materials and their characteristics

	Density /g·m^{-2}	Air permeability //m^3·m^{-2}·min^{-1}	Acid resistance	Alkali resistance	Temperature /°C
Glass fiber needle felt filter material	800 ~ 900	8 ~ 15	Excellent	Good	220 ~ 280
FMS9806 needle felt filter material	> 800	8 ~ 15	Excellent	Good	220 ~ 260
FMS9808 needle felt filter material	> 800	8 ~ 15	Excellent	Excellent	160 ~ 220
P84 needle filter felt filter material	500	1.8 ~ 3.6	Excellent	Medium	220 ~ 260
Nomex needle felt filter material	500	1.8 ~ 3.6	Medium	Excellent	160 ~ 240

5.4.4 Electrostatic Precipitator

1. Electrostatic precipitator

Electrostatic precipitators are widely used due to their convenience, ranging from the flue gases cleaner for the largest power plants to small household air cleaners, as shown in Fig. 5.11. The basic operating principle of the electrostatic precipitator is that an electric field is applied to the flue gas to force the particles to charge, then an electric field is used to migrate the particles to the oppositely charged electrode which is generally designed to be at right angles to the direction of the gas stream, as shown in Fig. 5.12. In electrostatic precipitation, external forces act directly on a single particle, while particle separation is indirectly exerted by the force applied to the whole gas flow. Particles are collected on the electrode. Particle charging is achieved by generating ions with corona developed by a highly charged electrode. The electrostatic field exists between the collecting electrode and that highly charged electrode. The device will be termed a single-stage electrostatic precipitator if the same pair of electrodes are used for the charging and collecting particles.

When the volumetric flux of gas is high, the ESPs are usually used for gas cleaning. This device has a wide range of applications. Not only can it be used to remove fly ash from power plant flue gas, but it is also widely used for the removal of particles and acid mists emitted from the metallurgical and chemical process industries. Because of the influence of the electrostatic field, charged particles can be trapped as they move to the dust collection plate. The dust layer will be formed with the accumulation of fly ash, then it will be wrapped into the ash hopper to remove the particles [34–37]. The operating temperature of conventional electrostatic precipitators is generally lower than 200 °C, and the application of these precipitators has formed a relatively large-scale application, making the dust removal efficiency close to 99.99%, and the dust emission concentration is lower than 50 mg/m^3 [38].

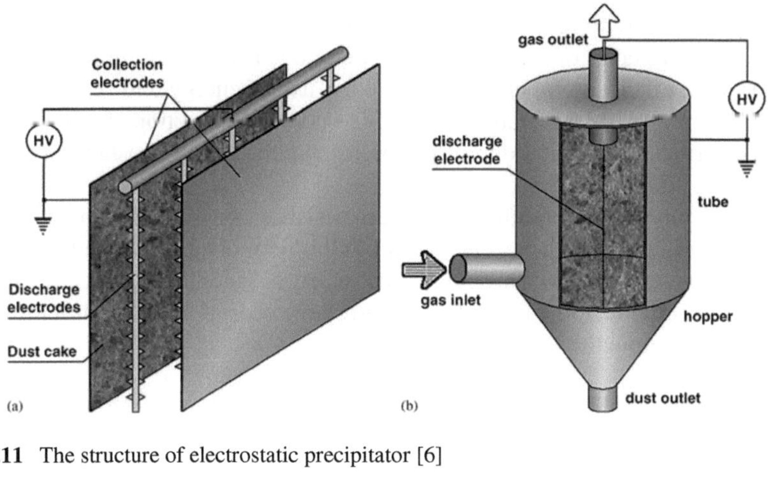

Fig. 5.11 The structure of electrostatic precipitator [6]

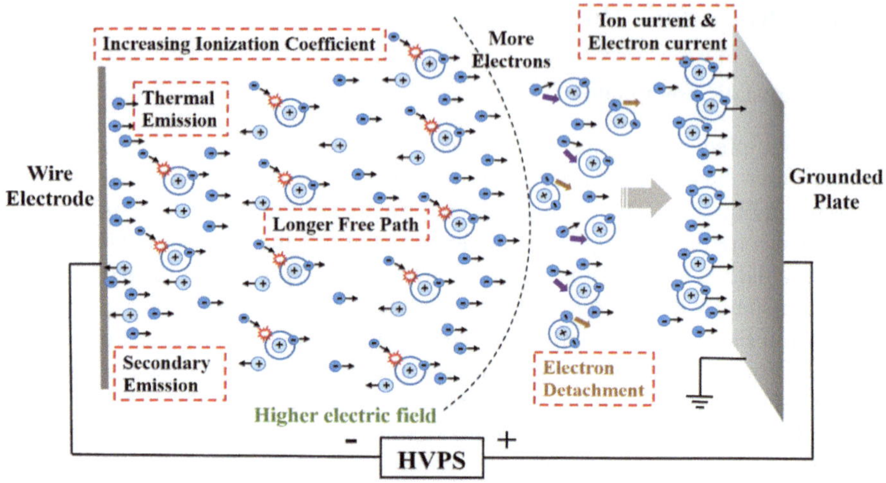

Fig. 5.12 Generation of corona in the electrostatic precipitator

2. High-temperature electrostatic precipitator

High-temperature electrostatic precipitator refers to the electrostatic precipitator whose operating temperature is higher than 350 °C, which is able to achieve the purification of high-temperature flue gas and also possesses substantial advantages such as high dust removal efficiency, small pressure loss, large flue gas disposal amount, great flue gas adaptability, simple operation, and low maintenance costs. It is, therefore, the most promising manner of high-temperature dust removal technology.

The operating principle of particulate removal technologies is roughly divided into two steps—first, reduce the high-temperature flue gas below 200 °C by waste heat boiler, and then use an electrostatic precipitator or baghouse to remove particles. For many process systems, the cleaned flue gas needs to improve the temperature to satisfy subsequent requirements. Therefore, this process not only wastes lots of heat, but also increases costs of investment and operation. Thus, it is of great significance for improving efficiency by effectively making use of energy and providing assurances to the process of production and stable operating conditions for the facilities. The superiorities and limitations of the high-temperature electrostatic precipitator are summarized in Table 5.5.

At high-temperature conditions, the characteristics of corona discharge including breakdown voltage, corona current, and the stability of discharging and characteristics of high-temperature dust particularly the variety of specific resistance are the most critical influencing factors of an efficient and stable operation of a high-temperature electrostatic precipitator.

Table 5.5 High-temperature electrostatic precipitator

Advantages	1. Low pressure loss
	2. Good capability of processing flue gas and great adaptability of varieties of flue gas
	3. Small operating energy consumption
	4. High dust removal efficiency and reliability
	5. Simple operation and maintenance; low cost of operation
Disadvantages	Dust-specific resistance influences the efficiency of dust removal——back-corona phenomenon will occur when it is higher and re-entrainment will appear when it is lower
	The investment in high-temperature steel is huge

5.5 Comparison of PM Control Technologies

5.5.1 4E Evaluation of PM Control Technologies

Based on the principles and technologies for PM control discussed in the above sections, here, we summarize the environmental, energy, engineering, and economic evaluation of these technologies for a more comprehensive understanding, as shown in Tablc 5.6.

The removal efficiency of electrostatic precipitators and filter dust collectors can reach as high as 99%. And they can remove particles with a particle diameter as low as 0.01 μm. The electrostatic precipitator has a high removal efficiency, low resistance, and a large amount of flue gas treatment, and the filter dust collector has a high removal efficiency, high resistance, and high removal efficiency of fine particles. While the mechanical collectors and wet dust collectors are usually used as pre-dusting equipment for their relatively lower removal efficiency. In the application field, the mechanical collector is usually used as a pre-dusting device for multi-stage dust removal in small industrial boilers and in some steel enterprises; the wet dust collector is usually applied for purification for high-temperature, flammable, explosive, and harmful gas, such as metallurgy and chemical industry; the electrostatic precipitators and filter dust collector are widely used in electric power, steel, cement, metallurgy, and other industries.

The economic cost consists of capital costs and operating costs, which vary according to the technology. The capital cost of the mechanical collector and wet dust collector is medium, for their simple structure and low resistance, while the capital cost of the electrostatic precipitator and filter dust collector is higher. As for the operating cost, the electrostatic precipitator is lower, and the filter dust collector is higher.

Table 5.6 4E evaluation of PM control technologies

4E Evaluation	Classification	Mechanical collector			Wet dust collector			Electrostatic precipitator		Filter dust collector		
		Gravity settlement	Inertial dust separator	Cyclone	Wet scrubber	Water-film dust collector	Venturi dust collector	Dry-ESP	Wet-ESP	Baghouse	Electrostatic-fabric integrated precipitator	High-temperature filtration
Environmental	Removal efficiency (%)	40–60	50–70	70–92	75–95	85–98	90–98	99–99.9	70–85	99–99.95	99–99.98	85–99.95
	Size of particle capture (μm)	≥ 50	≥ 20	≥ 5	≥ 5	≥ 1	≥ 0.01	≥ 0.01	≥ 0.01	≥ 0.01	≥ 0.01	≥ 0.01
Energy	Pressure drop (Pa)	50–130	300–800	500–2000	250–500	500–2000	3000–10,000	100–200	125–500	800–1500	600–1200	1000–5000
Engineering	Operating temperature (°C)	-	-	≤ 1000	150–370	~ 160	-500	90–600	≤ 60	≤ 250	≤ 250	≤ 600
	Gas flow velocity	1.5–2 m/s	15–20 m/s	20–30 m/s	1–2 m/s	10–20 m/s	60–90 m/s	0.8–1.5 m/s	2.5–4 m/s	0.6–1.2 m/min	0.8–1.5 m/min	≤ 0.8 m/min
	Capture mechanism	Gravity	Inertial force	Centrifugal force	Inertial impaction and condensation			Electrostatic attraction	Electrostatic attraction and inertial collision with droplet	Inertial impaction and interception	Charge and filter interception	Inertial impaction and interception
	Main features	Simple structure, low resistance, and low removal efficiency		Simple structure and high removal efficiency	Low resistance, high removal efficiency of large particles, and high water consumption		High resistance and high removal efficiency	High removal efficiency, low resistance, and large amount of flue gas treatment		High removal efficiency, high resistance, and high removal efficiency of fine particles	High removal efficiency of fine particles and high resistance	High removal efficiency, high resistance, and high removal efficiency of fine particles

(continued)

Table 5.6 (continued)

4E Evaluation	Classification	Mechanical collector			Wet dust collector			Electrostatic precipitator		Filter dust collector		
		Gravity settlement	Inertial dust separator	Cyclone	Wet scrubber	Water-film dust collector	Venturi dust collector	Dry-ESP	Wet-ESP	Baghouse	Electrostatic-fabric integrated precipitator	High-temperature filtration
	Application field	Small industrial boilers and some steel companies			Purification for high–temperature, flammable, explosive, and harmful gas			Widely used in electric power, steel, cement, and metallurgy		Widely used in electric power, metallurgy, cement, ceramics, and other industries		
Economic	Capital cost	Medium			Medium			High		High		
	Operating cost ($/t PM)	Medium			Medium			Low		High		

5.5.2 Technology Selection

To achieve the aims of ultra-low particulate emissions, coal power plants often adopt primary dust removal and secondary dust removal.

The efficient removal of particulate before desulfurization is called primary dust removal. The mainstream technology includes electrostatic precipitator technology, electrostatic-fabric integrated precipitator technology and baghouse technology. The electrostatic precipitator technology is the combination of high-efficiency power supply, advanced ash cleaning methods, and low-temperature electrostatic precipitator technology, which can achieve the goals of dust removal efficiency of no less than 99.85%. In addition, the dust removal efficiency of an electrostatic-fabric integrated precipitator and baghouse is no less than 99.9%.

The synergistic removal of particulate matter in the process of flue gas desulfurization or the removal of particulate matter in flue gas after desulfurization can be called secondary dust removal or deep dust removal. For the technology of limestone–gypsum wet desulfurization process in a composite tower, the synergistic dust removal efficiency is more than 70% generally by applying the high-efficiency mist eliminator or add WESP in a wet desulfurization tower.

Moreover, in consideration of the high concentration of particulate matter in the flue gas in the technology of semi-dry desulfurization, the electrostatic-fabric integrated precipitator and baghouse are applied. Furthermore, if the ultra-low emission requirements of particulate matter cannot be achieved, the wet electrostatic precipitator must be installed.

The selection of technology in the specific project needs to be combined with the actual situation of the project, and the principles, characteristics, applicability, influencing factors, energy consumption, economy, maturity, and other factors of various technologies should be taken into account. When selecting a primary dust removal measure, the characteristic of the coal and fly ash is obtained first to determine whether it is suitable to use an electrostatic precipitator. If not, the baghouse and the electrostatic-fabric integrated precipitator are preferred. If only using a primary dust removal to achieve ultra-low emissions (the emission concentration is less than 10 mg/m^3 or 5 mg/m^3), the ultra-clean electrostatic-fabric integrated precipitator would be preferred. When the outlet concentration of the primary dust removal is 30–50 mg/m^3, the secondary dust removal should use WESP; when the outlet concentration of the primary dust removal is 20–30 mg/m^3, the secondary dust removal should use WFGD synergistic removal or WESP; and when the outlet concentration of the primary dust removal is less than 20 mg/m^3, the secondary dust removal should be WFGD synergistic removal. With comprehensive consideration, the coal-fired power plant particles ultra-low emission technology route is provided in Table 5.7.

Table 5.7 Particles ultra-low emission technology route

Boiler type	Unit size (10 k kW)	Inlet dust concentration (g/m^3)	Primary dust removal			Secondary dust removal	
			ESP (≥99.85%)	Electrostatic-fabric integrated precipitator (≥99.95%)	Baghouse (≥99.9%)	WESP (≥70%)	WFGD (≥70%)
Pulverized coal furnace (tangential combustion, wall combustion)	≤20	≥30	Fair	Excellent	Excellent	Excellent	Fair
		20~30	Good	Good	Good	Good	Good
		≤ 20	Excellent	Fair	Fair	Fair	Excellent
	30	≥30	Fair	Excellent	Good	Excellent	Fair
		20~30	Good	Good	Fair	Good	Good
		≤20	Excellent	Fair	Fair	Fair	Excellent
	≥60	≥30	Fair	Excellent	Fair	Excellent	Fair
		20~30	Good	Good	Fair	Good	Good
		≤20	Excellent	Fair	Fair	Fair	Excellent
Pulverized powder furnace (W flame combustion)		≥30	Fair	Excellent	Good	Excellent	Fair
		20~30	Good	Excellent	Fair	Good	Good
		≤20	Excellent	Good	Fair	Fair	Excellent
CFB boiler			Fair	Excellent	Good	Excellent	Fair

5.5.3 Challenges and Perspectives

At present, PM control is gradually maturing. But it still needs further exploration and improvement.

1. The real pyrolysis gas contains reducing atmospheres such as H_2, CO, and CH_4. Therefore, in order to cope with practical industrial applications, it is necessary to study the carbon particle removal performance under a reduction atmosphere.
2. Additionally, the depositional morphology of fly ash of different resistivities on the dust collecting plate, the charge release process of the particles, the interaction between the particles, and the interaction between the particles and the plates are still not clear, and the dust-specific resistance is under study. The mechanism of the effect of resistivity on the electrostatic removal characteristics based on the force analysis of the dust layer remains to be explored.
3. Simultaneous removal of other pollutants in the PM removal process still needs to be addressed.

PM control technology was born on the basis of comprehensive multi-disciplinary, and PM control technology has great value for broad research and application prospects.

1. Innovation and reform in each direction can promote the development of a more energy-efficient, efficient and comprehensive direction.
2. Dust removal mechanism and theoretical research still have great potential, exploring new dust removal mechanisms and applying them to dust removal equipment to further improve the performance of the dust collector.
3. Computers, big data, artificial intelligence, etc., can be applied to dust removal equipment, making the research and application of dust removal equipment reach a new level of improvement.

It is believed that with the continuous advancement of science and technology, new high-efficiency and multi-purpose PM control equipment will continue to emerge, and the dust removal industry will surely move toward a more mature and more modern stage.

Exercise

1. Describe and distinguish the forces affecting the particulate matter removal in the dust collectors.
2. Briefly describe the principle of the dry electrostatic precipitator for PM removal.
3. Show the particle migration model and explain how the migration velocity affects the PM removal.
4. Please compare the advantages and disadvantages of the filter dust collector and ESP from the aspects of technology, economy, and application, respectively.
5. Assume that the two spherical particles in the ESP have two particles, their sizes are 0.1 μm and 0.2 μm, respectively, and the distance between them is 1 nm. Please calculate the Van der Waals force between them.

References

1. Guarnieri M, Balmes JR (2014) Outdoor air pollution and asthma. Lancet 383(9928):1581–1592
2. Kim KH, Kabir E, Kabir S (2015) A review on the human health impact of airborne particulate matter. Environ Int 74:136–143
3. Li H, ea, (2002) Current state of study on the human health effect of inhalable particulates. J Environ Health 19(1):85–87
4. Maffei T, Frassoldati A, Cuoci A et al (2013) Predictive one step kinetic model of coal pyrolysis for CFD applications. Proc Combust Inst 34(2):2401–2410
5. YlãTalo S, JukkaHautanen, (1998) Electrostatic precipitator penetration function for pulverized coal combustion. Aerosol Sci Technol 29(1):17–30
6. Jaworek A, Krupa A, Czech T (2007) Modern electrostatic devices and methods for exhaust gas cleaning: a brief review. J Electrostat 65(3):133–155
7. Smith RD, Campbell JA, Nielson KK (2016) Characterization and formation of submicron particles in coal-fired plants. Atmos Environ 13(5):607–617
8. Neville M, Quann RJ, Haynes BS et al (1981) Vaporization and condensation of mineral matter during pulverized coal combustion. Symp Combust 18(1):1267–1274
9. Friedlander SKS, Dust, Haze (2000) Fundamentals of aerosol dynamics, 2nd edn. Oxford University Press, New York
10. Verma V, Ning Z, Cho AK et al (2009) Redox activity of urban quasi-ultrafine particles from primary and secondary sources. Atmos Environ 43(40):6360–6368
11. Seames WS (2003) An initial study of the fine fragmentation fly ash particle mode generated during pulverized coal combustion. Fuel Process Technol 81(2):109–125
12. Buhre BJP, Hinkley JT, Gupta RP et al (2005) Submicron ash formation from coal combustion. Fuel 84(10):1206–1214
13. Buhre BJP, Hinkley JT, Gupta RP et al (2006) Fine ash formation during combustion of pulverised coal–coal property impacts. Fuel 85(2):185–193
14. Yan L, Gupta RP, Wall TF (2001) The implication of mineral coalescence behaviour on ash formation and ash deposition during pulverised coal combustion. Fuel 80(9):1333–1340
15. Buhre B, Hinkley J, Gupta R et al (2004) Coal quality and fine ash formation during combustion. In: Pittsburgh coal conference, Pittsburgh.
16. Ninomiya Y, Zhang L, Sato A et al (2004) Influence of coal particle size on particulate matter emission and its chemical species produced during coal combustion. Fuel Process Technol 85(8):1065–1088
17. Zhang L, Ito M, Sato A et al (2004) Combustibility of dried sewage sludge and its mineral transformation at different oxygen content in drop tube furnace. Fuel Process Technol 85(8–10):983–1011
18. Yan L, Gupta R, Wall T (2001) Fragmentation behavior of pyrite and calcite during high-temperature processing and mathematical simulation. Energy Fuels 15(2):389–394
19. Flagan RC, Taylor DD Laboratory studies of submicron particles from coal combustion. In: Symposium (international) on combustion, 1981, vol 1. Elsevier (1981), pp 1227–1237
20. Quann R, Sarofim A (1982) Vaporization of refractory oxides during pulverized coal combustion. In: Symposium (international) on combustion, 1982. vol 1. Elsevier, pp 1429–1440
21. Eddings EG, Sarofim AF, Lee CM et al (2001) Trends in predicting and controlling ash vaporization in coal-fired utility boilers. Fuel Process Technol 71(1–3):39–51
22. Lee CM, Davis KA, Heap MP et al (2000) Modeling the vaporization of ash constituents in a coal-fired boiler. Proc Combust Inst 28(2):2375–2382
23. Yao H, Naruse I (2005) Control of trace metal emissions by sorbents during sewage sludge combustion. Proc Combust Inst 30(2):3009–3016
24. Zheng C, Liu X, Yan P et al (2018) Measurement and prediction of fly ash resistivity over a wide range of temperature. Fuel 216:673–680
25. Markowski G et al (1980) A submicron aerosol mode in flue gas from a pulverized coal utility boiler. Environ Sci Technol 14(11):1400–1402

26. Chen Y, Shah N, FEH et al (2005) Transmission electron microscopy investigation of ultrafine coal fly ash particles. Environ Sci Technol 39(4):1144–1151
27. Chen Y, Shah N, Braun A et al (2005) Electron microscopy investigation of carbonaceous particulate matter generated by combustion of fossil fuels. Energy Fuels 19(4):1644–1651
28. Yu D, Xu M, Yao H et al (2008) A new method for identifying the modes of particulate matter from pulverized coal combustion. Powder Technol 183(1):105–114
29. Yu D, Xu M, Yao H et al (2007) Use of elemental size distributions in identifying particle formation modes. Proc Combust Inst 31(2):1921–1928
30. Yang Q, Zhang Y, Cai X, Jiang G, Xu X (2013) Chaotic behaviors and fractal self-similar analysis of particles transport properties in RIKEN mesoscopic device. Acta Physica Sinica 62:8
31. Fujii K, Hirota M, Shiromaru N, Andou H, Yanagida M, Hayatsu M (2008) Corrosion potential evaluation and materials selection for wet electrostatic precipitator applied for SO3 mist removal.1819
32. Allen M, Raabe O (1985) Slip correction measurements of spherical solid aerosol particles in an improved millikan apparatus. Aerosol Sci Technol 4(3):269–286
33. Cortés C, Gil A (2007) Modeling the gas and particle flow inside cyclone separators. Prog Energy Combust Sci 33(5):409–452
34. Chen DWK, Zeng X (2014) Present situation and development of flue gas dust removal technology for coal - fired boilers. Environ Eng 09:70–73
35. WC (2006) Mechanism modeling and fuzzy multi-objective optimization design of electrostatic precipitation equipment. Zhejiang University
36. Liu ZLQ (2011) Electrostatic and its application in dust removal system. Journal of Hengshui University 01:105–107
37. AM (2000) Electrostatic precipitation. IEEE Trans Dielectr Electr Insul 7(5):615–624
38. Dastoori KKM, Mallard C et al (2011) Electrostatic precipitation in a small scale wood combustion furnace. J Electrostat 69(5):466–472

Chapter 6
CO_2 Control

6.1 Introduction

In recent days, many metro cities are facing visibility issues due to the presence of smog, which is a prominent form of air pollution. In general, definition, any substance that individuals bring into the atmosphere with harmful consequences for living things and the environment is considered as an air pollutant. There are two major kinds of air pollution, namely visible and invisible, which caused global warming. Preliminary greenhouse gases, such as carbon dioxide (CO_2), methane, nitrogen oxide, and water vapor, are majorly responsible for global warming [1]. Among them, CO_2 is the preliminary invisible air pollution emitted by natural source decomposition, ocean release, respiration, and human sources including transportation, industrial process, and residential heating [2]. In addition to that, CO_2 is considered as the most long-lived greenhouse gas in the Earth's atmosphere. Not only is the presence of CO_2 limited to the gas state but CO_2 is also soluble in water to form carbonic acid, which is a weak acid and is called volatile acid.

In the past few decades, higher industrialization and civilization activities have pumped enough CO_2 into the atmosphere, which resulted in the Earth's climate change. Over 7,500 sizeable CO_2 emission sources have been identified till today, which are distributed globally. In order to limit the emission globally, the following agreements worldwide have to be implemented:

1 Kyoto Protocol (1997)—The reduction of global greenhouse emission by 5%.
2 Copenhagen Accord (2009)—To prevent dangerous anthropogenic interference with the climate system.
3 Cancun Agreement (2010)—Green Climate Fund for developing countries to control emission level.
4 Durban Agreement (2011)—Carbon Capture and Storage (CCS) technologies were discussed.
5 Paris Agreement (2016)—Put taxes on carbon emissions or higher taxes on gasoline.

P.-C. Chiang and X. Gao, *Air Pollution Control and Design*,
https://doi.org/10.1007/978-981-13-7488-3_6

Carbon capture, storage, and utilization have been considered as one of the options of carbon mitigation by the International Energy Agency. Carbon Capture and Storage (CCS) is a technology aiming at separating, transporting, and permanently storing carbon dioxide (CO_2) underground to prevent its emission into the atmosphere. Carbon Capture and Utilization (CCU) can use the captured CO_2 in different industrial applications. The International Energy Agency [7] identified five categories of utilization products such as fuels, chemicals, building materials from minerals, building materials from waste, and fertilizer of biological processes. In other words, both CCS and CCU support a circular carbon economy where carbon emissions are reduced, reused, recycled, and removed.

According to Special Report on Emissions Scenarios (SRES), the emission of global CO_2 could be in the range of 29.3–44.2 $GtCO_2$ (8–12 GtC) and 22.5–83.7 (6–23 GtC) $GtCO_2$ in the years 2020 and 2050, respectively [3]. It is thus concluded that global anthropogenic fossil CO_2 emissions representing over 75% of the total anthropogenic greenhouse gas (GHG) emissions [1] have continuously increased over the last decades, peaking at 37.9 Gt in 2018 [1]. The Paris Agreement Work Program, under the United Nations Framework Convention on Climate Change, aims to maintain the increase in global average temperature to below 2 °C, ideally below 1.5 °C, above pre-industrial levels. Most efforts to mitigate the effects of climate change have been directed toward reducing anthropogenic CO_2 emissions [4].

6.2 CO_2 Emission Sources

As a part of the emission source of CO_2, human sources can be divided into two sections called stationary emission sources and mobile emission sources.

1 Stationary sources: They include installations for fuel combustion (generation of heat, power, or energy), facilities for industrial processes, and other non-mobile activities such as storage and transfer. It is a single localized emitter, such as fossil power plants, oil refineries, and massive industrial parks [5].

2 Mobile sources: They include means of transport such as cars, aircraft, and ships. They also contain engineering machinery such as tractors, forklift trucks, and construction machinery [6].

6.2.1 *Stationary CO_2 Emission Sources*

The high-level stationary CO_2 emission sources (those emitting more than 1 $MtCO_2$ per year) listed in Table 6.1 account for above 80% of cumulative emission from a total stable source. Among the causes, power and industry sectors combined account for about 60% of total global CO_2 emission [3]. In the power sector, coal is the dominant fuel at the current stage and remains dominant with a projection of 36% in 2020 [7], while natural gas will become the second largest source. Thus, the power generated

Table 6.1 Profile of worldwide large CO_2 stationary sources emitting more than 0.1 Mt CO_2 per year [3, 8, 9]

Types of Industry	Process	CO_2 concentration in gas stream % by vol	Number of sources	Emission ($MtCO_2$)	% of total CO_2 emission	Average emissions/source ($MtCO_2$ per source)
CO_2 from fossil fuels or minerals	Power					
	Coal boiler	12–14	2025	7984	59.69	3.94
	Natural gas turbine	3–4	985	759	5.68	0.77
	Natural gas boiler		743	752	5.62	1.01
	Fuel oil boiler		515	654	NA	1.27
	Other fuels		79	61	NA	0.77
	Hydrogen		2	3	0.02	1.27
Natural gas sweetening		2–65	NA	50	0.37	NA
Cement production	Combined	14–33	1175	932	6.97	0.79
Refineries		3–13	638	798	5.97	1.25
Iron and steel industries	Integrated steel mills	15	180	630	4.71	3.50
	Other processes	NA	89	16	0.12	0.17
Petrochemical industries	Ethylene	8	240	258	1.93	1.08
	Ammonia production	18	194	113	0.84	0.58
	Hydrogen production	15–20			NA	NA
	Methanol production	10			NA	NA
	Ethylene oxide	100	17	3	0.02	0.15
CO_2 from biomass	Bioenergy	3–8	213	73		0.34
	Fermentation	100	90	17.6		0.20

from fossil fuel combustion is the most significant source for CO_2 emission, while substantial amounts of the CO_2 rise in the oil and gas processing industries. After coal and gas (17.5%), nuclear (16.8%), oil (9.8%), and other renewables (1.6%) are the sources of power generation [3, 7]. In addition, fuels have been utilized in large-scale industries such as refinery, blast furnaces of steel production, and chemical manufacturing industries. While considering fuel, biomass can be a leading source for the replacement of traditional fuels. However, the availability and producibilities are insufficient. Some different industrial processes, namely cement manufacturing, refineries, steel making industries, and petrochemical industries, also play important roles. In cement industries, nearly 900 kg of CO_2 is emitted for every 1000 kg of cement production [1, 3]. Meanwhile, refineries emitted CO_2 from gas-fired heaters and steam boilers. For the case of steel or iron making process, CO_2 is emitted from fossil fuels for heating the ores at a concentration of 20%. In the petrochemical industries, ethylene and ammonia productions are significant sources of CO_2 emission as compared to hydrogen, methanol, and ethylene oxide productions [1]. In common for all the above processes, the capacity buildup for capturing CO_2 is possible with appropriate approaches.

6.2.2 Mobile CO_2 Emission Sources

Mobile source of air pollution includes air pollution emitted by vehicles and engines or equipment that can be moved from one location to another. The combustion of fossil energy in the transport sector accounts for 14% of global greenhouse gas emissions including CO_2, CH_4, and N_2O. However, there is no practical way of capturing and storing these emissions directly. Among the significant greenhouse gases, emission of CO_2 is maximum. A primary source of emission is the incomplete combustion of fuels. Table 6.2 describes the categories of mobile sources with their fuel uses.

There are three general equations that can be used to calculate CO_2 emission for each type of fuel combusted [6].

$$\text{Emissions} = \text{Fuel} \cdot \text{EF}_1 \tag{6.1}$$

$$\text{Emissions} = \text{Fuel} \cdot \text{HHV} \cdot \text{EF}_2 \tag{6.2}$$

$$\text{Emission} = \text{Fuel} \cdot \text{CC} \cdot 44/12 \tag{6.3}$$

where emission means mass of CO_2 emitted; Fuel = mass or volume of fuel combusted; EF_1 = CO_2 emission factor per mass or volume unit; EF_2 = CO_2 emission factor per energy unit; CC = fuel carbon content; 44/12 = ratio of molecular weight of CO_2 and carbon.

Table 6.2 Categories of mobile transport with average energy use and average CO_2 emissions

Types of transport	Category	Primary fuel used [6]	Energy use (Btu/passenger mile) [10]	CO_2 Equivalent (g/passenger mile) [10]
On-road vehicles	Passenger Cars	Gasoline, Natural Gas, Electric Batteries	4983 (1 person) 3154 (4 persons)	371 (1 person) 234 (4 persons)
	Sport utility vehicles (SUV)	Diesel	2492	185
	Motor Coach	Diesel	749	56
	Van-Pool	Diesel	1354	101
	Car Pool (2 persons)	Gasoline	2492	185
	Trolley Bus	Gasoline	1321	233
	Transit Bus	Gasoline	4245	299
Non-road vehicles	Construction Equipment	Diesel Fuel		
	Agriculture Equipment	Gasoline		
	Forklifts	Gasoline		
Rail	Light Rail	Diesel Fuel	1146	202
	Heavy Rail	Diesel Fuel	889	156
	Commuter Rail	Electric	1608	177
	Intercity Rail	Electric	2091	179
Water Transport	Ships	Diesel Fuels	4590	331
	Boats	Residual fuel oil, Gasoline	10,987	818
Air Transport	Commercial Aircraft	Kerosene jet fuel	3260	243

CO_2 per passenger mile was calculated using the following equations:

$$\mathrm{CO_2\ per\ passenger\ mile}\ (g/\mathrm{passenger\ mile}) = \mathrm{Total\ CO_2}(g)/\mathrm{Annual\ passenger\ mile} \tag{6.4}$$

$$\begin{aligned}\mathrm{Total\ CO_2}(g) = {} & \mathrm{Sum}(\mathrm{CO_2}(g/\mathrm{gal}) \cdot \mathrm{Annual\ gallons}) \\ & + \mathrm{Electricity}\ (\mathrm{kW} \cdot h) \cdot 600.6\ \mathrm{g\ CO_2/kwh}\end{aligned} \tag{6.5}$$

$$\mathrm{CO_2}(g/\mathrm{gal}) = (44/12) \cdot 453.6g/\mathrm{lb} \cdot \mathrm{fuel\ density(lb/gal)} \cdot \mathrm{fuel\ wt.\%\ carbon} \tag{6.6}$$

CO_2 emission per passenger mile was calculated using Eq. 6.4 and total CO_2 emissions for each mode were calculated using Eq. 6.5. For all liquids and gaseous fuels, CO_2 emissions per gallon of fuel burned were calculated using Eq. 6.6. After calculating emissions from the above equations, data are presented in Table 6.2.

As shown in Table 6.2, motor coaches from on-road transportation use 749 btu/pass-mi and produce 56 g/pass-mi of CO_2 with the least amount of energy utilized and the lowest CO_2 emission per person mile. Van-pools on average produce twice as much CO_2 per passenger mile as motor coaches, car pools produce four times as much, and public buses deliver almost six times as much for the case of on-road transportation. Meanwhile, ferryboats produce the highest CO_2 with consumption of the highest energy (10,987 btu/pass-mi) as well. From above, it can be concluded that switching freight transport from ship to air, and domestic individual transport from van-pool to rail can decrease energy consumption and CO_2 emissions.

6.3 CO_2 Control Principles

6.3.1 Capture

At some point in time in the process, CO_2 will need to be separated, which is classified by the three major processes, namely post-combustion, pre-combustion, and oxygen combustion [11]. Pre-conversion capture refers to capturing CO_2 generated as co-product of an intermediate reaction. Post-combustion capture describes any carbon capture technique, which is applied to the separation of CO_2 from the flue gases of emitted sources (fossil fuel power plants, cement industries). Meanwhile, oxy-combustion refers to the process where fuel or coal is burned with pure oxygen to produce flue gas with high CO_2 concentrations and free from NO and NO_2. In any case, it is clear that the capture will consist of a combination of energy or fuel conversion steps and separation steps [12]. Table 6.3 shows the advantages, disadvantages, various technologies, and application areas for the CO_2 capture process. Among all the techniques, post-combustion capture technologies have been majorly accepted than other approaches to carbon capture, and have reached the primary stage of development in the front of commercially available methods. It can be used to remove CO_2 from various industries, including power plants, production of ethylene oxide, cement, steel, and biogas sweetening. Post-combustion capture methods include physicochemical adsorption by solid sorbents, physicochemical absorption, mineralization, cryogenic, membrane technologies, etc. In addition to these, the biological process can be used to convert CO_2 into biomass. It is particularly important to integrate the energy conversion technologies with the separation technologies for CO_2 and O_2 to achieve low-cost and high-efficient capture routes. However, the fundamental task of separating a relatively low concentration of CO_2 (<15%) is challenging to achieve without significant energy consumption [1, 13].

Table 6.3 Advantages, disadvantages, and available capture technologies for associated combustion technologies

CO_2 source (Direct air, industry, vehicle, power generation, and others)			
	Post-combustion	Pre-combustion	Oxygen combustion
Advantages	Can be retrofitted to existing plants allowing the continued operation of valuable resources Mature technology with the greatest potential for CO_2 reduction Standard Retrofit Higher thermal efficiency	Production of carbon-free fuel A wide range of hydrocarbon fuels Syngas (H_2) production	High purity CO_2 stream Flexibility in design and operation Substantial reduction of NOx Avoid complicated post-combustion separation Potentially higher generation efficiency
Disadvantages	Pretreatment of the flue gas Additional energy requirement Limited availability of ideal sorbents High cost for capture Low CO_2 concentration affects the capture efficiency	Complex chemical process Repowering Large capital investment Greater Risk	Oxygen separation Corrosion problems may arise High-efficiency drop and energy penalty A lot of risk and safety issues High capital cost
CO_2 Technologies	Physicochemical adsorption Physicochemical absorption Mineralization Bio-based system Chemical looping Cryogenic Membrane technology Gas hydrates Ionic liquids	Physicochemical adsorption Physicochemical absorption Gasification Separation technologies Advance turbine Crystallization Cryogenic	Cryogenic distillation Pressurized oxyfuel Oxyfuel gas turbines Membranes Chemical looping
Applications	Power plants, natural gas sweetening, iron and steel industries, oil refineries, and cement industries	Power plants, ammonia productions, gas separations	Power plants, iron and steel industries, cement industries, syngas production, and upgradation

6.3.2 *Sequestration*

CO_2 sequestration is the process involved in carbon capture and the long-term storage of captured CO_2. Before sequestration, captured CO_2 must be transported (onshore or offshore) from the point of capture to a storage site [11, 14]. The captured CO_2

can be transported as a solid gas, liquid, or supercritical fluid. In general, there are two main transport operations, i.e., Pipelines and Shipping.

The emission of millions of tons of CO_2 makes shipping ideas impractical, but transport of a small amount of CO_2 depends on rail or heavy goods vehicles. Thus, pipelines are considered the only practical option for onshore transportation when millions of tons of CO_2 will be stored annually. Figure 6.1 represents the CO_2 sequestration mechanisms which include Ocean sequestration and Geological sequestration.

The ocean is the most promising place to sequester carbon, which currently takes up 2 billion metric tons per year. The amount of carbon that would double the load in the atmosphere would increase the concentration in the deep ocean by two percent only. However, a higher level of CO_2 turned the ocean water to acidic, which affects aquatic life.

Geological storage is the second type of sequestration technique, which includes injecting CO_2 into the depth of 800–1000 m [15]. The adsorption rate depends on the reservoir pressure and temperature. The main concern with CO_2 storage is its possible leaks and escaping to the environment. The annual leakages are reported between 0.0001 and 1%, depending on the geological structure and its faults. Table 6.4 represents the major storage projects worldwide. The major source of CO_2 in

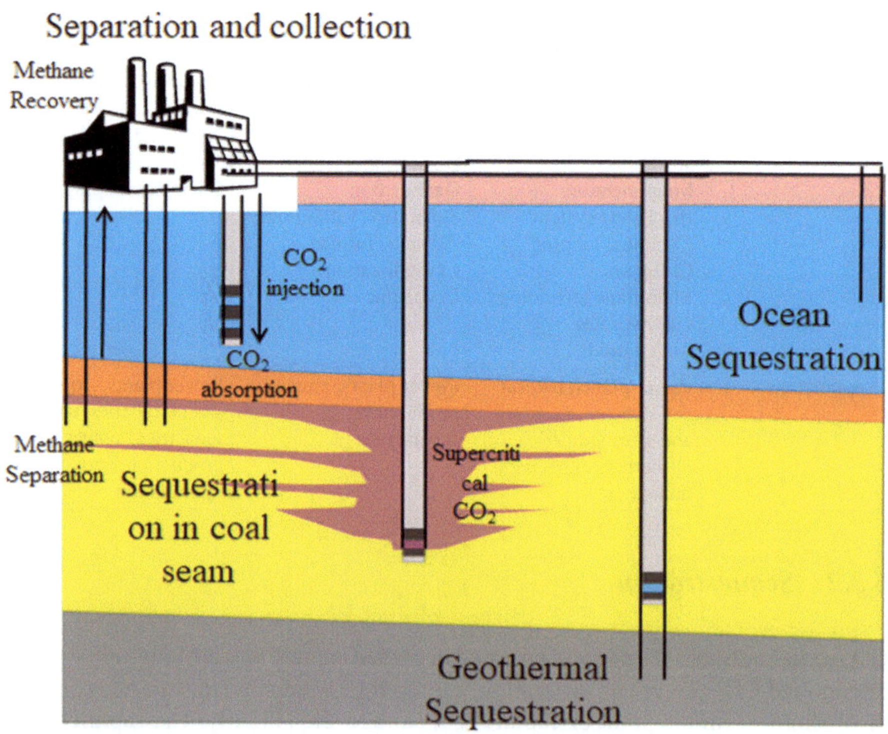

Fig. 6.1 Overview of CO_2 sequestration technology

Table 6.4 6.Major storage projects and proposed plant CCS projects [16]

Year of establishing	Storage country	CO_2 source/capture type	CO_2 storage type	CO_2 quantity (Mt/year)
1972	USA	Natural gas production/industrial separation	EOR	0.4–0.5
1982	USA	Fertilizer production/industrial separation	EOR	0.7
1986	USA	Natural gas production/industrial separation	EOR	7.0
1996	Norway	Natural gas production/industrial separation	EOR	1.0
2000	Canada	Synthetic natural gas/industrial separation	EOR	3.0
2008	Norway	Natural gas production/industrial separation	EOR	0.7
2010	USA	Natural gas production/industrial separation	EOR	8.4
2013	USA	Hydrogen gas production/industrial separation	EOR	1.0
2013	USA	Fertilizer production/industrial separation	EOR	1.0
2013	USA	Natural gas production/industrial separation	EOR	0.9
2013	Brazil	Natural gas production/industrial separation	EOR	1.0
2014	Canada	Power generation/post-combustion capture	EOR	1.0
2015	Canada	Hydrogen gas production/industrial separation	Geological storage	1.0
2015	Saudi Arabia	Natural gas production/industrial separation	EOR	0.8
2016	UAE	Iron and steel production/industrial separation	EOR	0.8
2017	USA	Ethanol production/industrial separation	Geothermal storage	1.0

(continued)

Table 6.4 (continued)

Year of establishing	Storage country	CO_2 source/capture type	CO_2 storage type	CO_2 quantity (Mt/year)
2017	USA	Power generation/post-combustion capture	EOR	1.4

EOR—Enhance oil recovery

listed projects is natural gas production with industrial separation capture types, while enhancing the oil recovery method is adopted for CO_2 storage.

6.3.3 *Utilization*

As an alternative to storage, captured CO_2 can be used as a commercial product for direct use. Utilization can be divided into chemical feedstock conversion, mineral carbonation, fuel production, and oil recovery enhancing applications [1]. In the utilization process, not only the CO_2 is used but fixed CO_2 would be reformed within a short time after CO_2 to chemical is used. In the direct utilization technology, CO_2 involves phase change (i.e., gas, liquid, solid, and supercritical fluids).

Table 6.5 presents the existing usage of CO_2 and the emerging usage of CO_2.

Table 6.5 Existing and emerging application areas for CO_2 utilization [17, 18]

Existing use of CO_2 (current demand (Mtpa))	Emerging usage of CO_2 (future demand (Mtpa))
EOR (30–300)	Enhanced coal bed methane (ECBM) recovery
Feedstock for urea production (5–30)	
Beverage carbonation (8)	
Prevent oxidation of wine (<1)	Polymer processing (5–30)
Food packaging (8.5)	Algal bio-fixation (>300)
Coffee decaffeination (1–5)	Improved geothermal power generation (<1)
In PCB as cleaning fluid	CO_2 as feedstock for the production of organic acid, alcohols, esters, and sugar
Welding (<1)	Baking soda (<1)
Fire suppression (<1)	Concrete curing (30–300)
Refrigerant gas (<1)	Renewable methanol (>300)
Pharmaceutical process (<1)	Solar fuel (<1)
Pneumatics hand tools (<1)	Formic acid, formaldehyde (>300)
Supercritical CO_2 as solvent (<1)	Fischer–Tropsch process
Pulp paper processing (<1)	Cultivate microalgae
Inserting gas (<1)	

Direct utilization of CO_2 is in the food industry as a carbonating agent, preservative, packaging gas, and as a solvent for the extraction of flavors and in the decaffeination process [19]. EOR and ECBM are other examples of direct utilization of CO_2 where it is used to extract crude oil from an oil field or natural gas. CO_2 can be utilized through carboxylation reaction where the CO_2 molecule is used as a precursor for organic compounds such as carbonates, acrylates, and polymers.

However, its conversion is energy-intensive as CO_2 is thermodynamically stable [20]. Microalgae is another approach to utilize fixed CO_2 from the waste to fuels [21]. However, the cost and energy associated with the harvesting stage make it less available in the market. Among all chemical processes, the mineral carbonation process with magnesium and calcium to form carbonate rocks is the simplest chemical process [1]. Overall carbonation reaction using serpentine is illustrated below

$$Mg_3Si_2O_5(OH)_4 + 3CO_2 - 3MgCO_3 + 2SiO_2 + 2H_2O \tag{6.7}$$

Meanwhile, multi-step carbonation consists of separating the metal from the mineral matrix in the presence of HCl or molten salt, followed by hydration reactions to obtain the metal in hydroxide form [19]. Finally, carbonation is taking place with captured CO_2. The reaction mechanism is illustrated below

$$Mg_3Si_2O_5(OH)_4 + 6HCl - 3MgCl_2 + 2SiO_2 + 5H_2O \tag{6.8}$$

$$MgCl_2.6H_2O - MgCl(OH) + HCl + 5H_2O \tag{6.9}$$

$$2MgCl(OH) - MgCl_2 + Mg(OH)_2 \tag{6.10}$$

$$Mg(OH)_2 + CO_2 - MgCO_3 + H_2O \tag{6.11}$$

6.3.4 *Conversion*

CO_2 can be considered as an important carbon source to convert valuable chemicals and fuel via CO_2 cycloaddition to epoxides, CO_2 hydrogenation, CO_2 reduction (electrocatalysis, biological, and photocatalysis method), CO_2 mineralization, and more [1, 22]. To convert to the product, additional energy inputs are essential due to the thermodynamical stability of the CO_2. Table 6.6 lists converted products from CO_2. At first sight, it is surprising that so many exothermic conversions exist but when considering Gibb's free energy, most of these reactions are associated with a positive number which is not thermodynamically favored [20]. In that case, many researchers have proposed heterogeneous and homogeneous catalysis systems to generate the product at mild conditions.

Table 6.6 Enthalpy and Gibbs free energy, reaction, and annual production of CO_2 converted products.

Product	Reactions	ΔH° (kJ/mol)	ΔG° (kJ/mol)	Annual production (tons) [18, 23]
Methanol	$CO_{2(g)} + 3H_{2(g)} – CH_3OH_{(l)} + H_2O_{(l)}$	– 137.8	– 10.7	100 millions
Formaldehyde	$CO_{2(g)} + 2H_{2(g)} – HCHO_{(g)} + H_2O_{(l)}$	– 11.7	46.6	9.7 million
Formic acid	$CO_{2(g)} + H_{2(g)} – HCOOH_{(l)} + H_2O_{(l)}$	– 31.0	34.3	700 thousand
Urea	$CO_{2(g)} + 2NH_{3(g)} – NH_2CONH_{2(aq)} + H_2O_{(l)}$	– 119.7		150 million
Dimethyl ether	$2CO_{2(g)} + 6H_{2(g)} – CH_3OCH_{3\,(g)} + H_2O_{(l)}$	– 264.9	– 38	
Dimethyl carbonate	$3CO_{2(g)} + 6H_{2(g)} – (CH3O)_2CO_{(l)} + H_2O_{(l)}$	– 28		10 million
Oxalic acid	$2CO_{2(g)} + 3H_{2(g)} – (COOH)_{2(l)}$	– 39.3	85.3	
Methane	$CO_{2(g)} + 4H_{2(g)} – CH_3OH_{(l)} + H_2O_{(l)}$	– 259.9	– 132.4	
Acetic acid	$CO_{2(g)} + CH_{4(g)} – CH_3COOH_{(l)}$	– 13.3	58.1	
Formamide	$CO_{2(g)} + H_{2(g)} + NH_{3(g)} – HCONH_{2(l)} + H_2O_{(l)}$	– 103	7.2	
Ethanol	$CO_{2(g)} + 3H_{2(g)} + CH_3OH_{(l)} – C_2H_5OH_{(l)} + 2H_2O_{(l)}$	– 221	– 88.9	
Acetaldehyde	$CO_{2(g)} + CH_{4(g)} + H_{2(g)} – CH_3CHO_{(l)} + H_2O_{(l)}$	– 14.6	74.4	
Acetone	$CO_{2(g)} + CH_{4(g)} + 2H_{2(g)} – (CH_3)_2CO_{(l)} + H_2O_{(l)}$	– 70.5	51.2	
Acrylic acid	$CO_{2(g)} + C_2H_{2(g)} + H_{2(g)} – CH_2 = CHCOOH_{(l)}$	– 223.6	– 115	
Benzoic acid	$CO_{2(g)} + C_6H_{6(l)} – C_6H_5COOH_{(l)}$	– 21.6	30.5	
Carbon monoxide	$CO_{2(g)} + C_{(s)} – 2CO_{(g)}$	172.6	119.9	
Ethane	$CO_{2(g)} + 2CH_{4(g)} – C_2H_{6(g)} + CO_{(g)} + H_2O_{(l)}$	58.8	88	
Ethylene	$2CO_{2(g)} + 2CH_{4(g)} – C_2H_{4(g)} + 2CO_{(g)} + 2H_2O_{(l)}$	189.7	208.3	
Acetaldehyde	$CO_{2(g)} + C_2H_{4(g)} – C_2H_4O_{(g)} + CO_{(g)}$	178	176	
Polycarbonate	$CO_{2(g)} + CH_3CHCH_2O –$ [–R–O–CO–O–CH_2-O–]$_n$			

Methanol, dimethyl carbonate, and dimethyl ether are promising materials for energy production, which reduce dependency on fossil fuels [13]. Apart from energy, methanol has been widely used in pharmaceuticals, dyes, paint, plastic, rubber, and fiber industries, while dimethyl ether is one of the derivatives of methanol that initially derived through the distillation of by-products produced under high-pressure methanol synthesis. It is widely used in electric power generation, and automotive fuel and domestic applications such as heating and cooking. In recent years, H_2 and solar fuels are produced using solar light as a photocatalytic light source. However, the conversion rate is very low, and the energy consumptions are very high. Thus, much attention is drawn toward system optimization for solar fuels conversion technologies. In recent years, incorporation of CO_2 into polymers such as urea–formaldehyde resin and melamine–formaldehyde resin have been found with 200 million tons of usage, while the mineralization of CO_2 or carbonation could be considered as a means of storage in geological formations. Another approach to CO_2 conversion via the biological method using fast-growing biomass is called enhanced biological fixation in which, microalgae have attracted considerable interest due to the conversion of biochemical or biofuels. This process is economical and energy perspective as compared to other reported techniques. After considering all associated techniques, control technology for system optimization has been developed.

6.3.5 Control Technologies

After considering all the above different aspects and principles, emission from a stationary source such as coal-fired power plants, natural gas thermal plants, natural gas turbine plants, iron, and cement industries are the primary sources of the mission. To control CO_2 by sequestration, utilization, and conversion, CO_2 capture is the initial step for supplying CO_2 to the next cycle. Thus, most of the technologies are designed based on CO_2 capture technology in its first step. According to "Electric Power Research Institute" post-combustion capture can be the leading technology for controlling CO_2 due to its higher efficiency for capturing CO_2. Figure 6.2 shows various CO_2 capture techniques.

6.3.6 Adsorption Technology

There are two kinds of adsorption, i.e., physical and chemical [25]. The main advantage of physical adsorption over chemical adsorption is its simple and energy-efficient operation. Pressure swing and temperature swing methods can achieve that. Activated carbon, metalorganic framework, clay, zeolite, mesopores silica, amine-based alkali earth metals, and Li-ion-based adsorbents have frequently been used [1, 26]. With the physical adsorption method, adsorbents selectively adsorbed CO_2 with the help of the Van der Waals force working between the molecules of fluid and the

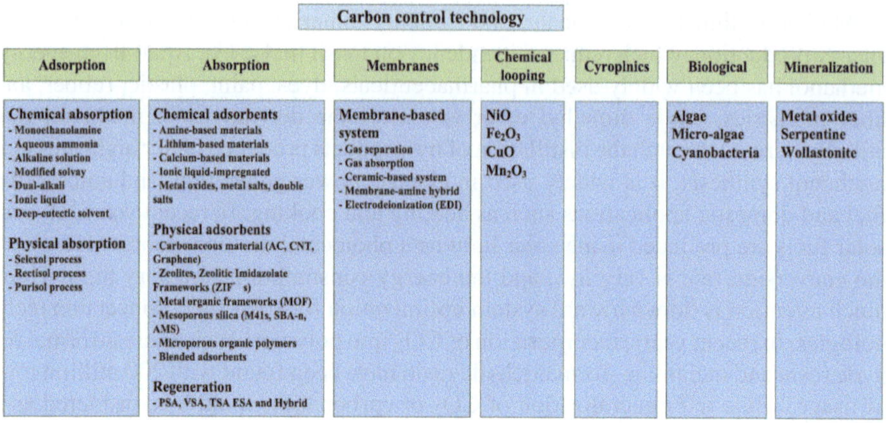

Fig. 6.2 Classification of CO_2 control techniques

surface of the adsorbents, and then release the adsorbed CO_2 under reduced pressure, consequently allowing CO_2 capture separation and recovery with high purity at high recovery rates [25, 26]. The first attempt to implement this technology for an adsorbed CO_2 from blast furnace gas was carried out in Japan Iron and Steel Federation with a project name "CO_2 Ultimate Reduction in Steelmaking process by innovative technology for cool Earth 50 (COURSE 50)" [27].

In the project, the "Advanced Separation systems by Carbon Oxides Adsorption" unit was constructed as described in Fig. 6.3. The actual blast furnace gas is compressed up to 300 kPa, which is cooled to 10 °C and removes water and sulfur compounds by adsorption with silica-alumina gel and activated carbon before feeding to the PSA unit. The CO_2 recovery rose up to 6.3 tons per day with a shorter cycle time [27].

In recent years, new processes called multilayers are being developed for dealing with humidity flue gas streams and impurities. Multilayered beds enable the use of adsorbents with high CO_2 selectivity, but it degrades significantly in the presence of water.

6.3.7 *Absorption Technology*

The absorption technique is used to remove CO_2 from gas streams by contacting the contaminated air with a liquid solvent. It has been classified as chemical absorption and physical absorption.

1. Chemical absorption

Chemical absorption involves one or more reversible chemical reactions between CO_2 and absorbent solutions, which mostly include alkanol amine. During the

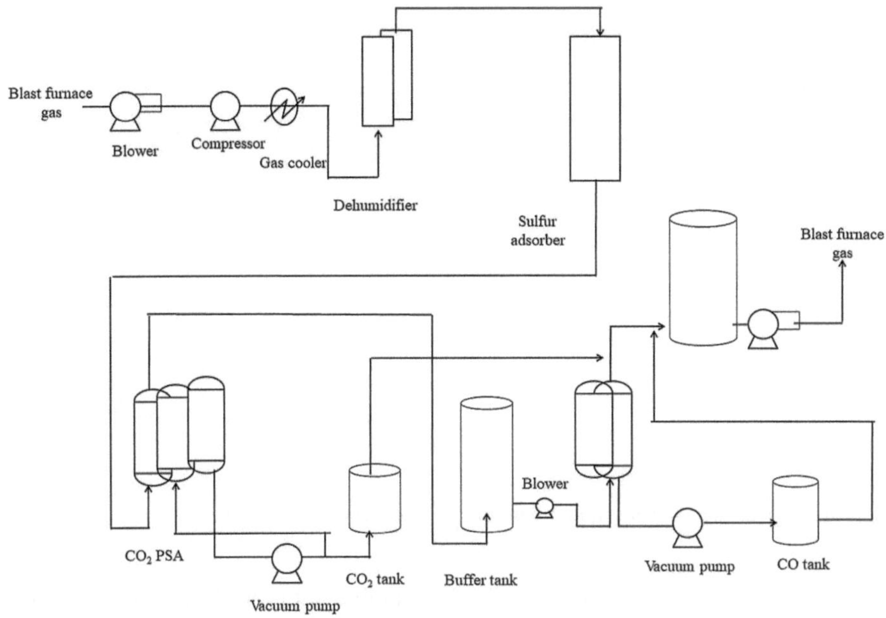

Fig. 6.3 Physical adsorption unit for CO_2 from Japan Iron and Steel Federation unit

reactions, the bond between absorbent and CO_2 is broken, and it will recover from flue gases. Recently, the amine-based process has been used commercially for the removal of acid gas impurities from mixed gas streams. In general, tertiary amines with three replacements of hydrogen groups through R (alkyl, aryl, or aralkyl) groups have been utilized for absorption. During the reaction, the -OH group serves to reduce the vapor pressure by increasing the water solubility, while the amino group provides the alkalinity in the water solution for absorbing acid gases [28, 29]. The diagram of a typical absorption system is shown in Fig. 6.4. During the process, the combustion flue gas (~230 °C) coming out of the stack, usually needs to be cooled below 50 °C before entering the absorber system. This is achieved by spraying cooling water in a cooler. After the cooling step, the blower is installed to give enough pressure to send flue gases to the absorption/desorption system. Conventionally, the waste gas to be scrubbed of CO_2 enters the absorber at the bottom, flows up, and leaves at the top, whereas the solvent enters the top of the absorber, flows down, and emerges at the bottom. In order to reduce the viscosity of the circulating fluid, dilution through water addition is performed. With higher viscosity, fluid would require more power to pump and form circulation. After dilution, the liquid amine solution containing the absorbed gas flows to the stripper unit, where it is heated, and the acid gases are liberated. The solvent regeneration can be carried out at low pressure to enhance the desorption of CO_2 from the liquid. Some amine solution is typically carried over into the acid gas stream from the regeneration step, and the amine solution is recovered using a

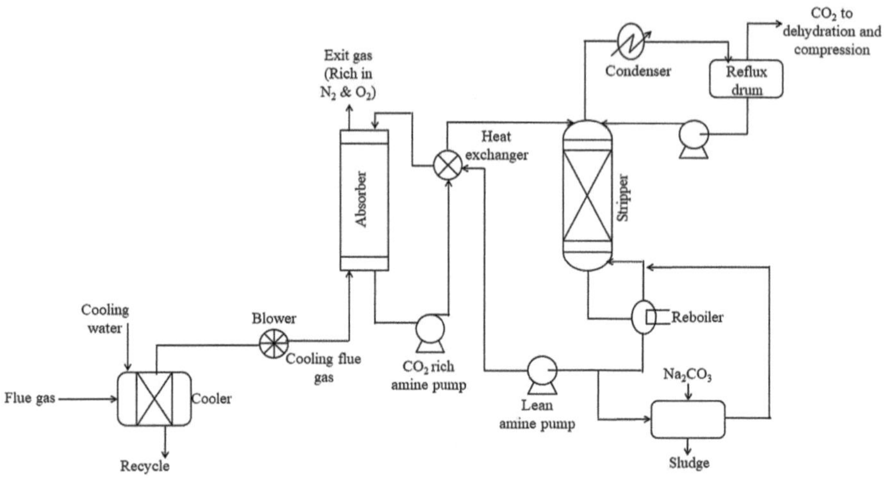

Fig. 6.4 Typical chemical absorption unit for CO_2 recovery from flue gas

condenser. After the condenser, the lean amine solution flows through a heat exchanger where it is contacted with the rich amine solution from the contact tower, and from there the lean amine solution is returned to the gas contact tower [30]. The above chemical absorption techniques have been adopted by Sumitomo Chemicals, Luzhou Natural Gas Chemicals, PROSINT, and Mitsubishi Heavy Industries worldwide.

2. Physical absorption

CO_2 is physically absorbed in a solvent according to Henry's law [29]. In most of the case, temperature and pressure are affecting the absorption properties. The advantage of this process is the lower energy consumption compared to others. The Selexol process, Rectisol process, Purisol process, and Fluor solvent process are commercially available processes for physical absorption [30]. The most popular process is the Selexol process for bulk CO_2 and H_2S removal. In general, the Selexol solvent is a mixture of dimethyl ether of polyethylene glycols. The typical Selexol process for acid gas removal is presented in Fig. 6.5. The first sulfur absorber column removes most of the H_2S from the feed syngas, followed by the second column for CO_2 removal. The rich solvent leaving the CO_2 absorption column is flashed in drums, from which relatively pure CO_2 is recovered. The solvent in the sulfur absorber column must be stripped in a column with a reboiler to remove the high H_2S content gases [31].

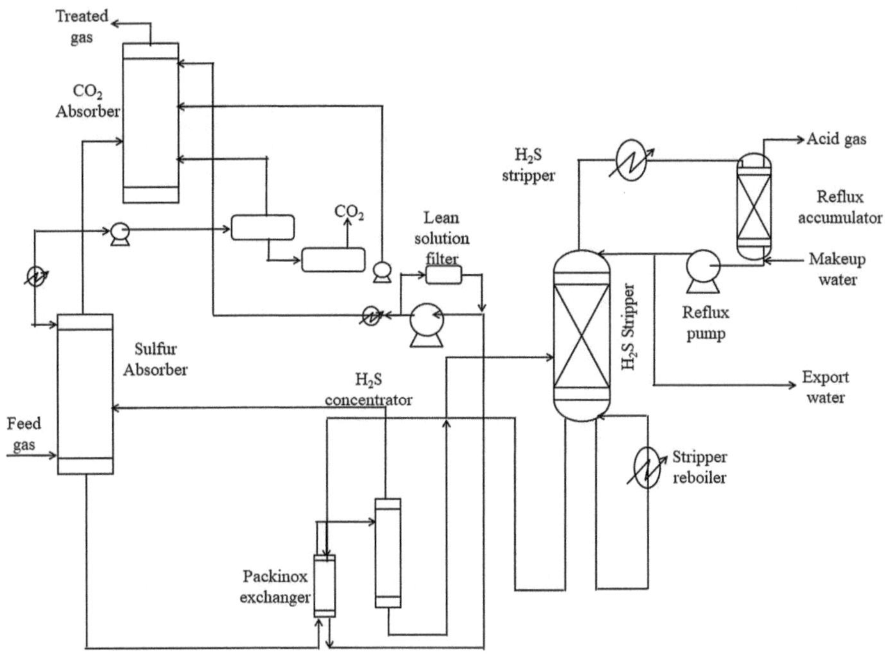

Fig. 6.5 Physical absorption (Selexol) unit for CO_2 and sulfur recovery from flue gas

6.3.8 Membrane Technology

Membrane technology has successfully been applied in various applications such as desalination by reverse osmosis, biogas purification up to CNG quality, removal of heavy metals from water, and CO_2 capture. Concerning CO_2 capture by membrane processes, progress can be observed in the last 10 years. Membranes with predominate polymer materials are thin barriers that allow selective permeation of certain gases. CO_2 selective polymer membranes offer an alternative to post-combustion. Primary research is focusing on developing membranes, which are sufficiently selective to produce high purity CO_2 in a single step. More functional active films are developed rather than high-pressure gradient films to pursue CO_2 separation more economically. Amine functional membrane systems are frequently used but are complex and expensive [32]. Membrane-based systems provide a simple, low-cost, compact solution for reducing the CO_2 content in mixed gas. These membranes are used for achieving pipeline CO_2 with specifications of 4 mol.% or less or for bulk CO_2 removal. These membranes are higher in module packaging capacity, lighter in weight, and have less space, making them be easily installed.

There are many limitations to single-step membrane technology such as cleaning, challenging to manage continuous processing, and separation performance. To solve these constraints, a two-stage system (Fig. 6.6) has been proposed where combustion air is used as a sweep gas to drive the second stage and increase the CO_2 concentration

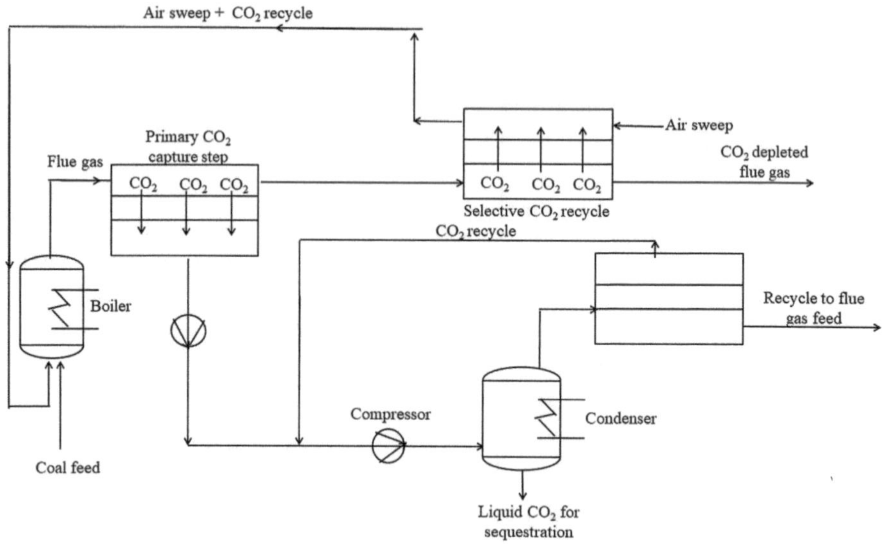

Fig. 6.6 Two-stage membrane process for CO_2 capture

at the first stage. In the second step, selective CO_2 recycles with the removal of air sweep [33], while recycling CO_2 passes to the boiler with decreasing temperature. However, the higher temperature of the boiler is never fully exploited by the steam cycle. To lower the cost, a high permeable membrane with advanced inorganic or organic materials with high selectivity needs to be developed.

It was thus concluded that membrane technology is potentially suitable for fuel gas purification and CO_2 enrichment, but there are still some obstacles, i.e., membrane resistance for humid feed stream, fouling, and long-term stability of thin selective layer should be overcome in the future.

6.3.9 Chemical Looping Technology

Chemical looping technology is a unique approach to carbon capture because it avoids the additional step of the gas separation process. Thus, it is capable of keeping the low energy cost of the process. By using substances which can be efficiently oxidized and reduced, oxygen may be inherently separated from the air [34]. In the process of chemical looping, a dual fluidized bed system has been deployed with the circulating system, where a metal oxide is employed as a bed material providing the oxygen for combustion in the fuel reactor [33]. The reduced metal is then transferred to the second bed (air reactor) and re-oxidized before being reintroduced back to the fuel reactor by completing the loop (Fig. 6.7).

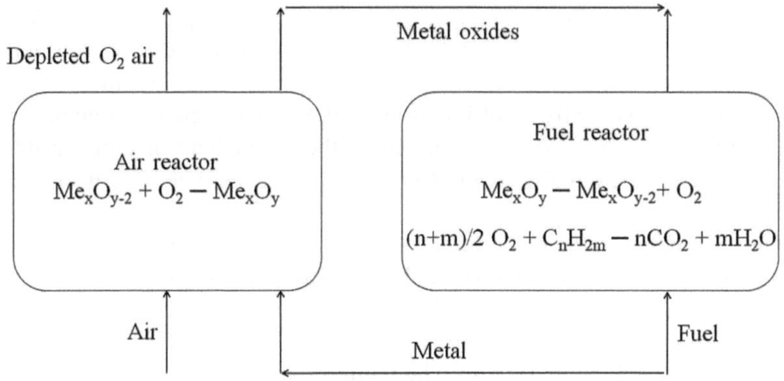

Fig. 6.7 Chemical looping technology for CO_2 capture

Chemical looping technology uses oxygen carrier materials, such as metal oxides or sulfates, to indirectly provide oxygen for full or partial oxidation of the fuel in the reducer tank. The oxygen-depleted oxygen carriers are then regenerated by a reaction with air in the oxidizer during the process of power generation. The stable oxygen carriers are circulated between the two reactors with transferring oxygen. In this process, solid–solid interactions are much slower. Thus, in situ gasification with fluidizing steam or CO_2 is required for the combustion of coal [35]. Oxygen carrier materials are the key player in determining the quality and efficiency of the process. Extensive research has been conducted to develop desirable oxygen carrier materials with enhanced redox reactivity, long-term stability, physical strength, and at the same time minimum cost and energy consumption. The chemical looping combustion pilot plant with 3 MW has been commissioned by Alstom (GE power) company in which, calcium sulfate as a carrier is used [30]. Meanwhile, Babcock & Wilcox have developed iron oxide-based chemical looping technology.

It is thus concluded that chemical looping has great potential for reducing the energy consumption and costs of CO_2 capture from fossil fuel-based power and chemical production while maintaining high efficiency. Further information on factors such as oxygen carrier durability, technical feasibility of downstream high-temperature valves and filters, and scale-up challenges will be required to select the best configuration. The chemical looping plant exhibits a promising decarburization technology by integrating renewable energy for further development [36].

6.3.10 *Cryogenic Technology*

The existing challenges that need to be overcome in cryogenic technology include cold energy sources, capture costs, impurities, etc. Sterling cooler consists of two isothermal and two isochoric heat transfer processes which may be determined by a high and a low temperature used for a novel cryogenic CO_2 capture system. Advanced cryogenic CO_2 capture processes can be easily retrofitted to the existing industrial emission facilities and avoid the challenges associated with chemical solvents or physical sorbents. Recently, heat integration of the developed cryogenic CO_2 gas separation process is employed under which the latent heat of condensate water, the sensible heat of residual gas, and CO_2 product are recovered by heat exchanger networks [37].

A conventional cryogenic distillation process was proposed by Holmes and Ryan for natural gas purification under which the feed gas is initially cooled by a precooler, and then chilled to low temperature by the heat exchanger. The cooled feed gas is sent to the distillation column, which contains a number of vapor–liquid contact devices (such as trays or packing materials). The methane separated at the top is withdrawn via a partial condenser. Condensed CO_2 is gathered at the bottom of the distillation column and then further separated to generate the purified CO_2 product extracted from the separator [38].

The CryoCell® process was developed by Cool Energy Ltd. which indicates that feed gas is initially dehydrated to low water specifications (5 ppm) for cryogenic operations. The dried gas then undergoes heat exchange with treated gas and cold CO_2 prior to cooling the CO_2 to its freezing point and then enters the CryoCell® separator as a three-phase mixture, as shown in Fig. 6.8. The solid CO_2 collected at the bottom of the separator is melted by a heater and separated as a liquid phase [37]. Table 6.7 shows the advantages and limitations of the existing low-temperature CO_2 capture technologies [38].

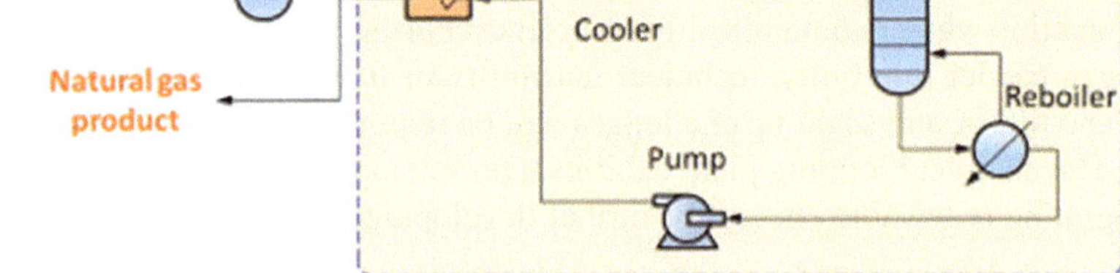

Fig. 6.8 Process flow diagram of the typical CryoCell® CO_2 capture process from natural gas [38]

Table 6.7 Advantages and limitations of the existing low-temperature CO_2 capture technologies.

Category	Cold energy source	Advantages	Limitations
Packed bed	Liquid nitrogen gas (LNG)	Atmosphere Simultaneous H_2O and CO_2 removal Avoiding high-pressure drop Surface area-to-volume ratio of the column	Depends on the availability of LNG Lab scale
AnSU	Liquefied natural gas (LNG)	Atmosphere Lower energy penalty than MEA absorption Pilot demonstration	Depends on the location of the natural gas station No H_2O adversely affects heat conduction Undesired mechanical stresses
CryoCell	Chiller	No process heating system required No corrosion potential No foaming potential Avoid compression cost	More suitable for high CO_2 concentration (higher than 20%) High compression power requirement
Distillation	Compressor and cooler	Avoid compression cost Easy to be pumped to a storage site Energy storage potential Water-saving potential Simultaneous removal of other pollutants (Hg, SO_X, NO_2, HCL, etc.)	Capital cost for the pressure difference High installation cost
Stirling cooler	Stirling cooler	Atmosphere Simultaneous H_2O and CO_2 removal Lower energy penalty than MEA absorption Energy storage potential	Exergy loss due to temperature difference Difficulty of frost layer scrapping Lab scale

6.3.11 Biological Treatment Technology

Biological treatment for CO_2 capture and conversion is a low-cost industrial technology. This method is performed by subjecting cultivation raceway tanks to natural or artificial light sources. However, this method cannot control cultivation conditions such as light intensity, evaporation of cultivation medium, and pH of the land [1, 39]. In that case, the tubular bioreactor is an alternative to this approach that can solve the limitation of light. Meanwhile, a photobioreactor is an efficient technique for CO_2 conversion. In this technique, photosynthetic microorganisms are suspended inside a medium containing a polluted air stream. These photosynthetic microorganisms (microalgae) can then use CO_2 as their carbon source [40]. The microalgae are available in various sizes in freshwater or marine environments. They utilize CO_2

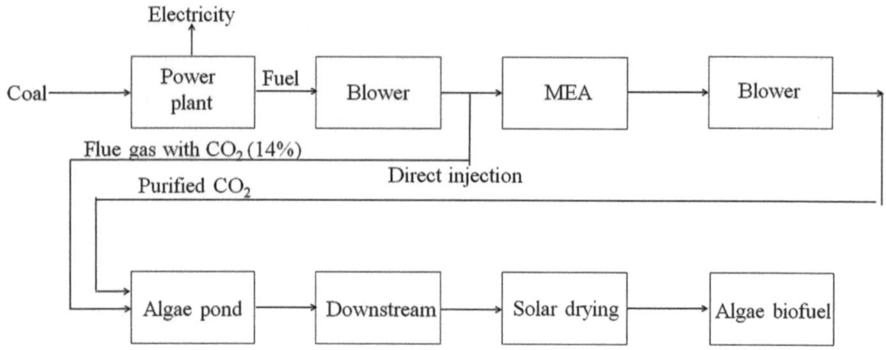

Fig. 6.9 Bio-algae based technology for CO_2 capture

as a source of energy for photoautotrophic growth and produce oxygen. Locating algae ponds near CO_2 point sources provides cost and energy-saving advantages. Integrated power plants with algae pond facilities would reduce the cost of CO_2 transportation, produce waste heat to warm ponds in winter, and give carbon credit to utilities [38]. Figure 6.9 describes a schematic diagram of an algae pond with the circular CO_2 economy [41].

There are two primary forms to grow algae, suspended and attached growth. In suspended growth systems such as photobioreactors and raceway ponds, the water flows around and goes through the free-floating algae [42]. These suspended growth reactors are cost-effective and are the secondary options among the CO_2 fixation methods, while the biological air treatment process uses attached growth in which the microorganisms, bacteria, or algae treating the waste are attached to the media of the reactor. Another technology for CO_2 fixations is bio-scrubber. The operation method of a bio-scrubber in biological treatment can be explained by the gas stream bubbled through an aqueous suspension, which contains active microalgae. The advantages and disadvantages of different types of algae reactors are mentioned in Table 6.8.

6.3.12 *Mineralization Technology*

CO_2 mineralization is a method that binds CO_2 permanently in a mineral form. In nature, this process is called weathering, in which, calcium and magnesium rocks tend to react with dissolved CO_2. This reaction results in various carbonate minerals that are inherently stable, preventing the release of CO_2 [43]. The process of weathering is prolonged. Thus, optimized pressure conditions have to be maintained to complete the reactions. Lackner and co-workers had studied the first gas–solid carbonation process (Eq. 6.12) [44]. Meanwhile, the enhanced CO_2 stored capacity in the presence of

Table 6.8 The advantages and disadvantages of different types of algae reactors

Type of reactor	Advantages	Disadvantages	Applications
Suspended growth	Easier controlling biomass No production of wastewater	Complex operations Plugging and drying require High-pressure drops	Wastewater treatment
Attached growth	Less expensive and less energy than suspended growth Simple operations Less equipment maintenance No inoculation	Large land required Odor issues Inability to operate with a high quantity of materials Less durability of natural matters Problems in treating acidic gases	Wastewater treatment Surfactant industries
Bio-scrubber	No drying Highly controllable units	Complex operations Expensive than other processes Discharge wastes need to be treated	Cigarette industries Pulp paper plant Wastewater treatment plant

moisture related to the fact that water vapor can be useful to convert oxides that may be present as hydroxide and then be carbonated as shown in Eqs. 6.13 and 6.14.

In that case, magnesium was selected for mineralization.

$$Mg_2SiO_4 + 2CO_2 - 2MgCO_3 + SiO_2 \tag{6.12}$$

$$MgO + H_2O - Mg(OH)_2 \tag{6.13}$$

$$Mg(OH)_2 + CO_2 - MgCO_3 \tag{6.14}$$

In order to optimize yield, the pretreatment step is to promote and accelerate carbonation reaction rates through enhancing the surface area step. There are two major processes, namely mechanical grinding and chemical leaching have been adopted [1, 45].

Figure 6.10 shows the magnesium silicate carbonation via $MgSO_4$ and $Mg(OH)_2$. In the schematic diagram, the first part is the mechanical grinding of magnesium silicate to increase the surface area. In the scheme, there are two by-product streams, one is for iron oxide, and another is for SiO_2. Carbonation of $Mg(OH)_2$, at 500 °C and pressurized conditions, gives pressurized steam as reaction products. The 30% of the total heat is generated only from the carbonation process, and it would be consumed by $Mg(OH)_2$ production process [45]. The use of a significant amount of steam to suppress calcination would complicate the final goals of the CO_2 sequestration. Thus, excess steam would be recycled in $Mg(OH)_2$ productions. The carbonation reaction is competing with calcination, which gives less reactive MgO.

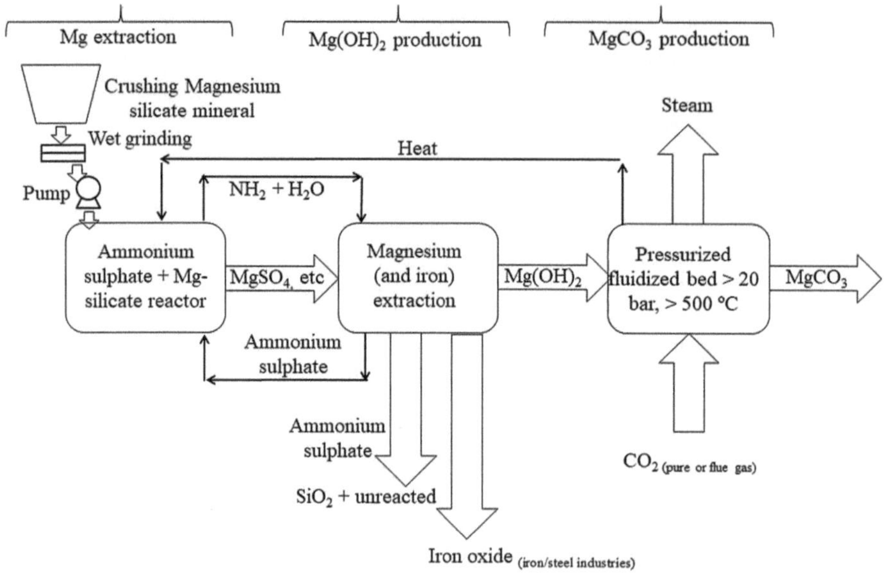

Fig. 6.10 CO_2 mineralization through magnesium silicate carbonation

On the other hand, $CaCO_3$ (Calcite) is another mineral which also has been considered as an alternate method for CO_2 mineralization and utilization. The following is the process chemistry of CO_2 capture by accelerated carbonation of alkaline wastes [46].

Step 1: calcium leaching from BOFS

$$Ca(OH)2(s) + H_2O(l) \rightarrow Ca^{2+}(aq) + 2OH^-(aq) + H_2O(l) \quad (6.15)$$

Step 2: gsdroud CO_2 dissolution into alkaline solution

$$CaO(s) + H_2O(l) \rightarrow Ca^{2+}(aq) + 2OH^-(aq) \quad (6.16)$$

$$CO_2(g) + 2OH^-(aq) \rightarrow CO_3^{2-}(aq) + H_2O(l)(\text{major reaction at pH} > 10.3) \quad (6.17)$$

$$CO_2(g) + OH^-(aq) \rightarrow HCO_3^-(aq)(\text{major reaction at } 6.3 < \text{pH} < 10.3) \quad (6.18)$$

Step 3: carbonate precipitation

$$Ca^{2+}(aq) + CO_3^{2-}(aq) \rightarrow CaCO_3(s)(\text{major reaction at pH} > 10.3) \quad (6.19)$$

$$Ca^{2+}(aq) + HCO_3^-(aq) \rightarrow CaCO_3(s) + H^+(aq)(\text{major reaction at } 6.3 < pH < 10.3) \quad (6.20)$$

According to the above observation, the process chemistry of accelerated carbonation is for BOFS. At first, the leaching of $Ca(OH)_2$ and CaO in BOFS would directly generate the Ca^{2+} and OH^- in the solution, as shown in Eqs. 6.15 and 6.16, respectively. Second, gaseous CO_2 can rapidly dissolve into the alkaline solution, where the predominant carbonate ions ($CO_3{}^{2-}$) could reduce the pH of the solution, as indicated in Eq. 6.17. Because the CO_2 continuously dissolved into the solution during the carbonation, the pH value would decrease gradually to Eq. 6.3, where the bicarbonate ions ($HCO_3{}^-$) were found to be dominated, as shown in Eq. 6.18. Finally, $CaCO_3$ would be formed by reacting the calcium ions with the carbonate ions ($CO_3{}^{2-}$) and bicarbonate ions ($HCO_3{}^-$) in the solution under a high pH and circum-neutral condition, respectively, as expressed by Eqs. 6.19 and 6.20 [46].

One kind of feedstock for accelerated carbonation is natural silicate minerals, such as wollastonite ($CaSiO_3$), serpentine ($Mg_3Si_2O_5(OH)_4$), olivine (Mg_2SiO_4), talcum ($Mg_3Si_4O_{10}(OH)_2$), pyroxene, and amphibole, which are rich in calcium or magnesium content. Although the CO_2 storage capacity of these natural Ca–Mg–silicate minerals is sufficient to fix the CO_2 emitted from the combustion of fossil fuels, the technological carbonation of these minerals is slow and energy-demanding. One way to avoid some of these drawbacks is to utilize alkaline waste residues. Table 6.9 summarizes the example of industrial alkaline solid wastes, e.g., coal- and oil shale-fired power plants, steelmaking slag, municipal solid waste incinerator (MSWI) ashes, air pollution control (APC) residues, cement kiln dust, red mud (RM), paper mill waste, and even concrete, as CO_2 adsorbents [47].

Figure 6.11 shows the main chemical composition of various alkaline wastes, e.g., steelmaking slag, cement, and fly ash in the literature [47].

6.4 Comparison of Technologies

It is necessary to understand all the factors affecting before adopting a particular CO_2 control technology. The following issues should be taken into consideration for the technology selection:

Basin suitability:

Different basin sites can be compared with the following criteria: Basin size and depth, faulting intensity, geothermal conditions, reservoir seal pairs, hydrocarbon potential, climate, location with onshore or offshore, industry maturity, etc. [48].

Table 6.9 Example of alkaline solid waste in the literature [47]

Alkaline Solid Waste	Example
Slag	Steelmaking slag (BOFS, EAFS) Coal slag Blast furnace slag (UF, FA, BFS)
Air pollution control (APC) residue	Municipal solid waste incinerator (MSWI) APC residue Cyclone dust
Fly ash	MSWI fly ash Coal fly ash Oil shale ash
Bottom ash	MSWI bottom ash
Cement wastes	Cement kiln dust (CKD) Cement bypass dust (CBD) Construction and demolition waste Cement/Concrete waste Blended hydraulic slag cement (BHC)
Mining and mineral processing waste	Asbestos tailings Nickel tailings Red mud (Bauxite)
Sludge (incinerator) ash	Sewage sludge incinerator ash (SSA) Steel wastewater sludge (SWS) Paper sludge incinerator ash (PSIA)
Paper pulping and mill waste	Paper mill waste (calcium mud) Green sludge dreg Lime mud Lime slaker grits

Storage capacity:

It must meet the volume requirements of neighboring currently identified by the CO_2 source.

Recycle potential:

The world is moving toward green science and technology. It is necessary to have possible materials, which can be recycled or reused. Especially when using membrane technology, it is essential to consider the outlook of recycling.

Cost and removal efficiency:

Direct capital costs and annual operating costs should be estimated. Each technology has its barriers, and thus it is necessary to consider each factor before pilot-scale production. For example, the algae method is low cost but needed a large number of land acquisitions.

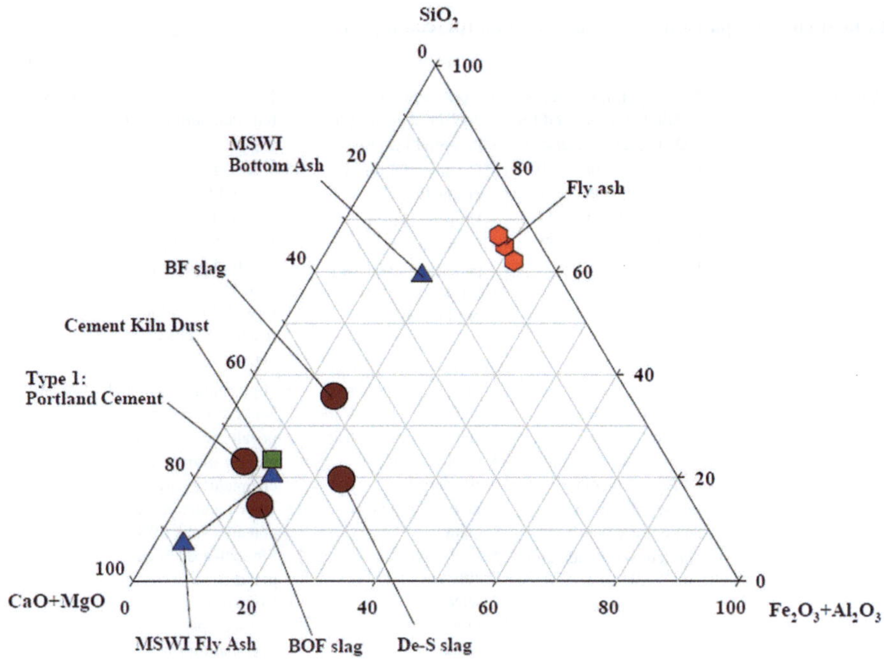

Fig. 6.11 The normalized CaO(MgO)–SiO_2–Al_2O_3(Fe_2O_3) phase diagram of various types of alkaline waste

Energy-efficient:

A lake of low-cost and energy-efficient capture technologies is there. Among them, low energy and near-zero-emission technology have to be certified for further scale-up. In that case, membrane technology has higher energy efficiency as compared to adsorption and absorption technologies.

Cost of transport:

The costs of transporting and storing CO_2 are mainly influenced by the local condition, i.e., if the CO_2 is used for EOR, the extra oil recovered can be worth more than the cost of the CO_2. However, if CO_2 capture works on-site with deep saline formations, the cost of transportation is saved.

Selectivity to cost of fuel:

Some of the technology depends on fuel or energy materials; the cost of fuel always varies with time. Thus, technology like mineralization has higher potential as compared to other technology as the low cost associated with the fuels.

Post-combustion CO_2 capture is the simplest technology that can be used to capture existing emission sources. Currently available CCS technologies are stated as follows: The chemical absorption using an amine-based solvent is regarded as the first CO_2 capture technology, with the Sleipner gas field project installed and

Table 6.10 Comparison of various CO_2 control techniques

Techniques	Features	Advantages	Disadvantages	References
Adsorption	The adsorption process is suitable to be used for CO_2 capture because it is reversible, has high adsorption capacity, less energy extensive, and low cost of adsorbent material	• Applicable to coal-fired power plants • Potential lower cost • Commercial for H_2 production • High efficiency • Low capture energy penalty (≈10–15%) • Large scale in H_2 production	• Novel materials for high-temperature CO_2 capture • High capital expenditure • Still in development • Insufficient large-scale H_2-fired power plant experience • Reusability of the material needs to be considered	[49, 50]
Absorption	Chemical absorption using amine-based solvent (Monoethanolamine (MEA)) is the first post-combustion technology that is applied worldwide on a large scale in several industrial processes	• Mature technology for natural gas processing and post-combustion • Capture-ready standard for carbon-fired power plant • High capture efficiency and selectivity • Low hydrocarbon losses • Adequacy via CO_2 partial pressure • Physical absorption having a low heat ratio for regeneration	• High capture ratio and heat ratio (energy penalty for solvent regeneration) for chemical absorption • High capture energy penalty (≈20–30%) for coal-fired power plants • Corrosion, emissions, and solvent degradation for chemical absorption • Stripping temperatures allowing the use of waste heat • Physical absorption having low selectivity and causing hydrocarbon losses • Requires high operating pressure	[49, 50]
Membrane	Membrane technology can be used to capture CO_2 due to it offering very high selectivity, high driving force, simple installation, low capital cost, and energy consumption in spite of the low partial pressure of CO_2	• Used in natural gas processing of large-scale floating production storage and offloading • No CO_2 regeneration • No chemicals • Low footprint • Adequacy via CO_2 partial pressure	• Very high operating pressure • Costly cleaning is an issue • Slow process • Demands compression of fed natural gas and permeates • Hydrocarbon losses • Trade-off permeability–selectivity • Low CO_2 partial pressure forbids it in post-combustion	

(continued)

Table 6.10 (continued)

Techniques	Features	Advantages	Disadvantages	References
Chemical looping	Chemical looping combustion is a unique approach to carbon capture which avoids any gas separation process and is therefore theoretically capable of achieving very low energy penalties which are mainly associated with CO_2 compression	• No cost of separation • Theoretically capable of achieving very low-energy penalties mainly associated with CO_2 compression • Potentially cost-effective CO_2-capture options with minimum energy losses • Largely minimizes No_x formation during the reaction • Combined with IGCC to produce syngas as a valuable by-product • Calcium-looping post-combustion process leading to an efficiency penalty of only 6–8%	• Still under development • Difficult to large-scale operations • Chemical looping combustion of coal requires in situ gasification with fluidizing steam or CO_2 • Inevitable loss of oxygen carrier with ash removal • Lack of efficient and stable oxygen-carrier materials	
Cryogenic distillation	Used commercially for high concentration and high-pressure gases Because of the high capital expenditure of cryogenic separation, this method is economically feasible only if the percentage of CO_2 is high in the stream	• Mature technology for processing natural gas with high CO_2 content • High selectivity • Low hydrocarbon losses • CO_2 obtained as liquid with benefits in CO_2 transport (no compressors needed, pumps used instead) • Appropriate for high CO_2 content	• Refrigeration energy penalties • Avoidance of CO_2 freeze-out required	
Biological treatment	Mitigation of CO_2 by biological treatment has been gaining momentum because it leads to the generation of energy from biomass grown by CO_2 fixation through photosynthesis. Photosynthetic CO_2 fixation is thought to be a feasible technology as it is energy-efficient, sustainable, and environmentally benign	• Microalgae growing rapidly and having high per-acre productivity • Water used to grow microalgae including various sources, such as wastewater and saline water • No further removal process of the CO_2 once the CO_2 is trapped with the microalgae • Capable to produce co-products, such as production of biofuels	• Uncertainty of efficiency • Production and downstream processing incurring huge costs	

(continued)

Table 6.10 (continued)

Techniques	Features	Advantages	Disadvantages	References
Mineralization	Mineral carbonation of CO_2 results in the production of stable mineral carbonates by treating CO_2 with metal oxides such as calcium and magnesium oxides that are naturally abundant in the form of mineral silicates	• Conversion to a solid material • Thermodynamically favored • Permanently fixing CO_2 • The exothermic nature of the mineralization reaction along with the geothermal gradient (up to 20 °c km^{-1}) contributing to a reduction in energy consumption • Flue gas capable to be used directly without removing impurities such as SO_X and NO_X • The presence of additives under mild conditions and short time periods • Diverse commodities such as magnesium/calcium carbonate, iron oxide, and silica compensating for the process cost	• Difficult for commercial scale • Still under development to get proper yield • Very slow kinetics despite injecting fluids with a higher concentration of CO_2 and increasing the temperature • Energy-intensive and high energy cost • High pressures (10.0–15.0 MPa) and temperatures (150–600 °C°C) to achieve a carbonation efficiency higher than 80% • Duration of the carbonation reaction is long (6–24 h), and the rocks should be mined (<37 μm) • Additives and other components with high-cost penalties • Complex process	

started in 1996 in Norway. Since amine solvents have the following excellent performance—(1) fast reaction rate of CO_2 absorption/desorption, (2) large capacity of CO_2, (3) low degradation, and (4) low energy consumption—it is currently the most promising approach and has become a benchmark for CO_2 capture technology.

Adsorption via porous materials such as carbonaceous materials and zeolites are considered as alternative materials in capturing CO_2 from flue gases as low energy consumption is needed to regenerate the adsorbent. As a result, adsorption may be an alternative process to capture CO_2 because of its low-cost raw materials, fast kinetics, low heat capacity, high CO_2 selectivity, and high CO_2 adsorption capacity.

Membrane technology can be used to capture CO_2 due to its high driving force, high selectivity, simple installation, low capital cost, and energy consumption. The disadvantages of membrane-based systems are the need for high-cost modules, and the relatively low CO_2 concentration and pressure weaken the driving force, hindering membranes to effectively function during the post-combustion CO_2 capture. Cryogenic CO_2 separation is applied commercially for high concentration and high-pressure gases. Due to the high capital cost and energy consumption of the cryogenic separation system, cryogenics is not an attractive approach [49] (Table 6.10).

Exercises

1. Define the "CCUS" and show the major technologies for CCS.
2. Explain how and why EOR and ECBM techniques can be successfully used for CO_2 utilization.
3. Compare the chemical and physical absorption for CO_2 control from the viewpoints of "Principles and Applications".
4. Show the process chemistry of CO_2 capture by accelerated carbonation of alkaline wastes.
5. Chemical looping technology has been considered as a newly developed method for CO_2 control. Please show advantages and disadvantages of this technology.

References

1. Chiang PC, Pan SY (2017) Carbon dioxide mineralizarion and utilization. Springer Nature, Berlin
2. Shahbazi A, Nasab BR (2016) Carbon capture and storage (CCS) and its impacts on climate change and global warming. J Pet Environ Biotechnol 7(4):10–4172
3. Lead C, Source of CO_2, IPCC special report on carbon dioxide capture and storage, Chapter 2, 75
4. Garcia-Garcia G, Fernandez MC, Armstrong K, Woolass S, Styring P (2021) Analytical review of life-cycle environmental impacts of carbon capture and utilization technologies. Chem Sus Chem 14: 995–1015
5. Greenhouse Gas Inventory Guidance, Direct Emission from Stationary Combustion Sources, US EPA, 2016
6. Greenhouse Gas Inventory Guidance, Direct Emission from Mobile Combustion Sources, US EPA, 2016
7. IPCC (2007) Climate change 2007: mitigation of climate change
8. IPCC (2002) Climate change and biodiversity
9. IPCC (2005) Carbon dioxide capture and storage
10. Comparison of Energy Use and CO_2 Emissions From Different Transportation Modes, American Bus Association May-2007
11. Leung DYC, Caramanna G, Valer MMM (2014) An overview of current ststus of carbon dioxide capture and storage technologies. Renew Sustain Energy Rev 39:426–443
12. Feron PHM, Hendriks CA (2005) CO_2 capture process principles and costs. Oil Gas Sci Technol 60(3):451–459
13. Huang CH, Tan CS, Review A (2014) CO_2 Utilization. Aerosol and Air Quality Res 14:480–499
14. Sood A, Vyas S (2017) Carbon capture and sequestration- a reviews. IOP conference series, earth and environmental science 83:1
15. Caprock Systems for CO_2 geological Storage, IEAGHG (2011)
16. Global CCS Institute, Project database, Large scale CCS facilities, https://www.globalccsinstitute.com/projects/large-scale-ccs-projects
17. Global CCS Institute, CO_2 reuse technology, https://hub.globalccsinstitute.com/publications/accelerating-uptake-ccs-industrial-use-captured-carbon-dioxide/1-co2-reuse-technologies.
18. Alper E, Orhan OY (2017) CO_2 utilization: developments in conversion process. Petroleum 3:109–126
19. Cuellar-France RM, Azapagic A (2015) Carbon capture, storage and utilization technologies: a critical analysis and comparison of their life cycle environmental impacts. J CO_2 Utiliz 9:82–102

20. Xiaoding X, Moulijn JA (1996) Mitigation of CO_2 by chemical conversion: plausible chemical reactions and promising products. Energy Fuels 10:305–325
21. Sayre R (2010) Microalgae: the potential for carbon capture. Bio Sci 60(9):722–727
22. Shah KJ, Imae T (2017) Photoinduces enzymatic conversion of CO_2 gas to solar fuel on functional cellulose nanofiber films. J. Mater. Chem. A 5:9691–9701
23. Aresta M (2010) Carbon dioxide as chemical feedstock, 1st edn. Wiley-VCH, New York
24. Assessment of post-combustion capture technology developments, Electric Power Research Institure Technical updates (2007)
25. Shah KJ, Imae T (2016) Analytical investigation of specific adsorption kinetics of CO2 gas on dendrimer loaded in organoclays. Chem Eng J 283:1366–1372
26. Shah KJ, Imae T, Shukla A (2015) Selective capture of CO_2 by poly(amido amine) dendrimer-loaded organoclays. RSC Adv 5:35985–35992
27. Saima H, Mogi Y, Haraoka T (2013) Development of PSA system for the recovery of carbon dioxide and carbon monoxide from blast furnace gas in steel works. Energy Proc 37:7152–7159
28. Yu CH, Huang CH, Tan CS (2012) A review of CO_2 capture by absorption and adsorption. Aerosol Air Quality Res 12:745–769
29. Huertas JI, Gomez MD, Giraldo N, Garzon J (2015) CO_2 absorbing capacity of MEA. J Chem, 1–7
30. Global CCS Institute, Module 2 CO_2 capture: Post combustion flue gas separation, https://hub.globalccsinstitute.com/publications/building-capacity-co2-capture-and-storage-apec-region-training-manual-policy-makers-and-practitioners/module-2-co2-capture-post-combustion-flue-gas-separation
31. Seloxol, Acid gas removal, National Energy Technology Laboratory, US department of Energy. https://www.netl.doe.gov/research/coal/energy-systems/gasification/gasifipedia/selexol
32. Shah KJ, Imae T (2016) Selective gas capture ability of gas-adsorbent-incorporated cellulose nanofiber films. Bio macromolecule 17:1653–1661
33. Lockwood T (2017) A compararitive review of next-generation carbon capture technologies for coal-fired power plant. Energy Proc 114:2658–2670
34. Li J, Zhang H, Fao Z, Fu J, Ao W, Dai J (2017) CO_2 capture with chemical looping combustion of gaseous fuel: an overview. Energy Fuels 31(4):3475–3524
35. Wang P, Means N, Shekhawat D, Berry D, Massoudi M (2015) Chemical looping combustion and gasification of coals and oxygen carrier development: a brief review. Energies 8:10605–10635
36. Osman M, Khan MN, Zaabout A, Cloete S, Amini S (2021) Review of pressurized chemical looping processes for power generation and chemical production with integrated CO_2 capture. Fuel Process Technol 214:10668
37. Song C, Liu Q, Ji N, Deng S, Zhao J, Kitamura Y (2017) Advanced cryogenic CO_2 capture process based on Stirling coolers by heat integration. Appl Thermal Eng 114, 887–895
38. Song C, Liu Q, Deng S, Li H, Kitamura Y (2019) Cryogenic-based CO_2 capture technologies: state-of-the-art developments and current challenges. Renew Sustain Energy Rev 101:265–278
39. Goli A, Shamiri A, Talaiekhozani A, Eshtiaghi N, Aghamohammadi N, Aroua MK (2016) An overview of biological processes and their potential for CO_2 capture. J Environ Manag 183:41–58
40. Fulazzaky MA, Talaiekhozani AR, Ponraj M, Abd Majid MZ, Hadibarata T, Goli A (2014) Biofilteration process as an ideal approach to remove pollutants from polluted air. Desalin Water Treat 52:3600–3615
41. Sen C (2012) Algae based carbon capture and utilization feasibility study - initial analysis of carbon capture effects based on Zhoushan case pre-study in China, Stockholm 2012
42. Gross M (2013) Development and optimization of algae cultivation system. Iowa State University
43. Pan SY, Shah KJ, Chen YH, Wang MH, Chiang PC (2017) Development of accelerated carbonation using alkaline solid waste for carbon mineralization and utilization towards a circular economy. ACS Sustain Chem Eng 5:6429–6437

44. Sanna A, Uibu M, Caramanna G, Kuusik R, Valer MMM (2014) A review of mineral caronation technologies to sequester CO_2. Chem Soc Rev 43:8049–8080
45. Geerlings H, Zevenhoven R (2013) CO_2 mineralization bridge between storage and utilization of CO_2. Ann. Rev 103–117
46. Pan S-Y, Chiang P-C, Chen Y-H, Chen C-D, Lin H-Y, Chang E-E (2013) Systematic approach to determination of maximum achievable capture capacity via leaching and carbonation processes for alkaline steelmaking wastes in a rotating packed bed. Environ Sci Technol 47:13677−13685
47. Pan S-Y, Chang EE, Chiang P-C (2012) CO_2 capture by accelerated carbonation of alkaline wastes: a review on its principles and applications. Aerosol Air Quality Res 12:770–791
48. Kaldi JG, Gibson-Poole CM, Payenberg THD (2009) Geological inputs to selection and evolution of CO_2 geosequestration site, Carbon dioxide sequestration in geological media- state of the science
49. Farihahusnah Hussin, Mohamed Kheireddine Aroua, Recent trends in the development of adsorption technologies for carbon dioxide capture: A brief literature and patent reviews (2014–2018), Journal of Cleaner Production, Vol.253, 2020.
50. de Queiroz O, Araújo F, Medeiros JL (2017) Carbon capture and storage technologies: present scenario and drivers of innovation. Curr Opin Chem Eng 17:22–34
51. Al-Mamoori A, Krishnamurthy A, Rownaghi AA, Rezaei F (2017) Carbon Capture and Utilization Update. Energ Technol. https://doi.org/10.1002/ente.201600747
52. Mondal M, Goswami S, Ghosh A, Oinam G, Tiwari ON, Das P, Gayen K, Mandal MK, Halder GN (2017) Production of biodiesel from microalgae through biological carbon capture: a review. 3 Biotech 7:99

Chapter 7
Thermodynamics

Nomenclature

c_P	The specific heat capacity at constant pressure
c_V	The specific heat capacity at constant volume
F	The force
g	The gravity
H	The enthalpy
L	The work
m	The mass
n	The number of moles
q	The heat
q_e, q	Adsorption capacity of activated carbon
R	The gas constant
T	The temperature
U	The internal energy
V	The volume
v	The velocity
W	The work
z	The height
η	The thermal efficiency

7.1 Introduction

7.1.1 Significance and Importance

The process of human industrialization is supported by the consumption of fossil fuels, and the whole stage is inseparable from the use and development of energy.

P.-C. Chiang and X. Gao, *Air Pollution Control and Design*,
https://doi.org/10.1007/978-981-13-7488-3_7

However, coal and petroleum-based fossil fuels are a real and potential threat to human health. Energy consumption is associated with economic activities and environmental damage. Economic growth must consume a certain scale of energy while sacrificing a certain environment. Energy is the lifeblood of China's economic development and social progress. Economic development is closely related to energy structure. Energy issues are related to people's livelihood, economic security, and sustainable development.

Therefore, it is very important to understand and master the research object, main contents, and methods of engineering thermodynamics, such as the two main ways of energy utilization (renewable and non-renewable), and working process of commonly used energy conversion devices. In consideration of engineering and technology, it is essential to study the thermodynamic properties of materials, the laws and methods of transforming thermal energy into mechanical energy, and the ways for utilizing energy effectively and reasonably.

7.1.2 Development of Thermodynamics

In the first half of the nineteenth century, as a great achievement of physics, the concept of "energy" appeared. People gradually realized that heat is only one of the many forms of energy that can be converted to each other. Scientists realized that the "Caloric Theory", which had dominated the scientific community for a century, was wrong. Thermodynamics came into being at the historic moment. In the mid-nineteenth century, the first law of thermodynamics and the second law of thermodynamics were summarized on the basis of the study of heat and work conversion, which solved the difference in quantity and quality between heat and mechanical energy conversion. From 1873 to 1878, Josiah Willard Gibbs (1839–1903) further summarized the relationship between thermodynamic functions describing the equilibrium of material systems and proposed the phase law. At the beginning of the twentieth century, Walther Hermann Nernst (1864–1941) proposed the thermal theorem, which made it possible to determine absolute entropy. In 1907, Gilbert Newton Lewis (1875–1946) proposed the concepts of fugacity and activity in order to use thermodynamic functions to deal with actual non-ideal systems.

7.1.3 Contents of This Chapter

In this chapter, the basic principles of thermodynamics will be introduced firstly, including the zero, first, second, and third laws of thermodynamics. Then the energy and environmental applications of thermodynamics will be discussed along with typical thermodynamic cycles, namely Rankine cycle, Brayton cycle, and piston engine cycle composed of Otto cycle and Diesel cycle. Finally, related applications and technologies will be briefly evaluated from the points of view of energy, environment, engineering, and economy.

7.2 Principles

7.2.1 Fundamentals

Thermodynamics is the study of macroscopic or volumetric mass systems and their relationships under steady-state conditions. Thermodynamic systems deal with a certain number of substances or specific areas in space [1]. The interaction between the system and its surroundings is determined by the boundary. The system can be clarified into two types, i.e., open or closed system. Open systems, also known as control volume, are selected areas in space. Closed systems, also known as control quantity, include a fixed quantity. Among them, when there is no energy exchange across the boundary, the system is called an isolated system.

When a system does not undergo any changes, attributes can be measured or calculated, thus, the state of the system could be fully described.

In thermodynamics, the process is defined by external constraints that exist during the changes [2]. The basic parameters, volume V, pressure P, and temperature T, are used to describe a simple compressible equilibrium system.

1. Heat and work

Force-induced displacement means work. The quantity of work is the product of force and displacement. Heat is the reduction of heat energy in the environment [3].

In thermodynamics, Q is used to represent heat. Heat is a process quantity, representing the energy that is emitted in the process of interacting with the surroundings. When the system absorbs heat, the value of heat is positive. The value of heat is negative when heat is released.

The work done by force F is defined as in Eq. (7.1):

$$L_{1-2} = \int_{x_1}^{x_2} F\,\mathrm{d}x \tag{7.1}$$

where

$L_{1-2} =$ the work, J;
F = the force, N;
$x_1 =$ the initial position, m;
$x_2 =$ the termination position, m.

In thermodynamics, when the work is done by the system, the value of the work is positive. When the work is done on the system, the value is negative.

Ideal gases

The ideal gas is a gas with its properties following a formulation as shown in Eq. (7.2):

$$PV = nRT \tag{7.2}$$

where

P = the pressure, Pa;
V = the volume, m^3;
n = the number of moles, mol;
R = the gas constant, 8.314 J/mol·K;
T = the temperature, K.

2. Heat engine

In the second law of thermodynamics, a hypothetical system is established, which has a relatively large thermal capacity and can supply or absorb a limited amount of heat without any temperature change. This system can be called a thermal energy storage, or just a storage.

From numerous observations, it was concluded that work can be directly and completely converted to heat, while the conversion of heat to work requires some specific devices, i.e., heat engines [4]. A typical diagram of a heat engine is shown in Fig. 7.1. A heat engine is defined as a device or machine that produces work from the heat in a cycle, which involves the absorption of heat Q_H at a high temperature T_H, and the release of heat Q_L to a low temperature T_L, to produce work L.

Thermal efficiency, which is the most important indicator of the performance of the heat engine, can be calculated as in Eq. (7.3):

$$\eta = \frac{L}{Q_H} \tag{7.3}$$

where

η = the thermal efficiency;
L = the produced work, J;
Q_H = the heat absorbed from high-temperature source, J.

According to the first law of thermodynamics, we can get Eq. (7.4):

$$L = Q_H - Q_L \tag{7.4}$$

where

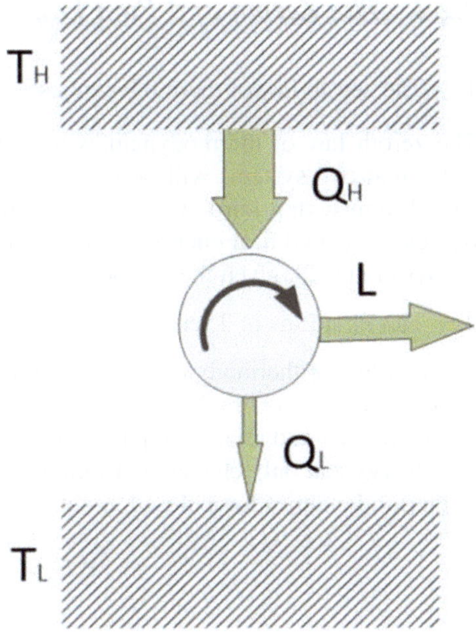

Fig. 7.1 A diagram of a heat engine

Q_L = the heat released to a low-temperature source, J.

Therefore, combined with Eqs. (7.3) and (7.4), thermal efficiency can be expressed as in Eq. (7.5):

$$\eta = \frac{Q_H - Q_L}{Q_H} = 1 - \frac{Q_L}{Q_H} < 1 \tag{7.5}$$

When the engine is in a reverse cycle, heat is absorbed at low temperature and released to a high-temperature heat source via external work. The device can be a refrigerator that absorbs heat from a low-temperature heat source for cooling, or a heat pump that releases heat to a high-temperature heat source for heating.

The performance of the refrigerator and heat pump can be expressed by the performance factor, as shown in Eqs. (7.6) and (7.7):

$$\text{Refrigerators:}\varepsilon_R = \frac{Q_L}{L} = \frac{Q_L}{Q_H - Q_L} = \frac{1}{Q_H/Q_L - 1} \tag{7.6}$$

$$\text{Heat pump:}\varepsilon_H = \frac{Q_H}{L} = \frac{Q_H}{Q_H - Q_L} = \frac{1}{1 - Q_L/Q_H} \tag{7.7}$$

When $Q_H = Q_L$, we can then obtain Eq. (7.8):

$$\varepsilon_H = \varepsilon_R + 1 \tag{7.8}$$

7.2.2 *The Zeroth and First Laws of Thermodynamics*

1. The zeroth law of thermodynamics

The zeroth law of thermodynamics provides the fundamentals of temperature. It states that two systems will be in thermal equilibrium if they are both in thermal equilibrium with a third system. The energy transferred by the random motion of particles is called heat energy. This energy transfers between the surroundings and the system [1, 2], given the symbol Q.

2. The First Law of Thermodynamics for a closed system

The first law of thermodynamics states that the energy of the universe is conserved in any process. The first law of thermodynamics provides a good basis for studying the interaction and relationship between various forms of energy. According to this law, energy can only change its form, not be created or destroyed [3]. The first law of thermodynamics for a closed system can be described as in Eq. (7.9):

$$\Delta E = q - W \tag{7.9}$$

where

q = the heat absorbed from outside the system, J;
ΔE = increase of internal energy, J;
W = work done by the system, J.

The enthalpy of a system, also called the heat content, is defined by Eq. (7.10):

$$H = E + PV \tag{7.10}$$

where

H = the enthalpy of a system, J.

Enthalpy is a static function. Since the absolute values of E cannot be obtained alone, the values of H are unknown. Therefore, it is important to consider the changes in enthalpy from the point of practical operation. At constant pressure, considering Eqs. (7.3) and (7.4) when PV work is performed, we can get Eq. (7.11) (Fig. 7.2):

$$q = (E_2 - E_1) + (P_2V_2 - P_1V_1) = (E_2 + P_2V_2) - (E_1 + P_1V_1) = H_2 - H_1 = \Delta H \tag{7.11}$$

3. The first law of thermodynamics for an open system

In open systems, the number of moles varies in a system. Thermodynamic functions depend on the numbers of moles in components, as well as on thermodynamic variables.

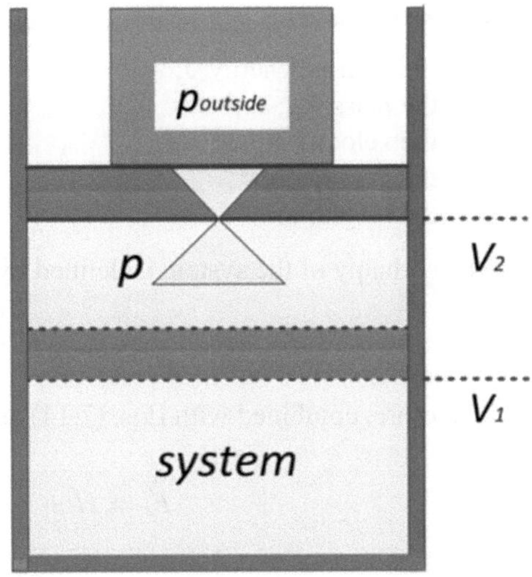

Fig. 7.2 A diagram of PV work

The law of conservation of mass is applied to a steady flow system, as expressed in Eq. (7.12):

$$\sum_i \dot{m}_{1i} - \sum_j \dot{m}_{2j} = 0 \tag{7.12}$$

where

$\dot{m}_{1i} =$ a variety of inflow of energy, kg/s;
$\dot{m}_{2j} =$ a variety of outflow of energy, kg/s.

When working fluid flows into an open system, the work W_f done by a unit mass of the fluid is calculated as in Eq. (7.13):

$$W_f = PV \tag{7.13}$$

where

P = the pressure, Pa;
V = the volume, m^3.

The internal energy of the working fluid is associated with kinetic energy and potential energy flowing into the system. The total energy E_t flowing into the system can be calculated as in Eq. (7.14):

$$E_t = U + PV + \frac{1}{2}mv^2 + mgz \tag{7.14}$$

where

U = the internal energy, J;
m = the mass, kg;
v = the velocity, m/s;
g = the gravity, m/s^2;
z = the height, m.

The enthalpy of the system is defined by Eq. (7.15):

$$H = U + PV \tag{7.15}$$

Therefore, combined with Eqs. (7.14) and (7.15), Eq. (7.16) is obtained:

$$E_t = H + \frac{1}{2}mv^2 + mgz \tag{7.16}$$

7.2.3 The Second and Third Laws of Thermodynamics

1. The Second Law of Thermodynamics for a closed system

There are several equivalent statements in the second law. For example, the statement proposed by Clausius is shown [5]: without an external mechanism, it is impossible to build a machine that can automatically transfer heat from a cryogenic object to a high-temperature object in a cycle.

The second law of thermodynamics is a hypothesis that does not derive from any previous principle like the first law. So far, except for the sub-micro phenomena, no exception has been found.

There are other ways to express the second law, such as the Kelvin–Planck statement: if a hypothetical machine is constructed to operate in a cycle and heat from the heat reservoir at a constant temperature, it is converted successfully without any changes in the heat reservoir or its environment. If this machine can be built, it will be "the second kind of perpetual motor [6]".

Based on Kelvin's statement, a heat engine that works with a single heat source cannot produce power outwards. It can be expressed as in Eq. (7.17):

$$\oint_{\text{1 heat source}} \delta L \leq 0 \tag{7.17}$$

When the cycle returns to the initial state, the change of the system energy is zero. Therefore, according to the first law of thermodynamics, we can obtain Eqs. (7.18) and (7.19):

$$\oint_{\text{Cycle}} \delta Q = \oint_{\text{Cycle}} \delta L \tag{7.18}$$

$$\oint_{\text{1 heat source}} \delta Q \leq 0 \tag{7.19}$$

Equation (7.19) is the second law of cyclic system thermodynamics for a single heat source.

Entropy, usually denoted as S, is a static function. It can be expressed as in Eq. (7.20):

$$\mathrm{d}S = \frac{\delta Q_{rev}}{T} \tag{7.20}$$

where

Q_{rev} = the heat in reversible cycle of the closed system, J;
T = the temperature, K.

The change of entropy from the initial state (state 1) to the final state (state 2) can be calculated as in Eq. (7.21):

$$S_2 - S_1 = \int_1^2 \frac{\delta Q_{rev}}{T} \tag{7.21}$$

In Eq. (7.21), since the temperature $T > 0$, entropy will increase when the system absorbs heat, while entropy will decrease when the system releases heat.

2. Entropy equilibrium for an open system

Most of the equipment in practical applications are open systems with inflow and outflow. Figure 7.3 illustrates a model of an open system in which mass and energy

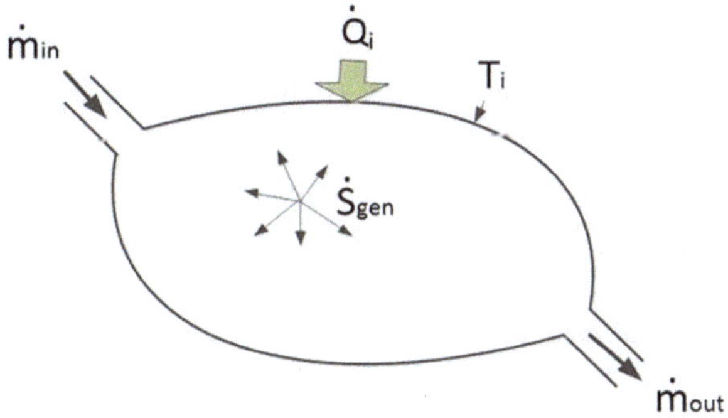

Fig. 7.3 A diagram of Entropy equilibrium for an open system

are exchanged with the surroundings.

Considering the input and output of heat and mass, as well as their changes over time, we can get the entropy equation as in Eq. (7.22):

$$\dot{S}_{gen} = \frac{dS}{dt} - \sum_i \frac{\dot{Q}_i}{T_i} + \sum_{out} \dot{m}s - \sum_{in} \dot{m}s \geq 0 \tag{7.22}$$

where

$\dot{S}_{gen} =$	the entropy generation per unit time;
$\frac{dS}{dt} =$	the entropy changes within the system;
$\sum_i \frac{\dot{Q}_i}{T_i} =$	the entropy flow for heat transfer;
$\sum_{in} \dot{m}s$ and $\sum_{out} \dot{m}s =$	the net entropy with the input and output mass.

For a closed system, $\sum_{in} \dot{m}s$ and $\sum_{out} \dot{m}s$ are equal to zero, therefore Eq. (7.22) can be converted to Eq. (7.23):

$$\dot{S}_{gen} = \frac{dS}{dt} - \sum_i \frac{\dot{Q}_i}{T_i} \geq 0 \tag{7.23}$$

When integrated with respect to time, Eq. (7.23) can be transformed to Eq. (7.24):

$$S_{gen} = (S_2 - S_1) - \int_1^2 \frac{\delta Q}{T} \geq 0 \tag{7.24}$$

For an isolated system without heat exchange from the surroundings, Eq. (7.25) is obtained:

$$S_{gen} = S_2 - S_1 \geq 0 \tag{7.25}$$

In terms of entropy, the second law states that, in any real process, there is net entropy increasing [7], expressed as in Eq. (7.26):

$$\Delta S_{\text{total}} \geq 0 \tag{7.26}$$

For any process, ΔS_{total} is always positive, approaching zero when the process becomes reversible.

3. The third law of thermodynamics

Regarding the properties of systems in equilibrium at absolute zero, the third law of thermodynamics states that the value of entropy for a perfect crystal is equal to 0 at absolute zero, as expressed in Eq. (7.27). The term perfect herein implies a single, pure substance.

$$\lim_{T \to 0} S = 0 \tag{7.27}$$

7.2.4 Environmental Consideration

7.2.4.1 Air Pollution Control

Particulate matter and gaseous pollutants are generally two kinds of air pollutants in industry. The removal of particles depends on physical forces, such as gravity, collision, electricity, and diffusion, in order to achieve efficient separation. Gaseous pollutants, on the other hand, are separated by chemical processes, including absorption, adsorption, and thermal processes. Reaction kinetics and mass transfer play a significant role in the selection and design of any method.

1. Adsorption

Adsorption technology is effective for removing pollutants in air or water. The process is carried out in a column with a powder adsorbent, such as activated carbon. When pollutants pass through a bed, they are adsorbed, leaving clean air or water in the exhaust. Once the bed is fully saturated, pollutants will reach the initial concentration as that of the feeding stream. At any time in bed, there are three distinct areas. Where pollutants flow in, the bed is saturated quickly. This is the so-called saturation zone, which is no longer effective for adsorption, while the front of the saturation zone is the most active region of adsorption, which is called the active adsorption zone. Far from the active adsorption zone are unused zones. If the feed rate remains unchanged, the zone of effective adsorption will slowly pass through the adsorption bed until the concentration of pollutants in the outlet reach the breakthrough level. The moving rate of the adsorption zone provides us with information about breakthrough time.

2. Thermal destruction

Thermal destruction, i.e., combustion of organic air pollutants, is a chemical process in which oxygen reacts rapidly with compounds to produce heat. Most fuels mainly consist of carbon and hydrogen, while other elements (e.g., N, P, S, and Cl) may also be included. Although the exact mechanism of combustion may be quite complicated, the whole process can be simplified using Eqs. (7.28) and (7.29):

$$C_nH_m + \left(\frac{n}{2} + \frac{m}{4}\right)O_2 \rightarrow nCO + \frac{m}{2}H_2O \tag{7.28}$$

$$nCO + \frac{n}{2}O_2 \rightarrow nCO_2 \tag{7.29}$$

Excess air is always involved in combustion, with the molar fraction of oxygen within the range of 0.1–0.15 for a hydrocarbon molar fraction of 0.001. These reaction rates indicate the reaction Eq. (7.30):

$$HC \rightarrow CO \rightarrow CO_2 \tag{7.30}$$

In a thermal incinerator, sufficient reaction time is required for completing the reactions. In addition, in the incinerator efficient mixing is also important. As a result, three important time parameters, i.e., residence time τ_{res}, chemical reaction time τ_{rxn}, and mixing time τ_{mix}, have to be taken into consideration for the incinerator design. Two dimensionless numbers can be used to represent the relative magnitudes of these time parameters: the Peclet number Pe, which is defined as $Pe = Z_s V_g / D$, and the Damköhler number Da, defined as $Da = Z_g k / V_g$,

where

$Z =$ the reaction zone length, m;
$V_g =$ the total reactor volume, m^3/s;
$D =$ the turbulent diffusivity, m^2/s.

In the case that Da is small and Pe is large, mixing will be the rate-determining step in the thermal incinerator. On the other hand, i.e., Pe is small and Da is large, the oxidation rate will be controlled by chemical kinetics. In most cases, mixing is not generally indeed rate-limiting as a reasonably moderate V_g is maintained. The overall efficiency for thermal destruction can be described as in Eq. (7.31):

$$\eta = 1 - \frac{[A]_{exit}}{[A]_{feed}} = 1 - e^{-k\tau_{res}} \tag{7.31}$$

7.2.4.2 Energy and Heat Recovery

An engine is a device or system that generates net power output. Engines always work in thermal cycles, which can be called power cycles. Refrigerators, air conditioners, and heat pumps are devices or systems that can produce refrigeration effects, known as refrigeration cycles [8]. According to the phase of fluid, the thermodynamic cycle can be divided into a gas cycle and a steam cycle. The working fluid in the gas cycle maintains the gas phase throughout the cycle. For the steam cycle, the working fluid will be steam for part of the cycle and liquid for the rest of the cycle.

Thermal efficiency is based on the theoretical thermal efficiency obtained from quasi-static processes in an ideal cycle. For a real thermal engine, there are many irreversible processes in the working fluid, so it is impossible to realize these quasi-static processes [9].

7.3 Applications

7.3.1 Combustion Modification

7.3.1.1 Formation of Thermal NOx

Nitrogen oxides in combustion products of coal-fired power plants are produced by complex chemical reactions. The design, operation, and fuel characteristics of boilers have an impact on NOx emissions. Usually, the NOx generated by coal combustion mainly comes from the oxidation of nitrogen compounds in fuel (fuel nitrogen, fuel N) and nitrogen in air (molecular nitrogen, N_2) used in combustion. According to different sources, the generated NOx could largely be divided into thermal NOx and fuel NOx.

The amount of fuel NOx depends on whether fuel N is oxidized to NOx or reduced to N_2 in a series of reactions. Because fuel N has been decomposed and oxidized below combustion temperature to produce NOx, the formation of fuel NOx has little relation with combustion temperature. Therefore, in this chapter, we will only discuss thermal NOx which is closely related to thermodynamics.

Two important reactions of NO and NO_2 occur at high temperatures:

$$\begin{aligned} N_2 + O_2 &\rightleftarrows 2NO \\ NO + \frac{1}{2}O_2 &\rightleftarrows NO_2 \end{aligned} \tag{7.32}$$

The chemical equilibrium of the above reactions is affected by the temperature and chemical composition of the reactants. When the temperature is less than 1000 K, the partial pressure of NO is very low, that is, the equilibrium constant of NO is very small. At a temperature >1000 K, considerable NO will be formed (Table 7.1).

At conventional combustion temperatures (>1500 K), considerable amounts of NO are generated.

Table 7.1 Effect of temperature and N_2/O_2 initial concentration ratio on equilibrium concentration of NO

T/K	Equilibrium concentration of NO/10^{-6}	
	$N_2/O_2 = 4$	$N_2/O_2 = 40$
1200	210	80
1500	1300	500
1800	4400	1650
2000	8000	2950
2200	13,100	4800
2400	19,800	7000

7.3.1.2 Measures to Reduce Thermal NOx

The amount of thermo NOx production mainly depends on the temperature level and the concentration of oxygen in the combustion zone, and the influencing factors are as follows.

Temperature has a decisive influence on the formation of thermo NOx. Because the activation energy (565 kJ) for NO formation is very large, the reaction rate is in a close relationship with the temperature change. When the combustion temperature is lower than 1500 K, the amount of thermal NOx is very small; when the combustion temperature exceeds 1500 K, the reaction rate will increase 6–17 times, and the amount of NOx will increase sharply as well. Therefore, the production of NOx is dramatically reduced by properly reducing the temperature in the combustion zone and avoiding local temperature peaks.

If the reaction mixture is detained for a long time, the concentration of NOx will also increase. Therefore, controlling the residence time for combustion products at high-temperature regions can reduce NOx formation.

There is a peak value in the relationship between NOx production and oxygen concentration. In theory, when the excess air coefficient $\alpha = 1$, the concentration of NOx is the highest. When $\alpha < 1$, the concentration of NOx increases with increasing oxygen concentration; when $\alpha > 1$, the concentration of NOx decreases with increasing oxygen concentration, which is due to the decrease of combustion temperature derived from dilution of oxygen.

7.3.2 Thermodynamic Cycles (Rankine Cycle, Brayton Cycle, and Piston Engine Cycle)

7.3.2.1 Rankine Cycle

The Rankine cycle is one of the simplest theoretical cycles in thermal power plants. The circulation system is mainly composed of boilers, steam turbines, condensers, deaerators, feed pumps, and economizers. The circulation process is as follows:

Water is used as a working fluid, with a deaerator used to get rid of the oxygen. After pressurization, it is pumped into the boiler economizer. This process is an adiabatic compression process.

After being preheated in the economizer, water is heated to saturated steam in the water wall, and then heated to superheated steam. This process is a constant pressure heating process. Superheated steam enters the turbine through steam pipes, which is an adiabatic expansion process.

The steam after the turbine is discharged into the condenser to cool down, which is a constant pressure exothermic process. At last, the condensed water is recycled to the boiler.

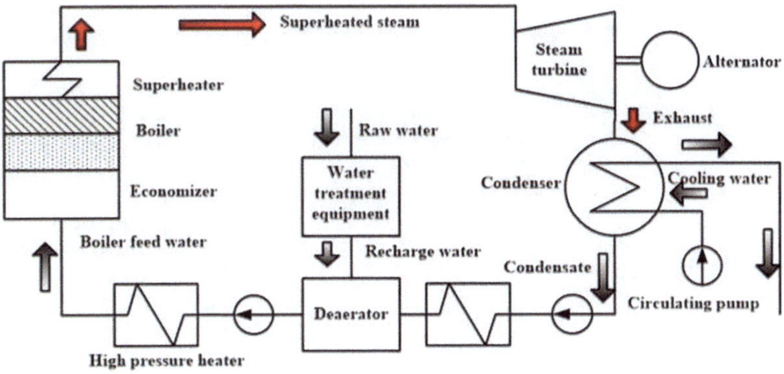

Fig. 7.4 A diagram of Rankine cycle

The realization of the Rankine cycle mainly depends on the working fluid to complete the four processes heat absorption, expansion, exothermic, and compression (Fig. 7.4).

The study of the steam power cycle must combine the characteristics of steam and thermal process. Because water and steam cannot burn, they can only absorb heat from the surroundings, they must be equipped with heat exchangers. Since combustion products do not participate in the cycle, not only a variety of conventional solid, liquid, gas, or nuclear fuels but also waste, inferior coal, as well as solar and geothermal energy could be used [10].

Increasing the initial steam temperature usually improves the thermal economy. It can be seen that the increase of initial gas temperature will lead to the decrease of the humidity of the exhaust gas, thus reducing the excessive loss of the humidity of the gas. It also increases the capacity of steam and reduces the amount of steam that accidentally leaks. If the final parameters are reduced, the Rankine cycle efficiency will be improved and exhaust humidity will increase.

Organic Rankine Cycle (ORC) is one type of the Rankine cycle using organic matter with a low boiling point as a working substance. It mainly consists of a waste heat boiler (or heat exchanger), turbine, condenser, and workflow pump, as shown in Fig. 7.5. ORC works by absorbing heat from waste heat flow in the heat exchanger and generating steam with certain pressure and temperature. And then the steam enters the turbomachinery for expansion work, thus driving the generator or other power machinery. Heat from the steam discharged from the turbine is released to the cooling water in the condenser. The steam condenses into a liquid state, and finally returns to the heat exchanger by means of a refrigerant pump. Solar energy is easy for utilization, with the characteristics of low radiation, low energy flow density, easy to convert into a low-temperature heat source. Combining solar energy with an organic Rankine cycle can form a solar-based low-temperature thermal power generation system.

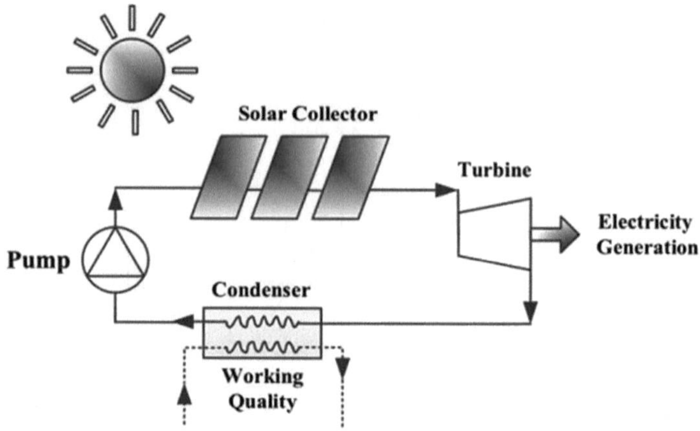

Fig. 7.5 Organic Rankine Cycle diagram

For low and medium enthalpy heat, ORC technology has quite a few advantages over the traditional steam Rankine cycle, mainly reflecting the higher efficiency of sensible heat recovery.

The power generation efficiency of the system is defined as $\eta = \frac{W_f - W_x}{Q}$. W_f is the power generated by the generator, W_x is the power consumed within the system, and Q is the heat absorbed by the working medium from the solar collector.

The main factors that restrict low-temperature Rankine cycle solar power generation are low thermal efficiency, high cost, and lack of suitable cycle refrigerants. The exhaust gas from the steam turbine directly enters the condenser, and a large amount of condensation heat is discharged into the atmosphere, which seriously affects the system's thermal efficiency. Therefore, the economy of the organic Rankine cycle depends directly on the thermodynamic properties of the workflow. It is yet important to develop solar Rankine cycle technology for utilizing the condensation heat of working fluid effectively and select a safe and reliable new type of working fluid.

The selection of organic working substances has a great influence on cycle efficiency. Good working substances should meet the requirements, such as low evaporation temperature, good heat transfer performance, stable chemical properties: non-toxic, non-polluting, and low price. The working fluids are mostly pure: R13, R123, R134a, Rl52a, R245fa, R600a, NH_3, and so on. Alkanes are low-priced, environmentally friendly but explosive, and have potential safety hazards. NH_3 is easy to produce ammonia water, which corrodes the pipe wall and machinery. Also, it has high critical pressure, requiring specific materials. The environmental protection characteristics of working fluids are mainly measured by the degree of ozone destruction: Global Warming Potential (GWP). Generally speaking, working fluids with more chlorine atoms are more destructive to the environment, while those with more fluorine atoms have stable chemical properties, but long life in the atmosphere. The ozone destruction potential (ODP) and greenhouse effect potential (GWP) of CFCs

Table 7.2 Characteristic parameters of commonly used refrigerants

Working quality	Molecular formula	Critical temperature/°C	Critical pressure /MPa	Boiling point/°C	ODP (R11 = 1)	GWP (100 years)
R113	$C_2Cl_3F_3$	214.1	3.39	47.6	0.9	5000
R123	$C_2HCl_3F_3$	183.7	3.67	27.7	0.012	120
R134a	CH_2FCF_3	101.1	4.07	−26.1	0	1300
R152a	$C_2H_4F_2$	113.5	4.52	−24.7	0	140
R245fa	$CF_3CH_2CHF_2$	256.9	3.65	15.3	0	950
R600a	$CH_3CH(CH_3)CH_3$	135.0	3.66	−11.8	0	20
NH_3	NH_3	132.4	11.2	−33.5	0	0

are very large. The characteristic parameters of common working fluids, including molecular formula, critical temperature and pressure, boiling point under standard condition, ODP, and GWP, are shown in Table 7.2.

Some scholars have proposed a set of binary mixtures suitable for the Rankine cycle with an evaporation temperature of about 100 °C. For example, 10% R152A and 90% R245fa are used as circulating fluids. At this point, the steam pressure is 0.9 MPa; the condensation pressure is 0.28 MPa; the expansion ratio is 3.54; the net output power is 26.2 kJ/kg; the Rankine cycle efficiency is 10.54%; the ODP is 0; and the GWP is 869. Therefore, ordinary solar collectors can be used. Moreover, the boiling points of mixed refrigerants are different, and there is a temperature slip during phase transformation, which effectively improves the thermal efficiency.

Organic Rankine Cycle (ORC) power generation technology can make full use of low temperature waste heat resources and convert heat into electricity, which is one of the efficient ways to improve energy efficiency. Its advantages are simple structure, wide temperature range of heat source, and high efficiency of waste heat recovery. ORC power generation technology has broad application prospects in the process of energy-saving and emission reduction [11].

7.3.2.2 Brayton Cycle

1. Traditional Brayton Cycle

A gas turbine is a kind of internal combustion power machine, in which the impeller is driven by the continuous flow gas to rotate at high speed, converting fuel into work, as shown in Fig. 7.6. It is a kind of rotary impeller thermodynamic engine. The main factors affecting the efficiency of a gas turbine are the initial gas compression ratio and the compression ratio of the light compressor. Increasing the initial gas temperature and the compression ratio can significantly improve the efficiency of a gas turbine. In the late 1970s, the highest compression ratio was 31; the highest initial gas temperature of industrial and marine gas turbines was about 1200 °C, and that of aviation gas turbines was higher than 1350 °C. At present, the thermal efficiency

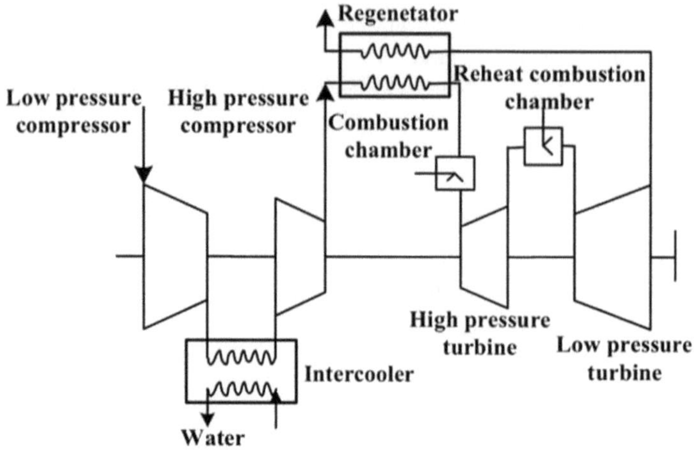

Fig. 7.6 Gas Turbine diagram

of a heavy gas turbine is 32–40% for a single cycle and 52–60% for a combined cycle. According to the different temperatures in front of the turbine, the thermal efficiency is different. The higher the turbine front temperature reaches, the higher is the thermal efficiency.

The structure of a gas turbine is simple, which can reflect a series of advantages like small size, fast start-up, lightweight, less cooling water, and no cooling water.

In the mainstream of air and gas, the simple cycle for gas turbine cycle mainly includes a combustor, a compressor, and a turbine. This simple cycle is widely applied in most gas turbines.

The compressor sucks air from the outside atmosphere, which is gradually compressed through the axial compressor and pressurized. The air temperature rises accordingly. Compressed air is pressed into the chamber and mixes with the fuel. As a result, high-temperature and high-pressure gases are produced, which then enter the combustion chamber. The turbine expands and drives the turbine-driven compressor. With the high-speed rotation of the external load rotor, the chemical energy of gas or liquid fuel is partly converted into mechanical energy and output electric energy.

Exhausts from turbines are released into the surroundings to release natural heat. The gas turbine in this way can convert chemical energy contained in the fuel to thermal energy, which is then converted into mechanical energy.

A compressor is usually driven by the gas turbine. About one-third to two-thirds of the mechanical power generated by turbines is utilized for driving compressors in a simple cycle, while the rest is used to drive generators. When starting a gas turbine, external power is needed first. When the mechanical power produced by the turbine is larger than that consumed by the compressor, the gas turbine can work independently.

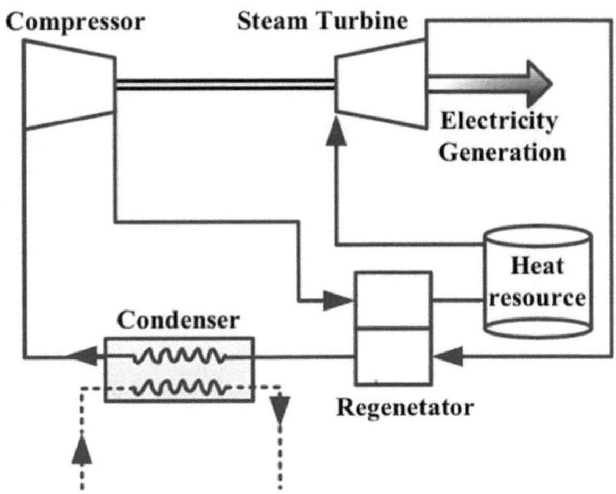

Fig. 7.7 The supercritical carbon dioxide Brayton cycle

2. Carbon Dioxide Supercritical Brayton Cycle

The supercritical carbon dioxide (SCO_2) Brayton cycle is based on SCO_2 as a circulating medium. CO_2 has relatively stable chemistry, good physical properties, reliable safety, and low price. In a supercritical state, special physical characteristics of low viscosity and high density make it have good fluidity, small specific volume, small compressibility, and high heat transfer efficiency. Therefore, SCO_2 is one of the most widely used supercritical fluids.

As shown in Fig. 7.7, based on the advantages of SCO_2, the SCO_2 Brayton cycle power generation system has the advantages of small space occupation, high efficiency, and good economy. Thus, thermal power, nuclear power, and other power generation systems try to use the SCO_2 Brayton cycle. Power generation system based on SCO_2 Brayton cycle mainly includes the heat source, the gas turbine, the generator, the compressor, the cooler, the regenerator, and so on.

In practical application, the high- and low-temperature sides of the regenerator will cause large heat transfer loss due to the large temperature difference, which will lead to the reduction of cycle efficiency [12]. Therefore, the regenerator is separated into a high-temperature regenerator and a low-temperature regenerator to improve the heat transfer efficiency. In addition, the recompression SCO_2 Brayton cycle also increases the thermodynamic processes such as recompression, reduces the temperature difference at the end of the cryogenic regenerator, and the heat taken away by the cooler, thus improving the safety and economic benefits of the system. Because of the characteristics of the SCO_2 Brayton cycle, it has a good application prospect in thermal power, nuclear power, and other power generation systems.

The cyclic efficiency of the three media at different temperatures is shown in Fig. 7.8.

In Fig. 7.8, water vapor, He, and CO_2 are used as working fluids. At 400 °C SCO_2 has obvious advantages. When the temperature reaches 550 °C, the efficiency can generally reach 45%. With the increase of temperature, the efficiency is also

Fig. 7.8 Cycle efficiency of three media at different temperatures

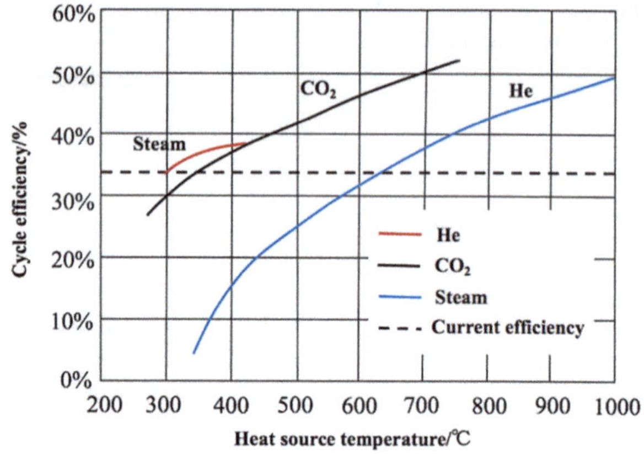

significantly improved. SCO_2 can achieve satisfactory conversion efficiency without a high cycle temperature. For the He cycle, to achieve 40% cycle efficiency, the cycle temperature must be above 750 °C, which poses a great challenge to the performance of component materials.

Due to the stable chemical properties of SCO_2, compared with high pressure and high-temperature water vapor, the corrosion rate of metal pipeline equipment is slower, so the requirements for materials are relatively low.

Because there is no water treatment system, a lot of water resources and water treatment agents are saved, and the initial investment is reduced.

Because of its small viscosity and high density, SCO_2 has typical advantages such as good fluidity, high heat transfer efficiency, and low compressibility. Therefore, the key components such as compressors and turbines are small in size and compact in structure.

Compared with the steam thermal cycle power generation system, the SCO_2 Brayton cycle power generation system has lower construction costs, lower operation and maintenance costs, longer service life, better economic benefits, and can reduce the leveling power cost by 8–15%.

At last, CO_2 is a greenhouse gas, which provides an effective way to recycle CO_2 resources.

At present, the research on the SCO_2 Brayton cycle system mainly focuses on the theoretical research of the thermal cycle process, cycle efficiency analysis and optimization, the selection and design of key components, and the construction of large-scale simulation units in a laboratory, but there is no formal large-scale generating unit.

7.3.2.3 Piston Engine Cycle

1. Gasoline Engine

A gasoline engine, as shown in Fig. 7.9, is an engine that converts internal energy into kinetic energy by using gasoline as fuel. Since gasoline has features of low viscosity and fast evaporation, it needs a gasoline injection system. After being compressed to a certain pressure and temperature, the gasoline can be ignited to make the gas expand. The thermal efficiency of a gasoline engine refers to the ratio of the mechanical power output of the engine to the chemical energy produced by burning fuel, that is, how much heat is converted into effective work. At present, the thermal efficiency of the engine is usually only a little more than 30%. The normal working temperature of an automobile engine is between 85 and 105 °C, which refers to the temperature inside the engine. Below 85 °C, gasoline atomization is not good besides lubricant fluidity, and the engine is easy to wear, showing poor performance; higher than 105 °C, the engine is easy to overheat. The advantages of gasoline engines include lightweight, simple structure, high speed, smooth operation, low cost, easy use, and maintenance, which make them widely applied in automobiles.

The working process of gasoline engine could be divided into the following phases:

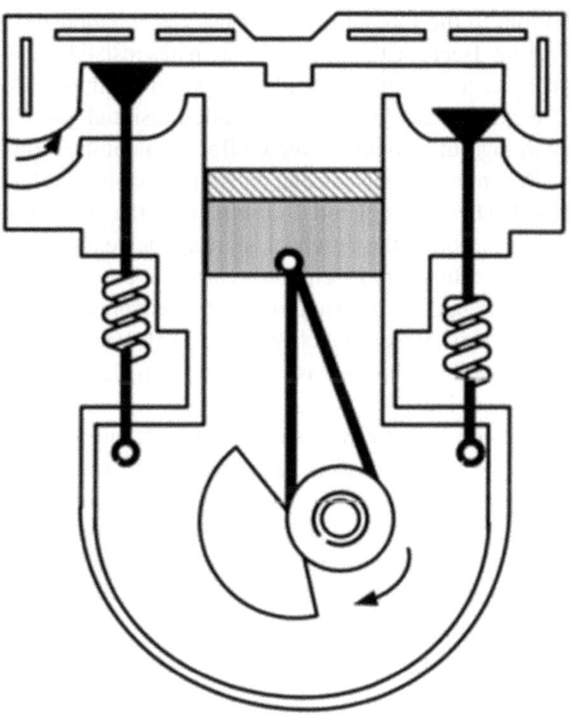

Fig. 7.9 Gasoline engine diagram

Intake stroke: At this point, the crankshaft drives the piston from the top to the bottom of the stop point, and while the intake valve opens, the exhaust valve closes. When the piston moves from top to bottom, the volume above the piston increases, and the gas pressure in the cylinder decreases, forming a certain degree of vacuum. Because the intake valve is opened, the cylinder is communicated with the intake pipe, and the mixture is sucked into the cylinder. When the piston moves to the bottom stop, the cylinder is filled with the fresh mixture and exhaust gas from the previous working cycle.

Compression stroke: The piston travels from the TDC to the BDC and the intake valve closes. The crankshaft rotates under the action of inertia force such as a flywheel, and the piston moves upward by connecting the rod. The gas volume in the cylinder decreases gradually, the gas is compressed, and the mixed gas pressure and temperature in the cylinder increase.

Work stroke: At this point, the intake and exhaust valves are closed at the same time, the spark plug ignites, the mixture burns violently, the temperature and pressure in the cylinder rise sharply, the high-pressure and high-temperature gas drives the piston downward, and the crankshaft rotates through the connecting rod. In the four strokes of the engine, only this stroke can achieve the conversion of thermal energy into mechanical energy, so the stroke can also be called the work stroke.

Exhaust stroke: At this point, the exhaust valve opens, the piston moves from the bottom to the top, and the exhaust gas is exhausted from the cylinder as the piston goes up. Because the exhaust system has resistance, and the combustion chamber also occupies a certain volume, at the end, it is impossible to exhaust the gas completely. The exhaust gas left behind is called residual gas. Residual gas not only affects gas charging but also has adverse effects on combustion.

When the exhaust stroke ends, the piston goes back to the TDC. A cycle of work has also been completed. Subsequently, the crankshaft continues to rotate depending on the inertia of the flywheel, starting the next cycle. Then the engine keeps running.

The technology of engine waste heat recovery can not only improve engine fuel economy but also improve engine fuel economy. Power density can effectively reduce engine emissions at the same time [13]. Vazaquez et al. predicted that if 6% of the energy in the exhaust was converted into electricity, 10% of the fuel would be saved for the engine.

The most direct means of automobile energy saving is to improve the efficiency of engine fuel utilization. In the coming decades, human beings will have to continue to rely mainly on oil and gas resources to drive cars [14]. Therefore, improving the efficiency of fuel engines and reducing fuel consumption are the most direct, realistic, mature, and indispensable means of automobile energy saving. For gasoline engines, "advanced inflatable technology" and "in-cylinder direct injection technology" are the two most feasible and efficient technologies recognized by the domestic and foreign industry at present [15].

2. Diesel Engine

Like gasoline engines, a four-stroke diesel engine consists of an intake stroke, a compression stroke, a work stroke, and an exhaust stroke, as shown in Fig. 7.10. Because diesel engine is fueled by diesel oil, its spontaneous combustion temperature is low and its viscosity is not easy to evaporate compared with gasoline, so compression ignition, also called compression ignition, is adopted in diesel engines. Its working process and system structure are different from that of gasoline engines.

The working process of a diesel engine could be divided into the following phases:

Intake stroke: Pure air first enters the cylinder. Due to the small resistance of the intake system of the diesel engine, the pressure at the inlet is higher than that of the gasoline engine. The inlet temperature is around 25–65 °C, lower than the gasoline engine.

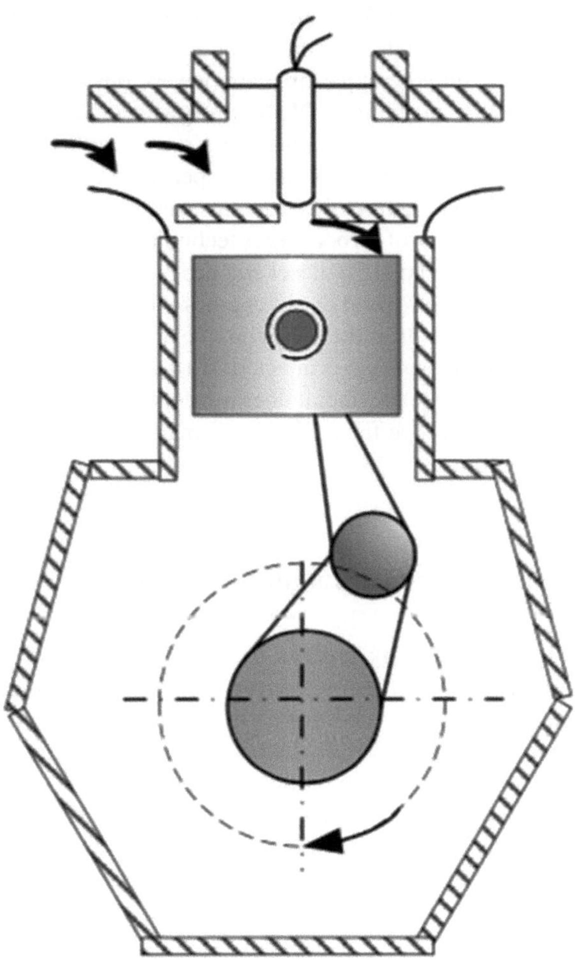

Fig. 7.10 Diesel engine diagram

Compression stroke: Because pure air is used as the compressed medium, the compression ratio of a diesel engine is generally higher than that of a gasoline engine. The compression end pressure is around 3000–5000 kPa, and the compression end temperature is around 750–1000 K, much higher than the spontaneous combustion temperature of diesel fuel (ca.520 K).

Working stroke: At the time that the compression stroke is close to the end, the diesel oil is injected into the cylinder combustor at ca. 10 MPa through the injector under the action of the high-pressure oil pump. After mixing with the air within a very short period of time, it ignites and burns itself quickly. The gas pressure in the cylinder rapidly increases, up to ca. 5000–9000 kPa and the highest temperature is 1800–2000 K. Hence, the diesel engine is a compression ignition engine since it is compressed and ignited by itself.

Exhaust stroke: Diesel engine exhaust is basically the same as gasoline engine exhaust, but the exhaust temperature of a gasoline engine is generally higher than that of a diesel engine [16]. The speed of a single-cylinder engine is uneven, the operation is unstable, and the vibration is large. This is due to the reason that only one stroke is working, and the other three are power-consuming strokes, which prepare to work. In order to solve this problem, the flywheel must have a sufficiently large moment of inertia, which in turn leads to an increase in the mass and size of the entire engine. Multi-cylinder engines could compensate for these deficiencies. Four-cylinder, six-cylinder, and eight-cylinder engines are generally used in modern cars.

The history of turbocharger technology is more than a hundred years. Large and medium-sized marine diesel engines basically use exhaust turbocharger technology, and small high-speed diesel engines with turbines are also developing rapidly [17]. The compressor of the diesel engine air system is driven by an exhaust gas turbine, and high-pressure gas is injected into the cylinder. A turbocharger is driven by exhaust energy of diesel engine; it does not consume the power of diesel engine itself, which helps to improve the economy of marine diesel engine and generally reduces fuel consumption.

7.3.3 Chemical Looping Process

The chemical ring process has the advantages of controllable capture of carbon dioxide and high energy conversion efficiency. It is a promising carbon fuel conversion technology. According to the products of chemical looping, generally, chemical looping technology is divided into two types: chemical looping full oxidation (CLFO) and chemical looping partial oxidation (CLPO). In CLFO, carbon-containing fuels are completely oxidized to carbon dioxide and water, usually through combustion, producing electricity. However, in CLPO, carbon-containing fuels are oxidized partially, usually through gasification, producing syngas or hydrogen. These two types of technologies have similar characteristics in operation, while the optimal process configuration and the oxygen carriers may differ a lot.

1. Fuel conversion

Because coal combustion produces many pollutants, it is the most carbon-intensive energy source. If pollution control is not implemented, the increase in coal utilization will lead to serious environmental problems. The main global concern is that fossil fuel combustion releases 270 billion tons of carbon dioxide annually. With the increase of coal consumption, in the absence of effective carbon dioxide emission reduction technology, the anthropogenic carbon dioxide emission rate may reach more than 400 billion tons in the next 20 years. By 2040, carbon fuels are expected to account for over 80% of energy consumption all over the world. The utilization of carbon fuels can also be divided into partial oxidation and full oxidation. Full oxidation, i.e., combustion, involves the reaction of hydrogen and carbon and atoms with oxygen to form water and carbon dioxide, respectively. The largest thermal energy release is through the full oxidation process, generating electricity and heat. On the other hand, partial oxidation minimizes the formation of carbon dioxide and water. Some oxidation products consist of syngas and hydrocarbons like olefins and aromatics. However, currently, the full and partial oxidation processes are not effective enough. In addition, the quantity and quality of carbon fuels are declining quickly. In order to maintain the raw materials, there is an urgent need for new technologies for fuel conversion with low emission and high energy efficiency. Chemical looping technology, as shown in Fig. 7.11, is expected to become a clean and efficient way to convert fuel into electricity, syngas, and hydrogen.

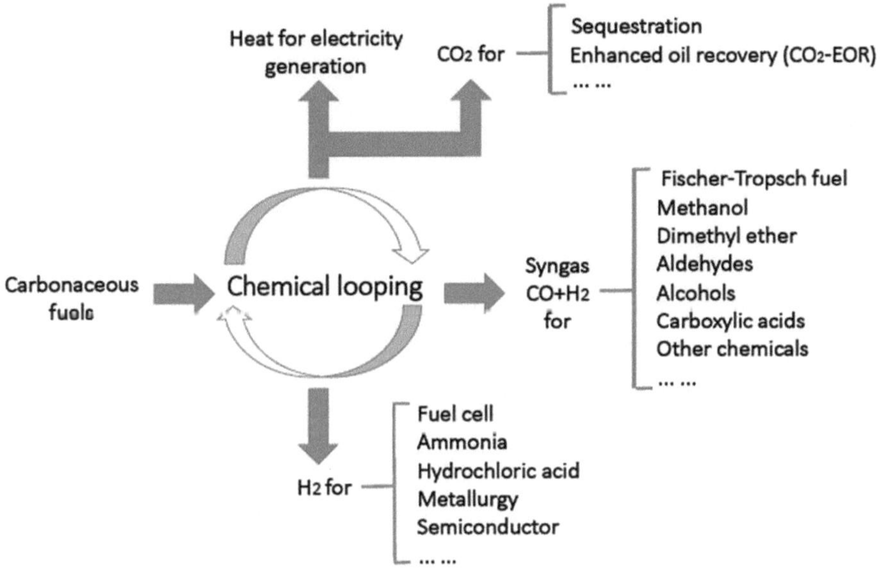

Fig. 7.11 Chemical looping processes for the conversion of carbonaceous fuels

2. CO_2 capture

The traditional combustion method is to mix the raw materials of carbon-containing fuel directly with air, and produce a kind of flue gas containing carbon dioxide, which is discharged into the environment. Currently, the concerns about global warming are related to carbon dioxide emissions. Therefore, it is very important to develop cost-effective carbon dioxide capture technology for the fossil fuel combustion process. Chemical looping combustion (CLC, as shown in Fig. 7.12) or CLFO was regarded as one of the most advanced technologies for power generation considering reducing CO_2 emission by US DOE.

CLC produces high efficiency for energy conversion because of the reduction of the irreversibility of the combustion reaction. The design of the reaction path can prevent immediate contact between fuel and air sources, leading to concentrated carbon dioxide flue gas without gas separation which is energy-intensive. The CLC process can capture more than 90% of carbon dioxide with an incremental power cost of less than 30%.

Biomass, the utilization of which is regarded as carbon neutral, is a kind of renewable energy with characteristics of high-water content, low energy density, and complicated ash composition. When biomass utilization technology is linked with carbon capture technology, even the total conversion of energy may be negative in terms of net carbon dioxide emissions, namely the so-called "bioenergy with carbon

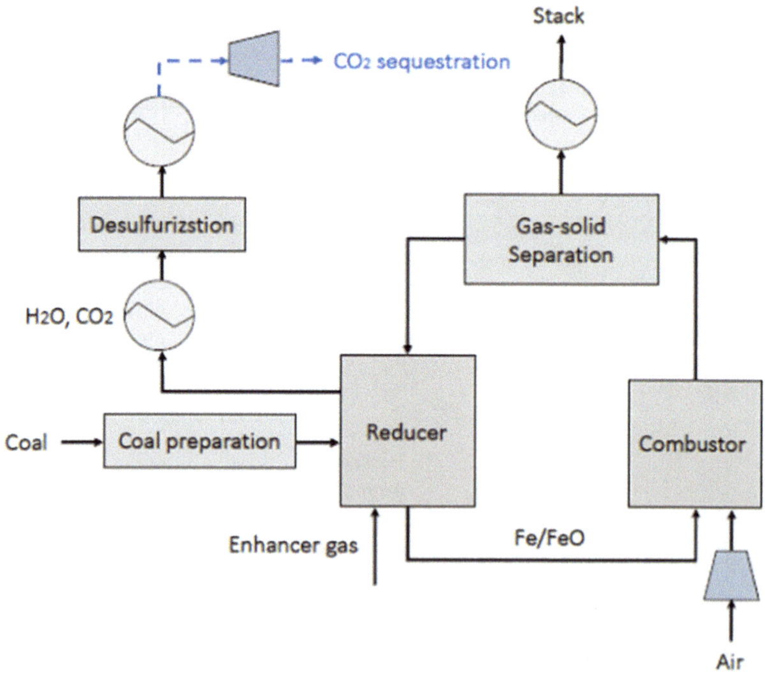

Fig. 7.12 Simplified process scheme of a chemical looping combustion system

capture and storage" (BECCS). At its beginning, the development of BECCS technology generally involved the application of a carbon dioxide capture device downstream of traditional thermochemical conversion processes. It has advantages of good adaption to existing technologies of energy conversion and carbon dioxide capture. However, it is limited in material and energy integration and systems engineering, likely resulting in real disruptive BECC technologies.

Because of its low exergy loss and inherent carbon dioxide separation capability, biomass-based chemical looping attracts much attention and has become a promising technology. When biomass is directly used in chemical cycles, no additional energy is needed to capture carbon dioxide, and the total exergy loss can be minimized. If a suitable carbon storage technology is combined with it, the net carbon balance can be negative. In addition, compared with coal utilization, volatile matter content is relatively high, biomass sulfur and ash content are relatively low, which can promote the operation of the chemical recycling process. Synthetic gas, hydrogen, and even carbon-based chemicals can be obtained via biomass-based chemical looping gasification (CLG) or calcium looping gasification (CALG) instead of thermal energy or electricity. The flexibility distinguishes biomass-based chemical looping processes (BCLPs) from other renewable energy sources, like wind or solar energy.

3. Syngas and H_2 generation

Partial oxidation of carbon fuels is a feasible method for the production of olefins, syngas, and other kinds of chemicals. Among them, hydrogen and syngas are regarded as the key intermediates for the production of other industrially important chemicals. At present, steam reforming of methane has become the main method for industrial production of hydrogen and syngas. In this traditional method, a large amount of steam is required in order to improve methane conversion. The extra amount of steam could facilitate water–gas shift reaction and produces syngas flow that is rich in hydrogen. This is helpful for the production of hydrogen, while it is not suitable for the production of methanol or Fischer–Tropsch synthesis. In addition, the thermal efficiency of steam methane reforming is significantly reduced by the energy loss during the processes of steam generation and condensation. Partial methane oxidation has now been industrialized for processes between gas and liquid, while these technologies are expensive because capital-intensive, high-purity oxygen is required.

In addition, chemical looping partial oxidation (CLPO) recently became a popular method for hydrogen and syngas from different kinds of fuel materials (like coal and natural gas), including chemical looping gasification (CLG, as shown in Fig. 7.13) and chemical looping reforming (CLR). In comparison with the requirement of steam methane reforming, CLPO needs a very small or no extra amount of steam. CLPO also has the advantage of generating syngas with high concentrations of H_2 and CO with an ideal CO/H_2 ratio.

For CLFO and CLPO, they are quite similar. When using carriers to transfer oxygen from air to fuel, both CLFO and CLPO include an oxidation (or air) and a reduction (or fuel) reactor. The oxygen carrier materials, usually metal oxides or sulfates, are used in the chemical looping process to provide oxygen indirectly for the partial or full oxidation of fuels in the reduction reactor. The oxygen-consuming

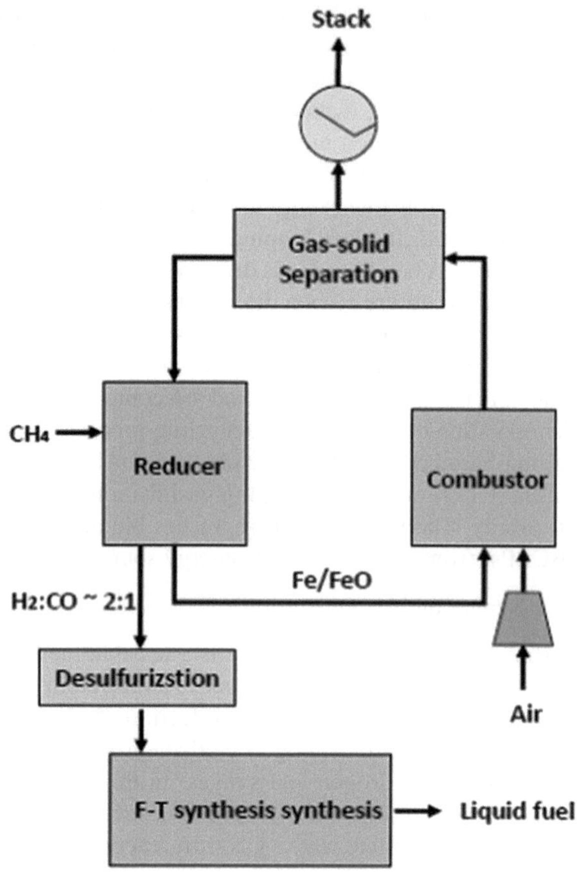

Fig. 7.13 Simplified process flow diagram of a chemical looping gasification system for liquid fuel synthesis

carrier is followed by regeneration by reacting with air in the oxidation reactor. At the same time, heat is released for process load and/or generation of power. The oxygen carriers circulate between the two reactors for transferring oxygen and releasing heat in order to separate the generated gases. Thus, carbon dioxide/syngas flow with high purity is generated from the reductant, while oxygen-consuming flue gas is generated from the oxidant. Through these reduction and oxidation processes, the energy consumption of the separation of high purity carbon dioxide or the production of syngas is minimal.

The oxygen carrier material is the key factor to determine the quality and process efficiency of chemical products. In order to develop ideal oxygen carrier materials, a large number of studies are carried out. In order to optimize the process, the key characteristics of oxygen carrier materials must be considered: physical strength, long-term stability, toxicity, redox reactivity, and cost. Quite a large number of oxygen carrier materials are investigated for the application of CLFO and/or CLPO. In the early stage, single metal oxides or sulfates were developed and used as active compounds. Recently, studies have led to the use of a variety of metal matrix composites to improve process performance.

7.3.4 *Good Engineering Practice*

7.3.4.1 Combustion Modification

The main characteristics of flameless oxidation can be summarized as follows:

(1) Even at high air preheating temperatures, flameless oxidation can be used for combustion.
(2) Combustor temperature, above 800 °C (depending on fuel and combustion devices).
(3) Flameless oxidation occurs under optimum conditions.
(4) No visible or audible appearance.

Flameless oxidation is possible in the adiabatic combustor and non-adiabatic combustor, which requires a high recycling rate. Compared with flame, there is no high gradient of temperature and species concentration. Flameless oxidation can be calculated by existing and available simulation software.

In addition to near-field burners, the temperature and flow conditions in the combustor of high-speed flame and flameless oxidation burners are similar. Due to the lack of high gradients found in flames, flameless oxidation is used in chemical kinetics data analysis and mathematical combustion model validation. So far, flameless oxidation technology has been applied in gas industrial furnaces. Achievable non-emission requirements require further research and exploration of other areas of use. The study of detecting active intermediates by laser technology should provide more information for the flameless oxidation process.

7.3.4.2 Chemical Looping Process

The process results are shown in Table 7.3, which compares the chemical ring syngas production system with the traditional self-heating and steam-reforming units. Compared with conventional systems, chemical cycle systems consume less steam per mole of treated carbon. Another advantage is that oxygen comes from iron oxide rather than from an air separation unit (ASU). This leads to higher selenium activity in natural gas oxidation. It can be seen that the chemical ring system has high selectivity for the equivalent syngas specification (H_2/CO ratio).

The higher the production efficiency of syngas, the less the natural gas will be consumed by syngas with the same output. The benchmark value of carbon utilization is selectivity plus low steam consumption and non-molecular oxygen demand, which makes chemical rings a promising and economical alternative to traditional systems.

Table 7.3 Carbon efficiency and overall production for the chemical looping case and the baseline DOE case

Process comparison		
Component	Base case	Chemical looping
H_2O/C (–)	0.68	0.2
H_2/CO (–)	2.19	2.18
Stoichiometric number (–)	1.59	1.84
Nat gas feed flow (kg/h)	354,365	334,427
Butane feed flow (kg/h)	18.843	18.843
Diesel fuel (bbl/day)	34,543	34,543
Gasoline (bbl/day)	15,460	15,460
Total liquids (bbl/day)	50,003	50,003
Carbon efficiency		
Natural gas to liquid fuels (%)	81.80	86.68

7.4 Challenges and Perspectives

7.4.1 Barriers to be Overcome

1. Institutional barrier

(1) Motorcycle gasoline engine

From the emission point of view, strictly speaking, all the countries in the world, including China, have not yet issued emission regulations for a motorcycle gasoline engine that can be applied to the actual operating conditions of motorcycles, and the research and establishment of this standard system will greatly promote the improvement of motorcycle emissions.

(2) General-purpose small gasoline engine

Since 2011, China has put forward emission control requirements for general small gasoline engines. The foundation of the establishment and implementation of this standard and regulation lags behind that of Europe and the United States.

The engine classification and emission limits of China's current emission standards for small gasoline engines (phase II) are the same as those of the EU's current regulations.

The current emission limits of handheld engines in China and the European Union are in line with those of the current sub-stage emission limits of the United States (EPA and CARB).

The emission limits of Non-Handheld engines in China and EU generally lag behind those of the third phase of the current emission limits of the United States (EPA and CARB).

The United States (EPA and CARB) has implemented evaporation emission standards, while the European Union and China have not yet implemented evaporation emission standards.

2. Regulatory barrier

At present, there are about 30 enterprises licensed to operate scrap SCR denitrification catalysts in China, of which 3–4 are catalyst manufacturers, 3–4 have regeneration and comprehensive utilization capabilities, and the remaining 20 only have regeneration capabilities.

Because of the vicious competition among the catalyst enterprises, the quality of catalysts has a serious downward trend. The proportion of regeneration of waste catalyst dismantled from power plants is very low (generally not more than 50%), and the price of new catalysts is low, which leads to the unwillingness of power plant users to regenerate. Instead, the waste catalyst is handed over to the waste catalyst disposal enterprise, which leads to only catalyst regeneration energy. The regenerated catalyst of the power plant has no market (except for a small amount of non-thermal denitrification projects), and the regenerated part acts as a new catalyst to return to the power plant, which will lead to a lower proportion of regenerated catalyst when the power plant is replaced next time.

For non-renewable catalysts, the best way is to remove the harmful substances to the environment and human body through harmless treatment, and then return them to the production of titanium–tungsten powder as raw materials for catalyst recycling. However, only the regenerative catalyst disposal enterprises have no way to deal with the non-regenerative part. Driven by the interests, some enterprises may transfer the harmless waste catalyst to the non-disposal qualified enterprises, including some non-disposal qualified catalyst manufacturers.

Therefore, it is imperative to strengthen supervision over the waste catalyst industry, especially on how to manage and dispose of the non-renewable parts of the catalyst disposal enterprises with only renewable capacity, so as to put an end to their illegal activities in circulation and disposal links. In this way, the number and flow direction of non-renewable catalysts can be accurately grasped, the market environment of denitrification catalysts can be purified, so that the regenerated catalysts can no longer flow to the market in the form of new catalysts, while ensuring the stability and reliability of the denitrification system, and the market space for the utilization of non-renewable parts of resources can be reserved.

3. Financial barrier

The denitrification cost mainly consists of several parts: financial cost, operation cost, investment cost, inspection and maintenance cost. Financial cost includes loan interest, value-added tax, and other financial expenditure which guarantees the fund raising; operation cost includes various expenditures of SCR denitrification device in coal-fired power plant, such as electricity, water, reducing agent, and others; the cost of inspection and maintenance refers to catalyst replacement and equipment check and repair.

Unit type, unit capacity, denitrification efficiency, reductant type, and inlet NOx concentration are the main factors affecting the cost of SCR denitrification of flue gas in coal-fired power plants. The specific contents are as follows: (1) Unit type. Practice has found that the higher the proportion between heating coal consumption and total coal consumption, the higher the operating cost of denitrification. The reason for the increase of denitrification operation cost of heating units is that in the process of calculation, the denitrification cost of heating units cannot be directly calculated because of various denitrification costs, which leads to the increase of the proportion between the required heating coal consumption and the total coal consumption in the denitrification process. (2) Unit capacity. There is an inversely proportional function between unit capacity and denitrification cost. When the unit capacity increases continuously during operation, the denitrification cost tends to decrease. (3) Denitrification efficiency. It is found that when the denitrification efficiency increases, the operation cost, inspection, and maintenance cost of denitrification will increase continuously, and the relationship between denitrification efficiency and denitrification cost is positive. In addition, the use of reducing agents may increase the efficiency of denitrification and increase the cost of denitrification. (4) Type of reductant. The data in Table 7.1 can explain the relationship between the type of reducing agent and the cost of denitrification. The denitrification cost of the liquid ammonia denitrification group is lower than that of the urea denitrification group, because the urea decomposition system needs electricity or fuel. Compared with liquid ammonia, the cost increases obviously. With the purpose of ensuring the normal operation of the denitrification unit, a part of the maintenance cost is needed. At the same time, it is necessary to increase the number of catalyst layers to ensure the efficiency of denitrification.

4. Technical barrier

At present, there are many problems in the research of the organic Rankine cycle of solar energy in China, such as theoretical research, programming and simulation with MATLAB software, and some state and equilibrium equations. For example, Feng Liming established the MATLAB/Simulink software calculation model of the piston expander by using the energy equation, continuity equation, gas flow equation, and heat transfer equation, and analyzed the influence of different design parameters and operating fluids on the output work, efficiency, and mass flow of the expander. The conclusion has laid a theoretical foundation for the design of the expander, but it has not been verified in practical application. There are few experimental studies on organic Rankine cycle power generation, which is also related to the difficulty of type selection and design of the experimental platform and the high price of the experimental equipment. On the one hand, the utilization of low- and medium-temperature industrial waste heat technology has not been popularized in China, and related enterprises have no awareness of using low-temperature waste heat to generate electricity. In addition, the low taste of low- and medium-temperature industrial waste heat makes it difficult to utilize, so most of them are abandoned. On the other hand, China's organic Rankine cycle low-temperature power generation market is not fully formed; there is no professional production of organic Rankine cycle generating units for commercial purposes. The above reasons mean the organic Rankine cycle technology in China is still in its infancy. Scientific researchers should conduct more experiments, compare and verify the experimental data with the theoretical analysis results, find out the problems, optimize the system and components, and combine theory with practice. At the same time, low-temperature waste heat power generation technology should be vigorously promoted and applied to actual production. At present, China's industrial waste heat recovery and utilization scale are not large, and the organic Rankine cycle power generation based on solar energy is just beginning. With the continuous attention to clean energy and the vigorous promotion of policies, it is believed that the organic Rankine cycle of solar energy power generation development space will be expanded.

5. Public acceptance

Urbanization and industrialization have been one of the most important activities of human beings for more than one hundred years. This process has brought abundant comfort to human beings, as well as the scarcity of resources and environmental pollution.

It is reported that air pollution is an environmental problem that troubles citizens' lives at present. 90% of urban residents attribute air pollution to automobile exhaust pollution, and 87.2% of the public believe that it should be treated.

In fact, the public's feelings and appeals are not unreasonable. Research shows that since the 1990s, the pollution of motor vehicles has increased significantly in Beijing, Shanghai, and Guangzhou. More than 80% of carbon monoxide and 40% of nitrogen oxide pollution come from motor vehicle emissions. Under certain weather conditions, some chemicals formed after chemical reactions directly stimulate human blood vessels. Besides, particulate matter emitted by motor vehicles seriously affects human health. These substances can reach the deep part of the stomach directly, and also contain a lot of toxic substances. Current research shows that 20%–30% of urban particulate pollutants come from motor vehicle emissions, and these motor vehicle emissions are located in the human respiratory belt, and their impact on the human body is more significant.

7.4.2 4E Evaluation

The 4E evaluation of the above-mentioned technologies is summarized in Table 7.4.

7.4.3 Integration for Innovation

Under the influence of the concept of saving energy and protecting the environment, it is necessary to optimize and reform the thermal system of power plants, improve the energy-saving and emission reduction capacity of the thermal system of power plants, and adopt the strategy of energy saving and emission reduction, so as to effectively improve operational efficiency. It improves the efficiency and greatly reduces the energy consumption of operations. By introducing advanced technology and equipment, optimizing the thermal system of power plants, and fully recovering

Table 7.4 4E evaluation of thermodynamics applications

4E evaluation	Classification	Rankine cycle			Brayton cycle	Auto cycle	Diesel cycle
		Thermal power plant	Solars power	Gas turbine	Thermal and nuclear power	Gasoline engine	Diesel engine
Energy	Energy resource	Coal	Solar energy	Natural gas	Fuel	Gasoline	Diesel oil
	Energy medium	Steam	stable chemical properties	Gas	Supercritical CO_2	Gas	Gas
	Power generation efficiency	35–70%	9–11%	32–60%	<45%	>30%	40–45%
Engineering	Operating temperature (°C)	400–570	– 300	1500–2000	>400	85–105	80–90
	Pressure (MPa)	2.45 –27	0.1–2.4	—	>7.38	0.6 –1.7	0.25–0.35
Economic	Capital cost	Low	High	Low	High	High	High
	Operating cost	Low	High	Low	High	High	High
Environmental	NOx	High	Almost no air pollutants	High	Almost no air pollutants	High	Low
	SO_2	High		—		—	—
	PM	High		—		Low	Low
	VOCs	High		—		High	Low
	CO	High		—		High	Low
	CO_2	High		—		High	Low

and utilizing waste heat and other energy, the concept of energy saving and emission reduction is implemented.

As an important part of solar thermal utilization, the performance of solar collectors directly affects the economy of the whole system. The choice of refrigerant directly affects the operation and life of the system. When choosing a refrigerant, we need to consider its thermophysical properties, heat source temperature, and the expansion process of the refrigerant in the expander. As a key component of the expander, the selection of the expander needs to consider many factors, such as output power, adiabatic efficiency, flow rate, and quality. Regeneration can improve the system efficiency, however, at the same time, the condition of the heat source and the stability of the system should be considered.

The research of ORC technology in China is still in its infancy. There are more theoretical studies and fewer experiments. In the practical application of engineering, there are fewer problems that need to be solved urgently. With the increasingly serious problems of energy shortage and environmental pollution, people begin to focus on the development and utilization of clean energy to enhance energy efficiency. This promotes the development of the organic Rankine cycle. The gradual improvement of ORC power generation technology will promote the use of low-grade heat energy, promote energy conservation and emission reduction, and optimize the energy structure.

At present, the research on supercritical carbon dioxide Brayton cycle abroad mainly focuses on nuclear reactors, including sodium-cooled reactor, lead-cooled reactor, and molten-salt reactor. In addition to the advantages of high efficiency and small volume, the supercritical carbon dioxide Brayton cycle has a great improvement on safety, in comparison with the steam system in the secondary circuit. According to the research plan of the DOE in the US and related progress, the application of supercritical carbon dioxide Brayton cycle in large nuclear power plants is expected to become a reality in the next 20 years. Coal, garbage, and biomass all provide heat in the form of combustion. For the application of supercritical carbon dioxide Brayton cycle, the International Energy Agency (IEA) in its report "Coal in the 21st Century" summarizes four future coal utilization technologies, taking coal as an example. These technologies represent significant progress in coal utilization. The closed Brayton cycle is one of them. The U.S. Department of Energy believes that the application of the supercritical CO_2 Brayton cycle can reduce the cost of solar power generation. The analysis shows that the supercritical CO_2 Brayton cycle can increase the efficiency of solar photothermal power generation by about 8%. The U.S. DOE plans to implement supercritical CO_2 Brayton cycle research to promote a substantial reduction of cost for solar photothermal power generation and improve its competitiveness. Basic research is the basis of technology development and application. The design and application of supercritical CO_2 Brayton cycle system depend on the mastery of the basic physical properties of supercritical carbon dioxide fluid, and the flow and heat transfer of supercritical CO_2 fluid in the channel (especially in the microchannel of heat exchanger). Meanwhile, reliable computer numerical simulation can effectively assist the design and verification of the system, reduce costs, and is also the focus of basic research. Strengthening basic research in these fields will help to promote the development of this technology.

At present, the main trend of internal combustion engine development is to widely apply high-tech integration for energy saving and emission reduction. First of all, if 40% of fuel consumption could be saved, then 40% of the development space will be expanded. Although electric vehicles, hydrogen fuel cell vehicles, and other vehicles are numerous, in many performance indicators, electric vehicles cannot replace gasoline vehicles. In the next 15 years, gasoline diesel locomotives will continue to dominate, and high-efficiency gasoline internal combustion engines will continue to assume the basic power of oil-electric hybrid vehicles. Currently, China remains one of the world's largest producers and sellers of automobiles, of which nearly 80% are gasoline diesel locomotives. The technical indicators of heavy trucks in China have been greatly improved. We must maintain the development advantages of this kind of internal combustion engine technology, continue to implement energy saving and emission reduction, and promote the stability and sustainable development of automotive internal combustion engines.

Exercises

1. Define the following terms and explain why they should be considered for air pollution control design: heat, work, enthalpy, and entropy.
2. Explain why the performance of the supercritical carbon dioxide Brayton cycle is higher than that of the traditional Brayton cycle.
3. Explain why the advanced Rankine cycle can be used for energy saving and emission reduction. What are its advantages and disadvantages?
4. If a system accepts 160 kJ of work from the environment and its thermodynamic energy increases by 200 kJ, how much heat will it absorb or emit?
5. The ideal gas of 1 mol is heated by isobaric heating to raise its temperature by 72 K, and the heat transferred to it is 1.6×10^3 J; please calculate (1) the work W done by the gas; (2) the increment of the internal energy ΔE of the gas.

References

1. G A (2012) Principles of thermodynamics. Springer
2. Baehr HD, Stephan K (2011) Heat and mass transfer. Springer
3. Hornstein J (1998) Atmospheric thermodynamics. Oxford University Press
4. Musgrave CB (1998) Solid state thermodynamics. Stanford University
5. Klotz IM, Rosenberg RM (2008) Chemical thermodynamics basic concepts and methods. Wiley
6. Y C (2016) Thermodynamics an engineering approach. Tsinghua University Press
7. Kaufman M, Kaufman M (2001) Principles of thermodynamics. Marcel Dekker Inc
8. Cao Q (2018) Attainability of the Carnot efficiency with real gases in the regenerator of the refrigeration cycle. Appl Energy 220:705–712. https://doi.org/10.1016/j.apenergy.2018.03.102
9. Mostowy M, Szablowski L (2018) Comparison of the Brayton-Brayton cycle with the Brayton-Diesel cycle. J Power Technol 98(1):97–105
10. Ding Y, Sheng D (2013) Optimization of start-up process in combined cycle steam turbine of a power plant. Proc CSEE,33(2):9–15

11. El-Shaarawi MAI (2014) Comparative analysis between constant pressure and constant temperature absorption processes for an intermittent solar refrigerator. Int J Refrig 41:103–112
12. Wright SA, Conboy TM (2012) High-temperature split-flow recompression brayton cycle initial test results. Office of Scientific & Technical Information Technical Reports
13. Taylor AMKP (2008) Science review of internal combustion engines. Energy Policy 36(12):4657–4667
14. Weerasinghe WMSR, Stobart RK, Hounsham SM (2010) Thermal efficiency improvement in high output diesel engines a comparison of a Rankine cycle with turbo-compounding. Appl Thermal Eng 30(14–15):2253–2256
15. Wang T, Zhang Y, Peng Z, Shu G (2011) A review of researches on thermal exhaust heat recovery with Rankine cycle. Renew Sustain Energy Rev 15(6):2862–2871
16. Wen S, Qu X, Zhu YL (2014) Study on recovery system for marine diesel engine waste heat. Adv Mater Res 912–914:795–798
17. Song J, Song Y, Gu CW (2015) Thermodynamic analysis and performance optimization of an Organic Rankine Cycle (orc) waste heat recovery system for marine diesel engines. Energy 82:976–985

Chapter 8
Heat Transfer

Nomenclature

A	The heat transfer area
a_m	The percentage of melting
c_p	The specific heat capacity, J/(kg·K)
c_{sp}	The average specific heat capacity of [T_i-T_m], kJ/(kg·K)
c_{tp}	The average specific heat capacity of [T_m-T_f], kJ/(kg·K)
$E_{\lambda,b}(\lambda, T)$	The spectral emissive power of the blackbody
h	The heat transfer coefficient, W/(m^2·K)
h_r	Radiation heat transfer coefficient
h_{tot}	Total heat transfer coefficient
Δh	Latent heat of fusion, J/kg
$I_{\lambda,b}(\lambda, T)$	The blackbody spectral intensity
k	The thermal conductivity, W/(m·K)
$q_{1,2}$	The amount of heat transfer between black surfaces 1 and 2
L	The thickness of the plate, m
Q, q	The heat rate
Q_{input}	The total heat input into the system, J
T	The temperature, K
T(x, y, z)	The scalar temperature field
T_1	The high temperature, K
T_2	The low temperature, K
T_f	The final temperature, K
T_i	The initial temperature, K
T_m	Melting point, K
ΔT_m	The logarithmic average temperature difference
$W_{cooling}$	The cooling output, J
$W_{electricity}$	The electrical power output, J
W_{heat}	The heat output, J
W_{total}	The total work output, J

P.-C. Chiang and X. Gao, *Air Pollution Control and Design*,
https://doi.org/10.1007/978-981-13-7488-3_8

$X_{1,2}$	The angular coefficient of surface 1 to surface 2
η_L	External heat loss coefficient, taken as 0.97–0.98
ε	Emissivity
σ	The Stefan–Boltzmann constant, 5.67×10^{-8} W/($m^2\cdot K^4$)

8.1 Introduction

8.1.1 Importance and Significance

In Chap. 7, we discussed thermodynamics, from which we know that energy can be transferred via work and heat. Heat is defined as the energy which goes through the boundary of a system because of the temperature difference between this system and its surroundings [1]. In thermodynamics, only the end states of the processes are considered. No information on the properties or rates of the processes is provided [2]. Heat transfer is a discipline that concerns the way of heat exchange. It is an integral part of heat-related science, technology, and industrial development, and plays a significant role in environmental science. It is especially important in energy and environmental engineering related to daily life.

Heat transfer mainly involves three modes: conduction, convection, and radiation. In the control of atmospheric pollutants, heat transfer is often applied to the design of condensers, which separates the condensable contaminants in the gas flow by cooling and condensing.

8.1.2 Technology Development

For more than 100 years, researchers have conducted extensive and in-depth investigations on heat transfer phenomena, yielding a large number of scientific papers, reports, and valuable academic monographs. The research results have been widely used in various fields such as industry, agriculture, space, and biotechnology, and have produced significant economic benefits in improving heat transfer efficiency, reducing material consumption, and product cost. Summarizing the existing work, including the fundamental laws of heat transfer, and pointing out the challenges and future development directions are of great significance. In short, heat transfer itself is the basis of cross-industry expertise, which is developed on the basis of mathematics (mainly differential equation theory), thermodynamics, fluid mechanics, and quantum mechanics, and it must be based on experiments. Hence, the development of heat transfer science relies on the progress of mathematics, thermodynamics, fluid mechanics, and quantum mechanics theory; on the other hand, it also needs continuous scientific measurement technology.

8.1.3 Contents of This Chapter

In this chapter, we will introduce the basic concepts of the heat transfer processes, including conduction, convection, and radiation. Additionally, the calculation method and application of the overall heat transfer coefficient are introduced. A brief introduction to heat exchangers is presented. Finally, an assessment of existing heat exchangers according to the evaluation of environment, energy, engineering, and economic (4E) indices are included. Challenges and perspectives of heat exchanger development are pointed out according to the current control situation.

8.2 Principles

Basically, heat transfer can be classified into three mechanisms: conduction, convection, and radiation. A medium is required for conduction, while only heat is diffused without the requirement of moving the medium. In convection, heat is transferred only via a flowing fluid medium like gas or liquid. In radiation, however, a medium is not required. Radiation can pass through even in a vacuum condition. Heat is actually transferred via electromagnetic waves. The actual three heat transfer modes can be performed simultaneously or in any two of them simultaneously.

8.2.1 Conduction

Conduction is the transfer of energy caused by a temperature gradient existing in a stationary medium, based on random atomic or molecular activity [3]. It can be viewed as energy transfer between the particles of a substance, from the particles which are more energetic to the less one.

1. Fourier's law

To quantify the conduction processes, we can use Fourier's law to calculate the conduction rate at a steady state. For a one-dimensional plane wall shown in Fig. 8.1, the heat flux q (W/m^2), which indicates the heat transfer rate per unit area in a direction x in Fig. 8.1, can be calculated as in Eq. (8.1):

$$q = -k\frac{\mathrm{d}T}{\mathrm{d}x} \tag{8.1}$$

where

k = the thermal conductivity, W/(m·K);

T = the temperature, K.

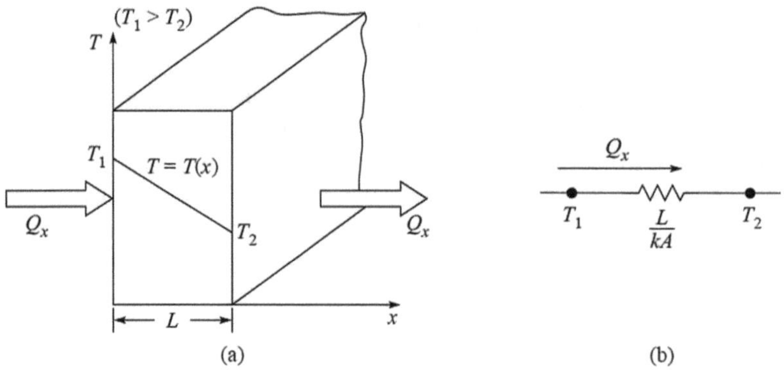

Fig. 8.1 **a** Heat conduction through a plane wall. **b** The analogous electrical circuit

Under a steady state with linear temperature distribution, as described in Fig. 8.1, the temperature gradient can be calculated using Eq. (8.2):

$$\frac{\mathrm{d}T}{\mathrm{d}x} = \frac{T_2 - T_1}{L} \tag{8.2}$$

where
T_1 = the high temperature, K;
T_2 = the low temperature, K;
L = the thickness of the plate, m.

Based on Eqs. (8.1) and (8.2), the heat flux can be calculated as in Eq. (8.3):

$$q = k\frac{T_1 - T_2}{L} = k\frac{\Delta T}{L} \tag{8.3}$$

The heat rate Q (W) through a plane of area A (m^2) can be calculated by Eq. (8.4):

$$Q = qA = \frac{T_1 - T_2}{L/(kA)} \tag{8.4}$$

where A represents the cross-sectional area of the wall which is perpendicular to the direction of heat flow. Equation (8.4) is similar to the current–voltage relation in a direct current (DC) electric circuit (Ohm's law). The term L/(kA) represents the thermal resistance of the wall through which heat flows at a rate of Q_x under a difference between temperatures (T_1–T_2). The analogous electrical circuit is shown in Fig. 8.1b.

Fourier's law indicates that heat flux is a quantity with a specific direction. Therefore, the expression for Fourier's law considering three-dimensional conduction heat transfer can be expressed as in Eq. (8.5):

$$q = -k\nabla T = -k\left(i\frac{\partial T}{\partial x} + j\frac{\partial T}{\partial y} + l\frac{\partial T}{\partial z}\right) \tag{8.5}$$

where
T(x, y, z) = the scalar temperature field.

For conduction heat transfer, Fourier's law is fundamental. It is an empirical equation, which indicates that the heat flux in the object is in proportion to the gradient of temperature. It should be noted that Fourier's law applies to all matter, no matter if it is gas, liquid, or solid.

2. Thermal conductivity

When Fourier's law is applied, we must know the thermal conductivity k of the substance. The value is determined by the matter's physical structure, which is related to its state.

According to Eq. (8.5), thermal conductivity in the x-direction can be defined by Eq. (8.6):

$$k_x \equiv -\frac{q}{(\partial T/\partial x)} \tag{8.6}$$

Similar definitions for thermal conductivities along the y- and z-directions can be given as well. It should be noted that the thermal conductivity at each direction of heat transfer is the same for an isotropic material, i.e., $k \equiv k_x = k_y = k_z$.

Generally, the thermal conductivity of a substance/material in its solid state is the largest, followed by that of a liquid, while thermal conductivity for gas is always the smallest. The thermal conductivity of a solid, like silver, or aluminum, may be 2 or 3 orders of magnitude higher than that of a liquid, and 3 or 4 orders of magnitude larger than that of gas, as shown in Fig. 8.2.

The thermal conductivities of some materials, including metallic solids and non-metallic solids which are of industrial importance, are listed in Table 8.1.

8.2.2 Convection

When a moving fluid and a stationary surface are at different temperatures, the heat transfer that occurs between them is called convection [4]. Generally, two categories of convection heat transfer exist: (1) forced convection, where the fluid is pushed against a surface by external methods; (2) free convection, where the movement of the fluid is due to density variation resulting from body forces [5]. The flow around a solid in a homogeneous inflow is called an external flow, whereas a fluid surrounded by a solid interface is called internal flow. In order to study the convection process, the concept of the convective boundary layer is introduced.

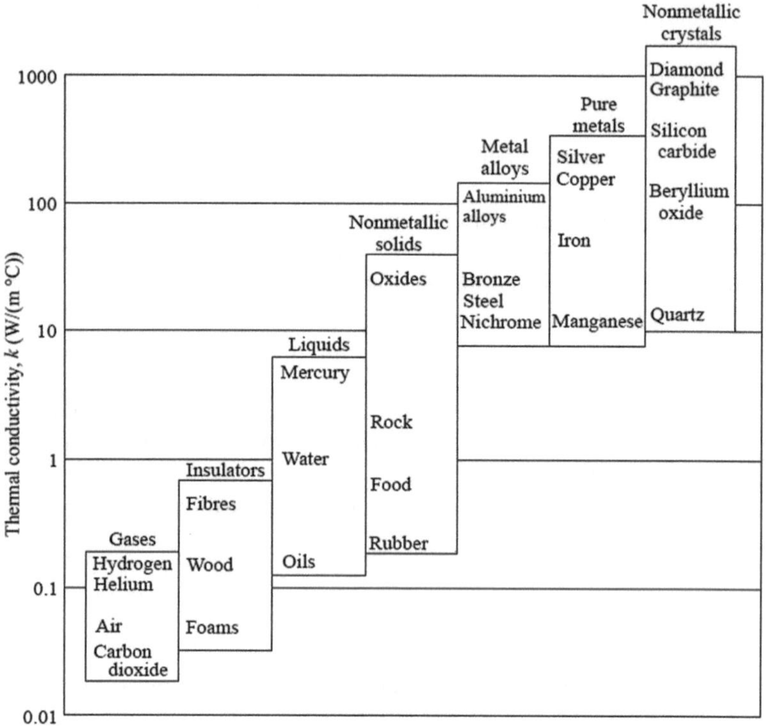

Fig. 8.2 Range of thermal conductivity at normal temperatures and pressures for a variety of states of matter

Table 8.1 The thermal conductivities of several materials at different temperatures

Material	Thermal conductivities [W/(m·K)] at different temperatures (K)							
	100	200	300	400	600	800	1000	1200
Pure Al	302	237	237	240	231	218	–	–
Pure Cu	482	413	401	393	379	366	352	339
Pure Fe	134	94.0	80.2	69.5	54.7	43.3	32.8	28.3
Plain carbon steels	–	–	60.5	56.7	48.0	39.2	30.0	–
Pure Ni	164	107	90.7	80.2	65.6	67.6	71.8	76.2
Pure Pt	77.5	72.6	71.6	71.8	73.2	75.6	78.7	82.6
Sapphire	450	82	46	32.4	18.9	13.0	10.5	–
Amorphous carbon	0.67	1.18	1.60	1.89	2.19	2.37	2.53	2.84
Diamond	10000	4000	2300	1540	–	–	–	–
Pyroceram	5.25	4.78	3.98	3.64	3.28	3.08	2.96	2.87

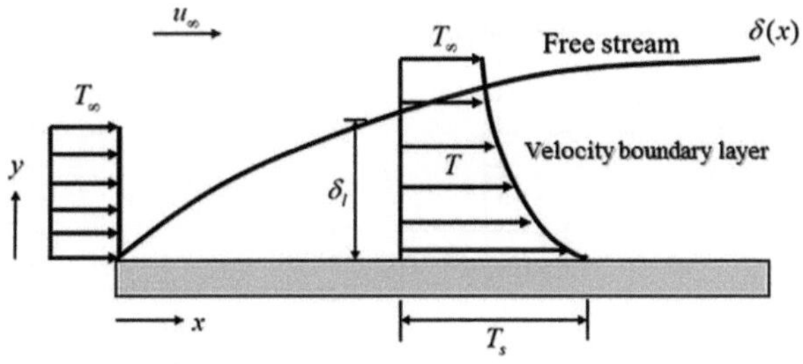

Fig. 8.3 Velocity boundary layer

The concept of the boundary layer is important for understanding the convection processes for both mass transfer and heat transfer between the fluid flow and surface. We introduce a velocity boundary layer as well as a thermal boundary layer to study the heat and mass transfer between the fluid flow and the surface through the surface.

1. Velocity boundary layer

Figure 8.3 shows the defined velocity boundary layer. When the fluid molecules are attached to the solid surface, their velocities are zero. These fluid particles hinder the movement of those in the adjacent fluid layer, which in turn hinders the movement of the upper layer. This effect can be ignored until the distance $y = \delta$ from the surface. Define δ as the thickness of the velocity boundary layer; usually, the δ value is the height where $u = 0.99u_\infty$, as shown in Fig. 8.3.

2. Thermal boundary layer

The defined thermal boundary layer is shown in Fig. 8.4. When considering the fluid flow on an isothermal plate, we assume that the temperature of the isothermal plate is T_s and the uniform temperature of the bulk fluid is T_∞. The fluid particles, which are in contact with the surface, will approach the surface temperature T_s when they exchange energy with the plate. These fluid particles will continue exchanging energy with those in the upper layer, again followed by reaching thermal equilibrium. As a result, a temperature distribution will present within this flow field, ranging from T_s (surface temperature of the plate) to T_∞ (the temperature far away from the plate surface). The thickness of the thermal boundary layer δ_t is described as the distance from the surface to the point where the difference of temperature $(T-T_s)$ equals to $0.99(T_\infty-T_s)$.

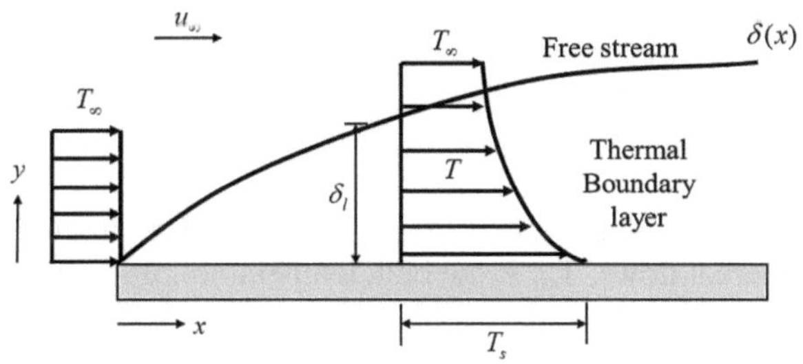

Fig. 8.4 Thermal boundary layer

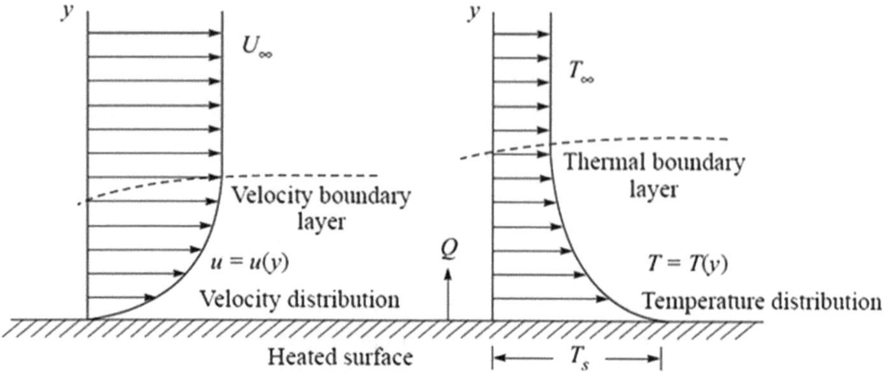

Fig. 8.5 Velocity and temperature distributions in boundary layers in convective heat transfer mode

Laminar flow is dominated by fluid viscosity, and the turbulent flow velocity changes with time. When the flow velocity and object size increase, the flow will transit into turbulence (Fig. 8.5).

According to the no-slip condition, the velocity of the fluid at the interface between the fluid and the surface is zero. Therefore, heat can be transferred from the surface to the fluid by conduction, which transfers kinetic energy between the molecules by their random motions. For laminar flow, the contribution of advection by bulk fluid motion arises with the growth of the boundary layer in the direction of flow. The heat being conducted to the boundary layer goes downstream and will finally be swept to the fluid outside this layer. On the other hand, in the case of turbulent flow, the transfer of heat will be based on the transfer of mass and momentum in the turbulent zone, but there is also a laminar flow zone. Therefore, knowledge of the boundary layer is essential for understanding the mechanistic aspects of convective heat transfer clearly.

Irrespective of the details of the mechanism, the heat transfer rate by convection (both free and forced) between a fluid and a solid surface is calculated from the relation, as shown in Eq. (8.7):

$$Q = hA\Delta T \tag{8.7}$$

where

Q is the rate of heat transfer by convection;

A is the heat transfer area;

$\Delta T = (T_s - T_\infty)$ is the difference between the surface temperature T_s and the temperature of the fluid T_∞ at the bulk flow region.

In the case of external flows, T_∞ is the free stream temperature (Fig. 8.4), and for internal flows, T_∞ is the bulk temperature of the fluid.

The basic rate equation used to quantify the convection, i.e., the calculation of convective heat flux q (W/m^2), is expressed as in Eq. (8.8), which is also known as Newton's law of cooling:

$$q = h(T_s - T_\infty) \tag{8.8}$$

where
h = the heat transfer coefficient, $W/(m^2 \cdot K)$.

Here, h is dependent on boundary layer conditions, affected mainly by the features of the flow, surface geometry, thermodynamics, and the transport properties of the flow. In every investigation of convective heat transfer, the goal will eventually become the determination of h. Table 8.2 lists several typical values of h for different processes.

Equation (8.8) can also be expressed in terms of heat flow, thermal potential, and thermal resistance for convective heat transfer as in Eq. (8.9):

$$q = \frac{T_s - T_\infty}{1/(hA)} \tag{8.9}$$

The analogous electrical circuit is shown in Fig. 8.6.

We can use the proportional analysis to further simplify the above equation to obtain the relationship between the heat transfer coefficient and the flow parameters of the laminar flow on the plate. This analysis is used to confirm the validity of analysis or numerical solution.

So, we define the heat transfer coefficient h and associate it with a dimensionless number $Nu = hL/k$, which is called the Nusselt number and means the ratio of convection to conduction. We can obtain this dimensionless number in different situations through the derivation of the boundary layer and the similarity principle.

Table 8.2 Typical values of the convective heat transfer coefficient

Forced convection	h [$W/(m^2 \cdot K)$]	Free convection	h [$W/(m^2 \cdot K)$]	Convection with phase change	h [$W/(m^2 \cdot K)$]
Gases	25–250	Gases	2–25	Boiling or condensation	2,500–100,000
Liquid	100–200,000	Liquid	50–1,000		

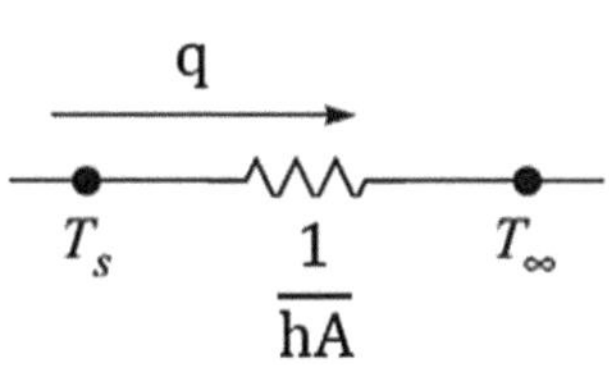

Fig. 8.6 The analogous electrical circuit of convective heat transfer problems

8.2.3 Radiation

Radiation is the phenomenon of electromagnetic waves transmitting energy. Electromagnetic radiation due to thermal heating is called thermal radiation. As long as the body temperature is above absolute zero, the object keeps releasing heat radiation, along with constantly absorbing the heat radiation projected on the surface. For example, the sun's energy travels to the earth in the form of electromagnetic waves, followed by being selectively absorbed and scattered as it goes through the atmospheric layers, before eventually heating up everything around us. Radiation allows us to see and sense everything in our surroundings. In this process, it is not only the strength of the radioactive energy that is important, but also how it interacts with matter and how it energizes all the living organisms.

Radioactive heat transfer is the result of phase-free radiation and absorption between objects. A heat transfer medium is not required for radiation heat transfer. Even in a vacuum, heat can be transmitted by electromagnetic waves. When thermal radiation is projected onto the surface of an object, three phenomena commonly occur, which are absorption, reflection, and penetration. As shown in Fig. 8.7, the energy Q from the outside can be divided into three parts: the energy Q_α absorbed by the object, Q_ρ reflected by the object, and the rest Q_τ through objects, as expressed in Eq. (8.10):

$$Q = Q_\alpha + Q_\rho + Q_\tau \tag{8.10}$$

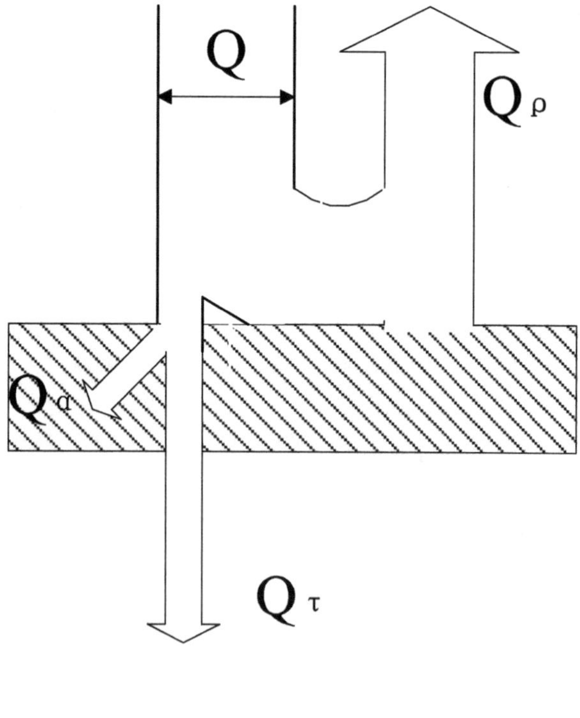

Fig. 8.7 The object absorption, reflection, and penetration of heat radiation

where Q_α/Q, Q_ρ/Q, Q_τ/Q are called the absorption ratio, reflectance, and penetration ratio of the object and denoted by α, ρ, and τ, respectively. Therefore, Eq. (8.11) can be obtained:

$$\alpha+\rho+\tau=1 \tag{8.11}$$

For practical problems, due to the great differences in the surface conditions of different materials, thermal radiation is difficult to study. Therefore, we will study thermal radiation from the ideal surface. What we refer to as an object which can absorb all the heat radiation energy being put into its surface is a scientific imaginary object that does not exist in real life. Blackbodies can be made artificially.

Radiation force is introduced to quantitatively describe the radiation energy to the outside of a unit blackbody surface at a specific temperature. The energy per area of an object that is emitted from the unit surface area to the hemispherical space is denoted as E (W/m^2). The relation between the radiation force of the blackbody and the thermodynamic temperature (K) is defined by the Stefan–Boltzmann law, as expressed in Eq. (8.12):

$$E_b = \sigma T^4 \tag{8.12}$$

where

σ = the Stefan–Boltzmann constant, 5.67×10^{-8} $W/(m^2 \cdot K^4)$.

Therefore from Eq. (8.12), the total radiant energy emitted from a blackbody in vacuum is only dependent on its temperature [6].

In practical applications, Eq. (8.13) is used to calculate the net radiation heat exchange:

$$q = h_r\left(T - T_{surf}\right) \tag{8.13}$$

where from Eq. (8.13), h_r is the radiation heat transfer coefficient, as described in Eq. (8.14):

$$h_r = \varepsilon\sigma\left(T + T_{surf}\right)\left(T^2 + T_{surf}^2\right) \tag{8.14}$$

As firstly determined by Planck, the blackbody spectral intensity I can be calculated as in Eq. (8.15):

$$I_{\lambda,b}(\lambda, T) = \frac{2hc_0^2}{\lambda^5\left[\exp(hc_0/(\lambda kT)) - 1\right]} \tag{8.15}$$

where

h = the universal Planck constant, 6.626×10^{-34} J·s;
k = the universal Boltzmann constant, 1.381×10^{-23} J/K;
c_0 = the speed of light in vacuum, 2.998×10^{8} m/s;
λ = the wavelength, m;
T = the absolute temperature of the blackbody, K;

The spectral emissive power of the blackbody, $E_{\lambda,b}$ unit (W/m^3), can be calculated as in Eq. (8.16):

$$E_{\lambda,b}(\lambda, T) = \pi I_{\lambda,b}(\lambda, T) = \frac{C_1}{\lambda^5\left[\exp(C_2/(\lambda T)) - 1\right]} \tag{8.16}$$

where

C_1 = first radiation constant, 3.742×10^{-16} W·m^2;
C_2 = second radiation constant, 1.4388×10^{-2} W·K.

Equation (8.16) is also called the Plank distribution, which describes the dependence of the spectral radiation force of the blackbody on the wavelength and temperature.

The relationship between the wavelength λ_m corresponding to the maximum spectral radiation force and the temperature T is described as in Eq. (8.17):

$$\lambda_{\max} T = C_3 = 2.8976 \times 10^{-3} \mathrm{m} \cdot \mathrm{K} \tag{8.17}$$

Equation (8.17) is also called Wien's displacement law for the blackbody, where C_3 is the third radiation constant. This law shows that the maximum value of the spectral emissive power shifts to shorter wavelengths when the temperature increases.

It has been pointed out that a blackbody is an ideal object for studying thermal radiation. Here, we will study radiation from real surfaces on this basis. The emission capacity of a blackbody is larger than that of a real surface at the same temperature. Therefore, we define the emissivity (also called blackness) ε as shown in Eq. (8.18):

$$\varepsilon = \frac{E}{E_b} = \frac{E}{\sigma T^4} = \frac{\int_0^\infty \varepsilon(\lambda) E_{\lambda,b} d\lambda}{\sigma T^4} \tag{8.18}$$

The above formula is only for direction and spectral averages, but in reality the emission capabilities of a real surface vary with direction and spectrum. Therefore, it is necessary to define the spectral emissivity for a specified direction (θ, φ) and wavelength λ, as described in Eq. (8.19):

$$\varepsilon_{\lambda,\theta}(\lambda, \theta, \varphi, \mathrm{T}) = \frac{I_\lambda(\lambda, \theta, \varphi, \mathrm{T})}{I_{\lambda,b}(\lambda, \mathrm{T})} \tag{8.19}$$

Three correction coefficients are introduced: emissivity ε, spectral emissivity $\varepsilon(\lambda)$, and directional emissivity $\varepsilon(\lambda)$.

Radiation between parallel surfaces. Radiation heat transfer between surfaces is related to several factors such as surface geometry, size, and the relative position of the surface. Those factors are expressed by the angular coefficient. The concept of the angular coefficient was put forward in the 1920s with the appearance and development of radiation heat transfer calculations based on solid surfaces. The most widely used name is the angular coefficient. It should be noted that angular coefficients apply only to diffuse surfaces, surface emissions, radiation, and the projected radiation uniformity [7].

The angular coefficient is defined in two surfaces, numbered 1 and 2, in which a transparent medium is applied. The angular coefficient of surface 1 to surface 2 $X_{1,2}$ is as follows: the surface 1 is projected directly onto the surface 2, accounting for surface 1 of the radiant energy on the surface, as shown in Eq. (8.20):

$$X_{1,2} = \frac{\text{The irradiated energy of surface 1 onto surface 2}}{\text{Effective radiation of surface 1}} \tag{8.20}$$

According to the definition and the analytic form of the angle coefficient, the algebraic properties of the angle coefficients can be derived into three parts, as expressed in Eqs. (8.21)–(8.23):

(1) Relativity

$$A_1 X_{1,2} = A_2 X_{2,1} \tag{8.21}$$

(2) Completeness

$$X_{1,1} + X_{1,2} + X_{1,3} + \cdots + X_{1,n} = \sum_{i=1}^{n} X_{1,i} = 1 \tag{8.22}$$

(3) Additive

$$X_{1,2} = \sum_{i=1}^{n} X_{1,2i} \tag{8.23}$$

Using the various properties of the angle coefficient, we can obtain a set of algebraic equations and obtain the angle coefficient. It is worth noting that (1) the premise of the method is that the system must be closed; (2) the number of concave surfaces must be equal to the number of invisible surfaces.

An example of two steady-state radiative heat transfers is given, that is, a surface-to-surface radiant heat transfer in a closed system consisting of two isothermal blackbodies or two isothermal gray bodies. The enclosed system is filled with a transparent medium that does not absorb any radiation. Figure 8.8 shows a blackbody surface. The amount of heat transfer between black surfaces 1 and 2 can be calculated as in Eq. (8.24):

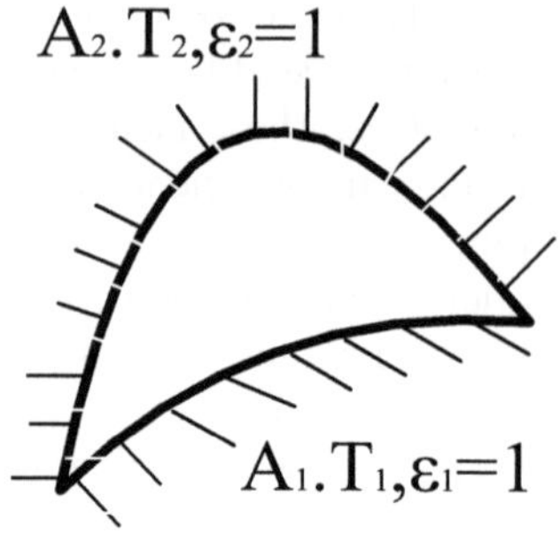

Fig. 8.8 Radiation heat transfer of blackbody system

$$q_{1,2} = A_1 E_{b1} X_{1,2} - A_2 E_{b2} X_{2,1} = A_1 X_{1,2}(E_{b1} - E_{b2}) \tag{8.24}$$

Multiple reflections between the gray bodies make it complicated to calculate the radiation heat transfer processes, and the concept of input radiation G and effective radiation J is required. In the following case, the relationship between J and the surface net heat transfer is investigated under the assumption that the surface physical properties and temperature are known. As shown in Fig. 8.8, for surface 1, the net radiation heat transfer q is expressed as in Eqs. (8.25) and (8.26):

$$q = J_1 - G_1 = E_1 - \alpha_1 G_1 = \varepsilon_1 E_{b1} - \alpha_1 G_1 \tag{8.25}$$

$$J_1 = E_{b1} - \left(\frac{1}{\varepsilon_1} - 1\right) q \tag{8.26}$$

The radiation heat transfer in two isothermal diffuse gray closed surfaces is analyzed. The net heat exchange rate of the two surfaces is calculated as in Eq. (8.27):

$$q_{1,2} = A_1 J_1 X_{1,2} - A_2 J_2 X_{2,1} \tag{8.27}$$

Conservation of energy can be described by Eq. (8.28):

$$q_{1,2} = \frac{E_{b1} - E_{b2}}{\frac{1-\varepsilon_1}{\varepsilon_1 A_1} + \frac{1}{A_1 X_{1,2}} + \frac{1-\varepsilon_2}{\varepsilon_2 A_2}} \tag{8.28}$$

System blackness (or system emissivity) is defined as in Eq. (8.29):

$$\varepsilon_s = \frac{1}{1 + X_{1,2}\left(\frac{1}{\varepsilon_1} - 1\right) + X_{2,1}\left(\frac{1}{\varepsilon_2} - 1\right)} \tag{8.29}$$

Then we can get Eq. (8.30):

$$q_{1,2} = \varepsilon_s A_1 X_{1,2}(E_{b1} - E_{b2}) \tag{8.30}$$

In comparison to blackbody radiation heat transfer, there exists an ε_s in the above equation, which is considered a factor that affects heat transfer due to the absorptions and reflections of the gray body system.

8.2.4 *Overall Heat Transfer*

1. Thermal resistance

To solve complex heat conduction problems, thermal resistance and an equivalent thermal network will be introduced here. Without internal heat source and physical properties, there is an analogy between thermal diffusion and charge diffusion. Just like the relationship between resistance and conduction, we define the thermal resistance to be the ratio of the driving potential to the thermal conductivity, as shown in Eq. (8.34):

$$R_{t,cond} \equiv \frac{T_{s,1} - T_{s,2}}{q_x} = \frac{L}{kA} \tag{8.34}$$

It is also possible to extend the definition of thermal resistance to thermal convection, which is known from Newton's law of cooling, as presented in Eq. (8.35):

$$R_{t,conv} \equiv \frac{T_s - T_\infty}{q} = \frac{1}{hA} \tag{8.35}$$

And Radiant thermal resistance can be described as in Eq. (8.36):

$$R_{t,\text{rad}} \equiv \frac{T_s - T_{sur}}{q_{rad}} = \frac{1}{h_r A} \tag{8.36}$$

2. Equivalent thermal network

Circuit notation is a useful tool for conceptualization and quantitative calculation of heat transfer problems. Heat transfer efficiency can be calculated by separately calculating each component in the network. The basic relationship between components are series and parallel (Fig. 8.9).

For systems in series, the components distribute the thermal drive potential distributed across the two ends, while the heat transfer rate of each component is equal, as shown in Eq. (8.37):

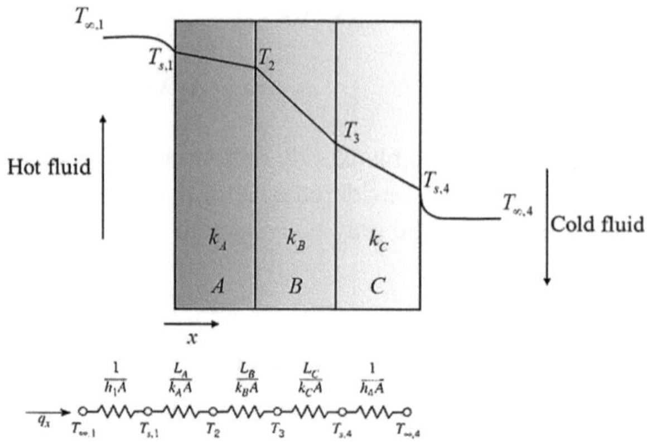

Fig. 8.9 The diagram of equivalent thermal network in series

$$q_x = \frac{T_{\infty,1} - T_{s,1}}{\frac{1}{h_1 A}} = \frac{T_{s,1} - T_{s,4}}{\frac{L}{kA}} = \frac{T_{s,4} - T_{\infty,4}}{\frac{1}{h_2 A}} = \frac{T_{\infty,1} - T_{\infty,4}}{1/(h_1 A) + L/(kA) + 1/(h_2 A)} \quad (8.37)$$

For a system in parallel, the thermal drive potentials at both ends of the component are the same, and the sum of the thermal conduction rates of all components is the thermal conductivity of the system, as presented in Eq. (8.38):

$$q_x = \frac{T_{\infty,1} - T_{\infty,2}}{1/(h_1 A)} + \frac{T_{\infty,1} - T_{\infty,2}}{L/(kA)} + \frac{T_{\infty,1} - T_{\infty,2}}{1/(h_2 A)} \quad (8.38)$$

Example A concrete wall is 200 mm thick with a thermal conductivity k = 0.8 W/(m·°C). This wall is exposed to air at 80 °C on one side and to air at 30 °C on the other side. The average convective heat transfer coefficients are 40 W/(m^2·°C) at the 80 °C side and 20 W/(m^2·°C) at the 30 °C side. Calculate the rate of heat transfer per unit surface area of the wall and the surface temperatures of the two sides.

Solution Figure 8.10 shows the equivalent electrical network system.

To calculate the rate of heat transfer Q, we use Eq. 8.36.

$$q_x = \frac{80 - 30}{1/40 + 0.2/0.8 + 1/20} = 153.85\,\mathrm{W/m^2}$$

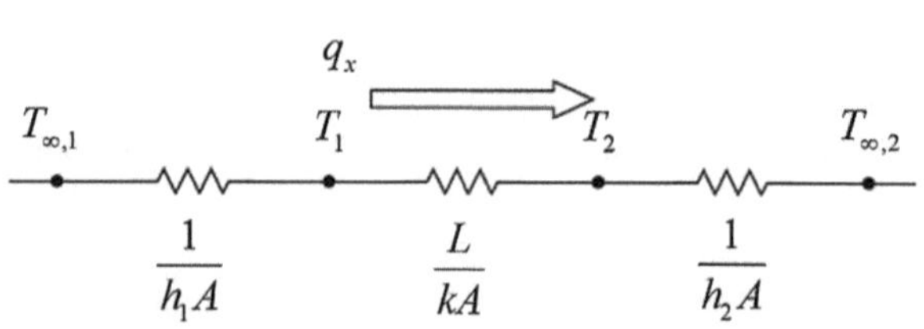

Fig. 8.10 The equivalent thermal network

Assume T_1 and T_2 to be the surface temperatures on the 80 °C-side and 30 °C-side, respectively.

$$T_1 = T_{\infty,1} - q_x \times (1/h_1 A) = 80 - 153.85/40 = 76.15\,^{\circ}\mathrm{C}$$

$$T_2 = T_{\infty,2} + q_x \times (1/h_2 A) = 30 + 153.85/20 = 37.69\,^{\circ}\mathrm{C}$$

8.3 Applications

8.3.1 Heat Exchanger: Recuperator and Regenerator

Heat exchangers are devices for transferring thermal energy (enthalpy) at different temperatures and in thermal contact between fluids and solid particles, between fluids and solid surfaces, or between two or more fluids. There usually exists no interaction between external heat and work. The applications of heat exchangers typically include cooling or heating fluid streams as well as condensing or evaporating fluid streams with one component or multiple components. In addition, it may also be the goal of a heat exchanger for heat recovery or rejection, or for pasteurization, sterilization, distillation, fractionating, crystallizing, concentrating, or controlling process fluids. The heat exchanger is usually classified according to the transfer process in the heat exchanger. In some of the heat exchangers, fluids are in direct contact when exchanging heat. Mostly, heat transfer between fluids occurs via the separation wall or in and out of the wall in a transient manner. Fluids are always separated by a surface. Ideally, fluids will not mix or leak. This type of heat exchanger is called a direct transfer heat exchanger, or simply a recuperator. On the other hand, a heat exchanger that has intermittent heat exchange between fluids with different temperatures, through storage or release of thermal energy via the surface or matrix of the exchanger, is called an indirect transfer heat exchanger, or simply a regenerator (Fig. 8.11).

1. Recuperator

In a recuperator, heat is continuously transferred from the high-temperature fluid to the low-temperature one via a separation wall. Although two (or more) fluids are required to flow simultaneously in the exchanger, the two (or more) fluids are not directly mixed since each of them flows in a separate fluid channel. Generally, no moving parts exist in most heat exchangers. This kind of heat exchanger is referred to as a recuperative heat exchanger (or a recuperator). Several examples of direct transfer heat exchangers include plate-type, tubular, and extended surface exchangers. It should be noted that the term "recuperator" is not commonly used for plate and shell-and-tube heat exchangers. They are subdivided into primary surface exchangers and extended-surface exchangers. Fins or extended surfaces are not used on any fluid side

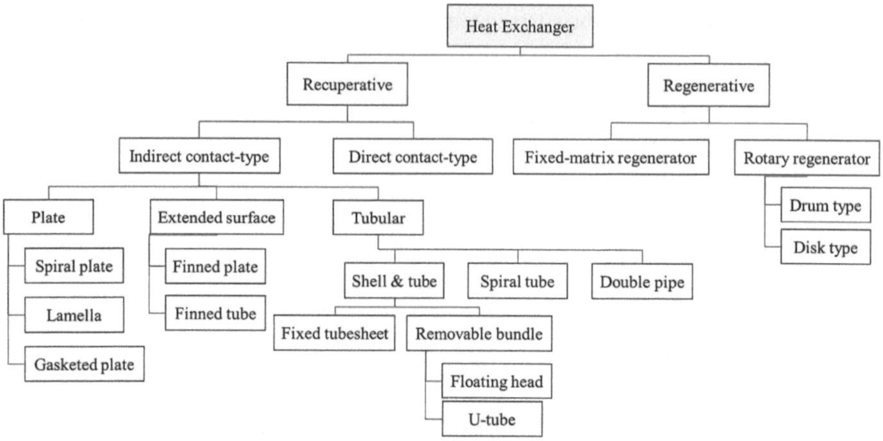

Fig. 8.11 Classification of recuperator and regenerator

in a primary surface exchanger, among which good examples include plain tubular exchangers, shell-and-tube exchangers with plain tubes, as well as plate exchangers. Of all the heat exchangers, recuperators account for the vast majority.

(1) Tubular heat exchanger

Tubular heat exchangers are usually composed of round tubes, despite that oval tubes, rectangular tubes, or round/flat twisted tubes are also sometimes applied. For the design of a tubular heat exchanger, it is very flexible since the core geometry can be easily changed by varying the length, diameter, and arrangement of the tube. It can also be designed for high-pressure differentials between environments and high-pressure fluids. It is mainly applied for heat transfer processes from liquid-to-liquid and from liquid-to-phase change (condensing or evaporating). When operation temperatures and/or pressures are extremely high, or fouling takes place on at least one fluid side, the tubular heat exchanger is mainly applied for gas-to-liquid and gas-to-gas heat transfer processes. The tubular heat exchangers can be divided into shell-and-tube exchangers, double-pipe exchangers, and spiral tube exchangers. All of them are main surface exchangers, except for finned heat exchanger outside/inside tubes (Table 8.3).

(2) Extended surface heat exchangers

The above-mentioned tube and plate heat exchangers are the main surface heat exchangers, except for a shell-and-tube one with low-finned tubes. The effectiveness over these exchangers is always 60% or below, with the surface area density of heat transfer always being less than 700 m^2/m^3. While much higher effectiveness (up to about 98%) is needed in some applications, along with limited box volume and mass, it requires a more compact surface. In addition, in the cases of heat exchangers with gases or several types of liquids, the overall coefficient of heat transfer at one or both fluid sides may be very low. Hence, a large surface area for heat transfer is

Table 8.3 Characteristics of shell-and-tube heat exchangers

Shell and tube exchangers	Typical TEMA code	Advantages	Disadvantages
Fixed tube sheet	BEM, AEM, NEN	Provides maximum heat transfer area for a given shell and tube diameter. Provides for single and multiple tube passes to assure proper velocity. Less costly than removable bundle designs	Shell side/outside of the tubes are inaccessible for mechanical cleaning. No provision to allow for differential thermal expansion developed between the tube and the shell side. This can be taken care of by providing an expansion joint on the shell side
Floating head	AEW, BEW, BEP, AEP, AES, BES	Floating tube sheet allows for differential thermal expansion between the shell and the tube bundle. Both the tube bundle and the shell side can be inspected and cleaned mechanically	To provide the floating head cover, it is necessary to bolt it to the tube sheet. The bolt circle requires the use of space where it would be possible to place a large number of tubes. Tubes cannot expand independently so that huge thermal shock applications should be avoided. Packing materials produce limits on design pressure and temperature
U-tube	BEU, AEU	U-tube design allows for differential thermal expansion between the shell and the tube bundle as well as for individual tubes. Both the tube bundle and the shell side can be inspected and cleaned mechanically. Less costly than floating head or packed floating head designs	Because of the U-bend, some tubes are omitted at the center of the tube bundle. Because of U-bend, tubes can be cleaned only by chemical methods. Due to U-tube nesting, an individual tube is difficult to replace. No single tube pass or true countercurrent flow is possible. Tube wall thickness at the U-bend is thinner than at the straight portion of the tubes. Draining of tube circuit is difficult when positioned with the vertical position with the head side upward

Fig. 8.12 An extended surface heat exchanger

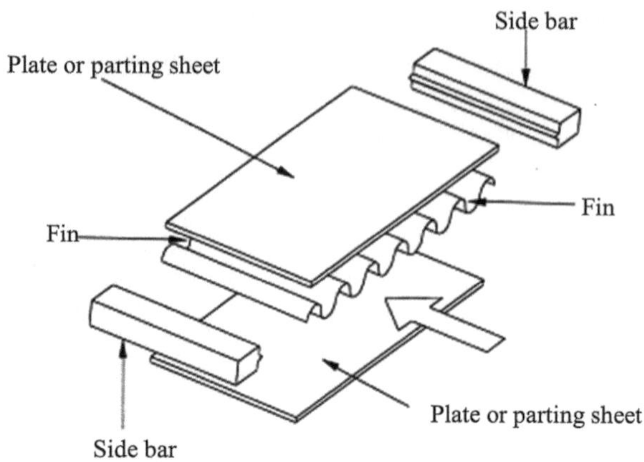

required. Depending on the design requirements, one of the most conventional ways for enhancing the surface area as well as the compactness of the heat exchanger is the addition of the extended surface (fins) on one or both fluid sides, with a fin density as high as possible. Depending on the design, adding fins can promote the surface area by 5–12 times. This type of heat exchanger is called an extended surface heat exchanger. By using thin gauge materials and proper core size adjustment, the flow area can be increased. The heat transfer coefficient of an extended surface can be either higher or lower than that of a surface without fins. Here, we take interrupted fins and internal fins in a tube as two typical examples. The former one can increase both the area and the heat transfer coefficient. On the other hand, the latter one can enhance the tube-side surface area while leading to a slight decline of the heat transfer coefficient, which is dependent on fin spacing. In general, the enhancement of fin density decreases the fin-related heat transfer coefficient. Interruptions by the flow could promote the heat transfer coefficient of the corresponding flat-fin surface by two–four times (Fig. 8.12).

(3) Plate-type exchangers

Thin plates are always used to build this type of heat exchanger. In general, this type of heat exchanger, however, does not work under very high temperatures, pressures, and temperature and pressure differences. They can further be divided into plate, thin plate, spiral plate, and plate coil exchangers. Among these heat exchangers, the most important type is the plate heat exchanger. Typically for a plate-and-frame heat exchanger, it has many thin rectangular corrugated or embossed metallic plates that are sealed around the edges and connected together in a frame. Plate sets with fixed and movable end caps are clamped by long bolts, consequently forming a seal. Between the two fluids, sealing is achieved through elastically molded gaskets. The most common flow configuration is one-way to one-way countercurrent, in which all the inlets and outlets are connected to a fixed end cap (Table 8.4).

Table 8.4 Plate and Frame Heat Exchanger Parameters

Main unit	
Maximum surface area, m^2	2500
Number of plates	3–700
Port size, mm	<400
Plates	
Thickness, mm	0.5–1.2
Size, m^2	0.03–3.6
Spacing, mm	1.5–5
Width, mm	70–1200
Length, m	0.6–5
Operation	
Temperature, °C	−40–260
Pressure, MPa	0.1–2.5
Channel flow rates, m^3/h	0.05–12.5
Maximum port velocity, m/s	6
Maximum unit flow rate, m^3/h	2500
Performance	
Heat exchanger effectiveness, %	<93%
Temperature approach	As low as 1
Heat transfer coefficients for water–water duties, $W/m^2 \cdot K$	3000–7000

2. Regenerator

Two fluids in the storage heat exchanger alternatively flow through the same flow channel resulting in continuous heat transfer. The surface for heat transfer surface is usually a porous structure called a matrix, or a permeable solid material called a packed bed. As the high-temperature gas flows through the heat transfer surface, the thermal energy carried by the gas can be stored in the matrix wall. As a result, the high-temperature gas is cooled during the heating period of the matrix. Later when a low temperature gas flows over the same surface, thermal energy will be resealed by the matrix wall, followed by being absorbed by the gas. Therefore, heat is not continuously transferred, while it is stored and released by the matrix wall. This kind of heat exchanger is called a regenerative heat exchanger, or a regenerator. In order to work continuously within the required temperature range, the gas flows and matrices are periodically switched. As a result, the same channel is periodically occupied by cold and hot gases. Hence, the real time for a high-temperature gas flowing through a low-temperature regenerator matrix is referred to as the hot period. Similarly, the actual time for a low-temperature gas flowing through a high-temperature regenerator matrix is referred to as the cold period. In order to operate successfully, the hot and cold periods do not need to be the same. After switching fluids, a small portion of the fluid is inevitably entrained into another fluid stream, which is called carry-over leakage. Moreover, if the pressures of the two fluids are different, leakage will take place from the high-pressure fluid to the low-pressure fluid that seals radially, peripherally, and axially or flows through the valve. This leakage is called pressure leakage. Due to the fact that these leaks are unavoidable, regenerators are applied for gas-to-gas transfers of heat and mass. In several other cases, moisture can be transferred using a regenerator from humid air to dry air up to about 5%.

(1) Rotary Regenerators

Figure 8.13 shows two types of rotary regenerators. Rotary regenerators, depending on their applications, can be called a thermal wheel, heat wheel, Munter wheel, or Ljungstrom wheel. It can also be called a laminar flow wheel if the gas flow is laminar. In this type of heat exchanger, the geometry of any flat-fin surface can be used in a matrix made of thin metal sheets. Discontinuous channel surfaces (like strip fins and shutter fins) are not used, because if the two fluids are at different pressures, a lateral (to main flow) flow leak will occur. Two fluids will be mixed, resulting in the contamination of fluid at low pressure and hence reducing exchanger efficiency. Therefore, generally a matrix has continuous flow channels. Fins are stacked by using flat or wavy gaskets. For these surfaces, the fluid remains unmixed in any cross-section. While for herringbone or diagonal channel matrices, gaskets are not needed to stack the fins. These types of surfaces are designed with Reynolds numbers ranging from 100 to 1,000. In the regenerator, the matrix is rotated using a peripheral ring gear drive or a hub shaft. Each matrix element is periodically transferred from the high-temperature fluid to the low-temperature fluid and back again. The time needed for the matrix to fully rotate is equal to the total period for the fixed-matrix regenerator. In rotary regenerators, the fixed radial seal position controls the leading edge area required for each fluid, as well as helping to minimize the major leakage.

(2) Fixed-matrix regenerator

This kind of regenerator is also known as a valved, fixed-bed, periodic-flow, or stationary regenerator. In order to operate continuously, there are at least two (usually

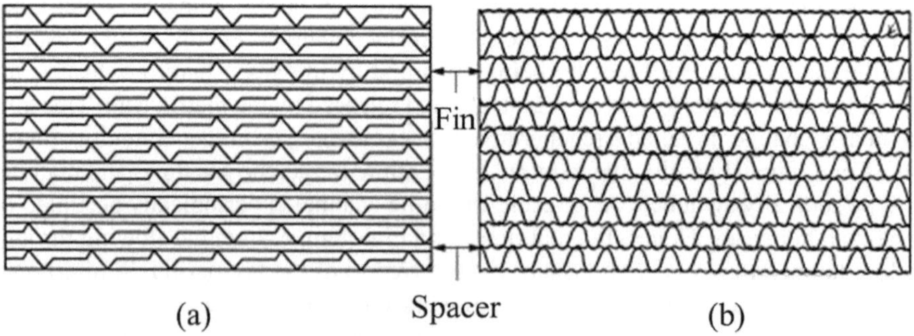

Fig. 8.13 Continuous-channel matrices for a rotary regenerator: **a** notched plate; **b** triangular passage

three or four) same matrices running in parallel for reducing variations of temperature of the heated cold gas at the outlet in high-temperature applications. On the other hand, one single matrix is enough for continuous operation in a rotary or rotating hood type of regenerator. There are two types of elements for heat transfer in a fixed-matrix regenerator: (1) checker work and (2) pebble beds. In the former one, there are two main types: (a) non-compact regenerators for applications with corrosive gases at high temperatures (typically 925–1600 °C); and (b) highly compact regenerators for applications from low to high temperatures. This regenerator is a key thermodynamic element in the Stirling engine cycle. While there is only one matrix in it, as a result it has no continuous fluid flows like other regenerators.

3. Design of recuperator and regenerator

Heat calculation is the basis of heat exchanger design. Heat (force) calculation of an interwall heat exchanger is introduced based on an interwall heat exchanger. Design the new heat exchanger and determine its area. However, the heat transfer area of the same size can adopt different construction sizes, and different construction sizes will affect the heat transfer coefficient, so it is generally conducted with the structure calculation.

Heat transfer equation

$$Q = \int_0^A h\Delta T dA \tag{8.39}$$

Q = Heat rate;
h = Heat transfer coefficient;
ΔT = Temperature difference on the microelement surface.

$$Q = h_{tot} A \Delta T_m \tag{8.40}$$

h_{tot}, Total heat transfer coefficient;
ΔT_m, Logarithmic average temperature difference.

If heat loss is not taken into account,

$$Q = M_1(i_1' - i_1'') = M_2(i_2'' - i_2') \tag{8.41}$$

Subscript 1 represents the thermal fluid and subscript 2 cold fluid. Supcrscript 1 prime means import, and superscript 2 prime means export.

If there is no phase transition,

$$Q = M_1 \int_{t_1''}^{t_1'} C_1 dt = M_2 \int_{t_1'}^{t_1''} C_2 dt \tag{8.42}$$

$$Q = M_1 c_1(t_1' - t_1'') = M_2 c_2\left(t_2'' - t_2'\right) \tag{8.43}$$

When heat loss is considered,

$$Q_1 \eta_L = Q_2 \tag{8.44}$$

η_L, external heat loss coefficient, taken as 0.97–0.98.
The heat transfer temperature difference,

$$\Delta t_m = \varphi \frac{\Delta t_{\max} - \Delta t_{\min}}{\ln(\Delta t_{\max}/\Delta t_{\min})} \tag{8.45}$$

Heat transfer efficiency,

$$\varepsilon \equiv \frac{\Phi}{\Phi_{\max}} = \frac{q_{m1}c_1(t_1' - t_1'')}{(q_m c)_{\min}(t_1' - t_2')} = \frac{q_{m2}c_2(t_2'' - t_2')}{(q_m c)_{\min}(t_1' - t_2')} \tag{8.46}$$

$$\text{NTU} = \frac{kA}{(q_m c)_{\min}} \tag{8.47}$$

$$R_c = \frac{(q_m c)_{\min}}{(q_m c)_{\max}} \tag{8.48}$$

The flow mode has a great influence on the rationality of the whole heat exchanger design. The following aspects should be paid attention to when selecting the flow mode:

(1) Under a given temperature condition, a large average temperature difference is ensured to reduce the consumption of metal or other materials by reducing the heat transfer area.
(2) Make the temperature change of the fluid as large as possible, so that the heat of the fluid can be used reasonably, reduce its consumption, and save the investment and energy consumption of pumps or fans.
(3) Make the heat transfer surface temperature as even as possible, and make it work at a lower temperature, so as to make the heat exchanger with cheaper materials.
(4) The best heat transfer condition should be provided, so that a higher heat transfer coefficient can also reduce the heat transfer surface.

8.3.2 Cogeneration and Trigeneration

In consideration of energy efficiency, air pollution, and cost, an efficient and economical way to reduce CO_2 emissions is cogeneration. It could reduce emissions at a minimal cost and provide a maximum amount of energy for use. Cogeneration reduces demand for fuel, as well as environmental and economic aspects for fuel transportation and storage.

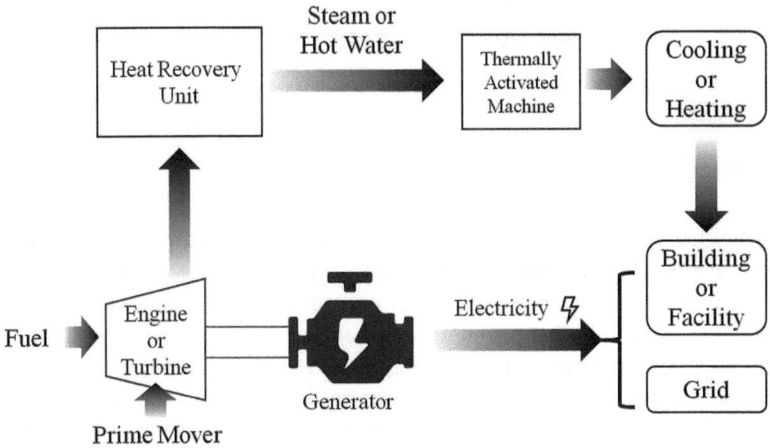

Fig. 8.14 A schematic of a typical cogeneration configuration

Cogeneration, which is also called combined heat and power (CHP), simultaneously produces heat and electricity. The heat can be used to produce process steam, cool, or be used directly for heating. The generation of electricity, heating, and cooling together is also called trigeneration. Cogeneration is a technology applied to meet user needs. Compared to the individual generation of electricity and thermal energy, cogeneration is more thermally efficient because in electricity production a fraction of energy must be rejected as waste heat, while in cogeneration the waste heat can be utilized.

A typical scheme of a cogeneration system is shown in Fig. 8.14. The prime mover is used to convert the chemical energy into mechanical shaft power, followed by being converted into electricity through a generator. The gases with high temperatures from the turbine enter the heat recovery unit to produce steam. This unit is called the heat recovery steam generator (HRSG). Then the steam is sent to a thermally activated machine, where the steam is processed for heating and/or cooling.

The thermal efficiency $\eta_{\text{Cogeneration}}$ can be calculated as in Eq. (8.49):

$$\eta_{\text{Cogeneration}} = \frac{W_{total}}{Q_{input}} = \frac{W_{electricity} + W_{heat}}{Q_{input}} \tag{8.49}$$

where

W_{total} = the total work output, J;

$W_{\text{electricity}}$ = the electrical power output, J;

W_{heat} = the heat output, J;

Q_{input} = the total heat input into the system, J.

For a trigeneration system where cooling is achieved at the same time, the thermal efficiency is calculated as in Eq. (8.50):

$$\eta_{\text{Trigeneration}} = \frac{W_{total}}{Q_{input}} = \frac{W_{electricity} + W_{heat} + W_{cooling}}{Q_{input}} \tag{8.50}$$

where

W_{cooling} = the cooling output, J;

For every heat engine, theoretical efficiency cannot exceed the Carnot cycle value, or the Rankine cycle value for steam turbine power plants, or the Brayton cycle value for gas turbines. The reason for the efficiency drop is mainly associated with the latent heat vaporization of steam which is not recovered. In a cogeneration system, the exhaust steam is at a higher temperature, making it useable for heating or cooling. Most of this heat is from condensed steam that creates latent heat. Therefore, for conventional separated generation of electricity and thermal energy, the overall efficiency is generally around 40–50%. For cogeneration systems, the overall efficiency can reach 70–80%.

Emission of air pollutants is always to be considered for fossil fuels or biomass combustion. Cogeneration is regarded as one of the most cost-efficient methods to reduce CO_2 emissions from heating systems in cold climates. According to the International Energy Agency, cogeneration can reduce CO_2 emissions from new generators in 2015 by more than 4% (170 Mt/year), while in 2030 this saving increases to more than 10% (950 Mt/year) [8].

However, pollution of water and air as well as their disposal cause problems for cogeneration. To meet environmental standards, additional pollutant treatment equipment, like air scrubbers and water treatment, are needed, which increases the overall cost of the project. If pollutant treatment facilities are already available in the commercial or industrial processes, the selection of cogeneration is economically a good option.

8.3.3 Phase Change Energy Storage

Energy storage refers to the transformation of energy into a more stable form under natural conditions. In the process of energy development, conversion, transportation, and utilization, there are always differences in quantity, form, and time between the supply and demand of energy. To make up for these differences and utilize energy more effectively, the process or technical means of storing and releasing energy is often called energy storage technology, which can be classified into chemical energy storage, mechanical energy storage, electromagnetic energy storage, wind energy storage, water power storage, thermal energy storage, etc. [9]. These technologies are used to prevent the automatic deterioration of energy quality, to improve the performance of the energy conversion process, to facilitate the economic use of energy, to reduce pollution, and to protect the environment. Therefore, these technologies are important for the rational, efficient, and clean use of energy.

The most widely used materials for thermal energy storage are phase change materials (PCM). PCM is a kind of material that changes the phase when the temperature changes, with the process of absorption or release of large amounts of latent heat [10]. In recent years, PCM have been widely applied in many fields, such as building, battery thermal management, and solar energy, due to their constant temperature and high energy storage capacity during the phase change process. PCM have become common energy storage materials in many new energy engineering fields.

The transformation of matter from one phase to another is called phase change. The phase change of matter usually exists as solid–gas, liquid–gas, liquid–solid, and solid–solid (from one crystalline form to another). The process of phase change is generally isothermal or nearly isothermal. The energy which is either absorbed or released within the melting-solidification cycle is regarded as the latent heat of fusion [11]. The amount of latent heat of phase change is generally large, so we can store and utilize the latent heat of phase change and the sensible heat of absorption. Therefore, the basic principle of phase change energy storage technology is that when the material changes phase, energy is stored by the latent heat of isothermal PCM dressed in components, and then released when needed.

The capacity of the latent heat storage system using PCM can be calculated using Eqs. (8.51) and (8.52):

$$Q = \int_{T_i}^{T_m} mc_p dT + ma_m \Delta h_m + \int_{T_m}^{T_f} mc_p dT \tag{8.51}$$

$$Q = m[c_{sp}(T_m - T_i) + a_m \Delta h_m + c_{tp}(T_f - T_m)] \tag{8.52}$$

where

Q = stored heat, J;
m = the mass of PCM, kg;
c_p = the specific heat capacity, J/(kg·K);
c_{sp} = the average specific heat capacity of [T_i–T_m], kJ/(kg·K);
c_{tp} = the average specific heat capacity of [T_m–T_f], kJ/(kg·K);
T_m = melting point, K;
T_i = the initial temperature, K;
T_f = the final temperature, K;
a_m = the percentage of melting;
Δh = latent heat of fusion, J/kg.

PCM are mainly divided into inorganic materials and materials according to their material types. Inorganic PCM mainly consist of crystalline hydrated salts, molten salts, metals, and alloys. Due to the limitation of the phase change temperature, crystalline hydrated salt is mostly used in building applications. The melting point ranges from several degree Celsius to more than 100 °C. Other widely applied materials are alkali and alkali metal halides, sulfates, nitrates, carbonates, acetates, etc. Crystalline hydrated salts store heat or release heat by absorbing or releasing crystalline water during freezing or melting. Crystalline hydrated salts are denoted by $AB{\cdot}mH_2O$. Their phase change energy storage mechanism is described as

$$AB \cdot mH_2O \overset{\square eating(T \gg T_m)cooling(T \ll T_m)}{\leftrightarrows} AB + mH_2O - Q \tag{8.53}$$

$$AB \cdot mH_2O \overset{\square eating(T \gg T_m)cooling(T \ll T_m)}{\leftrightarrows} AB \cdot pH_2O + (m-p)H_2O - Q \tag{8.54}$$

where
T_m = the melting point, °C;
Q = the fusion heat, kJ/kg.

Crystalline hydrated salt has the advantages of a wide application range, low price, high thermal conductivity, large melting heat, large volume heat storage density, and general neutrality. The disadvantages are the subcooling phenomenon during solidification, i.e., the temperature should be reduced to a certain temperature under the "solidification point" and then coagulation starts. Phase separation occurs when melting, i.e., the number of crystal water released by the temperature rising is not enough to dissolve solid insoluble salt dehydration while these dehydration salts cannot recrystallize. This results in irreversible phase change and leads to the decrease of energy storage capacity.

Organic phase change energy storage materials include paraffin, alkane, fatty acid, alcohol, etc. Usually, the phase change temperature and enthalpy of homologous organic compounds will increase with the increase of the carbon chain, so we can get a series of PCM of different temperatures. However, with the increase of carbon chain, the increase in the phase transition temperature will eventually tend to a fixed value. In order to get the appropriate phase change temperature and good performance of PCM, several organic phase change materials are usually used to form two or multiple phase change materials. Sometimes, the organic and inorganic materials are combined to overcome such shortcomings. The advantage of organic PCMs is their good formability in the solid state. It is difficult to appear supercooling and phase separation. The corrosion of the material is small with stable performance, low toxicity, and low cost. The disadvantages are the low thermal conductivity, low density, small specific energy storage capacity per unit volume, large volume variation during the phase change process, and low melting point which makes it not suitable for high-temperature applications. It is also volatile, highly flammable, even explosive, and easily aging when exposed to air.

The solid–solid PCM absorbs or releases heat due to the transformation of the crystal structure before and after the phase change. Because of the absence of the liquid phase in the phase change process, it presents small volume change before and after phase change, is non-toxic, is not corrosion, does not have high material requirements and production technology of the container, and has a long service life. The solid–solid PCM are mainly inorganic salts, polyols, and cross-linked high-density polyethylene (HDPE). Among them, the inorganic salt solid–solid phase change energy storage material has a higher phase transition temperature, and is suitable for storage and temperature control at high temperatures. At present, the layered perovskite, Li_2SO_4, and KHF_2 are widely applied. Polyalcohol solid–solid phase change materials are mainly pentaerythritol (PE) and neopentyl glycol (NPG), 2-amino-2-methyl-1, 3-propylene glycol (AMP), 3-hydroxymethyl ethane, and 3-hydroxymethyl amino methane. The phase change temperatures of them are within 40–200 °C, suitable for medium- and high-temperature energy storage. After radiation or chemical cross-coupling, the crystal transition temperature of HDPE is 120–135 °C, which is lower than its softening point, 150 °C. Hence, it can also be used as PCM. This material is of long service life, stable performance, no cooling and chromatography, good mechanical properties, and easy processing into a variety of shapes. So it has good potential for practical applications.

8.4 Challenges and Perspectives

8.4.1 Barriers to be Overcome

In many practical industrial applications, heat is exchanged between two fluids with different temperatures, separated by a solid wall. The device for implementing this process is a heat exchanger [12, 13], in which two different fluids are involved, one of which has a higher temperature and gives out heat while the other has a lower temperature and absorbs heat [14]. There are many applications for heat exchangers in industrial production, such as superheaters, economizers, air preheaters for boilers, condensers, deaerators, and feedwater heaters in power plant systems and evaporators in the refrigeration industry.

1. Institutional barrier

Recently, obstacle factors of the progress of heat transfer technology should be mainly attributed to institutional constraints. The establishment of market rules and system innovation of thermal recycling is more important than technology. The government and legislature must play a greater role, especially in formulating a series of complementary policies, laws, and regulations.

2. Regulatory barrier

At present, energy enterprises are facing endless technical barriers to trade, and they will even pay a lot of unnecessary costs in the process of coping with them. Therefore, how to solve regulatory barriers is a problem we must face. To this end, law experts, government regulatory authorities, and enterprises should work together to explore solutions.

3. Financial barrier

In the past 2 years, the engine of traditional industry growth has lost its power. Consumption growth is filling a part of the gap. Under the general trend of consumption upgrading, user behavior habits are changing dramatically. The business-driven mode of consumer finance has also changed from traditional capital-driven and scenario-driven to technology-driven. Therefore, finance is also a big challenge we have to face.

4. Technical barrier

With the progress of science and technology, whether in the Internet or finance, new ideas and technologies are emerging endlessly, and the speed of technological updating is becoming faster and faster. Technical barriers have gradually become a major challenge in technology promotion.

5. Public acceptance

As an important criterion for decision-making and evaluation of national key construction projects, public acceptance is a factor that we have to pay attention to.

8.5 4E Analysis

In the above sections, we discussed the principles and technologies for heat exchangers. Herein, we summarized the characteristics and costs of these technologies, from the point of view of environmental, energy, engineering, and economic aspects, representing the average values for coal-fired power plants, as shown in Table 8.5. It can be observed that the common heat exchanger types can be divided into the tubular heat exchanger and shell and tube heat exchanger. Relative to the tubular heat exchanger, the latter one has a narrower heat transfer type and it is more expensive to manufacture and operate. However, it has relatively high heat transfer efficiency, high flue gas flux, and suitable space size. Among them, more specific technology options can be adopted depending on the actual application environment, such as fixed shell and tube heat exchanger, and U-tube heat exchanger. When choosing a suitable heat exchanger, the following points should be considered: (1) high heat transfer efficiency, (2) suitable space size, and (3) high enough flue gas flux.

Exercises

1. Describe and distinguish the three basic types of heat transfer, i.e., conduction, convection, and radiation.
2. Please briefly describe the unique features of double-pipe-type heat exchangers.

Table 8.5 4E evaluation of heat exchanger technologies in coal-fired power plant

4E Evaluation	Classification	Water Wall	Superheater	Reheater	Economizer	Air preheater	Cooling apparatus	Deaerator	The cooling tower
Environmental	Coolant	Water	Superheated steam	Superheated steam	Water/Steam	Air	Water/air	Water	Water/Air
	Space size	Huge	Small	Small	Small	Medium	Small	Medium	Huge
	Thermal pollution	Small	Small	Small	Small	Small	Medium	Small	Huge
Energy	Heat transfer efficiency	Medium	High	High	High	Medium	Medium	Medium	Small
Engineering	Heat flow direction	Air–Water/steam	Air–steam	Air–steam	Air–water/steam	Air–Air	Water–air	Steam–water	Steam–air
	Heat exchange method	Radiation	Radiation/Convection	Convection	Convection	Convection	Convection/Conduction	Convection	Convection
Economic	Capital cost	Medium	Medium	Medium	Medium	High	High	High	Low
	Operating cost	Low	Low	Low	Low	Medium	Low	Medium	High

3. Analyze each heat exchanger and compare its advantages and disadvantages from the aspects of technology, economy, and application, respectively.
4. A concrete wall is 400 mm thick, and the thermal conductivity is k = 1.2 W/(m·°C). The wall is exposed to air at 90 °C on one side and to air at 30 °C on the other side. The average convective heat transfer coefficients are 60 W/(m^2·°C) on the 90 °C side and 20 W/(m^2·°C) on the 30 °C side. Calculate the overall heat transfer coefficient and determine the surface temperatures of the wall on both sides.
5. Cooling industrial exhaust gas with a double-pipe-type heat exchanger (counterflow). The flow rate of the exhaust gas in the inner tube (D_i = 30 mm) is 0.2 kg/s. The flow rate of water in the outer annular space (D_o = 50 mm) is 0.1 kg/s. The inlet temperatures of exhaust gas and water are 100 °C and 30 °C, respectively. If it is necessary to cool the flue gas to 50 °C, how long does the pipe need? (Assume that the physical properties of the exhaust gas are consistent with the air.)

References

1. Baehr HD, Stephan K (2011) Heat and mass transfer. Springer
2. Incropera FP, DeWitt DP, Bergman TL, Lavine AS (2007) Fundamentals of heat and mass transfer. Wiley
3. Liu L, Zheng LC, Liu FW, Zhang XX (2016) An improved heat conduction model with Riesz fractional Cattaneo-Christov flux. Int J Heat Mass Trans 103:1191–1197
4. Ghiaasiaan SM (2011) Convective heat and mass transfer. Cambridge University Press
5. Das D, Roy M, Basak T (2017) Studies on natural convection within enclosures of various (non-square) shapes—A review. Int J Heat Mass Trans 106:356–406
6. Howell JR, Siegel R, Mengüc MP (2010) Thermal radiation heat transfer. Taylor & Francis Group, LLC
7. Holman JP (1986) Heat transfer. McGraw-Hill Book Company
8. Agency IE (2008) Combined heat and power evaluating the benefits of greater global investment
9. Huggins RA (2016) Energy storage fundamentals, materials and applications. Springer
10. Raoux S, Wuttig M (2009) Phase change materials science and applications. Springer
11. Fleischer AS (2015) Thermal Energy storage using phase change materials fundamentals and applications. Springer
12. Bhuiyan AA, Islam A (2016) Thermal and hydraulic performance of finned-tube heat exchangers under different flow ranges: a review on modeling and experiment. International J Heat Mass Trans 101:38–59
13. Mobtil M, Bougeard D, Russeil S (2018) Experimental study of inverse identification of unsteady heat transfer coefficient in a fin and tube heat exchanger assembly. Int J Heat Mass Trans 125:17–31
14. Fsadni AM, Whitty JPM (2016) A review on the two-phase heat transfer characteristics in helically coiled tube heat exchangers. Int J Heat Mass Trans 95:551–565

Chapter 9
Advanced Combustion

Nomenclature

a	Oxygen to combustion mole ratio
E, e	The total energy
g	Gravitational acceleration
$\overline{g}^0$	One component Gibbs function
$G, \overline{g}$	Gibbs function or free energy
$h_f^0, \overline{g}_f^0$	Formation enthalpy
K_p	The equilibrium constant is dimensionless
LHV	Low calorific value
m	The quality
$\dot{m}$	Mass flow rate
N	Molar mass
Nu	Nusselt number
Pr	Prandtl number
R	Gas constant
$S, s, \overline{s}$	Entropy
t	Time
T	The temperature
$U, u, \overline{u}$	Internal energy
v	Speed
V	Volume
W, w	Work
x	The number of carbon atoms in a fuel
y	The number of oxygen atoms in a fuel
Y	The mass fraction
z	Highly
Z	Elements than
κ	Proportional constant
ρ	The density

P.-C. Chiang and X. Gao, *Air Pollution Control and Design*,
https://doi.org/10.1007/978-981-13-7488-3_9

Φ	Equivalent than
χ	The mole fraction
$*_{ad}$	Adiabatic
$*_A$	Air
$*_{CV}$	The control of body
$*_f$	Final or generated
$*_F$	Fuel
$*_g$	Gas
$*_i$	Component *i*
$*_l$	Liquid
$*_{ref}$	The reference state
$*_s$	Explicit
$*_{sat}$	Saturated state
$*_T$	At temperature T
$*^0$	Standard pressure

9.1 Introduction

9.1.1 Significance and Importance

So far, combustion is the most common first step in converting the energy stored in chemical bonds into expendable energy for human use. By reducing emissions, the harm to the environment from combustion has been reduced and will be reduced in the future. Emission reductions stem from two interrelated phenomena: efficiency gains and response pathway projects. Increasing the efficiency of combustion (the amount of fuel required per unit of energy service) simply reduces the amount of matter that the combustion system handles, thereby limiting the amount of potentially harmful pollutants. Adjusting the way reactants are converted into products can reduce the amount of substances that may be harmful to the environment in the chaotic combustion reaction zone. The efficiency of combustion usually depends on the way the reactants are converted into products. The process is complex, and the knowledge involved is growing rapidly. New sensing technologies have made it possible to monitor the progress of responses to unprecedented levels, and increased computational resources have made it possible to predict emissions from new sources even before they are tested.

It is as simple as striking a match, but extremely complex in its application. It is the main force that exists for a long time after industry but realizing its future potential will exceed the limit of science and technology. Combustion systems are used to generate steam and heat for important manufacturing processes, to heat various materials such as metals and chemical materials, and to change the mechanical and chemical properties of materials and products. The new technology can be used to develop more efficient, clean, fuel-flexible, and reliable combustion systems that can produce uniform, high-quality end products that meet environmental requirements with high productivity.

Given the huge potential for energy savings, almost every manufacturing sector is linked to improvements in combustion technology. Boilers, furnaces, and other heating equipment together account for two-thirds of the energy used in U.S. manufacturing. In several key sectors, the benefit ratio is even higher.

Environmental quality: the long-term goal of efficient boilers is to reduce nitrogen oxide (NO_x) emissions to 2 ppm (volume parts per million), 98% below the general uncontrolled level. Other goals include reducing emissions of carbon monoxide, particulates, and sulfur dioxide. With regard to fuel flexibility, future combustion systems will maximize the use of a wide range of fuels, including waste and renewable fuels. Related to cost-effectiveness and system reliability, lower lifecycle costs will be achieved by saving operating costs and fuel usage and improving system reliability: for example, one goal is to double the time between planned bailer outages.

9.1.2 Development of Combustion Technology

Partnerships between government and industry are essential to achieving this kind of innovation. The challenges of improving the combustion process are extremely complex, and the combustion equipment industry—a highly segmented, low-margin industry—lacks the resources to address these issues effectively on its own. By making selective investments in promising research and development projects, governments and businesses are laying the foundation for increased manufacturing competitiveness, improved environmental quality, and energy security.

Advances in computational science are revolutionizing our world, enabling us to study increasingly complex scientific and engineering problems, from predicting weather patterns to mapping the human genome. Computational modeling and simulation of the combustion process is one of the most complex physical and chemical processes. In fact, the combustion mathematical model was one of the first major challenges designated by the department of energy in USA under the high-performance computing act of 1991, and its solution was identified as one of nine key issues to meet the needs.

Currently, new combustion systems must be designed and "tuned" using a tried-and-tested approach, then customized for integration with other process steps and equipment for specific applications. Advanced modeling and system engineering capabilities will bring more predictability and repeatability, including more science—the art of combustion design, producing systems that can deliver thermal loads with unprecedented precision, reliability, efficiency, and cleanliness. The primary goal of future combustion systems is to achieve a low-cost solution to control the emission of priority pollutants while maintaining or improving process efficiency. In developing the vision and technology roadmap, industry participants expect NO_x emissions to be below 10 ppm in the short term and 2 ppm in the long term. The scientific basis for single-digit emissions is in place, but technical advances are needed to achieve acceptable levels of cost, operational performance, and reliability in practice. Ultra-low NOx burners—using combustion grading, premix combustion, and advanced control capabilities to reduce peak flame temperatures—are a promising low-cost way to reduce emissions.

9.1.3 Objective of This Chapter

In the following sections, the basic principles of combustion including the fundamental concepts, thermodynamics, reaction kinetics, and the combustion process will be discussed. The combustion applications and practices, including combustion modification and catalytic oxidation, will also be introduced. The developed technologies should be based on sound principles thereby solving the issues mentioned in the introduction section, mainly including (1) minimizing secondary pollutants; (2) achieving higher performance efficiency, and (3) maximizing heat and energy recovery. Numerous technologies have been developed for advanced combustion by cutting-edge technologies. Processing engineering does not necessarily choose the most popular mainstream technology but chooses the most appropriate technology according to the evaluation of engineering, environmental, energy, and economic (4E) aspects. The challenges and perspectives of advanced combustion are affordable. The clean combustion technology should improve the following operational parameters, including cost-effectiveness and system reliability, fuel flexibility, environmental quality, and energy efficiency.

9.2 Principles

9.2.1 Fundamentals

The combustion process is an exothermic reaction between the fuel and the oxidant, usually accompanied by the phenomenon of flame, luminescence, or smoke. The process requires a sufficient amount and concentration of combustible and combustion-supporting substances, a certain temperature and enough energy from ignition source, and a drastic chemical reaction between fuel and oxygen to convert the chemical energy of fuel into heat energy. The basic process of combustion is a chemical reaction, which is divided into a single-phase combustion chemical reaction and a heterogeneous combustion chemical reaction.

Conditions for complete combustion are affected by four factors, including

(1) Air conditions. Sufficient air is required for combustion. However, if the air volume is too large, the furnace temperature decreases along with increased heat loss.

(2) Temperature. Ignition temperature of the fuel should be reached for combustion.
(3) Time. The residence time of the fuel in the high-temperature area should exceed the time required for fuel combustion.
(4) Turbulence. Fuel and oxygen should be fully mixed for combustion.

One of the concepts that has to be mentioned in air conditions is the theoretical air volume, which refers to the minimum amount of air per unit of fuel required for complete combustion, also known as theoretical air requirement.

There are several basic theoretical assumptions that need to be made before a combustion chemical equation can be established.

(1) Air is only composed of nitrogen and oxygen, and its volume ratio is $V_{N2}/V_{O2} = 79.1/20.9 = 3.78$ (molar ratio, volume ratio);
(2) Fixed oxygen in fuel can be used for combustion;
(3) Sulfur in fuel is mainly oxidized to SO_2;
(4) Low nitrogen content in the fuel, a small amount of hot NO_x production, can be ignored in the calculation of theoretical air quantity;
(5) Nitrogen in the fuel is mainly converted into nitrogen N_2;
(6) The chemical formula of fuel is set as $C_xH_yS_zO_w$, where subscripts *x*, *y*, *z*, and *w* represent the atomic numbers of carbon, hydrogen, sulfur, and oxygen, respectively.

Based on the above assumptions, a chemical equation can be obtained for the combustion of a certain kind of fuel.

Combustion equation can be expressed as

$$\begin{aligned} C_xH_yS_zO_w + \left(x + \frac{y}{4} + z - \frac{w}{2}\right)O_2 + 3.76\left(x + \frac{y}{4} + z - \frac{w}{2}\right)N_2 \\ \rightarrow xCO_2 + \frac{y}{2}H_2O + zS_2O + 3.76\left(x + \frac{y}{4} + z - \frac{w}{2}\right)N_2 + Q \end{aligned} \tag{9.1}$$

The theoretical air volume for combustion can be calculated as

$$V_a^0 = \frac{22.4 \times 4.78 \times (x + y/4 + z - w/2)}{12x + 1.008y + 32z + 16w} \mathrm{m^3/kg} \tag{9.2}$$

For solid fuels,

$$V_{0,O_2} = \left(\frac{C}{12} + \frac{H}{4} + \frac{S}{32} - \frac{O}{32}\right) \times 22.4 \times 10^{-2} \tag{9.3}$$

$$V_{0,\mathrm{air}} = \frac{V_{0,O_2}}{0.21} \tag{9.4}$$

For gasoline,

$$V_{0,O_2} = \left[\frac{1}{2}CO + \frac{1}{2}H_2 + \frac{3}{2}H_2S + \sum\left(n + \frac{m}{4}\right)C_nH_m - O_2\right] \times 10^{-2} \quad (9.5)$$

$$V_{0,\mathrm{air}} = \frac{V_{0,O_2}}{0.21} \quad (9.6)$$

Theoretical smoke amount refers to the smoke amount generated by the complete combustion of unit fuel and theoretical air, which is specific to the above combustion equation,

$$V_{0,co_2} = \frac{22.4}{12} \times \frac{C}{100} \quad (9.7)$$

$$V_{0,SO_2} = \frac{22.4}{32} \times \frac{S}{100} \quad (9.8)$$

$$V_{0,N_2} = \frac{22.4}{28} \times \frac{N}{100} + 0.79V_{0,\mathrm{air}} \quad (9.9)$$

$$V_{0,N_2} = \frac{22.4}{18} \times \frac{W}{100} + \frac{22.4}{2} \times \frac{H}{100} \quad (9.10)$$

For gasoline,

$$V_{0,CO_2} = \left(CO + CO_2 + \sum nC_nH_m\right) \times 10^{-2} \quad (9.11)$$

$$V_{0,SO_2} = H_2S \times 10^{-2} \quad (9.12)$$

$$V_{0,H_2O} = \left(H_2 + H_2O + H_2S + \sum \frac{m}{2}C_nH_m\right) \times 10^{-2} \quad (9.13)$$

$$V_{0,N_2} = N_2 \times 10^{-2} + 0.79V_{0,\mathrm{air}} \quad (9.14)$$

Excess air coefficient refers to the superimposed coefficient when the actual air demand is usually greater than the theoretical air demand (Table 9.1).

$$V_{\alpha,\mathrm{air}} = \alpha V_{0,\mathrm{air}} \quad (9.15)$$

Table 9.1 Excess air ratio for various types of furnaces

Combustion type	Anthracite coal	Bituminous coal	The gas	Heavy oil
Hand stove and coal thrower stove	1.3–1.5	1.3–2.0	–	–
The chain furnace	1.3–1.4	1.3–1.5	–	–
The stove	1.2	1.25	1.15–1.2	1.05–1.1

where

a—excess air coefficient;
$a = 1$, the ratio of fuel to air is known as the chemical equivalent;
$a < 1$, the actual amount of air supplied is less than the theoretical amount. Incomplete combustion;
$a > 1$, the actual amount of air is greater than the theoretical amount to ensure complete combustion;
Gaseous fuel, $a = 1.02$–1.2;
Liquid fuel, $a = 1.1$–1.3;
Solid fuel, $a = 1.3$–1.7.

1. Three T's Principles

Although similar to oxidation, the reaction is much faster and requires a certain minimum temperature to initiate the reaction. This rapid chemical reaction is controllable, so under normal circumstances it would not explode. In order to burn effectively, the reaction element must be mixed with oxygen, kept at an appropriate temperature, and given sufficient time to react. Thus, the three T's that determine the success of a combustion process are combustion, turbulence, temperature, and time. The combustion process releases heat, so in general, once the ignition is complete, the temperature is kept at a high value. As the combustion proceeds, the concentration of oxygen and reactive elements decreases, and continuous turbulence is required to bring the unreacted particles into contact with oxygen. Therefore, sufficient time is required from initial ignition to full combustion.

2. Time

This reduces the amount of excess combustion air required to ensure that each drop of fuel is completely surrounded by oxygen and burns completely. As the amount of combustion air decreases, combustion efficiency increases. The O_2 reading in the manufacturer's specification indicates that the flame retention performance of the burner meets the design requirements. For gas systems, air and fuel are mixed in the burner. Also, increasing the time it takes for the fuel and air to fully mix helps ensure complete combustion. A common misconception is that the shutter on an atmospheric burner controls the combustion inlet. Although some degree of combustion air does enter through the blinds, ventilation remains unchanged. When the shutters are closed, more combustion air is sucked in through the "secondary" air intake, so little combustion air is provided. The adjustable intake louver on the atmospheric burner mainly controls the air velocity when the fuel/air mixture enters the throat or "mixing zone" of the burner. This can be achieved by adjusting the blinds while conducting a combustion test. The difference in oxygen/carbon dioxide readings is small unless the shutter is fully open or closed. With the increase of reaction time, the flue gas in the heat exchanger also increases the heat transfer time. In the manufacturer's specification, O_2 and stack temperature readings indicate that flue gas is passing through the heat exchanger at a rate that allows maximum heat transfer time, while still allowing the introduction of additional combustion air. Therefore, time plays a very important role in combustion efficiency. On the other hand, in a mathematical sense, 3Ts also refers to the characteristic time of the following three modeling problems as follows:

$$\tau_c = \frac{1}{k'}(\text{chemical time}) \tag{9.16}$$

$$\tau_r = \frac{V}{Q} = \frac{L}{u}(\text{residence time}) \tag{9.17}$$

$$\tau_m = \frac{L^2}{D_e}(\text{mixing time}) \tag{9.18}$$

where

V: Volume of the reaction zone, m^3;
Q: Flow rate, m^3/s;
L: Length of the reaction zone, m;
u: gas velocity, m/s;
D_e: effective (turbulent) diffusion coefficient, m^2/s.

3. Temperature

By reducing the amount of combustion air entering the combustion process to the absolutely necessary minimum, we increase the ΔT between the flame/flue gas and the distributed air or boiler water. During heat transfer, radiant energy is more effective than convection/conductive heat transfer. It is mainly caused by the flame itself and is directly related to the flame temperature. As excess air is reduced, higher flame temperatures produce more radiant heat. The rate of combustion is very sensitive to temperature. If the flame temperature is increased by 10%, the rate of combustion will increase by more than twice. Unfortunately, when there is enough oxygen, an increase in flame temperature will also increase the production of NOx gas by more than 10 times. According to the Arrhenius equation, temperature has a great influence on the reaction rate of any system, as shown in the equation below:

$$k = Ae^{\frac{-E_a}{RT}} = \frac{Z'Sy_{O_2}P}{R'}e^{\frac{-E_a}{RT}} \tag{9.19}$$

where

V: Volume of the reaction zone, m^3;
k: reaction rate of the reaction.

4. Turbulence

The agitation of the flue gas in the heat exchanger is to provide continuous circulation in the hotter flue gas in contact with the surface of the heat exchanger. Typically, the surface of a heat exchanger has a variety of irregular surfaces, including bumps, ridges, etc., to provide this effect. Boilers and domestic hot water heaters usually have turbulence to provide this mixing process. These surfaces also generate eddies, which reflux the flue gas, increasing the time the flue gas stays in the heat exchanger. These turbulent and irregular heat exchanger surfaces are designed to be most effective in the appliance's full combustion rate. In the process of combustion, the burner passes through the boiler with the least resistance to produce smaller flue gas, so that the flue gas is washed and mixed on the fire side of the heat exchanger.

9.2.2 Thermodynamics

The intensity of combustion between fuel and oxidant depends on their relative concentration. When their concentration ratio is chemically correct and all the reactants can be completely consumed in the reaction, the combustion intensity is close to the maximum, which can be called stoichiometric combustion [1]. The complete oxidation of simple hydrocarbon fuels produces carbon dioxide (CO_2) and water (H_2O) from all the carbon and hydrogen, hydrogen, that is, for hydrocarbon fuels whose general composition formula is C_nH_m,

$$C_nH_m + \left(n + \frac{m}{4}\right)O_2 \rightarrow nCO_2 + \frac{m}{2}H_2O \tag{9.20}$$

Even in the ideal case of full combustion, the calculation of all kinds of combustion emissions is not simply the measurement of carbon dioxide and water. Since the fuel is burned in air rather than pure oxygen, nitrogen in the air may participate in the combustion process to produce nitrogen oxides. In addition, many fuels contain elements other than carbon and hydrogen that may be converted during combustion. Finally, combustion is not always complete, and the expelled gas contains both unburned and partially burned products in addition to CO_2 and H_2O. Air consists of oxygen, nitrogen, a small amount of carbon dioxide, argon, and other trace substances. Since most of the diluents in the air are nitrogen, it is reasonable to think of air as a mixture of 20.9% (molar) O_2 and 79.1% (molar) N_2. Therefore, for every mole of oxygen required for combustion, 3.78 mol of nitrogen must also be introduced. Although nitrogen may not significantly change the oxygen balance, it does have a significant impact on the thermodynamics, chemical dynamics, and pollutant formation in combustion systems. Therefore, it is useful to carry "inert" materials in combustion calculations. The stoichiometric relationship for the complete oxidation of hydrocarbon fuel (C_nH_m) becomes

$$C_nH_m + \left(n + \frac{m}{4}\right)(O_2 + 3.78N_2) \rightarrow nCO_2 + \frac{m}{2}H_2O + 3.78\left(n + \frac{m}{4}\right)N_2 \quad (9.21)$$

Thus, for every mole of fuel burned, $4.78(n + m/4)$ moles of air are required and $4.78(n + m/4) + m/4$ moles of combustion products are generated. The molar fuel/air ratio for the stoichiometric combustion is $1{:}4.78(n + m/4)$. Gas compositions are generally reported in terms of mole fractions since the mole fraction does not vary with temperature or pressure as does the concentration (moles/unit volume). The product mole fractions for complete combustion of this hydrocarbon fuel are

$$y_{CO_2} = \frac{n}{4.78(n + m/4) + m/4} \quad (9.22)$$

$$y_{H_2O} = \frac{m/2}{4.78(n + m/4) + m/4} \quad (9.23)$$

$$y_{N_2} = \frac{3.78(n + m/4)}{4.78(n + m/4) + m/4} \quad (9.24)$$

The large quantity of nitrogen diluent substantially reduces the mole fractions of the combustion products from the values they would have in its absence.

Unlike other system properties that make no reference to any process, specific heats are system properties that are defined for the infinitesimally small processes. Thus, through heat interaction dq (kJ/kg), if the system temperature changes by dT, then the specific heat is defined as

$$c \equiv \frac{dq}{dT} \frac{\text{kJ}}{\text{kg·K}} \quad (9.25)$$

where

c: The specific heat, kJ/(kg·K);
q: Heat interaction, kJ/kg;

To facilitate engineering calculations, however, the values of c are documented for only two processes: a constant-volume process and a constant-pressure process.

For a constant-volume process, $dV = 0$. Therefore, we have

$$c_v \equiv \left.\frac{du}{dT}\right|_v \quad (9.26)$$

where

c_v: The constant-volume specific heat, kJ/(kg·K).

In thermodynamics, to facilitate a compact definition of specific heat at constant pressure Cp, a new property, called enthalpy, is defined and denoted by h (kJ/kg).

Thus,

$$h \equiv u + pv \tag{9.27}$$

where

h: The enthalpy, kJ/kg.

From the above definition, for an infinitesimally small process, $dh = du + pdv + vdp$. For a constant-pressure process, however, $dp = 0$.

$$c_P \equiv \left.\frac{dh}{dT}\right|_p \tag{9.28}$$

A material balance is a quantitative expression of the law of conservation of matter: What goes in comes out (unless it stays behind).

$$\text{input} = \text{output} + \text{accumulation} \tag{9.29}$$

This expression is always true for elements flowing through combustion systems (except for minor deviations in the unusual case where radioactive materials are involved). It is often not true for compounds participating in combustion reactions. Because of its fundamental character and intrinsic credibility, the material balance is one of the most useful tools for combustion system analysis.

All chemical reactions are accompanied by the absorption or evolution of energy, usually in the form of heat [2]. From the most basic principle, you can determine the amount of heat, and thus the temperature and composition of the product. Spectral data and statistical calculations enable us to determine the internal energy of a substance. The internal energy of a given substance depends on its temperature, pressure, and state, not on the method by which it is attained. Also, the change of internal energy, ΔE, any physical change, or the result of a chemical reaction system depends only on the initial and final states of the system. The total change in internal energy is the same regardless of whether energy evolves as heat, energy, or work. If the change in kinetic and potential energies of the flow reaction is negligible and no work is done except for the work required for the flow, then the added heat is equal to the increase in enthalpy of the system

$$Q = \Delta H \tag{9.30}$$

where Q is the heat added and H is the enthalpy. For a uniform flow reaction proceeding at constant pressure, the heat added is also equal to the gain in enthalpy

$$Q = \Delta H \tag{9.31}$$

if heat released,

$$Q = -\Delta H \tag{9.32}$$

Most thermochemical calculations are made for closed thermodynamic systems, and stoichiometry is most conveniently represented in terms of the molar quantities as determined from statistical calculations. In dealing with compressible flow problems in which it is essential to work with open thermodynamic systems, it is best to employ mass quantities. Throughout this text, uppercase symbols will be used for molar quantities and lowercase symbols for mass quantities.

The condition for equilibrium is determined from the combined form of the first and second laws of thermodynamics; that is,

$$dE = TdS - PdV \tag{9.33}$$

where S is the entropy. This condition applies to any change affecting a system of constant mass in the absence of gravitational, electrical, and surface forces. However, the energy content of the system can be changed by introducing more mass. Assume that the contribution to the energy of the system on adding one molecule i is μ_i. The introduction of a small number dn_i of the same type contributes to a gain in energy of the system of $\mu_i dn_i$. All the possible reversible increases in the energy of the system due to each type of molecule i can be summed to give Eq. (9.34):

$$dE = TdS - PdV + \sum\nolimits_i \mu_i dn_i \tag{9.34}$$

It is apparent from the definition of enthalpy H and the introduction of the concept of the Gibbs free energy G

$$G \equiv H - TS \tag{9.35}$$

that

$$dH = TdS - VdP + \sum\nolimits_i \mu_i dn_i \tag{9.36}$$

and

$$dG = -SdT - VdP + \sum\nolimits_i \mu_i dn_i \tag{9.37}$$

Recall that P and T are intensive properties that are independent of the size of the mass of the system, whereas E, H, G, and S (as well as V and n) are extensive properties that increase in proportion to mass or size. By writing the general relation for the total derivative of G with respect to the variables, it becomes

$$dG = \left(\frac{\partial G}{\partial T}\right)_{P,n_i} dT + \left(\frac{\partial G}{\partial P}\right)_{T,n_i} dP + \sum\nolimits_i \left(\frac{\partial G}{\partial n_i}\right)_{P,T,n_j(j\neq i)} dn_i \tag{9.38}$$

9.2.3 Heat and Mass Transfer

Pulsation is the power source to realize momentum, heat, and mass transfer in a turbulent flow field. The continuous generation and dissipation of turbulent energy are the two most basic characteristic processes of fluid turbulent motion. The speed of the fluid particle w and pressure p are varied with time constant τ, the irregular change, sometimes micelle fluid around the instantaneous axis without rules, and often influenced by disturbance vortex motion (vortex)—pulse (Table 9.2).

The momentum transfer includes the normal stress of turbulence

$$\rho \overline{w'^2} \tag{9.39}$$

Shear stress

$$\rho \overline{w'_i w'_j} \tag{9.40}$$

Turbulence (turbulence intensity)

$$E = \frac{\sqrt{\left(\overline{w'_x}^2 + \overline{w'_y}^2 + \overline{w'_z}^2\right)/3}}{w^*} \tag{9.41}$$

The heat transfer as the heat flux

$$c\rho \overline{T' w'_y} \tag{9.42}$$

Mass transfer as the mass transfer flux

$$\rho \overline{m' w'_y} \tag{9.43}$$

The turbulent kinetic energy transferred can be obtained

Table 9.2 A basic mathematical description of momentum, heat, and mass transfer

Transfer flux G		The characteristic parameters K of "3Ts", m^2/s	The transmission power Y of "3Ts"	Power supply
Characteristic	Unit			
Momentum transfer	N/m^2	Viscosity ν	$\rho \frac{dw}{dy}$	Speed difference Δw
Heat transfer	W/m^2 or $J/(m^2 \cdot s)$	Thermal diffusivity a	$C_P \rho \frac{dT}{dy}$	Temperature difference ΔT
Mass transfer	$kg/(m^2 \cdot s)$ or $mol/(m^2 \cdot s)$	Diffusion coefficient D	$\frac{dC}{dy}$	Concentration difference ΔC

$$\frac{1}{2}\rho\left(\overline{w_x^2 + w_y^2 + w_z^2}\right) \tag{9.44}$$

Turbulent dissipation

$$\nu\left\{\begin{matrix} 2\left[\left(\overline{\frac{\partial w_x'}{\partial x}}\right)^2 + \left(\overline{\frac{\partial w_y'}{\partial y}}\right)^2 + \left(\overline{\frac{\partial w_z'}{\partial z}}\right)^2\right] + \\ \left[\left(\overline{\frac{\partial w_x'}{\partial y}} + \overline{\frac{\partial w_y'}{\partial x}}\right)^2 + \left(\overline{\frac{\partial w_x'}{\partial z}} + \overline{\frac{\partial w_z'}{\partial x}}\right)^2 + \left(\overline{\frac{\partial w_y'}{\partial z}} + \overline{\frac{\partial w_z'}{\partial y}}\right)^2\right] \end{matrix}\right\} \tag{9.45}$$

According to the Navier–Stokes equation and homogenization of viscous incompressible fluid, the following equations were derived (Table 9.3):

$$\rho\overline{g}_x - \frac{\partial\overline{P}}{\partial x} + \frac{\partial}{\partial x}\left(\mu\frac{\partial\overline{w}_x}{\partial x} - \rho\overline{w'^2{}_x}\right) + \frac{\partial}{\partial y}\left(\mu\frac{\partial\overline{w}_x}{\partial y} - \rho\overline{w_y'w_x'}\right) + \frac{\partial}{\partial z}\left(\mu\frac{\partial\overline{w}_y}{\partial z} - \rho\overline{w_z'w_x'}\right)$$

Table 9.3 Analogy of heat and mass transfer

Item		Heat transfer Q		Mass transfer $\dot{m}$	
		Name	Equation and unit	Name	Equation and unit
Power supply		Temperature difference ΔT	ΔT	Concentration difference ΔC	ΔC
Basic rules	I	Fourier law of heat conduction	$Q = -\lambda\frac{dT}{dx}\cdot f$	Fick's law	$\dot{m} = -D\frac{dC}{dr}\cdot f$
	II	Newtonian convection heat transfer	$Q = \alpha\cdot\Delta T\cdot f$	Convective mass transfer	$\dot{m} = \alpha_{zl}\cdot\Delta C\cdot f$
Criterion number symbol		Nu	$Nu = \frac{aL}{\lambda} = f(\text{Re, Pr})$	Nu_{zl}	$Nu_{zl} = Sh = \frac{\alpha_{zl}\cdot L}{D} = f(\text{Re}, Sc)$
		Heat	$Q = qf$,W	Mass	$\dot{m}$, kg/s or mol/s
		Heat transfer	$q = \frac{Q}{f}$,W/m^2	Mass transfer	$\frac{\dot{m}}{f}$, kg/(m^2s) or mol/(m^2s)
		Heat conductivity	λ, W/(mK)	Mass diffusivity	D, m^2/s
		Temperature gradient	$\frac{dT}{dx}$, K/m	Concentration gradient	$\frac{dC}{dx,\text{kg/m}^4\text{ or mol/m}^4}$
		Heat emission coefficient	α, W/(m^2K)	Mass exchange coefficient	α_{zl}, m/s
		Temperature difference ΔT	ΔT, K	Concentration difference ΔC	Δc, kg/m^3 or mol/m^3
		Heat transfer area	f, m^2	Mass transfer area	f, m^2

$$= \rho\left[\frac{\partial}{\partial x}(\overline{w}_x\overline{w}_x) + \frac{\partial}{\partial y}(\overline{w}_y\overline{w}_x) + \frac{\partial}{\partial z}(\overline{w}_z\overline{w}_x)\right] \tag{9.46}$$

$$\rho\overline{g}_y - \frac{\partial\overline{P}}{\partial y} + \frac{\partial}{\partial x}\left(\mu\frac{\partial\overline{w}_y}{\partial x} - \rho\overline{w'_x w'_y}\right) + \frac{\partial}{\partial y}\left(\mu\frac{\partial\overline{w}_y}{\partial y} - \rho\overline{w'^2_y}\right) + \frac{\partial}{\partial z}\left(\mu\frac{\partial\overline{w}_y}{\partial z} - \rho\overline{w'_z w'_y}\right)$$

$$= \rho\left[\frac{\partial}{\partial x}(\overline{w}_x\overline{w}_y) + \frac{\partial}{\partial y}(\overline{w}_y\overline{w}_y) + \frac{\partial}{\partial z}(\overline{w}_z\overline{w}_y)\right] \tag{9.47}$$

$$\rho\overline{g}_z - \frac{\partial\overline{p}}{\partial z} + \frac{\partial}{\partial x}\left(\mu\frac{\partial\overline{w}_z}{\partial x} - \rho\overline{w'_x w'_z}\right) + \frac{\partial}{\partial y}\left(\mu\frac{\partial\overline{w}_z}{\partial y} - \rho\overline{w'_y w'_z}\right) + \frac{\partial}{\partial z}\left(\mu\frac{\partial\overline{w}_z}{\partial z} - \rho\overline{w_z'^2}\right)$$

$$= \rho\left[\frac{\partial}{\partial x}(\overline{w}_x\overline{w}_z) + \frac{\partial}{\partial y}(\overline{w}_y\overline{w}_z) + \frac{\partial}{\partial z}(\overline{w}_z\overline{w}_z)\right] \tag{9.48}$$

9.2.4 Fuel Property

The main components of fuel are carbon (C), hydrogen (H), oxygen (O), nitrogen (N), sulfur (S), ash (A), and water (M). Among them, carbon, hydrogen, and sulfur are combustible components, while water and ash are incombustible components.

Substances are made up of molecules. However, because of the different states of molecules, the existence state of substances is also different, which will also lead to different characteristics of solid, liquid and gas fuel. The following Table 9.4 compares the molecular structures of solid, liquid, and gas.

The volatile components of solid fuels are distilled and burned in the gaseous state, while the fixed components are burned in the solid state. The combustion rate is mainly controlled by the diffusion of oxygen molecules to the solid surface. Liquid fuel is burned in gaseous form after evaporation, and the combustion rate is mainly controlled by the evaporation process. Gaseous fuels can burn rapidly and their combustion state is basically controlled by the diffusion or mixing of air and fuel.

Table 9.4 The comparison of the molecular structures of solid, liquid and gas fuel

	Solid	Liquid	Gas
Molecular arrangement	Close	No fixed position	Extremely messy
Intermolecular distance	Very small	Smaller	A large
Intermolecular forces	Strong	Larger	Tiny
Molecular motion	Vibration between the balance point	Vibration within a certain range	Freedom of vibration
A certain volume	Yes	Yes	No
A certain shape	Yes	No	No
Liquidity	No	Yes	Yes

1. Solid fuel

Natural wood, peat, lignite, bituminous coal, anthracite, and oil shale are typical solid fuels. Solid fuel combustion is difficult to understand and analyze, less efficient, and generates more ash. Solid fuel can be directly used as a fuel source, and it also can be used to convert to liquid fuel and gas fuel. For example, there are three main types of reactions in the coal combustion process, including pyrolysis reaction, chemical bridge bond fracture, and formation of free radicals. In the secondary reaction process, some cracking reactions, dehydrogenation reactions, hydrogenation reactions, condensation reactions, and bridge bond decomposition occur. The proximate analysis of coal is given in terms of percentage moisture, volatile matter, fixed carbon (or char), and ash [3]. The volatiles burn rapidly compared to the solid carbonaceous char that remains after devolatilization [4]. Coal may be classified according to rank and grade. The rank of coal describes the progressive metamorphism of coal from lignite (low rank) to anthracite (high rank). Coal rank is based on heating value (HV). This value and percentage of fixed carbon increases as the rank increases and the volatile matter decreases from lignite to low volatile bituminous coal [5] (Fig. 9.1).

The thermal equilibrium equation is

$$m_p c_p \frac{dT_p}{dt} = Q_1 - aS_p(T_p - T_\infty) - \varepsilon_0 \delta_p S_p(T_p^4 - T_\infty^4) + Q_3 \qquad (9.49)$$

where

$m_p c_p \frac{dT_p}{dt}$, for the internal energy increases;
Q_1, for the heterogeneous heat of reaction;
$aS_p(T_p - T_\infty)$, for the convection heat;
$\varepsilon_0 \delta_p S_p(T_p^4 - T_\infty^4)$, for the radiation heat;
Q_3, for the external heating.

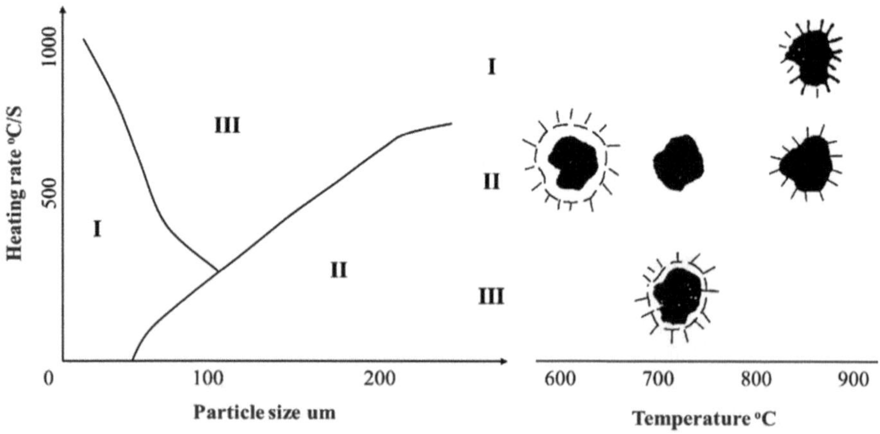

Fig. 9.1 Pyrolysis curves of typical coals and various coal particles firing at different temperatures

2. Liquid fuel

Liquid fuel is a liquid combustible substance used to generate heat and other forms of energy such as electricity. Liquid fuel contains mostly hydrocarbons or their mixtures. Examples of liquid fuel include natural petroleum and some processed products such as gasoline, kerosene, diesel, and fuel oil.

Liquid fuel atomization is the first step of the liquid fuel combustion process. Atomization is the process of breaking liquid fuel into small droplets. Liquid fuel atomization can increase the specific surface area of fuel, accelerate the vaporization of fuel, and facilitate the mixing of fuel and air, so as to ensure the rapid and complete combustion of fuel [3]. Meanwhile, the quality of atomization plays a decisive role in the combustion process of liquid fuel. Liquid atomization includes the following steps:

(1) Liquid flows out of the nozzle to form a liquid column or liquid film.
(2) Due to the initial turbulence of the liquid jet and the effect of surrounding gases on the jet (pulsation, friction, etc.), the liquid surface fluctuates and folds, and is ultimately separated into liquid fragments or filaments.
(3) Under the action of surface tension, liquid fragments or filaments shrink into spherical droplets.
(4) Under the action of aerodynamics, the large droplets will break up further.

Figure 9.2 illustrates the composition distribution and energy balance of a droplet in the evaporation–combustion process at static high temperature. Among them,

w_{xg} is the air mass fraction in air,
w_{lg} is the mass fraction of fuel vapor in air,
w_{xgs} is the air mass fraction of the surface of the droplet,
w_{lgs} is the fuel vapor mass fraction of the droplet surface.

Two important dimensionless terms are often used to describe the crushing process. Weber number: the ratio between the inertial force and the surface tension force.

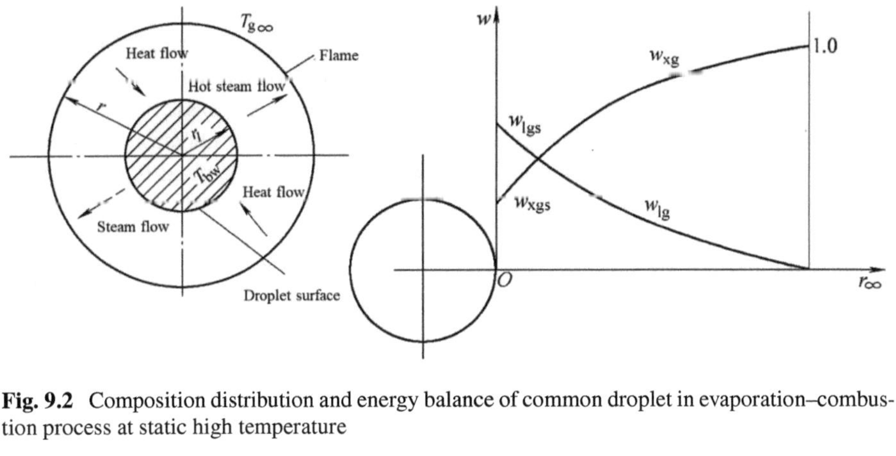

Fig. 9.2 Composition distribution and energy balance of common droplet in evaporation–combustion process at static high temperature

$$We_g = \frac{\rho_g d_l (v_l - v_g)^2}{\sigma} \tag{9.50}$$

Ohnesorge number: ratio of viscous force to surface tension:

$$Oh_d = \frac{\mu_l}{\sqrt{\rho_l d_l \sigma}} \tag{9.51}$$

As the radius increases, the content (mass fraction) decreases to infinity, $w_{lg\infty} = 0$. For air, the change in mass fraction is just the opposite, at infinity, $w_{xg\infty} = 1.0$, and it gradually reduced to the w_{xgs} value on the droplet surface. Obviously, at any radius, $w_{xg} + w_{lg} = 1.0$.

In fact, there is no macroscopic flow of air components. The real flow is caused by Stephan flow and the fuel vapor convection:

$$q_{ml,0} = v_{gs} \rho_{gs} 4\pi r_1^2 w_{lgs} \tag{9.52}$$

where

$q_{ml,0}$, outward flow of fuel vapor, kg/s;
v_{gs}, the velocity of the gas leaving the surface of the droplet, m/s;
ρ_{gs}, the density of the mixed gas on the surface of the droplet, kg/m^3;
r_1, droplet radius, m;
w_{lgs}, the fuel vapor mass fraction of the droplet surface.

When the droplet reaches a certain temperature, the heat generated by the droplet is exactly equal to the heat required for evaporation. This temperature is called the equilibrium temperature at which droplets evaporate.

3. Gaseous fuel

Gaseous fuel is a gaseous combustible substance that produces heat or power. Generally, it contains low molecular weight hydrocarbons, hydrogen, carbon dioxide, and other flammable gases, and often contains nitrogen and carbon dioxide, and other non-flammable gases. Combustion characteristics are affected by the characteristics, pressure, temperature, composition, inert gas content, additives, etc.

Gas fuel combustion is generally divided into three processes: (1) fuel and air mixing stage (formation of combustible mixture), (2) combustible mixture of heating and ignition stage, (3) completing the combustion chemical reaction stage. The normal flame propagation process is the local first ignition of the combustible mixture, and the ignition part transfers heat and active particles to the unburned part to make them catch fire successively. The process is concentrated in the flame surface layer, relying on the heat conduction, and the propagation speed is limited, rather than the whole combustible mixture at the same time (Fig. 9.3).

The thermal conductivity differential equation for a one-dimensional laminar combustion chamber system with chemical reactions can be derived as follows:

Continuity equation,

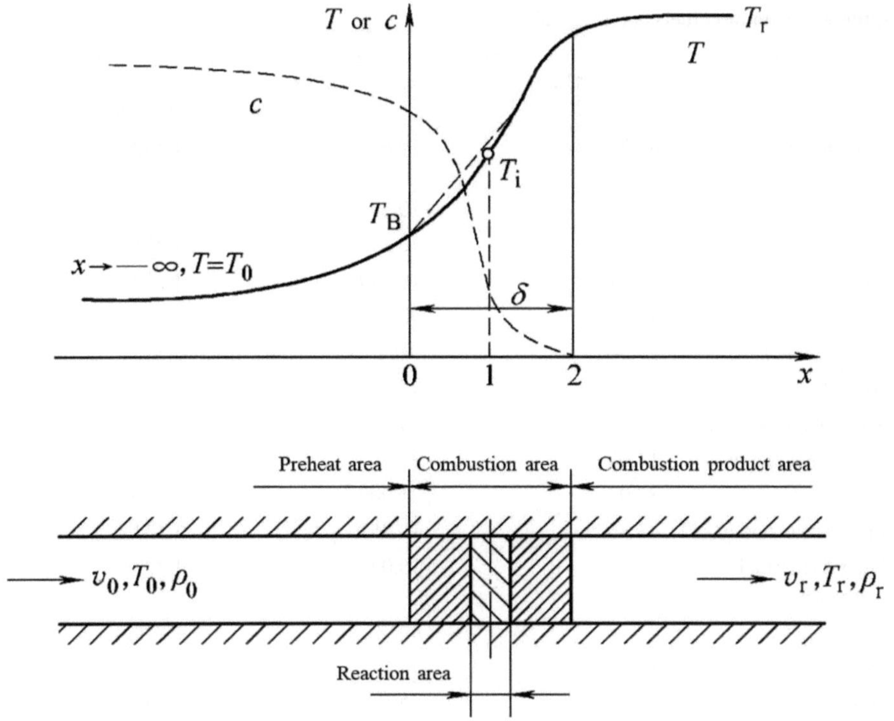

Fig. 9.3 Typical gas combustion models in one-dimensional laminar combustion chamber systems, where $-\infty < x \leq 0$ is the preheating zone of combustible mixture, $0 \leq x \leq \delta$ combustion area (delta width), $\delta \leq x < +\infty$ product area of combustion reaction

$$\rho_0 v_0 = \rho_x v_x \tag{9.53}$$

Thus,

$$\frac{d}{dx}\left(\lambda \frac{dT}{dx}\right) - c_p \rho_0 v_0 \frac{dT}{dx} + Qw = 0 \tag{9.54}$$

and no combustion reaction occurs in the preheating zone when $-\infty < x \leq 0$

$$\frac{d}{dx}\left(\lambda \frac{dT}{dx}\right) - c_p \rho_0 v_0 \frac{dT}{dx} = 0 \tag{9.55}$$

According to the boundary conditions,

$$x = -\infty, \frac{dT}{dx} = 0\, T = T_0 \tag{9.56}$$

Table 9.5 Common fuels and their calorific values

Liquid fuel name	Average low calorific value/(MJ/kg)	Average low calorific value/(kcal/kg)	Equivalent to the standard coal coefficient
Liquefied petroleum gas (liquid)	45.22	10,801	1.543
Gasoline	43.11	10,297	1.471
Kerosene	43.11	10,297	1.471
Diesel	42.71	10,199	1.457
Crude oil	41.03	9800	1.400
Heavy fuel oil	39.36	9401	1.343
Asphalt	37.69	9002	1.286
The water coke slurry	23.01	5495	0.785
Coal water slurry	14.65	3500	0.500
LPG	87.92	21,000	3.000
Field associated gas	45.46	10,857	1.551
Oil gas (thermal cracking)	42.17	10,073	1.439
Natural gas	36.22	8652	1.236
Oil gas (cracking)	18.85	4501	0.643
Biogas	18.85	4501	0.643
Coke oven gas	16.75	3997	0.571
Water gas	10.05	2401	0.343
Electricity	3.60	861	0.123

$$x = 0,\ T = T_B \tag{9.57}$$

Among them,

$$T = T_0 + (T_B + T_0)\exp\left(\frac{\rho_0 v_0 c_p}{\overline{\lambda}} x\right) \tag{9.58}$$

$\overline{\lambda}$ Is the average thermal conductivity of the gas medium at $-\infty < x \leq 0$ (Table 9.5).

9.2.5 Reaction Kinetics

Chemical reaction kinetics is an important method to study the chemical reaction process of combustion, which mainly includes the mechanism of the chemical reaction and the rate of the chemical reaction. The basic theories involve molecular collision theory and chain reaction theory. The main research object is a chemical reaction system with ideal mixing and uniform temperature.

Most combustion is a complex chemical reaction: it is not completed in one step but involves several intermediate products in successive intermediate reactions before the final product is formed. Each reaction that constitutes a complex reaction is called an elementary reaction. The total package reaction represents the chemical reaction mechanism of a specific process, which can be effectively used for the calculation of heat and mass balance of some combustion reactions.

Among them, the chain reaction theory is one of the two basic theories of a chemical reaction mechanism. Many chemical reactions are caused by the formation of extremely active components, leading to a series of continuous and competitive intermediate reactions with the formation of reactants into products. The intermediate reaction to produce a number of unstable free radicals or free atoms is known as the active center. At a very high rate, active centers react with the original reactant molecule, while themselves disappear and new ones are created, which keep the reaction going all the way to the final product. The activation center (ACTS) serves as an intermediate link. The reaction rate of branch chain reaction is very small at the beginning stage and increases sharply when the active center accumulates to a certain extent. The time from the start of the reaction to the time when the rate of the chemical reaction increases to a sensible level is called incubation time. After a period of incubation time, the reaction rate rises sharply until the concentration of the activation center reaches the maximum, forming the explosion phenomenon of branching reaction.

1. Molecular collision theory

Molecular collision theory is the theory of the rates and mechanisms by which one chemical changes into another. In order to provide basic background knowledge and make full use of the precious achievements of predecessors, it is necessary to introduce some languages of chemical dynamics. First, some basic terms: chemical reaction is the conversion of one substance to a different form of chemical substance; a combustion reaction is a chemical reaction in which a fuel combines with an oxidant (usually, oxygen in the air) to form a combustion product. Combustion reactions are described by chemical symbols and reaction equations, with the starting substances, reactants, and final products connected by arrows, which means "mutual reaction formation". Therefore, for the combustion of methane in air, the combustion reaction is

$$CH_4 + 2O_2 \rightarrow CO_2 + 2H_2O \quad (9.59)$$

The stoichiometric coefficients 2, 1, and 2 represent the number of moles of oxygen, carbon dioxide, and water that participate in the combustion of 1 mol of methane. At the molecular level, similar expressions are used to describe molecular events that cause observed changes. For example, hydrogen atoms attack oxygen molecules to form hydroxyl radicals and oxygen atoms as follows:

$$H + O_2 \rightarrow OH + O \tag{9.60}$$

Chemical transformations that are also real molecular events are called basic reactions or basic steps. Combustion reactions occur because a large number of different fundamental reactions are combined to convert fuel and oxidant into combustion products, as described in the chemical equation. The mechanism of combustion reaction can be described at the molecular level, which is a very difficult task, needing many years of hard work for researchers. The mechanism includes the choice of some elements, assembling into hydrogen, carbon monoxide, methane, methanol-air mixture and the reaction of different high hydrocarbon fuels. The whole reaction is carried out by highly reactive intermediates, in which one free radical produces two or more free radicals (such as $H + O_2 + OH + O$) [O], where the OH and H reactants are called chain branches. These reactions that result in the net disappearance of free radicals (such as $H + OH + M + H_2O + M$) are called chain termination reactions. The intermediates are chain propagation reactions (such as $OH + H + H_2O + H$); termination reactions are molecules in the third body M, which are necessary to remove the energy released by the binding.

2. Chain reaction theory

Chain reactions involve the formation of one free radical species and the subsequent generation of another free radical [6]. This theory involves knowing the rate of each fundamental reaction. Therefore, for the basic reaction $H + O_2 + OH + O$, the increase in the concentration of OH and O per unit time due to the reaction will be a measure of the reaction rate. Since the result of this reaction is a decrease in the concentration of H and O_2, the change in concentration caused by this reaction will be multiplied by (−1) to get the rate. Therefore, the encounter between OH and O may lead to the formation of H and O_2, which is called the reverse reaction; the net rare value R of a basic reaction is considered to be the difference between the forward and reverse rates:

$$R_n = \text{forward rate} - \text{reverse rate} \tag{9.61}$$

At any time, the net interest rate must be positive or negative because the termination reaction is less frequent than the bimolecular reaction, but plays an important role in chain termination and energy release, so the composition of the reaction system changes over time and the direction of the net reaction may change. To simulate the combustion reaction, as described in Chap. 2, two opposite directions will be considered separately. The higher the concentration of the reaction molecule, the faster the reaction will proceed. Since the rate in both directions is proportional to the concentration of each reactant molecule, the mass action rate law relates the reaction rate to the concentration of the reactant. Therefore, for $H + O_2 + OH + O$ basic reaction, the forward and reverse rates can be written as follows:

$$\text{forward rate} = k_f[\mathrm{H}][\mathrm{O}_2] \tag{9.62}$$

$$\text{reverse rate} = k_f[\mathrm{OH}][\mathrm{O}] \tag{9.63}$$

$$R_n = k_f[\mathrm{H}][\mathrm{O}_2] - k_f[\mathrm{OH}][\mathrm{O}] \tag{9.64}$$

$$or\ \mathrm{d}[\mathrm{OH}]/dt = d[\mathrm{H}][\mathrm{O}_2]/dt - d[\mathrm{OH}][\mathrm{O}]/dt \tag{9.65}$$

The proportionality constants k_t and k, are called the rate coefficients, rate parameters, or rate constants in the forward and reverse directions, and the brackets are used to denote concentrations. This rate is expressed as a function of temperature in the form of the Arrhenius expression:

$$k = A\exp(-E/RT) \tag{9.66}$$

where A is constant, and E is the activation energy. More details about the derivation of k will be given below.

3. Chemical equilibrium in combustion chemical reactions

When the state of the system does not change spontaneously with respect to time, it is said that it is in a state of equilibrium in the back zone of the flame, and many combustion products are in chemical equilibrium or may be in mobile equilibrium. The basic equations of chemical equilibrium will be discussed in the following section as well as the equilibrium program to calculate the adiabatic temperature and composition.

The chemical reactions shown below proceed in the forward and reverse directions concurrently:

$$aA + bB \leftrightarrow yY + zZ \tag{9.67}$$

When the forward and reverse rates are in balance, i.e., the concentrations of the constituents are constant with zero net rates of species formation or destruction, the reaction is said to be in a state of chemical equilibrium:

$$\text{i.e.}\ \ \mathrm{R}_f = \left[\frac{-1}{a}\frac{d[A]}{dt}\right]_f = R_r = \left[\frac{1}{a}\frac{d[A]}{dt}\right]_r \tag{9.68}$$

$$\text{or}\ \ k_f[A]_e^a[B]_e^b = k_r[Y]_e^y[Z]_e^z \tag{9.69}$$

where the suffix e denotes the equilibrium concentration. Thus, at equilibrium,

$$\frac{[Y]_e^y[Z]_e^z}{[A]_e^a[B]_e^b} = \left[\frac{k_f}{k_r}\right] = K_n \tag{9.70}$$

where K, the equilibrium constant, varies with temperature as do the rate constants. Note that since $(a + b)$ is not necessarily equal to $(y + z)$ in Eq. (9.67), the equilibrium constant is not necessarily dimensionless; thus, the suffix n is employed to indicate its definition in terms of molar densities.

First of all, we need to establish a reference to explain the amount of chemical heat released or absorbed during the transformation of different species. Therefore, for each substance, the heat of formation at constant volume (E) and the heat of formation at constant pressure (H) can be defined as the heat needed to generate a mole of a substance from the elements in their standard state. The reaction is carried out in a closed system. Whether at constant volume or at constant pressure, the chemical reaction representing this process is a positive reaction at the same initial temperature and the final temperature,

$$\sum_{j=1}^{L} v'_{i,j} M^{O}_{i,j} \rightarrow M_i \tag{9.71}$$

By definition, the heat of formation is zero for elements in their standard states. For a given substance, h is positive if heat is absorbed by the system, and negative if heat is released. As examples, we have

$$C(gr) + \frac{1}{2}O_2 \rightarrow CO,\ \overline{h}^o_i\left(T^o = 298.15\ \mathrm{K}\right) = -26.42\ \mathrm{kcal/mole\ of\ CO} \tag{9.72}$$

$$\frac{1}{2}\mathrm{H_2} + \frac{1}{2}\mathrm{I_2} \rightarrow \mathrm{HI},\ \overline{h}^o_i\left(T^o = 298.15\ \mathrm{K}\right) = 6.30\ \mathrm{kcal/mole\ of\ HI} \tag{9.73}$$

The negative and positive signs indicate that the formation reactions of CO and HI are exothermic and endothermic, respectively. We also note that the reaction $CO + 1/2O_2 \rightarrow CO_2$ evolves −67.63 kcal/mole of CO_2. This is not the heat of the formation of CO_2 because CO is not an element (Table 9.6).

In this section, we discuss theories specially developed for unimolecular reaction rates. A unimolecular reaction describes a chemical process in which a reactant undergoes an isomerization or decomposition process. The reaction

$$R \underset{\rightarrow}{k} P \tag{9.74}$$

indicates such a transformation in terms of R and P. Since the reactant of a unimolecular reaction must acquire sufficient energy before the reaction can take place, it requires intermolecular energy transfer through collision with another molecule. Consequently, a unimolecular reaction is really second order in nature, exhibiting a pseudo-first-order behavior only under certain conditions. To demonstrate this behavior, an arbitrary species M is added to the reaction equation, as in

Table 9.6 Standard enthalpy of formation, reaction entropy, and Gibbs free energy of some common substances

Material	$\Delta_f H_M{}^{\theta}$	$\Delta_f G_m{}^{\theta}$	$S_m{}^{\theta}$
C (graphite)	0.0	0.0	5.7
C (diamond)	1.9	2.9	2.34
C(g)	716.7	671.3	158.1
CO(g)	−110.5	−137.2	197.7
CO_2(g)	−393.5	−394.4	213.8
CH_4(g)	−74.6	−50.5	186.3
C_2H_2(g)	227.4	209.9	200.4
C_2H_4(g)	52.4	68.4	219.3
C_2H_6(g)	−84.0	−32.0	229.2
C_3H_8(g)	−103.8	−23.4	270.3
C_4H_6(g)	165.5	201.7	293.0
C_4H_8(g)	1.17	72.04	307.4
n-C_4H_{10}(g)	−124.73	−15.71	310.0
C_6H_6(g)	82.9	129.7	269.2
Cu(cr)	0.0	0.0	33.2
F_2(g)	0.0	0.0	202.8
Fe(cr)	0.0	0.0	27.3
H_2(g)	0.0	0.0	130.7
I_2(cr)	0.0	0.0	116.1
I_2(g)	62.4	19.3	260.7
H_2O(l)	−285.83	−237.13	69.91
H_2O(g)	−241.8	−228.6	188.8
H_2O_2(l)	−187.8	−120.4	109.6

$$R + M \rightleftharpoons P + M \tag{9.75}$$

It is useful for the incinerator designer to have a basic understanding of the principles involved in combustion reaction kinetics. At a minimum, one should understand the concepts of an ignition temperature, the situations that lead to "quenching" of combustion reactions, and the significance of oxygen depletion on burnout. Also, there is merit to understand specific kinetics relationships for common products of incomplete combustion (carbon monoxide and soot) and the oxidation and pyrolytic kinetics for a spectrum of important organic compounds.

9.2.6 Environmental Consideration

The fuel quickly completes the combustion reaction in the combustion chamber, releasing heat and possibly inhibiting the production of combustion-related pollutants. Therefore, various chemical reactions of pollutant formation are one of the research subjects of combustion science.

The mechanism of combustion is actually to study the mechanisms for the formations of CO, carbon black, sulfur oxide, nitrogen oxide, smoke, and other pollutants produced in the combustion process. Combustion related pollutants designated with environmental standards are mainly sulfur oxides, nitrogen oxides, CO, dust and hydrocarbons gases. CO_2, as a greenhouse gas produced by combustion, also faces great pressure to reduce emissions.

The coal combustion process will be used as an example to analyze and introduce the formation of pollutants in the combustion process.

1. Primary and Secondary Pollutants

Air pollutants can be simply divided into primary pollutants and secondary pollutants, as shown in Fig. 9.4. The first-level pollutant is the pollutant directly discharged into the atmosphere, and the second-level pollutant is the pollutant formed by the chemical or photochemical reaction after the first-level pollutant is discharged into the atmosphere and exposed to sunlight. Unburned hydrocarbons, NO_x, particulate matter and sulfur oxides are examples of major pollutants. Particles may be lead oxide produced by the oxidation of lead dioxide in automobiles, as well as the formation of various types of carbon. Nitrobenzoic acid and ozone are examples of secondary pollutants. Some pollutants are divided into two categories. Nitrogen dioxide, which is directly

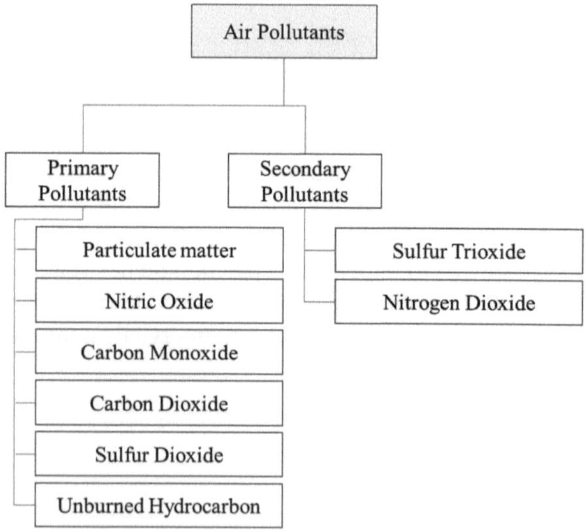

Fig. 9.4 Types of primary and secondary pollutants

emitted from automobile exhaust, is also formed in the atmosphere under the photochemical effect. Aldehydes released in automobile exhaust are also formed in the photochemical oxidation of hydrocarbons. Carbon dioxide mainly comes from automobiles and stationary sources, and is also the product of atmospheric hydrocarbon oxidation.

2. The Generation of NOx

The nitrogen oxides produced by combustion are mainly nitric oxide (NO) and nitrogen dioxide (NO_2), collectively known as NO_x. NO_x from fuel combustion is mostly NO, and only a small amount of NO is converted to NO_2 when a mixture of gas and air is burned at high temperatures with rapid cooling. However, a small amount of nitrous oxide (N_2O) is produced in the low-temperature combustion process (equivalent to the fluidized bed combustion condition) at 800 °C.

NO_x production and emissions are related to the combustion mode of the fuel, especially the combustion temperature and excess air coefficient. It is necessary to understand the NO_x formation mechanism in the combustion process, however, the NO_x formation mechanism in the combustion process is not very clear at the present time, and the research on NO_x formation mechanism and control technology is very active. NO_x is produced by combustion from two sources, nitrogen in the air and nitrogen in the fuel. The formation of NO_x can be divided into thermal NO_x, fuel NO_x, and fast NO_x (Fig. 9.5).

It has been well established that if a laboratory chamber containing NO, a trace of NO_2, and air is irradiated with ultraviolet light, the following reactions occur:

$$NO_2 + hv\left(3000\left(\frac{\pi}{2} - \theta\right)\text{Å} \le \lambda \le 4200\ \text{Å}\right) \rightarrow NO + O(^3P) \tag{9.76}$$

$$O + O_2 + M \rightarrow O_3 + M \tag{9.77}$$

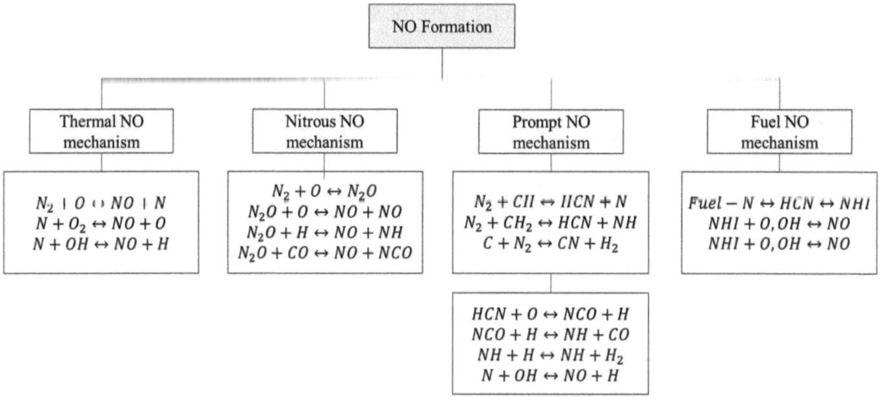

Fig. 9.5 Formation mechanism of NO

$$O_3 + NO \rightarrow O_2 + NO_2 \tag{9.78}$$

The net effect of irradiation on this inorganic system is to establish the dynamic equilibrium:

$$NO_2 + O_2 \overset{hv}{\leftrightarrow} NO + O_3 \tag{9.79}$$

Combustion of N in fuel (solid and liquid) and N in the input air produces NO_x. Generally, about 90% of NOx produced from combustion is NO, and the remaining nitrogen compound is mainly NO_2. The amount of NO_x produced during fuel combustion can be estimated by the following formula (Table 9.7):

$$G_{NO_x} = 1630 \times B \times (N \times \beta + 0.000938) \tag{9.80}$$

where

G_{NOx}, NOx emissions, kg;
B, the amount of coal (oil) consumed, T;
N, the amount of nitrogen in the fuel, %;
β, nitrogen conversion rate in fuel, %.

3. The Generation of SOx

All fuels contain small amounts of sulfur in the form of inorganic or organic sulfur. Gas fuel in general contains less sulfur than coal and oil. Gaseous fuels contain less sulfur, mainly H_2S, typically below 0.5%. H_2S in the gas phase is easier to remove from the fuel. Therefore, gaseous fuels are clean fuels, and the sulfur pollution caused by combustion is usually very small. A small part of the sulfur in liquid fuels is inorganic sulfur, and the majority is a complex compound combining sulfur with carbon, hydrogen, oxygen, and other elements (Fig. 9.6).

Ozone and SO_3 appear to be the only products of the photochemically induced reaction. The following essential steps were postulated:

Table 9.7 Nitrogen content and conversion rate of fuel

The name of the fuel coal	Nitrogen mass percentage (%)		The furnace	NOx conversion rate (%)
	Value	Average		
The inferior heavy oil	0.5–2.5	1.5	Layer combustion furnace	50
General heavy oil	0.2–0.4	0.2	Pulverized coal furnace	25
Inferior diesel oil	0.08–0.4	0.14	Oil burner	40
The name of the fuel	0.005–0.08	0.02		

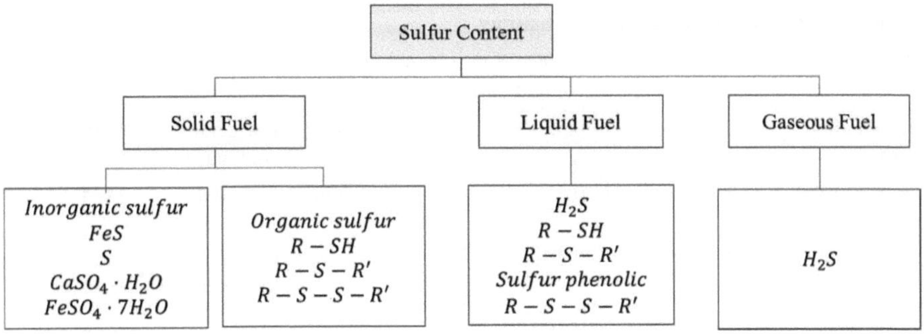

Fig. 9.6 The form of sulfide in fuel

$$SO_2 + hv \rightarrow SO_2^* \tag{9.81}$$

$$SO_2^* + O_2 \rightarrow SO_4 \tag{9.82}$$

$$SO_4 + O_2 \rightarrow SO_3 + O_3 \tag{9.83}$$

Nevertheless, this point might not be one of great concern since it is also possible to write the reactions as shown in Eqs. (9.84) and (9.85):

$$SO_2 + hv \rightarrow {}^1SO_2^* \tag{9.84}$$

$$^1SO_2^* + SO_2 \rightarrow {}^3SO_2^* + SO_2 \tag{9.85}$$

A secondary reaction route to produce SO_3 could be

$$SO_4 + SO_2 \rightarrow 2SO_3 \tag{9.86}$$

Sulfur in coal can be divided into combustible sulfur and noncombustible sulfur; combustible sulfur accounts for about 80% of the total sulfur. The formula to calculate the sulfur dioxide emissions from coal combustion is as follows (Table 9.8):

$$G_{SO_2} = 2 \times 80\% \times 1000 \times B \times S \times (1 - \text{Thedesulfurizationefficiency } \%) \tag{9.87}$$

where

G_{SO_2}, SO_2 emissions, kg;
B, coal consumption, t;
S, total sulfur content in coal burning, %.

Table 9.8 Total sulfur content of coal burning

Name of mine	The name of the coal	Total sulfur content (%)	Name of mine	The name of the coal	Total sulfur content (%)
Jixi mining area	1/3 coking coal	0.53	Shuangyashan mining area	Long flame coal, gas coal	0.30
Hegang mining area	1/3 coking coal	0.5	Qitaihe mining area	Gas coal	0.24
Shuangyashan mining area	Gas coal	0.25		Coking coal, lean coal	0.53
	Long flame coal, gas coke coal, lean coal	0.36		Lean coal	0.36

The calculation formula of sulfur dioxide emissions generated by fuel combustion is as follows:

$$\mathrm{G}_{\mathrm{SO_2}} = 2 \times 1000 \times B \times S \times (1 - \text{The desulfurization efficiency }\%) \quad (9.88)$$

where

$G_{\mathrm{SO_2}}$, emissions, kg;
B, fuel coal (oil) consumption, t;
S, total sulfur content in coal burning, %.

The amount of sulfur dioxide produced when hydrogen sulfide is burned in natural gas

$$G_{SO_2} = 2.857 \times V \times G_{H_2S} \quad (9.89)$$

where

$G_{\mathrm{SO_2}}$, SO_2 emissions, kg;
B, gasoline fuel consumption, L;
G_{H2S}, total sulfur content in coal burning, %.

4. The Generation of Soot

The soot produced by fuel combustion includes black smoke and fly ash. Black smoke is an incomplete combustion material, mainly free carbon (i.e., carbon black) and volatile substances, most of which are combustible substances. The particle size of black smoke is generally between 0.01 and 1 micron. Fly ash is a kind of noncombustible mineral ash particles in soot, whose particle size is generally more than 1micron, and their production amount is related to the fuel composition, equipment, and combustion condition.

(1) Gaseous carbon black

Gaseous carbon black is a kind of solid particle formed by thermal decomposition under the condition of insufficient air. Particle sizes are very small (0.02–0.05 micron) and increase significantly when they assemble into chains. Carbon black is produced in a flame, radiating force is increased, giving off light, form a luminous flame.

(2) Residual carbon black

Residual carbon black is the residue of solid particles, called putty or soot, from the burning of liquid fuel. The oil droplets are heated by the high temperature in the furnace and the flame around them, producing oil vapor. At the same time, the oil droplets undergo a condensation reaction, foaming fiercely on one side and solidifying on the other side, generating flocculent hollow microbeads with high porosity. The particle size is very large (10–300 micron), and the shape is approximately spherical. It is easy to form residual carbon black when heavy oil or residue is burned, but not easy to produce when gasoline and diesel fuel are burned.

(3) Snowflakes

Snowflake takes carbon black as the core. When the temperature of the flue gas is close to the dew point temperature, carbon blacks adsorb sulfuric acid in the flue gas, grow into snowflake shaped smoke, grow into snowflake shaped smoke, also known as acid smoke. Particles are in large size, often sink near chimneys, and are highly corrosive.

(4) Carbon deposit

Carbon deposition can be considered as a residual type of carbon black; its formation process is oil droplets attached to the wall of the burner and combustion chamber; oil droplets continue to gasify and the remaining material form carbon deposit. The amount of carbon deposition has a complicated relationship with the flame temperature, especially the wall temperature. When the temperature increases, the carbon deposition will increase or decrease, and the final result mainly depends on the temperature range. The volatilization, boiling point, and composition of fuel also have obvious effects on carbon deposition.

The formula of coal smoke calculation is shown as follows (Table 9.9).

$$G_{sd} = 1000 \times B \times A \times d_{fh} \times (1 - \eta)/\left(1 - C_{fh}\right) \tag{9.90}$$

where

G_{sd}, soot emissions, kg;
B, coal consumption, t;
A, ash content in coal, %;
d_{fh}, the percentage of soot in the smoke content, %;
η, dust removal system dust removal efficiency, %;
C_{fh}, the percentage of fuel in the smoke, %.

Table 9.9 Coefficient of coal-smoke calculation

Type	d_{fh} (%)	G_{sd}	Type	d_{fh} (%)	G_{sd}
The percentage of soot in the smoke content (d_{fh})					
Hand stove	25	117	The coal furnace	40	187
The chain furnace	25	117	Fluidized bed	60 (25%)	224
Reciprocating furnace	20	93	Pulverized coal furnace	85 (8%)	259
Vibration of the furnace	40	187			
Ash content in coal (A)					
Unit	Ash content (%)	Unit	Ash content (%)	Unit	Ash content (%)
Jixi	35.61	Hegang	24.90	Shuangyashan	22.49
Combustible content of smoke and dust (C_{fh})					
Type	C_{fh} (%)	Type	C_{fh} (%)	Type	C_{fh} (%)
Nomal	45	Pulverized coal furnace	8	Fluidized bed	25

5. The Generation of CO

CO is the most widely distributed and abundant pollutant in the atmosphere, and also one of the important pollutants generated in the process of fuel combustion. The main source of CO in the atmosphere is internal combustion engine exhaust, followed by the incomplete combustion of fossil fuels in boilers; CO can stay in the troposphere for about half a year.

CO is an intermediate product produced during the combustion of a carbon-containing fuel. CO can be formed from all the carbon originally present in the fuel. The thermal decomposition of fuel also produces CO. Therefore, CO mainly comes from incomplete combustion and pyrolysis of fuels. The causes of incomplete fuel combustion include insufficient oxygen and partial hypoxia, which produces large amounts of CO. The combustion area temperature is not high enough; there is a local low-temperature area and CO is in direct contact with the low-temperature wall surface, so CO cannot produce a combustion reaction. CO does not stay long enough in the combustion chamber, such as small combustion chamber volume, short circuit of gas flow, or ignition delay. The gas is not sufficiently mixed.

The carbon monoxide emissions from the combustion of the fuel are calculated as follows:

$$G_{CO} = 2330 \times B \times C \times Q \tag{9.91}$$

where

G_{co}, CO emissions, kg;
B, fuel consumption, t;
C, carbon content in fuel, %;
Q, fuel combustion incomplete value, %.

The carbon monoxide emission produced by burning gas is calculated as follows (Table 9.10):

$$G_{CO} = 1.25 \times V \times Q \times \left(V_{CO} + V_{CH_4} + 13V_{C_mH_n}\right) \quad (9.92)$$

where

G_{co}, CO emissions, kg;
V, gasoline fuel consumption, L;
Q, fuel combustion incomplete value, %;
V_{CO}, CO volume content, %;
V_{CH4}, CH_4 volume content, %;
V_{CmHn}, C_mH_n volume content, %.

Table 9.10 Coefficient of CO generation calculation

Fuel type	Q	C	Fuel type	Q	C
Incomplete fuel combustion values (Q)					
Wood	4	50	Coke	3	85
Peat, mixed coal	4	60	Heavy oil	2	90
Lignite	4	70	The artificial coal gas	2	20
Bituminous coal	3	80	Natural gas	2	75
Anthracite coal	3	90	Light oil	1	90
Volume content of main substances in gas (V)					
Gas species	Volume content (%)		Gas species	Volume content (%)	
Hayi Gas	CO:10, CH_4:25, $C_{III}H_{II}$:1		Natural Gas	CO:5, CH_4:95	
Carbon monoxide pollution coefficient (G_{co}) (kg/t)					
Fuel type	C(%)	G_{co}	Fuel type	C(%)	Gco
wood	50	46.6	Coke	85	59.4
Peat, mixed coal	60	55.9	Heavy oil	90	41.9
Lignite	70	65.2	The artificial coal gas	20	9.32
Bituminous coal	80	55.9	Natural gas	75	35.0
Anthracite coal	90	62.9			

9.3 Applications

9.3.1 Combustion Modification: Novel Burners

Burner is an important equipment inside the industrial oil-fired boiler and gas-fired boiler, which can ensure the stable ignition and complete combustion of fuel. A low nitrogen burner, by adjusting the combustion air and combustion head, can obtain the best combustion performance.

1. Low excess air ratio burners

The principle of low excess air ratio burners is to make one part of the fuel-burning too thick, the other part of the fuel-burning too light, but the overall air volume remains unchanged. By lowering the excess air coefficient, a reduced atmosphere can be formed in certain parts of the boiler, inhibiting the transformation of volatile N compounds into NO_x [7]. It should also be noted that NO_x can be reduced on the coke surface to form N_2 or N_2O [8]. This combustion is also known as deviation combustion or non-stoichiometry combustion.

(1) Pollution minimum (PM) low pollution burner

The key equipment to separate coal powder is a separator. Based on the centrifugal principle of pulverized coal passing through the elbow, it is composed of one elbow of pulverized coal and two sprinklers of one concentration and one light. Under the action of centrifugal force, the concentrated phase pulverized coal airflow enters the upper layer and the light phase pulverized coal airflow enters the lower layer. The two are fed into the furnace with independent nozzles. PM combustion nozzle angle can be adjusted within the 30°, at the same time the burning device also has the flue gas recirculation nozzle and reburning nozzle. With combustion air classification and flue gas recycling technology, low pollution burners have been applied in many domestic units. With the continuous maturity and improvement of the structure of the separation device, Mitsubishi began to pay attention to the precise adjustment of the separation device (Fig. 9.7).

(2) Wide Range (WR) burner

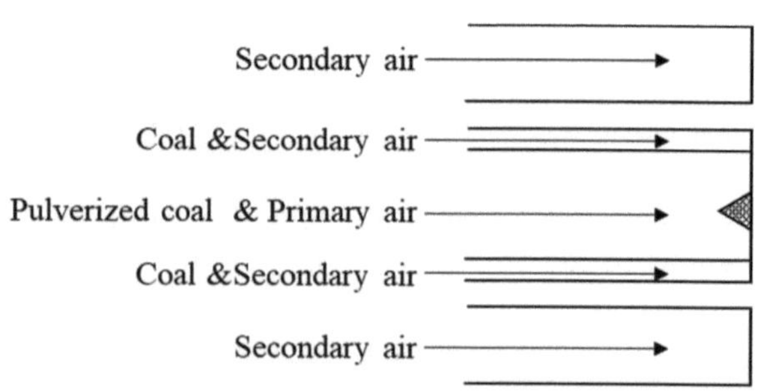

Fig. 9.7 Schematic diagram of the PM burner

WR burner makes use of 90° elbow to form two types of coal, heavy and light, achieving horizontal partition. The ratio of heavy and light coal can reach 7:3 until the burner outlet. An additional "V"-shaped obtuse body is added at the burner outlet to form a high-temperature reflux area. In addition, with the regulation of corrugated diffuser cone and circum-boundary wind, the reheat temperature is adjusted by an up–down swing at the first level, and the second-level combustion is formed by the top wind, which has a better performance of low NOx emission. Harbin Boiler Factory has improved on this basis by installing corrugated plates around the nozzle to form an improved WR burner (Fig. 9.8).

(3) Horizontal concentrated pulverized coal burners and radial swirl burners

In the tangential combustion pulverized coal furnace, a pulverized coal concentrator is installed on the primary wind pulverized coal pipeline to divide the primary wind into two streams of horizontal thick and light air. The high-concentration pulverized coal airflow is on the fire side, forming an imaginary cutting circle inside, while the low-concentration pulverized coal airflow is on the back-fire side, forming an imaginary cutting circle outside. The pulverized coal concentrator adopts a cyclone separator or shutter; the ratio of the concentration to dilution is 5–6. This type of burner is widely used in China and can be stably burned at a 50% low load. In four-corner burners all concentrated airflow sprayed into the combustion chamber from the fire side, which ensured timely ignition, combustion stability and effectively inhibited the NOx emission.

(4) Double grade low NOx burner

In the primary air channel, the concentration of coal powder is separated by separating the blunt body. The concentrated phase airflow is directly connected to the coal nozzle, while the light phase airflow is connected to the secondary air through the guiding pipeline and enters the furnace from the secondary air nozzle. The dilute phase air volume can also be adjusted by separating and adjusting the baffle opening degree. This unique structure connects the primary and secondary air ducts inside the burner and can realize fuel and air classification. At the same time, the arrangement is flexible, and the concentration of primary air pulverized coal can be adjusted without changing the original burner of the boiler (Fig. 9.9).

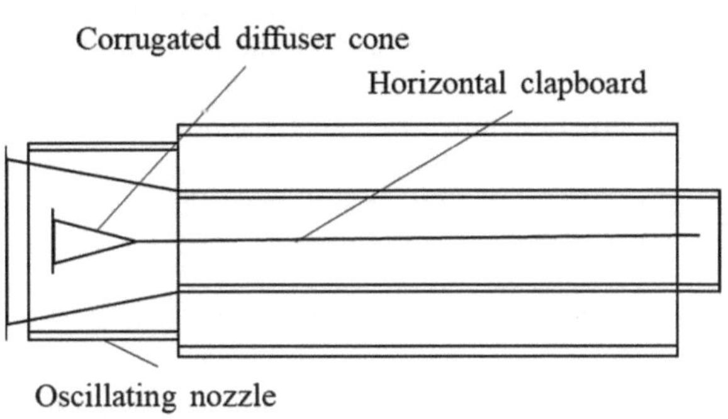

Fig. 9.8 Schematic diagram of the WR burner

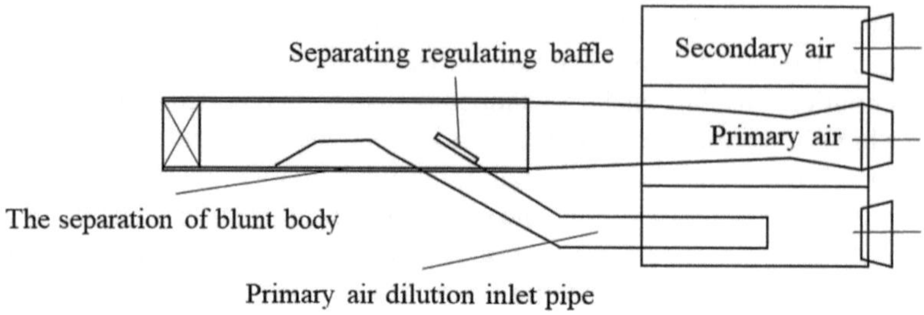

Fig. 9.9 Schematic diagram of double grade low NOx burner

2. Two-stage burners

The stage burner designed according to the principle of fractional combustion makes fuel and air mixed combustion in sections. As the combustion deviates from the theoretical equivalent ratio, NOx generation can be reduced. The basic principle is to divide the secondary wind into two strands and send them into the burning area in stages. In the primary stage of pulverized coal ignition, part of secondary air is added to continue the combustion of rich fuel for a certain distance to form the primary combustion area, while the other part of secondary air is sent downstream to form the secondary combustion area where the excess coefficient of air is >1.

(1) Low NOx concentric firing system (LNCFS) burner

A wedge diverter is installed in front of the divider, which can make the fuel flow diffuse when it leaves the nozzle outlet, leading to the reduction of the penetration ability of the fuel flow jet and the increase of surface area. In addition, bump blocks on the wedge diverter cause turbulent mixing of fuel flows and cause the ignition to occur near the burner. The volatiles of coal rapidly precipitate near the nozzle of pulverized coal and quickly ignite, which improves the flame stability and reduces the generation of NOx.

(2) NOx reduction (NR) series burner

Its core concept is "NOx reduction in flame". By rapidly igniting pulverized coal, a reducing environment can be formed under the condition of sufficient fuel, thereby reducing the generation of NOx. Pulverized coal concentrator is installed in a wind tunnel, via the radial separation the pulverized coal particles pass through stable-combustion ring in the end of the nozzle. Then, due to negative pressure induced by the structure, pulverized coal and hot flue gas fully mixed in the backflow zone. With diaphragm adding, the zone further expands, beneficial to fast ignition and decomposition of NOx in the flame. The deflector optimizes the secondary and tertiary air injection to further separate the secondary and tertiary air (Fig. 9.10).

(3) Double-regulated cyclone burner

The primary air outlet is equipped with a toothed flame stabilizer ring, which can delay the mixing of primary air and internal and secondary air, and the internal and

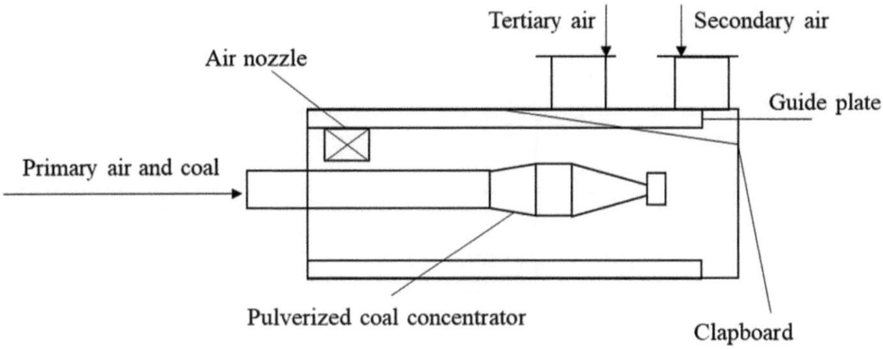

Fig. 9.10 Schematic diagram of the HT-NR3 burner

external secondary air enters axially. Adjustable rotating blades are set in the channel and a sliding gate valve is used to control the flow (Fig. 9.11).

(4) Double cone burner

The burner adopts the grading air distribution technology and flue gas recycling technology. 5% of the total air carries coal powder from the center tube into the combustion chamber, under the action of backflow caps reverse spraying into the combustion chamber, ensuring high concentration pulverized coal and secondary air contact before fully preheating. While secondary air, via tangential blade, rotates into the pre-combustion chamber. Then under the action of the reverse jet, secondary air forms backflow zone between cyclone and conical space, stabilizing the combustion (Fig. 9.12).

3. Low NOx burners

(1) Split flame burner

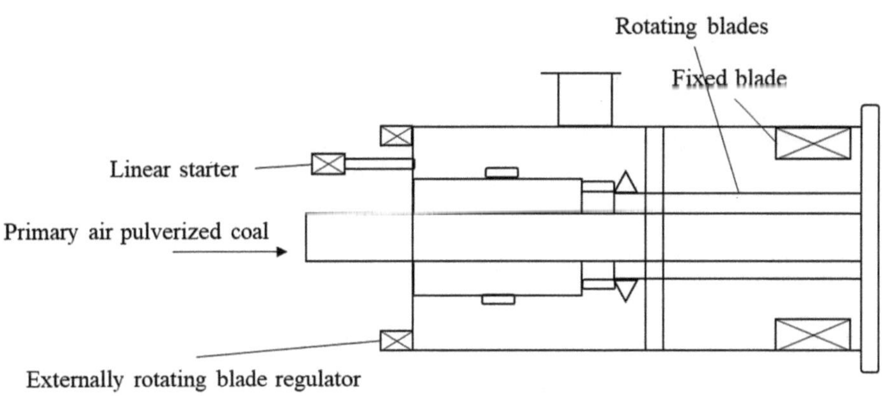

Fig. 9.11 Schematic diagram of the DRB-4Z burner

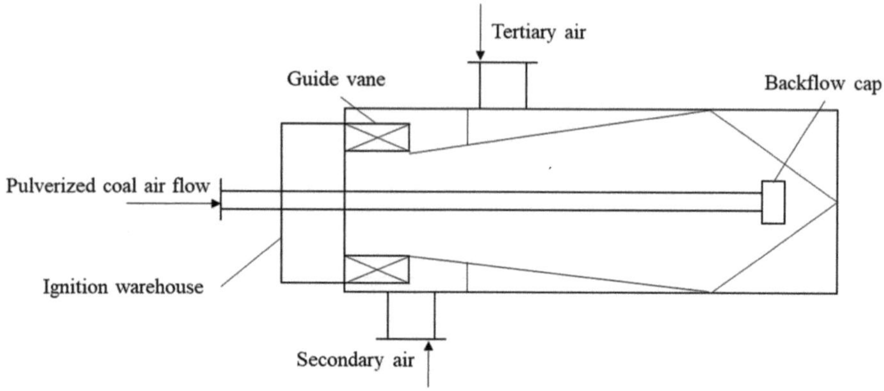

Fig. 9.12 Schematic diagram of the double cone burner

The principle is to divide the main flame into several small flames, because the small flame heat dissipation area is large, and the flame temperature is low, so that the "thermal reaction NO" decreases. In addition, the small flame shortens the residence time of oxygen, nitrogen, and other gases in the flame, and significantly inhibits the "thermal reaction NO" and "fuel NO".

(2) Mixed accelerator burner

The residence time of flue gas in the high-temperature area is one of the main factors affecting NOx production. By improving the mixing of combustion and air, the thickness of the flame surface can be reduced. When the combustion load remains unchanged, the residence time of flue gas in the flame surface, i.e., the high-temperature area, is shortened, thus reducing NOx production. The mixed-boost burner is designed based on this principle.

(3) Low NOx pre-combustion chamber burner

Research around pre-combustion chamber has been conducted in China in recent 10 years, gradually forming a high efficiency and low NOx combustion technology. Generally speaking, in a pre-combustion chamber the wind or secondary air and fuel mix together. Due to the lack of oxygen, only part of the fuel starts to fire. Poor oxygen content and low flame temperature reduce NOx formation.

(4) Recirculators

The recirculator incorporates the combustion air pressure head, part of the combustion smoke suction into the burner, and air mixed combustion. Due to the recirculation of flue gas, the heat capacity of combustion flue gas is large, the combustion temperature is reduced, and NOx is reduced. Another kind of self-recirculating burner is to recycle some flue gas directly in the burner and add the combustion process. This kind of burner has double effects of inhibiting nitrogen oxide and saving energy.

The SGR-type burner recirculates flue gas directly. The recirculating gas mixes with pulverized coal in primary air. Because of oxygen dilution, a reducing atmosphere forms, lowering combustion rate and temperature, inhibiting the formation of NOx (Fig. 9.13).

4. Design of burners

The guidelines and procedures included herein are appropriate for designing a typical direct-fired heater of the types found in refineries and gas plants, and are not expected to include specialty heaters or furnaces that include reaction in the tubes due to heat or catalysts. We will assume that for this discussion, the heater design will need to meet API standards. For this first example of selecting a heater design, we will assume the heater is a small unit of the type used in the petrochemical industry.

In the following, according to the typical pre-combustion chamber burner based on porous media combustion mechanism and the design requirements of industrial burners, the design of pre-combustion chamber burner is explained in accordance with the design guidelines of industrial-grade porous media burners.

(1) Selection of porous media materials and structures

In the process of porous media combustion, the combustion occurs inside the porous medium, where the material's temperature is higher, and in the process of burner start–stop, thermal stress caused by thermal expansion may cause damage to the porous medium material, at the same time, that caused through the working substance may be corrosive, thus higher requirements on the performance of the porous medium material (Table 9.11).

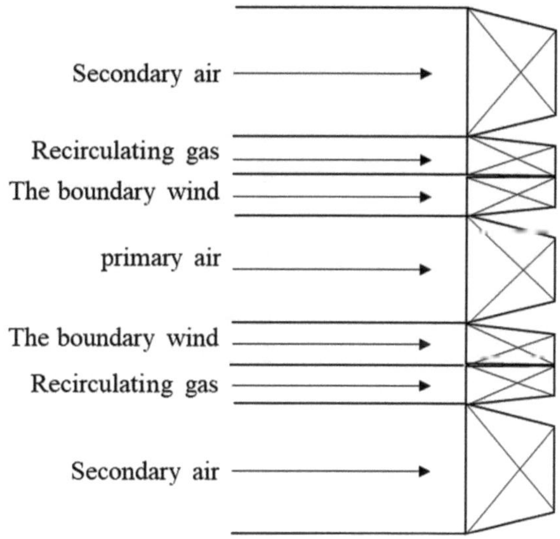

Fig. 9.13 Schematic diagram of the SGR low NOx burner

Table 9.11 Coefficient of CO generation calculation

Material	Maximum available temperature/°C	Coefficient of thermal expansion/(10^6/K)	Specific heat capacity/[J/(g K)]	Coefficient of thermal conductivity/[W/(m K)]	Emissivity	Thermal shock resistance
Alumina	1900	8	0.9–1	20–30	0.28	Fare
Silicon carbide	1600	4–5	0.7–0.8	80–150	0.9	Good
Zirconia	1800	10–13	0.5–0.6	2–5	0.31	Good
Mullite	1800	3.2	4.55	5.2	08–0.94	Fare
Cordierite	1450	1–2	0.92	1.97–2.5	0.5–0.7	Good
Quartz ceramics	1900	0.45	0.18	0.6–1.8	0.55	Good
Aluminum titanate	1550	0.78	0.92	0.96	0.5–0.7	Good

The porosity, specific surface area, volume density, thermal shock resistance, and resistance characteristics of different types of porous media are different. Therefore, the stable-combustion range, heat transfer characteristics, and temperature characteristics of gas during combustion are different (Table 9.12).

(2) Combustion head design

For the porous media burner with a given design power, the gas flow is

Table 9.12 Comparison of different porous media structures

Structure	Aperture/diameter	Porosity (%)	Advantages	Disadvantages
Particle packing type	3–30 mm	20–50	High volume density and high heat storage capacity. Thermal shock resistance, thermal expansion resistance, thermal impact resistance, wear resistance. Long life, easy to install and replace	The structure arrangement has strong randomness, and the actual arrangement structure is complex. The porosity is low, and the flow resistance loss is large. The runner clogs easily
Straight hole mesh type	–	50–70	High porosity and large specific surface area. Low thermal inertia. Small resistance loss. Not easy to accumulate ash and block	Poor thermal shock resistance, thermal expansion resistance, and thermal impact resistance
Fiber mesh pattern	–	>95	High porosity, strong permeability, and low flow resistance loss. High strength, stable structure	The production technique is difficult, the craft is complex, and the cost is high
The bubble type	8–120 PPI	70–90	Small density, large specific surface area, and uniform pore distribution. It has certain heat resistance, abrasion resistance, and corrosion resistance. Good flow characteristics, low flow resistance loss	High porosity reduces the mechanical properties of porous media structures and the thermal stress caused by thermal expansion easily destroys the structures

$$q_{v1} = \frac{p}{q} \tag{9.93}$$

where

q_{vl}, combustion gas flow, m^3/s;
p, burner design power, kW;
q, low gas calorific value, kJ/m^3.

The combustion airflow is

$$q_{v2} = q_{v2} \times \frac{1}{\varphi} \times \frac{2}{0.21} \tag{9.94}$$

where

q_{v2}, combustion airflow, m^3/s;
φ, design and operation equivalent ratio of burner.

The premixed gas flow rate is

$$q_v = q_{v2} + q_{v2} \tag{9.95}$$

The cross-sectional area of the porous medium is

$$A = \frac{q_v}{v_{in}} \tag{9.96}$$

where

A, the cross-sectional area of the porous medium, m^2;
v_{in}, the rate of premixed gas flow between the two layers of porous media when the flame surface can be stabilized, m/s.

Pressure loss calculation,

$$\frac{\Delta P}{L} = \frac{150(1-\varepsilon^2)}{\varepsilon^3}\frac{\mu V}{d_m^2} + \frac{1.75(1-\varepsilon)}{\varepsilon^3}\frac{\rho V}{d_m^2} \tag{9.97}$$

where

V, cross-section of the velocity, m/s;
ε, porosity of porous media;
μ, dynamic viscosity of fluid, Pa S;
ρ, fluid density, kg/m^3;
L, the effective length of fluid flow through porous media, m;
ΔP, pressure drop along the flow, Pa;

d_m, average pore size of porous media, m.

(3) External insulation design

Calculation of heat conduction flow of insulation layer

$$\phi = \frac{2\pi L(T_1 - T_3)}{\ln(d_2/d_1)/\lambda_1 + \ln(d_3/d_2)/\lambda_2} \tag{9.98}$$

The large space natural convection heat transfer model is adopted for the convection heat transfer between the metal shell and the atmosphere,

$$\phi_1 = 0.48\left(\frac{ga_v(T_3 - T_0)d_3^3}{\nu^2}\,\mathrm{Pr}\right)^{\frac{1}{4}} \frac{\lambda_3}{d_3}\pi d_3 l[T_3 - T_0] \tag{9.99}$$

The radiative heat exchange between the metal shell and the outside is

$$\phi_2 = \pi d_3 l \varepsilon \sigma \left(T_3^4 - T_0^4\right) \tag{9.100}$$

According to the heat balance,

$$\phi = \phi_1 + \phi_2 \tag{9.101}$$

9.3.2 Catalytic Oxidation

1. Regenerative thermal oxidation (RTO)

Direct oxidation is a method of direct combustion of VOC in gas flow under a high temperature of 800 °C or so. The direct oxidation method can effectively remove VOC gas with high concentrations. The main technological parameters that affect the VOC removal rate of this method are combustion chamber temperature, inlet concentration, and residence time. In the process of use, by adjusting these three parameters, the balance between treatment efficiency and operating cost can finally be achieved, and production resources can be effectively utilized (Table 9.13).

The regenerative thermal oxidizer is an energy-saving environmental protection device used to treat medium and low concentrations of organic waste gas [9]. Its core is to recycle the heat of burning VOC through the heat storage bed for preheating subsequent organic waste gas. When the organic waste gas preheats to a certain temperature, it enters the combustion chamber for incineration. The operation cost of this method is relatively low when dealing with organic waste gas, but the primary investment is high, and it is usually a fully automatic operation, leading to a lower maintenance cost (Table 9.14).

Table 9.13 Theoretical reactor temperatures required for 99.99% destruction by thermal incineration for a 1 s residence time

Compound	Temperature, °F
Acrylonitrile	1344
Allyl chloride	1276
Benzene	1350
Chlorobenzene	1407
1,2-dichloroethane	1368
Methyl chloride	1596
Toluene	1341
Vinyl chloride	1369

The recycling thermal incineration uses gas or fuel direct combustion with exhaust gas containing an organic solvent; under high temperature, the organic matter is decomposed into carbon dioxide and water; the cleaned flue gas passes through multi-stage heat exchangers, increasing the temperature of air or water, making full use of the heat generated, lowering the overall energy consumption. The advantages of this method are that the VOC decomposition rate is high, and heat generated in the VOC treatment process is effectively utilized. However, heat exchangers have limited life in intermittent production, which will increase the maintenance and operation cost of the equipment (Fig. 9.14).

2. Regenerative catalytic combustion (RCO)

There are many kinds of catalytic oxidation methods, including photocatalysis, catalytic ozone oxidation, and other methods, among which the most common and effective one is catalytic combustion (Table 9.15).

The regenerative catalytic combustion device is mainly used to treat high concentrations of organic waste gas. This method is suitable for all processes of the same production line, and the organic waste gas composition, concentration, and fluctuation are different due to different raw materials and production capacity. The heat storage principle of this method is similar to that of RTO. The main feature of this method is to make the surface porous structure of the catalyst in the process of catalytic oxidation and apply it to adsorb the reactant molecules so that they can be enriched on the surface of the catalyst to maximize the efficacy of the catalyst. The catalyst's action of reducing activation energy speeds up the reaction process, making the reactant molecules flame-free at lower temperatures, decomposing organic pollutants into water and carbon dioxide, and releasing large amounts of heat. Compared with RTO, this method can realize automatic operation with relatively low one-time investment, and the reaction can be carried out at a lower temperature due to the intervention of the catalyst, which greatly reduces the energy cost. The disadvantage of this method lies in the large area of the equipment used, and the catalyst failure phenomenon is easy to occur in the long-time running, leading to the increase of the cost of replacement catalyst (Table 9.16).

Table 9.14 The working process of the regenerative thermal oxidation

	Process I	Process II
Schematic diagram	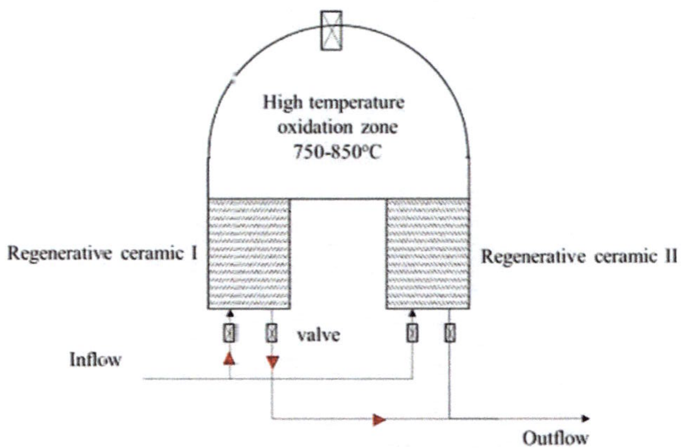	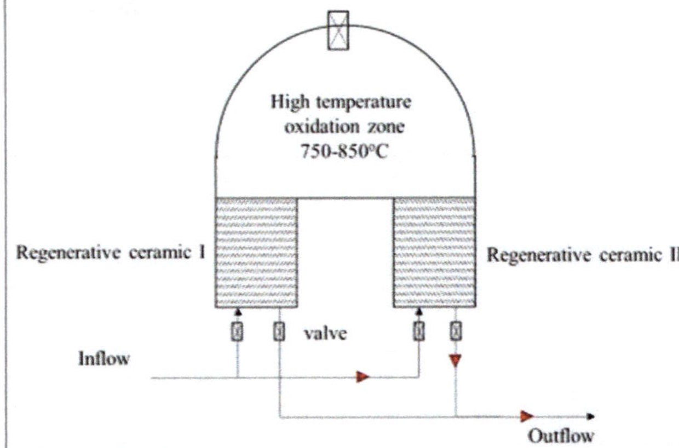
Regenerative ceramic I	Waste gas absorbs heat heat storage ceramic I is exothermic	Clean gas releases heat; heat storage ceramic I absorbs heat
Regenerative ceramic II	Clean gas releases heat: heat storage ceramic II absorbs heat	Waste gas absorbs heat; heat storage ceramic II is exothermic
The combustion chamber	High-temperature oxidation decomposition	

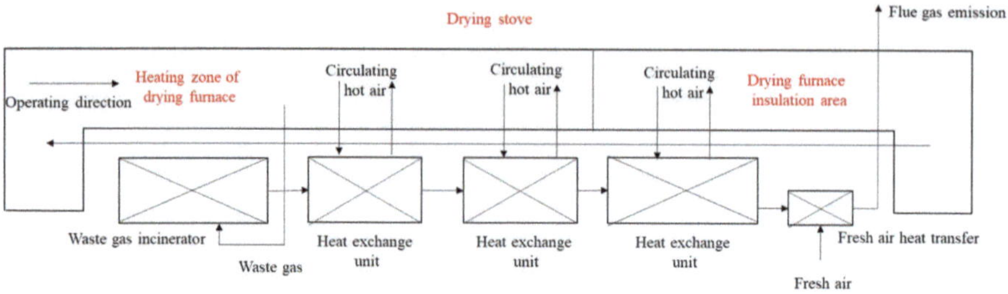

Fig. 9.14 Schematic diagram of recycling thermal incineration

Table 9.15 Catalyst temperature required for oxidizing 80% of inlet VOC to CO_2 of two catalysts

Compound	Temperature, °F	
	CO_3O_4	Pt–Honeycomb
Acrolein	382	294
n-butanol	413	440
n-propylamine	460	489
Toluene	476	373
n-butyric acid	517	451
1,1, 1-trichloroethane	661	>661
Dimethyl sulfide	–	512

3. Runner concentration + regenerative catalytic combustion

This method concentrates the VOC in the industrial waste gas with low concentration and large air volume and decomposes the concentrated polluted air with high concentration and small air volume by combustion. The rotary wheel is divided into adsorption area, desorption area, and cooling area. The polluted air first passes through the adsorption area and is adsorbed by the adsorbent. As the rotator runs, the VOC on the adsorbent is carried into the stripping section, under the action of high-temperature air; adsorbent finishes stripping, stripping down the VOC into regeneration in the air (the core of the process is that the adjustable air volume of the regeneration air, when the regeneration air volume for processing air volume 1/X, the regeneration of VOC concentration in the air for processing the VOC concentration in the air of X). The rotary wheel that completes the desorption process enters the cooling area and returns to the adsorption area after cooling. At this point, the recycling process of adsorption, desorption, and regeneration is completed, and the concentration process of polluted air is completed for further treatment. High concentration and small air volume polluted air after concentration is treated with RCO (Fig. 9.15).

4. Design of oxidizers

(1) General treatment of material and energy balances

In the sizing and costing of the incinerator and the calculation of the auxiliary fuel requirements, it is necessary to conduct the material and energy balances around the

Table 9.16 The working process of the regenerative catalytic combustion

	Process I	Process II
Schematic diagram	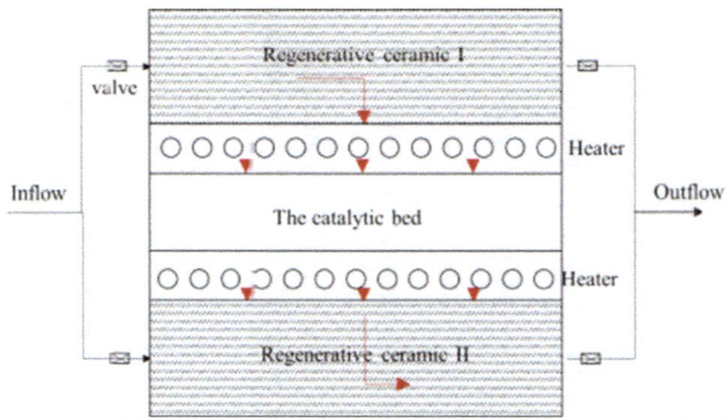	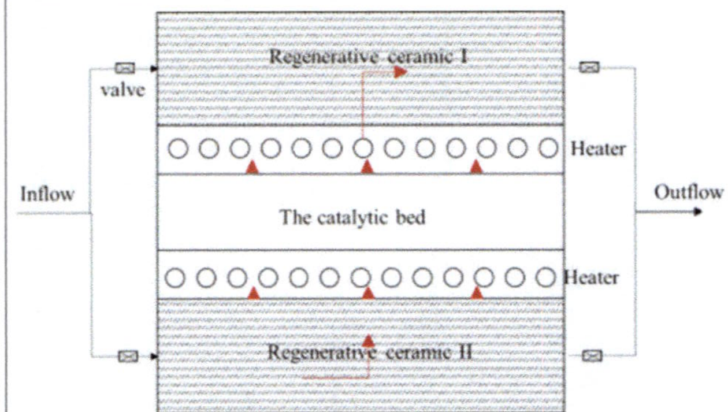
Regenerative ceramic I	Waste gas absorbs heat heat storage ceramic I is exothermic	Clean gas releases heat; heat storage ceramic I absorbs heat
Regenerative ceramic II	Clean gas releases heat heat storage ceramic II absorbs heat	Waste gas absorbs heat; heat storage ceramic II exothermic
The combustion chamber	Medium temperature catalytic decomposition	

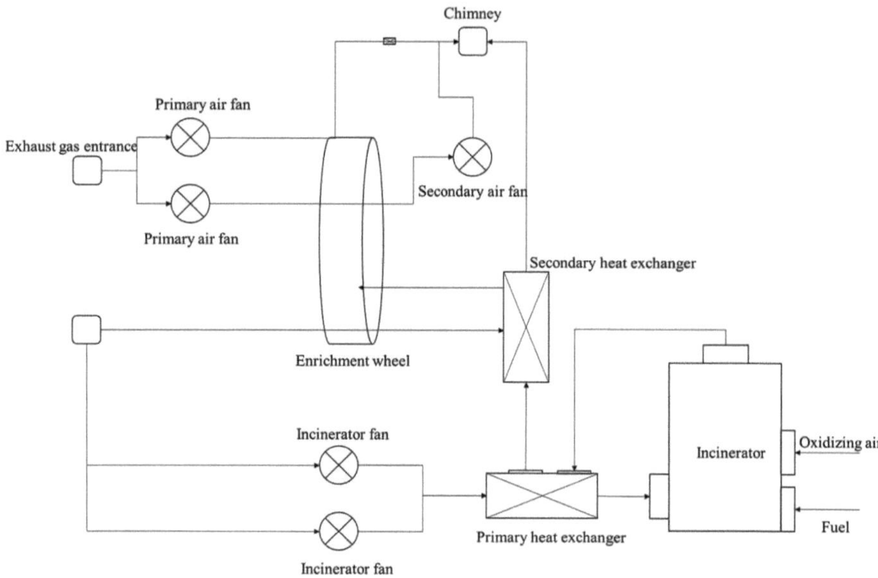

Fig. 9.15 Schematic diagram of runner concentration + regenerative catalytic combustion

entire incinerator unit and around certain parts of the unit, such as the combustion chamber or the preheater. They can be stated in general equation form as

$$\text{In} - \text{Out} + \text{Generation} = \text{Accumulation} \tag{9.102}$$

Because the incineration process is a steady-state process, the accumulation term is zero and the equation becomes

$$\text{In} - \text{Out} + \text{Generation} = 0 \tag{9.103}$$

For mass balances, it is useful to restrict the balances to be made on the mass of each atomic species so that for mass balances the generation term becomes zero. Thus, the simplified equations are

$$\text{In} - \text{Out} = 0,\ \text{for sterady} - \text{state mass balances} \tag{9.104}$$

$$\text{In} - \text{Out} + \text{Generation} = 0,\ \text{for sterady} - \text{state mass balances} \tag{9.105}$$

For the incineration process, the two terms In and Out are generally mass terms (for a mass balance) of the form,

$$\dot{m} = \rho q \tag{9.106}$$

where

ρ = density (mass per unit volume),
q = volumetric flow rate (volume per unit time),
or sensible heat terms (for an energy balance) of the form,

$$\dot{Q} = \dot{m}\Delta h = \rho q p\left(T - T_{ref}\right) \tag{9.107}$$

where

C_p = heat capacity;
T = temperature.

When the reference temperature is taken as zero, the sensible heat terms become

$$\dot{Q} = rQC_pT \tag{9.108}$$

For the incineration process, the generation term for energy balances accounts for the energy released through the combustion reactions. This term is generally of the form

$$\dot{Q} = rQ(-\Delta h_c) \tag{9.109}$$

where

$(-\Delta h_c)$ = heat of combustion.

(2) The oxygen content of the waste gas

There must be sufficient oxygen in the waste gas to support the combustion of the waste organics (including VOCs) and the auxiliary fuel, if auxiliary fuel is needed. It may be necessary to add auxiliary air if the oxygen content is less than about 20%. This example is based on air streams that contain >20% oxygen, as shown below for 1000 ppmv of benzene and methyl chloride:

$$\text{Air Content, Vol\%} = 100 - \frac{1000(\text{ppm})}{10^6} \times 100 - \frac{1000(\text{ppm})}{10^6} \times 100 = 98\% \tag{9.110}$$

This gives the oxygen content in percent as

$$\text{Oxygen Content, Vol\%} = 98.8\% \times 0.209 = 20.86\% \tag{9.111}$$

(3) The percent of the LEL of the gas mixture

The example chosen here is typical, in that there is more than one VOC component in the gas stream. An approximate method to calculate the LEL of a mixture of compounds, LEL_{mix}, is given as

$$\mathrm{LEL_{mix}} = \left[\sum\nolimits_{j=1}^{i} \frac{x_j}{\left(\sum\nolimits_{i=1}^{j} x_i \right) \mathrm{LEL}_j} \right]^{-1} \quad (9.112)$$

where

x_i = volume fraction of combustible component *I*;
LEL_j = lower explosive limits of combustible component *j* (ppmv);
n = number of combustible components in mixture.

(4) The volumetric heat of combustion of the waste gas streams

The energy content of the gas stream, expressed in terms of the heat of combustion, is calculated as follows:

$$\left(-\Delta h_{c_w} \right) = \sum_{i=1}^{n} (-\Delta h_{ci}) x_i \quad (9.113)$$

where

$(\&\Delta h_{cw})$ = heat of combustion of the waste stream (Btu/scf);
$(\&\Delta h_{ci})$ = volumetric heat of combustion of component *i* at 25 °C (Btu/scf);
x_i = volume fraction of component *i* in the waste gas;
n = number of combustible components in the waste gas.

With these assumptions, the equation for fractional energy recovery for a heat exchanger becomes

$$\text{Fractional Enery Recovery} = \frac{T_{w_0} - T_{w_i}}{T_{f_i} - T_{w_i}} \quad (9.114)$$

The auxiliary fuel requirement
Thus, if the reference temperature is near ambient conditions,

$$H_L = 0.1 \rho_{fi} Q_{fi} C_{pm_{fi}} \left(T_{fi} - T_{ref} \right) \quad (9.115)$$

With these assumptions, the mass and energy balance around the combustion chamber reduces to the following equation:

$$\text{Total Enery Input} = r_{fi} Q_{fi} C_{pm_{fi}} \left(T_{fi} - T_{ref} \right) \quad (9.116)$$

The total volumetric flow rate of gas through the incinerator is calculated below:
The increase in volumetric throughput due to an increase in the number of moles of gas as a result of combustion is neglected. Shown below is the sum of the inlet streams to the incinerator.

$$Q_{fi} = Q_{w_0} + Q_a + Q_{af} \tag{9.117}$$

9.4 Comparison of Control Technologies

9.4.1 4E Evaluation

The 4E evaluation of the above-mentioned technologies is summarized in Table 9.17. To address these technical and commercial limitations, industrial controls are implemented to achieve better full combustion. For this purpose, many alternative technologies are being evaluated, and the 4E assessment is proposed to evaluate different technologies. It includes environmental, energy, engineering, and economic indicators.

9.4.2 Affordable and Clean Combustion Technology

Through the improvement of the pollutant minimization process and combustion technology, combustion has become a safe and less harmful process to the environment. Emission reduction results from two interrelated phenomena: efficiency improvement and reaction pathway engineering to improve combustion efficiency (the amount of fuel required per unit of energy service) will only reduce the total amount of materials processed by the combined system, thus limiting the potentially harmful emissions. By adjusting the way that reactants are transformed into products, the amount of environmentally harmful substances that may be produced in the chaotic combustion reaction zone can be reduced.

Combustion is an applied science, which is of great significance in transportation, power generation, industrial process, and chemical engineering. Actually, combustion must be safe, effective, and clean at the same time. It has a long history from ancient times to the middle ages. Fire, soil, water, and gas are considered as one of the four basic elements of a poem. However, with the work of Antoine Lavoisier (1785), one of the founders of the chemical revolution and the discoverer of the law of conservation of mass, its importance has been re-recognized. From 1775 to 1777, Lavoisier first proposed that the key to combustion was oxygen. He realized that the newly separated components in the air (in 1772–1774, Joseph Priestley of England and Carl Scheler of Sweden) were elements. Then he named it and formulated a new definition of combustion, that is, the process of chemical reaction with oxygen. In the precise quantitative experiment, he established a new theory in a short period of time, and it has been widely recognized. At first, the number of scientific publications on combustion was very limited when combustion experiments were

Table 9.17 4E evaluation of combustion technologies

4E evaluation	Classification	Combustion modification				Catalytic oxidation	
		Stage burner	Self-recirculating burner	Thick-light burner	Split flame burner	Direct oxidizer	Catalytic oxidizer
Environmental	NOx	√	√	√	√	√	–
	SO_2	√	√	√	√	√	–
	PM	√	√	√	√	√	–
	VOCs	√	√	√	√	√	–
	Others	–	–	–	–	CO/HC	–
	Removal efficiency	40–50%	40–50%	5–15%	20–30%	95–99%	90–99%
Energy	Energy resource	All fuels	Coal	All fuels	Gasoline	–	–
	Combustion adjustment	Air	Recirculating Air	Air	Air	Air	Air
	Energy medium	Gas	Gas	Gas	Gas	Gas	Gas
	Heat efficiency	99%				70–85%	60–75%
Engineering	Operating temperature (°C)	–	–	–	–	700	300–500
	Pressure (MPa)	–	–	–	–	0.6–1.5	4–8 [10]
	System complexity	High	Low	Low	Low		
Economic	Capital cost	Low	Low	Low	High	High	High
	Operating cost	Low	High	Low	High	High	Low

conducted in chemical laboratories. From the beginning to now, chemistry has made a great contribution to the understanding of combustion at the molecular level.

Combustion efficiency usually depends on the way that reactants are converted into products. This process is complex. Knowledge about combustion is growing rapidly. New sensing technology makes the ability to monitor the combustion reaction process reach an unprecedented level. The increase of computing resources makes it possible to predict emissions of new strategies even before testing.

1. Opportunity fuels

The term "opportunity fuel" is usually defined by listing the combustible resources used in this case. The usual approach is "I know as soon as I see it." However, a more general approach can also be taken. Opportunity fuels can be defined as combustible resources that are not mainstream fuels for commercial use, but can be effectively used in industrial and commercial applications to generate electricity or improve process and space heat. In some cases, materials once defined as port fuels can become mainstream and very popular energy resources.

Since opportunistic fuels are not the mainstream of commercial fuels, the most common types of opportunistic fuels are residues from other processes or low-value products. These wastes include but are not limited to petroleum coke, sawdust, waste wood (mixture of sawdust, waste skin, shavings, and other solid wood product residues), pulping waste liquid, rice husk, oat husk, edible lard from turned plants, and agricultural activities. In addition to the solid residues, the moving wheat straw and other straw, unavailable hay, peat and a pile of other solid materials, blast furnace gas, coke oven gas, refinery tail gas, and other products can also be regarded as opportunity fuels, which are mainly used for industrial site processing, many of which enter the commercial field to some extent, while other fuels are strictly used on site. In this series of fuels, petroleum coke and wood waste are the most widely used opportunity fuels. Wood fuel alone provides about 3 EJ (or a quarter) for the U.S. economy and about 20 EJ worldwide.

The range of opportunity fuels, depending on the source, is as wide as that allowed by local opportunities and licensing agencies. Resources are based on the principle of recovery and reuse, not disposal; sources are also based on the maximum and complete value of each raw material extracted from the earth's natural resources. Some of these opportunity fuels are quite new. Other opportunity fuels, such as petroleum coke, wood waste, blast furnace gas, coke powder, and refinery tail gas, have a long history of use.

2. Combustion solution for tomorrow

To conserve energy, efficient use of energy resources is necessary. For industries striving toward energy efficiency and environmental stewardship, advanced combustion technologies and alternative fuels offer unique possibilities for cost reduction, productivity, and sustainability. The operating cost of biomass combustion is usually higher than that of coal. The most restrictive factor is the cost of fuel, which leads to a huge economic disadvantage for energy crops relative to residues. Even if the fuel is nominally free when it is produced (because there are many residues), its transportation, preparation, and on-site treatment will generally increase the effective cost of unit energy, making it compete with coal, sometimes even more than coal. A general conclusion is that biology Mass CO combustion is usually a little more expensive than dedicated coal systems, and if there is no incentive to reduce carbon dioxide emissions, it is difficult to justify CO combustion. However, compared with alternative renewable energy, biomass CO combustion power generation is generally much cheaper.

Due to the high steam parameters and many technical measures to improve the efficiency of the thermal power plant, the conversion efficiency (higher calorific value) can reach 30–38%, which is easy to exceed the efficiency of the dedicated biomass system. Compared with many estimated efficiencies in the future or beyond the advanced biomass system, adding biomass to the coal-fired boiler will not affect the overall power generation efficiency of the coal-fired power plant, or in the worst case. In this case, the power generation efficiency will be slightly reduced. Due to the preparation and treatment of more or less energy-intensive fuels, the efficiency may change slightly (whether positive or negative), while the generally increased moisture content in the fuel will slightly reduce the overall efficiency.

9.4.3 Integration for Innovation

Improving the efficiency of the internal combustion engine is a technically proven and cost-effective way to significantly improve the fuel economy of a national fleet in the near to medium term. The corresponding benefits are reducing our dependence on foreign oil, reducing carbon emissions by improving combustion process, reducing engine friction and other losses, reducing energy loss of emission control system, and promoting in order to improve the efficiency; it is necessary to comply with the exhaust emission regulations and require the use of post-treatment technology combined with the engine combustion method. The fuels considered include hydrocarbon-based fuels (petroleum- and non-petroleum-based and gas fuels such as natural gas). Due to its relatively low cost, high performance, and ability to utilize renewable fuels, internal combustion engines, including those in hybrid vehicles, will continue to be critical to our transportation infrastructure for decades.

The flow of people and the transportation of goods are the basic needs of modern society. Any way to meet these needs must meet sustainability standards, as well as customer recognition. Therefore, solutions and services must meet safety, energy efficiency and safety, environmental compatibility, and affordability standards. In order to meet these standards, the development of transportation system will be combined with the improvement and change of power systems and vehicle technology, and fuel and energy parties The development of the case is closely related to the electrification of the developed powertrain, which will continue to be promoted according to different vehicle development goals, while the need to improve energy security and reduce the use of fossil fuel and petroleum fuel will promote the wide use of renewable energy. Based on the global needs, the long-term trends related to the evolution of personal mobile demand and the long-term trends related to cargo transportation will be deployed at the local level according to the supply of raw materials and local policies. The market mix that will participate in the formation of transportation solutions is expected to have more people living in Metropolitan communities, who will continue to demand "clean" and efficient solutions scheme. This "clean" factor will not only be provided by the technological progress of vehicles but also by the increasing role of connectivity and the optimization of its use. Similarly, the long-distance transport system will benefit from connectivity, which will improve the conditions for autonomous driving, especially on intercity and highway routes, and benefit from the promotion of a broader multimodal transport model scheme.

Exercise

1. Explain 3Ts principle.
2. Describe the unique features of the burners for low NOx emission.
3. List NOx formation mechanisms while different oxygen-containing fuels are introduced for combustion.
4. Discuss the environmental and human impacts of nitric oxide, sulfur trioxide, unburned hydrocarbons, and particulate matter from diesel engines.
5. Compare the performances of RTO and RCO from the aspects of 4E.

References

1. Law CK (2006) Combustion physics. Cambridge University Press
2. Glassman I, Yetter RA (2008) Combustion. Elsevier
3. Date AW (2011) Analytic combustion with thermodynamics, chemical kinetics, and mass transfer. Cambridge University Press
4. Sajwan KS, Twardowska I, Punshon T, Alva AK (2006) Coal combustion byproducts and environmental issues. Springer
5. El-Mahallawy F, Habik SE-D (2002) Fundamentals and technology of combustion. Elsevier
6. Turns SR (2012) An introduction to combustion: concepts and applications. McGraw-Hill
7. Chen J (2017) Research and application on the low-NOx combustion of 130t/h CFB boiler. Zhejiang University, Hangzhou
8. Liao Z (2011) N_2O formation and control in circulation fluidized bed boilers. Zhejiang University, Hangzhou
9. Lewandowski D (2000) Design of thermal oxidation systems for volatile organic compounds. CRC Press, Boca Raton. https://doi.org/10.1201/9781315141060
10. Dent JC, Derham JA (1974) Air motion in a four-stroke direct injection diesel engine. Proc Inst Mech Eng 188(1):269–280. https://doi.org/10.1243/pime_proc_1974_188_030_02

Chapter 10
Renewable Energy

Nomenclature

A	Certain area in calculation
c	Specific heat capacity
C_{COP}^{carnot}	Coefficient of performance
E	Incident solar energy
E_{CO2}	CO_2 emission
Ef	CO_2 emission factor
ESC	Solar constant
Et	Electricity generation in the year t
EWi	Kinetic energy of the wind
Ft	Fuel expenditures in the year t
G	The annual power generation capacity of power plants
Gb	Sirect (beam) radiation
Gd	Siffused radiation
Gg	Global radiation
h_{util}	Utilizable head of a river or a stream
H	Net enthalpy gains of the biomass
It	Investment expenditures in the year t
LCOE	Levelized cost of electricity
m	Mass
Mt	Operations and maintenance expenditures in the year t
n	Life of the system
N	Total capacity of the power generation system
P	Output power
PWa	Work energy capacity of water
Q	Volumetric flow rate
r	Discount rate
S	Related surface area

P.-C. Chiang and X. Gao, *Air Pollution Control and Design*,
https://doi.org/10.1007/978-981-13-7488-3_10

T	Temperature
v	Velocity
η	Efficiency
ρ	Density
τG, τRD, τMD, τGA, τPA	Related factor

10.1 Introduction

10.1.1 Importance and Significance

Based on the renewable potential, natural energy resources can be generally divided into two categories: renewable energy, which can be replenished naturally and non-renewable energy, which cannot be regenerated or can only be regenerated very slowly. Non-renewable energy resources are mainly minerals and fossil fuels, for example, coal, crude oil, and natural gas. Comparing with renewables, non-renewables are approaching their depletion and normally produce more pollutions and emissions during energy consumption which accelerates global warming. Therefore, humans contemporarily are devoted to finding substitutional renewable energy to deal with these issues. Section 10.1.1 will state the pollution and global warming caused by non-renewables, and summarize the overall status of the development of renewable energy.

By the late eighteenth century and the first part of the nineteenth century, with the process of the First Industrial Revolution, coal became the main energy resource supporting human activities. Smog and soot resulting from burning coal had serious health impacts on the residents of increasing urban centers. In the Great Smog of 1952, pollutants from factories and home fireplaces mixed with air condensation killed at least 4,000 people in London, over the course of several days. In October 1948, 43% population of Donora, a small town in the USA, got sick with sore throat, runny nose, thirst, sore limbs, cough, chest tightness, vomiting, diarrhea, and other symptoms. It is estimated that during the event, the SO_2 concentration was several times higher than the normal value and dust particles were also found. The analysis suggests that the combination of SO_2 and its oxidation products with dust particles in the atmosphere is a causative factor. Public health was a serious concern after these big events, and finding substitute energy is of great importance (Fig. 10.1).

Apart from air pollution, global warming also turns into a big issue for human beings. Two pieces of convincing evidence proved that the global warming pace is accelerating. The first evidence is the Global Mean Surface Temperature (GMST) for 100 recent years, which was recorded by major measuring stations throughout the world, shown in Fig. 10.2a. Note that the rate of increase of GMST has increased over recent decades due to the increasing demand for global fossil fuels. The other evidence is the decreased extent of Arctic sea ice, which leads to increased sea level, shown in Fig. 10.2b. The main factor of global warming is the Greenhouse effect.

Fig. 10.1 The Great Smog of 1952 in London [1]

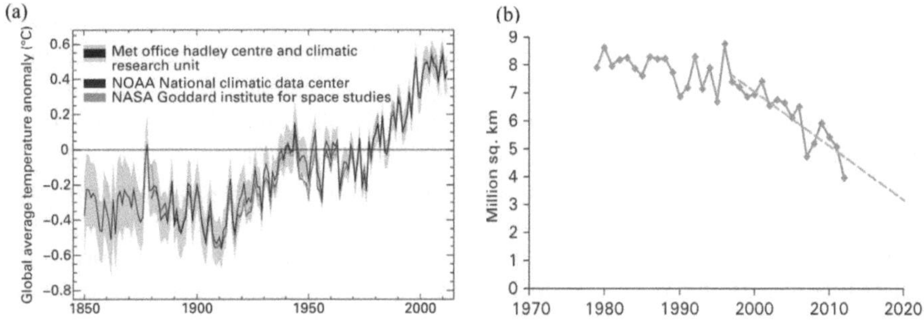

Fig. 10.2 **a** The global mean surface temperature (1850–2012), data source: from WMO (2013). **b** Decreased extent of Arctic sea ice in September from 1979–2012, data source: the US National Snow and Ice Data Center

The atmosphere of the earth acts like the glass of a greenhouse likewise (i) allows the incoming shortwave solar radiation to transmit through at daytime, (ii) absorbs the outcoming longwave radiation of grounds and decreases the connection between indoor and outdoor. Here, greenhouse gases (GHGs) are responsible for absorbing longwave radiation, notably carbon dioxide (CO_2), nitrous oxide (N_2O), and methane (CH_4). The global primary energy consumption in 2014 was 12.9 billion tons of oil equivalent (BTOE). During the combustion of fossil fuels, 10 billion tons of carbon was generated in 2014 [2]. The utilization of renewable energy can provide alternative solutions to global challenges on energy and the environment.

Renewable energy from natural processes that are continuously replenished and involve solar energy, wind energy, biomass energy, ocean energy, geothermal energy,

Fig. 10.3 Sustainable development goals. *Source* https://www.undp.org

and hydropower [2]. Renewable energy in the proportion of international energy consumption is growing rapidly. Research and development of renewable energy is the focus of the current international energy strategy. Developing renewable energy, adjusting the energy structure, and enhancing the integrated application rate of renewable energy, can help reduce the global energy demand for fossil fuels. Improving energy and economic security is of strategic significance for the promotion of economic and social development. According to the Sustainable Development Goals (SDGs), which act as a guide of UNDP policy and funding until 2030, affordable and clean energy is the 7th goal of the 17 sustainable development goals (Fig. 10.3). To achieve this, investing in solar, wind and thermal power, improving energy productivity, and the same as expanding infrastructure and upgrading technology provide clean and more efficient energy in all countries. Furthermore, another report named Better Business Better World [3] declares the 60 fastest-growing sustainable market opportunities sharing more than US$12 trillion by 2030 when delivering the SDGs. Energy and materials are listed as two major categories of 60 markets, among which expansion of renewables, energy efficiency, and energy access are relating to renewable energy.

Table 10.1 summarizes the renewable electric power capacity of the world, the European Union (EU), the United States (US), and China [4]. It can be observed that the power capacity of hydropower is the largest among all the renewable energy resources (1096 GW), which accounts for more than half of the global total renewable electric power capacity, followed by wind power (487 GW), solar PV (303 GW), and bio-power (112 GW). The total capacity for the rest energy types, including geothermal power, solar thermal power, and ocean power, is relatively small less than 20 GW. The development of renewable energy in China is very rapid. In 2016, the total renewable power capacity accounts for approximately 28% of the global total capacity, higher than that of the EU and the US. In China, hydropower, wind power, and solar PV mainly contribute to renewable power generation, while solar thermal power, ocean power, and geothermal power have a negligible contribution.

Table 10.1 Renewable electric power capacity of the World, EU, US, and China in 2016 (GW) [4]

Energy type	World	EU−28	US	China
Solar PV	303	106	41	77
Solar thermal power	4.8	2.3	1.7	~0
Wind power	487	154	82	169
Bio-power	112	37	16.8	12
Hydropower	1,096	127	80	305
Ocean power	0.5	0.3	~0	~0
Geothermal power	13.5	0.9	3.6	~0
Total capacity	2,017	428	225	564

10.1.2 Technology Development

Energy always plays an irreplaceable role in our lives. The development of human society is inseparable from the emergence of high-quality energy and the use of advanced energy. At the same time, in recent years, global climate problems have become increasingly serious. The greenhouse effect caused by excessive carbon dioxide emissions has begun to attract people's attention. Therefore, as an important part of the strategy, many countries have comprehensively developed clean, pollution-free, and renewable energy technologies to solve the problems of traditional energy consumption.

10.1.3 Contents of the Chapter

In this chapter, we will discuss the basic principles of renewable energy, including solar energy, wind energy, biomass energy, ocean energy, hydro energy, and geothermal energy. The fundamentals and estimation of the energy potential of these energy types will be introduced. The applications related to these renewable energy types will also be discussed. Finally, an evaluation of the technologies related to renewable energy utilization from the point of view of environmental, energy, engineering, and economic (4E) aspects is provided.

10.2 Principles

The fundamentals of energy, energy estimation, and energy utilization of each renewable will be discussed in this section. To further understand the conversion process, some examples will be demonstrated.

10.2.1 Principles of Solar Energy

It is obvious that solar energy arises from sun radiation, which is one of the earliest energy sources used by human beings. The sun is a hot gas ball with pressure of 3×10^{11} atm. The hydrogen atoms lose electrons while nuclei are at high temperatures, and only protons exist due to the thermal motion of the high-speed collision that occurred by thermonuclear reactions. Four protons polymerize into a helium nucleus and release a lot of heat. The 90% of solar radiation is produced in the sun core area (0–0.23R, R represents sun radius, 6.96×10^5 km), and the temperature is about (800–4000) $\times 10^4$ K; the radiation energy transmission area is between 0.23 and 0.7R, and the temperature drops to approximately13 $\times 10^4$ K; the convection zone is between 0.7 and 1.0R, and the temperature drops to approximately 5000 K. The surface of the sun, which is visible to the naked eye, is the outer layer of the convection region, referred to as the photosphere. It consists of ionized gases, which are capable of absorbing and emitting a continuous radiation spectrum. It is also the largest radiation source for the sun. The outer sphere is a reversed layer (luminous and almost transparent solar atmosphere), chromosphere, and corona in turn.

Solar radiation travels at a rate of 3×10^8 m/s, taking about 8 min traveling from the sun to the earth's atmosphere. When solar radiation passes through the earth's atmosphere, O_3, O_2, H_2O (g), and CO_2 in the atmosphere can absorb this radiation. On the other hand, it is absorbed by the scattering of air molecules and dust. These gases are selective for the absorption of solar radiation, and they can only absorb radiation energy at a certain wavelength range. For example, O_3 has strong absorption of ultraviolet light. O_3 and O_2 have a certain absorption of visible light, while H_2O (g) and CO_2 mainly absorb the radiation energy in the infrared region. Due to scattering, the total solar radiation which reaches the surface of the earth is the sum of direct radiation and scattered radiation. A typical picture of solar radiation and absorption and scattering of solar energy is shown in Fig. 10.4.

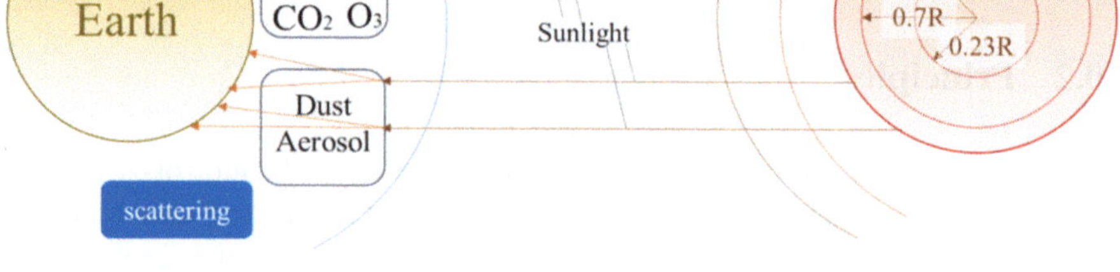

Fig. 10.4 Illustration of solar radiation

Solar energy is utilized mainly in two ways, namely, solar thermal conversion and solar electricity conversion. The solar thermal conversion system usually contains one or more materials to restore the heat from the sunshine. Direct utilization of solar energy includes directly utilizing heat such as a solar water heater, and indirectly utilizing heat such as water expansion to drive the turbine to generate electricity. These key energy conversion processes are heat conduction, heat convection, and heat radiation. However, the power of the sunlight reaching the ground of the earth per square meter is nearly 1 W, which means, to gain high temperature, we need a large area to absorb the sunlight. To solve this problem, humans invented different technologies including the solar mirror device. The next part will give an example to demonstrate the conversion theory.

Solar electricity conversion is based on the photovoltaic effect found by the French physical scientist Becquerel in the nineteenth century. It refers to the phenomenon that light causes a difference of potential between different parts of the semiconductor or semiconductor to which the semiconductor is bonded. This is the process to convert photons into electrons and subsequently from light energy to electrical energy. In addition, it is also a process to form voltage.

In general, solar energy provides a number of unique benefits:

(1) Abundant reserves. The amount of solar radiation to the surface of the Earth every year is almost the same as 13 billion tons of standard coal, or about 2×10^4 times of the current global total energy consumption.
(2) Permanence. The earth is supplied with solar radiation continuously. According to the current estimated nuclear rate produced by the sun, solar radiation energy can last for billions of years. While the history of life on earth is about millions of years, so solar energy can be regarded as inexhaustible.
(3) Universality. Solar radiation can be used locally in most areas of the earth, and it has great advantages in solving the energy supply problem in remote areas.
(4) Cleanness and safety. Solar energy produces energy with almost no pollution, it is far cleaner than conventional energy resources and safer than nuclear power.

However, solar energy has its weaknesses:

(1) Dispersion. Despite a large amount of solar radiation to the surface of the Earth, however, the density of this part of the energy is very low, which is only about 1 kW/m^2 at the ground.
(2) Discontinuity and instability. Solar radiation has not only the regular change with the diurnal, seasonal, latitudinal, and altitude factors, but also the random change with the weather.

Direct radiation is the main part of the absorption of solar radiation for most solar energy equipment. Hence, the solar radiation reaching the earth's surface is "weakened". Considering transmission factor, τ_G, which covers all weakening effects, the global radiation G_g can be calculated as shown in Eq. (10.1):

$$G_g = E_{SC}\tau_G \tag{10.1}$$

where
G_g = the global radiation;
E_{SC} = the solar constant;
τ_G = the transmission factor.

The transmission factor consists of the Rayleigh diffusion τ_{RD}, the Mie diffusion $\tau_{MD,}$ and the absorption within gases τ_{GA} as well as the absorption within particles τ_{PA}. Hence, Eq. (10.2) can be used to calculate the transmission factor

$$\tau_G = \tau_{RD}\tau_{MD}\tau_{GA}\tau_{PA} \tag{10.2}$$

where
τ_{RD} = the Rayleigh diffusion factor;
τ_{MD} = the Mie diffusion factor;
τ_{GA} = the absorption within gases;
τ_{PA} = the absorption within particles.

The diffusion mechanisms within the atmosphere cause diffuse and direct radiation to the incident on the surface of the earth. Direct radiation is the radiation incident on a particular spot, having traveled a straight path from the sun. In contrast, diffused radiation is the radiation that emerged by diffusion in the atmosphere and thus indirectly reaching a particular spot on the earth's surface. The sum of direct (beam) radiation G_b and diffused radiation G_d, which is related to the horizontal receiving surface, is called global radiation G_g. The diffused radiation G_d consists of the radiation diffused in the atmosphere, the atmospheric counter-radiation, and the radiation reflected by the neighborhood. Therefore, the global radiation can also be expressed as shown in Eq. (10.3)

$$G_g = G_b + G_d \tag{10.3}$$

It is noted that the global radiation insolation varies remarkably with seasons and weather conditions. Typically, both the maximum value of G_g and the length of the day are much smaller in winter than in summer. Furthermore, G_g of a clear day is larger than that of a cloudy day. With instrumental measurements and simulations, the amount of solar radiation can be measured and determined. To estimate one site total insolation, formulas given below are used

$$G_g = G_h^{max}\sin(\pi t/N) \tag{10.4}$$

Then integrating over the daylight period, the total insolation can be determined.

$$H_g = \int (G_b \cos\theta + G_d)dt = (2N/\pi)G_g^{max} \tag{10.5}$$

For example, for a clear day in the tropic area, $G_g^{max} \approx 950\ Wm^{-2}$, the daylight period keeps 12 h throughout the year, and $H_g \approx 26\ MJ\ m^{-2}\ day^{-1}$. For a cloudy day, the average measured values of H_g are typically 50–70% compared to the values for the clear sky value.

10.2.2 Principles of Wind Energy

Solar radiation received and emitted by longwave radiation is different on earth and in the atmosphere, so the temperature is different everywhere, which creates a difference in atmospheric pressure. The atmosphere flows from high pressure to low pressure. The atmospheric flow in the horizontal direction is the wind, which is caused by the difference in atmospheric pressure resulting in different air pressure. Figure 10.5 shows the pressure and wind band on the Earth.

Besides variations in the atmosphere, there are also considerable regional and local variations caused by geographical and environmental factors listed below.

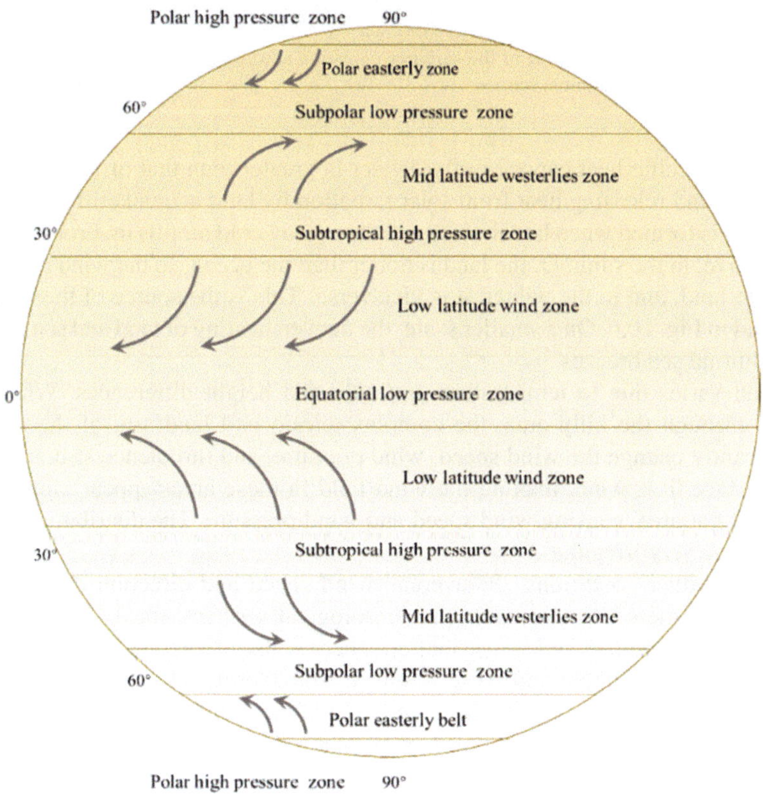

Fig. 10.5 Pressure and wind band on the earth

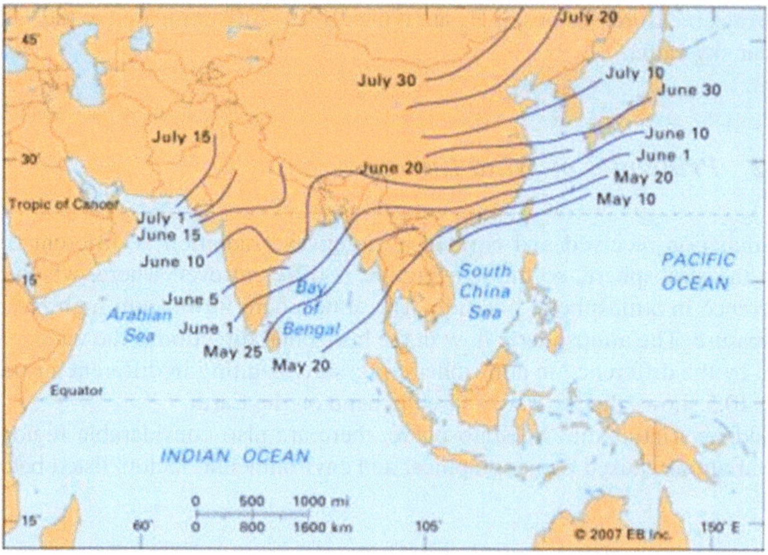

Fig. 10.6 Average date of onset of the summer monsoon across different regions of Asia (*Source* from Encyclopedia Britannica, Inc.)

As the specific heat capacity of seawater is greater than that of land, the rate of absorbing and releasing heat from solar radiation by land is quicker than the ocean. The wind is formed when hot air rises and the nearby cold air fills in. From a seasonal perspective, in the summer, the land is hotter than the ocean, so the wind blows from ocean to land, and in the winter, it is vice versa. This is the source of the monsoon, depicted in Fig. 10.6. On a smaller scale, the uneven heating of land and sea produces local diurnal sea breezes.

Wind varies due to temperature gratitude and height differences. When wind passes through the hilly area, the complex terrain and landform of the area will significantly change the wind speed, wind pressure, and turbulence structure of the near-surface flow wind, making the wind field in these areas appear more distinct from the flat area, causing wind speed and wind pressure. The distribution of wind is therefore very irregular.

Like weather conditions, the average wind speed and direction depend on the season and time of the day. Using meteorological services, local residents can be informed about the local wind distribution. There is also prediction nowadays on wind via computer simulation and AI technology. Considering all the factors, engineers can locate the best position for the installation of the wind power stations.

The energy of the wind is derived from solar radiation. The world has huge wind energy resources. It is estimated that the world's wind energy resources are as high as 53 trillion kW·h per year.

From the described context above, the air masses within the atmosphere are constantly moving. The kinetic energy E_{Wi} of these moving air masses can be calculated as shown in Eq. (10.6) [5]

$$E_{Wi} = \frac{1}{2} m_{Wi} v_{Wi}^2 \tag{10.6}$$

The air mass flow entering through a particular surface m_{Wi} is defined in Eq. (10.7) [6]

$$m_{Wi} = S\rho_{Wi} \frac{dx}{dt} = S\rho_{Wi} v_{Wi} \tag{10.7}$$

With the Eqs. (10.6) and (10.7), the kinetic wind capacity P_{Wi} for a free flow can hence be calculated as Eq. (10.8)

$$P_{Wi} = \frac{1}{2} m_{Wi} v_{Wi}^2 = \frac{1}{2} S\rho_{Wi} v_{Wi}^3 \tag{10.8}$$

According to Eq. (10.8), the wind power is proportional to the cube of the wind velocity; furthermore, it depends on the density of the air ρ_{wi} and the cross-sectional area S of the wind flow.

10.2.3 Principles of Biomass Energy

Biomass refers to all organic matter derived from plants and animals including their wastes and residues. Biomass energy is a form of energy based on biomass as the carrier and green plants using chlorophyll to convert solar energy into chemical energy stored in plants [7]. It is a unique renewable source of carbon and can be converted into solid, liquid, and gaseous fuels. Biomass can not only contribute energy, but also produce thousands of chemical products like coal and oil. While solar energy, wind energy, hydropower, and other renewable energy cannot produce fuel material by themselves. Biomass is the product of the photosynthesis of green plants. Figure 10.7 shows a typical process of photosynthesis.

Photosynthesis uses CO_2 in the air and water from the soil to convert absorbed solar energy into carbohydrates and O_2, as expressed in Eq. (10.9)

$$CO_2 + 2H_2O \xrightarrow{\text{light}} [CH_2O] + O_2 + H_2O \tag{10.9}$$

where

$[CH_2O]$ = a basic unit of carbohydrates.

It should be noted that the two oxygen atoms in the product O_2 are from the reactant H_2O rather than CO_2.

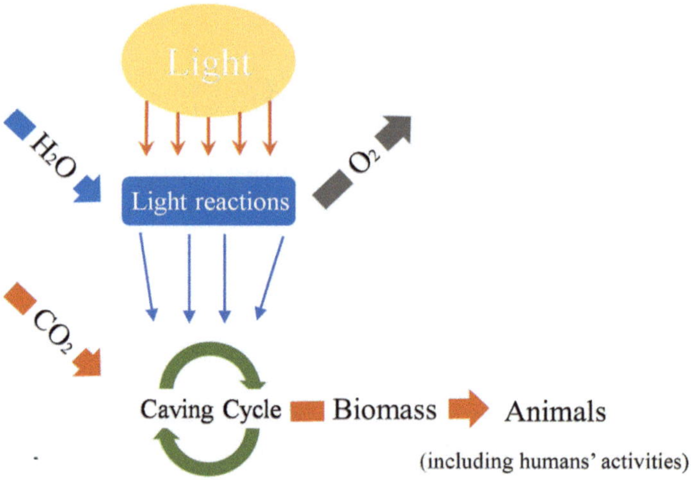

Fig. 10.7 Photosynthesis process

Photosynthesis can also occur by reducing CO_2 with compounds other than H_2O, as shown in Eq. (10.10) [8]

$$CO_2 + 2H_2X \rightarrow [CH_2O] + X_2 + H_2O \tag{10.10}$$

e.g., X may be sulfur (S), relating to certain photosynthetic bacteria that grow without oxygen.

Through the photosynthesis process, carbohydrates are produced. For humans, respiration and combustion are the two main ways to utilize biomass energy, demonstrated in Eq. (10.11)

$$[CH_2O] + H_2O + O_2 \xleftrightarrow{combustion/photosynthesis} CO_2 + 2H_2O + \Delta H \tag{10.11}$$

where

$\Delta H = 460$ kJ per mole C $\approx$16 MJ kg^{-1} of dry carbohydrate material.

Here, ΔH is the enthalpy change during human respiration and combustion activity. The difference is that respiration occurs at ~20 °C due to catalytic enzyme while combustion takes place over 400 °C.

Biomass energy has the following characteristics:

(1) The global biomass reserves are huge. The biomass used for energy derivation can be obtained by agricultural and forestry waste, deforestation, and special planting. The utility model has the advantages of wide variety and wide distribution. It is a convenient option for local utilization, and it has various forms and low energy density. The sources of biomass include the following:

Urban waste: industrial, living, and commercial waste.

Organic wastewater: industrial wastewater and domestic sewage.

Excrement: livestock, poultry, human excrement, and so on.

The forest biomass: firewood, branches, bark, roots, leaves, sawdust, wood shavings, etc.

Agricultural wastes: straw, husk, nut, corn cob, sugar cane, sugar beet pulp, the aquatic plants, algae, seaweed, duckweed, water hyacinth, reed, water hyacinth, etc.

The energy plant: trees shrubs and herbs with rapid growth and short rotation, such as cotton, sesame, peanut, soybean, etc.

In order to obtain biomass, deforestation will be limited by environmental protection, and the industry base of biomass energy may gradually be transferred to planting.

(2) The technology based on biomass energy is mature. Direct combustion has a long history using firewood and crop straw production by fermentation biogas for cooking and lighting is also common in rural areas. Sugarcane and corn are used to produce ethanol, replacing gasoline as fuel. A large scale application could be seen in the United States and Brazil.

(3) Energy-saving and environmental protection effects. The sulfur content and nitrogen content in biomass resources are lower than those of other fuel sources, and the ash fraction is also very small. As a result, SO_2, NO_x, and dust emissions after combustion are much lower than emissions from burning fossil fuels, which can prevent atmospheric acid rain. At the same time, the greatest contribution of biomass to the ecological environment is that, it has the characteristics "zero-emission" of CO_2 [9], which contributes to the prevention of the warming of the earth's surface. This is because the photosynthesis and biomass combustion form a cycle, where the consumption of CO_2 in biomass photosynthesis and the generation in combustion tend to achieve the goal of conservation.

Note that, not all the solar radiation is absorbed by the plant because the photosynthesis process requires an appropriate temperature. Plants have evolved to ensure the temperature, which is why leaves are seldom black. The photosynthesis efficiency is defined as Eq. (10.12) [8]

$$\eta = \frac{H/A}{E/A} \tag{10.12}$$

Data shows that one-year average global photosynthesis efficiency (radiation incident beneath the atmosphere onto all land and sea) is only around 0.1%. Not all the enthalpy gain of the biomass is stored as biomass available for reaction with oxygen for bioenergy. Gross primary production (GPP) is the initial rate at which plants capture energy. Net primary production (NPP) is the rate at which plants store chemical energy lesser than the energy used in their own respiration and growth, expressed in Eq. (10.13)

$$\text{NPP} = \text{GPP} - \text{energy of respiration} \tag{10.13}$$

Typically, NPP $\approx$ 0.5 GPP, though the ratio varies between plants and between ecosystems. Global terrestrial NPP can be estimated by combining satellite measurements of the amount of living plant matter (based on the spectral characteristics of chlorophyll), typically at a resolution of approximately 0.5 $\times$ 0.5 deg, calibrated against many land-based measurements. This yields average global NPP $\approx$ 50 Pg C(Pg $= 10^{15}$ g) in 2009, predicted by Potter et al. [10].

The processes of decreasing carbon dioxide from the atmosphere are called "carbon sinks". There are mainly two types of carbon sinks: one is through photosynthesis, in which, plants extract carbon in the form of CO_2 from the atmosphere to turn into carbohydrate material, which directly decreases the number of greenhouse gases. The other is that plant detritus in fallen leaves and branches form organic matter with the help of decomposers, which indirectly increases carbon concentration in the soil and decreases the greenhouse gases concentration in the atmosphere.

However, burning biofuels turns the carbon fixed in organic materials back to the atmosphere. Only if the plant material is burnt to replace (abate) specific use of fossil fuel, there will be a long-term benefit by preventing that fossil carbon from otherwise reaching the atmosphere. Developing high-efficient technology to produce methane from biomass is also significant. Even though it releases carbon dioxide, carbon dioxide is 23 times less of a greenhouse gas than methane. Generally, utilizing waste products avoids food versus fuel trade-off, and burning methane gas will reduce greenhouse gas emissions.

10.2.4 Principles of Ocean Energy

The total surface area of the ocean is 3.61 $\times$ 10^8 km^2, accounting for 71% of the total earth's surface area. The average depth of the ocean is about 380 m below sea level, and the volume of the whole sea is as much as 1.37 $\times$ 10^9 km^3. The ocean water not only provides shipping, fisheries, and abundant mineral resources for human beings, but also stores tremendous energy.

Ocean energy refers to the renewable energy generated from the sea. Oceans receive, store, and distribute energy through a variety of physical processes. All ocean energy sources, including tidal energy which is originated from the moon and the sun's gravitational pull, are derived from solar radiation. The energy generated from seawater includes heat energy, tide energy, tidal current energy; ocean current energy and wave energy is mechanical energy; and the salinity of seawater in the

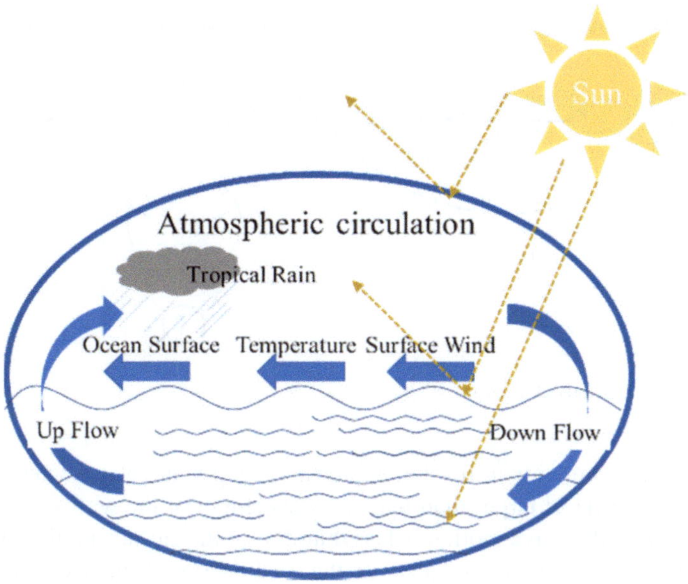

Fig. 10.8 *Source* of ocean energy

estuary area is chemical energy. Each ocean energy resource has a considerable energy flux: the tidal energy and salinity gradient can be about 2 TW, and wave energy is also at this order of magnitude, while ocean energy is at least two orders of magnitude larger than it. Figure 10.8 shows the source of ocean energy.

Generally, ocean energy has the following characteristics:

(1) Ocean energy is abundant in ocean water bodies, but the energy per unit volume, unit area, and unit length is relatively small. Therefore, the efficiency of deriving ocean energy is one of the limiting factors when considering energy utilization.
(2) Ocean energy comes from the gravitational attraction between solar radiation and celestial bodies. As long as the sun, moon, and other celestial bodies coexist with the earth, the ocean energy can be regenerated.
(3) The ocean energy can be divided into stable energy including temperature difference energy, salinity difference energy, and ocean current energy, and unstable energy including regular change (e.g., tidal energy and tidal energy) and irregular change (e.g., wave energy).
(4) Marine energy belongs to clean energy which has little effect on the environment after its exploitation.

Herein, two main physical types of ocean energy, wave energy and thermal energy, are introduced in detail in this section.

1. Wave energy

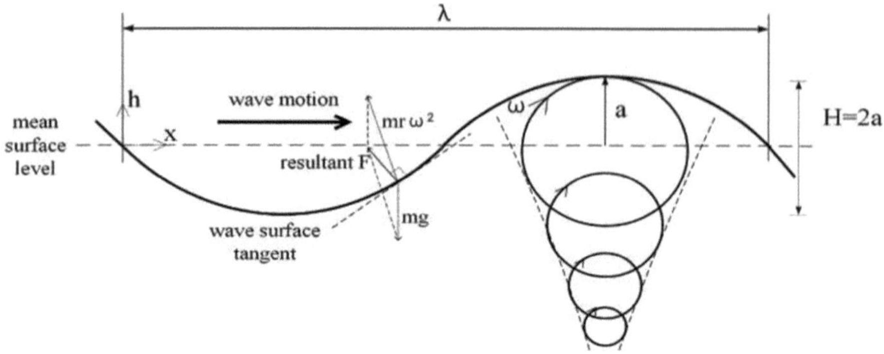

Fig. 10.9 Characteristics of waves

Wave energy is commonly utilized in the form of deep-water waves where the depth of the surface waves is larger than half the wavelength of the waves (D > λ/2). At the depth of λ/2, the wave motion is less than 5% of the surface motion, which can be considered as negligible. To calculate the total energy of the deep-water waves, we need to know the motion of the waves first.

Figure 10.9 shows the process of wave motion. There is no net movement for the water particles. In practice, water particles run in a circle and it means that the particles on the surface remain on the surface. Four main parameters that can be used to describe the wave include the amplitude a, the wavelength λ, the velocity c, and the period T. A particle of water in the surface has a circular motion of radius an equal to the amplitude of the wave and half of the height of waves on the surface, a = H/2. The radius r decreases with the depth of wave z, and the relationship is

$$r = ae^{kz} \tag{10.14}$$

where

k = wave constant number, equal to 2π/λ;

z = depth below the surface, a negative quantity.

For the motion of waves, the centrifugal force and the gravity force work together on wave particles. Thus, the water surface is perpendicular to resultant force F of gravitational and centrifugal force acting on an element of water. On the surface, the following equations can be derived:

$$\frac{dz}{dx} = \frac{a\omega^2 \sin\theta}{g + a\omega^2 \cos\theta} \approx \frac{a\omega^2 \sin\theta}{g} = \frac{a\omega^2 \cos\omega t}{g} \tag{10.15}$$

$$v_{vertical} = \frac{dz}{dt} = -a\omega \cos\omega t \tag{10.16}$$

The solution of the two equations above is:

Table 10.2 Parameters of waves

Parameter	Formula
Wavelength λ	$\lambda = \frac{2\pi g}{\omega^2}$
Wave period T	$T = \sqrt{\frac{2\pi\lambda}{g}}$
Velocity of a particle υ	$v = a\omega = a\sqrt{\frac{2\pi g}{\lambda}}$
Wave surface velocity c	$c = \frac{gT}{2\pi}$

$$z = a\sin(kx - \omega t) \tag{10.17}$$

This equation shows that the relationship between the wave position and the angular velocity. Other parameters' formulas are listed in Table 10.2.

The total energy of a certain wave can then be obtained.

$$\delta E_K = \frac{1}{2}\rho r^2\omega^2 dz = \frac{1}{2}\rho(ae^{kz})^2\omega^2 dz \tag{10.18}$$

where

δE_K = the kinetic energy of a unit length of the wave;

Hence

$$E_K = \int_{z=-\infty}^{z=0} \frac{1}{2}\rho(ae^{kz})^2\omega^2 dz = \frac{1}{4}\rho\frac{\omega^2 a^2}{k} = \frac{1}{4}\rho a^2\frac{\frac{2\pi g}{\lambda}}{\frac{2\pi}{\lambda}} = \frac{1}{4}\rho a^2 g \tag{10.19}$$

Since total energy $E = E_P + E_K$, so

$$E = E_P + E_K = \frac{1}{4}\rho a^2 g + \frac{1}{4}\rho a^2 g = \frac{1}{2}\rho a^2 g \tag{10.20}$$

This formula is of great importance, which means the potential energy of some waves can be calculated by just measuring the amplitude of the wave.

2. Thermal energy

Ocean thermal energy conversion (OTEC) is the conversion of ocean thermal energy into useful work to generate electricity. In an idealized OTEC system, the power input P_0 can be calculated using Eq. (10.21) [8]

$$P_0 = \rho c Q \Delta T \tag{10.21}$$

where

ρ = the density, for water is 1.0×10^3 kg/m^3;

c = specific heat capacity, for water is 4.2×10^3 J·kg^{-1}·K^{-1};

Q = the volume flow rate, m^3/s.

ΔT is the temperature difference and can be expressed as Eq. (10.22):

$$\Delta T = T_h - T_c \quad (10.22)$$

where

Th = the temperature when the flow enters the system, K;
Tc = the temperature when the flow leaves the system, K.

According to the second law of thermodynamics, the maximum output of work energy P_1 can be calculated using Eq. (10.23)

$$P_1 = \eta_{Carnot} P_0 \quad (10.23)$$

η_{Carnot} is the efficiency of an ideal Carnot engine, as expressed in Eq. (10.24)

$$\eta_{Carrot} = \frac{\Delta T}{T_h} \quad (10.24)$$

From Eqs. (10.21)–(10.24), the ideal mechanical output power is shown in Eq. (10.25)

$$P_1 = \frac{\rho c Q}{T_h}(\Delta T)^2 \quad (10.25)$$

Ocean energy is a type of renewable energy with characteristics such as sustainability, minimal chemical pollution, and natural variability.

However, scientists have no conclusion if the extraction and utilization of ocean energy at a very large scale may cause potential environmental effects, such as impact on marine ecology, which required further research.

10.2.5 Principles of Hydro Energy

Hydro energy includes kinetic and potential energy stored in the rivers and lakes. Among the total solar energy incident on earth, approximately 21% or 1.2×10^6 EJ/a are used to maintain the global water cycle (water evaporation and precipitation). However, only 0.02% or 200 EJ/a out of this amount of energy are finally available for utilization [11].

Due to gravitation, water flows within a stream or a river from a higher site to a lower one. Each side has its own pressure and velocity. A simplified 1D model can be used to describe the flow process and illustrate the exchange between potential and kinetic energy, assuming there is no friction and liquid is incompressible. Under the preconditions, the hydrodynamic Bernoulli pressure equation can be obtained as shown in Eq. (10.26)

$$\frac{P_1}{\rho g} + h_1 + \frac{v_1^2}{2g} = \frac{P_2}{\rho g} + h_2 + \frac{v_2^2}{2g} + h_{util} \tag{10.26}$$

where
h_{util} = the utilizable head of a river or a stream, m.

Therefore, if the volumetric flow rate Q and density of water ρ are defined, the work energy capacity of certain water can be expressed

$$P_{Wa} = \rho Q g h_{util} \tag{10.27}$$

where
P_{Wa} = work energy capacity of water, kg·m/s^3 or W.

Hydro energy has the following advantages:

(1) Environmentally friendly: hydropower is commonly considered to be environmentally friendly and pollution-free;
(2) Low cost: the operation cost of a hydropower plant is relatively low and the payback period of the plant is normally about 3–5 years;
(3) Additional benefits: a hydropower plant can prevent floods and adjust ship channels.

While the disadvantages are:

(1) Obvious limits: distribution of the hydro energy is limited by natural conditions such as hydrology, climate, and landforms;
(2) Ecological impact: construction of the hydropower plant brings some effects to the ecology system, and aggravating downstream soil erosion;
(3) Weather factor: in areas with large changes in precipitation season, there is less or even no power generation in the dry season.

10.2.6 Principles of Geothermal Energy

Geothermal energy is the heat from the subsurface of the earth, which comes from the decay of molten magma and radioactive material of the earth. The earth is a solid ellipsoid with an equivalent radius of 6378 km. It can be divided into crust, mantle, and centrosphere according to its structure from outside to inside. The underground radioactive material of the earth is constantly decaying and releasing a lot of heat via volcanic eruptions, intermittent fountains and hot springs, and other means. A steady flow of heat comes from the interior to the ground with thermal conduction, convection, and radiation.

An example of geothermal energy utilization is the heat pump system, which is driven by a power source and provides heating and/or cooling. The optimum theoretical performance of a heat pump is described by the Carnot cycle, the coefficient of performance C_{COP}^{Carnot} can be expressed as Eq. (10.28)

$$C_{COP}^{carnot} = \frac{P_{out}}{P_m} = \frac{T_{out}}{T_{out} - T_g} \tag{10.28}$$

where

P_m = the motor power, W;
P_{out} = the heat delivered by the heat pump, W;
T_g = input temperature in a Carnot cycle, K;
T_{out} = output temperature in a Carnot cycle, K.

However, it should be noted that the Carnot analysis assumes the process is infinitely slow, and it is reversible. Hence, the ideal C_{COP}^{Carnot} value is larger than the practical value, which is generally 3–5 for a commercial ground-sourced heat pump.

Geothermal energy has the following advantages:

(1) Geothermal energy is generally considered to be environmentally friendly which does not emit a significant amount of air pollution;
(2) The land area required to support a geothermal power plant is generally smaller than a fuel-fired plant (e.g., coal, nuclear, hydroelectric, solar, or wind);
(3) Geothermal resources are abundant. It is estimated that the worldwide geothermal potential is approximately 2 terawatts (TW);
(4) Geothermal energy is available everywhere, despite only some resources are profitably exploitable.

While the disadvantages are:

(1) Geothermal power plants may induce land subsidence, seismicity, landslides;
(2) Noise is generated during the development of a geothermal power project;
(3) Geothermal power plants may cause disturbances of wildlife habitat, vegetation, and scenic vistas;
(4) Geothermal power plants are very location-specific, where most resources are not cost-competitive.

10.3 Applications

10.3.1 Solar Energy

Due to the low energy density of solar radiation, a solar collector, which is also one of the most important devices for solar energy utilization, is required to absorb solar radiation. During the absorption of solar energy, the temperature of the collector will increase. If the collector is connected to a reservoir, energy will be transferred to it. While due to heat conduction and convection, the collector will also emit radiation and lose heat energy. Therefore, an efficient collector is to increase solar energy absorption and reduce the energy loss. Typical measures include operating the collector at temperatures not too high above ambient air temperature. When a higher temperature is required, appropriate transparent covers and insulation are needed to reduce the heat loss rates. A typical diagram of a solar collector is shown in Fig. 10.10.

The technologies for solar energy applications mainly include (1) solar photoelectric conversion technology [12, 13], (2) solar energy—heat exchange technology, and (3) solar—chemical energy conversion technology, as summarized in Table 10.3.

Photovoltaic power generation and solar thermal power generation are both widely used in the power generation industry. Differences between the two techniques are listed in Table 10.4.

The focal point of solar photoelectric conversion technology (PV technology) is the solar cells. The development of solar cells can be divided into three generations. The first generation is called the conventional and wafer-based cells made of crystalline silicon (c-Si). According to crystallinity and crystal size, the c-Si cells can be further classified into monocrystalline (mono-Si), polycrystalline (multi-Si), and ribbon silicon (ribbon-Si, no market-share now) cells. The second generation is promoted by thin-film technology and includes amorphous silicon (a-Si), CdTe, and CIGS cells. This generation is commercially significant in utility-scale photovoltaic power stations. The third generation of solar cells is often constructed by organometallic compounds and is currently at the development stage. Multi-junction (MJ) solar cells are another hotspot in research and can be defined as a hybrid type of the three kinds above. It contains several different semiconductor materials, typically using the metalorganic vapor phase epitaxy. Different materials have different

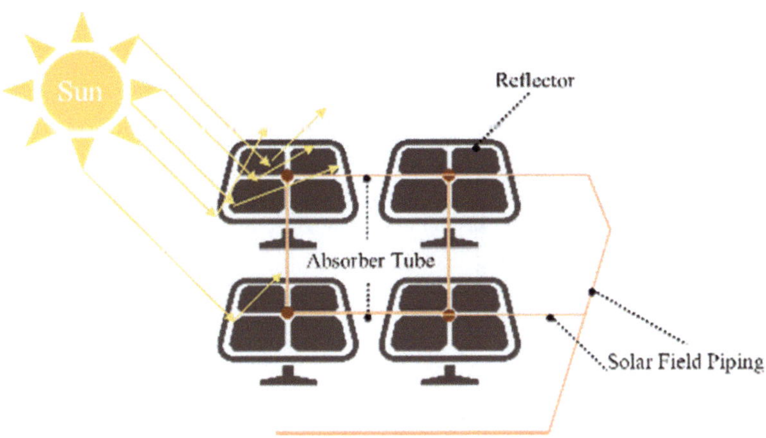

Fig. 10.10 Solar collector

Table 10.3 Technologies and applications of solar energy

Technology		Application	Diagram
Solar photoelectric conversion technology	Photovoltaic power generation	Semiconducting materials are used to convert sunlight directly into electricity. When the sunlight irradiates onto the solar cells, some of the photons from the light are absorbed along with inducing electron–hole pairs. Under the electric field of the cells, the photo-induced electrons and holes are separated. The flow of electrons generates an electric current	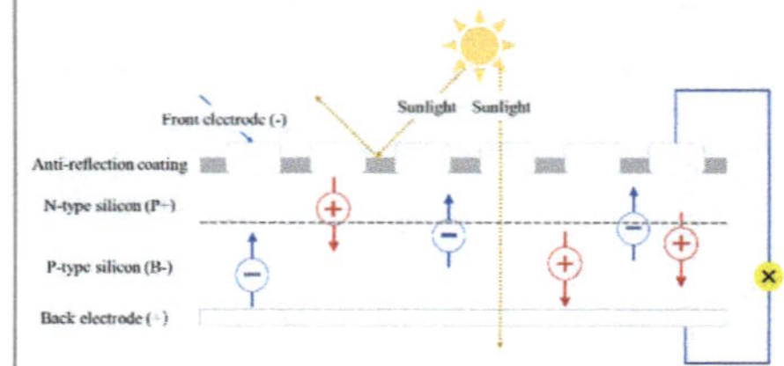
Solar energy-heat exchange technology	Solar thermal power generation	Heliostats are used to concentrate and reflect the sunlight to a collector, in which high temperatures are generated. The heat-transfer medium (liquid or gas) is then heated, followed by generating the steam, which then drives the turbine and generator to produce electricity. There are four main types of systems, as named by their collector type: 1) Parabolic trough, 2) Fresnel trough, 3) Solar tower, and 4) Dish/Stirling	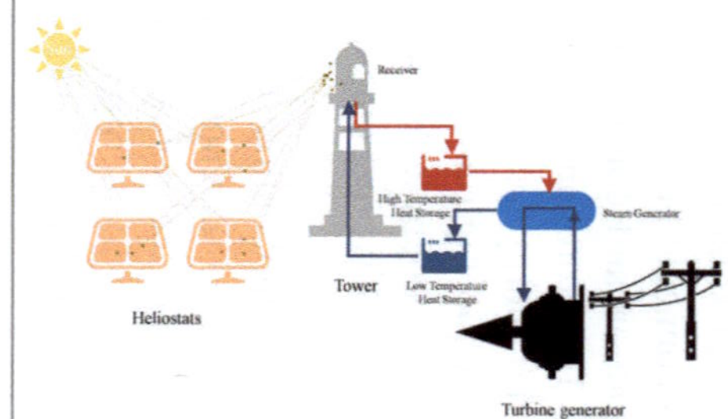
	Solar water heaters	The solar water heater is the device that converts solar energy into thermal energy for water heating. Solar radiation is absorbed by the collector and energy is then transferred to the fluid, followed by the heating of water. The hot water is stored in a hot water system for household use	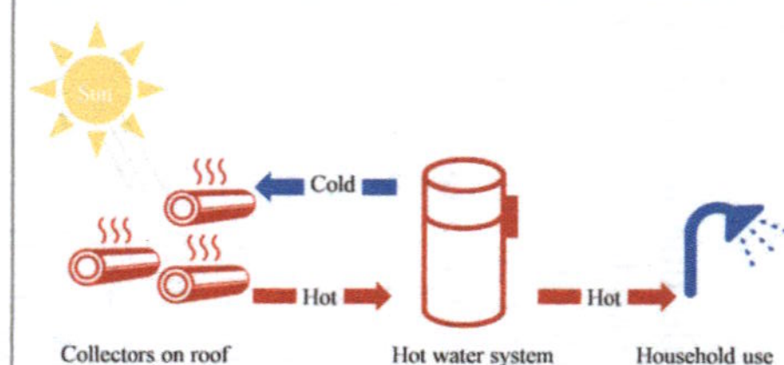

(continued)

Table 10.3 (continued)

Technology		Application	Diagram
	Solar refrigeration	Solar refrigeration is similar in principle to conventional refrigeration. There are two main categories: 1) Solar radiation is used as the driving energy directly. Absorption refrigeration, adsorption refrigeration, and jet refrigeration are the typical types 2) The solar-generated mechanical is used as the driving force. Herein compressed refrigeration, photoelectric refrigeration, and thermoelectric refrigeration are mainly involved	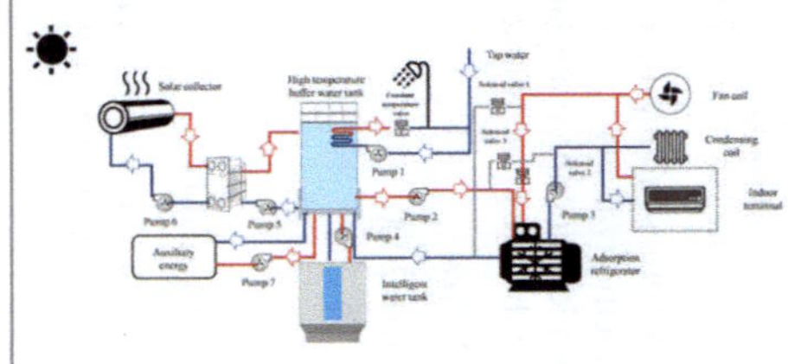
solar-chemical energy conversion technology		Solar photochemical conversion is the technology that converts solar energy into chemical energy, including photochemical effects, photosynthesis, and photoelectric conversion. Typical applications include light-assisted hydrogen production, green plant photosynthesis, thermochemical reaction to produce fuels	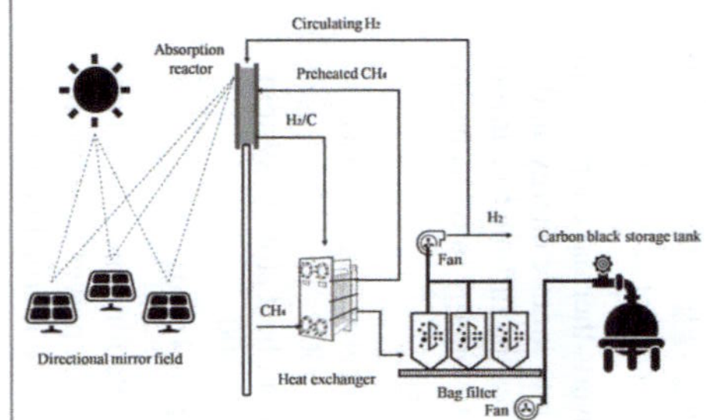

Table 10.4 Comparison between photovoltaic and solar thermal power generation

	Solar thermal power generation	Photovoltaic power generation
Principle	Converts thermal energy of the sun into kinetic energy, and utilizes the turbine to convert kinetic energy into electricity	Converts sunlight into electricity using semiconducting materials that exhibit the photovoltaic effect
Conversion efficiency	12–25%	10–20%
Electricity generation hours per year (h)	With heat storage system: 4000–5000 Without heat storage system: 1000–2000	1000–2000
Land need	35–40 m^2/MW	25–30 m^2/MW
Application situation	Suitable for concentrated and big-scale power generation, similar to thermal power generation	Suitable for distributed and small-scale power generation
Current operation cost ($/kWh)	0.12–0.3	0.2–0.35
Future operation cost ($/kWh)	0.04–0.06	0.04–0.10
Advantages	The cost of heat storage is low and the efficiency is high. It can also be combined properly with other power generation technology	Technology and industry are both mature now
Disadvantages	High demands for the land situation	The production and maintenance process causes pollution. And it needs higher stability

bandgaps, thus capable of absorbing a broader wavelength of sunlight, leading to high solar electricity conversion efficiency. MJ solar cells are often combined with light concentration technology to obtain higher efficiency and can reach the efficiency of 37.9% (without concentrator). Details of technology parameters are shown in Table 10.5.

Discontinuity and instability are the main disadvantages of solar photoelectric conversion technology. On a rainy day and a cloudy day, it may induce drastic fluctuation in the electricity grid. The solution is to link the battery group to the

Table 10.5 Technology parameters of different solar cells

Materials	Mono-Si (%)	Multi-Si (%)	A-Si (%)	CdTe (%)	CIGS (%)
Cell efficiency (lab)	27.6	21.9	14	22.1	22.6
Market share	32.2	60.8	0.3	2.3	1.9

PV system. Researches have been conducted on solar thermal power generation technology to moderate the problem above [8].

Due to the low energy intensity of sunlight, solar thermal technology is a suitable option for concentrated solar power (CSP) generation. There are mainly four types of CSP stations, parabolic trough power plant, Fresnel trough power plant, solar tower power plants, and dish/Stirling power plant. Detailed parameters of the four types are listed in Table 10.6.

10.3.2 Wind Energy

When the wind flows through the windmill blades, three types of forces are generated (Fig. 10.11), including the axial force (i.e., the drag force by the contact of air with the blade), the radial force (i.e., the lift force which gives the blade a tendency to move), and the tangential force. The first two forces are involved in power generation. The horizontal-axis wind turbines are mainly driven by the axial force, while the vertical shaft fan uses the radial force. Modern wind turbines mainly utilize the lift force, in which the direction is perpendicular to the wind direction. When the air flows through the blade surface, the flow velocity at the windward side of the blade is much higher than that of the leeward side, resulting in the pressure on the windward surface smaller than that of the leeward surface. Consequently, the lifting force is generated, which causes the rotor to rotate around the central axis.

Wind energy is used to drive wind turbines to generate electricity and it also could be converted into mechanical energy, see Table 10.7.

10.3.3 Biomass Energy

Biomass contains a large amount of amino acid., protein, and lignocellulose (like grasses, agricultural wastes, forestry residues, and woody materials). It is a renewable energy source and is an inexpensive feedstock to produce useful mineral materials, chemicals, fuels, and heat as well as power [15]. There is a large variety of biomass with different characteristics, making it more complicated for applications.

The application of biomass energy can generally be divided into three categories: (1) physical conversion, (2) chemical conversion, and (3) biological conversion, demonstrated in Fig. 10.12. Among them, the chemical conversion of biomass is widely used to produce biofuels, heat, power, etc. In addition, biomass can also be used for preparing energy materials, such as porous carbon materials for electrochemical energy storage [16].

Application of biomass energy may include ethanol production, biomass gasification, and combustion of biomass for power generation, as shown in Table 10.8.

Table 10.6 Parameters of four types of typical solar thermal power plant

Type	Focusing type	Efficiency[a] in %	Operation temperature	Concentration factor[b]	Capacity factor[c] in %	Typical capacity	Technology status[e]	Appearance
Parabolic trough	Line	10–23	300 °C	50–90	23–56	30–300 MW	++	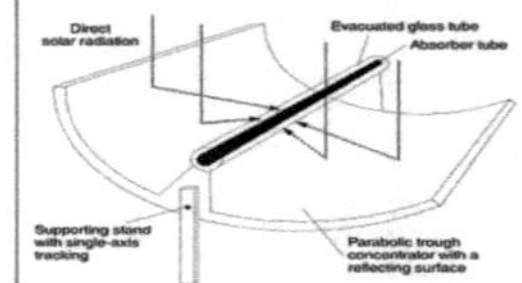
Fresnel trough	Line	9–17	390 °C	25–50	25–90	10–200 MW	0	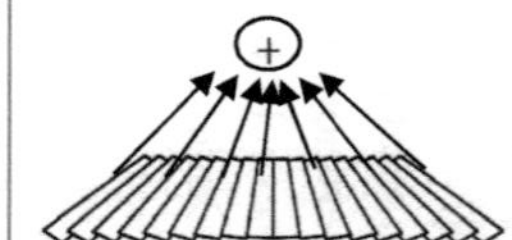
Solar tower	Point	10 – 28	575 °C	600–1,000	20–78	10–200 MW	+	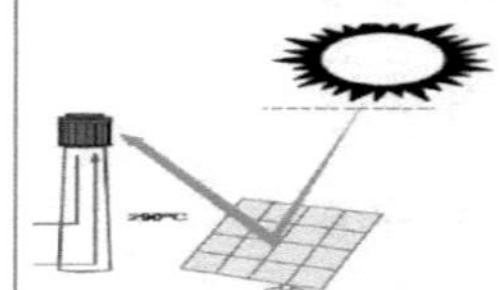

(continued)

Table 10.6 (continued)

Type	Focusing type	Efficiency[a] in %	Operation temperature	Concentration factor[b]	Capacity factor[c] in %	Typical capacity	Technology status[e]	Appearance
Dish/stirling	Point	15–25	750 °C	up to 3,000	24–25	%1.%2– 1[d] MW	+	

[a] conversion of radiation energy into electrical energy
[b] geometrically concentration factor (C_{geom}), describing the ratio of the solar aperture surface A_{ap} to the absorber surface A_{abs}
[c] real amount of electricity generation and yearly amount of electricity generation operating at rated power
[d] power of single Dish/Stirling power generation module, can be interconnected with others
[e] 0 successful operation of demonstration plants, + successful continuous operation of demonstration plants, ++ commercial plants in operation

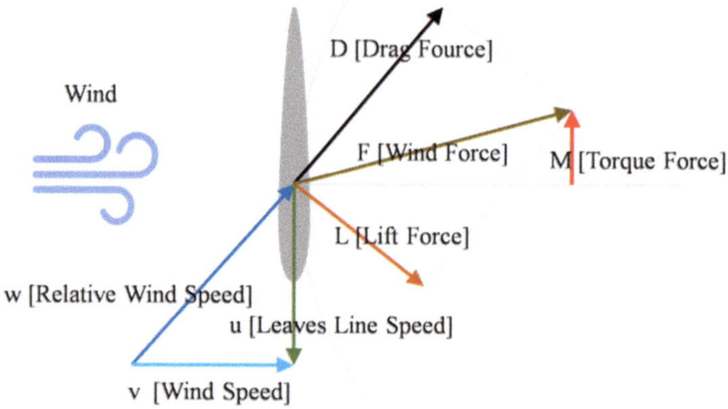

Fig. 10.11 Airflow and forces generated by wind

10.3.4 Ocean Energy

The utilizations and applications of tidal energy, wave energy, thermal gradient energy, and salinity gradient energy are summarized in Table 10.9.

10.3.5 Hydro Energy

The hydropower plants were constructed to utilize the power of flowing water. A typical hydropower plant usually includes dams, intake systems, penstock systems, factory buildings, and outlet systems. Figure 10.13 shows the procedure of power generation and energy conversion processes.

The total energy generated by the system can be calculated by Eq. (10.29).

$$N = 9.81\eta QH \tag{10.29}$$

where

N, the total capacity of the power generation system;

Q, the volumetric flow rate of the river or stream;

η, the efficiency of the turbine.

H, the difference between the height of the intake and the height of the turbine.

There are mainly two kinds of hydropower plants, including the water-storage type and no-storage type as shown in Table 10.10.

Table 10.7 Technologies and applications of wind energy

Technology	Application	Diagram	
Wind energy for electricity generation	Wind turbines are energy conversion devices that convert wind energy into mechanical, electrical or thermal energy. The devices are mainly divided into two types: (1) onshore wind power generation and (2) offshore wind power generation [14]. The former one is the basis of offshore wind power generation technology, which has the advantages of small land usages, high wind velocity, small turbulence intensity, large capacity of power generation, and negligible noise and vision impacts. Therefore, offshore wind power has become a new frontier for the development of wind power utilization	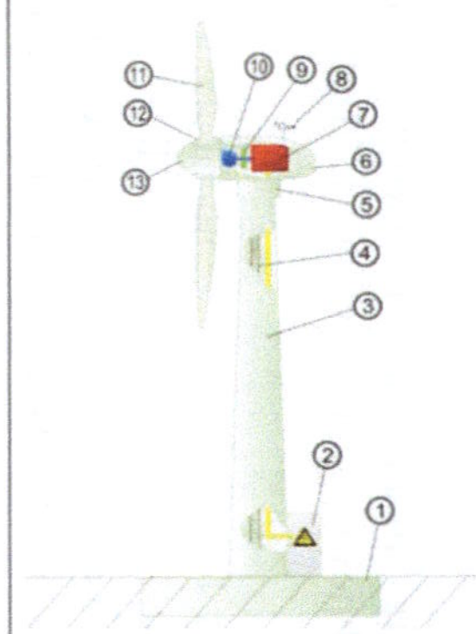	1. Foundation, 2. Connection to the electric grid, 3. Tower, 4. Access ladder, 5. Wind orientation control (Yaw control), 6. Nacelle, 7. Generator, 8. Anemometer, 9. Electric or Mechanical Brake, 10. Gearbox, 11. Rotor blade, 12. Blade pitch control, 13. Rotor hub
Wind energy—mechanical energy conversion	Herein wind energy is used to directly drive a variety of mechanical devices, such as water pumps. The advantages of this wind engine include (1) small investment, (2) high efficiency, and (3) durability	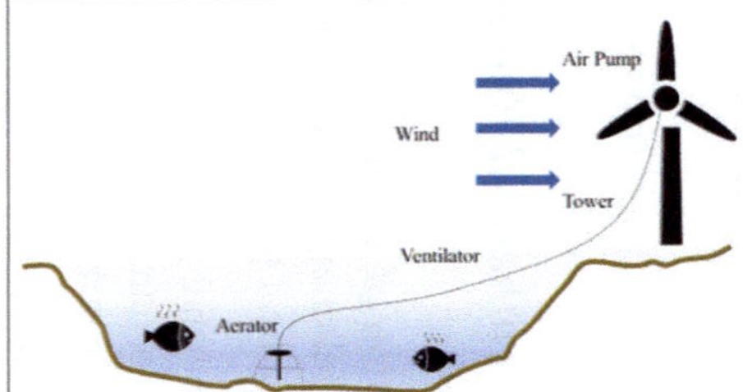 Fish Tank Aerator Diagram	

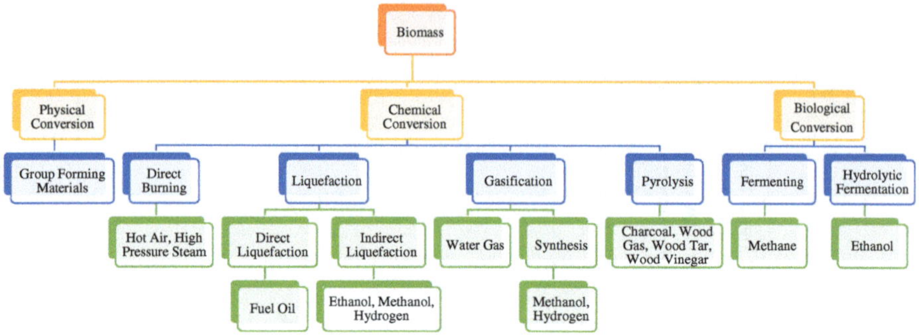

Fig. 10.12 Biomass energy conversion technology and its products

10.3.6 Geothermal Energy

The utilization of geothermal energy mainly includes two categories: conversion of geothermal energy for electrical power generation, and utilization of geothermal heat, such as heat pumps.

(1) Geothermal power generation. In this process, geothermal energy is converted into mechanical energy first and then electricity. To utilize geothermal energy, there must be a heat carrier to carry the underground heat to the ground. At present, the heat carriers that can be used by geothermal energy utilization devices mainly are groundwater and/or heated water.

(2) Heat pump. The heat from the near-surface ground, including soil, groundwater, or from rivers and/or lakes, is input to heat pumps, which can provide heat for use at increased temperature. A heat pump absorbs heat at a low temperature and releases it at a high-temperature level with a heat exchange device. Therefore, thermal energy from a low-temperature heat source can be withdrawn by a heat pump. Then the absorbed thermal energy can be used as thermal energy at higher temperatures. The technology above is now applied throughout the world and is one growing rapidly geothermal utilization technology.

Table 10.11 listed the typical use of geothermal energy including geothermal power generation and heat pump.

10.4 Challenges and Perspectives

10.4.1 Barriers to Be Overcome

The aforementioned paragraphs discuss mainly the technological aspects of different types of renewable energy. It is important to know that apart from technical aspects, the development and deployment of technologies are also affected by the political,

Table 10.8 Technologies and applications of biomass energy

Technology	Application	Diagram
Ethanol production from biomass	The biomass is pretreated using dilute sulfuric acid, followed by mixing with lime to increase the pH value and remove compounds that are harmful to the bioconversion device. The end product of ethanol is then purified by a two-column distillation and adsorption using molecular sieves. The residual solids are removed and fed to a power plant boiler. The liquid at the distillation bottom is concentrated by evaporation with the syrup, which is also fed into the boiler. The wastewater is treated by anaerobic digestion, a portion of the treated wastewater is recycled	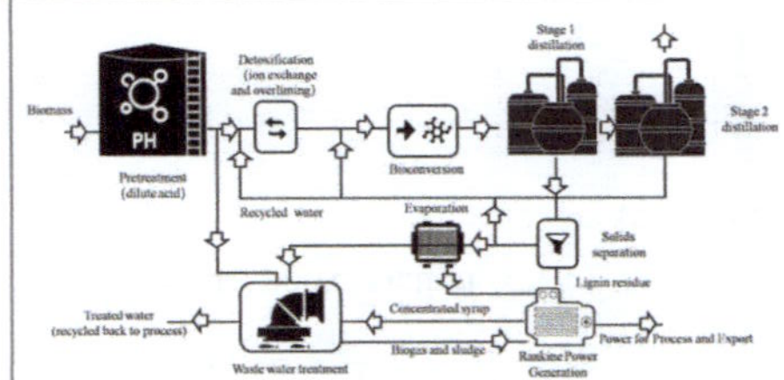
Biomass gasification	The biomass raw materials are added from the top of the furnace. In the drying layer, water in the raw materials evaporates. In the cracking process, the dry materials are cracked to produce combustible gases, including CO, CH_4, H_2, etc. The gas mixture containing combustible gases and CO_2 comes out from the cracking process for further use. The resulted carbon is reacted with the gasification agent (such as air, added from the bottom of the furnace) to form CO (incomplete combustion) and CO_2 (complete combustion), releasing a large amount of heat. The CO_2 in the reduction layer (no oxygen) can react with carbon to produce combustible CO, which then comes out from the outlet in the cracking process	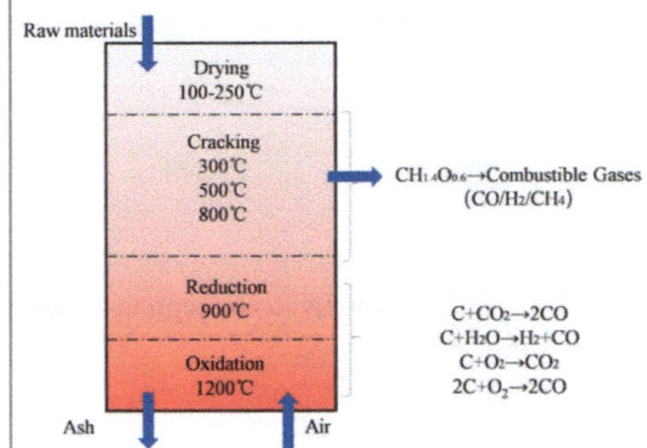
Biomass combustion for power generation	Biomass is burned in a boiler, releasing heat to produce superheated steam from water. The steam is then fed into a turbine generator, in which energy from the steam is converted into power. The outlet stream from the turbine is fed into a condenser and then goes to an economizer. The water undergoes several phase changes and returns to the original state	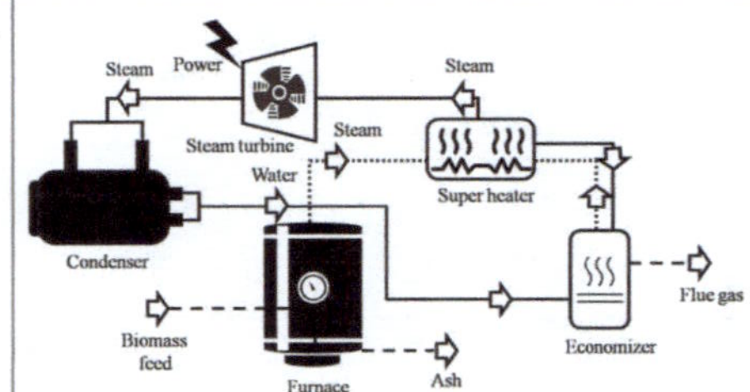

Table 10.9 Technologies and applications of ocean energy

Technology	Application	Diagram
Tidal power generation	The main utilization of tidal energy is tidal power generation, consisting of one-basin and two-basin tidal power stations In a one-basin system, a bay is insulated from the sea by a dam, while it is still connected to the sea using a floodgate and a turbine. Water can enter the bay only through the floodgate and exit the bay only via the turbine. The turbine is connected with an electrical generator to produce electricity For a two-basin system, another turbine is integrated between the two basins. This turbine is built either into a dam or into a canal located between the two basins, which are also connected to the sea via turbines. The inflow and outflow of water are controlled when water enters Basin 1 at high tide and exits via Basin 2 at low tide, along with having passed through the turbines	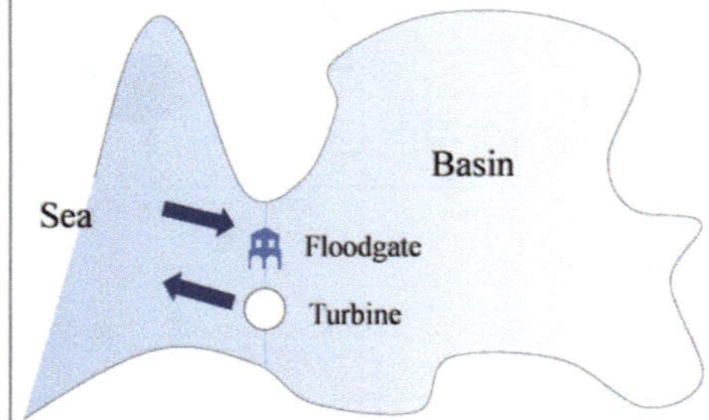 One-basin tidal power station 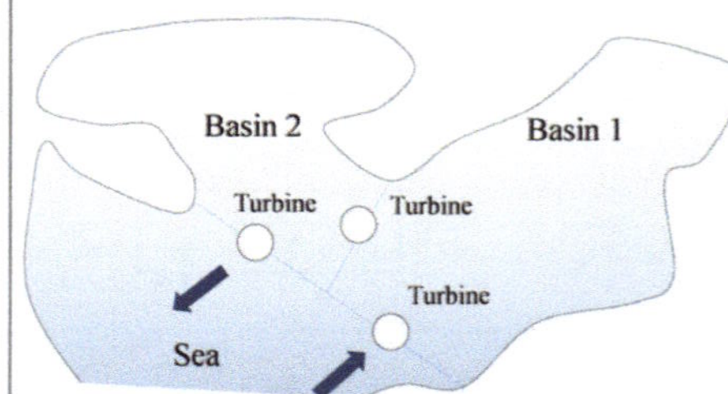 Two-basin tidal power station

(continued)

Table 10.9 (continued)

Technology	Application	Diagram
Wave power generation	When waves pass onto a partially submerged cavity, a water column fluctuates in the cavity. With the fluctuates waves, a reciprocating motion in the air is induced, which is linked to the atmosphere via a turbine connected with a generator. A Wells turbine is usually used. Once started, it continues to turn in the same direction. Therefore, a continuous power generation is produced regardless of the wave energy changes	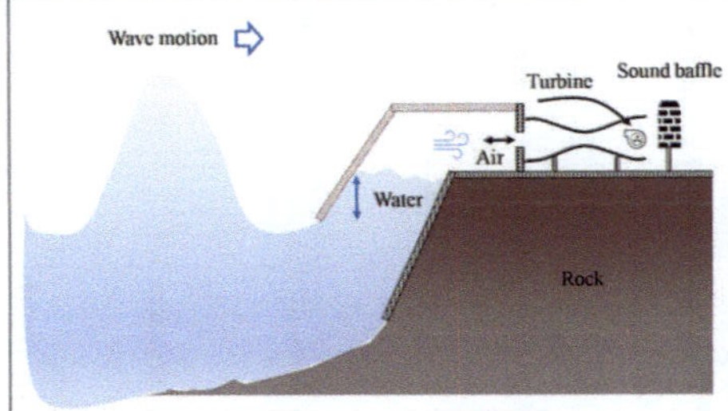
Thermal gradient power generation	When the warm surface seawater enters the evaporator, the working fluid boils (usually at 25–30 °C), driving the turbine and generator to produce electricity. The outlet vapor condenses in a condenser (usually at around 5 °C), through which cold deep seawater passes. The outlet water from the evaporator and the condenser can be used for other purposes	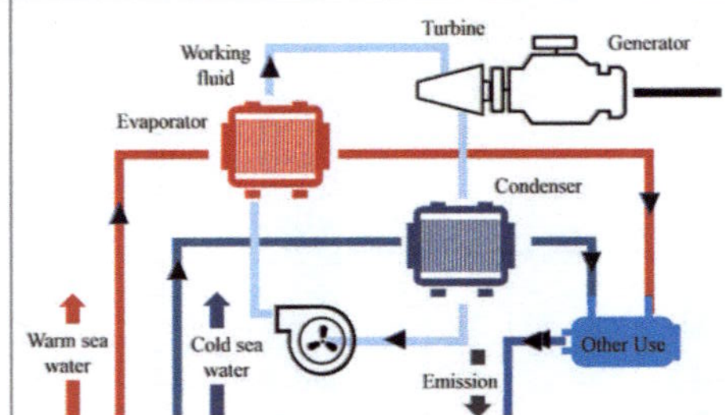
Salinity gradient power generation	When the seawater is separated from the pure water by membrane modules in the container, the pressure on the seawater side of the membrane is higher than that on the pure waterside. As a result, a part of the pure water will flow out of the water tank through the membrane, and generate a huge power that can push the water motor to operate, thereby generating electricity	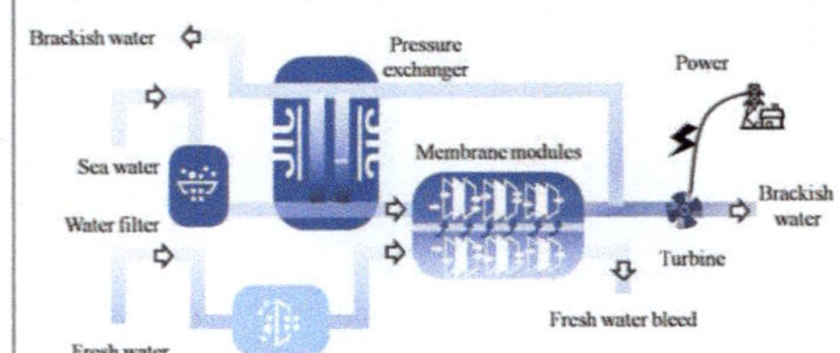

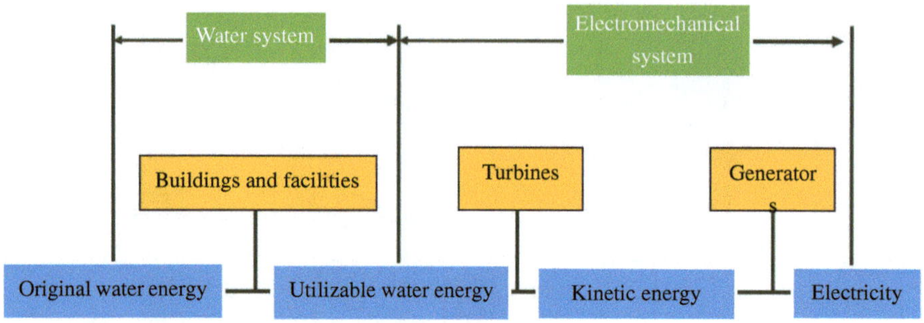

Fig. 10.13 The procedure of power generation and energy conversion

economic, and social factors, such as climate change mitigation targets set up by the governments, increasingly new players flocking in with higher revenues, job employment. However, there still exist multiple barriers that inhibit the energy transition process:

1. Institutional and administrative barriers

Institutional and administrative barriers include a lack of institutions and authorities dedicated to renewables; the absence of clearly defined responsibilities; complicated licensing procedures; difficulty in land acquisition and permission; inadequate planning guidelines; and complex, slow, lengthy or opaque permitting processes. Political or other resistance to renewable energy—such as institutional corruption and anti-renewables lobbying—can impede further development.

2. Regulatory and policy

Regulatory and policy include inappropriate policy, discontinuity of policies, perverse or split incentives, unfavorable or inconsistent policies, unclear agreements (such as power purchase agreements, feed-in tariffs or self-consumption), and a lack of transparency. Uncertainty and inconsistency about targets and policies, including retroactive changes, significantly hamper renewable energy expansion, as support schemes or procedures that are unclear lower confidence among investors and developers.

Table 10.10 Technologies and applications of hydro energy

Technology	Application	Diagram
No storage hydroelectricity	Most hydroelectric power stations use dams to regulate the water and to drive a water turbine. A hydroelectric power station normally consists of a dam, intake works, penstock, and in some cases a headrace, plus the powerhouse and tailrace. The generation power depends on the volume of the water flow and the height between the intake and the turbine. The height difference is called the head	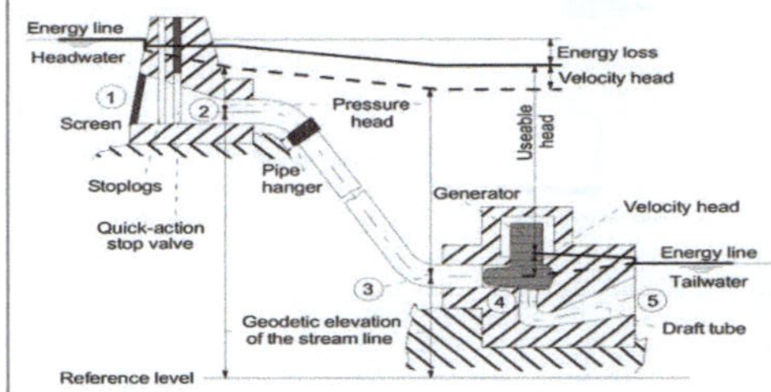
Pumped-storage hydroelectricity	This technology can compensate for the electricity shortage, and is also suitable for combining with other renewable energy to promote the stability of the electricity system. When there is low electrical demand, excess generation capacity is used to pump water into the upper reservoir. When there is higher demand, water is released back into the lower reservoir through a turbine, generating electricity. Reversible turbine/generator assemblies act as a combined pump and turbine generator unit (usually a Francis turbine design)	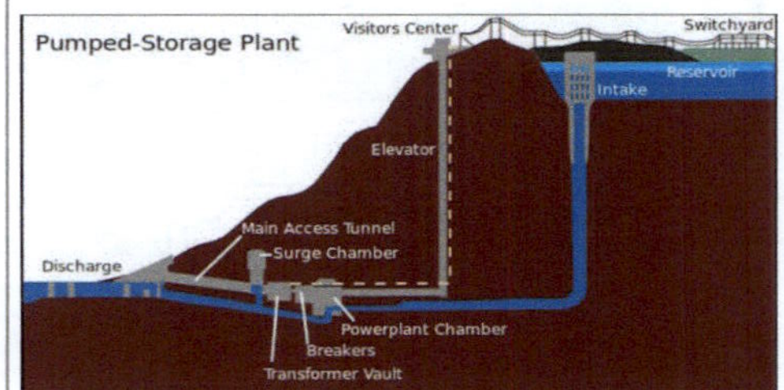

Table 10.11 Technologies and applications of geothermal energy

Technology	Application	Diagram
Geothermal power generation	Superheated water at high temperatures and pressures is stored in the geothermal reservoir. The steam, when the temperature of which is higher than 180 °C, can be directly passed from the ground into the turbine, operating with a Rankine cycle. The turbine starts to rotate, followed by driving the generator to produce electricity. When the water temperature is low (from 110 to 180 °C), a heat exchanger is used in which other working fluids are heated by the water. The working fluids are generally organics and boil at ~ 80 °C, hence generating pressurized vapors into the turbine, driving it and the generator to produce electricity	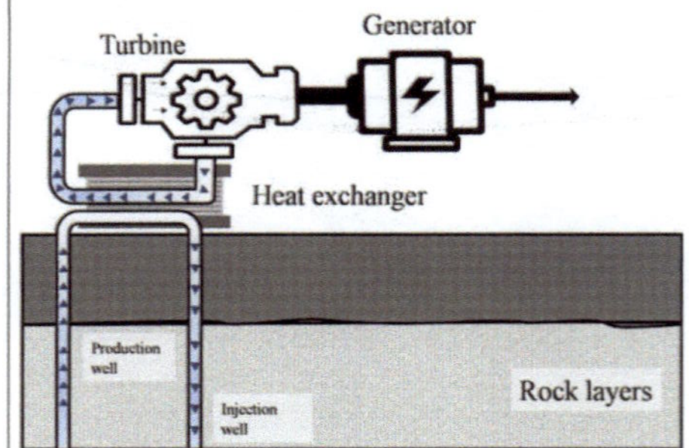
Heat pump	The working agent is evaporated in the evaporator by heat addition, which is made from geothermal energy. The vapor-phase working agent is suctioned and compressed by a compressor. Its temperature is increased. The working agent then flows to a condenser and is liquefied, followed by releasing heat to the heat utilization system. It subsequently flows to the low-temperature region through the expansion valve, completing a cycle	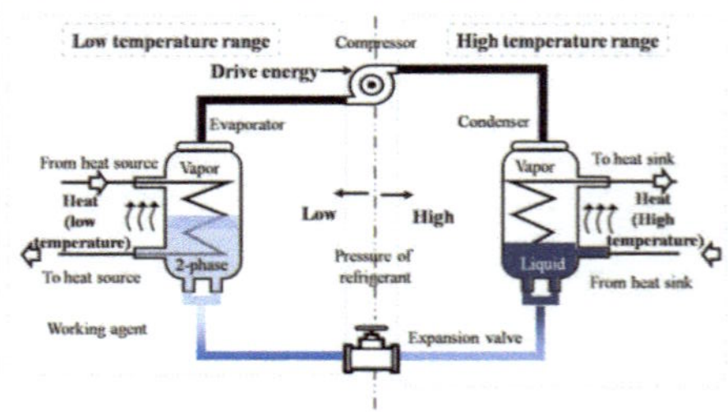

3. Financial barriers

Financial barriers pertain to the lack of adequate funding opportunities and financing products for renewables [17]. As the cost of renewables falling, many projects on renewable energy become commercially available in more and more countries. Nevertheless, government investment still stays lower than its potential. There are three main barriers in investment [18], including project start-up and development, investment risk management, and up-scaling investment. Technical assistance and grant funding in the start-up process, targeted non-financial interventions such as project initiation and facilitation tools, together with debt-based instruments are all needed to solve these financial barriers.

The usage of the large-scale renewable energy has a long way to go. Initially, safety and technical maturity need a guarantee. Secondly, renewable energy needs to be cost-competitive compared with other fuels when several energy options are offered. Thirdly, a lack of economies of scale can also contribute to higher costs for the system, particularly in the early stages of market growth. According to IEA (2016), the cost-effectiveness of renewable options has improved due to a combination of sustained policy support, technology progress, expansion into newer markets with better resources, and better financing conditions [19].

Market barriers also exist, including inconsistent pricing structures that lead to disadvantages for renewables, irregular pricing of renewable energy products, information asymmetries, distortions in market power, fossil fuel and nuclear subsidies, and a failure to incorporate social and environmental externalities into costs. Many countries have energy tariffs that are not fully cost-reflective, as well as fossil fuel and/or nuclear subsidies that inhibit the deployment of renewables. Low fossil fuel prices can similarly slow the pace of deployment, particularly for renewable heating and cooling and transport. Trade barriers in some countries—such as import duties—also make importing renewable energy products more expensive.

4. Infrastructure barriers

Infrastructure barriers pertain to the availability of needed infrastructure to incorporate renewable energy into the energy system, which includes problems linked to system flexibility and the ability of the power grid to integrate renewable energy. As the deployment of renewable power capacity is being developed, challenges to grid integration can arise, amplified by a weak grid infrastructure or a lack of required upgrades for transmission and distribution infrastructure. In some countries, these challenges can result in the curtailment of power from renewable sources. Restricted grid connection or access is another concern, particularly for distributed technologies and in cases of a vertically integrated power sector. In addition, a lack of district heating or adequate cooling infrastructure hinders progress in the heating and cooling sectors, and the absence of appropriate engines in vehicle fleets hampers the deployment of biofuels in the transport sector.

5. Public acceptance and environmental barriers

Public acceptance and environmental barriers constitute constraints that could lead to a renewable energy project being found unsuitable for a specific location. A lack of public acceptance alone can lead to higher costs, delays, and even cancelation of projects. Local planning and zoning regulations and environmental features can further hinder the deployment of renewable energy in certain areas.

There are also awareness and capacity barriers relating to insufficient skilled personnel and training programs, which usually happen in developing countries. Not sufficient awareness of renewables' technology and inability to carry out operation and maintenance activities, lead the renewable systems to fail after implementation. However, it is difficult to fulfill the request in a short time. Focusing on improving education, training more skilled workers, and upgrading local value chain are necessary tasks to implement.

10.4.2 4E Evaluation

To better understand the perspectives and challenges of renewable energy, a 4E evaluation (including energy, environment, economic, and engineering) should be performed. Among all these types of renewable energy, biomass energy has the highest energy efficiency with a relatively low cost. However, this traditional renewable energy only received modest investment support, while emerging solar energy and wind energy have largely developed with the highest investments. All solar, wind, and biomass energy are clean energies with no CO_2 emission, and these significant environmental benefits might be one of the reasons for their remarkable developments. Moreover, all these emerging technologies have created millions of direct and indirect jobs, which are essential for human society. The 4E evaluation of the technologies for renewable energy is shown in Table 10.12.

10.4.3 Integrated and Innovative Solutions

There are quite a few positive developments of renewable energy, which indicates the growth of renewable energy in global energy consumption, especially in the renewable power sector. Some of the renewable energy technologies have been widely accepted as a mainstream energy source with their competitive cost for a long time, including hydropower, bioenergy, and geothermal power. Some emerging technologies are now joining them with largely increasing install capacity globally, especially solar PV and wind power.

The importance of integrating renewable energy is widely and increasingly recognized, and the application of renewable energy sources is expected to keep increasing (Fig. 10.14). Governments and corporates around the world have shown great interest in renewable power and the investments show an optimist growth worldwide (Fig. 10.15).

Although more and more countries have established policies and regulations supporting renewable energy (Fig. 10.16), there are still sustainability debates and

Table 10.12 Evaluation of Renewable Energy (data obtained in 2017)

Classification key parameters		Hydropower	Solar energy		Wind energy	Biomass energy	Ocean energy	Geothermal energy
			Solar PV	CSP				
Energy	Installed capacity (GW) [20]	1,114	402	4.9	539	122	0.5	12.8
	Annually added capacity (GW) [20]	19	98	0.1	52	8	0.01	0.707
	Global power generation (TWh) [21]	–	442.6		1122.7	555	32.2	
Environment	CO_2 emission reduction (Mt/year)	–	441.27		1119.33	553.335	28.98	
	CO_2 reduction benefit (billion USD/year)	–	6.619		16.79	6.64	0.435	
Economic	Levelised cost of energy (USD/kWh) [20]	0.03–0.12	0.07–0.14	0.12–0.32	0.06–0.24	0.05–0.12	–	0.05–0.10
	New investment (billion USD) [20]	3.2	160.8		107.2	6.6	0.2	1.7
Engineering	Created job vacancy (thousand jobs) [20]	1,804	3,365	34	1,148	3,055	–	93

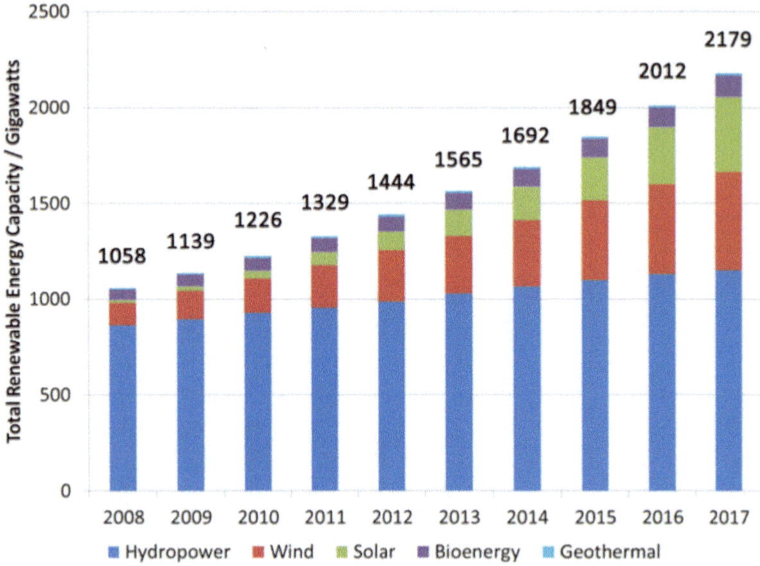

Fig. 10.14 Global renewable power capacity, 2008–2017 (data from IRENA: https://www.irena.org/)

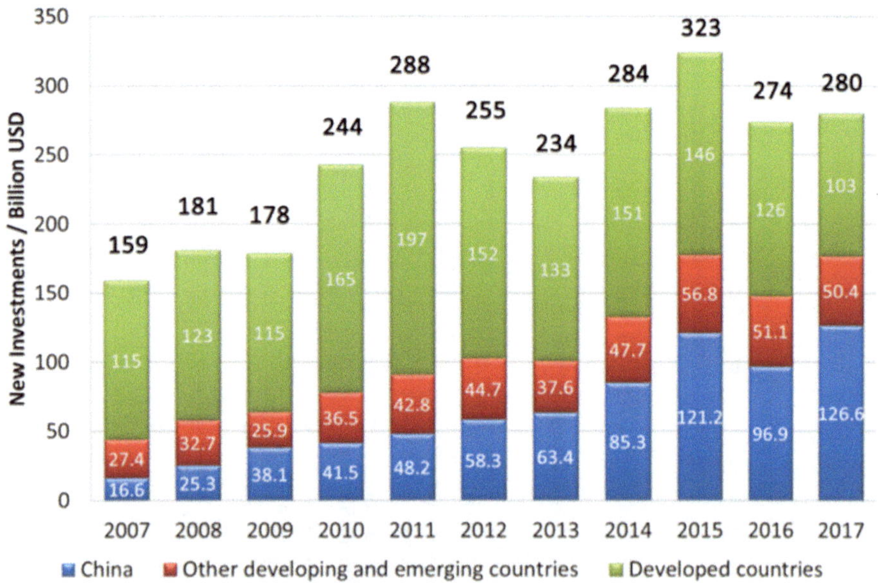

Fig. 10.15 Global new investment in renewable power and fuels, 2007–2017 [20]

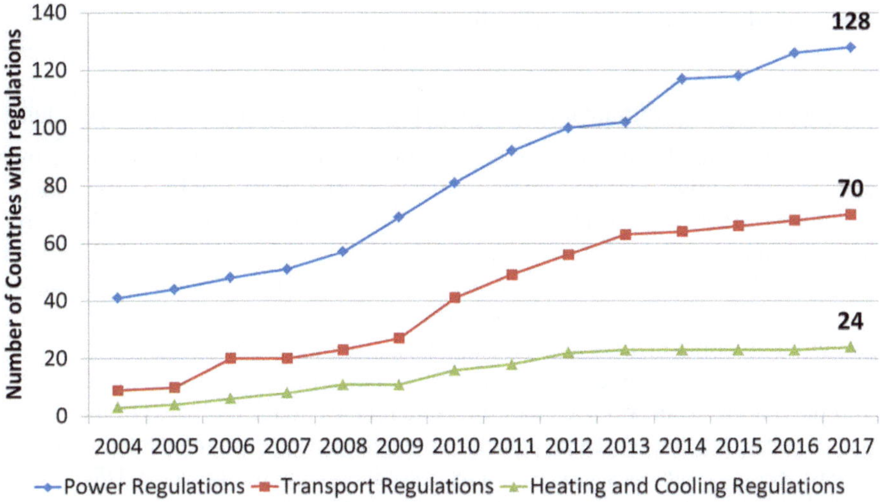

Fig. 10.16 Number of countries with renewable energy regulatory policies, 2004–2017 [20]

policy uncertainty in both the energy transport and the heating and cooling sector, and the advances are uneven across regions. The lack of policy landscape and slow applicable technology progress are the most critical challenges. Although 179 countries have established their renewable energy targets, the majority of them are focusing on the power sector. For example, in the transport sector, the policies mainly focused on promoting biofuels and electric mobility for road transport, while transport by rail, aviation, and shipping is minimal. In the heating and cooling sector, policies on energy efficiency have spread in the building sector, while it is still lacking in the industry.

Although the ratio of renewable energy and conventional energy is increasing, there still exist some unneglectable bottlenecks of Utility-Scale Renewable Energy. The affecting factors are listed as follows:

(1) Renewable energy power plants are always constructed in remote areas thus requiring the construction of new transmission infrastructure. Considering the instability and intermittence of renewable energy, for example, solar energy is sensitive to weather conditions, additional support equipment is needed to reduce the impacts on electric grids. This means more investment money.
(2) Some of the renewable energy has drawbacks due to characteristics or immature techniques. Carbon dioxide is generated during the production process of silicon solar cells, and people are focusing on reducing greenhouse gas emissions. The rotation of wind turbine blades produces low-frequency noise pollution, which is harmful to animal health. Giant dams of hydropower plants can result in local climate and ecosystem change.
(3) The challenge of constructing a renewable energy generation plant is the complicated regulatory process. An abundant amount of capital is needed by

related companies to support the projects and get them through the multi-year interagency process ranging from local, provinces, to the national level. Because of both uncertainty and taking longer time to obtain regulatory approval, investors are not so willing to invest in the renewable energy industry.

In general, the global application of renewable energy shows promising growth with its competitive cost and energy efficiency, and there would be increasingly remarkable progress if sufficient policies are established. It is expected that corporations' sourcing of renewable energy would continuously increase and they would play a significant role in spreading renewable energy to new regions in the future.

The following section sets out some key priorities and practical policy measures that will help policymakers to accelerate the energy transition.

Focus area 1. Combine energy efficiency and renewable energy measures (for example, public sector policies that integrate renewable technologies in the renovation of public buildings). Deploy technologies that promote renewable energy and increase energy efficiency (for example, combined heat and power (CHP) systems fueled by renewables to recover waste heat for use by industrial plants or commercial and residential buildings).

Focus area 2. Increase use of electricity in transport, buildings, and industry. This will help to unlock substantial efficiency gains and yield a wide range of other benefits, including the reduction of air pollution in cities. To achieve this objective, policymakers should:

(1) Set targets for the replacement of the conventional fossil fuel-based technologies by electric vehicles, heat pump systems, and electric stoves.
(2) Facilitate sector coupling between power and end-use sectors, to facilitate the integration of variable renewables in the power sector.
(3) Increase flexible electricity demand by means of demand side management, smart charging and vehicle-to-grid for electric vehicles, flexible heat pump heating and cooling, thermal storage fed by electricity, etc.
(4) Use information communication technology and digitalization, along with demand side management, to reduce peak electricity demand, lowering the need to invest in power capacity, and reduce operational costs.

Focus area 3. Foster system-wide innovation. Innovation has played a key role in the energy transition to date, particularly in solar and wind technologies. A critical but often underappreciated point is that innovation is far broader than technology research and development: it must cover a technology's life cycle, including demonstration, deployment (technology learning), and commercialization. Innovation ecosystems should also cover a wide range of activities, including new approaches to power system operation, market design, enabling technologies, and business models. Alongside innovative generation technologies, innovation is needed across the whole energy system to assist the system to integrate variable renewable energy generation and accelerate the widespread adoption and up-scale clean energy.

Focus area 4. Align socio-economic structures and investment with the transition. The success of the energy transition depends on how well it is connected to efforts to address the transition of the broader socio-economic system itself. Implementing the energy transition requires significant investments. More than that, climate boundaries require a rapid scale-up of investment in renewable energy and energy efficiency within a relatively short time window. The quicker the energy transition gets underway, the lower the climate change adaptation costs will be, and the smaller the socio-economic disruption. But the longer the needed scale-up in investment is delayed, the higher the capital cost needed to finance the transition. The financial system needs to be aligned with broader sustainability and energy transition requirements. A timely mobilization of investments requires addressing the barriers that conventional financing faces, as well as mobilizing additional investment streams.

Focus area 5. Ensure that transition costs and benefits are fairly distributed. A transition can be regarded as fair to the degree that it also seeks to reduce historical divergences in levels of energy access. Universal energy access is, in fact, a key component of a fair and just transition. Beyond energy access, huge disparities exist in the energy services available in different regions. The transition process will only be completed when energy services converge in all regions.

Exercises

1. Name at least 5 independent ultimately primary sources of energy.

2. Compare the principles and applications between Parabolic and Fresnel trough collectors utilized in solar power generation?

3. Analyze biomass utilization processes and compare their advantages and disadvantages from the aspects of technology, economy, and application, respectively.

4. Assume that the velocity of wind is 10 m/s and the wind turbine power generation efficiency is 30%. The surface area that the wind flows through the turbine is $10m^2$. Calculate the electric power of the wind turbine.

5. The flow of a river can be idealized into the form shown in the following figure. In region 1, before the weir, the stream velocity u1 is uniform with depth. In region 2, after the weir, the stream velocity increases with depth h in the water. Use Bernoulli's equation to calculate that the speed, uh, of the streamline passing over the weir at a depth h.

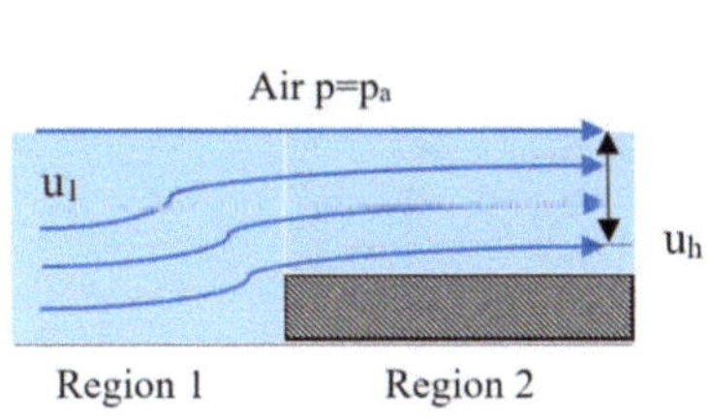

References

1. Thousands suffocate as Great Smog descends on London (1952)
2. Bhatia SC (2014) Advanced renewable energy systems
3. Commission BSD (2017) Better business better world
4. REN21 (2017) Renewables 2017 global status report. REN21 Secretariat, Paris
5. Nelson V (2009) Wind energy renewable energy and the environment. CRC Press Taylor & Francis Group, LLC
6. Fleming D (2016) Wind energy developments. Nova Science Publishers Inc., Potential and Challenges
7. Sørensen B (2004) Renewable energy Its physics, engineering, use, environmental impacts, economy and planning aspects. Elsevier
8. Twidell J, Weir T (2015) Renewable energy resources, 3rd edn.
9. Cheng J (2010) Biomass to renewable energy processes. CRC Press Taylor & Francis Group
10. Potter C, Klooster S, Genovese V (2012) Net primary production of terrestrial ecosystems from 2000 to 2009. Clim Change 115(2):365–378. https://doi.org/10.1007/s10584-012-0460-2
11. Kaltschmitt M (2007) Renewable energy. In: Basics of renewable energy supply
12. Carter JG (2017) Solar energy and solar panels systems. Nova Science Publishers Inc., Performance and Recent Developments
13. Heller P (2017) The performance of concentrated solar power (CSP) systems analysis. Elsevier, Measurement and Assessment
14. Ng C, Ran L (2016) Offshore wind farms technologies. Elsevier, Design and Operation
15. Pan S-Y, Lin YJ, Snyder SW, Ma H-W, Chiang P-C (2015) Development of low-carbon-driven bio-product technology using lignocellulosic substrates from agriculture: challenges and perspectives. Curr Sustain/Renew Energy Rep 2(4):145–154. https://doi.org/10.1007/s40518-015-0040-y
16. Gao M, Pan S-Y, Chen W-C, Chiang P-C (2018) A cross-disciplinary overview of naturally derived materials for electrochemical energy storage. Mater Today Energy 7:58–79. https://doi.org/10.1016/j.mtener.2017.12.005
17. IRENA IaR (2018) Renewable energy policies in a time of transition. IRENA, OECD/ IEA and REN21
18. IRENA (2016) Unlocking renewable energy investment: The role of risk mitigation and structured finance, 2016(b) edn.
19. Agency IE (2016) Medium-term renewable energy market report 2016. https://doi.org/10.1787/renewmar-2016-en
20. Renewables 2018 global status report (2018) REN21
21. BP statistics review of renewable energy (2018) BP

Chapter 11
Energy Materials

11.1 Introduction

Adopting novel and alternative energy is a vital means of inhibiting the adverse anthropogenic effects on the environment. The adoption of new energy significantly depends on the development of new materials, which could improve the efficiency and reduce the cost.

New energy materials refer to the materials for new energy conversion, storage, and utilization, which are developing along with the advancement of novel energy technologies and become the key issues. From the point of view of materials science, all new materials which could store energies and effectively utilize them could also be classified as new energy materials. The invention of new energy materials boosts the development of new energy system, improves the system efficiency, and directly affects the investment and operation cost. Therefore, the development and utilization of new energy materials play a role of great importance in the application of new energy technologies.

At present, new energy materials focus on battery materials (e.g., lithium-ion battery materials), fuel cell materials, and solar cell materials. They also include materials for nuclear reactors, energy storage materials, and other related ones, such as composite materials for high power wind turbine blades, biomass and bioenergy, capacitor materials, thermoelectric conversion materials.

In this chapter, the fundamentals and applications of typical new energy materials used for lithium-ion battery, fuel cell, solar cell, and energy storage, will be introduced. In the end, the environmental impact of these materials will also be discussed.

P.-C. Chiang and X. Gao, *Air Pollution Control and Design*,
https://doi.org/10.1007/978-981-13-7488-3_11

11.2 Principles

11.2.1 Lithium-Ion Battery Materials

Batteries utilizing lithium-ion electrodes are currently the most popular and highly developed rechargeable ones. This is mainly due to the high voltage of operation (greater than 3.6 V for ordinary ones and up to 5.0 V in novel high voltage electrolyte) and relatively high energy density. Some are also characterized by long life cycle with lower memory effect as compared to Ni-Cd or previous rechargeable batteries [1, 2]. Accordingly, lithium-ion batteries are widely used in civil and military fields, such as portable electronic devices and electric vehicles. Its working principle is shown in Fig. 11.1. Generally, when the Li-ion battery is fabricated, de-lithiation of the positive electrode material occurs during the initial charging process, which involves extraction of Li ions from the positive electrode, diffusion through the electrolyte, and intercalation into the negative electrode. During discharging, on the other hand, electrons removed from the negative electrode lead to de-intercalation of the Li ions, which then diffuse through the electrolyte back to the positive electrode for intercalation. The negative enthalpy of the discharging process thus leads to electrons been forced through the external circuit, thus providing energy to the external load. In practice, this first discharge process usually results in significant capacity fade in the cell, however, after few cycles, the energy becomes stable. The reversible transfer of Li ions between the positive and negative electrodes during charging/discharging is the reason for this mechanism to be termed as "rocking-chair".

For example, when $LiCoO_2$ is used as the cathode (positive electrode) material and graphite as the anode (negative electrode) material [3], the reactions during discharging/charging of Li-ion battery are shown in Eqs. (11.1)–(11.3).

Positive electrode (forward reaction during charging and vice versa)

$$\mathrm{LiCoO_2} \rightarrow \mathrm{Li}_{1-x}\mathrm{CoO_2} + x\mathrm{Li}^+ + xe^- \tag{11.1}$$

Negative electrode (forward reaction during charging and vice versa)

$$C + x\mathrm{Li}^+ + xe^- \rightarrow \mathrm{Li}_x C \tag{11.2}$$

Overall reaction:

$$\mathrm{LiCoO_2} + C \rightarrow \mathrm{Li}_{1-x}\mathrm{CoO_2} + \mathrm{Li}x\mathrm{C} \tag{11.3}$$

During the development of lithium-ion battery, the anode materials included lithium alloys and oxides, graphite materials, amorphous carbon materials, nitrides, silicon materials, tin-based materials, etc. At present, the commercial lithium-ion battery anode material is mainly graphitized carbon materials.

Lithium metal has the highest specific capacity as the anode material. However, lithium metal can react with many materials due to its high reactivity. During the

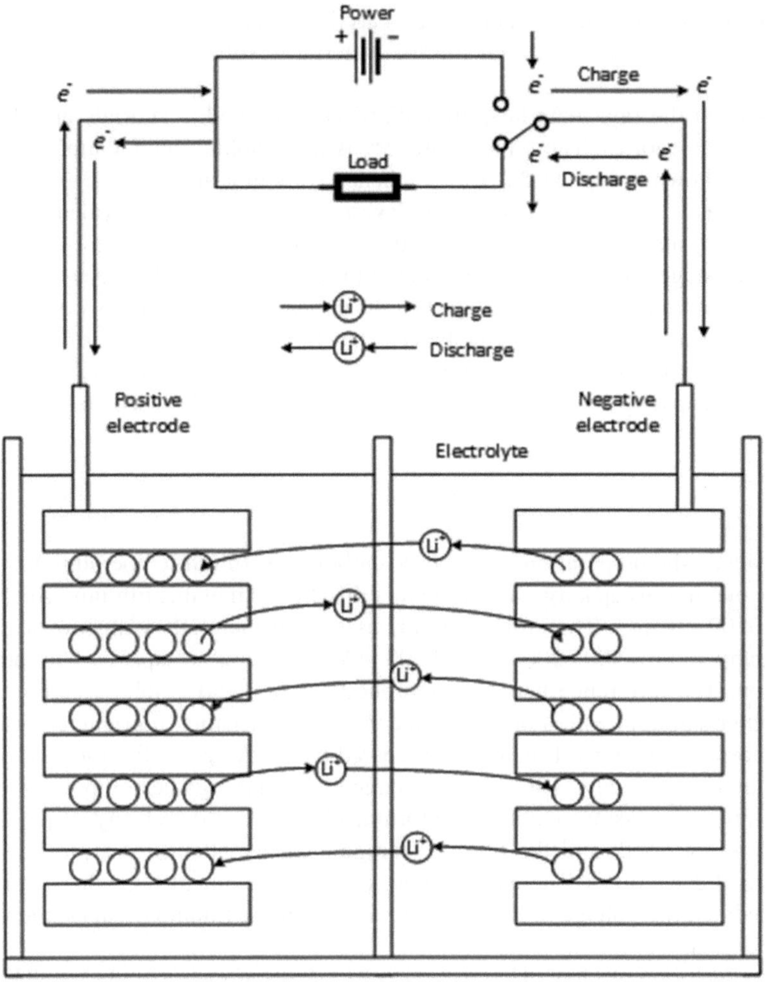

Fig. 11.1 Working principle of lithium-ion battery

charging process, the surface of the negative electrode is easy to generate dendritic crystallization, leading to short circuit of battery and associated safety problems and short battery life. Accordingly, it cannot be used in practice for rechargeable Li-ion battery systems, although it has been used in primary lithium batteries. In order to solve these problems, the research on lithium alloy as anode material has attracted much attention. The alloys mainly include Li-Al, Li-Si, Li-Sn, Li-Sb, Li-Bi, Li-Pb, Li-Cd, Li-Mg, and other alloys. The biggest problems for the alloy anode materials are the volume expansion and contraction caused by deep embedding and prolapse of lithium, which causes gradual powdering of the electrode material in repeated charging and discharging processes, followed by falling off and thereby shortening the lifetime of the battery. Therefore, it is proposed that the use of special structural design, such as using carbon- and silicon-based materials with stable electrochemical properties to fixed alloy elements in the material to solve the problems of volume change.

Based on the raw materials, graphitized carbon anode materials are classified into many categories. The typical category is the graphite intercalation compound (GIC), as shown in Fig. 11.2a. The graphene layers in graphite are stacked through van der Waals forces and atoms in the layers are connected by covalent bonds [4]. The graphite materials embedding lithium can form different "stages" of the GICs. Herein the stage is defined as the number of the graphite layers between two adjacent embedded atomic layers, e.g., 1 stage Li-GIC indicates that there is only one graphene layer between two adjacent Li embedded layers with the structure of -Li-C-Li-C- (LiC_6). As for the graphite anode material, it can reach LiC_6 at most, so its theoretical maximum capacity is 372 mA·h/g. Equation (11.4) can be used to describe this process

$$Li^+ + e^- + C_6 \rightarrow LiC_6 \tag{11.4}$$

However, the preparation of graphitic carbon needs high-temperature treatment, and the maximum capacity is much smaller than that of metallic lithium, which could reach 3800 mA·h/g. Therefore, amorphous carbon materials with low heat treatment temperature (500 –200 °C) and 900 mA·h/g reversible capacity have attracted attention. The difference between amorphous carbon and graphitic carbon is that the former's graphitization process is incomplete, consisting of graphite crystallites and amorphous regions. It is mainly prepared by cracking small molecular organic compounds over catalysts and cracking large molecular organic compounds directly or by other processing methods. Unfortunately, the cycling ability of amorphous carbon is not good, and the lithium storage capacity attenuates rapidly with the increase of cycle times. There is also hysteresis in the voltage, i.e., when the lithium is embedded, the voltage is below 0.3 V, and when the lithium escapes, the voltage is higher than 0.8 V.

From the point of view of composite material preparation, silicon-based lithium-ion battery anode materials are promising [5]. Silicon has two forms: crystalline

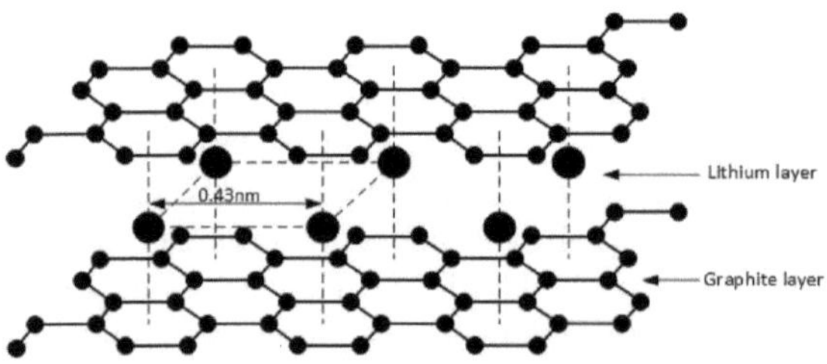

(a) Structure of LiC_6

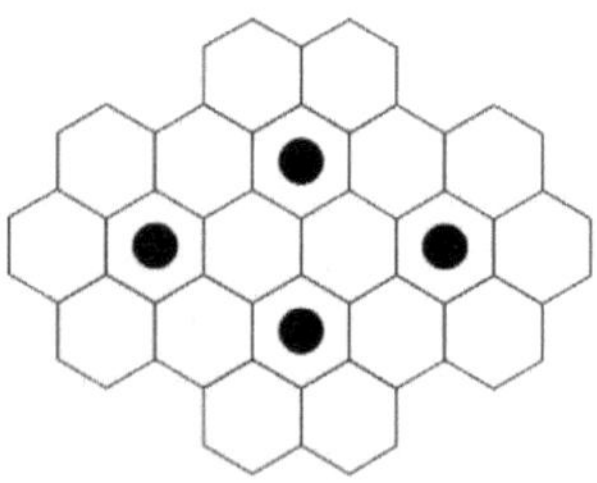

(b) Location of Li in LiC_6

Fig. 11.2 LiC_6 structure of anode material

and amorphous, where the amorphous silicon material has better performance when used as a negative electrode material. Therefore, some amorphous particles could be added to obtain amorphous silicon during the preparation procedure. The inserted Si and Li compounds could reach the level of Li_5Si with the initial reversible capacity at 0 – V being up to 800 mA·h/g, while the capacity decay is fast. SiB_n (n = 3.2 –6) is a typical compound of silicon and nonmetal combination, which could increase initial reversible capacity, and has high efficiency for the first time. When silicon and graphite are grounded to form $C_{0.8}Si_{0.2}$, the initial reversible capacity could also be enhanced. Among the new silicides formed by silicon doped metal, manganese silicide has the best performance. Its average discharge voltage is comparable to that of graphite, but its capacity and cycling performance are superior to that of natural graphite.

Due to the high ionic conductivity of Li_3N and the easy migration of Li ions, nitride lithium-ion battery materials are well studied. The nitride $Li_{3-x}M_xN$ negative electrode materials are obtained after doping Li_3N with transition metal elements such as Co, Ni, and Cu. The nitride structure is similar to the graphite structure with the hexatomic ring. As shown in Fig. 11.3, in the case of $Li_{2.5}Co_{0.5}$ N, it is composed of two layers: layer A is Li-N and layer B is Co-Li formed by the substitution of Li by Co in layer A. Half of the lithium in layer A and all the lithium in layer B can be reversibly escaped with a maximum discharge voltage of 1.4 V. When the voltage exceeds 1.4 V, layer A (Fig. 11.4) would break down, resulting in the destruction of the structure. In the process of lithium extraction, the nitride transforms from crystal to amorphous. Element rearrangement occurs and remains amorphous in all cycles. In this type of nitride, $Li_{3-x}Cu_xN$ has the best performance, with a reversible

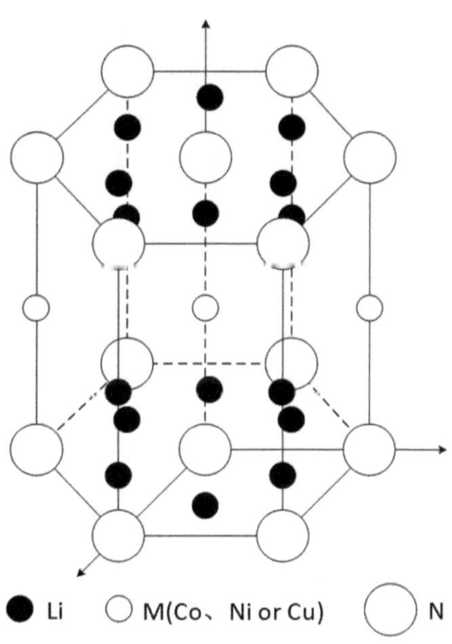

Fig. 11.3 Schematic of the $Li_{2.5}Co_{0.5}$ N structure

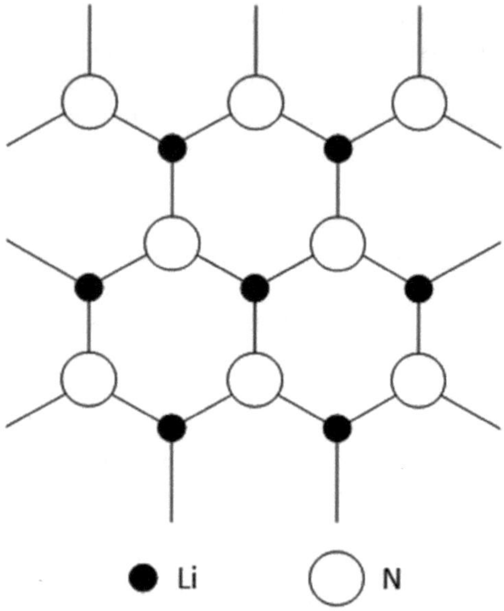

Fig. 11.4 Structure of Li-N layer (layer A) after half lithium protrusion

capacity of 650 mA·h/g, followed by $Li_{3-x}Co_xN$, whose capacity is 560 mA·h/g. The reversible capacities of other nitrides are lower than 372 mA·h/g, which is the theoretical maximum capacity of graphite. This type of nitride has a good cycling performance when the voltage is less than 1.4 V. But the synthesis conditions are harsh in which high temperature and pressure (30 MPa, 750 °C) are needed, limiting its application.

Tin-based anode materials include tin oxides, mixed oxides, tin salts, and other tin compounds. At present, there are two types of lithium storage mechanisms over tin oxides. One is called alloy type, in which Li and Sn oxide react to generate Li oxide and metallic tin, although with alloy formation by lithium and reduced tin. This process is described in Eqs. (11.5) and (11.6)

$$4Li + SnO_2 \rightarrow Sn + 2Li_2O \qquad (11.5)$$

$$Sn + xLi \rightarrow Li_xSn(x \leq 4.4) \qquad (11.6)$$

Another mechanism is called ion type, in which lithium is in the form of ions, without generating a separate Li_2O phase. The first charge and discharge efficiencies are relatively high. Equation (11.7) can be used to describe this process

$$xLi + SnO_2 \rightarrow Li_xSnO_2 \qquad (11.7)$$

When some metal or nonmetal elements, such as B, Al, P, Si, Ge, Ti, Zn, Mn, Fe, are introduced into the stannous oxide and stannic oxide with heat treatment, mixed oxides can be obtained. They can also be prepared by the mechanical grounding of SnO and B_2O_3, which can change the electrochemical properties of the material. In addition to oxides, tin salts can be used as negative electrode materials for lithium-ion secondary batteries, such as $SnSO_4$, whose maximum reversible capacity is up to 600 mA·h/g. The insertion of lithium over $SnSO_4$ are expressed as Eqs. (11.8) and (11.9)

$$Li + SnSO_4 \rightarrow Sn + Li_2SO_4 \quad (11.8)$$

$$4Li + Sn \rightarrow Li_4Sn \quad (11.9)$$

Other tin-based anode materials include tin silicon oxynitride, tin hydroxide, tin sulfide, nano metallic tin, etc. Tin-based anode materials have application potential for negative electrodes of lithium-ion secondary batteries. Some aspects, like the insertion mechanism of lithium in tin-based materials, remain to be investigated.

From the point of view of crystal structure, the cathode materials of lithium-ion battery mainly have three types: the layered $LiMO_2$ (M = Co, Ni, NiCo, NiCoMn, $LiMn_2O_4$), $LiMn_2O_4$ of spinel structure, and $LiMPO_4$ (M = Fe, Mn, Co) of olivine structure. The typical layered structure is $LiCoO_2$, whose crystal structure is α-$NaFeO_2$ type, which belongs to the hexagonal system, R3m space group. The oxygen atoms present ABCABC cubic close packing arrangement and between the adjacent oxygen atom layers lithium ions and cobalt ions alternately occupy the interlaminar eight body positions as shown in Fig. 11.5a. $LiCoO_2$ is a semiconductor, and the room temperature diffusivity of lithium ions in $LiCoO_2$ is 10^{-12} –10^{-11} cm^2·s. The theoretical specific capacity is 274 mA·h/g when lithium completely escapes. $LiMn_2O_4$ spinel structure belongs to the Fd3m space group as shown in Fig. 11.5b, in which oxygen atoms form the cubic close packed structures, in location of cell 32e manganese occupies half of octahedral voids 16d positions, and lithium occupies the 1/8 position of 8a tetrahedral. The empty tetrahedral and the octahedral voids are connected with each other by coplanar and common edges, forming a three-dimensional channel for the diffusion of lithium ions. The chemical diffusivity of lithium-ion in spinel is 10^{-14} –$10^{-12}$$cm^2$·s. The theoretical capacity of $LiMn_2O_4$ is 148 mA·h/g, and the actual capacity is about 120 mA·h/g. The olivine $LiMPO_4$ is an orthorhombic D162h-Pmnb space group as shown in Fig. 11.5c, in which each cell contains four units of $LiMPO_4$. Compared with the other lithium-ion cathode material, $LiFePO_4$ has low conductivity (10^{-12} –$10^{-9}$$cm^2$·s), and the chemical diffusion coefficient of Li^+ in $LiFePO_4$ is low as well, which limits its practical application. The theoretical specific capacity of Li^+ prolapse is about 120 mA·h/g. But due to its low cost, abundant resources, stable structure, and high thermal stability, $LiMPO_4$ is expected to be applied in power batteries and energy storage batteries.

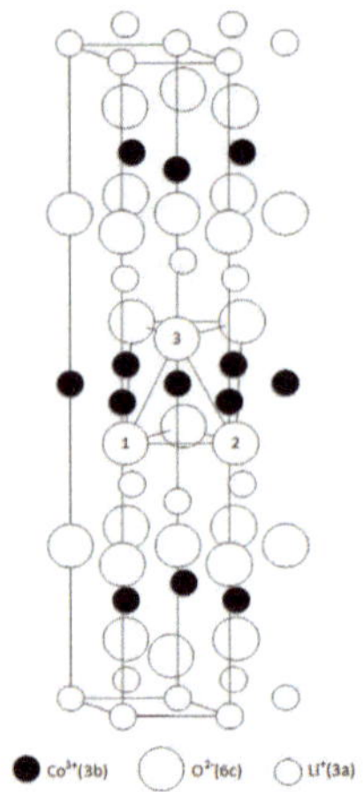

(a) Layered lithium cobalt oxide structure

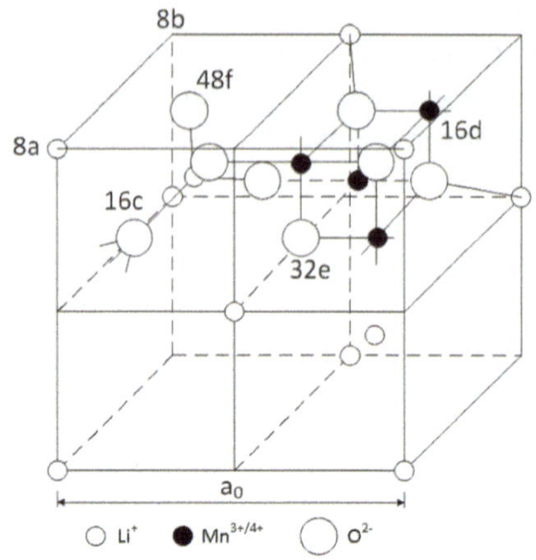

(b) $LiMnO_2$ spinel structure

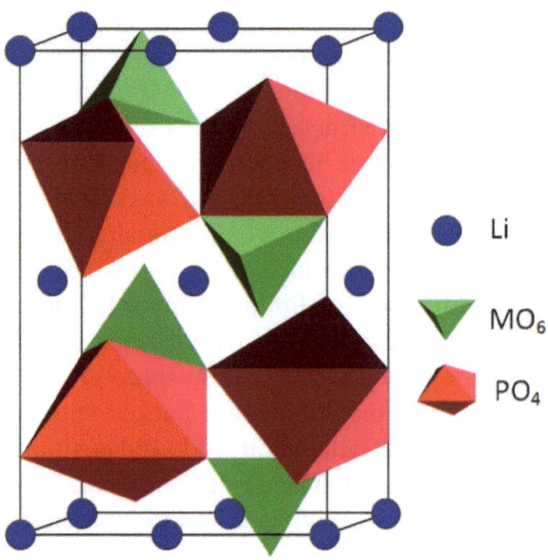

(c) $LiMPO_4$ olivine structure

Fig. 11.5 Crystal structure of cathode materials for lithium batteries

11.2.2 Fuel Cell Materials

The fuel cell is an energy conversion system that can be used to generate power through the conversion of chemical energy in fuels and oxidizers into electrical energy. Unlike batteries, the fuel cell does not require expensive metals and uses only cheap fuels for chemical reactions [6]. The fuel cell also has the advantages of zero noise pollution and high efficiency with low emissions [7]. The reaction process of a typical H_2-O_2 fuel cell is described as Eqs. (11.10)–(11.12)

Fuel electrode:

$$H_2 \rightarrow 2H^+ + 2e^- \quad (11.10)$$

Air electrode:

$$1/2O_2 + 2H^+ + 2e^- \rightarrow 2H_2O \quad (11.11)$$

Overall reaction:

$$H_2 + 1/2O_2 \rightarrow 2H_2O \quad (11.12)$$

The flow rate Q (mol/s) of hydrogen at current I (A) can be calculated as Eq. (11.13)

$$I = nFQ \quad (11.13)$$

When the chemical energy of a fuel cell is completely converted to electrical energy, the efficiency of the fuel cell is referred to as the theoretical efficiency (ε_{th}), as expressed in Eq. (11.14)

$$\varepsilon_{th} = \frac{\Delta G^{\ominus}}{\Delta H^{\ominus}_{298}} \quad (11.14)$$

where
$\Delta G^{\ominus}$ = the standard Gibbs energy change for the reaction, kJ/mol;
$\Delta H^{\ominus}_{298}$ = the standard enthalpy change at 298 K.

Under standard state, the theoretical potential ($E^{\ominus}$) can be expressed as Eq. (11.15)

$$E^{\ominus} = \frac{-\Delta G^{\ominus}}{nF} \quad (11.15)$$

where
n = the number of electrons participating in the fuel cell reaction;
F = Faraday constant, 96,485 C.

For each substance, the values of the standard enthalpy of formation and the standard Gibbs free energy can be found in the handbook of thermodynamic properties. Therefore, the theoretical efficiency and theoretical potential of the fuel cell can be calculated.

When the temperature and pressure are not under standard conditions, the relationship between the Gibbs free energy change (ΔG) and the temperature and pressure can be expressed as Eq. (11.16)

$$\Delta G = \Delta G^{\ominus} + RT \cdot \ln\left(\frac{p_{H_2O}}{p_{H_2} p_{O_2}^{1/2}}\right) \quad (11.16)$$

where

R = the gas constant, J/ (K·mol);
T = the temperature, K;
p_{H2} = the H_2 partial pressure in fuel side, atm;
p_{H2O} = the H_2O partial pressure in fuel side, atm;
p_{O2} = the O_2 pressure in air side, atm.

In the above equation, the first part, i.e., standard Gibbs free energy $\Delta G^{\ominus}$, is independent of the pressure and varies only with temperatures. The second part varies with pressures, gas compositions, and temperatures when the fuel cell operates.

At present, there are 6 types of fuel cells [8], i.e., alkaline fuel cell (AFC), phosphoric acid fuel cell (PAFC), molten carbonate fuel cell (MCFC), solid oxide fuel cell (SOFC), solid polymer fuel cell (SPFC, also known as proton exchange membrane fuel cell, PEMFC), and the biological fuel cell (BFC). The characteristics, materials, and reactions of the fuel cells are shown in Tables 11.1, 11.2, and 11.3.

11.2.3 Solar Cell Materials

Solar energy is a renewable energy source with the advantages of being abundant, safe, and environmentally friendly [9, 10]. There are three types for solar energy utilization: photo-thermal conversion, photoelectric conversion, and photochemical conversion [11, 12]. Among them, the photoelectric utilization of solar energy is growing rapidly in recent years. Solar cells are devices that use the interaction between solar energy and materials to generate electricity directly [13]. As the bandgap of the semiconductor material (0–3.0 eV) and visible light energy (1.5–3.0 eV) should be in correlation, light can be partially absorbed to a generate electromotive force when incident on a semiconductor, which is called the photovoltaic effect.

The working principle of a solar cell is shown in Fig. 11.6. The semiconductor p-n junction accepts sunlight to convert light energy into electrical energy according to the photovoltaic effect. The p-n junction is that when p and n type semiconductors join together, a special thin layer is formed in the interface between the two semiconductors. The p type of the interface is negatively charged, and the n type is positively charged. Since the p semiconductor is rich in holes, while n type semiconductor is rich in free electrons, therefore, concentration differences of electrons and holes generate. Electrons in the n region diffuse into the p region, and holes in the p region diffuse into the n region. Once the diffusion occurs, an internal electric field is formed pointing from the n region, preventing the diffusion and achieve equilibrium. A special thin layer is formed with a potential difference, i.e., p-n junction. When

Table 11.1 Comparison of characteristics of various fuel cell types

Cell		AFC	PAFC	MCFC	SOFC	PEMFC
Electrolytes	Type	Potassium hydroxide	Phosphoric acid	Li_2CO_3 Na_2CO_3	$ZrO_2 + Y_2O_3$	Cation-exchange membrane
	Conductive ion	OH^-	H^+	${CO_3}^{2-}$	O^{2-}	H^+
	Specific resistance /Ω·cm	~1	~1	~1	~1	~20
	working temperature /°C	50–150	190–200	600–700	About 1000	80–1000
	Corrosive	Medium	Strong	Strong	–	Medium
	Use form	Substrate impregnation	Substrate impregnation	Substrate impregnation or paste	Thin-film	membrane
Electrodes	Catalyst	Nickel and silver	Platinum	Unwanted	Unwanted	Platinum
	Fuel electrode	$H_2 + 2OH^- \rightarrow 2H_2O + 2e^-$	$H_2 \rightarrow 2H^+ + 2e^-$	$H_2 + {CO_3}^{2-} \rightarrow H_2O + CO_2 + 2e^-$	$H_2 + O_2\text{-} \rightarrow H_2O + 2e^-$	$H_2 \rightarrow 2H^+ + 2e^-$
	Air pole	$1/2O_2 + H_2O + 2e^- \rightarrow 2OH^-$	$1/2O_2 + 2H^+ + 2e^- \rightarrow H_2O$	$1/2O_2 + CO_2 + 2e^- \rightarrow {CO_3}^{2-}$	$1/2O_2 + 2e^- \rightarrow O^{2-}$	$1/2O_2 + 2H^+ + 2e^- \rightarrow H_2O$
Fuel (reactant)		Pure hydrogen (not containingCO_2)	Hydrogen (containing CO_2)	Hydrogen and carbon monoxide	Hydrogen and carbon monoxide	Hydrogen (containing CO_2)

Table 11.2 Structural materials for various fuel cells

Parts	PAFC	MCFC	SOFC	PEMFC
Electrolytes	H_3PO_4	Li_2CO_3 Na_2CO_3	$ZrO_2 + Y_2O_3$	Cation-exchange membrane Nafion
Substrates	SiC	γ-$LiCO_3$ powder Reinforcing fiber (Al_2O_3)	–	–
Fuel electrodes	Porous carbon plate Pt/C + PTFE	Ni-AlCr	Ni-YsZ Cermet	Porous carbon plate Pt/C + PTFE
Air poles	Porous carbon plate Pt/C + PTFE	NiO + Base rare earth element	$La_{1-x}Sr_xMnO_3$ ($x = 0.1$ –0.15)	Porous carbon plate Pt/C + PTFE
Composition materials	Membrane plate Carbon plate	membrane plate SUS310S/Ni overburden SUS310 + Al SUS316L	Bipolar junction plate: $LaCr_{1-x}Mg_xO_3$ Carrier: calcium oxide Stabilized zirconia	Membrane plate Carbon plate

Table 11.3 Reactions of various fuel cells

Type	Fuel electrode	Air pole	Total reactions
PAFC	$H_2 \rightarrow 2H^+ + 2e^-$	$1/2O_2 + 2H^+ + 2e^- \rightarrow H_2O$	$H_2 + 1/2O_2 \rightarrow H_2O$
PEMFC	$H_2 \rightarrow 2H^+ + 2e^-$ $H_2 + CO_3^{2-} \rightarrow 2H_2O + CO_2 + 2e^-$	$1/2O_2 + 2H^+ + 2e^- \rightarrow H_2O$ $1/2O_2 + CO_2 + 2e^- \rightarrow CO_3^{2-}$	$H_2 + 1/2O_2 \rightarrow H_2O$ $H_2 + 1/2O_2 \rightarrow H_2O$
MCFC	$CO + H_2O \rightarrow H_2 + CO_2$		
SOFC	$H_2 + O^{2-} \rightarrow H_2O + 2e^-$ or$CO + O^{2-} \rightarrow CO2 +$ 2e-	$1/2O_2 + 2e^- \rightarrow O^{2-}$ $1/2O_2 + 2e^- \rightarrow O^{2-}$	$H_2 + 1/2O_2 \rightarrow H_2O$ $CO + 1/2O_2 \rightarrow CO_2$

sunlight reaches the surface of the cell, photons with sufficient energy can excite electrons from the valence bands in p type silicon and n type silicon and generate electron hole pairs. Electrons and holes near the p-n junction will be separated from each other by the electric field of the space charge before recombination. Electrons move toward positively charged n regions and holes toward negatively charged p regions. Due to the charge separation of the p-n, an outward detectable voltage can be generated between the p region and the n region. The process is called the "photovoltaic effect" [14]. If the load circuit is on both sides of the cell, the separated electrons and holes are gathered through the grid lines on the surface of the solar cell, and the photocurrent is generated in the outer circuit.

Solar cells can be represented by a circuit consisting of p-n junction diode D, constant current source I_{ph}, series resistance $R_{s,}$ and shunt resistance R_{sh} equivalent

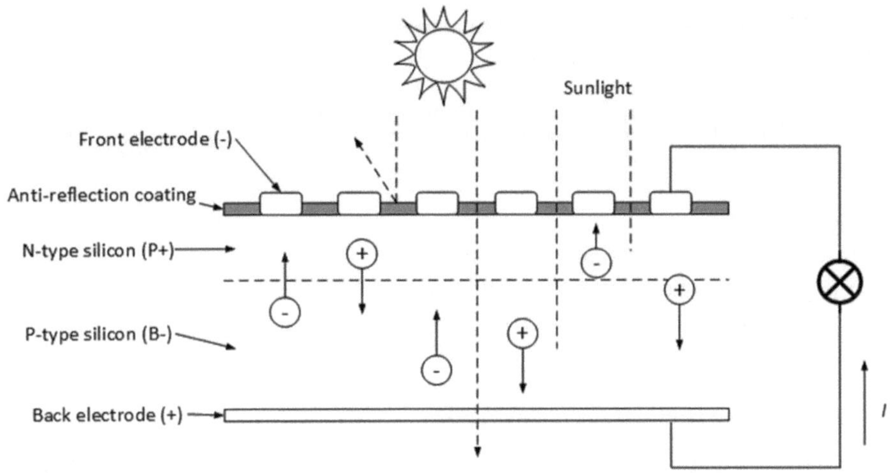

Fig. 11.6 Working principle of solar cell

to p-n junction leakage current, as shown in Fig. 11.7. The circuit is the equivalent circuit of the solar cell. Among them, I_{ph} is the photocurrent, and I_{ph} is proportional to the area of solar cells and the irradiance of incident light. I_D is the dark current, which is the single-phase current flowing through the p-n junction when the solar cell has an external voltage in the absence of illumination. I_L is the load current for solar cells. V_{OC} is the open circuit voltage of the battery. R_L is the external load resistance of the battery. R_s is a series resistance that is generally less than 1Ω. R_{sh} is a bypass resistor, usually several thousand ohms. R_s and R_{sh} are the intrinsic resistances of the solar cell itself. For an ideal solar cell, its R_s value in series is very small, and the parallel R_{sh} value is very large. Therefore, they can be neglected in the calculation of ideal circuits and the ideal equivalent circuit is equivalent to a constant current source of I_{ph} and a diode in parallel.

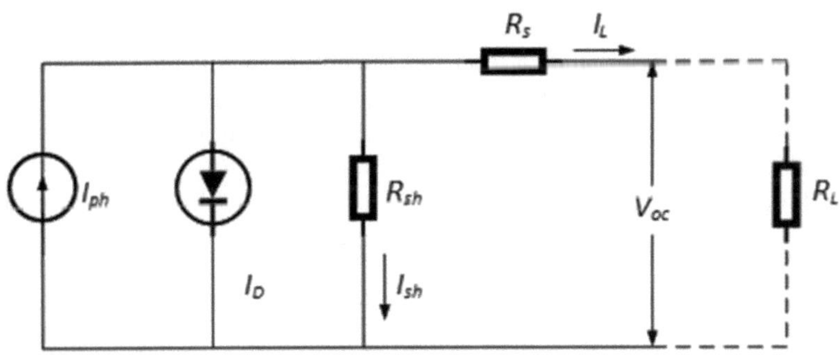

Fig. 11.7 Equivalent circuit of solar cells

The relationship between the current and the voltage at the two sites of the solar cell can be obtained by the equivalent circuit diagram, as expressed in Eq. (11.17)

$$I_L = I_{ph} - I_D - \frac{V_D}{R_{sh}} = I_{ph} - I_0\left[\frac{\exp(q(V_{oc} + I_L R_s))}{nkT} - 1\right] - \frac{V_D}{R_{sh}} \tag{11.17}$$

where

I_0 = the reverse saturation current, mA;
V_D = the terminal voltage of an equivalent diode, V;
q = an electron charge, 1.6×10^{-19}C;
K = a Boltzmann constant, 0.86×10^{-4} eV/K;
T = the temperature, K;
n = the curve constant for p–n junction.

I_0 is related to the material properties of the cell's component, and it is a reflection of the amount of photo-generated carriers that can be transported within the cell. It is generally a constant and is not affected by light intensity.

In order to make the solar cell generate more power, it is necessary to minimize the series resistance R_s and increase the parallel resistance R_{sh}. The equation of the equivalent circuit under ideal condition ($R_s \to 0$, $R_{sh} \to \infty$) is shown in Eq. (11.18)

$$I_L = I_{ph} - I_D - \frac{V_D}{R_{sh}} = I_{ph} - I_D \tag{11.18}$$

Depending on the materials, solar cells can be classified as silicon solar cells, multi-component thin-film solar cells, organic semiconductor solar cells, and nanocrystalline solar cells [15–17], as shown in Table 11.4. Among them, silicon is the most abundant and cheapest material, and its manufacturing technology is the most advanced [18]. Therefore, silicon-based solar cells are the widely applied [19].

The study of solar cell materials is in fact the study of semiconductor materials that can perform the photovoltaic conversion. The parameters of commonly used solar cells are shown in Table 11.5.

Figure 11.8 shows the input–output characteristic of solar cells, also known as voltage–current characteristics. It indicates the ability of solar cells to convert solar energy into electricity. The solid line in the figure is the voltage–current characteristic of the solar cell when it is irradiated by light. The phantom is a voltage–current characteristic of solar cells which is not illuminated by light.

The dark current is equivalent to the diffused current of the p-n junction, and the voltage–current characteristics can be expressed by Eq. (11.19)

$$I = I_0\left[exp\left(\frac{eV}{nkT}\right) - 1\right] \tag{11.19}$$

where

I_0 = an inverse saturation current, mA;

Table 11.4 Types of solar cells

Types of solar cells		Solar cell materials
Silicon solar cell	Crystalline state	Monocrystalline solar cell
		Polycrystalline silicon solar cell
		Polycrystalline thin-film solar cells
	Amorphous state	Amorphous solar cell
Multi-component thin-film solar cells	III-V group	GaAs, InP, GaInP
	II-VI group	CdS, CdTe, Cu_2S
	I-III-VI group	$CuInSe_2$, $CuInS_2$
Organic semiconductor solar cell	Pigment sensitization	DSSC
	Organic film	ITO/sexithiophene/Al, MEH-PPV/C60
Nanocrystalline solar cells	Thin-film	Dye sensitized TiO_2 nanocrystalline solar cells (DSSC) and dye sensitized ZnO nanocrystalline solar cells

V = the terminal voltage of solar cells for light exposure;
n = the diode factor;
K = Boltzmann constant;
T = the temperature, K.

When the p–n junction is irradiated by light, the carrier's direction of motion is opposite to the current in the upper equation, denoted as I_{sc}. The relationship between the solar cell terminal voltage V and the photocurrent density I_{ph} is shown in Eq. (11.20)

$$I_{ph} = I_0\left[\exp\left(\frac{eV}{nkT}\right) - 1\right] - I_{sc} \tag{11.20}$$

where I_{sc} is related to the intensity of the irradiated light, equivalent to the short current at the terminal of the solar cell. It is called short circuit photocurrent.

From the above equation, we can see that when the solar cell is in the open state, it will produce voltage, i.e., the open-circuit voltage V_{oc}. When the solar cell terminal is open, $I_{ph} = 0$.V_{oc} can be expressed as Eq. (11.21)

$$V_{oc} = \frac{nkT}{e}\ln\left[\frac{I_{sc}}{I_0} + 1\right] \tag{11.21}$$

When the solar cell connects with optimum load resistance, the best-fit point P is the intersection of maximum voltage V_m and the maximum current I_m in the voltage–current characteristics. The shaded area corresponds to the output power P_{out} of the solar cell, described as Eq. (11.22)

Table 11.5 Parameters of commonly used solar cell materials (dia stands for diamond structure, zb stands for sphalerite structure, and ch stands for chalcopyrite structure)

Materials	Crystal structures	Light absorption coefficient	Refractive index	Electron affinity/eV	Lattice constant/Å	Density/(g/cm^3)	Coefficient of thermal expansion/$\times 10^{-6}$ K	Melting point/K
c-Si	dia	11.9	3.97	4.05	5.431	2.328	2.6	1687
GaAs	zb	13.18	3.90	4.07	5.653	5.32	6.03	1510
InP	zb	12.56	3.60	4.38	5.869	4.787	4.55	1340
a-Si	–	About 11	3.32	–	–	–	–	–
Cd-Te	zb	10.2	2.89	4.28	6.477	6.2	4.9	1365
$CuInSe_2$	ch	–	–	4.58	–	–	6.6	– 1600
Al_xGa_{1-x} $0 \le x \le 0.45$	zb	13.18–3.12x	–	4.17–1.1x	5.653 + 0.0078x	5.36–1.6x	6.4–1.2x	–
Al_xGa_{1-x} $0.45 \le x \le 1$	zb	–	–	3.64–0.14x	–	–	–	–

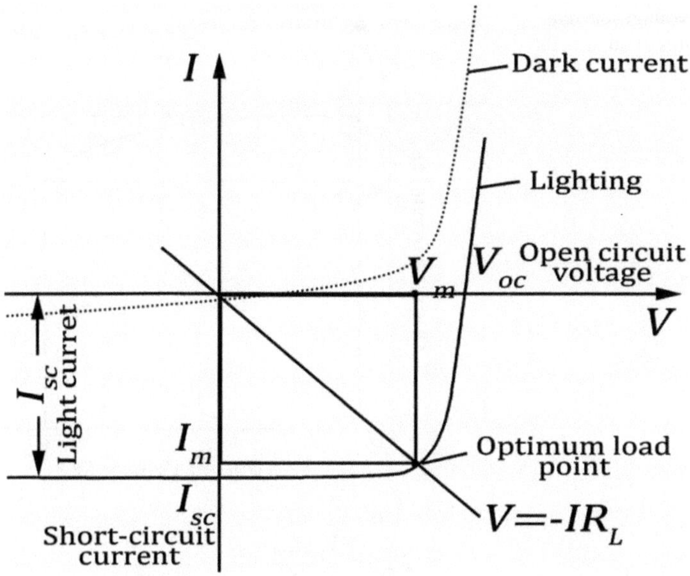

Fig. 11.8 Voltage–current characteristics of solar cells

$$P_{out} = VI = V\left[I_{sc} - I\left[\exp\left(\frac{eV}{nkT}\right) - 1\right]\right] \tag{11.22}$$

Since the output power at the optimum load point P is the maximum, the optimum operating voltage V_{op} and the optimum operating current I_{op} of the solar cell can be calculated as Eqs. (11.23)–(11.25)

$$\frac{dP_{out}}{dV} = 0 \tag{11.23}$$

$$\exp\left(\frac{eV_{op}}{nkT}\right)\left(1 + \frac{eV_{op}}{nkT}\right) = \frac{I_{sc}}{I_0} + 1 \tag{11.24}$$

$$I_{op} = \frac{(I_{sc} + I_0)eV_{op}/(nkT)}{1 + eV_{op}/(nkT)} \tag{11.25}$$

When the light is irradiated on the solar cell, the relationship between the voltage and the current of the solar cell can be simply represented as shown in Fig. 11.9. The V_{oc} is open-circuit voltage, I_{sc} is short-circuit current, V_{op} is the best operating voltage, and I_{op} is the best working current.

The voltage–current characteristic curve of the actual solar cell deviates from the rectangle, and the deviation degree is expressed by the fill factor F_F. The fill factor is the ratio of the rectangular area ($P_m = V_{op} \times I_{op}$) of the shaded part of the image to the area of the dotted part ($V_{oc} \times I_{sc}$), as shown in Eq. (11.26)

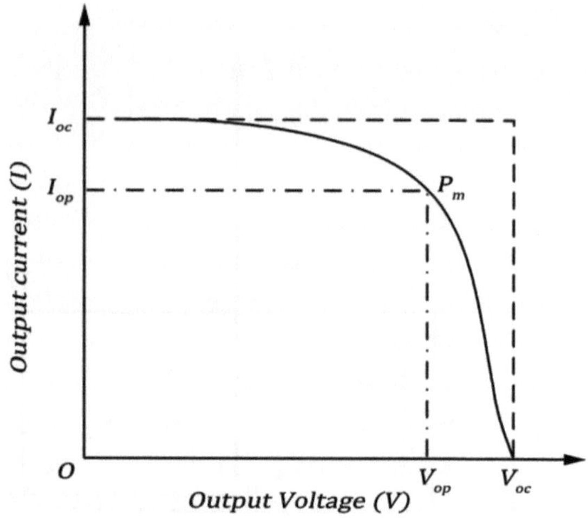

Fig. 11.9 Voltage–current characteristics of solar cells

$$F_F = \frac{V_{op} I_{op}}{V_{oc} I_{sc}} \tag{11.26}$$

The conversion efficiency of the solar cell is used to indicate the amount of light energy irradiated on the solar cell that is converted into electrical energy. The conversion efficiency of a solar cell is defined as the ratio of the maximum output power P_m of the solar cell to the total radiant energy P_{in}, shown in Eq. (11.27)

$$\eta = \frac{P_m}{P_{in}} \times 100\% = \frac{F_F V_{oc} I_{sc}}{P_{in}} \times 100\% \tag{11.27}$$

Among the various performance indicators of solar cells, the most important one is the conversion efficiency from light to electricity. In practical work, the constant power line of I × V = constant is often used in the voltage–current characteristic curve. If the input power is constant, it is the equal efficiency line, as shown in Fig. 11.10. When a voltage–current curve is tangent to an equal efficiency line, the efficiency line represents the photoelectric conversion efficiency of the p-n junction. Theoretical calculations show that the thermodynamic limit efficiency of solar cells is 32%. But the actual efficiency is lower than this value, e.g., monocrystalline silicon solar cell efficiency is about 29%. Due to the series resistance and surface reflection and other factors, the efficiency is reduced to 10–15%. Therefore, the efficiency of the average material calculated according to different parameters is about 20%.

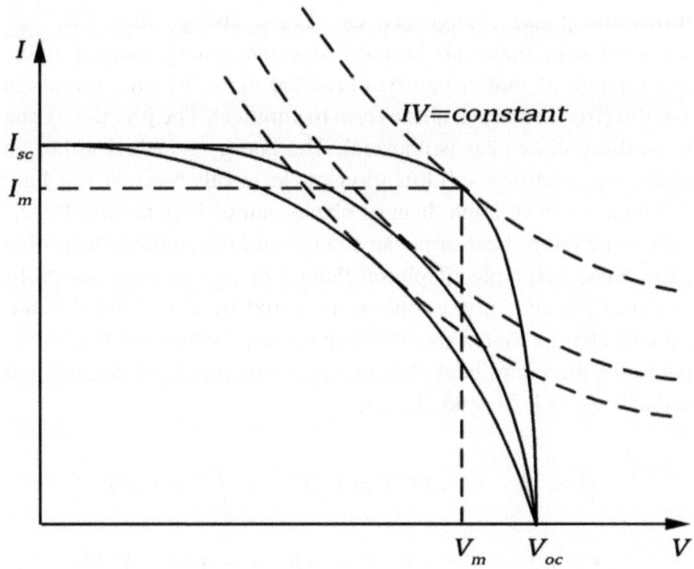

Fig. 11.10 Equivalent efficiency lines of solar cells (Isc = Im)

11.2.4 Energy Storage Materials

Energy storage refers to the transformation of energy into a more stable form under natural conditions. In the process of energy development, conversion, transportation, and utilization, there are always differences in quantity, form, and time between the supply and demand of energy. In order to make up for these differences and utilize energy effectively, the process or technical means of storing and releasing energy is often called energy storage technology. Energy storage can be classified into mechanical energy storage, chemical energy storage, electromagnetic energy storage, compressed air energy storage, water power storage, thermal storage, etc [20]. Energy storage technologies are used to prevent the automatic deterioration of energy quality, to improve the performance of the energy conversion process, to facilitate the economical use of energy, to reduce pollution, and to protect the environment. Therefore, energy storage technology is an important way for the rational, efficient, and clean use of energy.

The most widely used materials for energy storage are phase change energy storage materials. The transformation of matter from one phase to another is called phase change. Phase change energy storage material is a kind of material that changes the state when the temperature changes, with the process of absorption or release of large amounts of latent heat [21]. In recent years, phase change energy storage materials have been widely applied in many fields, such as building, battery thermal management, and solar energy, due to their constant temperature and high energy storage capacity during the phase change process. Phase change materials have become common energy storage materials in many new energy engineering fields.

The phase change of matter can be classified as: solid–gas, liquid–gas, liquid–solid, solid–solid (from one crystalline form to another). The process of phase change is generally isothermal or near isothermal. The energy which is either absorbed or released within the melting–solidification cycle is regarded as the latent heat of fusion [22]. The amount of latent heat of phase change is generally large, so we can store and utilize the latent heat of phase change and the sensible heat of absorption. Therefore, the basic principle of phase change energy storage technology is that when the material changes phase, energy is stored by absorbing the latent heat of isothermal phase change materials, and then released when needed.

The capacity of the latent heat storage system using phase change materials can be calculated as Eqs. (11.28) and (11.29)

$$Q = \int_{T_i}^{T_m} mc_p dT + ma_m \Delta h_m + \int_{T_m}^{T_f} mc_p dT \tag{11.28}$$

$$Q = m\left[c_{sp}(T_m - T_i) + a_m \Delta h_m + c_{tp}\left(T_f - T_m\right)\right] \tag{11.29}$$

where

Q = stored heat, J;
m = the heat storage medium quality, kg;
c_p = the specific heat capacity, J/(kg·K);
c_{sp} = the average specific heat capacity of $[T_i\text{-}T_m]$, kJ/(kg·K);
c_{tp} = the average specific heat capacity of $[T_m\text{-}T_f]$, kJ/(kg·K);
T_m = melting point, K;
T_i = the initial temperature, K;
T_f = the final temperature, K;
a_m = the percentage of melting;
Δh = heat of fusion in units, J/kg.

According to the types of phase change, the materials can be divided into four types: solid–liquid phase change material, solid–solid phase change material, solid–gas phase change material, and liquid–gas phase change material. Among them, solid–liquid and solid–solid phase change materials have better application potentials. The solid–liquid phase change materials can be divided into inorganic materials and organic materials. Inorganic phase change materials include crystalline hydrated salts, molten salts, metals, alloys, and other inorganic substances. One of the most widely applied is the crystallization of hydrated salts. The melting point ranges from several degrees Celsius to more than 100 °C. Other widely applied materials are alkali and alkali metal halides, sulfates, nitrates, carbonates and acetates, etc. Crystalline hydrated salts store heat or release heat by absorbing or releasing crystalline water during freezing or melting. Crystalline hydrated salts are denoted as by $AB{\cdot}mH_2O$. Their phase change energy storage mechanism is described as

$$\mathrm{B}\cdot\mathrm{mH_2O} \overset{heating(T\gg T_m)cooling(T\ll T_m)}{\leftrightarrows} \mathrm{AB}+\mathrm{mH_2O}-\mathrm{Q} \quad (11.30)$$

$$\mathrm{AB}\cdot\mathrm{mH_2O} \overset{heating(T\gg T_m)cooling(T\ll T_m)}{\leftrightarrows} \mathrm{AB}\cdot pH_2\mathrm{O}+(m-p)\mathrm{H_2O}-\mathrm{Q} \quad (11.31)$$

where

T_m = the melting point, °C;
Q = the fusion heat, kJ/kg.

Crystalline hydrated salt energy storage material has the advantages of wide application range, low price, large thermal conductivity, large melting heat, large volume heat storage density, and general neutrality. The disadvantages are the subcooling phenomenon during condensation, i.e., the temperature should be reduced to a certain temperature under the "condensation point" and then coagulation starts. Phase separation occurs when melting, i.e., some crystal water released with the rising temperature and the remaining is not enough to dissolve salt. The dehydration salt cannot be recrystallized. This results in irreversible phase change and leads to the decrease of energy storage capacity.

Organic phase change energy storage materials include paraffin, alkane, fatty acid, alcohol, etc. The phase change temperatures and enthalpies of homologous organic compounds usually increase with the elongation of carbon chain, so we can get a series of phase change energy storage materials of different temperatures. However, with the elongation of carbon chain, the phase transition temperature will eventually saturate to a fixed value. In order to get the appropriate phase change temperature and good performance of phase change materials, several organic phase change materials are usually used to form two or multiple phase change materials. Sometimes the organic and inorganic materials are combined to make up for the shortcomings. The advantage of organic phase change materials is their good formability in the solid state. It is difficult to appear supercooling and phase separation. The corrosion of the material is small with stable performance, low toxicity, and low cost. The disadvantages are the small thermal conductivity, low density, small specific energy storage capacity per unit volume, large volume variation in the phase change process, and low melting point which makes it not suitable for high- temperature applications. It is also volatile, highly flammable and even explosive, easily aging when exposed in the air.

The solid–solid phase change materials absorb or release heat due to the transformation of crystal structure before and after phase change. Because of the absence of liquid phase in the phase change process, it presents small volume change before and after phase change with additional advantages of non-toxicity, low corrosion, low requirements of materials and production technology for the container, long service life. The solid–solid phase change energy storage materials are mainly inorganic salts, polyols, and cross-linked high-density polyethylene. Among them, the inorganic salt solid–solid phase change energy storage material has a higher phase transition temperature and is suitable for storage and temperature control at high temperatures.

At present, the layered perovskite, Li_2SO_4, and KHF_2 are widely applied. Polyalcohol solid–solid phase change materials are mainly pentaerythritol (PE) and neopentyl glycol (NPG), 2-amino-2-methyl-1,3-propylene glycol (AMP), 3-hydroxymethyl ethane, 3-hydroxymethyl aminomethane. The phase change temperatures of them are at 40–200 °C, suitable for medium- and high-temperature energy storage. After radiation or chemical cross-coupling, the crystal transition temperature of HDPE is 120–135 °C, which is lower than its softening point, 150 °C. Hence, it can also be used as a phase change energy storage material. This material is of long service life, stable performance, good mechanical properties, easy processing into a variety of shapes, making it a good candidate for practical application.

11.3 Applications

11.3.1 Lithium-Ion Battery

Due to the high specific energy and high energy density, lithium-ion batteries have gained a dominant position in the rechargeable battery market for electronic devices. Since lithium-ion batteries are at present readily available with a decreasing cost, they are more likely to be selected for a wide range of applications.

1. Hybrid electric vehicle (HEV) applications

Hybrid electric vehicles were first launched into the world market by Toyota Motor Company in 1997. Due to the increasing public concerns on environmental issues and their good fuel economy, HEV became very popular quickly. As a consequence, car manufacturers all over the world put more and more effort into the development of HEV technology. Roughly HEV can be classified into two categories: (1) the series HEV (S-HEV) and (2) the parallel HEV (P-HEV). S-HEV is always powered by a motor, which is connected with the engine via a battery and a generator. P-HEV is mainly powered by the engine, with partly being assisted by a motor. At present, P-HEV is the major applied technology. There are two types of P-HEV: (1) high voltage systems (300–144 V) and (2) low voltage systems (42–14 V).

A lithium-manganese spinel is always used as the positive electrode active material, while for the negative electrode active material, hard carbon is used. A fuel tank is used to store the energy for driving a car. Hence, the battery should be a power battery with high power density, rather than an energy battery with high energy density. While in order to satisfy the demand for high power density performance, the most effective measure is to make a thinner electrode, which can increase the electrode area to reduce the current density and also make the interelectrode distance shorter, lowering the cell's internal resistance.

Table 11.6 shows the specifications of an HEV cell and the battery module [2]. The cell has a cylindrical shape with a length of 108 mm and a diameter of 40 mm.

Table 11.6 Specifications of the HEV cell and the battery module [2]

	Cell	Module
Dimensions (mm)	φ40 × 108	541 × 260 × 160
Weight (kg)	0.3	20.2
Nominal voltage (V)	3.6	173
Capacity (Ah)	3.6	3.6
Output power density at 50% SOC (W/kg)	2,000	1,350
Cooling system	–	Compartment air suction

Its capacity is 3.6 Ah in constant current-constant voltage charge up to 4.1 V. The output power density at 50% of charge is 2,000 W/kg.

2. Human Implantable Lithium Batteries

A quite unique class of land applications of lithium batteries is human implantable batteries. Surgery is always required when the battery needs to be replaced in the case of fully implantable devices. Therefore, the cells must be very small and can provide reliability to work sufficient time so as to gain the most out of the procedure for the cells to be implanted. In addition, reliability with no tolerance for the risk of failure is highly required. Sometimes the surgery is followed by a lengthy recovery period depending on the implant type. Moreover, the cell must be well-sealed because of the direct contact with human tissues.

The fully implantable devices can be undoubtedly regarded as a technical advancement. In the cases of various types of medical prostheses, they are expected to provide more opportunities to develop and apply advanced implantable power sources, such as high-end lithium batteries, as shown in Table 11.7 [3]. However, there are still some barriers against their wide commercialization: (1) A general anesthetic and ca. a 2-h operation are always required for the surgery to place the implants. (2) The price is very high (reported to be from $12,000.00 to $15,000.00 per unit) [3], in which the surgery and related medical fees are even not included. (3) The insurance coverage is problematic for fully implantable hearing aids.

11.3.2 Fuel Cell

As green sources of energy, fuel cell industries are growing rapidly. Fuel cells can cater to the following three areas of power consumption, making them very much popular in the market: (1) portable fuel cells, which are used for charging up products designed to be moved; (2) stationary fuel cells, which provide electricity and heat; and (3) transport fuel cells, which provide propulsive power to vehicles.

Table 11.7 Lithium-ion battery-powered medical implants [3]

Battery manufacturer	Product name	Implant function
Medtronic, Inc	RestoreUltra® (Implant brand name)	Spinal cord stimulation and muscular pain management
Greatbatch, Inc	Xcellion™ (Battery product line)	Implantableneuromodulation, orthopedic and cardiac rhythm management devices
Quallion, LLC	QL0003I, QL0020B, QL0200I-A (Batteries)	Medical implants, sensors: cochlear implants, spinal cord stimulators, glucose sensors, and cardiac rhythm management devices
EaglePicher medical power, LLC	CONTEGO, Micro battery (Battery product lines)	Neurostimulation, middle ear, and cochlear implants, drug pumps, and body fluid pumps, telemetry (monitoring blood pressure, temperature, etc.)

1. Portable fuel cells

A portable SOFC that can be operating over a variety of liquid fuels was developed by a Japanese company, Advanced Manufacturing Research Unit (AIST). It was developed for disaster and emergency situations. It is a microtubular SOFC technology, in which hydrocarbon fuels, including liquefied petroleum gas, can be directly used. This type of portable fuel cell is utilized as an emergency power supply to families, office buildings, and the municipal government.

2. Stationary fuel cells

Up to now, stationary fuel cells are regarded as the best performers among fuel cell technology. The applications of stationary fuel cells have been across all scales. Among them, micro-combined heat and power (micro-CHP) units for residential use is a typical application of stationary fuel cells. The fuel cell micro-CHP first came to market in 2012, which quickly outsold conventional engine-based micro-CHP. Two types of fuel cell technologies are developed rapidly: (1) the proton exchange membrane fuel cell (PFMFC) and (2) solid oxide fuel cell (SOFC). The latter one owns a higher maximum efficiency, while they require continuous operation, making the efficiency slightly reduced at night. In Germany, a series of successful installations is achieved by Ceramic Fuel Cells Limited (CFCL), under the European SOFT-PACT Project. Also under this project, CFCL collaborated with software control company HOMA and utility E.ON, heating system manufacturers, for the installation of 100 fuel cell micro-CHP systems in a variety of configurations in both the UK and Germany.

3. Transport fuel cells

In transport industries, one of the most important issues is to fulfill the stringent greenhouse gas (GHG) emission regulations. Road transportation has contributed mainly to GHG emission. Therefore, increasing the applications of public transport tools and the use of low (even zero) emission vehicles are regarded as the ideal solutions. As a consequence, fuel cells in transportation are getting popular.

In order to reduce the manufacturing cost, automakers tend to collaborate to pool their resources, hence facilitating the mass production of fuel cells. For example, BMW and Toyota made a joint announcement in 2013, proposing technology sharing along with co-development of fundamental fuel cell vehicle platform by 2020. In this joint announcement, both a fuel cell system and a hydrogen tank will be developed. In addition, Toyota has a lot of expertise and experience in developing fuel cell and battery-powered trains. It also signed an agreement with Japanese automakers, oil, and energy companies in 2011 to facilitate the development of fuel cell electric vehicles.

11.3.3 Solar Cell Materials

The solar energy is transformed into electricity by a photovoltaic (PV) cell due to the photovoltaic effect. A variety of materials with different structures is included in the solar cells so as to reduce the cost and to obtain maximum efficiency. Several types of solar cell materials exist: (1) single crystal silicon, (2) polycrystalline, silicon (3) amorphous silicon, (4) thin-film (non-silicon) material, and (5) other semiconductor absorbing layers [14].

1. Single crystal silicon cell

Single crystal silicon is generally regarded as the most efficient type of silicon cells. This is because that no grain boundary exists in the single crystals. In the crystal structure, grain boundaries are defects that could reduce the material's electrical and thermal conductivity. They can also be regarded as barriers to electron flow. The crystalline silicon cell, which is fabricated from a single crystal or polycrystalline silicon, is always the workhorse of PV industry [11]. On the market, the modules made from this kind of cell are the most mature. Depending on the quality of the material as well as the specific cell technology, the efficiencies of these cells are generally between 13 and 15%. Up to 20–25 years at 80% of nameplate rating can be guaranteed by reliable PV module manufacturers.

2. Polycrystalline silicon solar cell

Polycrystalline silicon has obvious grain boundaries. Hence, their conversion efficiencies are demonstrated to be slightly lower than those of single crystal cells, generally in the range from 13 to 15%. The grain sizes of polycrystalline cells are from 1 mm to 10 cm. Polycrystalline PV modules are generally guaranteed for 20 years by reliable manufacturers. About 85–90% of the global annual market for PV modules is based on wafer-based crystalline silicon [11]. Due to advantages of long lifetimes, reliable technology, and abundant primary resources, it is expected that crystalline silicon PV modules could remain a dominant PV technology in the global market. While one key challenge for these modules still exists, i.e., how to improve the efficiency and effectiveness of resource consumption.

3. Thin-film (non-silicon) material

Due to the high cost of Si wafers (typically 40–50% of the total cost of the modules), the price of crystalline silicon solar cell remains high. In order to reduce the cost of material, using lesser quantity of absorbing material will be a good option. Thin-film technology, therefore, came into the picture. Strong light absorbers are selected to be the desired materials, which only need ca. 1 μm thick. As a result, the costs for materials are obviously reduced. Several attractive materials for thin-film solar cells include the following: cadmium telluride, a-Si: H, copper indium diselenide, and copper indium gallium diselenide.

Thin layers of photosensitive materials are deposited onto a low-cost substrate (e.g., stainless steel, glass or plastic) to obtain the thin-film modules. The layers of active materials over the thin-film solar cells are only ca. 10 nm thick. However, compared with the efficiency of 13%–15% over crystalline silicon solar cell, the efficiency of commercial thin-film PV is always smaller than 10%. In laboratories, testing results over the fabricated thin-film showed promising efficiencies, but these high efficiencies and high yields over the laboratory-constructed thin films have not yet been magnified to mass production volumes.

Several advantages exist over the thin-film solar cells [11]: (1) relatively low consumption of raw materials, (2) good performance at high temperature, (3) ease of building integration, (4) high automation and production efficiency, and (5) reduced sensitivity to overheating. While the major drawbacks are: (1) relatively low efficiency and (2) the industry's limited experiences for lifetime performances.

11.3.4 Energy Storage

Energy storage through the phase change between liquid and solid is best suited for systems that require thermal energy storage for later use. Phase change materials (PCMs) are generally applied for thermal energy storage.

1. Thermal management of electronics

Many portable electronics, like smartphones, are always in standby mode in low-power situations, while with random bursts of activity, which would induce processing power to the peak. In these cases, the bursts of energy can be absorbed by PCMs and then be released when the cycle comes to end. Hence, PCMs are widely applied for the transient thermal management of various electronics. There are two primary advantages over PCMs applications: (1) good reliability and (2) suitability for low form factor systems. While a major challenge still remains: the low thermal conductivity for many of the PCMs with melting temperatures in this range.

2. Energy storage in building materials

For the selection of building materials, energy storage is always one important part. PCM-based gypsum board is the main emphasis for the application of PCM to store energy in building systems. In most of these systems, the temperature range of the transitions from solid to liquid over PCMs is from 20 °C to 30 °C. Up to date, several PCM gypsum boards are commercially available, many of which apply an encapsulated PCM, melting at 23 °C.

When embedded in building materials, PCMs are proved to be one of the best solutions to reduce energy costs. The most attractive issue when using PCM in building materials is that they can store energy without increasing the temperature. As a consequence, the temperature rise can be mitigated thus mitigating the loadings of heating, ventilation, and air conditioning (HVAC) within the building. Up to now, several products are commercially available. While currently, one key challenge is the cost and wide-scale availability.

11.4 Challenges and Perspectives

11.4.1 4E Evaluation

The 4E evaluation for energy storage materials can be linked to the various development in the sourcing of the precursors, and also to the method of design in the technology. In addition to the material properties, the design of the systems for the technologies is also significant. For example, consider Fig. 11.11 below, which shows the design of the systems and their method of operation. It is seen that the

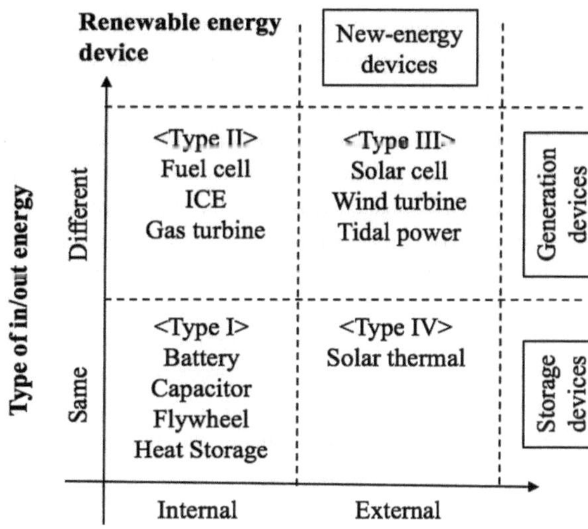

Fig. 11.11 The design of the renewable energy systems and their methods of operation

integration with the environment of the designed system can be linked not only to the precursor, but also to the system of integration. Furthermore, the technology selected for use with specific design considerations also relates to the systems and the mode of utilization. For example, fuel cells, internal combustion engines (ICE), and gas turbines, which are renewable energy generation systems, have different energy input and output compared to solar cells, wind turbines, and tidal power. Although for wind turbines and solar cells, the main environmental impact from the earlier can be linked to the materials, for example, mining the precious metals used for making the magnets. However, the decommissioning and construction aspects of the wind turbine would also have some environmental effects. For solar cells, on the other hand, their environmental impact compared to that of wind turbines are relatively smaller, however, in terms of energy density, the wind turbine has a relatively higher energy density though the sources of energy are different.

11.4.2 Social, Economic, and Environmental Benefits

Broadly speaking, graphite is the main material used for the negative electrodes of Li-ion batteries. Although synthetic graphite which is mainly produced in the United States is aimed at replacing the natural graphite, natural graphite is still used. Mostly produced in China, natural graphite mining involves the use of toxic chemicals such as hydrochloric acid, and the general impact on the lands is also an aspect that raises concern. Furthermore, since the water resources and land around mining sites, for example, in China have affected local populations in some cases, stringent government policies have thus been enacted to tackle these issues.

Although companies that use these materials in their production processes do not refer to the mining aspects, some such as Tesla, BMW, have made commitments to use materials with lower environmental impact through a detailed and critical understating of the life cycle of the materials obtained.

Some of the impacts of electrochemical energy storage systems are reviewed briefly below:

1. Lead-acid batteries: In general, Pb is toxic, and Pb mining usually creates massive land pollution in the regions where it is mined. Furthermore, the concentrated H_2SO_4 used is also highly corrosive, and most of the graphite components are usually fully corroded, such that they cannot be recycled in most cases.
2. Nickel batteries: there are various combinations for Ni batteries such as: Ni-metal hydrides, Ni-Cd (nickel–cadmium), etc. In general, some of these metals still pose immense environmental issues, however, since Ni-metal batteries can be recycled, some of these problems can be easily motivated. Moreover, the need for advanced resistant alloys such as chrome-steel, etc., which could make for the corrosiveness of these systems to be reduced, might increase the cost of producing new Ni batteries.

3. Sodium-sulphur batteries (Na-S) are operated in the molten state. Although these systems are usually safe, the high-temperature requirement coupled with the possibility of leakages and fumes, could result in potential environmental impacts of this technology.
4. Lithium-ion batteries: this is by far the most popular and comparatively the most advanced electrochemical energy storage system (EES) currently. Broadly speaking, the Li salts can be recycled, the positive (cathode) electrode materials such as $LiCoO_2$, $LiFePO_4$ can be recycled, and also the graphite negative electrode in most cases can still be recycled. However, the mining of cobalt, for example, has resulted in recent issues in the Congo such as environmental pollution and heavily invasive overhead mining methods. Furthermore, most of the materials required for Li-ion batteries are becoming more expensive because of mineral depletion. Thusly, this is the reason why alternative abundant health metal chemistries such as those involving Fe, or Mn are becoming more popular.
5. Supercapacitors: mostly use carbon based electrodes and conventional organic electrolytes such as acetonitrile, polycarbonates. Supercapacitors which are used for regenerative braking and high pulsed applications are usually recyclable. Moreover, the need to improve the design of these systems through the design of more complex chemistries involving Li such as Li-ion capacitors have implied that future technologies for supercapacitors would face similar environmental impact as those of Li-ion batteries.

11.4.3 Perspectives

Batteries and their related technologies have become more and more important for daily life [4]. For advanced Li-ion batteries, graphite is regarded as one of the most attractive anode materials. Via surface modification of graphitic anodes, their stability would be enhanced in prolonged cycling. In addition, to support the application of the advanced cathode materials, developing novel electrolyte systems with high anodic stability is very important. In this respect, ionic liquids based on derivatives of cyclic quaternary ammonium cations can be regarded as a promising candidate [5].

Considering bulk production and cost reduction, the fuel cells for vehicles or for stationary power are needed to achieve mass production, for which the advanced technology for automated mass production is of great significance, while most of the development on the preparation of membrane electrode assemblies (MEAs) is based on hands-on experiences. Therefore, automation of mass production technology for the preparation of MEAs should be taken seriously [8].

The rapid emergence of a new class of solar cells, i.e., perovskite-based solar cells, has been witnessed in the past several years [23]. In mid-2012, the first efficient solid-state perovskite cells were reported. The confirmed efficiency over the perovskite cells is rapidly increasing with effort worldwide. In addition, a variety of different fabrication techniques has been developed to further increase the efficiency, which also indicates that performance is still far from being fully optimized.

For the development of phase change materials (PCMs), there is a lot of work, i.e., to develop PCMs with higher specific heats, higher thermal conductivities, and higher latent heats [22]. In addition, the cost is also a key factor preventing the wide application of PCMs. While the application of PCMs for energy storage and thermal management has great potential. When the research and development sort out the above-mentioned concerns, the PCMs can be widely accepted.

Exercises

1. When $LiMn_2O_4$ is used as the cathode (positive electrode) material and graphite as the anode (negative electrode) material, provide the reactions during charging/discharging of the Li-ion battery.
2. Considering a fuel cell reaction given as follows. The change in Gibbs free energy is −698 kJ/mol. Determine the theoretical potential.

$$CH_3OH + 3/2O_2 \rightarrow 2H_2O + CO_2$$

3. When the overall solar cell current I_{ph} is 0 and the diode factor n is 1, determine the voltage V.
4. Describe the basic principle of phase change materials in energy storage applications.

References

1. Mikolajczak C, Kahn M, White K et al (2011) Lithium-ion batteries hazard and use assessment. Springer
2. Yoshio M, Brodd RJ, Kozawa A (2009) Lithium-ion batteries: Science and Technologies, Springer
3. Gulbinska MK (2014) Lithium-ion battery materials and engineering current topics and problems from the manufacturing perspective. Springer
4. De Las CC, Li W (2012) A review of application of carbon nanotubes for lithium ion battery anode material. J Power Sources 208:74–85
5. Marom R, Amalraj SF, Leifer N et al (2011) A review of advanced and practical lithium battery materials. J Mater Chem 21:9938–9954
6. Sammes N (2006) Fuel cell technology reaching towards commercialization. Springer
7. Kaur G (2016) Solid oxide fuel cell components interfacial compatibility of SOFC glass seals. Springer
8. Basu S (2007) Recent trends in fuel cell science and technology. Springer
9. Richter C, Lincot D, Gueymard CA (2013) Solar energy. Springer
10. Yang N (2017) The preparation of nano composites and their applications in solar energy conversion. In: Institute of process engineering. Chinese Academy of Sciences, Beijing
11. Mohanty P, Muneer T, Kolhe M (2016) Solar photovoltaic system applications a guidebook for off-grid electrification. Springer
12. Sugiyama M, Fujii K, Nakamura S (2016) Solar to chemical energy conversion theory and application. Springer
13. Schuster CS (2017) Diffractive optics for thin-film silicon solar cells. Department of Physics, University of York

14. Tiwari GN, Tiwari A, Shyam, (2016) Handbook of solar energy theory. Springer, Analysis and Applications
15. Fraas LM (2014) Low-cost solar electric power. Springer
16. Tress W (2014) Organic solar cells theory, experiment, and device simulation. Springer
17. Wu B, Mathews N, Sum T-C (2017) Plasmonic organic solar cells charge generation and recombination. Springer
18. Wang X, Wang ZM (2014) High-efficiency solar cells physics, materials, and devices. Springer
19. Korkin A, Goodnick S, Nemanich R (2015) Nanoscale materials and devices for electronics. Springer, Photonics and Solar Energy
20. Huggins RA (2016) Energy storage fundamentals. Springer, Materials and Applications
21. Raoux S, Wuttig M (2009) Phase change materials science and applications. Springer
22. Fleischer AS (2015) Thermal energy storage using phase change materials fundamentals and applications. Springer
23. Green MA, Ho-Baillie A, Snaith HJ (2014) The emergence of perovskite solar cells. Nat Photonics 8:506

Chapter 12
Adsorption

Nomenclature

a	$\frac{1}{H_d}$, ft^{-1}
a_p	The specific surface area
a_v	External area of solid particles, $\frac{ft^2}{ft^3}$
ax	Thickness modulus, dimensionless
A	The cross-sectional area of the adsorber
AC_t	The adsorption capacity at time t
A_{min}	The minimum cross-sectional area
[AS]	The amount of adsorbed phase in adsorbent
b	$\frac{G \cdot c}{\rho_B \cdot H_d}$, min^{-1}
b_τ	Time modulus
c	$\frac{y_0}{w_0} = const$
C	The molecular concentration of adsorbate in the gas phase
C_b	The outlet concentration of a pollutant at a breakthrough point or a point determined by regulations
C_i	Virial coefficients
D	The diameter of the vessel
D_G	Diffusion coefficient for gas, $\frac{ft^2}{h}$
D_L	The diffusion coefficient
D_p	Effective diameter of a particle in a packed bed, *ft*
e	The bed void space
G	Mass velocity of adsorbate-free fluid, $\frac{1b}{ft^2-h}$
ΔG	The change in Gibbs free energy
H_d	The height of a mass transfer unit
ΔH	The change of entropy
I_0	Bessel function of the first kind and of order zero
j_d	Chilton-Colburn analogy factor for mass transfer
k	The Boltzmann constant
k_a	Rate constants for forward reactions

P.-C. Chiang and X. Gao, *Air Pollution Control and Design*,
https://doi.org/10.1007/978-981-13-7488-3_12

k_1	Rate constants of adsorption process
k_2	Rate constants of desorption process
K_a	The adsorption rate
k_d	Rate constants for backward reactions
K_H	Henry's constant
L	The length of adsorption bed
m	The mass of the adsorbate molecule
M_c	The mass of the adsorbent for meeting service time requirements
$m_{sorbent}$	The mass of adsorbate
MW	The molecular weight of adsorbate
n^*	The amounts of adsorbate that are adsorbed at equilibrium state
$\overline{n}$	The total amount of adsorbate in the adsorber
N	The number of adsorption reactors which are in parallel mode
P	The partial pressure of adsorbate in bulk gas
$P_{(A)}$	The concentration of adsorbate in the gas stream
P_i	The partial pressure of adsorbate at the fluid-adsorbent interface
P_s	Saturation concentration of adsorbate
ΔP_s	The total system pressure drop
q	The adsorption capacity
q_m	The quantity of adsorbate necessary to cover the unit mass of the adsorbent with a monolayer
Q	The adsorption heat
Q_g	Gas flow rate
R	Gas constant
SAT	The saturation capacity of adsorbent
SAT_i	The saturation capacity of adsorbate i
Sc_G	Schmidt number of gas, $\frac{\mu_G}{\rho_G \cdot D_G}$
ΔS	The change of enthalpy
t_{reg}	The time for regeneration
Δt	The retention time of laden gas within the unit
T	Temperature in Kelvin unit
u	Gas velocity
ν	Pre-exponential factor of the desorption rate coefficient
ΔV	The gas volume in each unit
w	Adsorbate content of solid, $\frac{1b}{1b}$ of solid (adsorbate-free basis)
w_0	Adsorbate content of solid in equilibrium with y_0
W	The adsorbent weight
x	Distance in bed measured in the direction of flow, ft
y	Adsorbate content of fluid stream, $\frac{Ib}{Ib}$ of fluid (adsorbate-free basis)
y_0	Inlet adsorbate content of fluid stream, $\frac{Ib}{Ib}$ of fluid (adsorbate-free basis)
Z	The bed height
α	Dummy variable
β	Velocity
ε	The void fraction of the bed

θ	Surface coverage
μ_G	Gas viscosity, $\frac{Ib}{ft-h}$
π	Const = 3.14...
ρ	The density of the adsorbent
ρ_B	Bulk density of packed bed, $\frac{Ib}{ft^3}$
ρ_g	The density of the gas
ρ_G	Density of gas, $\frac{Ib}{ft^3}$
σ	The condensation coefficient
τ	Time, min

12.1 Introduction

12.1.1 Features of Adsorption Process

Adsorption is usually used for removing gaseous pollutants and/or collecting valuable materials from an exhausted gas stream. During adsorption, gaseous components (adsorbate) are removed from the flue gas stream by adhering to the surface of a solid (adsorbent). The mechanism behind adsorption is somewhat like condensation since the van der Waals forces that hold the gas molecules to the surface of the adsorbent are like vapor molecules, which cause condensation.

Popular adsorbents are activated carbon, silica gels, and aluminum salts which are usually highly porous particles. Adsorption occurs primarily on the internal surface of these particles. Meanwhile, the pollutant gas is stored in the pores of the adsorbent. In most cases, the adsorbent is regenerated and recycled so that its lifetime is extended while costs are minimized.

However, adsorption is not a final control process since it cannot destroy or transform the pollutant into nonhazardous forms. After saturation is achieved, the adsorbent must either be disposed of or desorbed by vapors. The desorbed vapors are highly concentrated, which makes the recovery or incineration (disposal) of pollutants easier and more economically feasible than before. Therefore, adsorption provides an efficient and economical method to reduce emissions of diluted gaseous pollutants.

12.1.2 Applications of Adsorption

Adsorption is used principally to control the emission of VOC where concentrations of VOC decrease from between 400 and 2,000 ppm to under 50 ppm. High-performance adsorptions are feasible to deal with VOC with concentration from 20 ppm to 25% of the Lower Explosive Limit (LEL), which is determined from safety concerns. If the concentration is lower than the lower edge of this range, pollution control may be difficult or uneconomical. Incineration, membrane separation, and condensation would be more economically practical and more technologically feasible when the concentration goes beyond the upper edge.

Adsorption makes recovery of the VOC possible in a non-destructive way. In some cases, adsorbate is valuable, therefore, recovery of this kind of VOC would drastically reduce the costs. Furthermore, adsorption could be implemented as a concentrating technology which reduces energy consumption in subsequent control measures such as incineration, membrane or condenser.

As a conclusion, adsorption is typically applied when (1) the pollutant gas is noncombustible or difficult to burn; (2) the pollutant gas is sufficiently valuable to warrant recovery; and (3) the pollutant gas concentration is diluted in the exhaust system.

12.2 Principles

12.2.1 Thermodynamics in Adsorption Process

When the adsorbent is exposed, the adsorbate, driven by a pressure gradient, will move toward it spontaneously. By measuring the partial pressure of adsorbate in bulk gas (p_0) and in the adsorbent pores (p), we have

$$\Delta G = \Delta H - \mathrm{T}\Delta S = \mathrm{RTln}(p/p_0) \tag{12.1}$$

where ΔG (J/mole) is the change in Gibbs free energy, ΔH (J/mol) is the change of entropy, ΔS (J/K·mol) is the change of enthalpy, R (J/K · mol) is gas constant, and T (K) is the temperature in Kelvin unit. When adsorption is in progress, p/p_0 will constantly be smaller than 1, leading to a negative ΔG. Therefore, if a pressure gradient exists, adsorption will occur spontaneously. Ultimately, when $p/p_0 = 1$ and $\Delta \mathrm{G} = 0$, the dynamic equilibrium is established. Currently, the adsorbent becomes saturated with an adsorbate.

An argument against the spontaneity of adsorption is that as adsorption proceeds, the adsorbate moves from bulk gas and attaches to the surface of the adsorbent. Simultaneously, the degrees of freedom, i.e., entropy for adsorbate shifts downward. In other words, the entropy of the adsorbate keeps descending, which violates the second law of thermodynamics, which states that the total entropy of an isolated system can only increase over time, i.e., $\Delta S > 0$. In most cases, adsorption is exothermic and heat is released, leading to a negative enthalpy change, i.e., $\Delta \mathrm{H}_{\mathrm{adsorbate}} < 0$. A popular explanation for this phenomenon is that the reduction in enthalpy is much greater than that of entropy, which makes Gibbs free energy ΔG a negative value and adsorption a spontaneous process.

As an exception, adsorption of herbicide diuron by activated carbon could occur endothermically, which could be due to the enhanced adsorbent-adsorbate interactions [1]. Such endothermic adsorption is spontaneous as long as it is in a suitable environment. Therefore, according to thermodynamics, entropy change should be negative in this case, as the formation of adsorbate molecules would be altered to planar, leading to improved accessibility into adsorbent microspores. It was also suggested that the endothermicity of this process is a result of the competition between diuron dehydration (endothermic) and adsorption (exothermic) reactions.

According to the amount of heat released, even though not precise, adsorption can be categorized into either physisorption or chemisorption. Physisorption happens because of intermolecular attractive forces, mainly van der Waals forces and sometimes electrostatic force, between adsorbent and adsorbate. For this reason, physisorption is always exothermic while adsorption heat depends on the magnitude of the attractive force. The heat liberated is in the general range of 2–20 kJ/(g·mol) which is similar to the enthalpy of vapor condensation. In addition, due to the long effective distance of Van der Waals forces, more than one layer of adsorbate can be adsorbed. One of the most significant features of physisorption is reversibility. By lowering the pressure of the adsorbate in the gas stream or by raising the temperature, the adsorbate is readily desorbed without a change in chemical composition (in most cases, but not all of them). The reversibility feature is extremely important if recovery of the adsorbed gas or recovery of the adsorbent for reuse is economically desirable.

Chemisorption results from the chemical interaction between the adsorbate and the adsorbing medium. The bonding force associated with this type of adsorption is much stronger than that of physical adsorption. Therefore, the heat liberated during chemisorption is much larger than that liberated during physical adsorption, e.g., on the order of an enthalpy (heat) reaction that may be in the range of 20–400 kJ/(g·mol). The energy (or activation) required for the chemisorbed molecules to react with another molecular species may be considerably less than the energy required when the two species are in the gas phase. This lower energy requirement is the basis of one explanation of the catalytic effect of solid surfaces in enhancing the rate of a chemical reaction. Unlike physisorption, the valence forces holding the adsorbate to the adsorbent are extremely short, such that only a monomolecular layer of adsorbate appears on the adsorbing medium. Moreover, chemisorption is frequently irreversible as the chemical nature of the original adsorbate undergoes a change when desorption is deployed (see Table 12.1).

12.2.2 Mass Transfer in Adsorption Process

Figure 12.1 presents distribution profiles of the adsorbate in gas and the adsorbent where the number of the transfer unit (NTU) and height of transfer unit (HTU) are used by dividing the adsorption bed into several units in series. According to the mass balance principle, the decreased adsorbate of laden gas equals the effects of

Table 12.1 Comparison of physisorption and chemisorption

Parameter	Physisorption	Chemisorption
Heat of Adsorption	8–20 kJ/mol	40–800 kJ/mol
Activation Energy, E	~0	Usually small
Temperature of Occurrence	Depends on the boiling point, but usually low	Depends on E, but is usually larger than the boiling point
No. of Layers Adsorbed	Multilayer	Monolayer
Force of Adsorption	Van der Waals force	Chemical bond
Rate of Adsorption	Fast	Slow
Specificity	Not Specific	Highly Specific
Reversibility	Yes	Not necessary
Example	Adsorption of water vapor by silica gel	Adsorption of H_2S by iron oxide

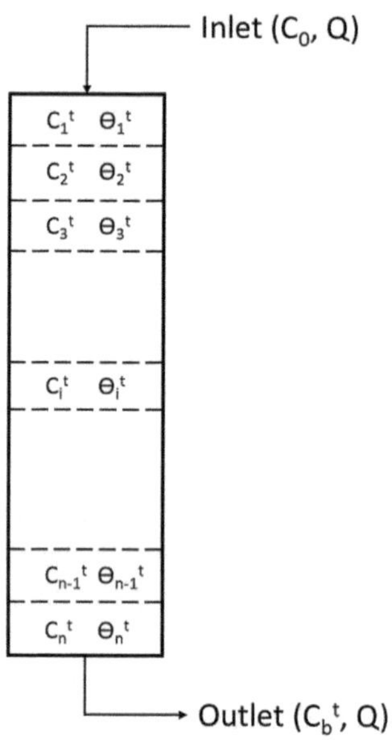

Fig. 12.1 Schematic diagram of the concentration profile in an adsorption reactor

diffusion, advection, and adsorption [2, 3], which is

$$\frac{\partial C}{\partial t} = \underbrace{D_L \frac{\partial^2 C}{\partial L^2}}_{\text{Diffusion}} - \underbrace{u \frac{\partial C}{\partial L}}_{\text{Advection}} - \underbrace{\frac{1-\varepsilon}{\varepsilon} \rho \frac{\partial q}{\partial t}}_{\text{Adsorption}} \tag{12.2}$$

where C (mol/m^3) is the molecular concentration of adsorbate in the gas phase, D_L (m^2/s) is the diffusion coefficient, L (m) is the length of adsorption bed, u (m/s) is the gas velocity, ε (–) the void fraction of the bed, ρ is the density of the adsorbent, and q (g/g) is the adsorption capacity.

Thereafter, we need to consider the equilibrium of adsorption which is achieved by balancing the forward and backward reactions:

$$\begin{cases} R_a = k_a(1-\theta)C \cdots \text{Forwad reaction rate} \\ R_d = k_d\theta \cdots \text{Backward reaction rate} \\ R = k_a(1-\theta)C - k_d\theta \cdots \text{Net reaction rate} \end{cases} \tag{12.3}$$

where k_a (m^3/s·mol) and k_d (1/s) are rate constants for forward and backward reactions, while θ (–) represents the surface coverage ratio of the adsorbent. It is worth noting that $\frac{1-\varepsilon}{\varepsilon}$ in Eq. (12.2) represents the ratio of the solid phase volume to the gas phase volume. Therefore, by signing the adsorbent weight W, and the cross-sectional area of the adsorber A, Eq. (12.2) can be rewritten as

$$\frac{\partial C}{\partial t} = -u\frac{\partial C}{\partial L} - \mathrm{S}\frac{\partial \theta}{\partial t} \tag{12.4}$$

where $\mathrm{S} = \frac{q_0 W}{LA - W/\rho}$. As a result, we can estimate the adsorption process in the ith section of the adsorption bed with Eq. (12.5) and (12.6)

$$\theta_i^{t+\Delta t} = \theta_i^t + \left[k_a\left(1-\theta_i^t\right)C_i^t - k_d\theta_i^t\right]\Delta t \tag{12.5}$$

$$C_i^{t+\Delta t} = C_i^t + \frac{Q\Delta t}{\Delta V}\left(C_i^t - C_{i-1}^t\right) - \left[k_a\left(1-\theta_i^t\right)C_i^t - k_d\theta_i^t\right]\mathrm{S}\Delta t \tag{12.6}$$

where ΔV (m^3) is the gas volume in each unit while Δt (s) represents the retention time of laden gas within the unit.

The adsorption capacity of the adsorbent is usually defined as the ratio of the amount of adsorbate that is adsorbed at the equilibrium state to the amount of adsorbent introduced. In practice, the capacity can be determined by a breakthrough curve via experiment, in which the gas containing the adsorbate with an inlet concentration $C_{i,\,it}$ keeps flowing through the adsorption bed until the outlet concentration reaches the specified value $C_{i,\,out}$, i.e., a breakthrough. Then the adsorption capacity, AC, can be calculated by (12.7)

$$\mathrm{AC} = \frac{m_i}{m_{sorbent}} = \int_{t=0}^{t=t_B} \frac{(C_{i,in} - C_{i,out}) \times Q_g}{m_{sorbent}}\,dt \tag{12.7}$$

where m_i and $_{sorbent}$ are the mass of adsorbate and adsorbent, respectively, Q_g is the gas flow rate, and t is the time. While in the experiment, where discrete measurement is performed, the adsorption capacity can be calculated by Eq. (12.8)

$$\mathrm{AC} = \frac{\mathrm{m_i}}{\mathrm{m_{sorbent}}} = \sum_{\mathrm{t}=0}^{\mathrm{t}=t_B} \frac{(\mathrm{C_{i,in}} - \mathrm{C_{i,out}}) \times \mathrm{Q_g}}{\mathrm{m_{sorbent}}} \times \Delta \mathrm{t} \tag{12.8}$$

The experimental setup for the determination of adsorption capacity is shown in Fig. 12.2. As the adsorbate containing fluid continuously passes through the adsorption bed, the adsorbent near the inlet interacts with adsorbate and ultimately becomes saturated. Then the frontier of the saturated zone would push toward the outlet until the entire adsorption process is finished. Conclusively, the shaded area of the breakthrough curve equals the adsorption capacity.

Mass transfer, which can be evaluated with the mass transfer coefficient, dominates the performance of the adsorption process and the utilization ratio of adsorbent to a large extent. However, the determination of the mass transfer coefficient is sophisticated. In this case, the linear driving force (LDF) model was developed to offer a simplified method for describing the mass transfer phenomenon in adsorption [4]. As indicated by Eq. (12.9), the coefficient k_{f} (1/s) can be expressed by assuming that the movement of adsorbate is in a linear driving force $(n^* - n)$

$$\left[\frac{\partial n}{\partial t}\right]_z = k_{\mathrm{f}}(n^* - n) \tag{12.9}$$

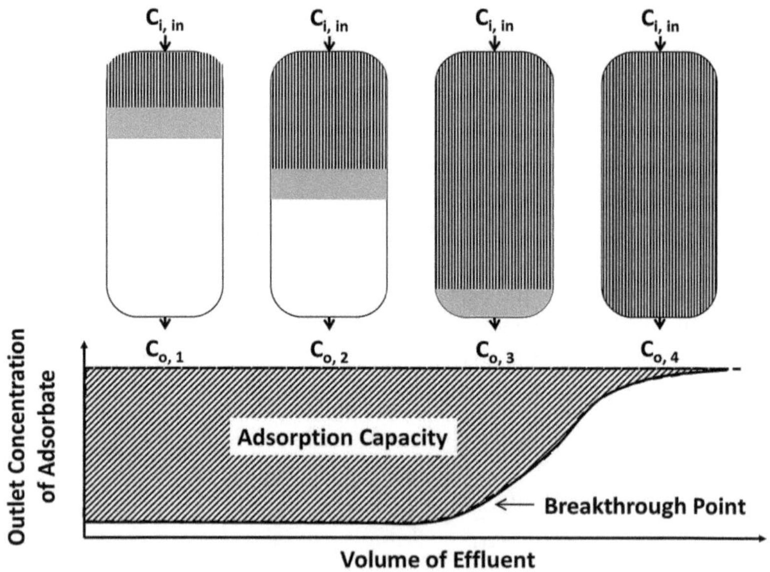

Fig. 12.2 Stages of adsorption and resulting outlet concentrations

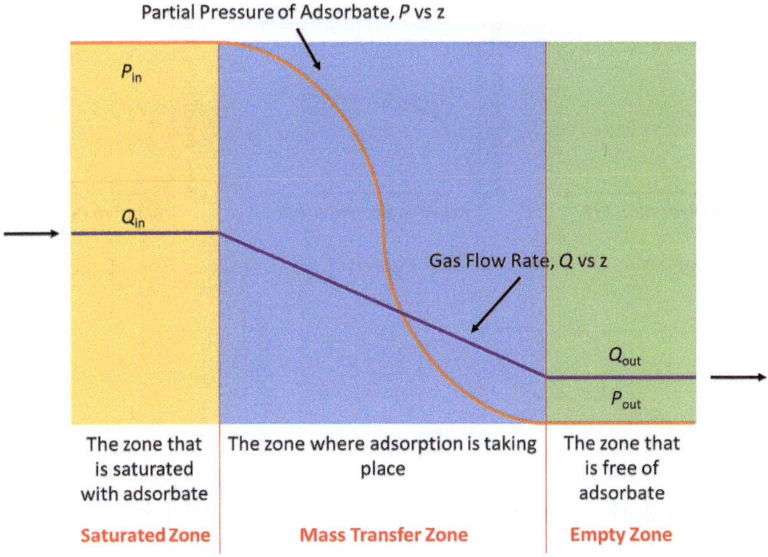

Fig. 12.3 Schematic diagram of gas flow regime and mass distribution in MTZ [5]

where n and n^* (mol/g) are the amounts of adsorbate that are adsorbed at (z, t) and at equilibrium state, respectively. The mass transfer coefficient k_f can be obtained by integrating Eq. (12.9) for which the profile of $n(z, t)$ is needed.

Figure 12.3 presents the gas flow profile and the mass distribution of adsorbate in MTZ at a specific depth (z) and time (t). Gas flow and mass distribution can be expressed as the functions of z and t, i.e., Q (z, t) and $P(z, t)$ with respect to partial pressure in the gas phase. In practice, the concentration or partial pressure of adsorbate is very low, contributing a negligible amount to the total gas flow rate. For this reason, we can assume that the gas flow rate is a constant, independent of z and t. Specifically, at z and t, the mass balance of adsorbate is given by Eq. (12.10)

$$\left[\frac{\partial \overline{n}}{\partial t}\right]_z = -\left[\frac{\partial QP}{\partial z}\right]_t = -Q\left[\frac{\partial P}{\partial z}\right]_t \tag{12.10}$$

where $\overline{n}$ (mol/com^3) is the total amount of adsorbate in the adsorber.

As adsorption proceeds, *MTZ* moves toward the outlet of the adsorber and the velocity β can be mathematically evaluated. M in the formula can be any characteristic property describing the adsorption process, i.e., P, n, $\overline{n}$.

$$\beta = -\left(\frac{\partial M}{\partial t}\right)_z / \left(\frac{\partial M}{\partial z}\right)_t = \left(\frac{\partial (QP)}{\partial \overline{n}}\right)_z \tag{12.11}$$

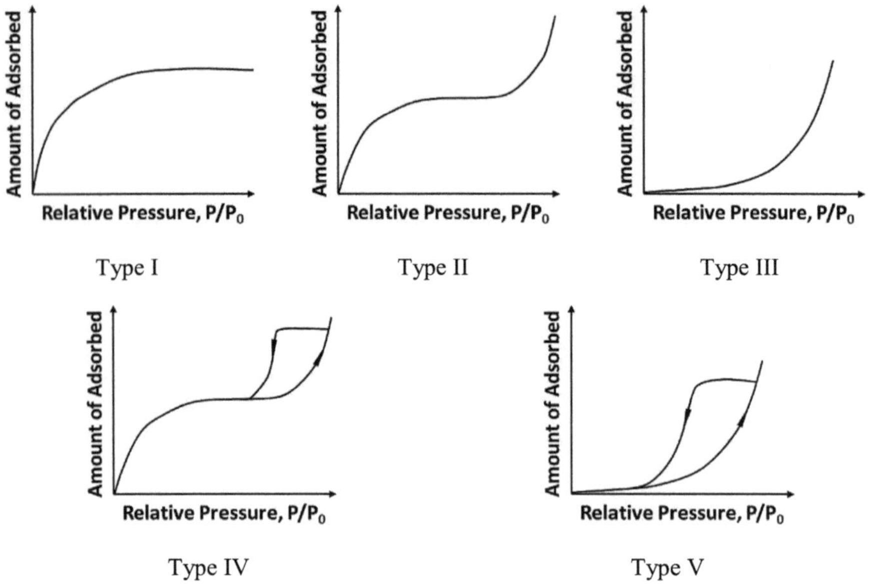

Fig. 12.4 Images of 5 types of adsorption profiles

Equation (12.11) can be integrated to get the relationships between P, $\overline{n}$, and Q, which are

$$\beta = \frac{Q \times P_{in}}{\overline{n}_{in}}; \overline{n} = \frac{Q \times P}{\beta} \tag{12.12}$$

Setting lower θ_{Ψ} and upper $\theta_{1-\Psi}$ limits for partial pressure or concentrations of adsorbate, effective mass transfer coefficient, k_{f}, can be obtained via integration of Eq. (12.9).

$$k_{\mathrm{f}} = \frac{\int_{\theta_{\Psi}}^{\theta_{1-\Psi}} \frac{d\theta}{(\theta^{*}-\theta)}}{t_{1-\Psi} - t_{\Psi}}; \theta = \frac{n}{SAT}, \theta^{*} = \frac{n^{*}}{SAT} \tag{12.13}$$

12.2.3 Adsorption Isotherm

An adsorption system involves the interaction between gaseous adsorbate A and active sites S in the adsorbent. For convenience, we assume the interaction would produce adsorption products, namely, AS. Simultaneously, desorption takes place, disintegrating AS and releasing adsorbate A back into the bulk solution. The processes are shown by Eq. (12.14).

$$A + S \overset{k_1/k_2}{\Longleftrightarrow} AS \tag{12.14}$$

If the partial pressure of adsorbent A is designated as $P_{(A)}$, the mole concentration of active site S is designated [S], and the mole concentration of adsorption product AS is [AS], then the adsorption and desorption rate can be readily calculated with Eqs. (12.15) and (12.16)

$$R_{\text{adsorption}} = k_1 \cdot P_{(A)} \cdot [S] \tag{12.15}$$

$$R_{\text{desorption}} = k_2 \cdot [AS] \tag{12.16}$$

where k_1 and k_2 are rate constants of adsorption and desorption process, respectively. When it reaches equilibrium, i.e., $R_{\text{adsorption}} = R_{\text{desorption}}$, we have equilibrium constant K as shown in Eq. (12.17)

$$K = \frac{[AS]}{P_{(A)} \cdot [S]} = \frac{k_1}{k_2} \tag{12.17}$$

Adsorption is always interpreted by the adsorption isotherm which relates the adsorbed phase to the gaseous phase of adsorbate, which is shown by Eq. (12.18).

$$P_{(A)} = \frac{[AS]}{K_H}\left[\frac{SAT}{SAT - [AS]}\right] e^{C_1 n + C_2 n^2 + C_3 n^3 + \cdots} \tag{12.18}$$

where P (Pa) is the partial pressure of adsorbate in bulk gas, [AS] ($mol \cdot kg^{-1}$) is the amount of adsorbed phase in adsorbent, K_H is the Henry's constant, SAT ($mol \cdot kg^{-1}$) is the saturation capacity of adsorbent which is the sum of [S] and [AS], and C_i are virial coefficients.

Physical adsorption can be divided into five categories, according to their adsorption isotherms [6]. When pore size is not much greater than adsorbate molecules, the Langmuir isotherm is observed which can be classified as type I and features a definite and distinct saturation limit. Type I adsorption occurs when the interaction between adsorbate molecules is completely overwhelmed by that between adsorbate and adsorbent. The occurrence of type II and type III isotherms indicates multilayer adsorption where the adsorbate load rises and adsorption shifts from a monolayer to a multilayer. Further, condensation in the capillary would happen even under higher pressures, leading to an increase in adsorption capacity. Type IV isotherm is usually observed in two-layer adsorption. Type V is another monolayer adsorption isotherm. Compared with type I, type V isotherm is observed when intermolecular attractions are non-neglectable.

Langmuir Isotherm

The Langmuir isotherm was developed based on the assumptions that the adsorption was taking place at active sites on the surface of the adsorbent. In addition, it also assumes that for each active site, only one molecule can be irreversibly adsorbed (monolayer adsorption) and that no intermolecular interactions occur. Mathematically, the Langmuir equation is derived by assigning zero to all virial coefficients, i.e., $C_1 = C_2 = C_3 = \cdots = 0$, in Eq. (12.18). An important concept in the Langmuir isotherm is surface coverage θ, which is defined as the number of occupied sites divided by a total number of sites.

$$\theta = \frac{[\mathrm{AS}]}{[\mathrm{S}] + [\mathrm{AS}]} = \frac{[\mathrm{AS}]}{\mathrm{SAT}} = \frac{\mathrm{K} \cdot P_{(A)}}{1 + \mathrm{K} \cdot P_{(A)}} \tag{12.19}$$

Equation (12.19) is the most popular form of Langmuir formula in use, where the equilibrium constant K can be derived from kinetic theory, which is

$$\frac{1}{K} = \frac{\nu}{\sigma}(2\pi mkT)^{1/2} e^{-\frac{Q}{RT}} \tag{12.20}$$

where ν is the pre-exponential factor of the desorption rate coefficient, σ is the condensation coefficient, m is the mass of the adsorbate molecule, k is the Boltzmann constant, and Q is the adsorption heat.

Based on the Langmuir isotherm, it is explicit that as $P_{(A)}$ approaches infinity, the adsorbent becomes saturated, i.e., $\theta \to 1$. On the other hand, when $P_{(A)}$ is at low levels, the adsorption process is dominated by Henry's law, shown by Eq. (12.21).

$$\lim_{\mathrm{P_{(A)}} \to 0} \left(\frac{[\mathrm{AS}]}{\mathrm{P_{(A)}}} \right) = \lim_{\mathrm{P_{(A)}} \to 0} \left(\frac{\mathrm{K} \cdot \mathrm{SAT}}{1 + \mathrm{K} \cdot \mathrm{P_{(A)}}} \right) = \mathrm{K} \cdot \mathrm{SAT} = \mathrm{K_H} \tag{12.21}$$

Equation (12.19) is usually rearranged to make it feasible for experimental data fitting.

$$\frac{P_{(A)}}{[AS]} = \frac{1}{K \cdot SAT} + \frac{P_{(A)}}{SAT} \tag{12.22}$$

or

$$\frac{1}{[AS]} = \frac{1}{SAT} + \frac{1}{K \cdot SAT} \cdot \frac{1}{P_{(A)}} \tag{12.23}$$

Freundlich isotherm

The Freundlich model assumes heterogeneous surface energies for adsorption which are distributed exponentially, leading to Eq. (12.24).

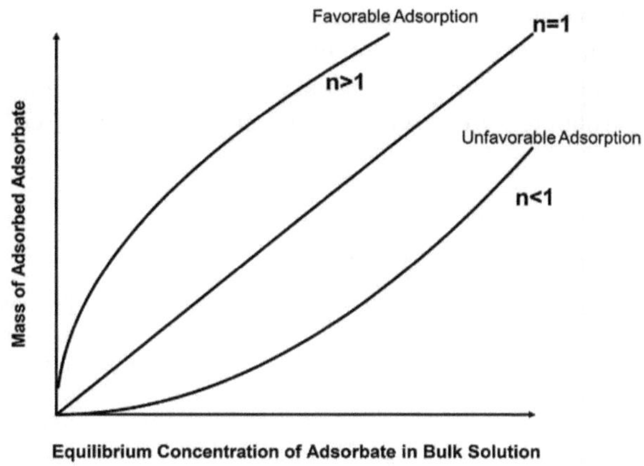

Fig. 12.5 Graphic presentation of Freundlich isotherm for the three cases, n > 1, n = 1, and n < 1

$$\ln\theta = \frac{RT}{Q_0}\ln P_{(A)} + \text{constant} \tag{12.24}$$

for small surface coverage, θ, Q_o is a constant. Equation (12.24) can be modified into the Freundlich isotherm

$$\theta = K_F P_{(A)}{}^{1/n} \tag{12.25}$$

The Freundlich isotherm does not indicate the saturation limit and it can be regarded as an expression of the Langmuir isotherm when $0<\theta<1$. Instead, according to the n value, we can tell whether the adsorption is favorable or not. A diagram of the Freundlich isotherm with three cases is shown in Fig. 12.5.

BET isotherm

BET was firstly developed by Brunauer, Emmett, and Teller to specify the relationship between adsorption capacity and the specific surface of the adsorbent, considering chemisorption and multilayer physisorption. Therefore, the stoichiometric formula that is applied to the Langmuir and Freundlich isotherms is not feasible in BET development. BET isotherm assumes that (1) Every layer of adsorbed molecules constitutes a base for the adsorption of the molecules of the second layer, which, in turn, acts as a base for the third layer, and so on; (2) The Langmuir isotherm can be applied to every adsorption layer; (3) Interactions between the adsorbed molecules are neglected; and (4) Adsorption energies of each layer are equal.

$$\frac{P_{(A)}}{q(P_s - P_{(A)})} = \frac{1}{q_m K} + \frac{(K-1)}{q_m K}\frac{P_{(A)}}{P_s} \tag{12.26}$$

where $P_{(A)}$ is the concentration of adsorbate in the gas stream, q is adsorption capacity at $P_{(A)}$, q_m is the quantity of adsorbate necessary to cover the unit mass of the adsorbent with a monolayer, P_s is the saturation concentration of adsorbate, and K is the constant. However, the BET model cannot describe the adsorption process for wide partial pressures of the adsorbate. As a result, BET should be used when the partial pressure is set between 0.05 and 0.35 [7] (see Table 12.2).

Table 12.2 Summary of the adsorption isotherms

Isotherm	Expression	Feature	Feasibility
Langmuir	$\theta = \frac{K \cdot P_{(A)}}{1+K \cdot P_{(A)}}$	• Feasible for monolayer adsorption • Indicate saturation limit	Type I
Freundlich	$\theta = K_F P_{(A)}{}^{1/n}$	• Feasible when 0 < θ<1 • Useful to determine the feasibility of adsorbent	
BET	$\frac{P_{(A)}}{q(P_s - P_{(A)})} = \frac{1}{q_m K} + \frac{(K-1)}{q_m K} \frac{P_{(A)}}{P_s}$	• Feasible for multilayer adsorption • BET is popular for determining the surface area of the adsorbent	Type II and Type V
D – R [8]	$W = W_0 \times \exp\left[-\left(\frac{A}{\beta E_0}\right)^2\right]$	• Based on potential theory • Postulate that adsorption in micropores is motivated by pore filling	Adsorption of an organic compound on microporous adsorbents
Polarization Theory	$\ln\left[\ln\left(\frac{P_0}{P}\right)\right] = \ln\left[\left(\frac{\varepsilon_0}{RT}\right) - \ln\left(\frac{d^3}{a}\right)^2\right] \times \frac{V}{V_0}$		
Harkins–Jura Theory [9]	$\ln\left(\frac{P}{P_0}\right) = B - \frac{a\sum}{2RT} \times \frac{1}{V^2}$		Provide a simple method to determine surface area of the adsorbent particle

12.3 Design Methods

12.3.1 Bed-Depth Service-Time (BDST) Approach

It is critical to have an adsorption reactor that is large enough to ensure long operating cycles. There is a fixed adsorption bed in which laden gas flows at a steady velocity v, so adsorbate gets removed. Taking a unit height, dz, of adsorption bed into the analysis, thus the partial pressure in this unit is denoted as dP. The schematic illustration is shown in Fig. 12.6.

Assuming the mass transfer between adsorbate and adsorbent is controlled by the fluid film, the mass transfer coefficient is denoted as k_f, and the mass balance of the adsorbate within dz is acquired

$$veA\mathrm{d}P = -k_a A(1-e)a_p(P-P_i)\mathrm{d}z \tag{12.27}$$

where A is the cross-sectional area of adsorption on the bed, a_p is the specific surface area, e is the bed void space, and P_i is the partial pressure of adsorbate at the fluid-adsorbent interface. Integration of Eq. (12.27) is usually difficult as P_i is not readily measured or calculated. Instead, we use the equilibrium partial pressure P^* and the overall mass transfer coefficient. By setting initial and boundary conditions, Eq. (12.27) can be integrated and yield the total bed depth (Z)

$$Z = \frac{ve}{(1-e)k_a a_p}\int_{P_i}^{P_o}\frac{\mathrm{d}P}{P-P*} \tag{12.28}$$

where P_i and P_o are partial pressures of adsorbate in bulk solution at the inlet and outlet of the adsorption bed, respectively.

The BDST approach, which is used to establish the relationship between bed depth and breakthrough time, is based on mass transfer within the adsorption reactor [10]. In the BDST model, it is assumed that adsorption is dominated by the surface reaction between adsorbate and the used active site on the adsorbent [11].

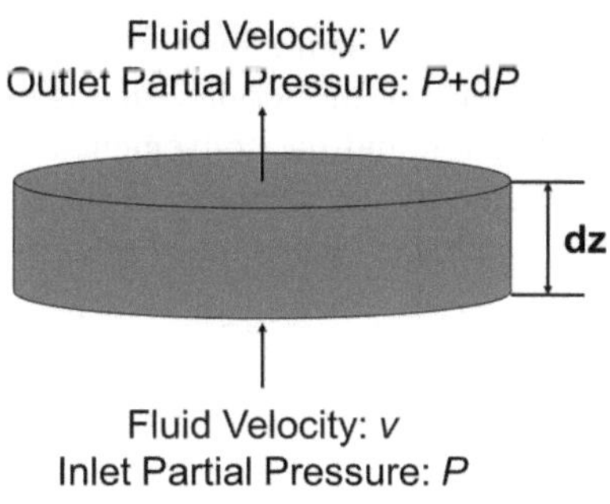

Fig. 12.6 Mass balance of adsorbate within a unit height of adsorption bed

$$\ln\left(\frac{C_{\text{in}}}{C_b} - 1\right) = \ln\left(e^{K_a \cdot AC_t \cdot Z \cdot Q} - 1\right) - K_a \cdot C_{\text{in}} \cdot t \tag{12.29}$$

where C_b is the outlet concentration of a pollutant at a breakthrough point or a point determined by regulations, K_a is the adsorption rate, AC_t is the adsorption capacity at time t, and Z represents the bed height (m). Since $e^{K_a \cdot AC_t \cdot Z \cdot Q}$ is much greater than 1 [12], Eq. (12.29) is thus simplified into Eq. (12.30)

$$\ln\left(\frac{C_{\text{in}}}{C_b} - 1\right) = \ln e^{K_a \cdot AC_t \cdot Z \cdot Q} - K \cdot C_{\text{in}} \cdot t \tag{12.30}$$

which further yields a linear relationship between bed depth and service time as shown in Eq. (12.31)

$$t = \frac{AC_t}{C_{\text{in}} \cdot v} Z - \frac{1}{K_a \cdot C_{\text{in}}} \ln\left(\frac{C_{\text{in}}}{C_b} - 1\right) \tag{12.31}$$

As a result, we can tell that service time is not only proportional to bed depth, but is also related to operation conditions including inlet concentration and fluid flow rate, etc. Note that by letting $t = 0$, we have

$$MTZ = \frac{v}{K_a \cdot AC_t} \ln\left(\frac{C_{\text{in}}}{C_b} - 1\right) \tag{12.32}$$

which is the critical bed depth or *MTZ* that prevents the outlet concentration from going beyond the breakthrough concentration.

However, the linear relationship depicted by Eq. (12.31) only holds finite time intervals since AC_t would rise as adsorption proceeds. In this case, Lee and Porter [13] proposed a modification which clarified the evolution of AC_t along operation time

$$t = \frac{AC_0\left(1 - e^{-a\sqrt{t}}\right)}{C_{\text{in}} \cdot v} Z - \frac{1}{K_a \cdot C_{\text{in}}} \ln\left(\frac{C_{\text{in}}}{C_b} - 1\right) \tag{12.33}$$

where AC_0 is saturation capacity.

In general, the service time is usually predetermined by the requirements of the application and the convenience of maintenance. In addition, the fluid velocity is restricted by the mechanical strength of the adsorbent. As a result, the appropriate bed depth can be estimated by operation conditions and application requirements.

12.3.2 Design Criteria

Variables that need to be determined in the adsorption process include inlet concentration, gas velocity, dimensions of the adsorption reactor, and operating temperatures.

Inlet concentration could drastically influence operating reliability and safety. As concentration increases, the mass transfer zone (MTZ) is suppressed so service time is extended. When combustible adsorbate is dealt with, concentration should be controlled below the lower explosive limit (LEL) for safety. For many VOCs, the LEL ranges from 1 to 1.5% of the flue gas volume which is equivalent to 2,500–3,750 ppm.

Gas velocity should be set by confirming the adsorbent is intact during the process. On the other hand, the greater the velocity is, the more significant *MTZ* will be.

Laden gas should be uniformly distributed throughout the adsorption reactor otherwise the adsorption bed may be partially penetrated (channeling), compromising adsorption capacity. For this reason, the cross-sectional area of the reactor should be narrowed while additional space should be added on both sides of the bed to regulate gas flow. In addition, the adsorption reactor should support and protect the adsorbent from pressure peak disturbances.

Since the adsorption process depends heavily on the thermodynamic situation, the temperature is a decisive factor. Adsorption is an exothermal process, and the temperature can increase due to the heat release, resulting in compromised performance.

In conclusion, the rule of thumb regarding adsorber design suggests that the superficial velocity should range in 20–100 ft/min. Considering that length of the mass transfer zone is typically 3 inches, the bed depth should be 1.5–3.0 ft [14].

12.3.3 Good Engineering Practice

Good engineering practices are built based on further consideration of operation and maintenance related issues. For instance, it is recommended that the adsorbent should not exceed 1/3 of the vessel volume [15] or

$$Z_{\max} \approx \frac{\pi D}{12} \tag{12.34}$$

where D is the diameter of the vessel.

Vessel dimensions are dependent on the carbon requirement and the laden gas flow rate. For a vessel where the carbon bulk density is 30 lb/ft^3, thus the diameter and the length can be, respectively, estimated by Eqs. (12.35) and (12.36)

$$D = \frac{0.127 M_c \cdot v}{Q} \tag{12.35}$$

$$L = \frac{7.87}{M_c}\left(\frac{Q}{v}\right)^2 \tag{12.36}$$

where M_c is the mass of the adsorbent for meeting service time requirements.

If the gas flow rate is less than 9,000 scfm, a vertical vessel is recommended. In this case, the diameter can also be estimated by [15]

$$D = \left(\frac{4Q}{\pi v}\right)^{1/2} \tag{12.37}$$

The gas velocity should conform to product specifications otherwise the adsorbent might get crushed. Meanwhile, the length of the vessel is the sum of bed depth plus the additional vacuum that is used for better distribution of gas streams. As a rule of thumb, 1 to 3/2 feet is recommended for smaller-diameter vessels. For longer vessels, 2 to 3 feet at each side of the bed is usually provided.

12.4 Design Procedures

An overview of the design procedure for an adsorption system is described in Fig. 12.7. The design starts from background investigations including emission regulation, the composition of the flue gas, and physio-chemical properties of the target substance. Subsequently, a pilot test simulating the practical situation is recommended where the various adsorbent and reactors are tested. According to the breakthrough curve obtained from the pilot test, some critical parameters such as mass transfer rate, adsorption capacity, etc., can be estimated. Based on them, the best available adsorbent and reactor can be proposed. Afterward, the dimensions of the reactor can be calculated. In this step, not only is the adsorption performance important, but other practical conditions such as space availability and ease of operation should also be taken into consideration. If the design can meet the requirements, a standardized operation and maintenance system should be established to ensure reliability.

12.4.1 Step 1: To Select the Appropriate Adsorption Reactor

Generally, there are five kinds of common adsorbers in use which include the fixed bed, drum, traveling bed, fluid bed, and chromatographic baghouse [16]. Among them, the fixed bed and drum are the most commonly used adsorbers for air pollution control, especially with VOCs.

Drums are a simple solution for all kinds of industries because they are easy to install, operate, and maintain. Drums are limited to controlling lower stream flow

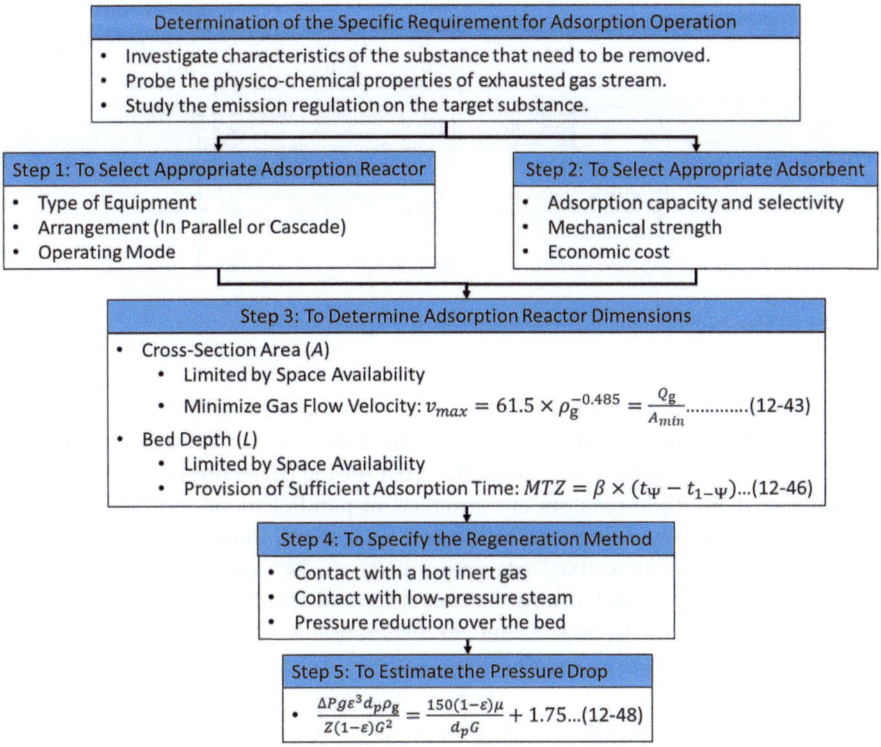

Fig. 12.7 Procedures for designing an adsorption system

Fig. 12.8 NIXTOX XP GAC Polyethylene Drum Adsorber (Courtesy: TIGG [17])

rates in intermittent operating modes. Drums are usually provided by vendors who are also responsible for regeneration and maintenance. As the adsorbent is saturated, the drum is switched off and then sent to a designated facility for regeneration.

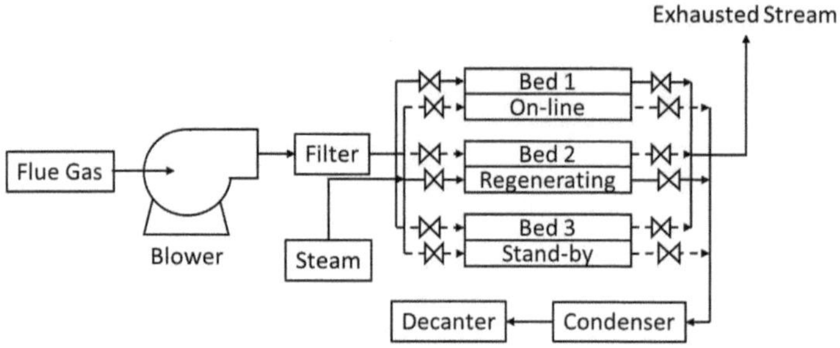

Fig. 12.9 Three-bed, continuously operated fixed adsorption system

A fixed bed is another kind of reactor used to manage air pollution emissions with a large flow rate. Compared to drums, the fixed bed can be regenerated where it is operated, which reduces the economic cost for the transportation of the reactor. In addition, if multiple reactors are arranged in parallel mode, adsorption can be operated continuously. Figure 12.9 depicts a typical continuous adsorption system that is comprised of three fixed adsorption beds. As shown in the figure, flue gas after pretreatment is directed into the operating bed (Bed 1) to remove the target pollutant. Simultaneously, bed 2 is under regeneration with steam which is collected and treated by a condenser and decanter. Afterward, bed 2 is put on stand-by like bed 3 is.

A more advanced design of the regenerable fixed-bed system is the rotating adsorber. A rotary adsorber is comprised of many discrete units which are installed on a rotary table. During operation, the adsorber rotates at a proper speed. Target substances in the flue gas are concentrated in the adsorption zone and are then desorbed when in contact with hot gas. The desorbed adsorbate flows into the combustion chamber for energy generation in the form of hot steam and/or electricity. The steam can be recycled and used for regeneration and cooling. Due to the limited contact time between the adsorption unit and flue gas, bed depth is usually shallow, resulting in a minimized pressure drop.

12.4.2 *Step 2: To Select the Appropriate Adsorbent*

Activated carbon is a popular adsorbent because it is cheap and is easy to manufacture. The feedstock of activated carbons includes coconut shells, bamboo, and many other bio-organic materials. Recently, activated carbon originating from coconuts has had the most attention due to their superior mechanical strength that can tolerate the impact of higher gas flows.

It is feasible for activated carbon to remove volatile pollutants with proper molecular weight and volatility at low concentrations. High molecular weight compounds are usually characterized by low volatility and stick to the adsorbent, which makes it difficult to remove them by regeneration. For this reason, adsorption is not feasible for such compounds. On the other hand, highly volatile materials are not readily adsorbed by carbon, which compromises the removal efficiency.

In most cases, activated carbon is used as a carrier so that functional groups can attach to its surface, enhancing the selectivity toward specific substances. After coating it with various kinds of functional groups, the selectivity of activated carbon is modified. Table 12.3 lists popular functional groups which are popular in air pollution control.

The second most popular adsorbent is zeolite (alumina-silica mineral) whose unique micro-structure makes it an excellent molecule separating agent. Unlike activated carbon, the pore size of zeolite is uniform and is dominated by its crystal structure. In addition, by adjusting the alumina-silica ratio, pore size and affinity to the polar molecules of zeolite can be controlled manually. Specifically, zeolite is originally usually hydrophilic due to the relatively high aluminum content, which is not desired for non-polar molecule adsorption. Therefore, dealumination is performed to transform zeolite into a hydrophobic adsorbent by replacing aluminum with silicon. In this case, zeolite becomes feasible for VOC control.

The most important property of an adsorbent is the adsorption capacity which is revealed by equilibrium isotherm under expected operating conditions including the temperature and pressure ranges of interest. Such information may be available by experimental tests or acquired from the adsorbent vendor. Apart from the breakthrough capacity (BRK), there are two important terms used to describe the adsorption capacity of the adsorbent bed. One of them is the saturation capacity (SAT) which represents the maximum amount of adsorbate that can be adsorbed by each unit mass of adsorbent. The exact amount of pollutants fixed by adsorption is termed as working charge (WC), which is usually 0.1 to 0.7 times the saturation capacity. After several adsorption-generation cycles, the capacity deteriorates since the residual adsorbate and vapor left on the bed causes the performance loss, which is termed as a heel. As a result, the relationship between breakthrough capacity, saturation capacity, working charge, and heel is presented in Eqs. (12.38) and (12.39)

$$BRK = [a \cdot SAT \cdot MTZ + SAT \cdot (Z - MTZ)]/Z \tag{12.38}$$

Table 12.3 Enhancement of activated carbon by functional groups [18–21]

Functional group	Applications
Oxygen	Gas purification from mercury
Sulfuric acid	Gas purification from ammonia, amine, mercury
Potassium carbonate	Gas purification from acid gases: HCl, HF, SO_2, NO_2
Manganese IV oxide	Gas purification from aldehyde
KOH/CaO	Gas Purification from H_2S
Phosphoric acid	Gas Purification from toluene, isopropanol

$$WC = BRK - HEEL \tag{12.39}$$

where a is the constant which is dependent on the breakthrough concentration set in adsorption.

If there is more than one adsorbate in the system (sometimes water vapor may exist) competition may happen, resulting in unsatisfactory performance. For this reason, the selectivity of the target substance is especially important. The working charge of a multiple-adsorbate system is expressed as

$$WC = \frac{1}{\sum_{i=1}^{n} (w_i / SAT_i)} \tag{12.40}$$

where n is the number of adsorbates in the system, w_i is the molar fraction of i in the adsorbates, and SAT_i represents the saturation capacity of adsorbate i. For a binary-adsorbate system, Eq. (12.40) simplifies to

$$WC = \frac{(SAT_A)(SAT_B)}{(w_A)(SAT_A) + (w_B)(SAT_B)} \tag{12.41}$$

In some cases where information on adsorption capacity is not available, we can estimate it by pore size distribution. Pore size distribution can be categorized as macropores (with diameter > 500 Å), mesopores (with diameter from 50 to 500 Å), and micropores (with diameter < 50 Å). Pore size is a decisive factor that affects the kinetics and the performance of the adsorption process. In general, macropores are related to the mass transfer and diffusion of the adsorbate molecules into the pores. Meanwhile, mesopores and micropores are responsible for the adsorption of medium and small size molecules, respectively.

Furthermore, the possibility of catalytic effects causing adverse chemical reactions in the gas stream and the formation of solid polymerizates on the adsorbent bed causes deterioration. Other characteristics include particle size, robustness, and resistance to the contaminant. In regeneration, retentivity, indicating the easiness of desorption, as well as contaminated effluent should be considered.

12.4.3 Step 3: To Determine Adsorption Reactor Dimensions

Dimensions of the adsorption reactor include the height and cross-sectional area, which depend on the mass of the adsorbent that is used. In general, the adsorbent requirement is dominated by operation duration, inlet concentration, and characteristics of the adsorbent and adsorbate.

$$M_C = N \cdot t \cdot Q_g \cdot (C_{\text{in}} - C_{\text{out}}) \cdot MW / WC \tag{12.42}$$

where N is the number of adsorption reactors that are in parallel mode, t is the duration of the adsorption cycle, and MW is the molecular weight of adsorbate. Considering the capacity loss due to *heel*, safety factors should be incorporated.

The cross-sectional area, A, is dependent on the gas velocity which is restricted by the mechanical strength of the adsorbent and the significance of the mass transfer zone. For the adsorbent with a diameter of 1/8 inch, the maximum gas velocity can be estimated with the following equation [22]

$$v_{max} = 61.5 \times \rho_g^{-0.485} = \frac{Q_g}{A_{min}} \tag{12.43}$$

where v (ft/min) is gas velocity, ρ_g (lb/ft^3) is the density of the gas, and A_{min} (ft^2) is the minimum cross-sectional area. Also, detailed information on crushing velocity is usually available from the manufacturer.

In general, superficial velocity is usually 20–100 ft/min. By setting a reasonable superficial velocity which should be lower than v_{max}, we can estimate the cross-sectional area and the bed depth according to Eqs. (12.44) and (12.45)

$$A = \frac{Q_g}{v} \tag{12.44}$$

$$Z = \frac{M_c/\rho_c}{A} \tag{12.45}$$

where ρ_c is the bulk density of the adsorbent.

Bed depth can be determined by comparing it with the length of the mass transfer zone. As shown in Eq. (12.32), the moving speed of *MTZ* is directly proportional to the gas flow rate. Note that gas flow and concentration of adsorbate measured at the outlet are mirror images of Fig. 12.3, therefore, the moving speed of *MTZ* can be evaluated using the change of characteristics in flow regime. An experimental determination of *MTZ* can be performed as follows:

$$MTZ = \beta \times (t_{\Psi} - t_{1-\Psi}) \tag{12.46}$$

Finally, vessel dimensions can be estimated according to GEP protocols or previous experience.

12.4.4 Step 4: To Specify the Regeneration Method

For the sake of economic performance, regeneration is usually implemented to reactivate adsorbent and recover valuable substances [23]. The mechanism of regeneration is shifting the thermodynamic equilibrium established during the adsorption cycle, which can be performed by increasing the temperature, reducing the pressure, or reducing the concentration of the gas stream out of the adsorbent [24]. Generally, desorption by increasing temperature can be accomplished with hot inert gas or steam. Steam is usually more effective and it has been reported that no more than 1 to 4 pounds of steam are required for recovering 1 pound of solvent. Inert gas is used when water vapor may be toxic to the adsorbent. Even though thermal desorption is effective in adsorbent recovery, it is not feasible when adsorbate, e.g., halogenated VOC, pyrolyzes. In this case, low-pressure desorption is recommended. A summary of the most commonly used desorption methods is listed in Table 12.4.

Regeneration affects the performance of an adsorption system to a very large extent. In general, a continuous and rapid operation is desired, otherwise, the performance is impaired. Therefore, it is suggested that (1) time required for regeneration

Table 12.4 Features and feasibilities of commonly used regeneration process [25, 26]

Method	Advantages	Disadvantages	Application
Thermal swing	• Good for strongly held species • Desorbed species recovered at high concentrations	• Thermal aging of adsorbents • Heat loss leads to thermal inefficiency • Long cycle times mean inefficient use of adsorbent • High latent heat for liquids	• Drying
Pressure swing	• Good for weakly held species • Desorbed species recovered at high concentrations	• Very low pressure may be required • Mechanical energy is more expensive than heat	• Dying • Hydrogen recovery • Air separation
Vacuum[a]	• Rapid cycling gives efficient use of adsorbents	• Desorbate recovered at low purity	• Separation of linear paraffins
Displacement	• Good for strongly held species • Avoid risk of cracking reactions during regeneration • Avoid thermal aging of adsorbent	• Product separation and recovery needed	• Separation of linear from branched and cyclic paraffins
Purge gas stripping	• Essentially at constant temperature and pressure	• Only for weakly held species • Purge flow is high • Not normally used when adsorbate needs to be recovered	• Relatively uncommon without thermal swing since purging alone is only suitable for weakly adsorbed species

[a] Special case of pressure swing

should be minimized; (2) regeneration steam should be provided at a higher flow rate, opposite to adsorption gas; (3) a larger fraction of heat should be provided to enable a more efficient heat transfer rate; and (4) low-retentivity carbon should be used so that desorption efficiency will increase. It is worth noting that desorption cannot totally remove adsorbate out of the adsorbent because the substance might be fixed in micropores and/or chemically adsorbed.

In the case of thermal swing, steam is required to raise the temperature of the adsorbent to a critical level so that desorption occurs. The amount of steam required for regeneration is estimated in Eq. (12.47)

$$Q_{\text{steam}} = St \cdot M_c / \left(t_{\text{reg}} - t_{\text{cool}}\right) \tag{12.47}$$

where Q_{steam} is the flow rate of steam, St is the amount of steam required by each unit mass of adsorbent which is usually 0.3 lb/lb, and t_{reg} and t_{cool} represent the time for regeneration and cooling cycles.

12.4.5 Step 5: To Estimate the Pressure Drop

Pressure drop, just like other air pollution control equipment is a big issue that can cause unnecessary power consumption for stream delivery is calculated by following [27]:

$$\frac{\Delta P g \varepsilon^3 d_p \rho_g}{Z(1-\varepsilon)G^2} = \frac{150(1-\varepsilon)\mu}{d_p G} + 1.75 \tag{12.48}$$

where ΔP ($\text{lb}_\text{f}/\text{ft}^2$) is the pressure drop, g (4.17×10^8 $\text{lb}_\text{m}\cdot\text{ft}/(\text{lb}_\text{f}\cdot\text{hr}^2)$) is the gravitational constant, d_p (ft) is the particle size of adsorbent, ρ_g ($\text{lb}_\text{m}/\text{ft}^3$) is the density of fluid, Z (ft) is the bed depth, G ($\text{lb}_\text{m}/(\text{ft}^2\cdot\text{hr})$) is the superficial mass flux, and μ ($\text{lb}_\text{m}/(\text{ft}\cdot\text{hr})$) is the gas viscosity.

For canister adsorber, the estimation of pressure drop is given as

$$Q_g = 0.0471 Q_g + 9.29 \times 10^{-4} Q_g \tag{12.49}$$

where Q_g is in the unit of inch of water and Q_g is in acfm.

12.4.6 A Rapid Method for Design of Adsorption Equipment

Mathematical Simplification

As suggested by Hougen and Marshall1 [28], the breakthrough curve during the adsorption can be expressed by the following dimensionless form:

$$\frac{y}{y_0} = 1 - e^{-b\tau} \int_0^{ax} e^{-\alpha} I_0\left(2\sqrt{b\tau \cdot \alpha}\right) d\alpha \tag{12.50}$$

$$\frac{w}{w_0} = e^{-ax} \int_0^{b\tau} e^{-\beta} I_0\left(2\sqrt{ax \cdot \beta}\right) d\beta \tag{12.51}$$

These two equations should link with the adsorption equilibrium curve as follows:

$$\text{y} = \text{cw} + \frac{c}{b}\frac{\partial w}{\partial \tau} \tag{12.52}$$

Using y = cw as the linear approximation for the adsorption equilibrium curve and substituting c = yo/wo, one obtains

$$\text{y} = y_0 \frac{w}{w_0} + \frac{y_0}{w_0 \cdot b}\frac{\partial_w}{\partial_\tau} \text{ or } \frac{y}{y_0} = \frac{w}{w_0} + \frac{\partial\left(\frac{w}{w_0}\right)}{\partial(b\tau)} \tag{12.53}$$

$$\frac{y}{y_o} = e^{-ax} \int_0^{b\tau} e^{-\beta} I_0\left(2\sqrt{ax\beta}\right) d\beta + e^{-ax} \cdot e^{-b\tau} \cdot I_0\left(2\sqrt{axb\tau}\right) \tag{12.54}$$

The values for the Bessel functions were obtained from the IMSL (International Mathematical and Statistical Libraries, Inc.) [29] software package, i.e., MMBSIO4. For values of ax greater than 350, the Bessel functions were approximated as

$$I_0(x) \cong \frac{1}{\sqrt{2\pi x}} e^x \tag{12.55}$$

The breakthrough curves obtained were plotted on a Calcomp plotter and are presented in Figs. 12.10 and 12.11.

$$\text{a} = \frac{1}{H_d} \tag{12.56}$$

$$\text{b} = \frac{G \cdot c}{\rho_B \cdot H_d} \tag{12.57}$$

If the mass transfer in the gas film is the limiting factor to adsorption, then the height of a mass transfer unit can be determined as

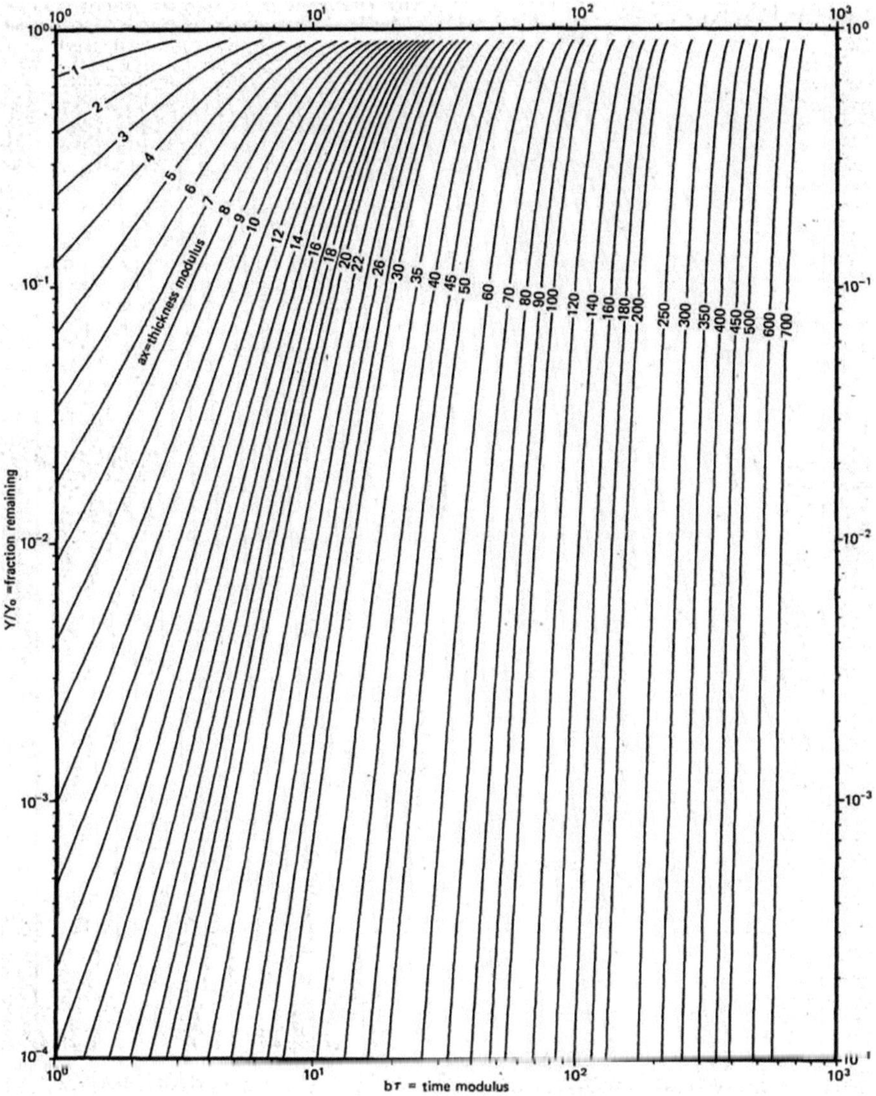

Fig. 12.10 Fraction of a component remaining in the fluid stream after passing through a stationary granular bed [32]

$$H_d = \frac{0.55}{a_v}\left(\frac{D_p \cdot G}{\mu}\right)^{0.51}\left(\frac{\mu}{\rho \cdot D_v}\right)^{2/3} \tag{12.58}$$

Fair [30] presents another approach to estimate Hd for mass transfer to a wetted solid from a gas stream as shown in Eq. (12.59).

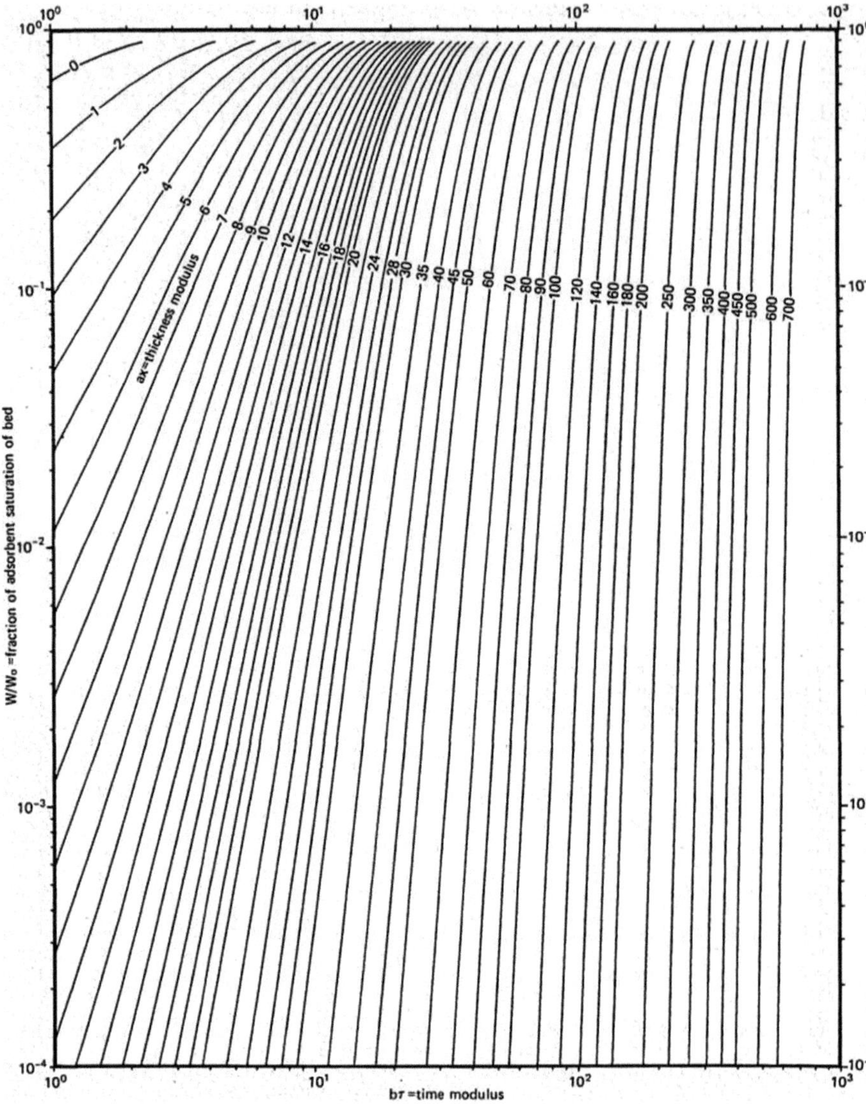

Fig. 12.11 Fraction of adsorbent saturation at any time and position in the bed [32]

$$H_d = \frac{(Sc_G)^{\frac{2}{3}}}{j_d \cdot a_v} \tag{12.59}$$

For large packings, jd can be calculated by the Hobson and Thodos [31] formula

$$j_d = 1.30\left(\frac{D_p G}{\mu_G}\right)^{-0.45} \tag{12.60}$$

In Eq. (12.57), c could be obtained from experimental equilibrium curves or by adsorption isotherm model.

12.5 Performance Evaluation and Enhancement

12.5.1 Performance Evaluation

Adsorption is generally applied to dilute mixtures of VOC and air. Removal efficiencies of 95–99% can be achieved. The maximum practical inlet concentration is usually about 10,000 ppm.

Capital costs for an adsorption reactor include the vessel, adsorbent, connections, and other auxiliaries. During the operation, the adsorbent needs routine replacement. A typical lifetime for carbon is about 5 years which may be influenced by operating conditions. If the adsorbate is difficult to desorb, the lifetime is expected to be shortened.

Energy requirements for adsorption include steam for regeneration, the ventilation system, and the pumping system for the delivery of cooling water.

$$\begin{cases} \text{Electricity}_{\text{fan}}(kW) = \frac{2.50\times10^{-4}\cdot Q\cdot\Delta P_s}{\varepsilon} \\ \text{Electricity}_{\text{pump}} = \frac{2.52\times10^{-4}\cdot q_{\text{cw}}\cdot H\cdot\rho}{\varepsilon} \end{cases} \quad (12.61)$$

where ΔP_s is the total system pressure drop which is equivalent to the bed pressure drop plus an additional extra inch of the water column, q_{cw} is the cooling water flow rate, H is the required water head, ρ the specific gravity, and ε presents the combined mechanical efficiency.

Adsorption barely follows the isotherm. Instead, as adsorption proceeds, temperature increases due to the heat release which is one of the most important problems that compromises the capacity of the adsorption bed. The rise of temperature, ΔT(°F), due to heat release can be calculated by Eq. (12.62)

$$\Delta T = \frac{6.1}{\frac{C_P}{C_i}\times10^5 + 0.51\frac{C_P}{AC_{\text{st}}}} \quad (12.62)$$

where C_P (BTU/ft^3/°F) is the heat capacity of adsorbent, C_i (ppm) is the inlet concentration of pollutants, and C_{st} is the capacity of the bed at $T + \Delta T$.

Bed capacity will seldom exceed 30 to 40% under actual operating conditions. Pilot-scale breakthrough curves must be determined. Water vapor in the gas stream can be captured by the adsorbent, leading to a deterioration of adsorption capacity. Therefore, care needs to be taken regarding relative humidity as it may adversely affect the adsorption processes [33]. Quantitatively, You and Lin [34] researched the

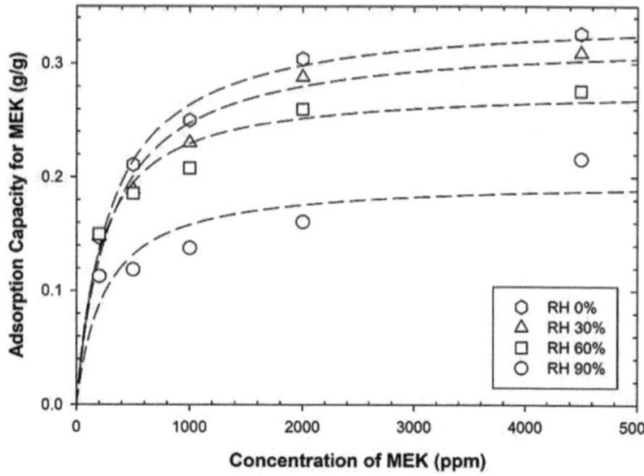

Fig. 12.12 Effects of relative humidity on adsorption for MEK. The data have been fitted with the Langmuir isotherm (*Data source* You and Lin [34])

adsorption of methyl-ethyl-ketone (MEK) in a RH 90% environment and they found that up to 47.0% of adsorption capacity was comprised as shown in Fig. 12.12.

12.5.2 Performance Improvement

Utilizing adsorption as an air pollution control technology has proved to be effective and efficient by following proper design and operating procedures. General requirements must be met in the design or selection of suitable adsorption equipment: (1) sufficient dwelling time, (2) good distribution of gas flow through the bed, (3) pretreatment to remove high concentrations of competing substances, (4) pretreatment to remove non-adsorbable substances that would impair the operation of the adsorption bed, and (5) renewing or regenerating the adsorbent after saturation.

Having sufficient dwelling time and standardizing gas flow is usually achieved by proper duct and vent design. As mentioned before, airflow below 100 ft/min through thicker carbon beds will provide adequate retention time for adsorption to occur. The excessive flow will reduce carbon efficiency and allow volatile compounds to escape into the atmosphere. Excessive airflow is also detrimental to process operations where unnecessary solvents are evaporated and delivered to carbon beds, imposing an additional load on the system. Uniform air distribution is important to the adsorption and desorption processes, otherwise tunneling effects may occur, reducing effective adsorption capacity. If multiple air flows are directed into the same system, adjustments of individual slide dampers may be required to obtain the correct airflow in each branch duct. Airflow switches should be installed in each branch to sense airflow and allow the process to operate when the airflow is adequate.

As pores and function groups play important roles in adsorption, it is critical to remove particulate matter and toxic substances from the flue gas. For this reason, adsorption is usually placed at the end of the entire treatment process. In addition, water vapor should be strictly monitored to prevent capacity loss. The aging of adsorbent may occur after several adsorption–desorption cycles. Therefore, a routine inspection is required to refresh the obsolete adsorbent. Personal protection should be carefully enforced during maintenance as residual adsorbate may be toxic to human health.

12.5.3 Innovative Technologies

PROTECT Adsorption Unit, Calgon Carbon [35]

Calgon Carbon Corporation's PROTECT High Flow VS (HFVS) carbon adsorber canisters are able to effectively control air emissions because they can contain up to 3,000 pounds of activated carbon. The HFVS system is compatible with Calgon Carbon's wide variety of activated carbon products. The most popular carbon product in use is 4 mm pelletized activated carbon with a Carbon Tetrachloride Number of 60 (Fig. 12.13).

Rotorsorb, CTP [36]

Rotorsorb provides a continuous air pollution control solution at room temperature whose benefits include low economic costs, no pressure peaks, and recirculation of desorption pollutant (Fig. 12.14).

Rotorsorb systems can achieve satisfactory removal efficiency (>97%) at a relatively low cost (pressure drop from 5 to 8 mbar). In addition, rotorsorb offers great operation flexibility because the system is able to combine with various auxiliaries (including combustion chambers and aerosol precipitators) which allow for the recovery of valuable substances or heat from the combustion of desorbed pollutant.

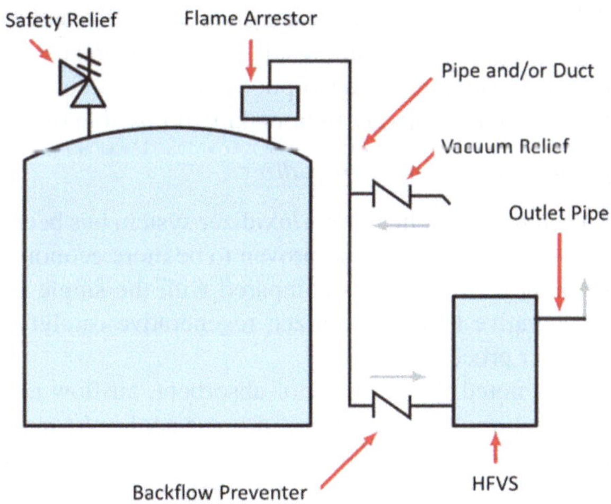

Fig. 12.13 System layout of HFVS adsorption unit (Courtesy: Calgon Carbon [35])

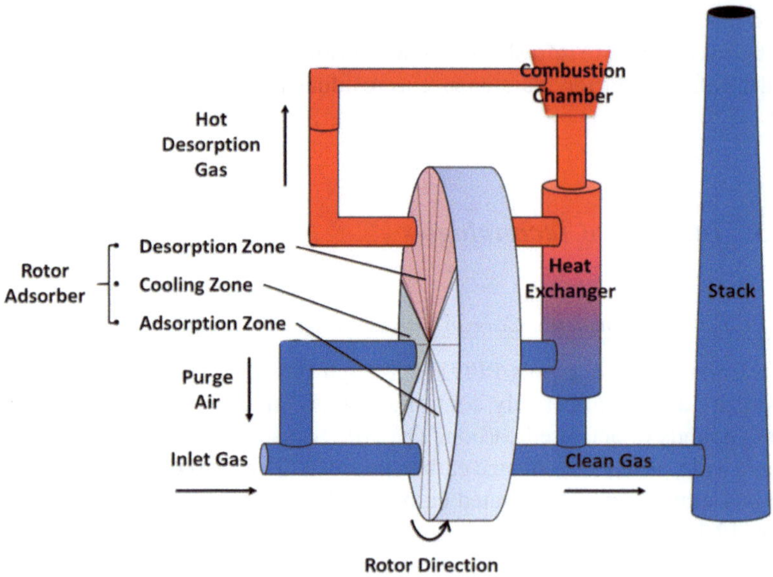

Fig. 12.14 Continuous rotary adsorber (Courtesy: CTP Air Pollution Control [36])

Fluidized Bed, TKS Industry [37]

The TKS Fluidized Bed Carbon (FBC) Abatement system consists of an adsorber, desorber, and control device. During the desorption cycle, hot gas is introduced to strip solvent from carbon beads, which is then ducted to a small oxidizer (the final control device) as shown in Fig. 12.15. System optimization is completed by reducing the oxidizer size and minimizing costs and energy use.

The TKS fluidized adsorption bed exhibits great engineering performance. Removal efficiency can reach more than 95% while the pollutant can be concentrated 50–1000 fold. As a result, energy use and carbon footprint are minimized, and CO_2 emissions in the adsorption process are controlled. In addition, the weight of the system is relatively light making it possible to be mounted on roofs.

Combined concentrator/oxidizer

The combined concentrator/oxidizer system has been proposed as an effective physical–chemical option and proven to be more economical, efficient, and yields lower secondary emissions as compared with the single oxidation processes such as the regenerative thermal oxidizer, regenerative catalytic oxidizer, or catalytic thermal oxidizer processes [38].

It is noted that the type of absorbent, air flow rate, rotating speed, regeneration temperature, the area ratio of process/cooling/regeneration zone, the flow rate ratio of the process flow to the regeneration flow, are the key parameters for design and control of the combined system. Table 12.5 shows the performance evaluation of RO, RTO, and RCO from the aspects of destruction efficiency, thermal recovery

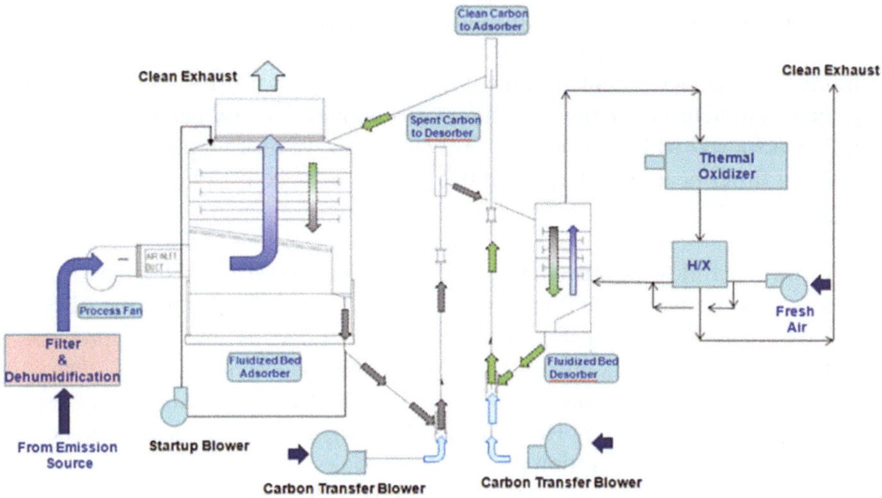

Fig. 12.15 Layout of Fluidized Adsorption Bed (Courtesy: TKS Industry [37])

Table 12.5 Comparisons among different oxidizers

	RO	RTO	RCO
Waste gas volume (m^3/h)	3600–49800	9600–43200	9600–43200
Operating temperature (°C)	700–750	760–810	300–350
Destruction efficiency (%)	99	95–99	95–98
Thermal recovery efficiency (%)	65–70	80–95	40–70
Pressure loss (Pa)	300–400	600–700	>700
Heat exchange type	Recuperative	Regenerative	Regenerative
Capital cost	Average	High	High
Operation cost	Average	Average	High
Maintenance	Low	Average	Average

efficiency, and cost effectiveness [38]. In general, the RO has a lower cost compared with RCO and RTO.

Microwave-heated adsorbent-reactor

A microwave-heated adsorbent-reactor system has been used for the continuous cleaning of air streams containing n-hexane at low concentrations. Both, a single catalytic bed (PtY zeolite) and a double (adsorptive DAY zeolite + catalytic PtY zeolite) fixed-bed reactor configurations were developed [39]. The zeolites were selectively heated by short periodic microwave pulses that caused the desorption of n-hexane and its subsequent catalytic combustion. The double bed configuration exhibits the same performance with only half of the catalyst load. The operation was especially efficient under realistic humid gas conditions, continuous gas cleaning could be achieved with short (3 min, 30 W) microwave heating pulses every 5 min, producing faster heating of the adsorptive and catalytic beds [39].

Exercises

1. Define the following terms and explain why they should be considered for design use: mass transfer zone, Langmuir isotherm, breakthrough capacity, BDST.
2. Please describe the general procedures for designing as adsorption process.
3. List the major factors affecting the design of an adsorption process.
4. Briefly describe the concept of a rapid method for the design of adsorption equipment suggested by Hougen and Marshall, and explain how it can be used for the design of adsorption process.
5. Determine the maximum mole fraction of H_2S that can be dissolved in Henry's law constant, $H_{H2S} = 483$ atm/mol fraction (1 atm, 6°8F); Partial pressure of $H_2S = 0.01$ atm.

References

1. Fontecha-Cámara MA, López-Ramón MV, Álvarez-Merino MA, Moreno-Castilla C (2006) About the endothermic nature of the adsorption of the herbicide diuron from aqueous solutions on activated carbon fiber. Carbon 44:2335–2338
2. Shaverdi G (2012) Developing a model for mass transfer in adsorption packed-bed filters. Concordia University
3. Chuang C, Chiang P, Chang E, Huang C (2003) Adsorption-desorption rate of nonpolar volatile organic compounds onto activated carbon exemplified by C_6 H_6 and CCl_4. Pr period Hazard Toxic Radioact Waste Manag 7:148–155
4. Sircar S, Kumar R (1983) Adiabatic adsorption of bulk binary gas mixtures. Analysis by constant pattern model. Ind Eng Chem Process Des Dev 22:271–280
5. Wu C-W, Kothare MV, Sircar S (2014) Column dynamic study of mass transfer of pure N_2 and O_2 into small particles of pelletized LiLSX zeolite. Ind Eng Chem Res 53:17806–17810
6. Brunauer S, Deming LS, Deming WE, Teller E (1940) On a theory of the van der waals adsorption of gases. J Am Chem Soc 62:1723–1732
7. de Boer JH, Lippens BC, Linsen BG, Broekhoff JCP, van den Heuvel A, Osinga TJ (1966) Thet-curve of multimolecular N2-adsorption. J Colloid Interface Sci 21:405–414
8. Hutson ND, Yang RT (1997) Theoretical basis for the Dubinin-Radushkevitch (DR) adsorption isotherm equation. Adsorption 3:189–195
9. Harkins WD, Jura G (1944) Surfaces of solids. XIII. A vapor adsorption method for the determination of the area of a solid without the assumption of a molecular area, and the areas occupied by nitrogen and other molecules on the surface of a solid. J Am Chem Soc 66:1366–1373
10. Bohart G, Adams E (1920) Behaviour of charcoal towards chlorine. J Chem Soc 42:523–544
11. Goel J, Kadirvelu K, Rajagopal C, Garg VK (2005) Removal of lead (II) by adsorption using treated granular activated carbon: batch and column studies. J Hazard Mater 125:211–220
12. Hutchins R (1973) Designing fixed bed adsorbers for wastewater treatment. Am J Chem Eng. 80:133–138

13. Lee V, Porter J, McKay G (2000) Development of fixed-bed adsorber correlation models. Ind Eng Chem Res 39:2427–2433
14. Joseph G, Beachler D, Course A (1981) 415: control of gaseous emissions. Student Manunal, US EPA, Research Triangle Park, NC
15. Vatavuk WM, Klotz WL, Stallings R (1999) Carbon adsorbers. VOC Controls
16. Calvert S, Englund HM (1984) Handbook of air pollution technology
17. TIGG (2015) NIXTOX XP | GAC Polyethylene Drum Adsorbers
18. Li YH, Lee CW, Gullett BK (2003) Importance of activated carbon's oxygen surface functional groups on elemental mercury adsorption☆. Fuel 82:451–457
19. Henning KD, Schäfer S (1993) Impregnated activated carbon for environmental protection. Gas Sep Purif 7:235–240
20. Isik-Gulsac I (2016) Investigation of impregnated activated carbon properties used in hydrogen sulfide fine removal. Braz J Chem Eng 33:1021–1030
21. Kang YH, Shiue A, Hu SC, Huang CY, Chen HT (2010) Using phosphoric acid-impregnated activated carbon to improve the efficiency of chemical filters for the removal of airborne molecular contaminants (AMCs) in the make-up air unit (MAU) of a cleanroom. Build Environ 45:929–935
22. McKetta JJ (1999) Encyclopedia of chemical processing and design: volume 67-water and wastewater treatment: protective coating systems to zeolite. Taylor & Francis
23. Lata S, Singh P, Samadder S (2015) Regeneration of adsorbents and recovery of heavy metals: a review. Int J Environ Sci Technol 12:1461–1478
24. Zhang W, Liu H, Sun Y, Cakstins J, Sun C, Snape CE (2016) Parametric study on the regeneration heat requirement of an amine-based solid adsorbent process for post-combustion carbon capture. Appl Energy 168:394–405
25. Ruthven DM (1984) Principles of adsorption and adsorption processes. Wiley
26. Crittenden B (1992) Selective adsorption-a maturing but poorly understood technology. The R&D Clearing House, pp 4–17
27. Ergun S (1952) Mass-transfer rate in packed columns-its analogy to pressure loss. Chem Eng Prog 48:227–236
28. Hougen OA, Marshall WR Jr (1947) Adsorption from a fluid stream flowing through a stationary granular bed. Chem Eng Prog 43:197
29. "The IMSL Library" (1979) International Mathematical and Statistical Libraries, Inc., vol 2, MMBSIO-1
30. Fair JR (July 14, 1969) Sorption processes for gas separation. Chem Eng 76:90
31. Hobson M, Thodos G (1951) Mass transfer: laminar flow of gases through granular beds. Chem Eng Prog 47:370
32. Ufrecht RH, Sommerfeld JT, Clay Lewis H (1980) Design of Adsorption Equipment in Air Pollution Control. J Air Pollut Control Assoc 30(12):1348–1352. https://doi.org/10.1080/00022470.1980.10465196
33. Guo J, Shah DB, Talu O (2000) Effect of water loading on effective diffusivity in industrial adsorbents
34. You J-H, Lin H-E, Wu H-M (2005) Effects of relative humidity on the adsorption characteristics for methyl-ethyl-ketone vapor on the activated carbon. J Chin Inst Environ Eng 15:47–56
35. Carbon C (2015) PROTECT high flow VS carbon adsorption canisters
36. Control CAP (2017) RotorSorb
37. Industrial T (2014) Fluidized bed carbon VOC abatement
38. Yang Ji, Chen Y, Cao L, Guo Y, Jia J (2012) Development and field-scale optimization of a honeycomb zeolite rotor concentrator/recuperative oxidizer for the abatement of volatile organic carbons from semiconductor industry. Environ Sci Technol 46:441–446
39. Nigar H, Julián I, Mallada R, Santamaría J (2018) Microwave-assisted catalytic combustion for the efficient continuous cleaning of VOC-containing air streams. Environ Sci Technol 52:5892–5901

Chapter 13
Absorption

Nomenclature

N_A	Molecular fluxes of component A
a	Interface area between gas and liquid phases
a_c	Centrifugal acceleration
a_p	Specific surface area of the packing
A	Surface area of the droplet
AF	Absorption factor
c	Molecular concentration of the mixture
C_A	Concentration in liquid phase
C_F	Correction factor dependent on the type of plates
d_e	Equivalent diameter
D_{AB}	Diffusivity of component A through B
f	Volumetric liquid holdup in the unit volume of the absorption column
fd	Distributing factor of liquid in the void space
F	Fraction factor of the flooding
F_p	Packing factor
$\overrightarrow{F}$	Force per unit area exerted by the surroundings
g_c	Gravitational constant
$\overrightarrow{g}$	Gravity vector
G	Flow rate for gas
G^E	Excess Gibbs energy
h	Packing axial height
H_T	Total column height
$k_y a$	Volumetric mass transfer coefficient
K	Ratio of kinetic energy of liquid jet to that of gas
K_L	Coefficient in liquid phase
K_G	Coefficient in gas phase
L	Flow rate for liquid
$\overline{L}_{min}$	Minimum liquid requirement based on MWR

P.-C. Chiang and X. Gao, *Air Pollution Control and Design*,
https://doi.org/10.1007/978-981-13-7488-3_13

m	Slop of the equilibrium curve
MWR	Minimum wetting rate
$\overrightarrow{n}$	Unit normal vector indicating the fundamental orientation
N_{Az}	Molecular flux of component A along z
P	Pumping pressure
P_A	Partial pressure in gas phase
P_T	System pressure during the absorption
r	Radius within the packing
r_{in}	Inner radius
r_{out}	Outer radius
R	Volumetric chemical reaction rate
S	Cross-section area
SdH	Liquid volume in the column
$u_{L,jet}$	Jet velocity of liquid at the nozzle
$u_{G,sup,flood}$	Superficial gas velocity at flooding point
v	Liquid viscosity
$\overrightarrow{v}$	Velocity vector
V_p	Incompressible droplet with a volume
V_F	Gas volumetric flux at the flooding point
$\overrightarrow{w}$	Boundary velocity vector
x_A	Mole fraction of component A
$(x)_{lm}$	Logarithmic-mean mole fraction
y	Mole fraction of pollutant in gas phase
Z_T	Height of the column
θ	Contact angle between the liquid and the packing material
μ_L	Solvent viscosity
ξ	Volume fraction
ρ	Density of the fluid
ρ_L	Solvent density
ϕ	Magnitude of increment in liquid-side mass transfer coefficient
ε	Porosity of packing
ψ	Relative density of solvent

13.1 Introduction

13.1.1 Features of Absorption Process

Absorption is a kind of unit process where soluble components in the flue gas are dissolved into the aqueous phase through a packed scrubber or tower. The process is driven by the difference between the actual concentration and the equilibrium concentration while the rate is dependent on the properties of the species in the

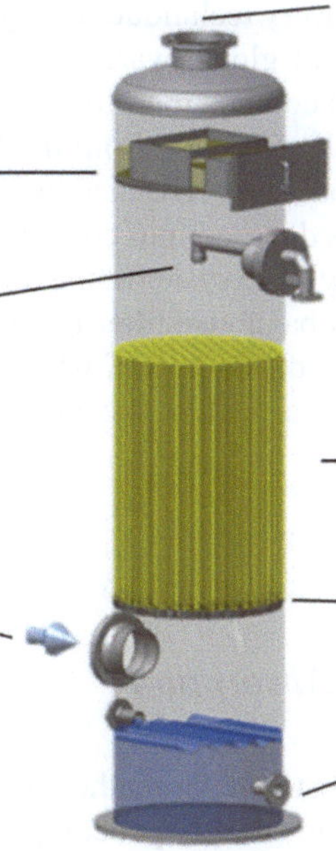

Fig. 13.1 Typical profile of absorption column (Courtesy: GEA Group [1])

system and the operating conditions. Based on liquid used or not, they can be categorized as wet, semi-dry, or dry absorptions. In addition, reactors are classified as countercurrent, cross-current, and co-current by distinct fluid flows. Furthermore, the internal in the reactor can either be packed or plate forms which are selected based on desired efficiencies and pressure drops.

The most popular reactor for absorption is the countercurrent tower shown in Fig. 13.1 where the solvent enters into the column from the top of the column while the gas from the bottom. After entering the tower, the gas stream rises upward through a packing zone wetted with solvent and gets purified. Meanwhile, the solvent flows downwardly throughout the packing zone which increases the contacting area between gas and liquid by filling packing materials.

13.1.2 Applications of Absorption Process

Absorption is primarily used for the removal of inorganic gases such as ammonia from fertilizer production, hydrogen fluoride from glass production, soluble organic compounds from chemical industries, odors from rendering plants, and sulfur dioxide from power generation [2, 3]. If wet type is implemented, simultaneous removal of particulate PM_{10} and $PM_{2.5}$ is achievable. Furthermore, absorption also provides raw material and/or product recovery technique for high-concentrations VOC reclamation. Recently in the context of global warming, absorption is also effective in reducing CO_2 emission from the combustion process [4].

In particular, removing of sulfur dioxide from the flue gas is termed as flue gas desulfurization or FGD. FGD could be performed by either wet or dry scrubbing. In the wet FGD process, flue gas is directed into contact with the solvent which is the slurry of lime or limestone. In dry FGD systems, the flue gas collided with solid-state solvent which are captured by a baghouse filter together. In comparison, dry FGD requires lower costs and energy consumptions, whereas its utilization is limited to the type of fuels (low-sulfur coal) as well as the amount of available solvent.

13.2 Principles

13.2.1 Mass Transfer of Absorption

As indicated by Fig. 13.2, the most popular mathematical model to describe absorption is film theory, where the removal of target compound by absorption consists of three steps that occur in series: (1) diffusion of the pollutant molecules through the gas to the surface of the solvent, (2) dissolution into the solvent at the gas–liquid interface, and (3) diffusion of the dissolved species inwards to the bulk solution. Diffusion dominates the mass transfer to a significantly large extent, which can be estimated by Fick's Law.

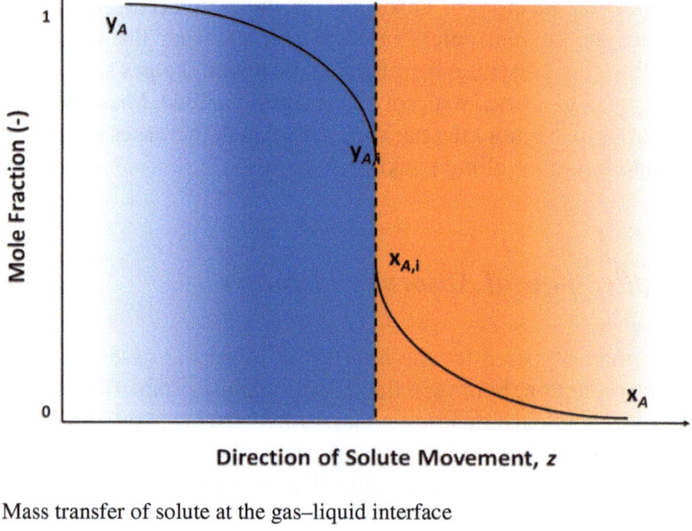

Fig. 13.2 Mass transfer of solute at the gas–liquid interface

The flux of solute A in a binary mixture system can be calculated by the following Eq. (13.1):

$$N_A = x_A(N_A + N_B) - cD_{AB}\nabla x_A \tag{13.1}$$

where N_A and N_B (mol m^{-2} s^{-1}) are molecular fluxes of component A and B, x_A (−) is the mole fraction of component A, c (g-mol m^{-3}) is the molecular concentration of the mixture, and D_{AB} (m^2/s) is the diffusivity of component A through B. It is clear that the first term on the right-hand side of Eq. (13.1) represents the molecular flux of component A contributed by convection, while the second one represents diffusion contributed flux.

In air pollution control, components in B are usually O_2 and N_2 which are not soluble to solvent. Therefore, N_B is negligible. In addition, assign z to the orientation from gas to liquid in which the mass flow can contribute to flux. In this case, Eq. (13.1) transfers into Eq. (13.2).

$$N_{AZ} = -cD_{AB}\frac{dx}{dz} + N_{AZ}x \tag{13.2}$$

Rearranging, we have

$$\int_{x_{A,i}}^{x_A} \frac{dx}{x-1} = \frac{N_{Az}}{cD_{AB}} \int_0^L dz \tag{13.3}$$

Finally, the molecular flux N_{Az} can be worked out by Eq. (13.4).

$$N_{AZ} = \frac{-cD_{AB}}{L} \ln\frac{1 - x_{A,i}}{1 - x_A} \tag{13.4}$$

In practice, X_A is often in the form of partial pressure, P_A, for gas phase or concentration, C_A, for liquid phase.

Mass Transfer without Chemical Reactions

Introducing logarithmic-mean mole fraction, $(x)_{lm}$

$$(x)_{lm} = \frac{(1-x)_1 - (1-x)_0}{\ln[(1-x)_1/(1-x)_0]} \tag{13.5}$$

where the subscript 0 and 1 represent the boundaries of interfacial film gas side and liquid side, respectively. Then we can rewrite Eq. (13.4) as

$$N_{AZ} = \frac{-cD_{AB}}{L(x)_{lm}}\left(x_A - x_{A,i}\right) \tag{13.6}$$

At steady state, we can assume the fluxes of component A along z in gas and liquid phases are equal. Therefore, $\frac{-c}{L(x)_{lm}}$ becomes a constant and the expression of N_{AZ} turns to

$$N_{AZ} = k_G(P_A - P_{A,i}) = k_L(C_A - C_{A,i}) \tag{13.7}$$

In practice, it is impossible to precisely determine the coefficients k_G and k_L as a result of the lack of awareness of film thickness. As an alternative method, Eq. (13.7) is usually transferred as

$$N_{AZ} = K_G(P_A - P_A^*) = K_L(C_A - C_A^*) \tag{13.8}$$

where P_A^* is the partial pressure of A over a solution of A having the bulk concentration C_A, and C_A^* is the concentration of the solution that would equilibrium with the partial pressure P_A in the gas phase.

As indicated by Fig. 13.3, the above transformation from Eqs. (13.7) and (13.8) was attributed to the changing of the driving force, i.e., pressure and concentration gradients. In the plot, the equilibrium line is determined by Henry's Law, that is

$$P^* = H \cdot C \tag{13.9}$$

Initially, the driving force in gas and liquid phases are $(P_A - P_{A,i})$ and $(C_A - C_{A,i})$, respectively. However, neither $P_{A,i}$ nor $C_{A,i}$ is easy to measure. For this reason, we use $(P_A - P_A^*)$ and $(C_A - C_A^*)$ instead for explicit estimation.

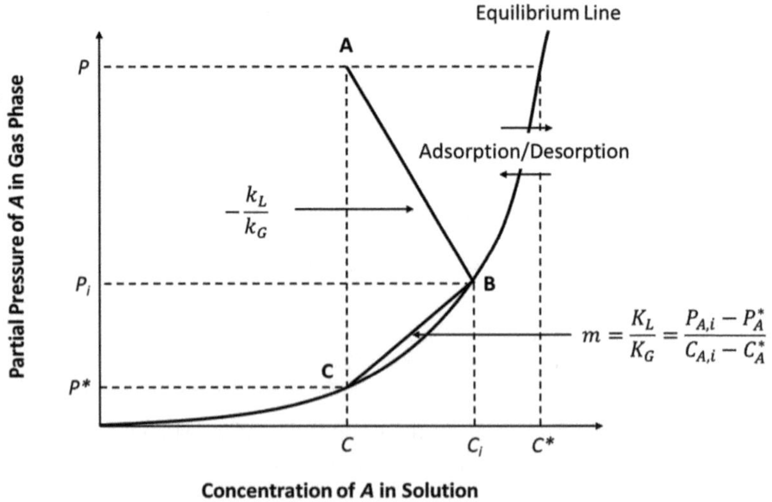

Fig. 13.3 Schematic diagram of the driving force, AB, in the two-film theory

Combining Henry's Law, Eqs. (13.7) and (13.8), we can establish the relationships among K_G, K_L, k_G, and k_L, which are

$$\frac{1}{K_G} = \frac{1}{k_G} + \frac{H_A}{k_L} \tag{13.10}$$

$$\frac{1}{K_L} = \frac{1}{k_L} + \frac{1}{H_A k_G} \tag{13.11}$$

In addition, Fig. 13.3 also suggests that in a specified absorption process, i.e., determined material system with inlet concentrations, the overall gaseous mass transfer coefficient is directly proportional to the aqueous one, that is

$$m = \frac{K_L}{K_G} = \frac{P_{A,i} - P_A^*}{C_{A,i} - C_A} \tag{13.12}$$

According to the relative magnitude between K_G and k_G, as well as K_L and k_L, we can determine the resistance process in absorption, which is specified as the following: (1) If $H_A \ll 1$, species A is very soluble, $K_G \approx k_G$, indicating a gas film controlled process; (2) If $H_A \gg 1$, species A is barely soluble, $K_L \approx k_L$, indicating a liquid film controlled process.

In an absorption reactor, due to the limited wetting rate, not all of the packing surface area is merged by the liquid. For this reason, the effective interfacial mass-transfer area a, should be considered in a column adsorption process whose cross-section area is defined as S. Then the mass balance can be expressed as

$$-d(GSy) = N_A a dV = N_A a S dH \tag{13.13}$$

For convenient use in engineering practice, the mass transfer efficiency is usually presented in the form of overall mass transfer coefficient, which is

$$K_G a = \frac{N_A}{H_T S P_T (y)_{lm}^*} \tag{13.14}$$

$$K_L a = \frac{N_A}{H_T S P_T (x)_{lm}^*} \tag{13.15}$$

where H_T is the total column height, and P_T is the system pressure during absorption.

Mass Transfer Associated with Chemical Reactions

A well-designed absorption technology shall consider the chemical reaction between solute and solvent to facilitate the overall efficiency while minimizing the size of the reactor.

Figure 13.4 presents the concentration profile of solute A at the gas–liquid interface. As the chemical reaction takes place in the liquid phase, k_L will be drastically

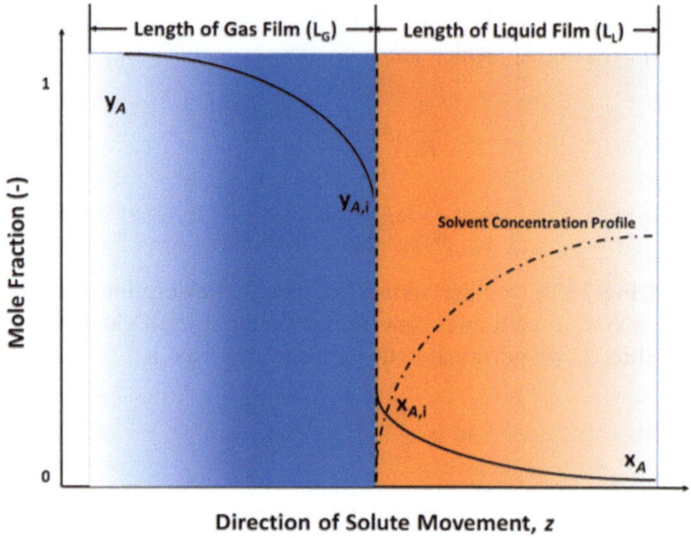

Fig. 13.4 Mass transfer of solute at the gas–liquid interface associated with chemical reaction

enlarged while k_G remains unchanged. The magnitude of increment in liquid-side mass transfer coefficient is defined as

$$\phi = k_L^{'}/k_L \geq 1 \tag{13.16}$$

As a result, the concentration of solute A at the interface would be suppressed. In conclusion, absorption associated with chemical reaction will be a gas side-controlled process.

In the absorption process where the chemical reaction is involved, the mass balance is given by Eq. (13.17)

$$dN_A = RfSdH = \left(\frac{K_L a}{\rho_L}\right)(x_i - x)SdH \tag{13.17}$$

where R is the volumetric chemical reaction rate, and f is the volumetric liquid holdup in the unit volume, SdH, of the absorption column. Therefore, fSdH indicates the liquid volume in the column.

13.2.2 Process Intensification by High-Gravity (HiGee) Rotating Packed Bed (RPB)

Mechanism of Process Intensification

From the above statements, we can tell that the performance of the absorption process is somewhat dominated by the process where the air pollutant moving from the gas to liquid phases. On the one hand, we can use chemical solvents to reduce the interfacial concentration hence increasing the driving force. On the other hand, modifications to the reactor could also impose a profound influence on the absorption performance.

Process intensification is developed to reduce economic costs, minimize the size of the reactor with better performance and energy efficiency [5]. Let's consider an incompressible droplet with a volume, V_p, then the mass flux can be expressed in the form of linear conservative momentum equation [6]

$$\underbrace{\frac{d}{dt}\left(\oint \rho\vec{v}dV_p\right)}_{\substack{\text{Linear}\\ \text{Momentum}}} + \underbrace{\oint \rho\cdot\vec{v}\cdot(\vec{v}-\vec{w})\cdot\vec{n}\,dA}_{\substack{\text{Rotational}\\ \text{Momentum}}} = \underbrace{\oint \rho\vec{g}dV_p}_{\substack{\text{Gravitational}\\ \text{Force}}} + \underbrace{\oint \vec{F}\,dA}_{\substack{\text{External}\\ \text{Force}}} \tag{13.18}$$

where ρ is the density of the fluid, $\vec{v}$ is the velocity vector, $\vec{w}$ is the boundary velocity vector, $\vec{n}$ is the unit normal vector indicating the fundamental orientation, A is the surface area of the droplet, $\vec{g}$ is the gravity vector, and $\vec{F}$ is the force per unit area exerted by the surroundings. Here, we can tell that the source of momentum includes the gravitational force and other forces from the outside of the system. From this perspective, the mass transfer can be intensified by enhancing the external forces, $\vec{F}$ to suppress the surface tension force that prohibits the mass from transferring into the liquid phase.

Generally, common pathways include heating, pressurizing, stirring, and vibrating, which are energy intensive and/or inefficient. Traditionally, process intensification could be achieved by heating, pressurizing, supersonic et al.

Process Intensification through a Rotating Packed Bed (HiGee)

Rotating packed bed was developed where gravitational force was significantly enhanced by several orders of magnitude via high-speed rotation, producing an intensified mass transfer performance [7]. As an innovative reactor for process intensification, rotating packed bed has drawn great attention since its development. As the principle, rotating packed bed generates an intensified gravitational field to level up the term $\oint \rho\vec{g}dV_p$. In addition, as the droplets move outwardly in the reactor, they would collide with a packing, exerting massive shearing force, thereby increasing the external-force term $\oint \vec{F}dA$.

Figure 13.5 presents three types of rotating packed beds in common use which

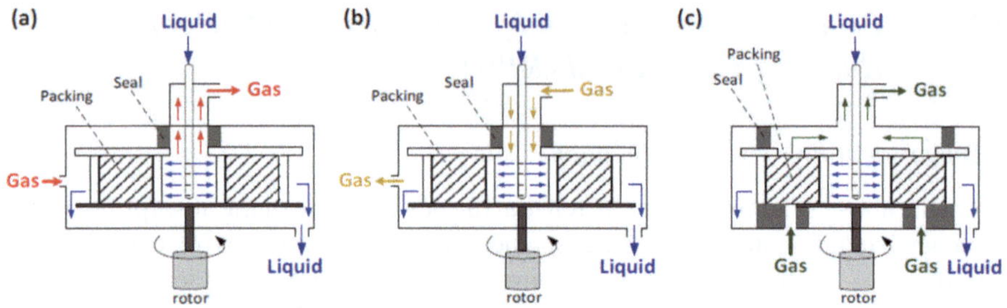

Fig. 13.5 Various types of HiGee reactors for the reaction between gas and liquid phases: **a** countercurrent flow, **b** co-current flow, and **c** cross flow [8]

are countercurrent, co-current, and cross-current devices. In general, countercurrent rotating packed beds exhibit higher mass transfer performances, however, which come at higher pressure drops. In contrast, co-current rotating packed beds are relatively energy-saving but can offer lower process intensification performance.

In general, the mass balance over a thin shell of fluid within the rotating packed bed is shown as Eq. (13.19) [9]

$$\frac{dQ_G}{2\pi rhdr} = K_G a_e \left(y^* - y\right) \tag{13.19}$$

where r is the radius within the packing, and h is the packing axial height. The integral of Eq. (13.19) leads to the overall mass transfer coefficient, which is

$$K_G a_e = \frac{Q_G}{h\pi \left(r_{out}^2 - r_{in}^2\right)} \ln\left(\frac{y_{in}}{y_{out}}\right) \tag{13.20}$$

where $\ln\left(\frac{y_{in}}{y_{out}}\right)$ represents the number of transfer unit (NTU) of the packing. According to the calculation mentioned above, we have the height of transfer unit (HTU) as

$$HTU = \frac{r_{out} - r_{in}}{NTU} = \frac{r_{out} - r_{in}}{\ln\left(\frac{y_{in}}{y_{out}}\right)} \tag{13.21}$$

and the area of transfer unit (ATU) as

$$ATU = \frac{\pi\left(r_{out}^2 - r_{in}^2\right)}{NTU} = \frac{\pi\left(r_{out}^2 - r_{in}^2\right)}{\ln\left(\frac{y_{in}}{y_{out}}\right)} \tag{13.22}$$

The attention of most gas–liquid or liquid–liquid process research is concentrated on the mass transfer process. Popularization of this intensification technology needs to be implemented to maximize the capabilities of RPB on an industrial scale. Enhanced micromixing has been the epitome of mass transfer in RPB and micromixing is directly related to characteristics of flow and mass transfer efficiency as well as the heat and momentum transfer. A trade-off will exist between implementing fundamental research and obtaining the maximum capabilities of RPB. Therefore, applications in the environmental process make it possible to make a contribution to integrated air pollution control, i.e., flue gas cleaning and particulate matters as well as CO_2 capture. In an emerging industry, preparation of nanoparticles and reaction of water gas shift using RPB has been illustrated to bring new life to chemical process [10].

13.3 Design Methods

13.3.1 Design Criteria

The first priority of an absorption column is that the effluent gas should meet the emission requirement which is either the pollutant concentration (majorly for conventional air pollutant) or the reduction rate (for hazardous air pollutant). In addition, the liquid distribution should be as uniform as possible to prevent a short-circuit from happening. For this reason, packed columns are usually sectioned and the liquid is re-distributed at the interface of sections. Since dissolved pollutants such as VOCs might cause environmental deterioration, attention should be paid to the prevention of secondary pollution.

Here, we introduce some rule of thumbs that help to guide the absorption design. For packed columns, it is recommended that: (1) the packed diameter should be no greater than one-eighth of the column diameter to prevent the poor liquid distribution; (2) the thickness of each packed layer is recommended from 0.3–0.7 mm; (3) tower diameter is recommended 15 times the packing diameter; and (4) the intervals between liquid distributors should be 5–10 tower diameters along with the tower height for pall rings or every 6.5 m for other packings [11].

For plate column: (1) the minimum tray number can be estimated by Fenske equation which is $N_{min} = \frac{\log\left[\left(\frac{X_d}{1-X_d}\right)\left(\frac{1-X_b}{X_b}\right)\right]}{\log\alpha_{avg}}$ [12]; (2) the minimum tray spacing should be 30 cm for the column diameter less than 1.5 m or 45 cm for column diameter greater than 300 cm [13]; and (3) the pressure drop for each tray should be 0.1 psi.

13.3.2 Theoretical Approach

Equilibrium approach is performed based on Henry's law. Assuming a proposed absorption column is going to deal with a contaminated flue gas with an inflow rate, G_{in}, whose inlet and regulated outlet pollutant concentrations are denoted as y_{in}, and $y_{out,}$ respectively. When the liquid flow rate, L, has been determined by the aforementioned methods, the operating line is established which can be derived from Eq. (13.23):

$$y = \frac{L}{G}x + \frac{(G_{out}y_{out} - L_{in}x_{in})}{G} \tag{13.23}$$

Explicitly, the operating line is straight under idealized situation and is shown in Fig. 13.6. Combined with the equilibrium, we can figure out the column height, z, by mathematical solution.

Let's consider the mass balance situation within a unit height of the column, dz, with respect to the gas phase.

$$d(Gy) = Gy|_{z+dz} - Gy|_z = k_y a(y - y_i)dz \tag{13.24}$$

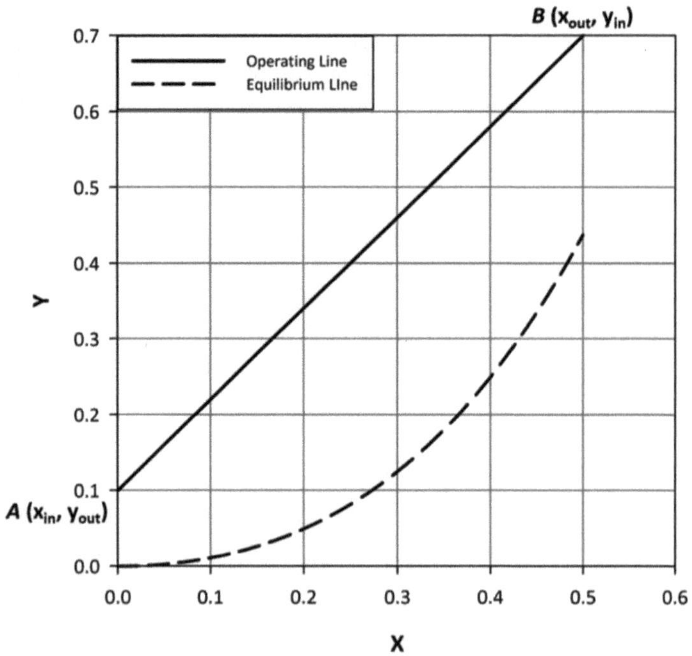

Fig. 13.6 Schematic diagram of equilibrium and operating lines

where a is the interface area between gas and liquid phases, and $k_y a$ is the volumetric mass transfer coefficient. Rearranged, we have

$$\frac{d(Gy)}{dz} = k_y a(y - y_i) \tag{13.25}$$

In practice, as the gas flows through the column, the flow rate, G, may vary with the traveling distance, making it become the function of the mole fraction of pollutant in the gas phase, y. On the other hand, we assume flow rates of inert compositions such as N_2, O_2, maintain constant throughout the column, which are denoted as G'. The relationship between G and y can be described as

$$G = \frac{G'}{(1-y)} \tag{13.26}$$

Then we have

$$d(Gy) = G'd\left(\frac{y}{1-y}\right) = G'\frac{dy}{(1-y)^2} = G\left(\frac{dy}{1-y}\right) \tag{13.27}$$

Therefore, the total height of the absorption column could be derived by integrating Eq. (13.25)

$$\int_0^{z_T} dz = \int_{y_{out}}^{y_{in}} \frac{G}{k_y a} \frac{dy}{(1-y)(1-y_i)} = \overline{\left(\frac{G}{K_y a}\right)} \int_{y_{out}}^{y_{in}} \frac{dy}{(1-y)(1-y^*)} \tag{13.28}$$

Since G and $k_y a$ would decrease from the inlet to the outlet, here we use an average value instead, denoted as $\overline{\left(\frac{G}{K_y a}\right)}$ which is also termed as the height of transfer unit (HTU). In addition, $\int_{y_{out}}^{y_{in}} \frac{dy}{(1-y)(1-y^*)}$ is termed as the number of transfer unit (NTU).

13.3.3 Good Engineering Practice

Since the performance of the absorption process is dependent on the internals including solvent distributor and packing material and structures to a large extent, the vendors could provide valuable information that is helpful for design.

For instance, as a pioneer in the field of absorption equipment, Raschig USA Inc. dedicates itself to improve absorption column, enhancing mass transfer while reducing energy penalties during the operation. Figure 13.7 presents the test results of No. 2 Raschig super-ring regarding pressure drop, transfer efficiency, height of transfer unit, and liquid holdup by which we can conclude a holistic good engineering practice protocol on the process design [10]. In detail, (a) and (b) give the information

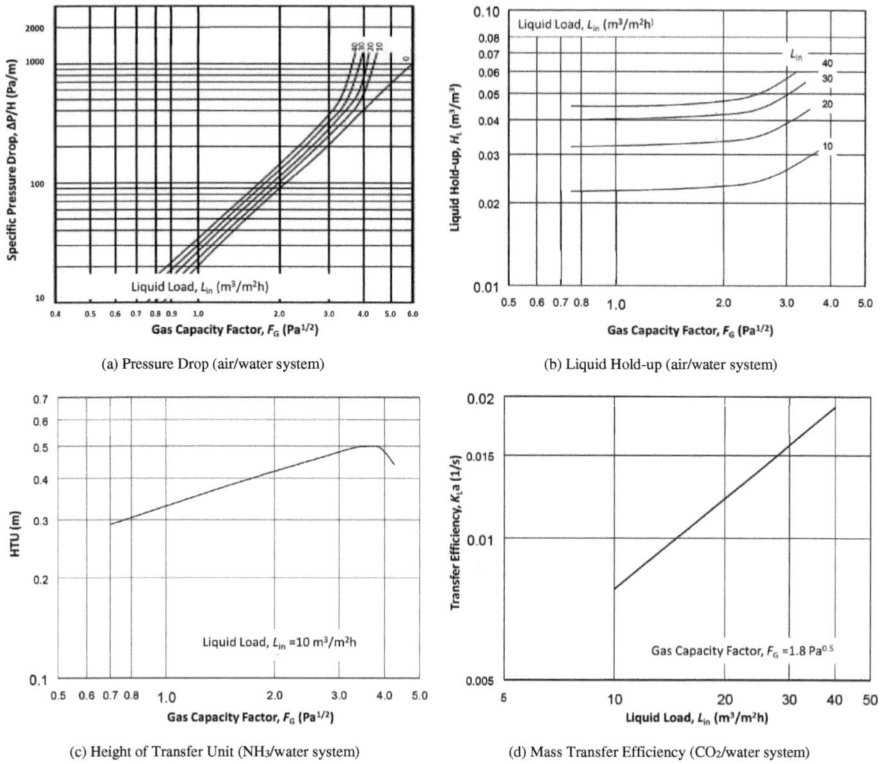

Fig. 13.7 Performance regarding pressure drop, transfer efficiency, height of transfer unit, and liquid holdup of No. 2 Raschig super-ring ®. Tests were performed in a column with a diameter of 0.288 m and a height of 2.0 m. Gas capacity factor is defined as $F_G = u_G \cdot \sqrt{\rho_G}$

on pressure drop and liquid holdup within the absorption column under different gas and liquid loadings. It is found that the pressure drop increases along with the increase in gas loading. In particular, a sudden significant increase in pressure drop is observed when the gas loading approaches and exceeds above about 3.0 Pa$^{1/2}$ which can be ascribed to the massive liquid holdup as indicated in (b). In (c), the height of transfer unit arises and reaches the peak value as the gas loading increases, which indicates the compromise of mass transfer performance. Figure 13.7d indicates the possibility of compensation of mass transfer performance by increasing the liquid feed.

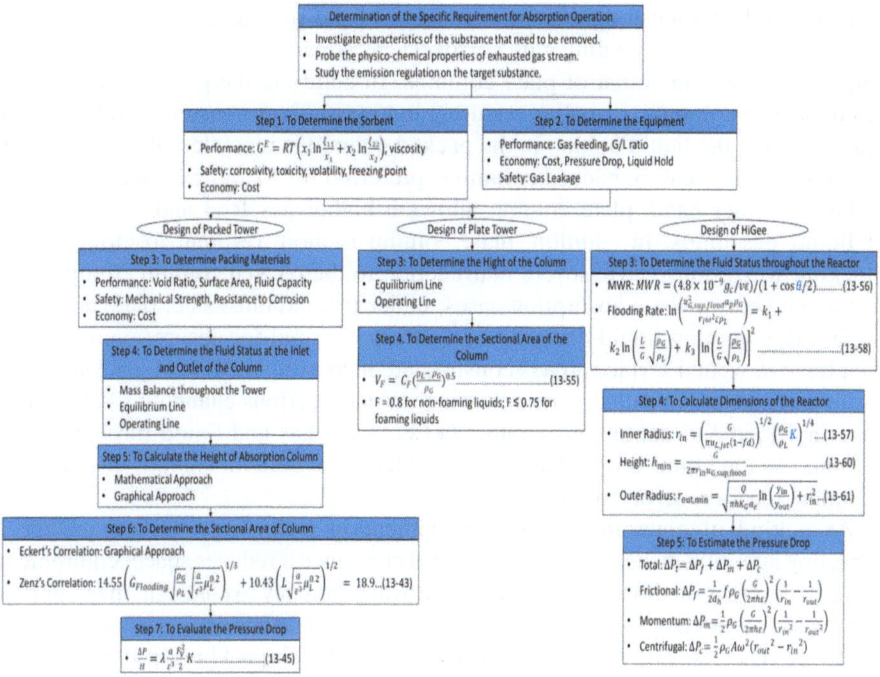

Fig. 13.8 Framework of absorption process design

13.4 Design Procedures

The function design of an absorption column should cover the solvent requirement, construction material, dimensions of the column, and the pressure drops. Designing a HiGee reactor is relatively more empirical where the dimensions including inner and outer radius and the packing height are largely dependent on the mass transfer coefficient obtained previously. In order to minimize secondary pollution, wastewater treatment facility should be carefully considered. In some cases where the fume at the stack exit is prohibited, a mist eliminator should be incorporated (Fig. 13.8).

13.4.1 *To Select the Appropriate Equipment and Absorbent (Step 1 and 2)*

Absorption Reactor

Conventional absorption reactor includes a packed column, plate column, and tray column. All of these types of reactors are designed to intensify the interactions between gas and liquid phases while reducing the pressure drop.

In general, the most significant advantage of the packed column is the higher benefit-cost ratios. For dealing with corrosive gas streams, the capital cost of packed columns is lower than that of plate columns. In addition, the pressure drop of the gas in packed columns is smaller, resulting in lower energy consumption for the gas draft. Further, the liquid holdup in the packed column is much less than in the plate column. For this reason, packed column is preferred for foaming liquids.

On the other hand, plate column allows a higher gas feed and a wider range of liquid–gas ratios. In addition, plate column is more resilient to clogging and blocking which may cause severe malfunction to the packed column. Temperature rise may occur in some absorption cases, under this circumstance, plate column is preferred. Optimum liquid distribution is always desired for absorption, otherwise the effective wetted surface area is constrained, increasing the risk of gas leakage. A well-designed and installed plate column eliminates serious channeling difficulties, ensuring good and continuous contact between the gas and liquid throughout the column. As the liquid is distributed at each plate, plat column is more favorable for chemical reaction-involved absorptions.

As a kind of emerging process-intensification technology, HiGee reactor is attracting attention in the air pollution control area since it reduces space requirement, minimizes retention time, and improves removal efficiency. At present, many derivative packing types have been developed to achieve the optimal engineering feasibility of HiGee reactors. The simplest packing is single-block packing. In detail, single-block rotating packed bed can be designed in the form of countercurrent or cross-current, depending on the distinct flow regimes of fluids. In a single-block rotating packed bed, mass transfer performance throughout the reactor is not uniform, and the highest mass transfer rate could be observed at the inner edge of the packing area, due to the most intensified gas-packing collision and the most rapid refresh rate of liquid [14]. One of the most significant disadvantages of block packing is the inherent higher pressure drop, as an alternative, blade packing is proposed. However, mass transfer performance is limited by its lower surface area [15].

Absorbent Selection

Absorbent is another key factor that dominates the performance of the absorption process where the prior consideration should be the solubility of the target compound. Such information is usually available from the vendors. In general, the feasibility of a solvent is dependent on several factors including solubility (especially for organic pollutants) and neutralization, which are exemplified in Table 13.1. From the safety perspective, sorbent is required to have low corrosivity and low toxicity. Other important consideration includes viscosity, volatility, freezing point, and cost.

Zheng [16] suggested a theoretical approach to the selection of sorbent which is conducted by evaluating the excess Gibbs energy

$$G^E = RT\left(x_1 \ln\frac{\xi_{11}}{x_1} + x_2 \ln\frac{\xi_{22}}{x_2}\right) \tag{13.29}$$

Table 13.1 Selection of suitable solvent for air pollution control

Solvent	Applicable pollutant
Calcium hydroxide, $Ca(OH)_2$	Sulfur dioxide, SO_2
Water, H_2O	Hydrogen chloride, HCl
	Hydrogen fluoride, HF
	Hydrogen bromide, HBr
	Phosgene, $COCl_2$
Sodium sulfite, Na_2SO_3	Sulfur dioxide, SO_2
	Chlorine, Cl_2
Sodium hydroxide, NaOH	Sulfur dioxide, SO_2
	Hydrogen sulfide, HS
	Organic acids, R-COOH
	Phenols, 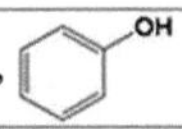
	Phosgene, $COCl_2$
	Chlorine, Cl_2
	Mercaptans, CH_3SH
Sodium thiosulfite, NaS_2O_2	Chlorine, Cl_2
Nitric acid, HNO_3	Ammonia, NH_3
Sulfuric acid, H_2SO_4	Ammonia, NH_3
	Amines, R_3N
Ammonia, NH_3	Formaldehyde, HCHO

where G^E (J/mol) is the excess Gibbs energy, x (−) is the molecular fraction, and ξ (−) is the volume fraction. In addition, the subscript 1 and 2 are indicating solute and absorbent, respectively.

13.4.2 The Design of Packed Tower

Step 3: To Determine Packing Materials

Packing plays a central role in the absorption process which serves to extend the effective contact area between liquid and gas phases. Factors that should be carefully considered include void ratio, available surface area, and fluid capacity: In addition, for the safety and durability concern, mechanical strength and the resistance to chemical corrosion should be compared among several kinds of technically feasible packings. Ultimately, economic consideration is important to the users. In this case, packing should be cheap and lightweight to reduce capital and maintenance costs, respectively.

As a rule of thumb, a satisfactory packing should be: (1) high wetted area per unit volume; (2) lightweight, (3) sufficient chemical resistance, (4) low liquid holdup, (5) low-pressure holdup, (6) and low cost (Table 13.2).

Table 13.2 Summary of some popular types of packings in common use

Packing types	Profiles	Features
Rasching ring	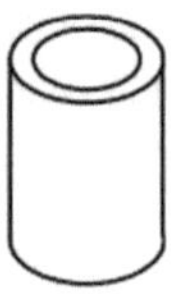	• Advantages – The most popular type due to the relatively lower cost – Great Feasibility and versatility – Highly structural durability • Disadvantages – Unsatisfactory efficiency than other packings – Poor liquid distribution due to internal liquid channeling and side-wall thrust
Lessing ring		• Advantages – Higher efficiency and durability than Rasching ring • Disadvantages – Higher pressure drops – High side-wall thrust
Berl saddle		• Advantages – Higher efficiency than Rasching ring – Lower potential of side-wall thrust – Lower unit pressure drops allowing higher flooding point than Rasching rings • Disadvantages – Relatively more expensive – Fragile than Rasching Rings
Intalox saddle		• Advantages – High efficiency – Little tendency or potential of nest – Higher uniformity of liquid distribution – Higher flooding limits and lower pressure drops • Disadvantages – Relatively more expensive – Fragile than Rasching Rings
Pall ring		• Advantages – Higher flooding limits and lower pressure drops – Good liquid distribution – High capacity • Disadvantages – High side-wall thrust
Tellerette		• Advantages – Higher flooding limits and lower pressure drops – Lightweight – Low side-wall thrust • Disadvantages – Relatively more expensive

Step 4: To Determine the Fluid Status at the Inlet and Outlet of the Column

Basically, absorption is a kind of process involving the interaction between gas and liquid phases, therefore, liquid or solvent distribution is a decisive factor in the performance of the absorption process. For this reason, it is important to determine the solvent flow rate throughout the column. Given a specific gas flow rate, solvent feeding falls into the range between the minimum possible liquid rate and the flooding rate.

The minimum liquid rate can be determined based on the mass balance within the absorption column as shown in Fig. 13.9. In an ideal absorption column, the amount of solute absorbed from the gas phase should equal to the added weight in the liquid phase which can be described as

$$y_{out}G_{out} + x_{out}L_{out} = y_{in}G_{in} + x_{in}L_{in} \tag{13.30}$$

where y and x are concentrations of solute in gas and liquid phases, respectively; G and L are flow rates for gas and liquid, respectively; the subscript out and in indicates the inlet and outlet. In practice, y_{out} and x_{in} are usually low such that $y_{out}G_{out}$ and $x_{in}L_{in}$ approximate to zero. Therefore, Eq. (13.30) can be simplified as

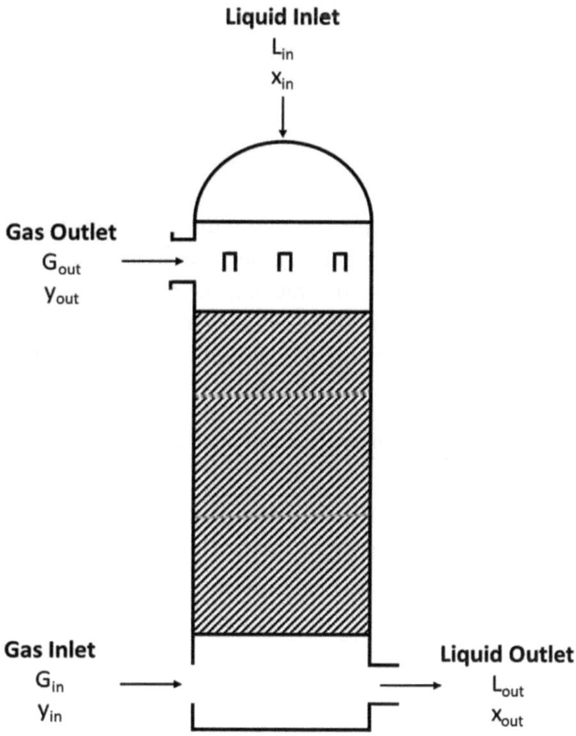

Fig. 13.9 Schematic diagram of mass balance within the absorption column

$$x_{out}L_{out} \approx y_{in}G_{in} \tag{13.31}$$

Therefore, the minimum liquid rate would be roughly estimated by

$$L_{min} = \frac{y_{in}G}{x_{out}} \approx \frac{my_{in}G}{y^*} \tag{13.32}$$

However, in some packed tower where the gases are very soluble, the minimum liquid rate determined above may not be sufficient to keep the packing surface totally wet, leading to a poor distribution of the liquid stream and a leak of contaminated gas stream. For this reason, we have to consider the minimum wetting rate to ensure the appropriate liquid distribution, which is described below.

Another approach to determining the minimum liquid rate is measuring the minimum wetting rate (MWR) which is defined as Eq. (13.33)

$$\text{MWR} = \frac{\text{Liquid Rate, ft}^3/\text{hr}\cdot\text{ft}^2\text{ cross section}}{\text{Packing Surface Area per Column Volume, ft}^2/\text{ft}^3} \tag{13.33}$$

The minimum wetting rate is highly dependent on the packing types and materials. Table 13.3 presents the MWR for different packing types and materials and it suggested that the packing in the form of rings needs the least quantity of liquid to get wet.

Therefore, the minimum liquid requirement based on MWR, i.e., $\overline{L}_{min}$, is calculated as the following equation:

$$\overline{L}_{min} = (\text{MWR})(\rho_L)(a) \tag{13.34}$$

where MWR (ft^2/h) is the minimum wetting rate, ρ_L (lb/ft^3) is the solvent density, and a (ft^2/ft^3) represents the surface area to volume ratio of the packing which is given in Table 13.4. In practice, the actual liquid flow rate is usually greater than $\overline{L}_{min}$ as much as 25–100%. However, besides the economic considerations, there is an upper limit for the practical liquid flow rate, which is the flooding point.

Table 13.3 Minimum wetting rate for different packing types and materials [17]

Packing	MWR
Rings (Raschig, Lessing, etc.), through 3-in. diameter	0.85
Grid type (wooden, etc.) (pitch: 2 in)	0.85
All packings larger than 3-in. diameter	1.30
Polished metal packings and poor wetting surfaces (some plastics, glazed porcelain, etc.)	1.30–2.50

Table 13.4 Packing factors for different types of packings [19, 22]

Packing type	Material	Nominal diameter	F_p	a
Raschig rings	Ceramic	1/2	640	111
		5/8	380	100
		3/4	255	80
		1	160	58
		3/2	95	38
		2	65	28
		3	37	
	Metal	1/2	410	118
		5/8	290	
		3/4	230	72
		1	137	57
		3/2	83	41
		2	57	31
		3	32	21
Pall rings	Metal	5/8	70	131
		1	48	66
		3/2	28	48
		2	20	36
		7/2	16	
	Polypropylene	5/8	97	110
		1	52	63
		3/2	32	39
		2	25	31
Berl saddles	Ceramic	1/2	240	142
		3/4	170	82
		1	110	76
		3/2	65	44
		2	45	32
Intalox saddles	Ceramic	1/2	200	190
		3/4	145	102
		1	98	78
		3/2	52	60
		2	40	36
		3	22	

Step 5: To Calculate the Height of Absorption Column

In general, the height of an absorption column is dependent on: (1) the equilibria between liquid and gas phases; (2) the specified outlet pollutant concentration or removal efficiency; and (3) the mass transfer efficiency of the column. Collectively, the height of the absorption column is the product of the number of transfer unit (NTU) and the height of transfer unit (HTU).

Integration of Eq. (13.28) is not mathematically practical, but by introducing some reasonable assumption, NTU can be calculated in simple ways:

(1) In the most idealized condition, if the operating and equilibrium curves are straight lines and parallel with each other, then

$$\mathrm{NTU} = \frac{y_{in} - y_{out}}{y - y*} \tag{13.35}$$

(2) If the curves are just straight

$$\mathrm{NTU} = \frac{y_{in} - y_{out}}{(y - y*)_{ln}} \tag{13.36}$$

where

$$(y - y*)_{ln} = \frac{(y - y*)_{in} - (y - y*)_{out}}{\ln\left[\frac{(y-y*)_{in}}{(y-y*)_{out}}\right]} \tag{13.37}$$

(3) If either is curved, then we have [18]

$$\mathrm{NTU} = \frac{\ln\left[\frac{(y_{in} - mx_{in})}{(y_{out} - mx_{in})}\left(1 - \frac{1}{AF}\right) + \frac{1}{AF}\right]}{1 - \frac{1}{AF}} \tag{13.38}$$

where m is the slope of the equilibrium curve, while AF is the absorption factor which is defined as

$$AF = \frac{L}{mG} \tag{13.39}$$

In general, AF is recommended to be 1.5–2.5 for higher economic efficiency [19]. If m approximates zero, indicating a difficult dissolution condition, then Eq. (13.38) would be transformed into

$$\mathrm{NTU} = \ln\left(\frac{y_{in}}{y_{out}}\right) \tag{13.40}$$

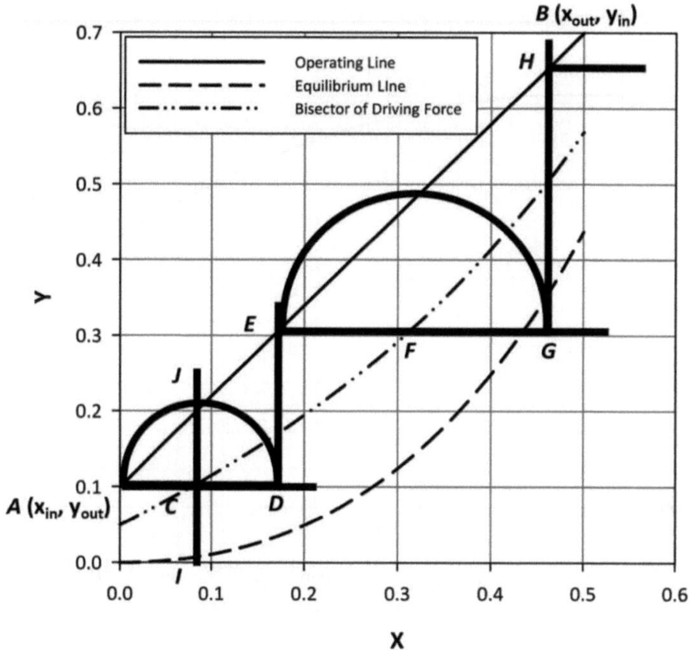

Fig. 13.10 Determination of N_{OG} by Baker's Method (Baker [20])

and (4) Graphic method provides an easy solution. As shown in Fig. 13.10, we have the operating and equilibrium curves, then

(1) Derive the bisector driving force by $\frac{y-y*}{2}$;
(2) Starting from A draw a horizontal line parallel to the abscissa and cross it with the bisector driving force line at C. Note that AC = CD;
(3) Draw a vertical line from C parallel to the ordinate and cross it with the operating line at J and equilibrium line at I;
(4) From D draw a line parallel to the ordinate axis cutting the operating line at E.

$$N_{OG} = \frac{DE}{(Y - Y^*)_{atC}} = \frac{DE}{JI} \tag{13.41}$$

Therefore, step ADE represents one transfer unit.

It is worth noting that an absorber should consider an additional height to provide the capability to meet future requirements. The additional column height would cause higher capital cost, but it could help to meet the future requirement without intensively modifying the absorber or building a new absorber.

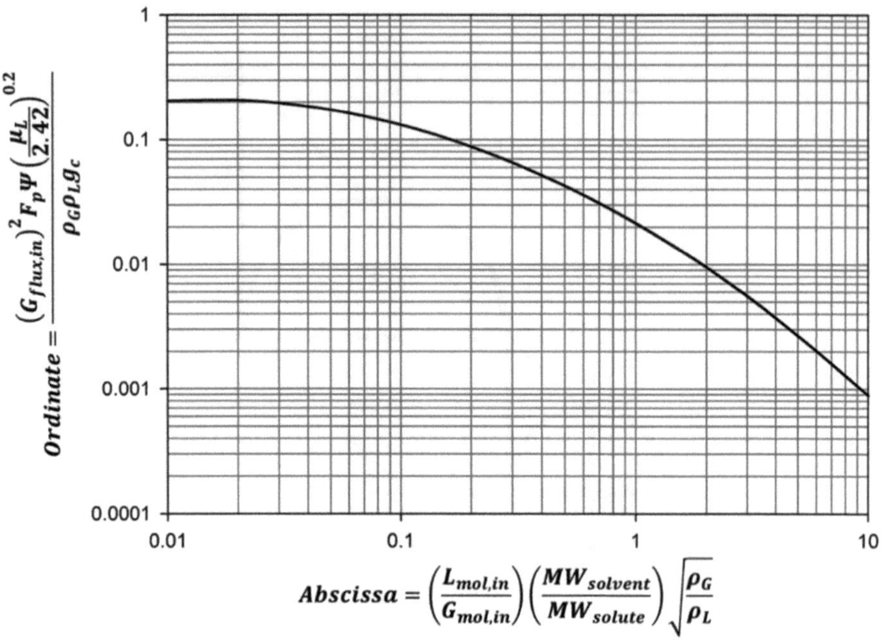

Fig. 13.11 Eckert's generalized pressure drop correlation

Step 6: To Determine the Sectional Area of Column.

The minimum sectional area of the column is determined by flooding rate, and the usual gas velocity falls from 60 to 75% of the flooding point [19]. The Eckert's generalized pressure drop correlation presented in Fig. 13.11 provides an indirect but simple method to determine the cross-sectional surface [21] (Fig. 13.11).

Where F_p (−) is the packing factor which can be referred to Table 13.4, ψ (−) is the relative density of solvent, and μ_L (lb/ft-h) represents the solvent viscosity.

The first step is calculating the abscissa, that is $\left(\frac{L}{G}\right)\left(\frac{\rho_G}{\rho_L}\right)^{0.5}$. Then proceed to the flooding line and locate the design parameter in the ordinate. Thereafter the maximum allowable gas flux rate, G_{max}, in terms of mass flow or volumetric flow per square feet is solved by setting up a fraction factor of the flooding (F, which is usually 0.6–0.75). Finally, we have the cross-sectional area as

$$S = \frac{G}{F \cdot G_{max}} \tag{13.42}$$

where V_F is the gas volumetric flux at the flooding point.

Zenz [23] modified the correlation proposed by Sherwood [21] and rearranged it in a simple form

$$14.55\left(G_{Flooding}\sqrt{\frac{\rho_G}{\rho_L}}\sqrt{\frac{a}{\varepsilon^3}\mu_L^{0.2}}\right)^{1/3} + 10.43\left(L\sqrt{\frac{a}{\varepsilon^3}\mu_L^{0.2}}\right)^{1/2} = 18.9 \quad (13.43)$$

Zenz [23] correlation provides a simple way to calculate all the combinations of flooding rates in which resorting to a chart is avoided once such condition has been established. For this reason, it requires no trial-and-error procedure to calculate the flooding gas rate. In addition, for a given gas–liquid system, Eq. (13.43) simplifies to Eq. (13.44) (Table 13.5).

$$(G_{Flooding})^{\frac{1}{3}} + (L)^{\frac{1}{2}} = \text{constant} \quad (13.44)$$

Table 13.5 Summary of flooding rates correlations

Authors	Correlation formula	Feasibility
Elgin and Weiss [24]	• $\log(Y) = 0.338(\log(X))^4 - 0.591(\log(X))^3 - 0.782(\log(X))^2 - 0.583\log(X) - 0.898$ • $X = \sqrt{\frac{G_{Flooding}}{L}}$ • $Y = \frac{a \cdot L^2}{g(\varepsilon - h_S)}$	$0.4 \le X \le 40$
Bertetti [25]	• $4.85\mu_L^{0.04}\left(\frac{\rho_G}{\rho_L}\right)^{0.33} G_{FLooding}^{0.67} + 4.1\mu_L^{0.1}L^{0.67} = \frac{6.4\varepsilon}{a^{0.33}}$	
Leva [26]	• $\log(Y) = -0.270(\log(X))^2 - 1.076\log(X) - 1.705$ • $X = \frac{L}{G_{Flooding}}\sqrt{\frac{\rho_L}{\rho_G}}$ • $Y = \frac{G_{Floooding}^2}{g}\left(\frac{\rho_G}{\rho_L}\right)\left(\frac{\rho_W}{\rho_L}\right)^2\left(\frac{a}{\varepsilon^3}\right)\mu_L^{0.2}$	$0.03 \le X \le 6$
Takahashi et al. [27]	• $G_{Flooding}^{1/2}\left(\frac{\rho_G\mu_L^{0.2}a}{\rho_L\varepsilon^3 g}\right)^{1/4} + L^{1/2}\left(\frac{\mu_L^{0.2}a}{\varepsilon^3 g}\right)^{1/4} = 0.552$	
Billet and Schultes [28]	• $G_{Flooding} = \frac{(\varepsilon - h_{T,Flooding})^{3/2}}{\varepsilon^{1/2}}\sqrt{\frac{2g}{\psi_L}}\sqrt{\frac{h_{T,Flooding}}{a}}\sqrt{\frac{\rho_L}{\rho_G}}$	
Kuźniewska-lach [29]	• $G_{Flooding} = A \cdot \varepsilon\left(\frac{a}{\varepsilon^3}\right)^B$ • $A =$ $\left(1608932L^3 - 7846.1L^2 + 237.3L + 15.2\right)\frac{\left(\frac{\rho_a\rho_L}{\rho_G\rho_W}\right)^{0.5}}{\left(\frac{\sigma_L}{\sigma_W}\right)^{0.25}\left(\frac{\mu_L}{\mu_W}\right)^{0.02}}$ • $B = \left(-144.6L^2 - 9.5L - 0.273\right)\frac{\left(\frac{\rho_W\mu_L}{\rho_L\mu_W}\right)^{0.02}}{\left(\frac{\sigma_L}{\sigma_W}\right)^{0.15}\left(\frac{\mu_G}{\mu_\theta}\right)^{0.1}}$	

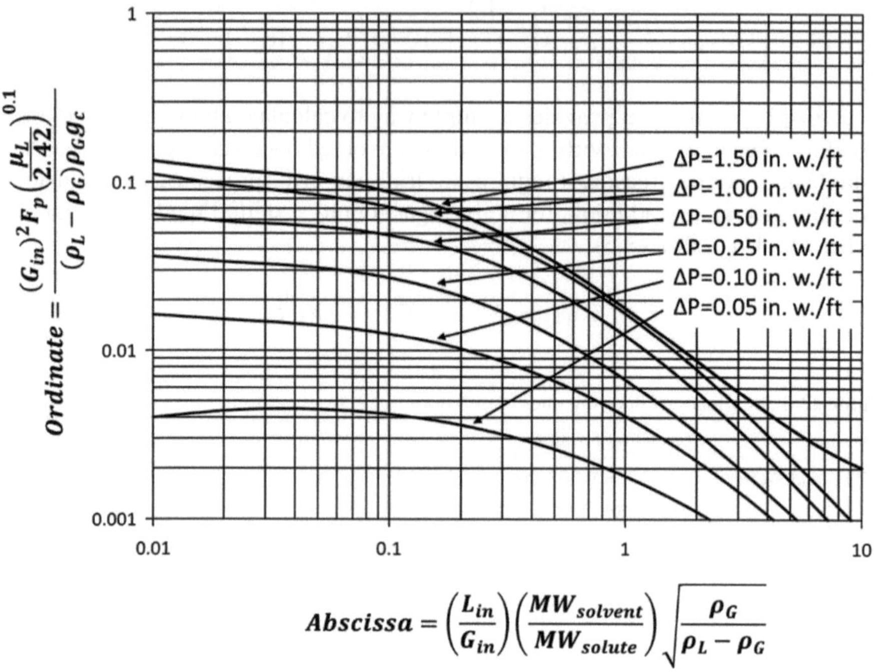

Fig. 13.12 Generalized pressure drop correlations

An alternative approach is based on setting a specific pressure drop while we try to determine a cross-sectional area that can secure the desired pressure drop. Similarly, we first need to determine the abscissa and the desired pressure drop in Fig. 13.12, then estimate the maximum gas flow rate from the ordinate. Eventually, the cross-sectional area is readily obtained.

Step 7: To Estimate the Pressure Drop

In general, pressure drop of the packed column is the function of dimensions of the column, packing, and operating conditions. Olujić [30] studied the effects of column diameter on pressure drop and concluded that the pressure drop of packed column correlates with column height, porosity, effective interfacial area, and wall factor

$$\frac{\Delta P}{H} = \lambda \frac{a}{\varepsilon^3} \frac{F_V^2}{2} K \tag{13.45}$$

where K (−) is the wall factor which is defined as

$$K = 1 + \frac{2}{3} \frac{1}{1 - \varepsilon} \frac{d_e}{d} \tag{13.46}$$

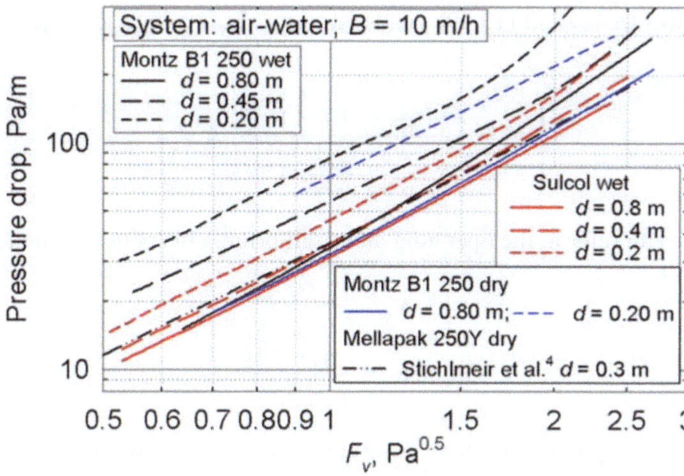

Fig. 13.13 Effects of column diameter on pressure drop [30]

d_e (m) represents the equivalent diameter which is calculated as (Fig. 13.13)

$$d_e = 6\frac{1-\varepsilon}{a_p} \quad (13.47)$$

13.4.3 The Design of Plate Tower

Prior to sizing a plate tower, we also need to acquire information about the phase equilibrium and establish an operating line. Subsequently, we can build a plate column again based on: (1) phase equilibrium, (2) desired removal efficiency, and (3) mass transfer.

Step 3: To Determine the Height of the Column

Analogous to the packed column, a plate column consists of several number of plate (NTU) with a specific height (HTU). As the equipment is used for air pollution control, the concentration of the target compound is very low, then we have

$$\frac{(y_{in} - y_{out})}{(y_{in} - y^*_{out})} = \frac{(AF^{NTU+1} - AF)}{(AF^{NTU+1} - 1)} \quad (13.48)$$

When the absorption factor AF = 1, the equation becomes to

$$\frac{(y_{in} - y_{out})}{(y_{in} - y^*_{out})} = \frac{NTU}{NTU + 1} \quad (13.49)$$

Equations (13.48) and (13.49) yields Kremser–Brown–Souders equation which is [18]

$$\begin{cases} N = \dfrac{\log\left[\left(\frac{y_{in}-mx_{in}}{y_{out}-mx_{in}}\right)\left(1-\frac{1}{A}\right)+\frac{1}{A}\right]}{\log A}, & A \neq 1 \\ N = \dfrac{y_{in}-mx_{in}}{y_{out}-mx_{in}}, & A = 1 \end{cases} \tag{13.50}$$

Also, we can refer to the operating and equilibrium curves to obtain the ordinate of Fig. 13.14

$$\text{Ordinate} = \frac{y_{in} - mx_{in}}{y_{out} - mx_{out}} \tag{13.51}$$

Then, according to the different absorption factor values, we have the number of plate from the abscissa.

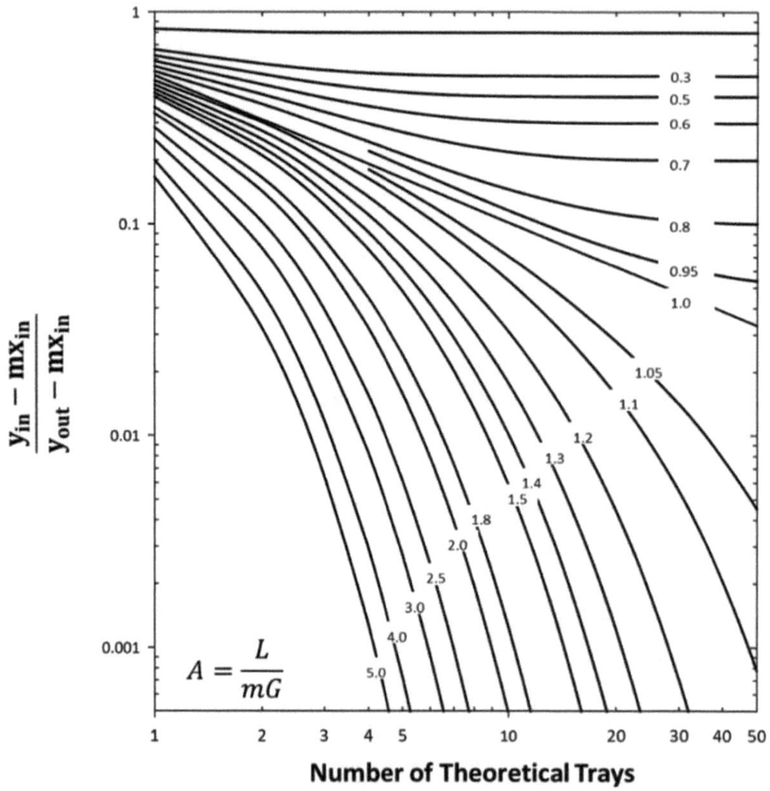

Fig. 13.14 Number of theoretical trays in the countercurrent absorption column

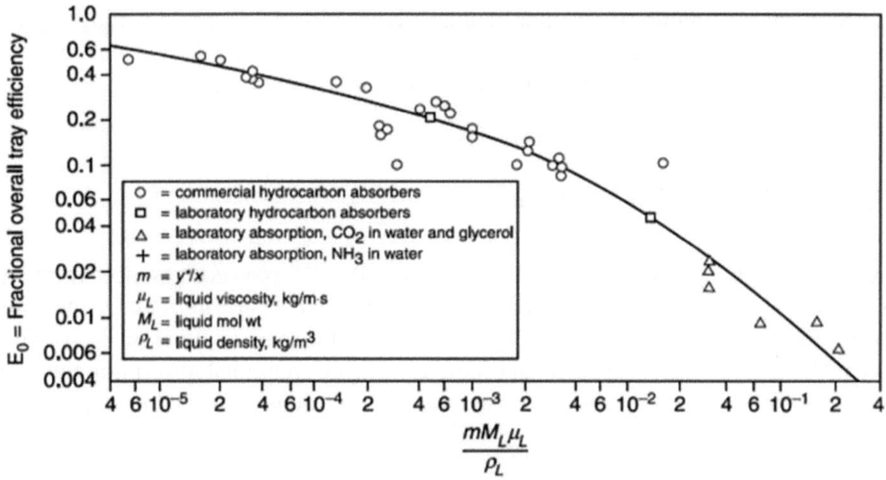

Fig. 13.15 Overall tray efficiencies of bubble-cap tray absorbers [31]

However, the NTU calculated above is theoretical, and we have to introduce the tray efficiency described in Eq. (13.52) or Fig. 13.15 to obtain the actual NTU

$$E_0 = \frac{\text{Equilibrium Trays}}{\text{Actual Trays}} = \frac{\log\left(1 + \frac{y_N - y_{N+1}}{y_N^* - y_{N+1}}\right)\left(\frac{1}{AF} - 1\right)}{\log\left(\frac{1}{AF}\right)} \tag{13.52}$$

Therefore, the actual NTU is

$$NTU_{act} = \frac{N}{E_O} \tag{13.53}$$

And the height of the column can be readily obtained

$$Z_T = HTU \cdot NTU_{act} \tag{13.54}$$

Step 4: To Determine the Sectional Area of the Column

Just like a packed column, the cross-sectional area should be large enough to prevent flooding from happening. Empirically, the maximum allowable superficial for a specific type of plate is given by

$$V_F = C_F\left(\frac{\rho_L - \rho_G}{\rho_G}\right)^{0.5} \tag{13.55}$$

where C_F is the correction factor dependent on the type of plates. Therefore, the cross-sectional area of the plat column can be obtained from Eq. (13.42).

As a rule of thumb, F for plate column is usually 0.8 in non-foaming liquids and less than 0.75 in foaming liquids. In addition, the determination of this factor is subject to the entrainment prevention and pressure drop optimization.

13.4.4 The Design of HiGee

As the gravitational force is replaced by high centrifugal force via high-speed rotation, the removal efficiency is determined by the diameter of the packing area. On the other hand, the loading capacity of a RPB is constrained by the axial height of packing area. Correspondingly, the removal efficiency of a conventional absorption tower is determined by its height while the cross-section area defines its loading capacity.

Step 3: To Determine the Fluid Status throughout the Reactor

Just like the case of a packed tower, we need to determine the appropriate liquid flow rate which ranges from the minimum wetting rate to the flooding point.

The minimum wetting rate can be obtained by following Eq. (13.56) [32]

$$\mathrm{MWR} = \left(4.8 \times 10^{-9} g_c / v\varepsilon\right) / (1 + \cos\theta / 2) \tag{13.56}$$

where g_c is the gravitational constant, v is the liquid viscosity, ε is the porosity of packing, and θ is the contact angle between the liquid and the packing material. The MWR is dependent on the rotating speed and the packing material. It was reported that the MWR for polypropylene and stainless steel packing were 0.08672 and 0.14360, respectively [33].

Apart from the hydraulic loading, flooding in a HiGee reactor also depends on the inner radius as well as the rotating speed, that is, flooding is more likely to occur in the reactor with a smaller inner radius running at lower speeds.

Therefore, inner radius, r_{in}, of the packing area should be predetermined in the design procedure. Determination of the inner radius is the result of competition between engineering performance and the maximum allowable gas flow, that is, an increase in inner radius enables the reactor to deal with the larger volumetric gas flow but requires more significant installation space. Eq. (13.57) presents the relationship between the inner radius and other factors [34]

$$r_{in} = \left(\frac{G}{\pi u_{L,jet}(1 - fd)}\right)^{1/2} \left(\frac{\rho_G}{\rho_L} K\right)^{1/4} \tag{13.57}$$

where $u_{L,jet}$ is the jet velocity of the liquid at the nozzle, fd is the distributing factor of liquid in the void space, and K is the ratio of the kinetic energy of liquid jet to that of gas. The calculation result should be carefully checked with the space requirement of nozzle, rotor, and other mechanical components.

Thereafter, Sudhoff et al. [35] suggested that it should be estimated in a way similar to the distillation column

$$\ln\left(\frac{u^2_{G,sup,flood}a_p\rho_G}{r_i\omega^2\varepsilon\rho_L}\right) = k_1 + k_2\ln\left(\frac{L}{G}\sqrt{\frac{\rho_G}{\rho_L}}\right) + k_3\left[\ln\left(\frac{L}{G}\sqrt{\frac{\rho_G}{\rho_L}}\right)\right]^2 \quad (13.58)$$

where $u_{G,sup,flood}$ is the superficial gas velocity at flooding point following the Eq. (13.59) [36]:

$$u^{1/2}_{G,sup,flood}\left\{\left(\frac{\rho_G}{\rho_L-\rho_G}\right)^{1/4} + k_1\left[\left(\frac{\rho_G}{\rho_L-\rho_G}\right)^{1/2}\frac{L}{G}\right]^{1/2}\right\} = k_5 a_p^{k_6}\left(\frac{a_c}{g_c}\right)^{k_7} \quad (13.59)$$

a_p is the specific surface area of the packing and a_c is the centrifugal acceleration. Similar to a packed tower, flooding line is feasible for the HiGee reactor (Fig. 13.16).

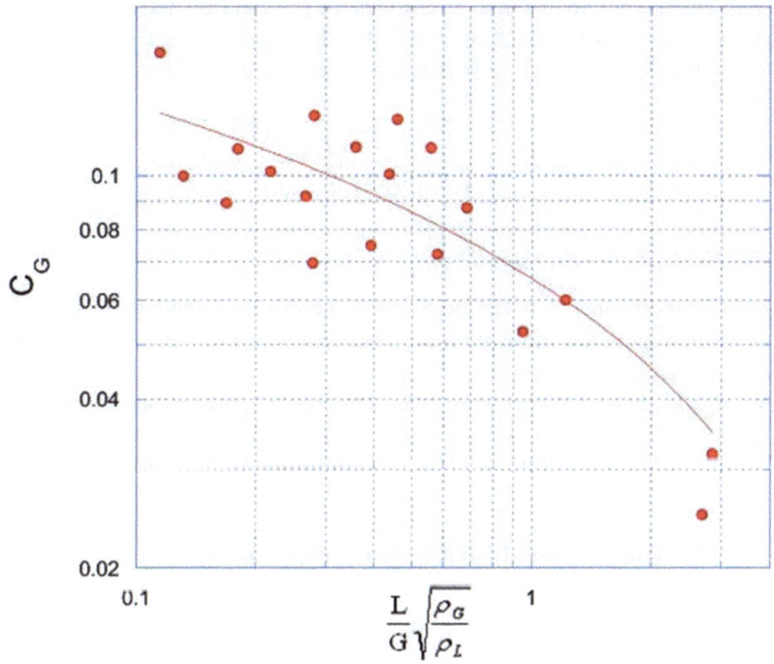

Fig. 13.16 Sherwood–Wallis flooding plot [37]

Table 13.6 Pressure drop expressions for HiGee reactor [34]

Pressure drop	Expression
Frictional	$\Delta P_f = \frac{1}{2d_h} f \rho_G \left(\frac{G}{2\pi h\varepsilon}\right)^2 \left(\frac{1}{r_{in}} - \frac{1}{r_{out}}\right)$
Momentum	$\Delta P_m = \frac{1}{2} \rho_G \left(\frac{G}{2\pi h\varepsilon}\right)^2 \left(\frac{1}{r_{in}^2} - \frac{1}{r_{out}^2}\right)$
Centrifugal	$\Delta P_c = \frac{1}{2} \rho_G A \omega^2 \left({r_{out}}^2 - {r_{in}}^2\right)$
Total	$\Delta P_t = \Delta P_f + \Delta P_m + \Delta P_c$

Step 4: To Calculate Dimensions of the Reactor

Height of packing should be higher than the minimum value to prevent the occurrence of flooding at the inner radius. Based on the superficial gas velocity mentioned before, the minimum packing height is

$$h_{min} = \frac{G}{2\pi r_{in} u_{G,sup,flood}} \tag{13.60}$$

Since we have calculated inner radius (r_{in}), height (h) of the packing, outer radius (r_{out}) becomes the only factor that determines the gas loading and size of the reactor. In order to ensure that the packing can fulfill the expected performance, the outer radius should be large enough to secure the necessary mass transfer coefficient

$$r_{out,min} = \sqrt{\frac{Q}{\pi h K_G a_e} \ln\left(\frac{y_{in}}{y_{out}}\right) + r_{in}^2} \tag{13.61}$$

Step 5: To Evaluate the Pressure Drop

Finally, the dimensions of a HiGee reactor should be amended by checking the pressure drop. In general, there are three contributing factors that should be considered when it comes to the pressure drop of a HiGee reactor, which are frictional, momentum, and centrifugal pressure drop. Theoretically, the expressions of all three pressure drop categories are listed in Table 13.6.

13.5 Performance Evaluation and Improvement

13.5.1 Performance Evaluation

Absorption is a kind versatile air pollution control device that is capable of removing soluble gaseous pollutants and particulate matters from the flue gas. It has been reported that the removal efficiencies of inorganic gases can be up to 99% and those of organic gases are typically 70–99%. In addition, more than 50% of the particulate matter can be removed where the process is termed as wet scrubber [38].

Economic cost of the absorption process varies drastically with different materials, structures, and operating modes. Specifically, stainless steel would cause the highest price, followed by ceramic and polypropylene [39]. On the other hand, Tellerette, Berl saddle, and Intalox saddle in the packed column would cost more if the same material is used. Operation and maintenance costs majorly come from solvent and energy consumption. In practice, lime and/or limestone are usually used for FGD for the relatively lower costs which can be compensated by the sale of gypsum in some cases.

Energy is consumed in the absorption process by pumping the solvent and drafting the flue gas which can be estimated by Eq. (13.62)

$$\begin{cases} \text{Electricity}_{\text{fan}}(\text{kW}) = \frac{1.17\times10^{4}\cdot G_{\text{in}}\cdot\Delta P}{\varepsilon} \\ \text{Electricity}_{\text{pump}} = \frac{0.746\times2.52\times10^{-1}\cdot L_{\text{in}}\cdot P}{\varepsilon} \end{cases} \tag{13.62}$$

where ε (−) is the mechanical efficiencies of fans or pumps, P (feet of water) is the pumping pressure [39].

An engineering-environmental-economic assessment for integrated air pollutants reduction, CO_2 capture, and utilization exemplified by the high-gravity process was presented by Tse-Lun Chen et. al. [40]. Table 13.7 shows Key performance indicators (KPIs) regarding engineering performance, environmental impacts, and economic cost. More detailed descriptions of the selected KPIs were stated as follows:

Engineering performance evaluation

The functional unit of the HiGee for the 4E analysis was one ton of PCFA input. The engineering performance was calculated using three KPIs, i.e., EP_1, EP_2, and EP_3, representing the CO_2 capture capacity, NO_x removal, and production of supplementary cementitious materials, respectively. The removal efficiency (η) of CO_2 and NO_x was evaluated by Eq. (13.63).

$$\eta(\%) = \frac{Q_i \times \rho_i \times C_i - Q_o \times \rho_o \times C_0}{Q_i \times \rho_i \times C_i \times 100\%} \tag{13.63}$$

where η (%) is the removal efficiency of a specific pollutant, ρ (kg/m^3) is the density of the flue gas, C_i (wt %/ppm) and C_o (wt %/ppm) are the concentrations of pollutant in the inflow and outflow gasses, respectively, and Q_i (m^3/min) and Q_o (m^3/min) are the flow rate of inflow and outflow gasses, respectively. The daily loading of the HiGee process was determined by Eq. (13.64)

$$Daily\ Loading = Q_i \times C_i \times \eta \times t \times 1440 \tag{13.64}$$

Table 13.7 Key performance indicators (KPIs) regarding engineering performance, environmental impacts, and economic cost

Aspect	Key performance indicators (KPIs)		Unit
Engineering performance	EP_1	CO_2 capture capacity	kg CO_2/t-PCFA
	EP_2	NO removal	kg NO/t-PCFA
	EP_3	Production of supplementary cementitious material	t/t-PCFA
Environmental impacts	EI_1	Global warming potential reduction	Kg CO_2/t-PCFA
	EI_2	Fossil depletion	Kg oil/t-PCFA
	EI_3	Particulate matter formation reduction	Kg PM_{10}/t-PCFA
	EI_4	Land occupation	m^2/t-PCFA
	EI_5	Ecosystem quality improvement	Points/t-PCFA
	EI_6	Risk of human health reduction	Points/t-PCFA
Economic cost	EC_1	Save in air pollution fees	USD/t-PCFA
	EC_2	Earning of carbon credit	USD/t-PCFA
	EC_3	Product sale	USD/t-PCFA
	EC_4	Operating cost	USD/t-PCFA

Environmental impact assessment

The environmental impacts from the HiGee process were assessed by the LCA using the ReCiPe midpoint/endpoint. For the BAU scenario, PCFA was disposed of in a landfill, while blowdown wastewater was directly discharged. Exhausted steam was emitted into the atmosphere through conventional air pollution control equipment (i.e., SCR-ESP-FGD). In the case of the HiGee process, PCFA and blowdown wastewater were mixed and then pumped into the reactor, where the PCFA slurry reacted with CO_2 in the flue gas. A total of six environmental impact indicators were determined: global warming potential reduction (EI_1), fossil depletion (EI_2), PM formation reduction (EI_3), land occupation (EI_4), ecosystem quality improvement (EI_5), and risk of human health reduction (EI_6).

Cost and benefit analysis

To determine the economic performance, a cost-benefit analysis was performed by calculating the differences between marginal costs and revenues under various daily capture scales. As shown in Fig. S1, the marginal cost (MC, with a unit of USD/t-CO_2) is defined as the additional cost for "producing one more unit of product" in Eq. (13.65)

$$MC = \frac{d(\text{Total operating costs})}{d(\text{Plant capture scale})} = \frac{d(\text{EC} + \text{CU} + \cdots)}{d(\text{CS})} \tag{13.65}$$

where EC (USD) is the cost for electricity consumption, CU (USD) is the cost for chemical use, and CS (t-CO_2/day) is the daily capture scale of CO_2. The major cost of the HiGee process was electricity consumption, where the electricity price was around 0.1 USD/kWh. The marginal benefit, with a unit of USD/t-CO_2, can be calculated by Eq. (13.66),

$$MB = \frac{d(\text{Total benefit})}{d(\text{Plant capture scale})} = \frac{d(\text{CC} + \text{PS} + \text{APF} + \cdots)}{d(\text{CS})} \tag{13.66}$$

where CC (USD/t-CO_2) is the carbon credit, PS (USD/t-CO_2) is the return from product sale, and APF (USD/t-CO_2) is the saved air pollution fee along with the CO_2 capture process.

13.5.2 Performance Improvement

An ideal absorption reactor exhibits characteristics including low liquid requirement, high gas treatment capability, low gas pumping cost, reduced internal complexity, low installation and operating cost, easier installation in confined spaces, and high absorption efficiencies. However, some problems may cause malfunction in the scrubber. For instance, in poor sorbent distribution and/or packing arrangement, gas leakage may occur. As a result, the operator need to increase the liquid flow rate and modify the locations of liquid distributors. In particular, care should be taken to the concentration of effective chemical composition. For instance, pH is an indicator of the concentration of alkaline solvent in SO_2 absorption which should be maintained above 6. For this reason, addition of fresh solvent and bleed-off of spent one is required. Attention should be paid to the treatment of discharged solution to prevent secondary pollution. If the absorption is designed to handle the organic pollutant, it is recommended that the solvent be regenerable or be less toxic.

The absorption reactor, especially packed column is vulnerable to clogging resulting from the particles in the flue gas or the insoluble substance produced by a chemical reaction between pollutants and solvents. For this reason, periodic inspection should be performed to evaluate the status of packing and to remove the precipitated solid. In general, removal of clogging is achieved by recirculating a chemical solvent that facilitates the dissolution of solids. Other practical options include high-pressure steam or hot water. In addition, periodic checks of the fan, pumps, chemical feed system, piping, duct, and liquid distributor are necessary to ensure effective operation. To sustain a full operation during cleaning or maintenance, a backup system including additional strainers or pumps should be considered.

13.5.3 *Innovative Technologies*

Wellman–Lord Process for Flue Gas Desulfurization [40, 41]

In general, wet flue gas desulfurization processes use lime as the reactant so that the final product would be gypsum which can be recovered as construction material. However, the process as such would generate a great mass of sludge and the gypsum purity fails to meet the relevant requirement, which makes it unfavorable for implementation.

Wellman–Lord is a regenerable process established on a closing-loop without throwaway sludges. At the startup stage, alkaline absorbent is injected into the system to react with SO_2, generating SO_3^{2-} which is the major species responsible for SO_2 removal afterward. A detailed example is given below by assuming the makeups are NaOH and Na_2CO_3 (Fig. 13.17).

The major reaction occurring in the absorption reactor is sulfur dioxide (SO_2) and sodium sulfite (Na_2SO_3) to produce sodium bisulfite ($NaHSO_3$). Since the existence of oxygen or other oxidant agent, sodium bisulfite may get oxidized and transform into sodium sulfate. The absorbent then flows into a surge tank for regeneration

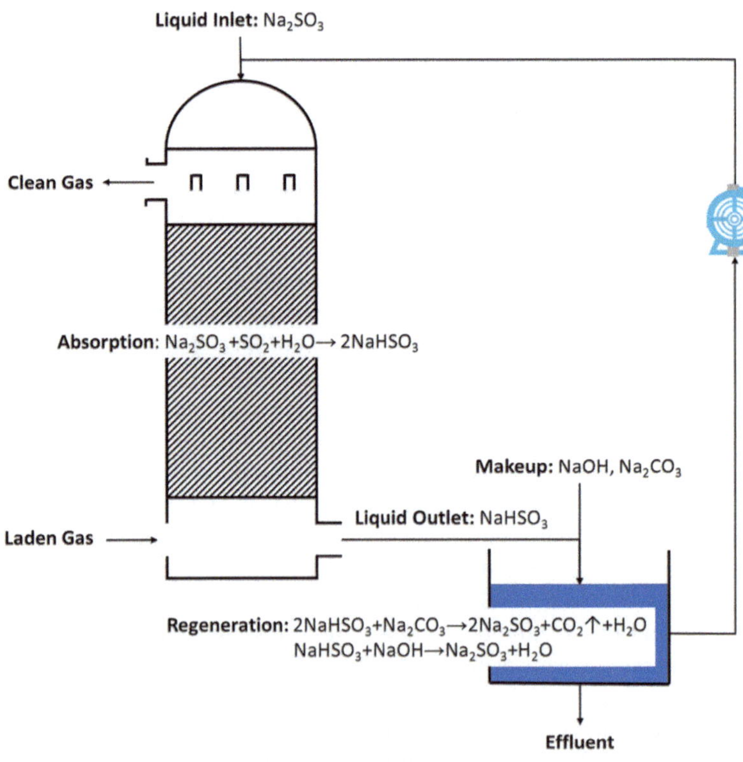

Fig. 13.17 Schematic diagram of Wellman–Lord process

where sodium hydroxide (NaOH) and sodium carbonate (Na_2CO_3) are added. The regenerated absorbent is partially pumped back into the system while the other is discharged to maintain the balance of the system.

The advantage of Wellman–Lord process is that it can drastically reduce sludge production and sodium sulfite can be recovered for industrial uses. However, the price of sodium hydroxide is much higher than limestone. For this reason, sodium carbonate is sometimes used as an alternative.

13.5.4 Higee Enhanced Absorption

Jassim et al. [42] used a HiGee reactor to perform carbon capture with MEA absorbent. Compared with conventional absorbers, they found that there was an around tenfold increase in mass transfer coefficient when the RPB was introduced to replace the conventional absorber. For this reason, HTU of the RPB was about 94% smaller than that of the conventional absorber (Table 13.8).

Recently, a favorable mass transfer fate for multiple air pollutant abatement via a high-gravity rotating packed bed was developed as shown in Fig. 13.18 [43].

In the experimental set up, the key operating dimensionless parameters including gas to liquid ratio (GLR), high-gravity factor (β) (i.e., estimating from the rotating speed), and liquid to solid ratio (LSR) were designed for the field tests. In order to generate a green solvent, the fly ash was added into water and mixed for 30 min and then settled down. The leachate was gained from the supernatant of the solvent with high alkalinity due to the chemical properties of fly ash. In addition, in order

Table 13.8 Performance comparison between conventional absorber and RPB for carbon capture

Item (Unit)		Conventional absorber [42]	HiGee reactor
Equipment specification	• Packing diameter (m)	4.42	0.40 (r_{out}) 0.16 (r_{in})
	• Packing height (m)	14.02	0.03
	• Packing volume (m^3)	215.01	<0.01
Operating parameters	• Pressure (atm)	1.00	1.00
	• Temperature (°C)	40	20–40
	• MEA concentration (%)	10–18	30
	• L/G (–)	4.45	16.30–30.75
Engineering performance	• CO_2 removal (%)	97.0–99.3	96.3–99.3
	• HTU_{OG} (m)	3.40	0.14–0.27
	• K_Ga (1/s)	0.068	0.64 –1.22

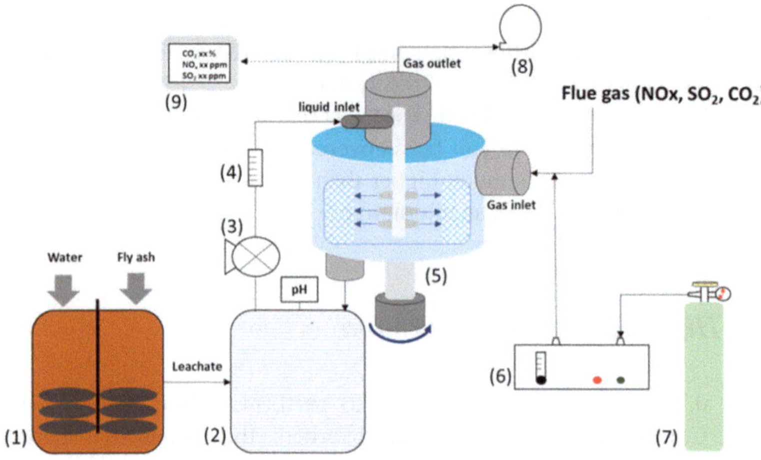

Fig. 13.18 Schematic diagram of Higee system for multiple air pollution control: (1) mixing tank; (2) solvent tank; (3) pump; (4) flow meter; (5) rotating packed bed; (6) O_3 generator; (7) oxygen cylinder; (8) blower; (9) gas analyzer

to overcome the lower solubility of NO, a corona ozone generator was prepared to oxidize NO with the injection of O_3 gas at a constant flow rate of 1 L/min with an O_3 concentration of 0.0838 g/min.

Figure 13.19 shows the determination of the favorable enhancement factor for NO_x–SO_2–CO_2 removal in a RPB using a systematical graphical presentation. The result indicated that the available energy consumptions of 0.103 kwh per t-NO_x, 0.047 kwh per t-SO_2, and 134.62 kwh per t-CO_2 were found at the β of 233.8. The estimated K_Ga values of NO_x, SO_2, and CO_2 up to 1.25, 1.08, and 1.7 could be determined, respectively, by the vertical arrows, and the enhancement factor of NO_x, SO_2, and CO_2 were exhibited as 1.8, 5.9, and 1.6, respectively.

Exercises

1. Define the following terms and explain why they should be considered for design use: number of transfer unit, height of transfer unit, minimum wetting rate, Baber's method.
2. Please describe the general procedures for designing an absorption process.
3. List the major factors affecting the design of an absorption process.
4. Describe the unique features and advantages of using HiGee for air pollution control.

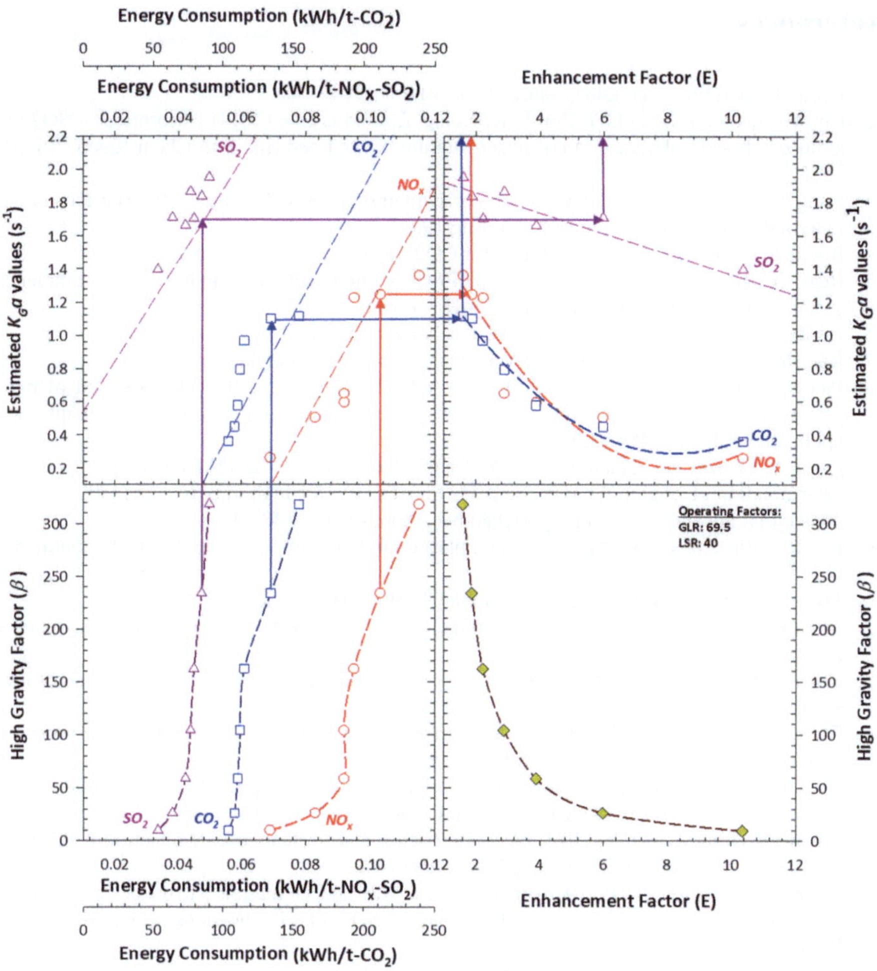

Fig. 13.19 Systematic approach to determine the favorable enhancement factors for simultaneous NO_x–SO_2–CO_2 removal in an RPB

5. Determine the liquid flow rate in lbmol/h·ft^2 for a packed countercurrent gas scrubber. The mole fractions of pollutant in inlet and outlet gas are 0.08 and 0.002, respectively. The mole fractions of pollutant in inlet and outlet liquid are 0.001 and 0.05, respectively. The gas flow rate is 18 lbmol/h·ft^2.

References

1. Group G. Scrubber technology: absorption columns 2018
2. Chu G-W, Luo Y, Shan C-Y, Zou H-K, Xiang Y, Shao L et al (2014) Absorption of SO2with ammonia-based solution in a cocurrent rotating packed bed. Ind Eng Chem Res 53:15731–15737
3. Jiang X, Liu Y, Gu M (2011) Absorption of sulphur dioxide with sodium citrate buffer solution in a rotating packed bed. Chin J Chem Eng 19:687–692
4. Rochelle GT (2009) Amine scrubbing for CO2 capture. Science 325:1652
5. Reay D, Ramshaw C, Harvey A (2013) Process intensification: engineering for efficiency, sustainability and flexibility. Butterworth-Heinemann
6. Whitaker S (1992) Introduction to fluid mechanics. Krieger
7. Ramshaw C, Mallinson RH (1981) Mass transfer process. Google Patents
8. Pan S-Y, Wang P, Chen Q, Jiang W, Chu Y-H, Chiang P-C (2017) Development of high-gravity technology for removing particulate and gaseous pollutant emissions: principles and applications. J Clean Prod 149:540–556
9. Pan S-Y, Eleazar EG, Chang EE, Lin Y-P, Kim H, Chiang P-C (2015) Systematic approach to determination of optimum gas-phase mass transfer rate for high-gravity carbonation process of steelmaking slags in a rotating packed bed. Appl Energy 148:23–31
10. Inc. RU. Raschig super-ring® metal and plastic random dump packing. Product bulletin 200. 2013
11. Jaya A (2011) Engineering design guideline. Packing Hydraulic R
12. Jones DS, Pujadó PP (2006) Handbook of petroleum processing. Springer Science & Business Media
13. Chuang K, Nandakumar K (2000) Tray columns: design. In: Encyclopedia of separation science, pp 1135–1140
14. Chen Y-S, Lin C-C, Liu H-S (2005) Mass transfer in a rotating packed bed with various radii of the bed. Ind Eng Chem Res 44:7868–7875
15. Lin C-C, Chu C-R (2015) Feasibility of carbon dioxide absorption by NaOH solution in a rotating packed bed with blade packings. Int J Greenhouse Gas Control 42:117–123
16. Zheng D, Dong L, Wu X (2013) New approach for absorbent species selection with excess gibbs function. Ind Eng Chem Res 52:9480–9489
17. Morris GA, Jackson J (1953) Absorption towers: with special reference to the design of packed towers for absorption and stripping. Butterworths Scientific Publications in Association with Imperial Chemical Industries
18. Treybal RE (1980) Mass transfer operations. New York
19. Perry JH (1950) Chemical engineers' handbook. ACS Publications
20. Baker T, Chilton T, Vernon H (1935) The course of liquor flow in packed towers. Trans AIChE. 31:296
21. Sherwood T, Shipley G, Holloway F (1938) Flooding velocities in packed columns. Ind Eng Chem 30:765–769
22. Coker A (1991) Understand the basics of packed-column design. Chem Eng Prog 87:93–99
23. Zenz F (1953) What every engineer should know about packed tower operations. Chem Eng 60:176–184
24. Elgin J, Weiss F (1939) Liquid holdup and flooding in packed towers. Ind Eng Chem 31:435–445
25. Bertetti JW (1942) Theoretical flooding velocities in packed columns. Trans AIChE, 38
26. Leva M (1992) Reconsider packed-tower pressure-drop correlations. Chem Eng Prog 88:65–72
27. Takahashi T, Akagi Y, Fujita K, Kishimoto T (1974) Flooding velocities of gas-liquid and liquid-liquid contactors of column type. J Chem Eng Jpn 7:223–225
28. Billet R, Schultes M (1995) Fluid dynamics and mass transfer in the total capacity range of packed columns up to the flood point. Chem Eng Technol 18:371–379
29. Kuźniewska-lach I (1999) Estimation of phase velocities at flooding point in packed columns for any gas/liquid system. Canad J Chem Eng 77:439–446

30. Olujić Ž (1999) Effect of column diameter on pressure drop of a corrugated sheet structured packing. Chem Eng Res Des 77:505–510
31. Theodore L, Buonicore AJ (1988) Air pollution control equipment
32. Jiang H, Lin J, Chow W (2002) Time-resolved photoluminescence studies of III-nitrides. III-Nitride semiconductors: optical properties, vol 1, pp 9
33. Jiao WZ, Liu YZ, Qi GS (2010) Gas pressure drop and mass transfer characteristics in a cross-flow rotating packed bed with porous plate packing. Ind Eng Chem Res 49:3732–3740
34. Agarwal L, Pavani V, Rao DP, Kaistha N (2010) Process intensification in HiGee absorption and distillation: design procedure and applications. Ind Eng Chem Res 49:10046–10058
35. Sudhoff D, Leimbrink M, Schleinitz M, Górak A, Lutze P (2015) Modelling, design and flexibility analysis of rotating packed beds for distillation. Chem Eng Res Des 94:72–89
36. Rajan S, Kumar M, Ansari MJ, Rao DP, Kaistha N (2011) Limiting gas liquid flows and mass transfer in a novel rotating packed bed (HiGee). Ind Eng Chem Res 50:986–997
37. Lockett M (1995) Flooding of rotating structured packing and its application to conventional packed-columns. Chem Eng Res Des 73:379–384
38. USEPA (2003) Air pollution control technology fact sheet. Packed Bed
39. Barbour W, Oommen R, Shareef GS (1995) Wet scrubber for acid gas
40. Adams RC, Lutz SJ, Mulligan SW (1979) Demonstration of Wellman-Lord/Allied chemical FGD technology: acceptance test results. TRW, Inc., Durham, NC (USA). (p. Medium: X; Size, Pages: 129)
41. Adams RC, Cotter J, Milligan SW (1979) Demonstration of Wellman-Lord/Allied Chemical FGD technology: demonstration test first year results. Annual report Sep 78-Aug 79. TRW, Inc., Durham, NC (USA). (p. Medium: X; Size: Pages: 99)
42. Jassim MS, Rochelle G, Eimer D, Ramshaw C (2007) Carbon dioxide absorption and desorption in aqueous monoethanolamine solutions in a rotating packed bed. Ind Eng Chem Res 46:2823–2833
43. Chen T-L, Chen Y-H, Chiang P-C (2020) Enhanced performance on simultaneous removal of NOx-SO2-CO2 using a high-gravity rotating packed bed and alkaline wastes towards green process intensification. Chem Eng J 393:124678

Chapter 14
Combustion Control

Nomenclature

a	Surface area per unit volume of bed, m^{-1}
acfm	Exhaust gas flow rate to incinerator (at average incinerator operating temperature), actual ft^3/min
A	Pre-exponential factor
B_f	Btu/scf of the fuel, Btu
B_v	Btu/scf of the vent stream, Btu
C_p	Heat capacity over temperature range from T to T_{ref}
Cp	Average heat capacity of composite air and waste gas stream, Btu/lb- °F
d	Effective diameter of the channel, m
D	Diffusivity, m^2/s
D_e	Effective (turbulent) diffusion coefficient, m^2/s
D_{min}	Minimum flare tip diameter
E_a	Activation energy, $E_a = -0.00966(\text{MW}) + 46.1$
f	Fanning friction factor
g_c	32.17 lb-ft/lb_f-h-s^2
h_e	Heat content of flue gas, Btu/lb
h_f	Lower heating value of natural gas, Btu/lb
Δh_c	Heat of combustion
$(\&\Delta h_{cw})$	Heat of combustion of the waste stream (Btu/scf)
$(\&\Delta h_{ci})$	Volumetric heat of combustion of component *i* at 25 °C(Btu/scf)
ΔH	Enthalpy
K	Orifice factor, (≒0.9)
L	Length of the reaction zone, m
LEL_j	Lower explosive limits of combustible component j
Lm	Length of one mass transfer unit
$\dot{m}$	Mass flow rate of gases
M	Average molecular weight
MFR	Momentum flux ratio

P.-C. Chiang and X. Gao, *Air Pollution Control and Design*,
https://doi.org/10.1007/978-981-13-7488-3_14

n	Number of combustible components in mixture
P	Absolute pressure, atm
Q	Volumetric flow rate (volume per unit time)
Q_e	Emission stream flow rate, scfm
Q_f	Natural gas flow rate, scfm
Q_{fi}	The flue gas flow rate
r	Density of gas
R	1546 ft-lb_f/R-lb-mole
R'	Gas constant, 0.08206 L-atm/mol-K
S	Steric factor = 16/MW
Sc	Schmidt number [$\mu/\rho D$]
t	Total average residence time, sec
T_{wi}	Waste gas inlet temperature
T_{fi}	Incinerator operating temperature
T_{he}	Emission stream temperature after heat recovery, °F
T_r	Reference temperature, °F
T_{ref}	Reference temperature (usually take to be zero, the temperature of a convenient stream, or the inlet gas stream)
ΔT	Temperature rise required to bring waste gas stream from temperature upstream of incinerator to required combustion temperature, °F
u	Gas velocity, m/s
V	Volume of the reaction zone, m^3
V_a	Acoustic velocity, or velocity of sound in the fluid, ft/s
V_c	The volume of the combustion chamber
V_{cat}	Overall bulk volume of the catalyst bed
V_g	The volumetric gas flow rate (based on cold entrance conditions)
$[VOC]_L$	Conc. of the VOC at length L
$[VOC]_0$	Inlet conc. of the VOC
x_i	Volume fraction of combustible component i
y_{O2}	Mole fraction of O_2 in the afterburner
Z'	Collision rate factor (from Fig. 11.5 for alkanes, alkenes, and aromatics)
ρ	Density (mass per unit volume)
τ	Fraction of heat intensity transmitter
Ø	Space velocity

14.1 Introduction

14.1.1 Features of Combustion Process

Combustion simply represents a chemical process in which a substance reacts rapidly with oxygen and releases heat. During combustion, new chemical substances which are called exhaust are created from the fuel and the oxidizer. The original substance is called fuel, and the proper amount of oxygen is called the oxidizer. The fuel can be a solid, liquid, or typically gas phase. In general, these fuels contain carbon, hydrogen, nitrogen, and sulfur that, when burned, produce other substances that are harmless than original ones. In the air pollution control, combustion which is also referred to oxidation or incineration is often used to treat the emissions of organic compounds with high removal destruction efficiencies. Combustion system particularly consists of burners which ignite the fuel and organic compound, and a chamber which provides appropriate residence time for the oxidation process.

A thermal incinerator (also known as thermal oxidizer, or afterburner) is a process unit for air pollution control in many chemical plants that decomposes hazardous gases at a high temperature and releases them into the atmosphere. The simplest technology of thermal oxidation is direct-fired thermal oxidizer. The heart of the thermal incinerator is a nozzle-stabilized flame maintained by a combination of auxiliary fuel, waste gas compounds, and supplemental air added when necessary. By integration with different auxiliary equipment, thermal incineration also has these categories:

Thermal incinerator is comprised only of the combustion chamber as shown in Fig. 14.1 [1]. The waste gas preheater and the secondary energy recovery heat exchanger are energy recovery devices and are not included as part of the direct thermal incinerator.

Recuperative incinerator is comprised of the combustion chamber, the waste gas preheater, and, if appropriate, the steam boiler and energy recovery heat exchanger as indicated in Fig. 14.2.

Regenerative incinerator uses direct contact heat exchangers constructed of a ceramic material that can tolerate the high temperatures needed to achieve ignition of the waste stream as shown in Fig. 14.3 [2].

Catalytic incinerators employ a bed of active material (catalyst) that facilitates the overall combustion reaction. The catalyst has the effect of increasing the reaction rate, and enabling the conversion at lower reaction temperatures than in thermal

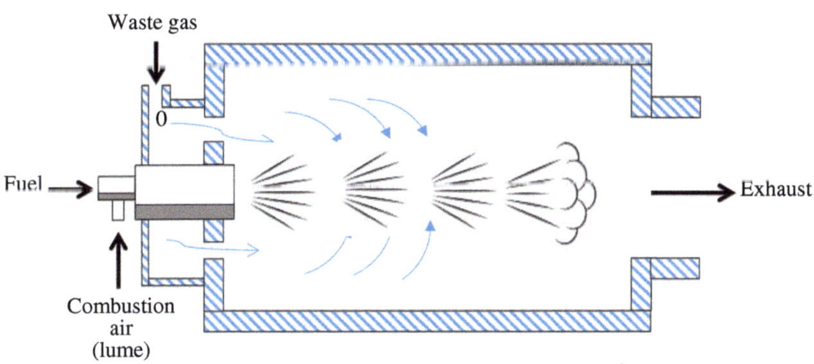

Fig. 14.1 Thermal incinerator schematic [1]

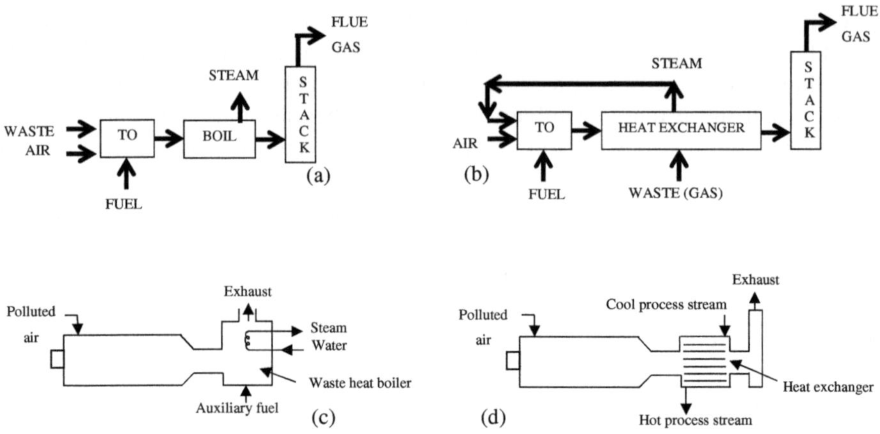

Fig. 14.2 Recuperative incinerator schematic (**a** and **c** with steam boilers, **b** and **d** with heat exchangers)

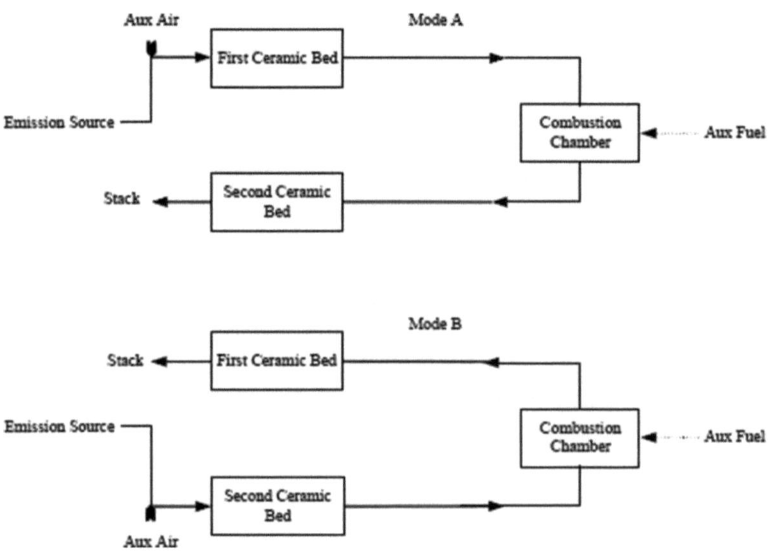

Fig. 14.3 Regenerable-type/thermal oxidizer [2]

incinerator units. Nevertheless, the waste stream must be preheated to a temperature sufficiently high (usually from 300 to 900 °F) to initiate the oxidation reactions. The waste stream is preheated either directly in a preheater combustion chamber or indirectly by the heat exchanger with the incinerator's effluent or other process heat or both and then passed over the catalyst bed. The chemical reaction (combustion) between the oxygen in the gas stream and the gaseous pollutants takes place at the catalyst surface [3] which can be divided into two types.

Fixed bed catalytic incinerator may use a monolith catalyst or a packed-bed catalyst for VOCs destruction. The catalyst is a porous solid block containing parallel, non-intersecting channels aligned in the direction of the gas flow shown in Fig. 14.4 [1].

Fluidized-bed catalytic incinerator the advantage of very high mass transfer rates, although the overall pressure drop is somewhat higher than in a monolith catalyst bed. This higher heat transfer rate to heat transfer tubes immersed in the bed may allow waste gas with higher heating values to be processed without exceeding maximum permissible temperatures in the catalyst bed shown in Fig. 14.5 [1].

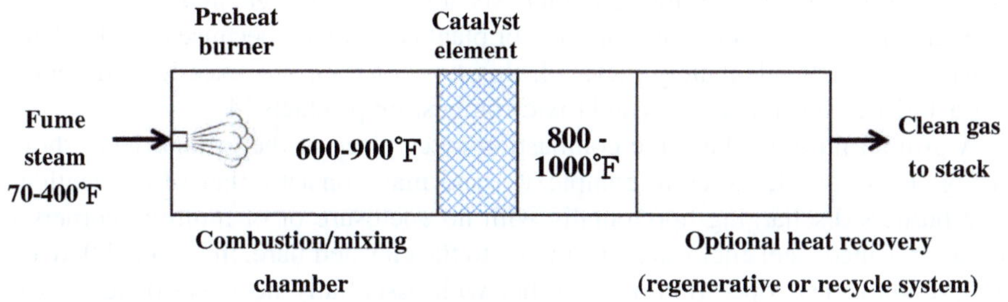

Fig. 14.4 Catalytic incinerator schematic [1]

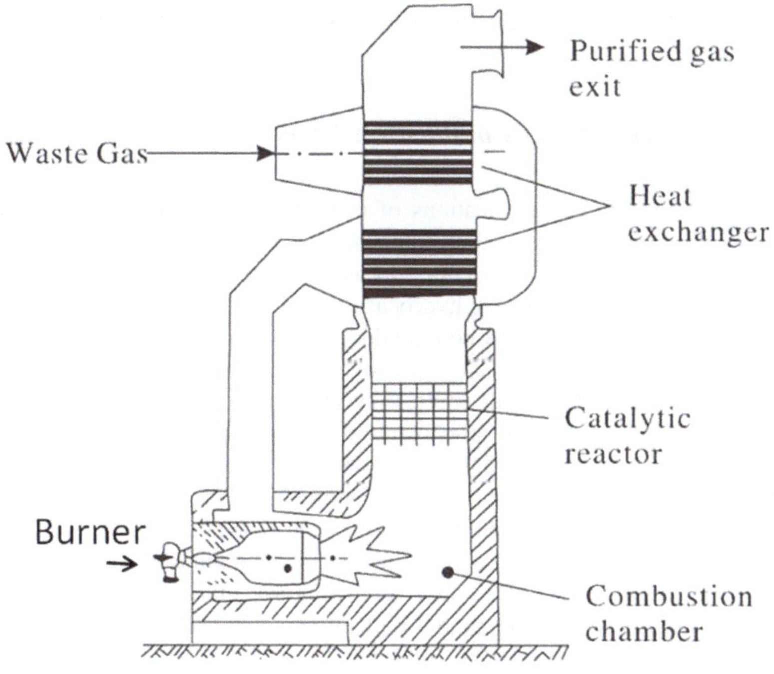

Fig. 14.5 Catalytic afterburner with array of ceramic cylinders coated with active catalytic material

The flare is utilized for pollution and disaster prevention during the emergency release system in a refinery or chemical plant. It uses to dispose of purged and wasted products from refineries, unrecoverable gases emerging with oil from oil wells, vented gases from blast furnaces, unused gases from coke ovens, and gaseous water from chemical industries.

Elevated flare is the most commonly used type in refineries and chemical plants as shown in Fig. 14.6a [4]. The waste gas stream is fed through a stack from 32ft to over 320 ft tall and is combusted at the tip of the stack. The elevated flare can utilize steam/air injection to make smokeless burning and with low luminosity up to about 20% of maximum flaring load. The disadvantage of steam/air injection is that it causes a noise pollution. Capital costs are relatively high, and an appreciable plant area may be rendered unavailable for plant equipment, because of radiant heat considerations. If adequately elevated, this type of flare has excellent dispersion characteristics for malodorous and toxic combustion products [4].

A ground flare is where the combustion takes place at the ground level shown in Fig. 14.6b [4]. It varies in complexity, and may consist either of conventional flare burners discharging horizontally with no enclosure or of multiple burners in refractory-lined steel enclosures. Compare to the elevated flare, the ground flare can achieve smokeless operation as well, but with essentially no noise or luminosity problems, provided that the design gas rate to the flare is not exceeded. However, it has poor dispersion of combustion product because the stack is near to ground. This may result in severe air pollution or hazard, if the combustion products are toxic or in the event of flameout. Capital, operating, and maintenance costs are high [4].

14.1.2 Applications of Combustion Process

Nowadays, a diversity of applications of combustion to control the emission of gaseous or fine particles are either combustible or thermally decompose at high temperature. It is a common practice to control odor and reduce the capacity of plumes, can emissions of reactive hydrocarbons (HCs). In normal combustion, this odorant is converted to water, carbon dioxide, and sulfur dioxide.

However, thermal oxidizers are typically used to destroy hazardous air pollutants (HAPs) and volatile organic compounds (VOCs) from industrial air streams. A process stream with hazardous gases is introduced into a firing box through or near the burner and enough residence time is provided to achieve the desired destruction removal efficiency (DRE) of the VOCs. Most direct-fired thermal oxidizers operate at temperature levels between 980 °C (1,800 °F) and 1,200 °C (2,190 °F) with air flow rates of 0.24–24 standard cubic meters per second. The ignition temperature varies with different compounds. The shorter the residence time, the higher the reactor temperature must be. Residence time is typically not longer than 1 s to the waste gas with typical temperatures of 1,200–2,000 °F.

Thermal incinerators have several advantages including complete removal of the organic pollutants without waste water or solid waste problems, good adaptability

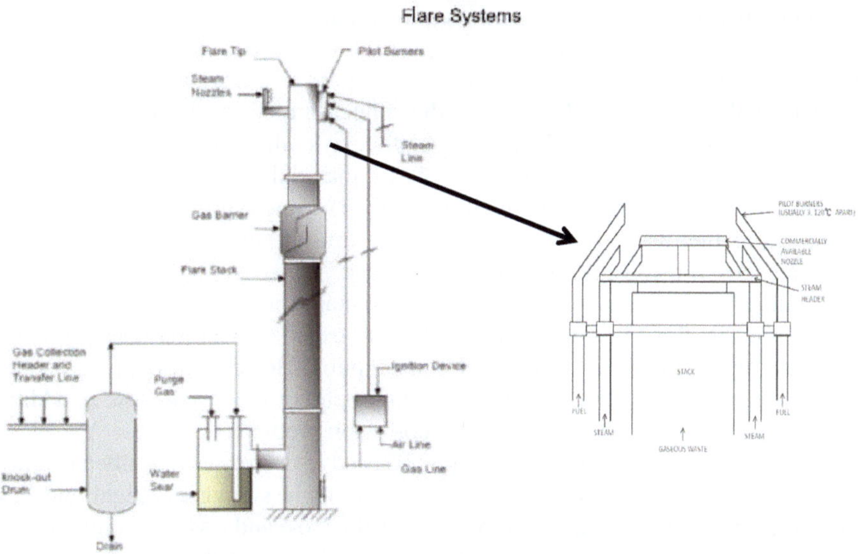

(a) Elevated flare equipped with mixing nozzle

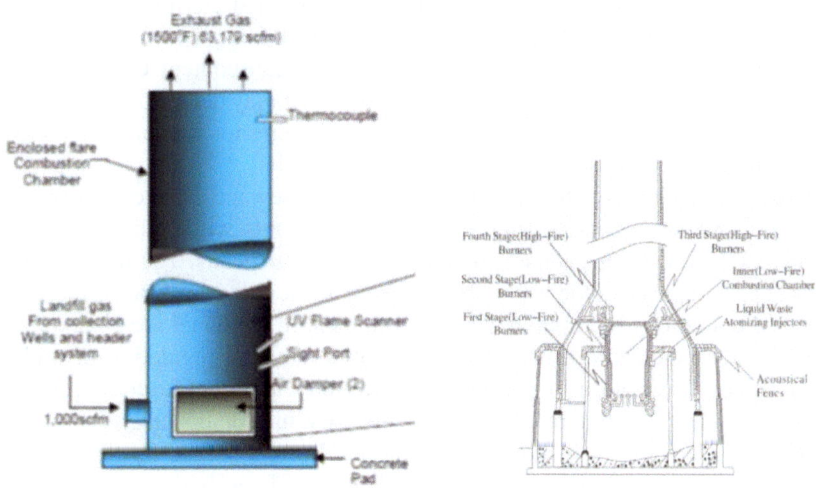

(b) Ground flare

Fig. 14.6 Elevated and ground flares schematics [4]

to changes in waste gas flow rate and composition, relatively simple design [5]. However, they also have disadvantages regarding the operating cost for injecting the supplementary fuel is rather high and constructing a steel outer shell lined with refractory material.

The purposes of the refractory lining are prepared to protect the steel shell from direct exposure to the effects of high temperature and corrosive materials and to improve the thermal efficiency. It also can be served as a thermal insulator to retain the steel shell's strength and to prevent the condensation of water vapor.

Refractory consists of various types of fireclays which are available in several forms including bricks and castable or plastic material. Plastic refractory material is normally used only to cover small area, for patching work, or to form intricate shape. More fuel usage required for longer warm-up periods is one of the major disadvantages of using a refractory-lined furnace.

Ceramic blanker insulation has several advantages including lightweight saving shipment costs, flexibility for field-installation, excellent thermal insulation short, and warm-up period for reducing fuel consumption. However, it requires the stud-and-washer system which is the most expensive part of the installation. In addition, it is very pervious to fumes may cause a corrosion problem.

Catalytic incineration can, in principle, be used to destroy essentially any oxidizable compound in an air stream with smaller units less oxygen weight and lower operating temperature, resulted in lower operation cost and less formation of smog-producing oxides of nitrogen. Consequently, the replacement of catalyst bed on a periodic basis is required to prevent the catalyst from suppressant or poison. However, there are practical limits to the types of compounds that can be oxidized due to poisoning and inhibition effects on the catalyst. In addition, most configurations require a low heating value of the inlet gas and a particulate content. Gases containing compounds with chlorine, sulfur, and other atoms that may deactivate the supported noble metal catalysts often used for VOC control were not suitably controlled by catalytic oxidation systems.

Flares are also used for burning waste gases from sewage digesters process, coal gasification, rocket engine testing, nuclear power plants with sodium, water heat exchangers, heavy water plants, and ammonia fertilizer plants. The flare provides a means of safe disposal of the vapor streams from its facilities, by burning them under controlled conditions such that adjacent equipment or personnel are not exposed to hazards, and at the same time meeting the environmental regulation of pollution control and public relations requirements. The chemical process used for flaring is a high temperature oxidation reaction to burn combustible components, mostly hydrocarbons, or waste gases from industrial operations. In combustion, the gaseous hydrocarbon (natural gas, propane, ethylene, propylene, butadiene, butane, etc.) reacts with atmospheric oxygen to form carbon dioxide (CO_2) and water with an efficiency over 98%. Several byproducts formed will be carbon monoxide, hydrogen, and others dependent upon what is being burned.

14.2 Principles

14.2.1 Combustion Stoichiometry

Combustion is a chemical reaction between substances, and oxygen is significant for high temperature combustion. To assure all materials are converted completely into harmless compounds before emitting to the atmosphere, a sufficient supply of oxygen must be present, so called stoichiometric or theoretical oxygen demand.

Combustion stoichiometry relates to the fuel constitution with carbon dioxide and water are mostly reaction products. In general, the combustible material are carbon and hydrogen and the supporting gas as air is approximately 20.9% O_2 and 79.1% N_2 by volume, which are expressed as below

$$C_nH_m + (n + m/4)(O_2 + 3.78N_2) \rightarrow nCO_2 + m/2H_2O + 3.78(n + m/4)N_2 \quad (14.1)$$

$$x_{CO_2} = \frac{n}{4.78(n + m/4) + m/4}$$

$$x_{H_2O} = \frac{m/2}{4.78(n + m/4) + m/4}$$

$$x_{N_2} = \frac{3.78(n + m/4)}{4.78(n + m/4) + m/4}$$

In some cases, the nitrogen exists in compounds being combusted normally enters the atmosphere partly as N_2, NO, or NO_2. Combustion technology is not a good method to treat these materials. However, some organic chemicals contain small amounts of organic nitrogen that will pass through the combustion process to form NO or NO_2, etc.

$$3(CH_3)_3N + 11\,^1\!/_2\, O_2 \rightarrow 6\, CO_2 + 9\, H_2O + 2\, NO \quad (14.2)$$

Focusing on equation written above, the stoichiometric amount of oxygen is identified by balancing chemical equation. The air-fuel equivalence ratio, λ (lambda) is an important parameter to classify if the mixture is rich or lean. Rich mixture is referred to as there is not enough oxygen to combine with all the fuel and waste, so that incomplete combustion presents. If more than stoichiometric amount of oxygen is applied, the mixture is referred to as "lean". The extra of oxygen then does not contribute to the oxidation reaction and passes through the incinerator. For example, if there is a compound consist of C, H, N, Cl, O, S, the chemical equation will be:

$$C_mH_nN_jCl_kO_oS_s + \left(m + \frac{n}{4} + j + s - \frac{o}{2} - \frac{k}{4}\right)O_2 \rightarrow m\, CO_2 + 0.5(n - k)H_2O + j\, NO_2 + k\, HCl + s\, SO_2 \quad (14.3)$$

14.2.2 Material, Heat, and Energy Balance

Since combustion is a very complicated thermal process it has a variety of physical and chemical reactions, and material, heat, and energy balance utilized for sizing and costing of the specific incinerator, as well as auxiliary fuel requirements. These balances are based on the law of conservation of mass and energy, which can be expressed as a general equation as follows:

$$\text{In} - \text{Out} + \text{Generation} = \text{Accumulation} \tag{14.4}$$

Because the incineration process is a steady-state process, the accumulation term is zero and the equation becomes

$$\text{In} - \text{Out} + \text{Generation} = 0 \tag{14.5}$$

For mass balances, it is useful to restrict the balances to be made on the mass of each atomic species so that for mass balances the generation term becomes zero. However, because the combustion reaction shifts mass and energy balances around equipments where combustion takes place would include a generation term.

$$\begin{aligned} \text{In} &= \text{Out (steady state for mass and heat balance);} \\ \text{In} + \text{Generation} &= \text{Out (steady state for energy balance).} \end{aligned}$$

For the mass balance,

$$\dot{m} = \rho Q \tag{14.6}$$

where

$\dot{m}$ mass flow rate of gases;
ρ density (mass per unit volume);
Q volumetric flow rate (volume per unit time).

Since the flame temperature is determined by a heat balance including the energy produced by combustion[15], the heat required has to be calculated to raise process to a specific temperature by using the law of conservation of energy in steady state [6]. The change of enthalpy is computed by following equation:

$$\Delta H = C_p\left(T - T_{ref}\right) \tag{14.7}$$

where

ΔH enthalpy
C_p heat capacity over temperature range from T to T_{ref}

T_{ref} reference temperature (usually take to be zero, the temperature of a convenient stream, or the inlet gas stream).

For energy balance, the reference temperature tends to be neglected and the sensible heat become

$$\dot{Q} = \dot{m}\Delta H = rQC_pT \tag{14.8}$$

where

r density of gas

In the incineration process, the generation term of energy balance accounts for the energy released through the combustion reactions as shown below

$$\dot{Q} = rQ(\Delta h_c) \tag{14.9}$$

where

Δh_c heat of combustion.

14.2.3 3T's Principle

The process of combustion could be considered as the most important method for air pollution control device. Through combustion, the energy contained within the fossil fuels is released and then used to generate steam or electricity as the process demands heat from fuel and make it available for further use. It is essential that the combustion is complete so that there are no toxic and hazardous emissions and amount of heat and energy recovery for further use. The combustion process is complex in nature and multiple parameters affect the combustion efficiency. Among them, time, turbulence, and temperature are referred to as 3T's of combustion. It is very important to control and optimize these 3T's to achieve the best performance of combustion.

1. Time

When a fuel is being burned, it is important that sufficient time is available so that the fuel burns completely. 100% combustion means that the fuel is fully oxidized and full oxidation of the carbon, hydrogen, and other combustible elements has taken place. If fuel remains in the combustion zone for a time lesser than necessary, it will be partially burned which increases the unburned losses. On the other hand, if it remains for a time higher than the required, the power output of the boiler will drop as new fuel will not be able to come in and get burned. Ideally, the fuel should reside sufficiently for the complete combustion and then replaced by the fresh fuel. Thus, the time plays a very important role in determining the combustion efficiency. On the other hand, in mathematical sense, 3T's are also referred to three characteristic times for modeling concerns as follows [6]:

$$\tau_C = \frac{1}{k}, \text{(chemical time)} \tag{14.10}$$

$$\tau_r = \frac{V}{Q} = \frac{L}{u} \text{(residence time)} \tag{14.11}$$

$$\tau_m = \frac{L^2}{D_e} \text{(mixing time)} \tag{14.12}$$

where

V Volume of the reaction zone, m^3;
Q Volumetric flow rate, m^3/s;
L Length of the reaction zone, m;
u Gas velocity, m/s;
D_e Effective (turbulent) diffusion coefficient, m^2/s.

2. Turbulence

Oxygen makes an essential part of the process of combustion. While burning the fuel, it is essential that it is broken down in small particles thereby increasing the surface area of the fuel and ensuring that sufficient air, i.e., oxygen is made available. Turbulence ensures a well mixing of the air and fuel. If turbulence is not maintained, certain part of the fuel will have excess oxygen for the combustion while the remaining having not sufficient oxygen resulted in incomplete combustion of carbon forming carbon monoxide instead of carbon dioxide. If proper turbulence is not maintained, part of the fuel and incomplete combustion by products will emit from the chimney which will increase the unburned losses.

3. Temperature

During the combustion, if the temperature is not sufficiently high, the fuel will take some time to ignite thus increasing the time of the combustion. Hence, it is very important to maintain the appropriate temperature which ensures that the fuel is quickly burned releasing the complete energy. The temperature greatly influences the reaction rate for any system by Arrhenius equation as shown below

$$\mathrm{k} = \mathrm{A}e^{\frac{-E_a}{RT}} = \frac{Z' S y_{O_2} P}{R'} e^{\frac{-E_a}{RT}} \tag{14.13}$$

where

A pre-exponential factor;
E_a activation energy, $E_a = -0.00966(\mathrm{MW}) + 46.1$;
Z' collision rate factor (from Fig. 11.5 for alkanes, alkenes, and aromatics);
S steric factor = 16/MW;
y_{O2} mole fraction of O_2 in the afterburner;

P absolute pressure, atm;
R' gas constant, 0.08206 L-atm/mol-K.

14.3 Design Methods

14.3.1 Combustion Kinetics

Combustion kinetics provided the rationale to design the combustion chamber. Despite related information are scarce, complex, and costly obtained by previous studies, there are some predicting VOCs kinetic models developed in the past. Ross [7] summarized the older methods by suggesting T_{design} be set several hundred degrees (F) above the VOCs auto-ignition temperature, at which combustible material of VOCs in air will ignite without any external source as shown in Table 14.1.

Another predicting models is Lee and coworker's approach, [8] a purely statistical model to predict the T_s required to give various levels of destruction in an isothermal plug flow afterburner based on their experimental studies. In order to reach 99.9% destruction efficiency, temperature can be predicted through following equation:

$$T_{99.9} = 594 - 12.2W_1 + 117W_2 + 71.6W_3 + 80.2W_4 + 0.592W_5 - 20.2W_6 - 420.3W_7 + 87.1W_8 - 66.8W_9 + 62.8W_{10} - 75.3W_{11} \quad (14.14)$$

where

$T_{99.9}$ temperature where destruction efficiency is 99.9%, °F;
W_1 number of carbon atoms;
W_2 aromatic carbon flag (0 = No, 1 = Yes);
W_3 C-C double bond flag (0 = No, 1 = Yes);
W_4 number of N atoms;
W_5 auto-ignition Temperature, F;
W_6 number of O atoms;

Table 14.1 List of auto-ignition temperature of volatile organic compounds

Substance	Auto-ignition temperature (°F)
Acetone	1000
Ethanol	799
Cyclohexane	514
Phenol	1319
Acrolein	453
Hydrogen sulfide	500
Styrene	915
Toluene	1026

W_7 number of S atoms;
W_8 hydrogen/carbon ratio;
W_9 allyl (2-propenyl) compound flag (0 = no,1 = yes);
W_{10} carbon-double-bond-chlorine interaction (0 = no,1 = yes);
W_{11} natural log of residence time.

In general, Cooper, Alley, and Overcamp's Approach [9] is widely used to describe combustion kinetics of VOCs. An integration of collision theory with empirical data and proposed a method to predict an effective first order rate constant k for hydrocarbon incineration over the range from 940 to 1,140 °K. Certain method depends on molar weight and the type of the hydrocarbon. T_{design} can be described by Arrhenius equation mentioned in Eq 14.13. The concentration of VOC to be burned is much less than the concentration of oxygen in the contaminated air stream can be described by the true kinetic expression:

$$\text{Decrease in conc. of VOC per time} = \frac{-d_{c_{VOC}}}{dt} = r = kc_{VOC}c_{O2} \qquad (14.15)$$

$$\text{Destruction efficiency} = 1 - \frac{HC_{out}}{HC_{in}} = 1 - e^{-k\tau_r} \qquad (14.16)$$

14.3.2 Good Engineering Practice

Combustion technology is widely applied to the industrial process to control the mission of gaseous or small particulate air pollutions which are either combustible or thermally decomposed at high temperature. As mentioned earlier, combustion technology is generally classified into three types including thermal incineration, catalytic incineration, and flares. An appropriate equipment selection is based on types of pollutants, preferred destruction efficiency and adapted exhausting requirement. Not only the concentration of combustible in the waste stream, but also inlet and outlet temperature, mixing behavior (turbulence), rate of heat loss, and combustion chamber geometry and its material construction may affect the combustion efficiency of equipment [10]. The design criterial for these three technologies is shown in Table 14.2.

The mixture continues to react as it flows through the combustion chamber. The required level of VOC control of the waste gas that must be achieved within the time that it spends in the thermal combustion chamber dictates the reactor temperature. Once the unit is designed and built, the residence time is not easily changed, so that the required reaction temperature becomes a function of the particular gaseous species and the desired level of control. Simple design with a flexible and reliable adaptation is one of the advantages of thermal incinerator. Beside of that, it contributes to a complete removal of organic pollutants without waste water, and solid waste which might cause adverse impacts to human health and ecological system [5].

Table 14.2 Design criterial of three combustion methods [10]

	Thermal incineration	Catalytic incineration	Flares (direct flame combustion)
Specific	The most flexible technique	Catalysts are easily poisoned	The most effective technique
Air volume	Sufficient	Sufficient	Low
Flame	Yes	No	Yes
Pollutant conc.	<25%LEL	<25%LEL	>UEL(50 BTU/SCF)
Supplemental fuel	Need	Need	Need
Catalyst	No	Yes	No
Reaction rate	Lower	Higher	Lower
Temperature	1200–1600 °F	500–1200 °F	2000–2500 °F(>1200 °F)
Minimum residence time	0.3 s (0.3–1.0)	0.03 s	0.3 s
Gas velocity (ft/sec)	30(single or double nozzle tunnel burners) 60–120(line-type)	≦8	≦15
Heat produced per 1ft^3 chamber volume	15000–20000 BTU/h		15000 BTU/h
Energy requirement	950 BTU/SCF(nature gas)	10 BTU/SCF(simple) 4 BTU/SCF(preheat)	50 BTU/SCF

In contrast, it still has disadvantages associated with operating cost for injecting high supplementary fuel. And, generating heat loss might reduce the utilization of supplementary energy.

1. Thermal incinerators

The reasons of poor performance for thermal incinerator are nonuniform temperatures and flows. There should be no dead spaces within the combustion chamber. It would be more economical to increase residence time and reduce operating temperature. In order to achieve efficient mixing, the residence time (t) should be more than 1 s.

(1) Determine the residence

$$t = \frac{V_c}{\dot{V}_g} \geq 1\text{s} \tag{14.17}$$

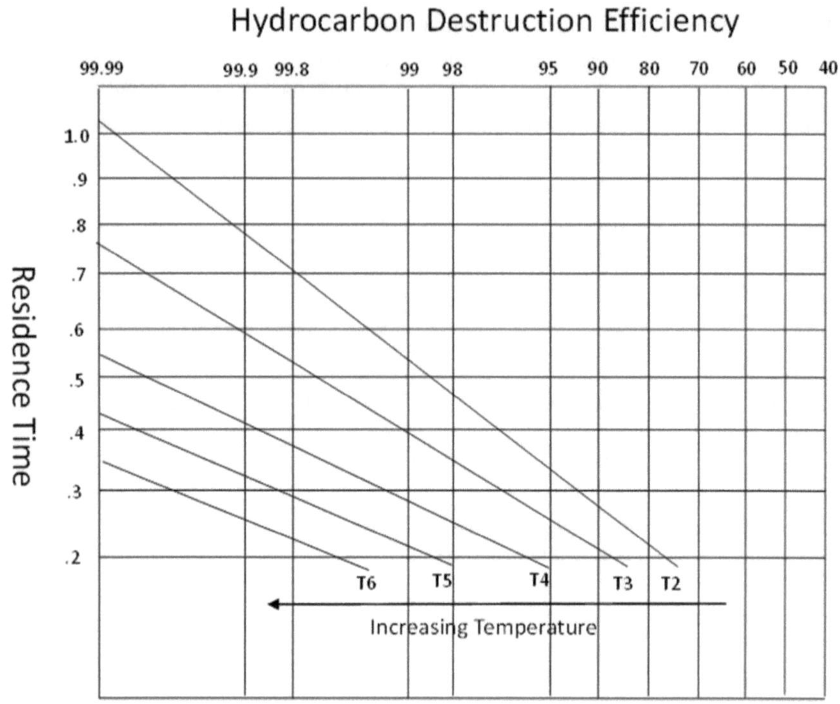

Fig. 14.7 Destruction efficiency

where

V_c the volume of the combustion chamber;
V_g the volumetric gas flow rate (based on cold entrance conditions)

Determine the appropriate residence time to achieve the anticipated pollution destruction as shown in Fig. 14.7.

(2) Determine the combustion chamber volume

$$V = \left(\frac{acfm}{60}\right)t \tag{14.18}$$

where

V = combustion chamber volume (following burner and prior to heat exchangers, dilution, etc.), ft^3;

acfm = exhaust gas flow rate to incinerator (at average incinerator operating temperature), actual ft^3/min;

t = total average residence time, sec

(3) Determine the fuel required to sustain combustion

The heat load (Btu/min) required to raise the temperature of the waste gas stream to the desired combustion temperature is

$$Q = mC_p\Delta T \tag{14.19}$$

where

Q = heat load, Btu/min;
m = mass flow rate of waste gas stream into incinerator, lb/min;
ΔT = temperature rise required to bring waste gas stream from temperature upstream of incinerator to required combustion temperature, °K;
Cp = average heat capacity of composite air and waste gas stream, Btu/lb- °K

2. Catalytic incinerators:

The overall rate of catalytic oxidation is most often limited by mass transfer rather than by reaction kinetics [11].

$$\frac{[\mathrm{VOC}]_L}{[VOC]_0} = e^{-L/L_m} \tag{14.20}$$

where

$[\mathrm{VOC}]_{\mathrm{L}}$ = conc. of the VOC at length L;
$[\mathrm{VOC}]_0$ = inlet conc. of the VOC;
L = length of the catalyst bed;
Lm = length of one mass transfer unit.

For laminar flow:

$$L_m = \frac{ud^2}{17.6D} \tag{14.21}$$

where

u = linear velocity in the channel, m/s;
d = effective diameter of the channel, m;
D = diffusivity, m^2/s.

For turbulent flow:

$$L_m = \frac{2}{fa}(Sc)^{\frac{2}{3}} \tag{14.22}$$

where

f = fanning friction factor;
Sc = Schmidt number [μ/ρD];
a = surface area per unit volume of bed, m^{-1}

Wark and Warner [12] have suggested a range of 0.2–0.5 ft^2 catalyst surface area per scfm of waste gas.

Volume of catalyst required:

$$V = \frac{exhaust\ flowrate\ (scfm)}{space\ velocity\left(h^{-1}\right)} \tag{14.23}$$

Snape [13] recommends 50000–100000 h^{-1} for honeycomb supports and 30000 h^{-1} for pelletized supports.

Hawthorn [14] reports space velocities of 500–2000 min^{-1} are typically used to achieve 85–95% destruction of VOCs.

3. Flares

The parameters including mass flow rate, discharge conditions, the type of flame holder used by the manufacturer, and density and heat-capacity ratios should be available to determine flare burner diameter. The pressure drop (limited to a pressure drop of 3 psig, 60 in. water) is the controlling factor in burner design.

The burner diameter (orifice equation) can be determined by the maximum discharge rate as an equivalent volume of air in scfh.

$$V_e = 1656 K A \sqrt{\Delta P} \tag{14.24}$$

where

V_e = volume equivalent flow rate, scfh;
K = orifice factor, (≒0.9);
A = area of flare burner tip, ft^2;
ΔP = allowable pressure drop at tip, in. water

Assume the gas behaves as a perfect gas, the maximum discharge velocity can be determined as

$$V_a = \frac{g_c K R T}{M} \tag{14.25}$$

where

V_a = acoustic velocity, or velocity of sound in the fluid, ft/s;
$g_c = 32.17$ lb-ft/lb$_f$-h-s^2;
K = Cp/Cv;
R = 1546 ft-lb$_f$/R-lb-mole;
M = average molecular weight

14.4 Design Procedure

See Fig. (14.8).

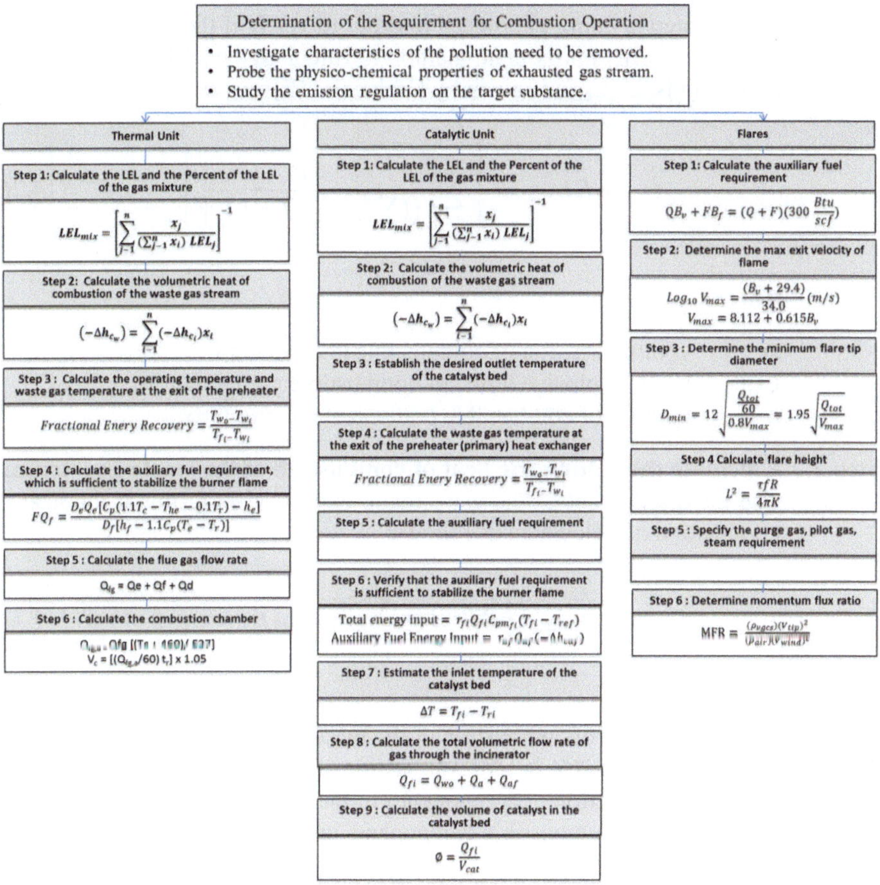

Fig. 14.8 General design flowchart of combustion device

14.4.1 Thermal Unit

To determine the specifications of the incinerator and the waste gas to be processed, the following parameters of the waste gas stream at the emission source, i.e., volumetric flow rate (scfm), temperature, oxygen content, chemical composition of the combustible, inert content, heating value, particulate content, desired control efficiency, and combustion chamber outlet temperature must be available. Confirmation of oxygen content of the waste gas has to exceed 20% for incineration design. There must be sufficient oxygen in the waste gas to support the combustion of the waste organics (including VOCs) and the auxiliary fuel, if auxiliary fuel is needed. It may be necessary to add auxiliary air if the oxygen content is less than about 20% [2].

Step 1 Calculate the LEL and the Percent of the LEL of the gas mixture

Certain step is going to proceed in order to verify if the dilution of the air is required. An approximate method to calculate the LEL of a mixture of compounds, LEL_{mix}, is given by Grelecki: [15]

$$LEL_{mix} = \left[\sum_{j-1}^{x} \frac{x_j}{\left(\left(\sum_{i-1}^{n} x_i \right) (LEL)_j \right]} \right]^{-1} \tag{14.26}$$

where

x_i volume fraction of combustible component i;
LEL_j lower explosive limits of combustible component j;
n number of combustible components in mixture

Step 2 Calculate the volumetric heat of combustion of the waste gas stream

The energy content of the gas stream, expressed in terms of the heat of combustion, is calculated as follows:

$$\left(-\Delta h_{c_w}\right) = \sum_{i-1}^{n} \left(-\Delta h_{c_i}\right) x_i \tag{14.27}$$

where
$(\&\Delta h_{cw})$ = heat of combustion of the waste stream (Btu/scf);
$(\&\Delta h_{ci})$ = volumetric heat of combustion of component *i* at 25 °C (Btu/scf);
x_i = volume fraction of component *i* in the waste gas;
n = number of combustible components in the waste gas

Step 3 Calculate the operating temperature and waste gas temperature at the exit of the preheater

Combustion temperature and residence time is achieved with a given destruction efficiency (DE). Studies by Mascone show that 98% DE can be met in a thermal incinerator operated at a temperature, T_{fi}, of 1600 °F and a residence time of 0.75 s.

As the VOC stream temperature leaving the heat exchanger, T_{wo}, increases, the auxiliary fuel requirement decreases, but at the expense of a larger heat exchanger. There is some limit on T_{wo}, such as T_{wo} must not be close to the ignition temperature of the organic-containing gas to prevent damaging temperature excursions inside the heat exchanger should the gas ignite. In the other hand, for gases containing halogens, sulfur, and phosphorous (or other acid-forming atoms), the flue gas temperature after the heat exchanger, T_{fo}, must not drop below the acid dew point. If there has heat exchanger following incinerator, temperature T_{wo} can be designed based on preferred fractional energy recovery:

$$Fraction\ Energy\ Recovery = \frac{T_{w_o} - T_{w_i}}{T_{f_i} - T_{w_i}} \tag{14.28}$$

where

T_{wi} waste gas inlet temperature;

T_{fi} incinerator operating temperature

Step 4 Calculate the auxiliary fuel requirement, which is sufficient to stabilize the burner flame

Auxiliary fuel will almost invariably be needed for startup of the unit.

$$Q_f = \frac{D_e Q_e\left[C_p(1.1T_c - T_{he} - 0.1T_r) - h_e\right]}{D_f\left[h_f - 1.1C_p(T_e - T_r)\right]} \tag{14.29}$$

where

Q_f natural gas flow rate, scfm;

Q_e emission stream flow rate, scfm;

D_e density of flue gas stream, lb/scf;

D_f density of fuel gas, lb/scf;

C_p mean heat capacity of air between T_c and T_r, Btu/lb-°F;

T_{he} : emission stream temperature after heat recovery, °F;

T_r reference temperature, °F;

h_e heat content of flue gas, Btu/lb;

h_f lower heating value of natural gas, Btu/lb

The factor 1.1 represents for an additional heat loss of 10% in the incinerator A small amount of auxiliary fuel (<5% of the total energy input) is needed to stabilize the burner flame. This check is made by calculating 5% of the total energy input to the incinerator and comparing it with auxiliary fuel energy input. The total energy input is calculated by

$$\text{Total energy input} = r_{fi} Q_{fi} C_{pm_{fi}}\left(T_{fi} - T_{ref}\right) \tag{14.30}$$

$$\text{Auxiliary Fuel Energy Input} = r_{af}\,Q_{af}\left(-\Delta h_{caf}\right) \tag{14.31}$$

The auxiliary fuel used in the design, Q_{af} should be the larger of 5% of the total energy input (28,900 Btu/min), and the auxiliary fuel energy input (146,500 Btu/min). The auxiliary fuel used easily meets this criterion.

Step 5 Calculate the flue gas flow rate

Flue gas is generated as a result of the combustion of supplementary fuel and the emission stream. Use the following equation to calculate flue gas flow rate:

$$Q_{fg} = Q_e + Q_f + Q_d \tag{14.32}$$

where

Q_{fg} flue gas flow rate, scfm;
Q_e emission stream flow rate, scfm;
Q_f natural gas flow rate, scfm;
Q_d dilution air requirement, scfm

Step 6 Calculate the combustion chamber

To obtain an indication of the residence time, the combustion chamber volume (V_c) should be calculated.

$$Q_{fg,a} = Q_{fg}[(T_c + 460)/537] \tag{14.33}$$

where

$Q_{fg,a}$ actual flue gas flow rate, scfm;
Q_{fg} flue gas flow rate, scfm

The combustion chamber volume, V_c, is determined by the following equation.

$$V_c = \left[\left(Q_{fg,a}/60\right)\,t_r\right] \text{ x } 1.05 \tag{14.34}$$

Note that 1.05 is used to account for minor fluctuations in the flow rate, and follows industry practice.

14.4.2 Catalytic Unit

Step 1–2 Similar to the thermal unit (Sect. 14.4.1)

Step 3 Establish the desired outlet temperature of the catalyst bed

The energy released by the oxidation of the VOCs in the catalyst bed will raise the temperature of the gases by an amount, ΔT, as the gases pass through the catalyst bed. Guidelines given by Combustion

Engineering indicate that values from 300 to 900 °F result in destruction efficiency between 90 and 95%. To prevent deactivation of the catalyst a maximum bed temperature of 1200 °F should not be exceeded.

Step 4 Calculate the waste gas temperature at the exit of the preheater (primary) heat exchanger

The waste gas temperature at the exit of the primary heat exchanger is estimated in the same manner as for the thermal incinerator.

$$Fraction\ Energy\ Recovery = \frac{T_{w_o} - T_{w_i}}{T_{f_i} - T_{w_i}} \tag{14.35}$$

where

T_{wi} waste gas inlet temperature;

T_{fi} incinerator operating temperature

Step 5 Calculate the auxiliary fuel requirement

The auxiliary fuel requirement must be positive. A negative fuel requirement indicates that the heat of combustion of the waste gas is too high for the fractional energy recovery in the primary heat exchanger that was selected.

Step 6 Verify that the auxiliary fuel requirement is sufficient to stabilize the burner flame

A small amount of auxiliary fuel (<5% of the total energy input) is needed to stabilize the burner flame. This check is made by calculating 5% of the total energy input to the incinerator and comparing it with auxiliary fuel energy input. The total energy input is calculated by

$$\text{Total energy input} = r_{fi} Q_{fi} C_{pm_{fi}} \left(T_{fi} - T_{ref}\right) \tag{14.36}$$

$$\text{Auxiliary Fuel Energy Input} = r_{af} Q_{af} \left(-\Delta h_{caf}\right) \tag{14.37}$$

Step 7 Estimate the inlet temperature of the catalyst bed

The inlet temperature to the catalyst bed must be calculated to ensure that the inlet temperature is above that necessary to ignite the combustible organic compounds in the catalyst that was selected for use.

The inlet temperature to the catalyst bed, T_{ri}, should be such that, when the temperature rise through the catalyst bed, T, is added to it, the resulting temperature is T_{fi}, 900 °F.

$$\Delta T = T_{fi} - T_{ri} \tag{14.38}$$

Step 8 Calculate the total volumetric flow rate of gas through the incinerator

Q_{fi} The total volumetric flow rate of gas leaving the incinerator is referred to as the flue gas flow rate, Q_{fi}, and is the gas rate on which the incinerator sizing and cost correlations are based.

$$Q_{fi} = Q_{wo} + Q_a + Q_{af} \tag{14.39}$$

Step 9 Calculate the volume of catalyst in the catalyst bed

For the purposes of cost estimation, the space velocity, in reciprocal time units, necessary to achieve the required level of destruction can be used to approximate the catalyst volume requirement. The space velocity is defined as

$$\emptyset = \frac{Q_{fi}}{V_{cat}} \tag{14.40}$$

V_{cat}: overall bulk volume of the catalyst bed, including interparticle voids (ft^3)

14.4.3 Flares

Step 1 Calculate the auxiliary fuel requirement

Some flares are provided with auxiliary fuel to combust hydrocarbon vapors when a lean flare gas stream falls below the flammability range or heating value necessary to sustain a stable flame. The amount of fuel required, F, is calculated based on maintaining the vent gas stream net heating value as the minimum of 300 Btu/scf is required by rules defined in Federal Register.

$$QB_v + FB_f = (Q + F)\left(300\frac{Btu}{scf}\right) \tag{14.41}$$

where

Q the vent stream flow rate, scfm;
B_v Btu/scf of the vent stream, Btu;
B_f Btu/scf of the fuel, Btu

Step 2 Determine the max exit velocity of flame

At too high an exit velocity, the flame can lift off the tip and flameout, while at too low a velocity, it can burn back into the tip or down the sides of the stack. The requirements are for steam-assisted, air-assisted, and non-assisted flares.

In steam-assisted flares and non-assisted flares, an exit velocity at the flare tip of less than 60 ft/s for 300 Btu/scf gas streams and less than 400 ft/s for >1000 Btu/scf gas streams. For gas streams between 300 and 1000 Btu/scf, the maximum permitted velocity is determined by the following equation [16].

$$Log_{10} V_{max} = \frac{(B_v + 29.4)}{34.0} (m/s) \tag{14.42}$$

For air-assisted flares, follows the equation below

$$V_{max} = 8.112 + 0.615 B_v \tag{14.43}$$

Step 3 **Determine the minimum flare tip diameter**
y determining the maximum allowed velocity, V_{max} (ft/s), and the knowing the total volumetric flow rate, Q_{tot} (acfm), including vent stream and auxiliary fuel gas, a minimum flare tip diameter, D_{min} (in), can be calculated. It is standard practice to size the flare so that the design velocity of flow rate (Q_{tot}) is 80% of V_{max}, i.e.,

$$D_{min} = 12\sqrt{\frac{\frac{Q_{tot}}{60}}{0.8 V_{max}}} = 1.95\sqrt{\frac{Q_{tot}}{V_{max}}} \tag{14.44}$$

Step 4 **Calculate flare height**
The height of a flare is determined based on the ground level limitations of thermal radiation intensity, luminosity, noise, height of surrounding structures, and the dispersion of the exhaust gases. Flare height may also be determined by the need to safely disperse the vent gas in case of flameout. The minimum flare height normally used is 30 feet. Equation 14.45 by Hajek and Ludwig may be used to determine the minimum distance, L, required from the center of the flare flame and a point of exposure where thermal radiation must be limited [17].

$$L^2 = \frac{\tau f R}{4\pi K} \tag{14.45}$$

where
τ fraction of heat intensity transmitter;
f fraction of hear radiated;
R net heat release;
K allowable radiation

Step 5 **Specify the purge gas, pilot gas, steam requirement**
Purge gas, typically natural gas, N_2 or CO_2, is used to maintain a minimum required positive flow through the system, to ensure a positive flow through upstream point in the flare transport piping.
The number of pilot burners required depends on flare size, gas composition, and wind conditions. Pilot gas usage is a function of the number of pilot burners required to ensure positive ignition of the flared gas, of the design of the pilots, and of the mode of operation.

The steam requirement depends on the composition of the vent gas being flared, the steam velocity from the injection nozzle, and the flare tip diameter. Although some gases can be flared smokeless without any steam, typically 0.01–0.6 pound of steam per pound of flare gas is required. The ratio is usually estimated from the molecular weight of the gas, the carbon-to-hydrogen ratio of the gas, or whether the gas is saturated or unsaturated.

Step 6 Determine momentum flux ratio

The data available indicate that the wake-dominated region begins at a momentum flux ratio (MFR) of 3 or greater. The MFR considers whether there is enough flare vent gas and center steam (if applicable) exit velocity (momentum) to offset crosswind velocity.

$$\mathrm{MFR} = \frac{\left(\rho_{vgcs}\right)\left(V_{tip}\right)^2}{\left(\rho_{air}\right)\left(V_{wind}\right)^2} \tag{14.46}$$

14.4.4 A Rapid Method for Sizing a Thermal Unit

As shown in Fig. 14.9, it suggests that the appropriate residence time and incineration temperature should be maintained for achieving a specific destruction ratio to meet the emission standard. For instance, a 1300 °K temperature with 1.0 s of residence

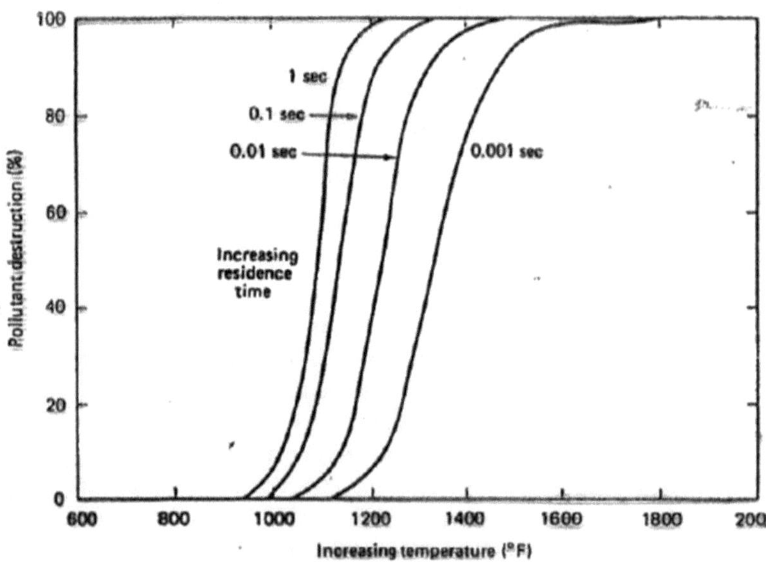

Fig. 14.9 Coupled effects of temperature and time on rate of pollutant oxidation [1]

Table 14.3 Design criterial for acceptable performance in various types of pollutant

Types of pollutant	Residence time (s)	Temperature (°F)
Hydrocarbons + CO (90% + destruction of HC + CO, as in LAPCD rule 66)	0.3–0.5	1250–1500
Odor		
(90–99% destruction)	0.3–0.5	1100–1300
(99% + destruction)	0.3–0.5	1200–1500
Smokes and plumes		
White smoke (liquid mist)	0.3–0.5	800–1000
Plume abatement (90% + destruction of HC + CO)	0.3–0.5	1250–1500

time is required to achieve a 99% destruction of VOC which was fairly consistent with the design criterial good engineering practice (GEP) as shown in Table 14.3 [1].

Air emissions with a flow rate of 5,000 cfm at 90 °F needs to raise the temperature to 1,300 °F achieving 99% destination. The natural gas requirements and the thermal incinerator size are determined as follows:

1. Determine the required heat to raise waste gas to 1,300 °F

   (1) Density of the exhaust air
   (29)*(492)/(460 + 90)/360 = 0.072 lb/ft^3
   (2) Mass flow rate of the air
   5,000cfm * 60 min/hr * 0.072 lb/ft^3 = 21,600 lb/hr
   (3) The heat rate to achieve 1,300 °F is
   21,600 * 0.26 * (1,300-90) = 7,000,000 Btu/hr
   (4) With a 10% heat loss the required rate is
   1.1 * 7,000,000 = 7,700,000 Btu/hr
   (5) Fuel requirement – 7,700,000/950 = 7,900 ft^3/hr

2. Determine the total gas flow rate at 1,300 °F

   (1) The volume rate of air at 1,300 °F
   5,000 *(460 + 1,300)/(460 + 90)/60 = 166 ft^3/s
   (2) The combustion products rate at 1,300 °F
   7,900 * 11.5 * (460 + 1,300)/(460 + 90)/3,600 = 86ft^3/s
   (3) The total gas flow rate at 1,300 °F is
   266 + 86 = 352 ft^3/s

3. Determine the size of the combustion chamber
   For single or double nozzle burner chamber:
   The combustion chamber's diameter for a velocity of 12 ft/sec
   For L/D = 2, → L = 12.8 ft
   Residence time 12.8/12 = 1.0 s > 1.0 s

14.5 Performance Evaluation and Enhancement

14.5.1 Performance Evaluation

Since the actual combustion process is much more complicated than the theoretical estimation, both the residence time and incineration temperature have to be high enough to ignite waste/fuel mixture, and an amount of air (oxygen), and waste/fuel mixture also have to be well-mixing in turbulent flow in the reaction chamber. The combination of three mentioned independent parameters are related to "3T's Principle", i.e., temperature, time, and turbulence.

Unfortunately, when these conditions are not proper to proceed a complete combustion process, there will lead to many of intermediate products, for example, chlorinated organics (Dioxins), polycyclic aromatic hydrocarbons (PAHs), hydrochloric acid, hydrofluoric acid, phosgene, and the formation of soot. Therefore, all combustion devices have to ensure that combustion is as complete as practical and emissions of products of incomplete combustion are as small as possible. Selecting a proper combustion chamber design depends on many factors, including the characteristic of hazardous compounds and combustibles in the waste gas stream, process flow rate, control requirement, and heat recovery. Many industries have to work on the cost-effectiveness analysis for the combustion chamber, a higher efficiency with lower fuel and energy consumption system would be expected to be built-in. In addition, the concentration of combustible gases should keep at some percentage below the lower flammability limit to ensure safe operation.

Pollution System Company located in Texas USA [18] provides many onsite service for air pollution systems such as system evaluation/relocation, and equipment maintenance/repairs inspections. For instance, the VOC Thermal Oxidizer Model CEF-5 is a trailer mounted, natural gas or propane thermal oxidizer designed to treat hydrocarbon contaminated air streams. For instance, low VOC contaminated streams 50% LEL. The oxidizer is designed to operate completely independent of the process and is equipped with an enable/disable signal interface in the system control panel. The unit and controls can be converted to operate with a catalyst module insert. The technical specifications regarding the thermal oxidizer are shown in Table 14.4.

The Anguil Company's regenerative catalytic oxidizer (RCO) [19] can destroy hazardous air pollutants (HAPs), volatile organic compounds (VOCs), and odorous emissions from industrial processes. The RCO achieves emission destruction converting the pollutants to carbon dioxide and water vapor while reusing the thermal energy generated to reduce operating costs through the process of thermal and catalytic oxidation. VOC and HAP laden process gas enters the oxidizer through an inlet manifold to flow control, poppet valves that direct this gas into energy recovery chambers where it is preheated. The process gas and contaminants are progressively heated in the ceramic media beds as they move toward the combustion chamber. In general, the design philosophy of RCO is to develop energy efficient and innovative products to meet the increased environmental demands enforced on by regulatory

Table 14.4 Technical specifications regarding the thermal oxidizer

Maximum air flow capacity	3050 scfm
Minimum air flow	800 scfm
Maximum inlet temperature from source	300°F
Burner capacity	4,800,000 BTU/h
Inlet connection	11" Dia.
Stack height	18'7" Above Ground
Stack diameter	30"
Maximum thermal operating temperature	1800°F
Optional catalyst module	Precious Metal Monolith
Catalyst operating temperature	650°F
Maximum exit catalyst temperature	1200°F
Maximum hydrocarbon throughput	
High VOC mode	~MMBTUH or 750 scfm
Thermal mode	<50% LEL
Catalytic mode	<25% LEL
Oxidizer process blowers	7.5 HP
Time to reach operating temperature	15 Min from cold start
Power requirements	460/3/60, 26 FLA
Trailer	16,000 lb Dual Axle

agencies, while also enhancing productivity and profitability. The specifications of Regenerative Catalytic Oxidizer are indicated below (Table 14.5).

According to technical report of NESTEC Inc., [20] a whole regenerative system is very complicated, not only RTO system, but also auxiliary components such as manifold, exhaust fan, control room, system control design, etc. As mentioned before, combustion technology has some important parts: burner, combustion chamber, stack, etc. In certain case, the burner's model is designed as RatioMatic burner size as 5 mm Btu/hr, NOx and CO output as less than 0.1 lb/mmBtu, 60 ppmv (max), respectively. The exhaust stack has 42" diameter, and 62' height, made by mild steel material.

Table 14.5 The specifications of regenerative catalytic oxidizer

Temperature	600–800 °F
Material	Multiple catalyst and ceramic media
Air flow rate	Up to 30,000 scfm, and single unit system up to 75,000 scfm
Maximum hydrocarbon throughput	2–3% LEL
System	• Programmable logic controller (PLC)-based controls with digital data recorder remote telemetry. • Variable frequency drive (VFD) allows for high volumetric turn-down during idle or low process conditions • Gas trains are designed to meet FM Global or any international certifications such as CSA, TSSA, EN, CGA, and ATEX • Low NOX burner designed to maintain temperature during full flow, no VOC process conditions

14.5.2 Performance Improvement

There are various types of combustion technology used to treat VOCs and gaseous pollutants. Combustion technology design is very complicated, not only the combustion chamber, but also auxiliary components such as heat exchanger, exhaust stack, control system and hydraulic systems, etc., as shown in Fig. 14.10. Each component has specific size, constructed material, function to serve specific air pollutant flow rate and type of pollutant.

Most of air pollution control design material mentioned about thermal, catalytic oxidizer, and flares as three typical equipment in combustion chamber design. However, with the fast acceleration of engineering and technology development, there exits enormous new combustion technology such as recuperative catalytic oxidizer (RCO), regenerative thermal oxidizer (RTO), or combination of different combustion chamber with auxiliary equipment to enhance combustion efficiency, destruction efficiency, heat recovery efficiency, and fuel and energy-saving of device. The most important step is determining waste gas stream specification, type of air

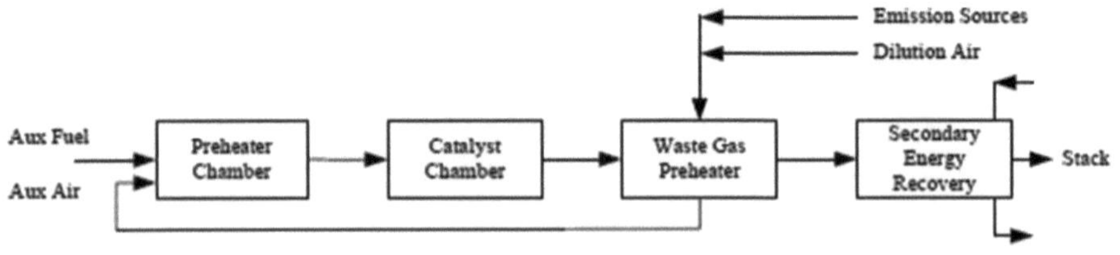

Fig. 14.10 Catalytic incineration with heat recovery [2]

pollutants, then apply appropriate device to treat specific air flow, also consider auxiliary devices need to be added into system in order to transport air flow and assure whole system is safe to worker.

14.5.3 Innovation Technologies

Table 14.6 shows comparisons of noble-metal and transition- metal-based catalysts for VOCs destruction [21]. The noble metal phases are usually supported by simple and low-cost oxides such as SiO_2 and Al_2O_3 but also by molecular sieves and transition metal oxides (TMOs) to increase the dispersion of active sites and aid reactant adsorption which can reduce the required noble metal load lowering cost. The transition metal oxides materials are generally less active than the single-metal-oxide catalysts (SNMCs), but they possess advantages including low cost, excellent reducibility and thermal stability, and resistance to poisoning. Generally, the destruction efficiency of single-metal-oxide catalysts is lower than that of precious metal catalysts for the removal of VOCs. The performance of metal-oxide catalysts can be improved by combining two or more oxides to impart a synergistic effect.

Table 14.6 Comparisons of noble-metal and transition-metal-based catalysts for VOCs destruction [21]

Catalysts		VOC type	Temp	Conversion
Noble-metal-based catalysts	Pt/Activated carbon	Benzene	112	100
		Toluene	109	100
		Ethylbenzene	106	100
		Xylene	104	100
	Pt/Al_2O_3	Benzene	165	90
		Toluene	160	90
		n-Hexane	190	90
	Pd/TiO_2	o-Xylene	140	90
		HCHO	Room Temp.	90
	Au/Co_3O_4	Toluene	138	90
Transition-metal-based catalysts	$MnOx/TiO_2$	Chlorobenzene	177	90
	Rod-like α-MnO_2	HCHO	250	90
	$MnOx/ZrO_2$	Toluene	250	90
	Cu-Ce	Chlorobenzene	328	99
	Ce-Zr	1-2 Dichloroethane	120	90
	$MnOx$-TiO_2-$SnOx$	Chlorobenzene	177	90
	$Mn_{0.5}Ce_{0.5}$-O_2	Formaldehyde	270	100

Catalysts and porous membranes can be combined to form a extractor, distributor, and contactor membrane reactor thereby providing an intimate contact between the VOC molecules and the wall of the pores, thus minimizing any diffusive resistance. The configuration of the reactor can have better performance of catalysts in VOC combustion due to the high contact efficiency between the propene and O_2 molecules [22].

Since much of the reactor volume is occupied by the catalyst support, rather than the active catalytic species itself. Nanoparticle fluidized-bed reactor exhibiting a low pressure drop, good dispersion of the active species, uniform temperature distribution inside the catalyst bed, and absence of intra-particle mass-transfer barriers was reported by He et al. [22]. It was evident that the lack of a rigid porous structure in the fluidized NiO nanoparticles facilitated the desorption of ethyl radicals from the surface of the NiO nanoparticles into the gas phase ultimately led to an enhancement in the reaction rate over the NiO nanoparticles.

Exercise

1. Define the following terms and explain why they should be considered for design use: combustion stoichiometry, destruction efficiency, lower explosive limit, fraction energy recovery.
2. Please describe the general procedures for designing a combustion process.
3. List the major factors affecting the design of a combustion process.
4. Compare the performance of thermal and catalytic incinerator from 3E aspects.
5. Given the following information, determine the outlet temperature of the incinerator:

   Heat input = 18.7 × 106 Btu/hr
   Total gas flow = 72,000/lb/hr
   Inlet air temperature = 200 °F
   Average cp = 0.26 Btu/lb·F.

References

1. Theodore L (2008) Air pollution control equipment. Wiley
2. Sorrels JL, Baynham A, Randall D, Hancy C (2017) Chapter 2 Incinerators and oxidizers: air economics group, OAQPS U.S. Environmental Protection Agency, November 2017
3. Hutten IM (2016) Federal law and regulatory activity that may relate to nonwoven filter media. Elsevier BV
4. KLM Technology Group (2007), Flare selection and sizing handbook
5. Brauer H, Varma YBG (1981) Air pollution control equipment. Springer Science and Business Media LLC
6. Cooper CD, Alley FC (1986) Air pollution control: a design approach. PWS Engineering, Boston
7. Ross RD (1977) Thermal incineration chapter 17 in air pollution control and design handbook. In: Cheremisinoff PN, Young RA (eds). Dekker, New York

8. Lee KC, Hansen JL, Macauley DC (1979) Predictive model of the time-temperature requirements for thermal destruction of dilute organic vapors, a paper presented at the 72nd Annual Meeting of the Air pollution control Association, Cincinnati, OH, June 24–29
9. Cooper CD, Cooper CD, Alley FC, Overcamp TJ (1982) Hydrocarbon vapor incineration kinetics. Environmental Progress 1(2), May 1982
10. Theodore L (2008) Air pollution control equipment calculations. Wiley
11. Reallick WB (1981) Design of transfer-limited catalytic incinerators. Chemical Engineering, p 123 (January 12, 1981)
12. Wark K, Warner CE (1981) Air pollution-its origin and control, 2nd edn. Harper & Row, New York
13. Snape TH (1977) Catalytic incineration. Chapter 21. In: Cheremisinoff PN, Young RA (eds) Air pollution control and design handbook. Dekker, New York
14. Hawthorn RD (1974) Afterburner catalysts-effects of heat and mass transfer between gas and catalyst surface. American institute of chemical engineers symposium series, vol 70, no 173
15. Grelecki C (1976) Fundamentals of fire and explosion hazards evaluation. AIChE Today Series, New York
16. Stone DK, Lynch SK, Pandullo RF, Evans LB, Vatavuk WM (2012) Flares. PartI: flaring technologies for controlling VOC-containing waste streams. J Air Waste Manag Assoc
17. Evans LB, Vatavuk WM (2000) Chapter 1 Flares: Office of Air Quality Planning and Standards U. S. Environmental Protection Agency, September 2000
18. Pollution Systems, Thermal Oxidizer and Recuperative Catalytic Oxidizer Specification Sheet
19. Anguil Environmental Systems, Inc., Regenerative Catalytic Oxidizer (RCO) Specification Sheet
20. Nestec Inc, Technical Specifications of Regenerative Thermal Oxidizer System
21. Kamal MS, Razzak SA, Hossain MM (2016) Catalytic oxidation of volatile organic compounds (VOCs)-A review. Atmos Environ
22. He C, Cheng J, Zhang X, Douthwaite M, Pattisson S, Hao Z (2019) Recent advances in the catalytic oxidation of volatile organic compounds: a review based on pollutant sorts and sources. Chem Rev

Chapter 15
Stack

Nomenclature

ρ	Density of a parcel of air
C_{ij}	Measured concentration of ith compound in the jth sample
C_P	Constant-pressure heat capacity of the gas
d	Stack exit diameter
d^1	Outer diameter of chimney
E_{ij}	Residual error which represents the difference between the measured concentrations and the modeled ones
f	Weighing factor which is used to determine the plume status
f_s	Compressive stress
f_w	Seismic forces
F	Natural frequency of the stack
F_{ip}	Concentration of *i*th compound in the *p*th source
h	Height of steel chimney above the section XX in meters
Δh	Plume rise
H	Height of the terrain popout adjacent to the emission source
H_c	Height above which the kinetic energy is enough for the fluid parcel climbing to the receptor height
$H\,(m)$	Height of nearby structures
$H_g\,(m)$	Stack height from the GEP protocol
H_W	Wake height
H_{WC}	Cloud depth under the worst case
I	Moment of inertia of circular ring section of the steel chimney
K	Shape factor
L	Lesser dimension of that popout
$L\,(m)$	Lesser dimension
m	Mass of air
N	Brunt–Vaiala frequency
P	Pressure of the environment

P.-C. Chiang and X. Gao, *Air Pollution Control and Design*,
https://doi.org/10.1007/978-981-13-7488-3_15

P_1	Intensity of wind pressure
P_a	Atmospheric pressure
Q	Emission rate of air pollutant
Q_h	Heat emission rate
R_S	Stack radius
S_{pj}	Contribution to the *j*th sample made by *p*th source
SP_s	Stack draft
St	Strouhal number
t	Thickness of steel plates in meters
T	Temperature of the environment
T_a	Temperatures of the atmosphere
T_{amb}	Temperature of ambient air
T_r	Reference temperature at which the flue gas would have the same density as the ambient air does
T_s	Temperatures of the stack gas
T_{sa}	Temperature of flue gas
u	Wind speed at stack exit
U	Wind speed under the worst case
v_s	Stack gas exit velocity
v_w	Mean wind speed which is a function of the elevation and atmospheric stability
V	Volume of air
W_L	Weight of Lining
W_s	Self-weight of the chimney
W_{sa}	Stack exit velocity
W_{WC}	Worst case cloud width
z_e	Relative height of the receptor above the terrain
Z	Modulus section of chimney ring section with no flue opening
θ	Potential temperature
ρ	Unit weight of steel
ρ_1	Unit weight of brick lining
ρ_{amb}	Density of ambient air
ρ_{sa}	Density of flue gas
σ	Standard deviation of concentration profiles
χ	Concentration profile

15.1 Introduction

Since the industrial revolution, ambient air pollution has escalated due to the burning of fossil fuels. With technology development and regulations evolution, emissions of air pollutants have been effectively controlled. However, air pollutant cannot be completely eliminated, dispersion enhancement is worth considering and especially due to emerging response and contingency plan. In other words, a stack is the end of an air pollution control process which is used to disperse air pollutants efficiently into the atmosphere so that ground-level concentration is well controlled.

Steel stack is made of steel plates and supported on foundation. It is made cylindrical in shape. The lower portion of the stack is widened or flared, in order to provide large base and greater stability. It is used to escape and disperse flue burned gases coming from furnaces of boiler to such a height that the gases do not contaminate the surrounding atmosphere. The cross-sectional area of stack depends on the type and quantity of fuel to be used in a plant, available draft for carrying the burned gases up the stack and losses due to friction within the stack.

The height of stack is kept to provide the required draft, which is defined as the difference between absolute gas pressures at any point in the duct or stack and the ambient atmospheric pressure. The draft depends on the height of stack above sea level, the type of fuel to be burned, the type of furnaces, and the temperature of the burned gases. Before structural designing the stack, it is necessary to decide the proper height and diameter of stack at the top, i.e., diameter and height are known data before designing the various components of the stack.

Well-designed stacks can offer other environmental and economic benefits including aesthetic values, heat recovery, etc.

15.2 Principles

15.2.1 Fundamentals of Pollutant Dispersion

Atmospheric Dispersion

As briefed earlier, a stack is designed and implemented to reduce pollutant concentration at ground level by facilitating pollutant dispersion in the atmosphere. A fundamental equation is therefore introduced to describe the fate of pollutants in a specific site:

$$(\text{Flowin}) + (\text{Creation}) = (\text{Flowout}) + (\text{Destruction}) + (\text{Accumulation}) \quad (15.1)$$

Creation and destruction are functions of temperature and the concentration of pollutants, while flow in and flow out are influenced by atmospheric conditions, topography, and the distribution of emission sources.

According to the ideal gas formula, the density of a parcel of air, ρ, can be expressed by Eq. (15.2):

$$\rho = \frac{m}{V} = \frac{MP}{RT} \quad (15.2)$$

where m and V are mass and volume of air, respectively. P and T are the pressure and temperature of the environment where the air is located. R is a constant. In addition, the pressure of fluids is a function of depth, which is

$$\frac{dP}{dz} = -\rho g = -\frac{gMP}{RT} \tag{15.3}$$

Assuming a parcel of gas undergoes a reversible, adiabatic process where no energy exchange occurs between the parcel and the atmosphere:

$$\frac{dP}{P} = \frac{C_P}{R}\frac{dT}{T} \tag{15.4}$$

where C_P is the constant-pressure heat capacity of the gas. Combining Eq. (15.3) and (15.4), we have the change of temperature with the elevation of altitude in a reversible, adiabatic process:

$$\Gamma = \frac{dT}{dz} = -\frac{gM}{C_P} = -\frac{9.8\,\text{m/s} \times 29\,\text{g/mol}}{3.5 \times 8.314\,\text{m}^3 \cdot \text{Pa/mol} \cdot \text{K}} = -9.78\,^{\circ}\text{C/km} \tag{15.5}$$

The temperature gradient, i.e., Γ, obtained from Eq. (15.5) is termed as the adiabatic lapse rate. When the actual temperature gradient is less than −9.78 °C/km, as indicated by (a) in Fig. 15.2, we call it superadiabatic. Under this condition, if the air parcel is moved away from original elevation due to natural disturbance, it would get accelerated as the result of the enlarged density difference between the parcel and the surrounding atmosphere. In (b), the lapse is the same as the adiabatic one under which the parcel moves at a constant speed. If the actual temperature gradient is higher than the adiabatic lapse, i.e., situation (c), the moving parcel would decelerate and finally return to original altitude. Situation (d) depicts a scenario where the temperature rises along the altitude, thereby producing an inversion layer, i.e., quite stable situation.

Momentum and Buoyancy Fluxes

The movement of an air parcel in the atmosphere can be regarded as the combination of movements in horizontal and vertical directions. The horizontal movement is due to the air advection such as the wind, whereas the vertical one is the result of momentum and buoyance forces. The momentum flux of the flue gas is generated due to the temperature difference which is expressed in Eq. (15.6):

$$F_{\text{m}} = {W_{\text{sa}}}^2{R_{\text{s}}}^2\left(\frac{\rho_{\text{sa}}}{\rho_{\text{amb}}}\right) = {W_s}^2{R_s}^2\left(\frac{T_{\text{sa}}}{T_{\text{amb}}}\right) \tag{15.6}$$

where W_{sa} (m/s) is the stack exit velocity; R_{s} (m) is the stack radius; ρ_{sa} and ρ_{amb} (kg/m^3) are densities of flue gas and the ambient air; and T_{sa} and T_{amb} (K) are the corresponding temperatures. On the other hand, the buoyancy flux is expressed as

Fig. 15.1 Revolving restaurant on the stack of Beitou Incineration plant [1]

$$F_m = gW_{sa}R_s^2\left(\frac{\rho_{sa} - \rho_{amb}}{\rho_{amb}}\right) = gW_{sa}R_s^2\left(\frac{T_{sa} - T_{amb}}{T_{amb}}\right) \quad (15.7)$$

As a result, we can see that flue gas should possess the sufficient heat and velocity to disperse efficiently. Here, we define potential temperature, θ, as the temperature that a parcel of air needs to compress adiabatically to a specific pressure. With the help of this factor, we can estimate the dispersion profile of plumes in the atmosphere:

$$\frac{d\theta}{dz} = \frac{dT}{dz} + \Gamma \quad (15.8)$$

If $\frac{d\theta}{dz} > 0$ or $\frac{dT}{dz} > 9.78\,°C/km$, i.e., the atmosphere is stable, then buoyancy of the plume decreases along the altitude as the hot plume elevates into a warmer region. In this scenario, atmospheric turbulence is too weak to affect the plume rise. If the thickness of stable layer is greater than plume's elevation distance, it would lose buoyancy and sink back to the original height. If $\frac{d\theta}{dz} = 0$ or $\frac{dT}{dz} = 9.78\,°C/km$, the atmosphere is neutral, and the buoyancy of the plume remains constant. In this situation, the atmosphere is usually turbulent, which suppresses the vertical momentum of the plume. $\frac{d\theta}{dz} < 0$ or $\frac{dT}{dz} < 9.78\,°C/km$ is desired when planning a stack since the atmosphere would be very unstable, meaning the pollutant would disperse efficiently.

Topography and Terrain Effects

In reality, the dispersion of airborne pollutant is influenced by structures and terrain. As a result, the surface friction together with pressure gradients interferes with atmospheric flow, generating a highly turbulent region thereby prohibiting the spread of

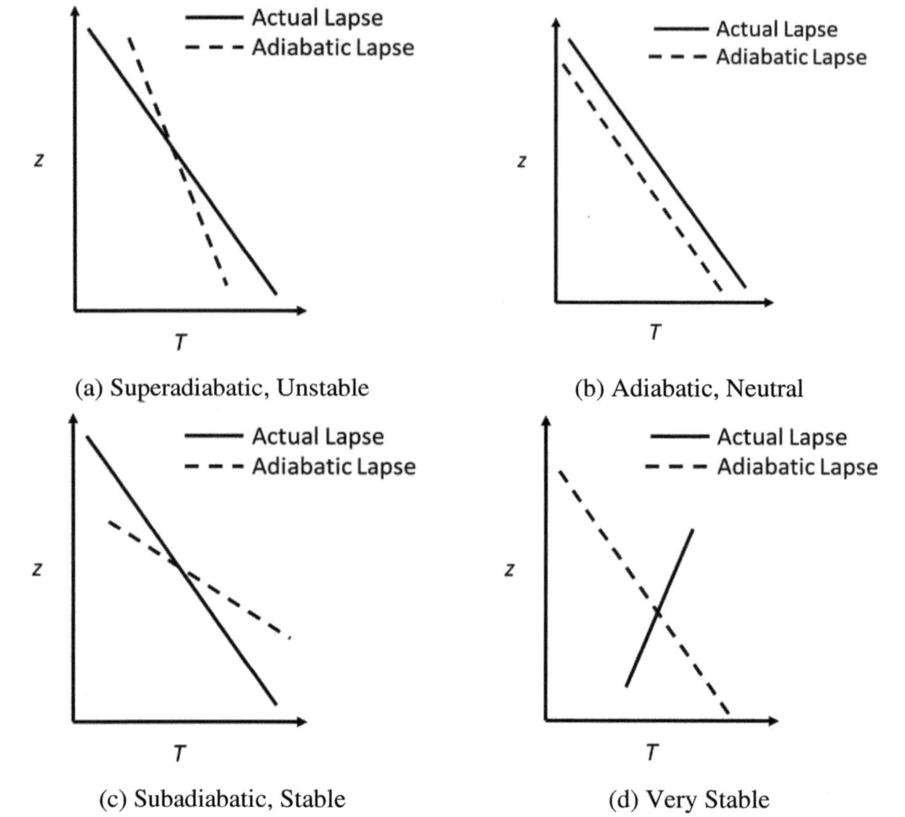

Fig. 15.2 Graphic presentation of atmospheric stability

air pollutants under lower elevations. This turbulent region can enhance the vertical dispersion, i.e., increase σ_z, resulted in reducing the effective stack height.

As shown in Fig. 15.3, terrain morphology can affect the dispersion profile by changing the streamline of atmospheric flow. When wind approaches raised terrain such as a mountain or building it will get elevated and accelerated. According to Bernoulli's law, the pressure near the terrain is lower compared to higher altitudes. As a result, the plume dispersion would be suppressed near the surface (wake region and cavity) which increases ground-level concentrations and creates downwash effects. Downwash effects would be so much significant if the wind speed is high and the source in interest is located within 5 times the lesser dimension (either height or width) of a structure downwind from the trailing edge of the structure. In this case, the profile of the plume is nearly horizontal. The quantitative estimation of the wake height is performed by defining it as the minimum centerline height which is not affected by the popout terrains and is given by Eq. (15.9) [2–4]:

$$H_W = H + 1.5L \tag{15.9}$$

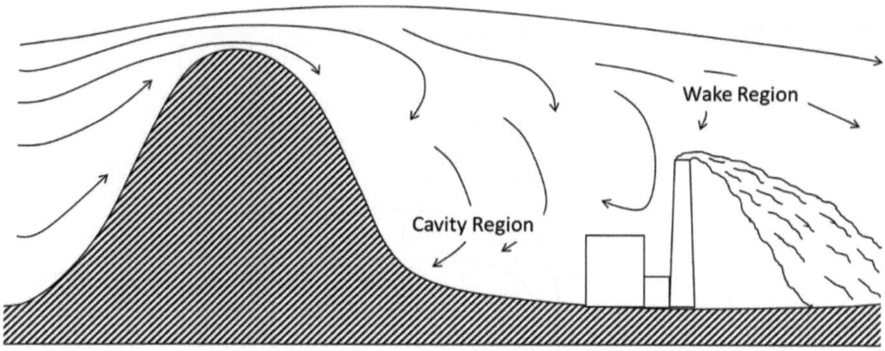

Fig. 15.3 Schematic diagram of effects of terrain on atmospheric dispersion

where H_W (m) is the wake height, H (m) is the height of the terrain popout adjacent to the emission source, and L (m) is the lesser dimension of that popout.

Cavity refer to a negative pressure zones on the top and all other sides except for the upstream side of the popout [5]. The height of the cavity is relatively lower than that of wake region and is given as [6, 7]:

$$H_C = H + 0.5L \tag{15.10}$$

15.3 Plume Rise

Figure 15.4 presents the abstract of the plume rise which is divided into three phases including initial, transition, and diffusion ones. In particular, the initial phase is characterized by its significantly strong internal turbulence and is further subdivided into jet, bent-over jet, and thermal sections. The jet section is close to the exit of the stack at which the upwardly flow of the plume is maintained particularly when the exit velocity is greater than 4 times the wind speed. As the plume keeps rising, it enters into the bent-over jet section where the entrainment of the cross flow is rapid as the result of the vortices. Thermal section is the end of the initial phase, characterized by its turbulence, which determines the growth of plume.

Since the heat flux and momentum get lost gradually after the plume is emitted into the atmosphere, the internal turbulence levels drop significantly so that the atmospheric eddies or energy-containing eddies of atmospheric turbulence deteriorate, which determine the ultimate height of the plume.

Fig. 15.4 Graphical presentation of plume flow

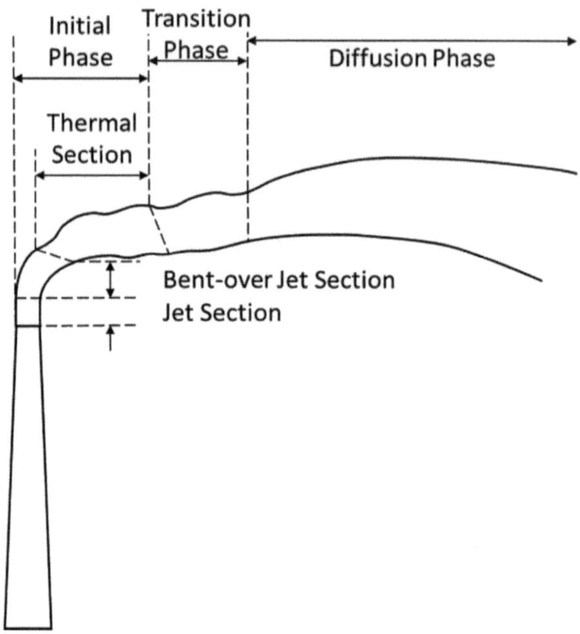

15.3.1 Dispersion Model

<u>*Gaussian Dispersion Model*</u>

The dispersion of air pollutant is also influenced by atmospheric conditions. Pasquill [8] categorized atmospheric stability into 6 discrete classifications based on measured wind speed, cloudiness, and insolation. As a result, the dispersion of air pollutant under different climate conditions can be evaluated (see Table 15.1).

Before further descriptions of pollutant dispersion, we need to establish the coordinate system. In general, the vertical direction is z, the direction of the prevailing wind is x, and the direction perpendicular to the prevailing wind is y. We assume that the pollutant does decompose and the plume is dispersing in Gaussian profile [8] along vertical and horizontal planes:

$$\underbrace{\chi(x, y, z, H_e)}_{\text{Pollutant concentration}} = \frac{Q}{2\pi \cdot \sigma_y \cdot \sigma_z \cdot v_w} \underbrace{\exp\left[-\frac{1}{2}\left(\frac{y}{\sigma_y}\right)^2\right]}_{\text{Cross - wind distribution}} \underbrace{\left\{\exp\left[-\frac{1}{2}\left(\frac{z-H_e}{\sigma_z}\right)^2\right] + \exp\left[-\frac{1}{2}\left(\frac{z+H_e}{\sigma_z}\right)^2\right]\right\}}_{\text{Vertical distribution}} \tag{15.11}$$

Table 15.1 Classifications of atmospheric stability [9]

Surface wind speed[a], m/s	Day solar insolation			Night cloudiness[e]	
	Strong[b]	Moderate[c]	Slight[d]	Cloudy (≥4/8)	Clear (≤3/8)
<2	A	A~B[f]	B	E	F
2~3	A~B	B	C	E	F
3~5	B	B~C	C	D	E
5~6	C	C~D	D	D	D
6	C	D	D	D	D

[a]Surface wind speed is measured at 10 m above the ground
[b]Corresponds to clear summer day with sun higher than 60° above the horizon
[c]Corresponds to a summer day with a few broken clouds, or a clear day with sun 35–60° above the horizon
[d]Corresponds to a fall afternoon or a cloudy summer day, or clear summer day with the sun 15–35°
[e]Cloudiness is defined as the fraction of sky covered by clouds
[f]For A~B, B~C, or C~D conditions, average the values obtained for each:
*A = Very unstable; B = Moderately unstable; C = Slightly unstable; D = Neutral; E = Slightly stable; F = Stable
Regardless of wind speed, Class D should be assumed for overcast conditions, day or night

where $\chi(x, y, z, H_e)$ (μg/m^3) represents the concentration profile at (x, y, z) with the effective stack height H_e; Q (μg/s) is the emission rate of air pollutant; σ_y and σ_z (m) are the standard deviation of concentration profiles in horizontal and vertical directions, respectively; v_w (m/s) is the mean wind speed which is a function of the elevation and atmospheric stability:

$$v_{w,z} = v_{\mathrm{w,ref}}\left(\frac{z}{z_{\mathrm{ref}}}\right)^{P} \tag{15.12}$$

where z is the elevation, v is the wind speed at corresponding height z, and P is the exponent listed in Table 15.2. Subscript ref indicates the reference elevation which is usually 10 m.

Table 15.2 Exponents for wind profile model for rough surface

Pasquill stability	Exponent	
	Urban	Rural
A	0.15	0.07
B	0.15	0.07
C	0.20	0.10
D	0.25	0.15
E	0.30	0.35
F	0.30	0.55

As a result, in an ideal situation with no external disturbance, the concentration profile is symmetrical to the wind direction, as shown in Fig. 15.5.

It is thus concluded that the mean concentration downwind of a point source under the worst case is given by Eq. (15.13) [10]:

$$C_{wc} = \frac{10^9 Q}{U H_{wc} W_{wc}} \tag{15.13}$$

where Q (kg/s) is emission intensity at the source; U (μg/m^3) is the wind speed at height $z = 10$ m under the worst case; H_{wc} (m) is the cloud depth under the worst case; W_{wc} (m) is the worst case cloud width.

Ambient Air Quality Model

Many ambient air quality models were developed to meet specific missions. Therefore, it is important to select on appropriate that meets the requirements of a specific air quality target. This section introduces the typical models which are commonly used for ambient air quality management.

1. Screening

A screening model is typically a steady-state Gaussian plume model that needs to rapidly estimate maximum downwind concentrations from an emissions source. In general, the predication must be conservative and estimate relatively higher concentrations than that by a refined model.

Typical screening models are

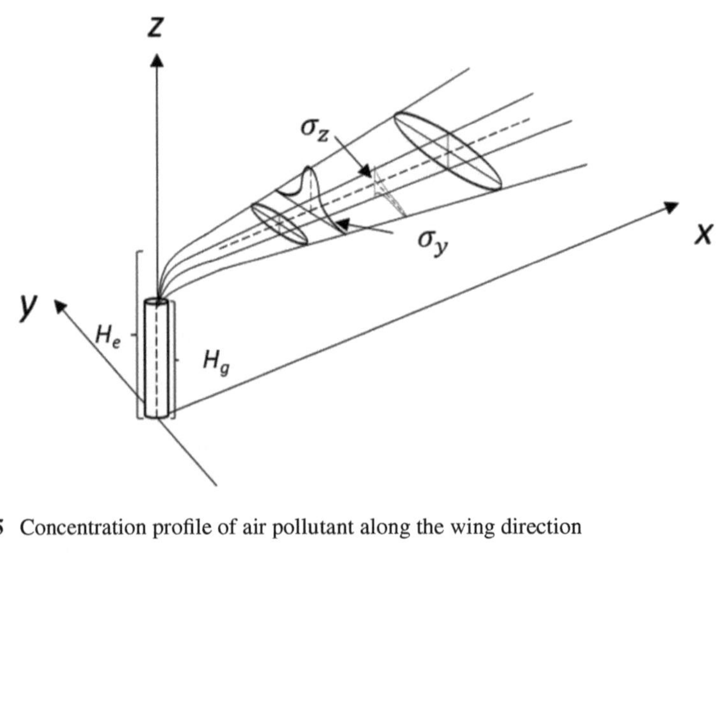

Fig. 15.5 Concentration profile of air pollutant along the wing direction

(1) SCREEN3 referred to the U.S. EPA (www.epa.gov/scram001/dispersion_screening.htm#screen3).
(2) AERSCREEN referred to the U.S. EPA (www.epa.gov/scram001/dispersion_screening.htm#aerscreen).

2. Local Scale Modeling

Local scale modeling using Gaussian principles is commonly utilized for new industrial sources, large industrial facilities, and construction projects.

By considering effects of buildings close to sources on plume rise, the model may predict isopleth diagrams for a specific grid of receptors or discrete receptor locations.

Typical local scale models:

(1) AERMOD referred to the U.S. EPA (www.epa.gov/ttn/scram/dispersion_prefrec.htm#aermod).
(2) Industrial Source Complex (ISC) referred to the U.S. EPA (www.epa.gov/scram001/dispersion_alt.htm#isc3).
(3) AUSPLUME referred to Australia EPA Victoria (www.epa.vic.gov.au/air/epa/ausplume-pub391.asp).
(4) AUSPUFF referred to Australia's Commonwealth Scientific and Industrial Research Organisation (CSIRO).
(5) Atmospheric Dispersion Modelling System (ADMS 4) referred to Cambridge Environmental Research Consultants (CERC) www.cerc.co.uk/software/adms4.htm.
(6) AirWare referred to Environmental Software and Services of Austria (www.ess.co.at/AIRWARE/product.html).

3. Source-Apportionment

A source-apportionment model based on linear algebra principles is introduced to estimate the contribution of sources at a specific location (i.e., a receptor). Pollution phsico-chemical characteristics of gases and particles need to be measured at the source and receptor for quantify source contributions to receptor concentrations.

Typical source-apportionment models are

(1) Chemical Mass Balance model (CMB) referred to the U.S. EPA.
(2) Unmix referred to the U.S. EPA (www.epa.gov/heasd/products/unmix/unmix.htm).
(3) Positive Matrix Factorization (PMF) referred to the U.S. EPA (www.epa.gov/heasd/products/pmf/pmf.htm).

4. Long-range Transport

Long-range transport models, i.e., lagrangian or Eulerian models are used when receptors are more than 50 miles from the source or when the plume transports through, i.e., complex mountains and valleys terrain and coastal regions with sea/land breezes.

Typical long-range transport models are

(1) CALPUFF referred to http://www.epa.gov/scram001/dispersion_prefrec.htm#calpuff.
(2) Lagrangian Atmospheric Dispersion Model (LADM) referred to www.dar.sciro.au/ladm/index.html.

5. Photochemical Modeling

Photochemical air quality models simulate the changes of pollutant concentrations in the atmosphere using a set of theoretical and empirical equations to assess the effectiveness of ambient air quality management. Relevant pollutants of concern include tropospheric ozone, i.e., volatile organic compounds (VOCs) and NOx and particulate matter, i.e., PM10 and PM2.5.

Typical photochemical models are

(1) Community Multiscale Air Quality Model (CMAQ) referred to http://www.epa.gov/asmdnerl/CMAQ/index.html.
(2) Comprehensive Air Quality Model with extensions (CAMx) referred to http://www.camx.com/.
(3) MUSE referred to www.ess.co.at/ISIREMM/a53.html.
(4) The Air Pollution Model (TAPM) referred to www.cmar.csiro.au/research/tapm/index.html.

15.4 Design Methods

In order to ensure the ground-level concentrations are in compliance with the ambient air quality regulations and standards, the stack height and the gas velocity are vital to stack design which dominate the dispersion extent of emitted pollutants. Both the theoretical approach and practical consideration in determining the appropriate stack height and its relevant requirements are stated as following:

15.4.1 Design Criteria

A well-designed stack is helpful in improving dispersion of air pollutant, saving human health while sustaining energy utilization. While designing a stack, relevant information including the local meteorological conditions (wind speed, prevailing wind direction), topographical conditions, specific air pollutant, and vulnerable populations all need to be gathered and analyzed.

Guidelines for a stack design include (1) the physical height should be 2.5 times higher than the surrounding buildings (1.5 times for flat country); (2) the ejection velocity should be greater than 60 ft/sec or 1.5 times the wind speed to prevent down washing; (3) The ground-level concentrations vary inversely as the square of effective height while the maximum concentration can be observed at 5–10 stack heights from the stack; and (4) Ground-level concentrations can be reduced to as low as 0.001% of stack-level.

15.4.2 Modeling Approach

Mathematical approach refers to using mathematical models which incorporates information on meteorology, terrain, and emission source to evaluate the effect of stack dimensions on ground-level concentrations. The modern science of air pollution modeling was originated in the 1920s which was used to estimate the dispersion of toxic chemicals released in the battlefield under various conditions [11].

As mentioned earlier, Gaussian plume model fails to predict the dispersion profile in complex scenarios. However, it provides a solid base on which we are able to estimate the ground-level concentrations by introducing reasonable boundary equations. Among the modified Gaussian model, the Industrial Sources Complex (ISC) model developed by the US EPA [12] has been widely used for estimating the concentrations of nonreacting pollutants within 10 miles of the stack, and the cross-wind distribution, σ_y, can be calculated by Eq. (15.14):

$$\begin{cases} \sigma_y = 465.11628x \cdot \tan\Theta \\ \Theta = 0.017453293(a - b\ln(x)) \end{cases} \tag{15.14}$$

where a and b (—) are constants related to Pasquill stability, which refers to Table 15.3.

A graphical presentation of the cross-wind distribution is shown in Fig. 15.6 which reveals that the standard deviation increases as distance from the stack increases. In addition, it also reveals that as the atmospheric stability increases from A to F, the standard deviation also rises, indicating a wider horizontal dispersion profile.

A similar distribution profile is also presented in vertical dispersion which is more susceptible to atmospheric stability. It has been widely suggested that vertical dispersion follows the power law as following:

$$\sigma_z = cx^d \tag{15.15}$$

Table 15.3 Constants for determination of cross-wind distribution deviations

Pasquill stability	a	b
A	24.1670	2.5334
B	18.3330	1.8096
C	12.5000	1.0857
D	8.3330	0.72382
E	6.2500	0.54287
F	4.1667	0.36191

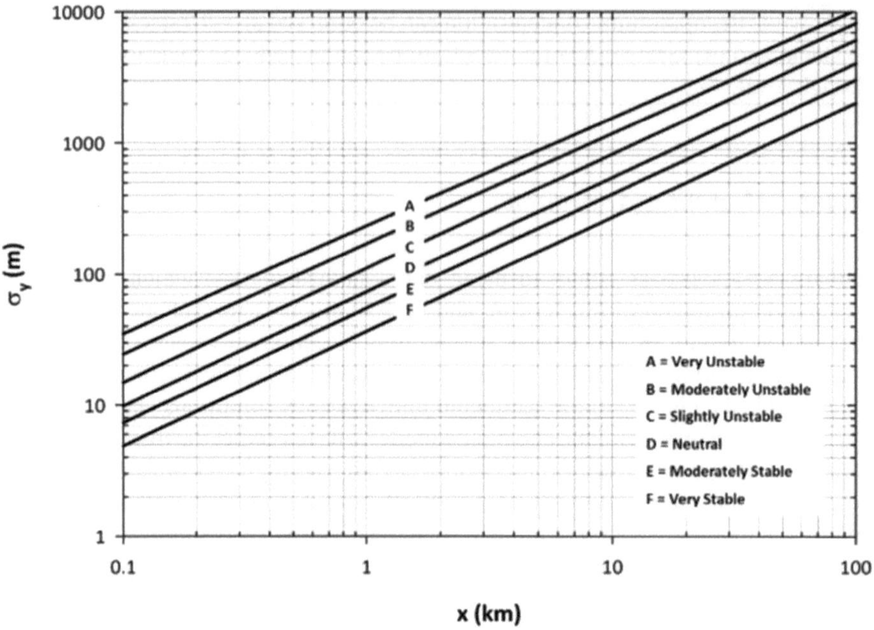

Fig. 15.6 Calculations of cross-wind distribution, σ_y, based on the Pasquill stability classes A-F

where *c* and *d* (−) are listed in Table 15.4.

Figure 15.7 presents the vertical distribution along the wind direction which indicates that the vertical distribution varies more with the atmospheric stability in comparison with cross-wind distribution. The wider dispersion is due to the greater temperature gradient and the mixing intensity of elevation changes.

15.4.3 Good Engineering Practice

It seems that the stack should be higher to promote the dispersion such that the ground-level concentration could be effectively controlled. However, too high stack may cause atmospheric loading thus is prohibited by Sect. 123, Clean Air Act. Meanwhile, the mathematical model is too sophisticated to be implemented case-by-case. As an alternative, US EPA proposed good engineering practice guiding the stack design with respect to the stack height [13]. The finalized GEP stack height protocol comprises three categories where the stack height equals (1) 65 m which is sufficient to avoid normal atmospheric effects and to prevent the atmospheric loading [14]; (2) 2.5 times the height (H) of the source; and (3) the height (H) of the source or a nearby structure plus one-and-one-half (1.5) times the lesser dimension (the height or width) of that structure (L), expressed by Eq. (15.16):

Table 15.4 Constants for determination of vertical distribution deviations

Pasquill stability	*x* (km)	*c*	*d*
A	<0.10	122.800	0.94470
	0.10–0.15	158.080	1.05420
	0.16–0.20	170.220	1.09320
	0.21–0.25	179.520	1.12620
	0.26–0.30	217.410	1.26440
	0.31–0.40	258.890	1.40940
	0.41–0.50	346.750	1.72830
	0.51–3.11	453.850	2.11660
	> 3.11	$\sigma_z = 5000$m	
B	< 0.20	90.673	0.93198
	0.21–0.40	98.483	0.98332
	> 0.41	109.300	1.09710
C		61.141	0.91465
D	< 0.30	34.459	0.86974
	0.31–1.00	32.093	0.81066
	1.01–3.00	32.093	0.64403
	3.01–10.00	33.504	0.60486
	10.01–30.00	36.650	0.56589
	> 30.00	44.053	0.51179
E	< 0.10	24.260	0.83660
	0.10–0.30	23.331	0.81956
	0.31–1.00	21.628	0.75660
	1.01–2.00	21.628	0.63077
	2.01–4.00	22.534	0.57154
	4.01–10.00	24.703	0.50524
	10.01–20.00	26.970	0.46713
	20.01–40.00	35.420	0.37615
	> 40.00	47.618	0.29592
F	< 0.20	15.209	0.81558
	0.21–0.70	14.457	0.78407
	0.71–1.00	13.953	0.68465
	1.01–2.00	13.953	0.63227
	2.01–3.00	14.823	0.54503
	3.01–7.00	16.187	0.46490
	7.01–15.00	17.836	0.41507
	15.01–30.00	22.651	0.32681
	30.01–60.00	27.074	0.27436
	> 60.00	34.219	0.21716

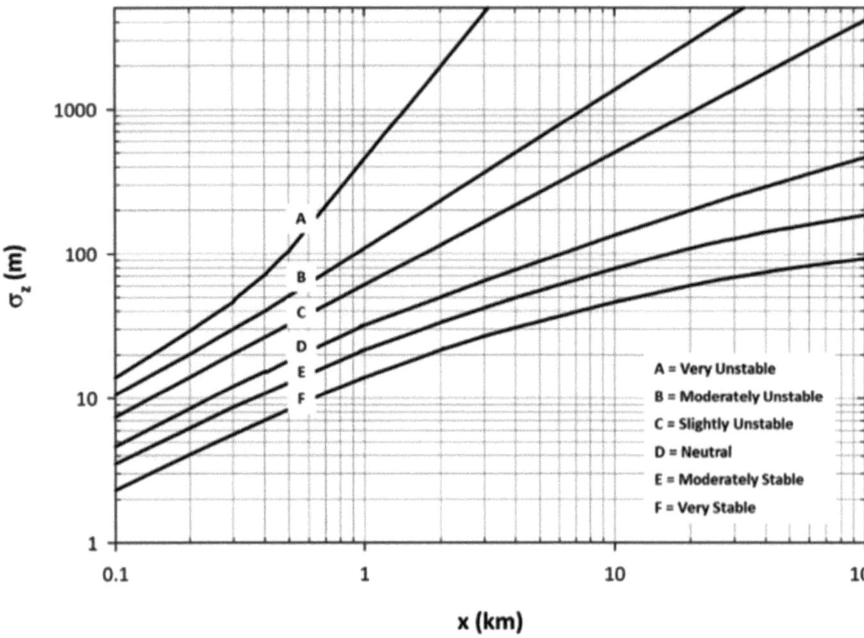

Fig. 15.7 Calculations of vertical distribution, σ_y, based on the Pasquill stability classes A-F

$$H_g = H + 1.5L \tag{15.16}$$

where H_g (m) is the stack height from the GEP protocol, H (m) is the height of nearby structures, and L (m) is the lesser dimension, either height (H) or projected width (W), of nearby structures. It can be seen that the GEP height in Eq. (15.16) is established to minimize the downwash effects in the wake region. The detailed definitions of "*nearby*" will be described in the following section.

15.5 Design Procedures

It has been widely acknowledged that the effective stack height, which is the sum of the physical height and plume rise, determines the dispersion of airborne pollutants. Since physical height is mainly determined by compliance to the GEP protocol, plume rise becomes an important factor in stack design. In general, higher plume rise is the result of greater emission velocity and temperature. A rigorous design procedure is presented in Fig. 15.8 which includes information gathering, dimension determination, draft calculation, and economic cost assessment.

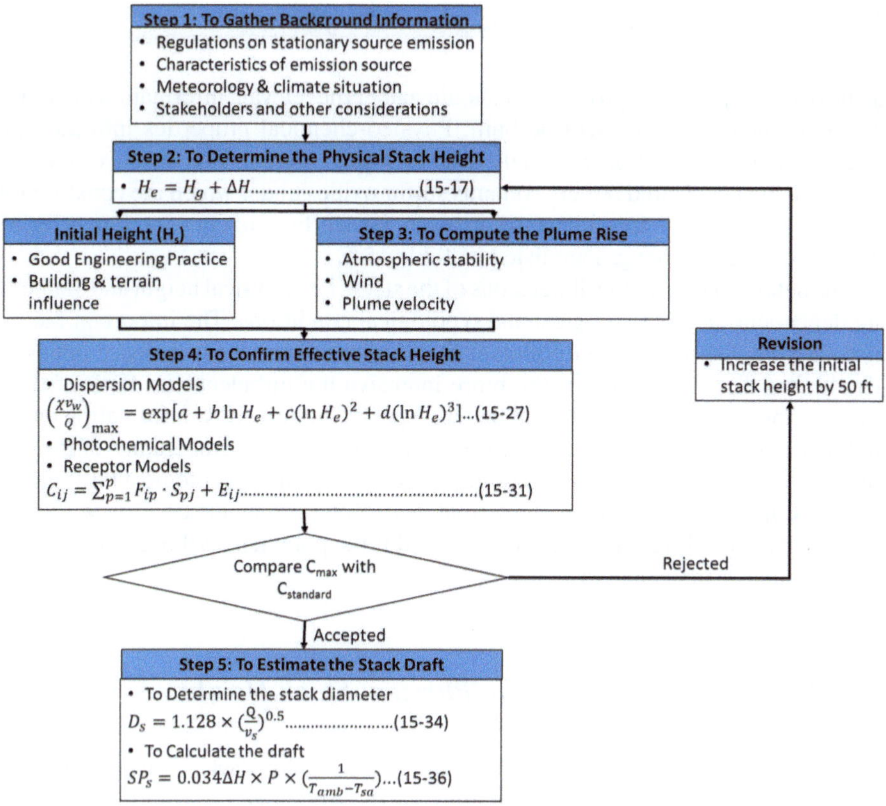

Fig. 15.8 Procedure of designing a stack

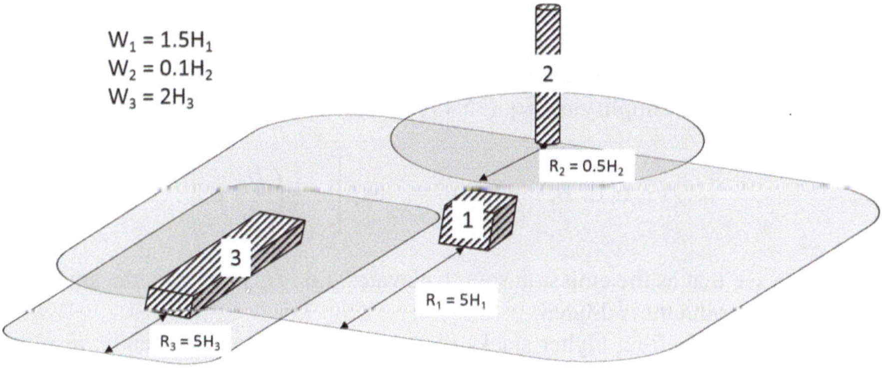

Fig. 15.8 Graphical information for determining the initial stack height

15.5.1 Step 1: To Gather Background Information

In the beginning, the design engineer should gather the relevant information including the area where the stack will be built. Physico-chemical properties influence the structure and construction materials to a very large extent. For instance, the flue gas from most coal-fired boilers contains sulfur oxides under which the construction material and structure are subject to corrosion. Feasibilities of construction materials subject to different flue gas are listed in Table 15.5.

The determination of the dimensions of the stack, i.e., physical height and diameter are dependent on meteorological and typological conditions. The important factors affecting the dispersion of air pollutant are wind speed and the turbulence intensity. The higher the wind speed is, the more intensive the turbulence is. Both of them are favoring mixing that resulted in better disperse. Additionally, information on the ambient temperature is crucial, i.e., we should increase the temperature of the flue gas to ensure a emission velocity for plume rise. Furthermore, geological information should be investigated which is a principal factor determining the construction material. For instance, brick stacks are not allowed if the potential earthquake magnitude is higher than 9 in China [15].

15.5.2 Step 2: To Suggest the Physical Stack Height

Stack height usually refers to the physical height of the stack. However, it is the effective height that matters in determining the dispersion profile of the air pollutant. Effective stack height is the sum of physical height and the plume rise, which is

$$H_e = H_g + \Delta h \tag{15.17}$$

Here we are interested in the ground-level pollutant concentration which can be readily obtained by simplifying Eq. (15.11):

$$\chi(x, 0, 0, H_e) = \frac{1}{\pi \cdot \sigma_y \cdot \sigma_z}\exp\left[-\frac{1}{2}\left(\frac{H_e}{\sigma_z}\right)^2\right] \tag{15.18}$$

We can see that as the emission source elevates, i.e., H_e increases, the amount of ground-level pollutant is diluted.

As mentioned before, higher stacks disperse flue gas more efficiently, resulting in a smaller impact on ground level concentrations. This section would detail the implementation of GEP in stack design by which we can estimate an appropriate stack height with good dispersion efficiency and acceptable construction cost.

The preceding section has specified the relationship between the GEP stack height with the dimensions of the nearby structures. The term "*nearby*" is defined as the area within 5 times the lesser dimension downwind from the trailing edge of the

Table 15.5 Feasibilities of construction materials subject to different kinds of flue gases [15]

Stack types			Gas properties				
			Dry			Humid	Wet
			Low corrosive	Medium corrosive	High corrosive		
Brick stack			Poor	Fair	Excellent	Excellent	Excellent
Single-tube reinforced concrete stack			Poor	Fair	Good	Good	Excellent
Tube-in-tube or multi-flue stack	Brick tube		Fair	Poor	Poor	Fair	Excellent
	Steel tube	Anticorrosion metal lining	Good	Good	Fair	Fair	Poor
		Light-weight anticorrosion brick lining	Good	Good	Fair	Fair	Poor
		Anticorrosion coating lining	Fair	Fair	Fair	Fair	Fair
		Acid resisting concrete lining	Fair	Fair	Fair	Good	Excellent
	Glass fiber lining		Good	Good	Fair	Fair	Poor

structure. In other words, GEP stack height only takes the upwind structures into considerations. Also, "*height*" and "*width*" are determined based on the projection plane perpendicular to the wind direction.

As shown in Fig. 15.8, the radius of the "nearby region" of building 1 is 5 times the height since the height is the lesser dimension compared to the width. As for building 2, the width is the lesser dimension which results in a radius of 0.5 H_2. For the same reason, the radius of the area for building 3 is 5 H_3. As a result, the GEP height for stacks located behind the building 1, 2, and 3 are 2.5 H_1, 1.15 H_2, and 2.5 H_3, respectively. It should be noted that adjacent structures should be regarded as one structure if their projection planes that lie upwind from the stack overlap. For instance, if a source is located downwind of building 1 and 3, the physical stack height should be 2.5 H_1 based on GEP. In other words, the initial stack height could be suggested by GEP and then for further confirmation by dispersion and/or receptor models ad stated below.

15.5.3 *Step 3: To Compute the Plume Rise*

The initial stack height should be revised by plume rise which is the result of velocity and density rises. The maximum plume rise can be estimated by Bosanquet and Pearson [16] as follows:

$$(\Delta h_v)_{\max} = \frac{4.77}{1 + 0.43(v_w/v_s)}\sqrt{\frac{Q \cdot v_s}{v_w}} \tag{15.19}$$

where v_w and v_s (ft/s) are velocities of wind and the flue gas at the stack exit, Q (scfs) represents the volumetric rate of the flue gas. Based on this, we can further estimate the actual plume rise from the stack:

$$(\Delta h_v)_{\mathrm{x}} = (\Delta h_v)_{\max}\left[1 - 0.8\frac{(\Delta h_v)_{\max}}{x}\right] \tag{15.20}$$

Note that Eq. (15.20) is only valid for $x > (\Delta h_v)_{\max}$.

As for density rise, we can estimate it with Eq. (15.21):

$$(\Delta h_d)_{\mathrm{x}} = \frac{6.37g \cdot Q_l \cdot (T_{\mathrm{sa}} - T_r)Z}{{v_w}^3/T_r} \tag{15.21}$$

where T_{sa} (K) is the flue gas temperature at the stack exit, T_r (K) is the reference temperature at which the flue gas would have the same density as the ambient air does, Z is a constant related to the distance from the stack:

$$aZ + bZ^2 = \frac{v_w \cdot x}{3.57\sqrt{Q_l \cdot v_s}} \tag{15.22}$$

A simpler solution for determination of the plume rise was proposed by Moses, Strom [17]:

$$\Delta h = -0.029\frac{\upsilon_s d}{u} + 2.26\frac{\sqrt{Q_h}}{u} \tag{15.23}$$

where Δh (m) is the plume rise, υ_s (m/s) is the stack gas exit velocity, d (m) is stack exit diameter, u (m/s) is the wind speed at stack exit, Q_h (kJ/s) is the heat emission rate which is

$$Q_h = \dot{m}c_p(T_{\text{sa}} - T_{\text{amb}}) = \frac{\pi d^2 \upsilon_s P}{4RT_s}c_p(T_{\text{sa}} - T_{\text{amb}}) \tag{15.24}$$

where C_p is the constant-pressure specific heat of the stack gas, T_s and T_a (K) are temperatures of the stack gas and atmosphere, respectively.

Moses's formula is limited to small stacks with diameters of 1.7–4.3 m and temperatures from 82 to 204 °C. However, Holland [18] developed an empirical formula applicable to large stacks:

$$\Delta h = \frac{\upsilon_s d}{u}[1.5 + 2.68 \times 10^{-3} P_a(\frac{T_s - T_a}{T_s})d] \tag{15.25}$$

or

$$\Delta h = \frac{\upsilon_s d}{u}(1.5 + 0.0096\frac{Q_h}{\upsilon_s d}) \tag{15.26}$$

where P_a (millibar) is the atmospheric pressure.

15.5.4 Step 4: To Confirm the Effective Stack Height

The calculated effective stack height should be validated by comparing the maximum ground-level contaminant concentrations with ambient air quality standards. If the ground-level concentration cannot meet the requirements, another 50 m in height is needed. This step will estimate the impact of the emission source on specific receptors by using different models which include dispersion models, photochemical models, and receptor models.

Dispersion Models

The basic dispersion model is the Gaussian model that was stated before as Eq. (15.11). Generally, the maximum ground-level concentration is observed at the central line of the flume, as indicated by Eq. (15.18). Since the standard deviation of dispersion, i.e., σ_y and σ_z, is a function of atmospheric stability and distance from the stack, the maximum ground-level concentration is dependent on atmospheric stability, distance from the stack, and effective height. Such relationships are depicted in Fig. 15.9 which indicates that a higher effective stack height can facilitate

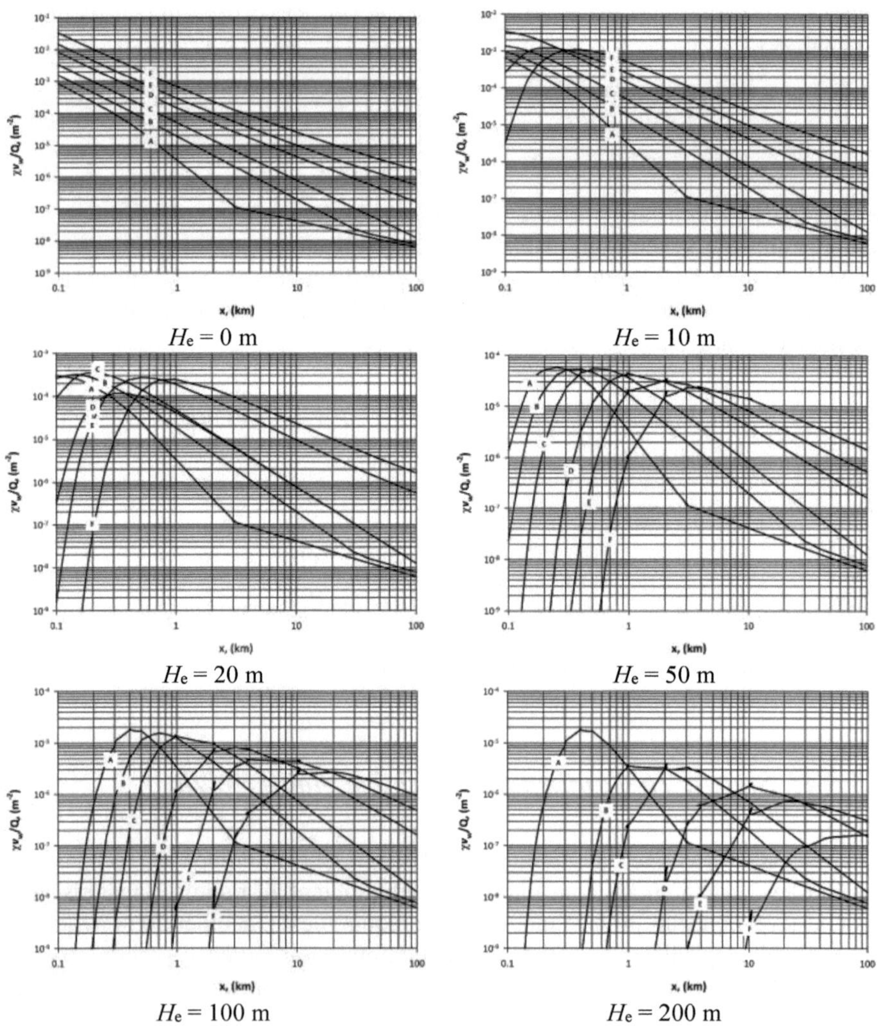

Fig. 15.9 Variation of maximum ground-level concentrations against the distance from the stack and the effective height under different atmospheric stabilities

Table 15.6 Coefficients for stability categories [19]

Pasquill stability	Coefficients			
	A	*b*	*c*	*d*
A	−1.0563	−2.7153	0.2161	0
B	−1.8060	−2.1912	0.0389	0
C	−1.9748	−1.9980	0	0
D	−2.5302	−1.5610	−0.0934	0
E	−1.4496	−2.5910	0.2181	−0.0343
F	−1.0488	−3.2252	0.4977	−0.0765

dispersion of the air pollutant, reducing ground-level concentrations.

With the help of these figures, we can readily estimate the maximum ground-level concentration once the distance from the stack, effective height, atmospheric stability, and emission status is given.

Ranchoux [19] simplified the relationship into

$$\left(\frac{\chi v_w}{Q}\right)_{\max} = \exp\left[a + b\ln H_e + c(\ln H_e)^2 + d(\ln H_e)^3\right] \tag{15.27}$$

where *a*, *b*, *c*, and *d* (−) are constants which relate to the atmospheric stabilities and given by Table 15.6.

When considering the effects of terrain on pollutant dispersion, US EPA introduced AERMOD model for further assessment [20]:

$$\underbrace{\chi_T(x, y, z, H_e)}_{\substack{\text{Pollutant}\\ \text{concentration}}} = \underbrace{f \cdot \chi_f(x, y, z, H_e)}_{\substack{\text{Contribution of}\\ \text{the horizontal plume}}} + \underbrace{(1 - f) \cdot \chi_f(x, y, z_e, H_e)}_{\substack{\text{Contribution of the}\\ \text{terrain following plume}}} \tag{15.28}$$

where f is the weighing factor which is used to determine the plume status, and z_e is the relative height of the receptor above the terrain.

Now we need to determine the weighing factor which should be related to the fraction of the plume below H_c:

$$f = \frac{1}{2}\left(1 + \frac{\int_0^{H_c} \chi_f(x, y, z, H_e)dz}{\int_0^{\infty} \chi_f(x, y, z, H_e)dz}\right) \tag{15.29}$$

where H_c is defined as the height above which the kinetic energy is enough for the fluid parcel climbing to the receptor height. Estimation of H_c can be performed with Eq. (15.30):

$$H_c = z - \frac{u(H_c)}{N} \tag{15.30}$$

where N is the Brunt–Vaiala frequency.

Photochemical Models

Photochemical models are drawing attention since they also take chemical processes into consideration. Among them, the Community Multiscale Air Quality (CMAQ) model is the most popular one. The CMAQ model establishes the connection between emission profiles and ground-level concentrations by incorporating information about emissions and properties of the pollutants. As a result, users acquired a spatial distribution as well as the composition of the mixture of pollutants. CMAQ consists of a large number of governing equations and computational algorithms. For detailed inspections of these components, please refer to Byun and Schere [21].

In order to simplify the calculation and analysis process, the US EPA developed a suite of software and have updated it to version 5.2.1 in March 2018. This software platform is comprised of four components, which are initial conditions processor (ICON), boundary conditions processor (BCON), meteorology-chemistry interface processor (MCIP), and CMAQ chemistry-transport model (CCTM).

Receptor Models

Receptor models were developed to determine the extent of air pollution that is contributed by a specific source. And chemical mass balance (CMB) is one of the two most popular receptor models in use which was developed based on the mass balance of chemical compounds

$$C_{ij} = \sum_{p=1}^{p} F_{ip} \cdot S_{pj} + E_{ij} \tag{15.31}$$

where C_{ij} is the measured concentration of ith compound in the jth sample, F_{ip} is the concentration of ith compound in the pth source, S_{pj} is the contribution to the jth sample made by pth source, and E_{ij} is the residual error which represents the difference between the measured concentrations and the modeled ones.

The use of CBM requires [22]: (1) identification of the contributing source types; (2) selection of chemical species or other properties to be included in the calculation; (3) knowledge of the fraction of each of the chemical species which is contained in each source type; (4) estimation of the uncertainty in both ambient concentrations and source profiles; and (5) solutions to the chemical mass balance equations.

The effective-variance least squares method is recommended by the US EPA to solve Eq. (15.31) where a Chi-square, χ^2, is developed in the first place [23]:

$$\chi^2 = \sum_{i=1}^{n} \left\{ \frac{[C_i - C_i(S_i)]^2}{{\sigma_{C_i}}^2 + \sum_{j=1}^{p} \left(\sigma_{F_{ij}} S_j\right)^2} \right\} \tag{15.32}$$

where σ_{C_i} and $\sigma_{F_{ij}}$ are errors resulting from the measurement and the apparatus. Equation (15.32) leads us to the contribution of air pollution by the source which is in the form of $S = \left(F^T wF\right)^{-1} F^T wC$ where w is the matrix of weighing factors:

$$w = \sigma_{C_i}{}^2 + \sum_{j=1}^{p} \left(\sigma_{F_{ij}} S_j\right)^2 \tag{15.33}$$

For simplicity in the use of CBM, US EPA has developed a CBM software platform and updated it to v8.2 in 2004.

Comparison of CMB and ISC Models

As suggested above, the major methods to determine the relationships between pollution sources and ambient air quality monitoring stations (receptors) are either dispersion or receptor models. The chemical mass balance (CMB) model was used to determine the amount of pollutants contributed by the emission source, and the industrial source complex (ISC) model was used to analyze the effect of plant emission on ambient air quality. Finally, results from the CMB and ISC models were used for comparison as shown in Fig. 15.11.

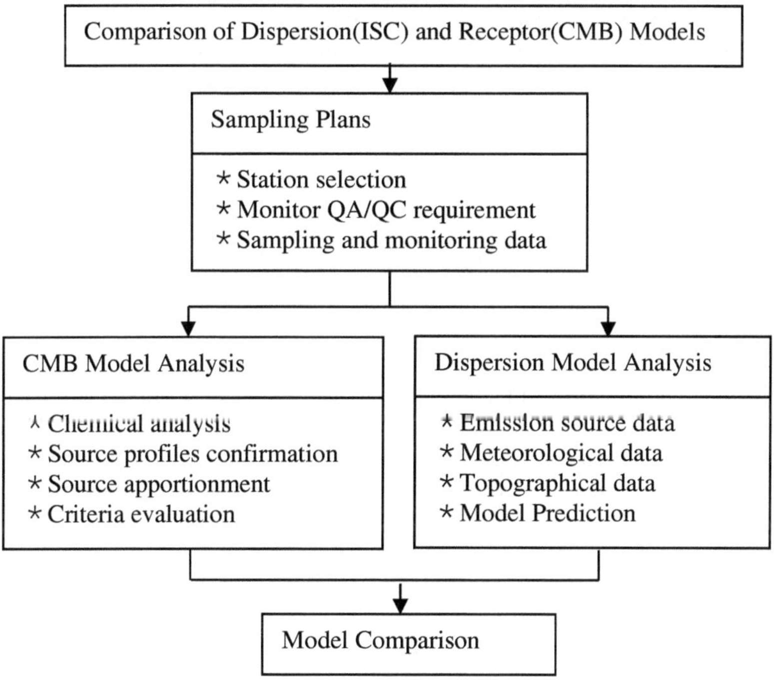

Fig. 15.11 Systematic approach to compare the levels of TSP emitted from a specific pollution source by dispersion and receptor Models

15.5.5 Step 5: To Estimate the Stack Draft

Plume rise is dependent on the exit velocity to a large extent which is a function of the temperature differential between stack gas and the ambient air. GEP suggested that the emission velocity of gas should be 1.5 times the wind speed so that the emitted gas possesses momentum passing through the wake region. Typically, exit velocities of 3,000–4,000 ft/min are adequate.

To Determine the Stack Diameter

The diameter of the stack is a function of flow rate and exit velocity of the flue gas which affect the plume height greatly. The plume height is expected to be greater under higher exit velocities so that the dispersion of flue gas is enhanced. The diameter of the stack can be calculated by the following Eq. (15.34).

$$D_s = 1.128 \times \left(\frac{Q}{v_s}\right)^{0.5} \tag{15.34}$$

It is noted that the diameter calculated in Eq. (15.34) herein is specifically the exit diameter. This is because, in most cases, the diameter of a stack gradually shrinks from the bottom to the top for structural stability. As a result, the gas velocity would be higher at the top of the stack than that at the bottom.

Gas flow rate, Q, is determined by previous processes, however gas pressure and/or the temperature should vary during the dispersion, the gas flow would be adjusted by the following Eq. (15.35):

$$Q_{\text{adj}} = Q'\left(\frac{T_2}{T_1}\right)\left(\frac{P_1}{P_2}\right) \tag{15.35}$$

To Calculate the draft

In general, the floating force that helps elevate flue gas is proportional to the temperature difference between flue gas and ambient air. When wind speed is about 5 m/s and gas temperature is between 100 and 200 °C, 1.5 m of elevation is obtained with 1 K of temperature difference between the stack gas (T_{sa}) and the ambient atmosphere (T_{amb}) [24].

However, gas loses static pressure and heat when it flows through hoods and ductwork, resulting in less momentum to pass the wake region. For this reason, an induced fan system is required to increase gas stream momentum so the gas flows smoother. The required draft is determined by the stack gas-ambient air temperature difference expressed as Eq. (15.36).

$$SP_s = 0.034\Delta H \times P \times \left(\frac{1}{T_{\text{amb}} - T_{\text{sa}}}\right) \tag{15.36}$$

where SP_s (inch of water column) is the stack draft, ΔH (ft) is the height difference, P (inch of water column) is the barometric pressure, T_{amb} (°R) and T_{sa} (°R) are temperature of atmosphere and stack gas, respectively.

The higher the gas-atmosphere temperature difference is, the less draft is required. In practice, engineers usually combine several individual exhausts into one stack, i.e., multi-flue stack, which reduces the temperature deterioration of the flue gas.

15.6 Safety Considerations

15.6.1 Forces Acting on Steel Stack

A self-supporting steel chimney or stack is made of steel plates and supported on foundations. When the lateral forces (wind or seismic forces) are transmitted to the foundation by the cantilever action of the stock, then the stock is known as self-supporting. The stock together with the foundation remains stable under all working conditions, without any additional support. In general, the self-supporting steel stock are made up to 10 m diameter and from 50 to 100 m in height. The various forces acting on the self-supporting steel stack are as follows:

Vertical Forces

(1) **Self-weight of the Chimney**

The self-weight of chimney is given by

$$W_s = \rho \cdot (\pi d)(t)(h) \tag{15.37}$$

where

ρ = unit weight of steel = 79 kN/m^3;

d = diameter of chimney in meters;

t = thickness of steel plates in meters;

h = height of steel chimney above the section XX in meters;

$\therefore W_s = 79\ (\pi dt \cdot h)$ kN.

The compressive stress in the steel plates at the section due to the self-weight of stack is given by

$$f_{s1} = \left[\frac{W_s}{\pi dt}\right] = \left[\frac{79\pi \cdot dt \cdot h}{\pi dt}\right] \text{kN/m}^2 = 0.079 \text{ h N/mm}^2 \tag{15.38}$$

(2) **Weight of Lining**

The weight of the lining in the steel stack also acts in the vertical toward direction. Assuming the thickness of the lining as 100 mm. The weight of lining

$$W_L = \rho_1 \cdot \pi d \cdot (0.1) \cdot h \tag{15.39}$$

where

ρ_1 = unit weight of brick lining = 20 kN/m^3.

$\therefore W_L = (2\pi \cdot d \cdot h)$ kN.

The compressive stress in the steel plates at the section X-X due to weight of lining

$$f_{s2} = \left[\frac{W_L}{\pi \cdot d \cdot t}\right] = \left[\frac{2\pi d \cdot h}{\pi \cdot dt}\right] \text{kN/m}^2 = 0.002\left[\frac{h}{t}\right] \text{kN/m}^2 \tag{15.40}$$

(3) **Wind Pressure**

The wind pressure acts horizontally and depends on the shape of the structure, i.e., width and height, location of the structure and climate condition can be expressed as

$$\text{P} = \text{K}.P_1 \times (\text{Projectareaofchimney}) \tag{15.41}$$

where

K = shape factor.

It accounts for the shape of the structure. p e shape factor for the cylindrical portion is 0.7

$$\begin{aligned} P_1 &= \text{Intensity of wind pressure} \\ \therefore \text{P} &= 0.7 \times \left(P_1 \times d^1 \times h\right) kN \end{aligned} \tag{15.42}$$

where

d^1 = outer diameter of chimney.

(4) **Seismic Forces**

The seismic forces also act horizontally and should be taken into consideration when the structure is located in the seismic areas. The following load combinations for calculation of stresses at any point of steel chimney are considered.

(i) Dead load + Wind load
(ii) Dead load + Earthquake load
(iii) Dead load + Earthquake Seismic load + Temperature effect + Live load.

The worst combination out of these three forces is considered for stack design.

PERMISSIBLE STRESSES

The windward side of the steel stack is subjected to tensile stress due to combined effect of the wind and the weight of steel stack. The leeward side of the steel stack is subjected to compressive stress due to combined effect of wind, weight of steel stack, and weight of lining. In order to control buckling in steel stack, the compressive stress caused by the combination of extreme stresses due to bending and direct load should not exceed the values given in Tables 15.7 and 15.8.

The permissible stresses in steel chimney in axial tension, shear, and bearing shall be adopted as specified in 18:800–1984. The allowable stress in axial compression and in bending for circular chimneys depends upon the effective height of the steel chimney h, the radius of gyration, k (0.707 × radius), the diameter and the thickness 't' of the chimney to be considered for different types of steel chimneys are shown in the Table 15.6.

In case, the increase in stresses is solely due to wind forces, the allowable stresses may be exceeded by 25%. However, the thickness of steel plates shall not be less than that needed if the wind forces were neglected. Similarly, when the effect of seismic load is considered, the allowable stresses may be exceeded by 25%. The thickness of steel plate shall not be less than that required if the seismic forces were not taken into account. The permissible stresses in rivets are adopted as per IS: 800–1984.

15.6.2 Safety Concerns

When subject to higher wind speeds, the stack vibrates to generate the alternating low-pressure zones under the downwind side of the stack which is known as the "vortex shedding". As wind speed increases to a critical value where the vibration frequency is close to the stack natural frequency resulting in a resonance. Critical wind speed can be calculated by Eq. (15.43):

$$v_{w,\text{critical}} = \frac{F \cdot D_s}{St} \tag{15.43}$$

where F (Hz) represents the natural frequency of the stack, and St (-) is the Strouhal number which is 0.2 for round stack in a laminar flow.

In some cases where a tube-in-tube stack or multi-tube stack is used, care should be taken to provide sufficient space among the tubes and outer shell. Otherwise, heat accumulates within the narrow space imposing heat stress on the structure. In multi-tube stacks, space among tubes and the outer shell should be larger than 750 mm. In

Table 15.7 Allowable stresses in axial compression for circular steel chimneys

h_l/k ration	Allowable stress in N/mm^2													
	(d/t) ratio													
	100 and less	125	150	175	200	225	250	300	350	400	450	500	550	600
0	125	125												
10	125													
20	125													
30	125	125												
40	125	125	125	125										
50	125		125	125	125	125	125	125						
60	125				125	125	125	125	124					
70	116								116	107				
80	103.5									103.5	92.8			
90	90.6				[b]						90.6	85.8		
100	78.6											78.6	78.3	71.8
110	68												68	68
120	58.9													58.9
140	45													45
150	39.7													39.7
160	35.3													35.3
170	31.5													31.5
180	28.4						[b]							28.4
190	25.5													25.5
200	23.2													23.2
210	21.1													21.1
220	19.2													19.2
230	17.6													17.6
240	16.2													16.2
250	15													15
300	10.6													10.6
350	7.72													7.72
130	51.3													51.3

Note
[a]For ratios of d/t and h_l/k in zone above zig-zag line, the stresses from the top line may be noted
[b]For ratios of d/t and h_l/k in zone below zig-zag line, the stresses from the column headed 100 and less may be noted
[c]Above values of stresses have been converted into SI Units from those given in m.k.s. Units

Table 15.8 Allowable stresses in bending for circular steel chimneys

h_1/k ration	Allowable stress in N/mm^2													
	(d/t) ratio													
	100 and Less	125	150	175	200	225	250	300	350	400	450	500	550	600
Upto 130	165	157.5	153	149.5	146.5	145	143.5	140	137					
140	159	157.5	153											
150	151	151	151	149.5	146.5	145	143.5	140						
160	145			145	145	138	138.5	138.5	137					
170	138								132					
180	132								132					
190	126				[b]				118	116.5				
200	118									113.5				
210	116.5									112				
220	113.5									108				
230	112			[c]						107				
240	108.5									105				
250	107									104				
260	105.5									102.5				
270	104									104				
280	102.5									102.5				
290	101										101			
300	99.2										99.2	94.5	86.6	78.17

Note For Tables 15.7 and 15.8

[a]For ratios of d/t and h_1/k in zone above zig-zag line, the stresses from the top line may be noted

[b]For ratios of d/t and h_1/k in zone below zig-zag line, the stresses from the column headed 100 and less may be noted

[c]Above values of stresses have been converted into SI Units from those given in m.k.s. Units

addition, the tube should be higher than the outer shell by 3 m or by 1 time the tube diameter, whichever is larger. For a tube-in-tube stack, the radius of the outer shell should be larger than the tube by 1000 mm while the additional height by which the tube taller than outer shell should range from 1 time the tube diameter to 2 times the tube-shell space.

15.6.3 Auxiliaries

In some cases, the stack is close to the boiler so the gas temperature is critically high. Under this circumstance, a heat recovery incorporated with an economizer should be introduced to reduce fuel consumption while minimizing environmental impacts. In general, a stack economizer is a heat exchanger that transfers the hot flue gas to a cold solvent which is usually water [25].

Other auxiliaries include emission monitoring station, inspection elevator, warning lamp (for aviation safety), lightning rod, etc., which are installed to comply with relevant regulations.

Exercises

1. Define the following terms and explain why they should be considered for design use: atmospheric stability, plume rise, wake height, physical stack height.
2. Please describe the general procedures for designing a stack.
3. List the major factors affecting the design of a stack.
4. How do you integrate CMB and ISC models for determining of the appropriate stack height.
5. For a coal-fired power plant given the following information, determine the wake height and the effective height of source:

   Stack height = 100 m;
   Internal diameter of stack outlet = 5 m;
   Velocity of outlet flue gas = 12.7 m/s;
   Temperature of outlet flue gas = 140 °C;
   Flow rate = 250 m^3/s;
   Average wind velocity at stack outlet = 4 m/s;
   Ambient temperature = 20 °C;
   Local atmospheric pressure = 978.4 hPa.

References

1. Protection TDoE (2018) Beitou incineration plant
2. Jensen M, Franck N (1963) Model-scale tests in turbulent wind. Danish Technical Press
3. Snyder WH, Lawson RE (1985) Fluid-modeling demonstration of good-engineering-practice stack height in complex terrain. Environmental Protection Agency, Research Triangle Park, NC (USA). Atmospheric Sciences Research Lab
4. Sherlock R, Stalker E (1940) The control of gases in the wake of smoke stacks. Mech Eng 52:455–458
5. Clarke JH (1967) Air flow around buildings. Heat Piping Air Cond 39:145–154
6. Evans BH (1957) Natural air flow around buildings. Texas Engineering Experiment Station
7. Lord GR, Baines W, Leutheusser HJ (1964) On the minimum height of roof mounted chimneys: results of an exploratory wind tunnel study. University of Toronto, Department of Mechanical Engineering
8. Pasquill F (1961) The estimation of the dispersion of windborne material. Meteorol Mag 90:33–49
9. Turner DB (1994) Workbook of atmospheric dispersion estimates: an introduction to dispersion modeling. CRC Press
10. Hanna SR (2010) Guidelines for use of vapor cloud dispersion models. Wiley
11. Sutton OG (1952) Micrometeorology
12. USEPA (1992) User's guide for the industrial source complex (ISC3) dispersion models volume II: description of model algorithms
13. EPA U (1977) Stack heights
14. USEPA (1985) Guideline for determination of good engineering practice stack height (Technical support document for the stack height regulations)
15. Niu C. Code for Design of Chimneys. 2013.
16. Bosanquet C, Pearson J (1936) The spread of smoke and gases from chimneys. Trans Faraday Soc 32:1249–1263
17. Moses H, Strom G, Carson J (1964) Effects of meteorological and engineering factors on stack plume rise. Nucl Safety 6
18. Holland J (1953) A meteorological survey of the Oak Ridge area. Rapport technique ORO-99, US AEC, Oak Ridge, Tennessee 71
19. Ranchoux RJP (1976) Determination of maximum ground level concentration. J Air Pollut Control Assoc 26:1088–1089
20. Venkatram A, Brode R, Cimorelli A, Lee R, Paine R, Perry S et al (2001) A complex terrain dispersion model for regulatory applications. Atmos Environ 35:4211–4221
21. Byun D, Schere KL (2006) Review of the governing equations, computational algorithms, and other components of the models-3 community multiscale air quality (CMAQ) modeling system. Appl Mech Rev 59:51–77
22. Coulter CT (2004) EPA-CMB8.2 Users Manual
23. USEPA (2016) Chemical mass balance (CMB) model
24. Carlton-Jones D, Schneider H (1968) Tall chimneys. Chem Eng 14:166
25. Energy Udo (2012) Advanced manufacturing program
26. Belevi H, Moench H (2000) Factor determining the element behavior in municipal solid waste incinerators. 1. field studies. Environ Sci Technol 34:2501–2506
27. Chang MB, Huang CK, Wu HT, Lin JJ, Chang SH (2000) Characteristics of heavy metals on particles with different size from municipal solid waste incineration. J Hazard Mater A79:229–239
28. Chiang PC (1997) Strategies on total suspended particulate control in Taiwan, EPA-86-F42–09–19. Environmental Protection Agency of ROC
29. Elliott SJ (1998) A comparative analysis of public concern over solid waste incinerator. Waste Manag Res 16:351–364
30. Greenberg MS, Zoller WH (1978) Composition and size distribution of particles released in refuse incineration. Environ Sci Tech 12:566–577

31. Hasselriis F, Licata A (1996) Analysis of heavy metal emission data from municipal waste combustion. J Hazard Mater 47:77–102
32. Kaupp H, McLachlan MS (1999) Gas/Particle partitioning of PCDD/Fs, PCBs, PCNs and PAHs. Chemosphere 38:3411–3421
33. Lorber M, Pinsky P, Gehring P, Braverman C, Winters D, Sovocool W (1998) Relationships between dioxins in soil, air, ash, and emissions from a municipal solid waste incinerator emitting large amounts of dioxin. Chemosphere 37:2173–2197
34. Maenhaut M, Francois F (1996) Size-fractionated aerosol composition at Gent, Belgium. Results from a one-year study. Nucl Inst Methods Phys Res B 109:476–481
35. Morselli L, Militreno S (1993) The presence and distribution of heavy metals in municipal solid waste incinerators. Toxicol Environ Chem 37:139–145
36. Schumacher M, Granero S, Rivera J, Müller L, Llobet JM, Domingo JL (2000) Atmospheric deposition of PCCD/Fs near an old municipal solid waste incinerator: levels in soil and vegetation. Chemosphere 40:593–600
37. U.S. EPA, 1981 (1981) Receptor model technical series, 11, July, 1981.
38. U.S. EPA, 1989 (1989) Receptor model technical series, Volume III (1989 Revision) CMB 7 user's manual; By Watson JG, Robinson NF, Chow JC, Henry RC, Kim B, Nguyen QT, Meyer EL, Pace TG, EPA-450/4-90-004, Environmental protection agency, Office of air quality, Washington DC

Chapter 16
Cyclone

Nomenclature

C_D	The Cunningham correction constant which is equal to $24\mu/d_p u$ for Stokes particles
D_c	The body diameter
D_p	The diameter of the particle
F_c	The centrifugal force
F_D	The dragging force on the particle
H_c	The height of the entrance
$H_1(X)$	The log-normal function for the fractional efficiency
$H_2(X)$	The log-normal function for the particle size distribution
k	The design number for cyclone which reflects the shape of the equipment
K_c	A function of the proportions of various dimensions
ℓ	The bottom edge of the gas outlet
N	The number of gas revolution
N_H	The inlet velocity
Q_{in}	The gas flow rate
u_{rel}	The relative particle velocity
V_ℓ	The annular volume around the central core zone from the bottom of the exit duct down to the natural length
V_s	The annular volume around the exit duct from the bottom duct up to half of the inlet
X	The particle diameter
η_i	Collection efficiency
θ_hn	The angle at which the particle hit the wall
μ	The viscosity of the gas
ρ_p	Particle density
ρ_G	Density of gas stream
σ_{gc}	The geometric standard deviation.

P.-C. Chiang and X. Gao, *Air Pollution Control and Design*,
https://doi.org/10.1007/978-981-13-7488-3_16

16.1 Introduction

16.1.1 Features of Cyclone

Cyclone dust collectors are widely used for dust-removal. In a cyclone, dust flows into a cylindrical or conical chamber tangentially and out of it through a central opening. Fundamentally, a cyclone is a settling chamber where centrifugal force is generated. While encountering the intensified centrifugal force, particles with a larger mass will be separated from the stream trajectory and hit the wall of the collector.

Cyclone dust collectors provide an affordable and reliable dust-collection solution. However, cyclones cannot handle particles smaller than 10 μm; Brownian diffusion rather than inertial interception would dominate particle collection with particle size less than 10 μm. Therefore, cyclones can be used as preclearance equipment.

16.1.2 Applications of Cyclone

Since the particle removal mechanism is analogous to settling chambers, cyclones have the same advantages, one of which is the moderate capital cost. In addition, operations and maintenance are easy and other relevant costs are reasonable, since no moving components are needed in the cyclone separator and pressure drop is acceptable. Furthermore, with the substitution of gravitational force with centrifugal force, the space requirement is less than a settling chamber. Last but not least, cyclones exhibit higher adaptability to the fluctuating dust load.

On the other hand, limited by the removal mechanism, cyclone separators have some inherent defects. The first one is the unsatisfactory removal efficiency for fine particles whose inertia momentum cannot be significantly enhanced by the centrifugal force. In addition, plugging, abrasion, and leakage may adversely affect collection performance.

As a result, cyclones are usually utilized as pretreatment equipment prior to a more efficient collector such as an electro-precipitator (EP), baghouse, or wet scrubber. In addition, cyclone collectors can be arranged in series or in parallel to enhance removal efficiency or reduce the dust load and pressure drop.

16.2 Principles

16.2.1 Fundamentals of Cyclone Collectors

As mentioned before, cyclones can drastically increase the magnitude of gravitational force, resulting in an intensified inertial impaction. Quantitatively, assuming that gas flow velocity is 10 m/s and radius of the cyclone is 0.5 m, the magnitude of centrifugal force is about 20 times greater than the gravitational force. When the particle can no longer maintain the trajectory of the stream line, it collides onto the outer wall of the cyclone and then is collected.

A particle in a spinning gas stream is subject to at least three forces, namely gravitational, centrifugal, and dragging forces. Among them, centrifugal is the dominant force in particle collection. According to Newton's law, the dynamics of the particle in radial direction follow Eq. (16.1):

$$m\frac{du_r}{dt} = m\frac{u_\theta^2}{r} - F_D, \quad (16.1)$$

where m is the mass of the particle in interest, $\frac{du_r}{dt}$ and $\frac{u_\theta^2}{r}$ represent the axial acceleration in the radial direction, and F_D is the dragging force on the particle. Experimental results indicate the variation of tangential velocity on different radii can be expressed by Eq. (16.2):

$$u_{\theta,1}r_1^{\mathrm{n}} = u_{\theta,2}r_2^{\mathrm{n}}. \quad (16.2)$$

Furthermore, n is a function of temperature for a given cyclone diameter which has been specified by Licht [1] and is expressed as follows:

$$n = 1 - \left(1 - 0.67{D_c}^{0.14}\right)\left(\frac{T}{283}\right)^{0.3}, \quad (16.3)$$

where D_c (m) is the body diameter and T (K) represents the gas temperature. Therefore, Eq. (16.1) can be re-arranged as Eq. (16.4) by introducing $u_r = \frac{dr}{dt}$:

$$\frac{d^2r}{dt^2} = \frac{u_{\theta,\mathrm{in}}^2 r_{\mathrm{in}}^{2\mathrm{n}}}{r^{2\mathrm{n}+1}} - \frac{3}{4}\frac{C_D}{d_p}\frac{\rho_G}{\rho_p}\left(\frac{dr}{dt}\right)^2, \quad (16.4)$$

where C_D is the Cunningham correction constant which is equal to $24\mu/d_p u$ for Stokes particles. Therefore, Eq. (16.4) can then be simplified as:

$$\frac{d^2r}{dt^2} = \frac{u_{\theta,\mathrm{in}}^2 r_{\mathrm{in}}^{2\mathrm{n}}}{r^{2\mathrm{n}+1}} - \frac{dr/dt}{\tau}, \quad (16.5)$$

where $\tau = \rho_p d_p^2/18\mu$ is the relaxation time which represents the time that is needed for the particle to decelerate to 1/e of its initial velocity. By neglecting the second order term, $\frac{d^2r}{dt^2}$, the initial condition, i.e., $r|_{t=0} = r_{\mathrm{in}}$, Eq. (16.5) can be integrated to be as follows:

$$t = \frac{r_{\text{in}}^2}{2(\text{n}+1)\tau u_{\theta,\text{in}}^2}\left[\left(\frac{r}{r_{\text{in}}}\right)^{2\text{n}+2} - 1\right]. \tag{16.6}$$

As a result, it is easy to estimate the elapsed time t for a particle entering the cyclone at r.

16.2.2 Laminar Flow Cyclone

Figure 16.1 illustrates a laminar flow cyclone. The velocity of the particle can be divided into the tangential u_θ and u_r radial terms which can be expressed by Eqs. (16.7) and (16.8), respectively,

$$u_\theta = \frac{Q_{\text{in}}}{H_c r \ln(r_2/r_1)}, \tag{16.7}$$

$$u_r = \frac{F_c}{3\pi\mu D_p}, \tag{16.8}$$

where Q_{in} is the gas flow rate, H_c is the height of the entrance, μ is the viscosity of the gas, D_p is the diameter of the particle, and F_c is the centrifugal force expressed as follows:

$$F_c = \frac{m_p u_\theta^2}{r} = \frac{\pi}{6}\rho_p D_p^3 \frac{Q_{\text{in}}^2}{{H_c}^2 r^3 [\ln(r_2/r_1)]^2}. \tag{16.9}$$

Based on the above equations, in a finite time duration, the particle would travel in the θ-direction expressed by

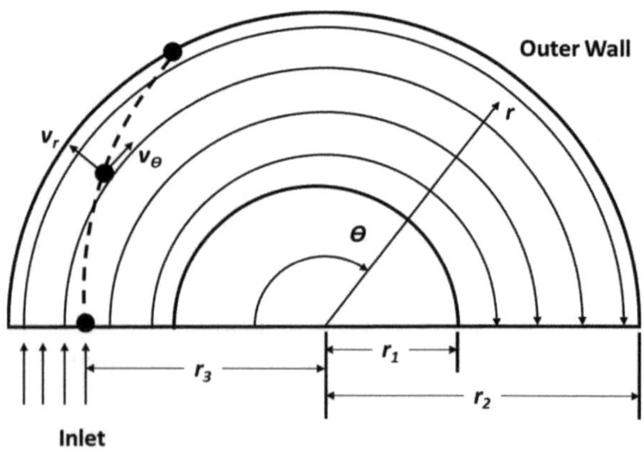

Fig. 16.1 Schematic diagram of the trajectory of a particle in a laminar flow cyclone collector [2]

$$u_\theta dt = rd\theta. \tag{16.10}$$

Similarly, it would travel in the r-direction expressed by

$$dr = u_r dt. \tag{16.11}$$

Then, we have

$$\frac{d\theta}{dr} = \frac{u_\theta}{ru_r} = \frac{18\mu H_c \ln(r_2/r_1) r}{\rho_p Q_{\text{in}} D_p^2}. \tag{16.12}$$

For particles entering the cyclone at $r = r_3$, the angle at which the particle hit the wall (θ_h) should follow the relationship below:

$$\theta_h = \frac{18\mu H_c \ln(r_2/r_1)}{\rho_p Q_{\text{in}} D_p^2}\left(r_2^2 - r_3^2\right). \tag{16.13}$$

Assuming that the gas flow rate and the particle concentration are uniform across the cross section of the entrance, it is reasonable to postulate that all particles entering the cyclone at $r \geq r_3$ can be collected. Further, we can extrapolate that the collection efficiency is proportionate to the fraction of the particles hitting the outer wall before $\theta = \theta_h$ as follows:

$$\eta(D_p) = \frac{1 - \left[1 - \frac{\rho_p Q_{\text{in}} D_p^2 \theta_h}{9\mu H_c r_2^2 \ln(r_2/r_1)}\right]^{1/2}}{1 - r_1/r_2}. \tag{16.14}$$

16.2.3 Turbulent Flow Cyclone

In a turbulent flow cyclone, the concentration of the particles is assumed to be uniform due to the turbulent flow regime under which the particle might be collected only when it travels to a thin layer, dr, at the outer wall as shown in Fig. 16.2. The fractional of particles removed over $d\theta$ is just the fraction of particles that are in the boundary layer, which is as follows:

$$N|_\theta - N|_{\theta + d\theta} = \frac{2r_2 dr}{\left(r_2^2 - r_1^2\right)} N|_\theta. \tag{16.15}$$

Note that $dr = u_{r2} r_2 d\theta / u_\theta$, then Eq. (16.15) is rearranged as follows:

$$\frac{dN}{d\theta} = -\frac{u_{r_2}}{u_{\theta_2}} \frac{2r_2^2}{r_2^2 - r_1^2} N. \tag{16.16}$$

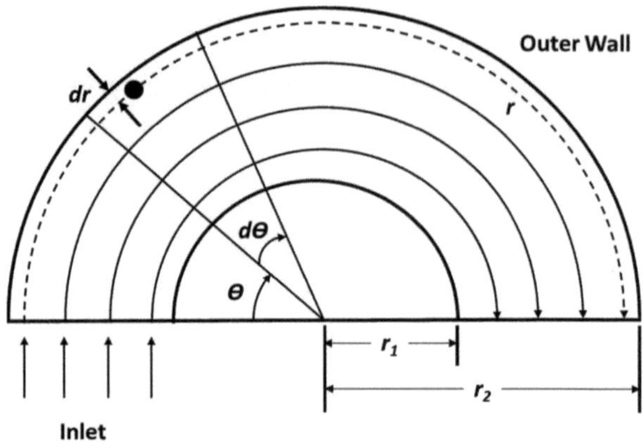

Fig. 16.2 Schematic diagram of the trajectory of a particle in a turbulent flow cyclone collector [2]

Introducing the initial condition that $N|_{\theta=0} = N_0$, we can obtain the following equation:

$$N|_{\theta} = N_0 \exp\left(-\frac{u_{r_2}}{u_{\theta_2}} \frac{2r_2^2}{r_2^2 - r_1^2}\theta\right). \tag{16.17}$$

The collection efficiency can be derived as:

$$\eta(D_p) = 1 - \frac{N|_{\theta}}{N_0} = 1 - \exp\left[-\frac{u_{r_2}}{u_{\theta_2}} \frac{2r_2^2}{r_2^2 - r_1^2}\theta_h\right]. \tag{16.18}$$

Similarly, the two velocity components can be expressed as follows:

$$u_{r_2} = \frac{\rho_p Q_{in}^2 D_p^2}{18\mu r_2^3 H_c{}^2 [\ln(r_2/r_1)]^2}, \tag{16.19}$$

and

$$u_{\theta_2} = \frac{Q_{in}}{H_c r_2 \ln(r_2/r_1)}. \tag{16.20}$$

The collection efficiency can then be expressed in terms of dimensions of the cyclone, showing the relationship between the impact angle θ_h and the desired collection efficiency.

$$\eta(D_p) = 1 - \exp\left[-\frac{\rho_p Q_{in} D_p^2 \theta_h}{9\mu H_c (r_2^2 - r_1^2)\ln(r_2/r_1)}\right]. \tag{16.21}$$

16.3 Design Methods

The purpose of cyclone design is to provide strong centrifugal force and retention time to divert particles from the gas streamline. Figure 16.3 shows the schematic diagram of a cyclone collector with principal dimensions which can determine the cyclone performance. As indicated by previous statements, the cyclone is limited to be effective to collect the particles larger than 5 μm [3]. More specifically, Eqs. (16.20) and (16.21) suggest that the particle removal efficiency of a cyclone increases

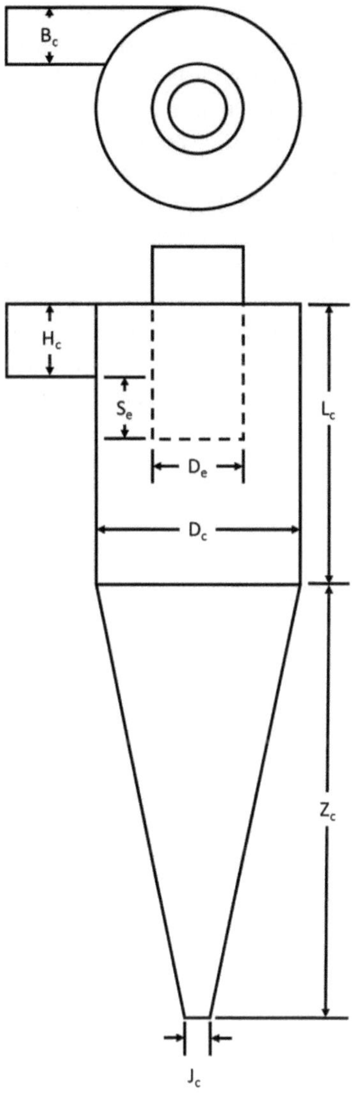

Fig. 16.3 Characteristics of a cyclone collector

along with the increase in particle density (ρ_p), particle diameter (D_p), particle concentration, gas flow rate (Q_{in}), body length ($L_c + Z_c$), the number of gas revolution (N), and the ratio of cyclone diameter to the outer diameter.

16.3.1 Classical Cyclone Design (CCD)

Classical cyclone design was firstly proposed by Shephered and Lapple [4] in the 1950s and was refined by Cooper and Alley [5]. It is suggested that the design through the CCD procedure should be based on the knowledge including the flow conditions, the dust load and particle size distribution (PSD), and the type of cyclone to be designed. Note that the PSD must be presented in the form of the mass fraction against the aerodynamic equivalent diameter of the PM. The cyclone type, including high efficiency, conventional, or high throughput, will provide all principal dimensions based on proportions of the dimensions to the cyclone barrel diameter (D_c).

The CCD process provides a standardized approach for the cyclone design. However, this process is limited in (1) presenting the relationship between the inlet velocity and cyclone dimensions; (2) predicting the correct number of gas revolution for different type cyclones, especially for 1D3Dt type; (3) establishing an accurate fractional efficiency curve which is used to calculate the overall collection efficiency [6]; and (4) estimating the pressure drop (see Eq. (16.22)) since it neglects the contribution made by vertical dimensions.

$$\begin{cases} \Delta P = \frac{1}{2}\rho_g {v_i}^2 H_V, \\ H_V = K\frac{HW}{{D_e}^2}. \end{cases} \tag{16.22}$$

16.3.2 Texas A&M Cyclone Design (TCD)

TCD was developed [7] to solve the limitations associated with CCD procedure. This method is based on the assumption that there is an optimal inlet velocity that enhances the overall collection efficiency. Specifically, the optimal inlet velocities for 1D3D, 2D2D, and 1D2D are 16 ± 2 m/s, 15 ± 2 m/s, and 12 ± 2 m/s, respectively. The barrel diameter (D_c) can be calculated as:

$$D_c = \sqrt{\frac{8Q}{v_\mathrm{i}}}. \tag{16.23}$$

TCD also provides a different and more precise equation to calculate the pressure drop (ΔP) throughout the collector which is given as Eq. (16.24):

$$\Delta P = K(VP_i + VP_o), \tag{16.24}$$

where VP_i and VP_o (Pa) are velocity pressures at the inlet and outlet, respectively; K (-) is a constant related to the cyclone type which equals to 5.1, 4.7, and 3.4 for 1D3D, 2D2D, and 1D2D cyclones, respectively.

16.3.3 Design Criteria

Cyclones are feasible for high-dust loading applications. The typical dust concentration ranges from 2.3 to 230 g/m^3, and the maximum concentration can reach up to 16,000 g/sm^3 [8]. The cost-effective flow rate for a single cyclone typically ranges from 0.5 to 12 sm^3/s; otherwise, the pressure drop throughout the system would become unacceptable and a multi-cyclone arranged in parallel is recommended.

16.3.4 Good Engineering Practice (GEP)

The most classic model for cyclone design was proposed by Lapple [9]. Upon which cyclone separators can be classified into three categories, i.e., high efficiency, conventional, and high throughput. All the dimensions of the above cyclones are related to the body diameter (D_c). Table 16.1 summarizes the relationships among dimensions

Table 16.1 Standard cyclone dimensions

Dimensions	High efficiency		Conventional		High throughout	
	[10]	[11]	[9]	[11]	[12]	[13]
Body diameter D_c/D_c	1.0	1.0	1.0	1.0	1.0	1.0
Inlet height H_c/D_c	0.5	0.44	0.5	0.5	0.75	0.8
Inlet width B_c/D_c	0.2	0.21	0.25	0.25	0.375	0.35
Outlet diameter D_e/D_c	0.5	0.4	0.5	0.5	0.75	0.75
Body length L_c/D_c	1.5	1.4	2.0	1.75	1.5	1.7
Cone length Z_c/D_c	2.5	2.5	2.0	2.0	2.5	2.0
Dust exit diameter J_c/D_c	0.375	0.4	0.25	0.4	0.375	0.4

with the body diameter which indicates that a cyclone with higher efficiency should have smaller inlet area and longer cone.

The long-lasting design evolution ultimately leads to the condensation of GEP cyclones which are the aforementioned 1D3D, 2D2D, and 1D2D types. The numbers proceeding Ds indicate the ratio of the barrel and cone lengths to the barrel diameter, that is, 1D3D means that the cyclone has barrel length equal to the barrel diameter and cone length three times of the barrel diameter.

Among the classical cyclone designs, 1D3D and 2D2D are the most efficient cyclone collectors for fine dust smaller than 100 μm. However, Michalslei [14] reported that the collection efficiency of these cyclones would affect by the "cycling lint" which is the product of the lint in the gas stream and diverts to the clean air exit stream. Simpson [15] proposed a new low-pressure cyclone which is also known as a high-through cyclone to solve the cycling-lint problem in high-efficiency cyclones, called the 1D2D cyclone. The 1D2D cyclone reduces the occurrence potential of "cycling lint" which makes it feasible in agricultural industry usage such as the cotton ginning process [16].

16.4 Design Procedures

Figure 16.4 presents the general design procedures of a cyclone which starts with the determination of the design parameters including gas flow condition, dust load, particle size distribution, and proposed type of equipment. According to the theory and practice, the body diameter (D_c) together with other dimensions can be estimated but need to be examined based on their removal efficiency and pressure drop.

16.4.1 Step 1. To Determine the Inlet Velocity

As suggested by Lapple [9], cyclone dust collectors can be classified into three categories, i.e., high efficiency, conventional, and high throughput. Prior to designing a cyclone via the Lapple model, the designer should acquire knowledge of (1) flow conditions, (2) dust load and particle size distribution, and (3) the type of cyclone. All dimensions are related to the body diameter (D_c). Table 16.2 summarizes the relationships among dimensions with the body diameter. It is revealed that cyclones with higher efficiencies should have smaller inlet areas and longer cones.

Cyclones remove particles with centrifugal force so it is important to determine the inlet velocity. If velocity is too low, then radial velocity will not be high enough to make particles collide with the wall. On the other hand, re-entrainment may occur under high velocities resulting in undesirably low removal efficiency. Kalen and Zenz [17] developed a mathematical method to predict the maximum allowable gas velocity (u_{max}), shown in Eq. (16.25):

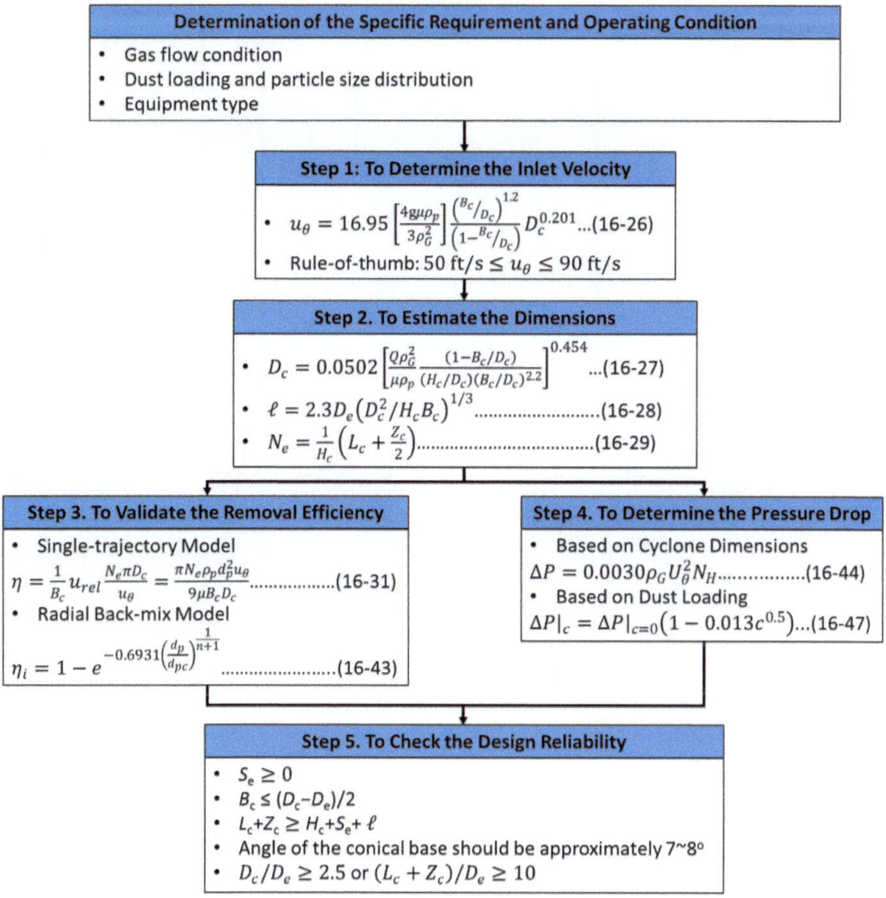

Fig. 16.4 Design framework of a cyclone

$$u_{\max} = 2.055\left[\frac{4g\mu\rho_p}{3\rho_G^2}\right]^{1/3}\left[\frac{(B_c/D_c)^{0.4}}{(1-B_c/D_c)^{1/3}}\right]D_c^{0.067}u_\theta^{2/3}. \tag{16.25}$$

Note that $u_\theta \approx \frac{Q}{H_c \times B_c}$, this equation uses imperial units. As a rule of thumb, u_θ should be in the range of 50–90 ft/s. Kalen and Zenz [17] concluded that the maximum available collection efficiency occurs when $u_{\max}/u_0 \approx 1.25$. Under this gas velocity, Eq. (16.25) becomes:

$$u_\theta = 16.95\left[\frac{4g\mu\rho_p}{3\rho_G^2}\right]\frac{(B_c/D_c)^{1.2}}{(1-B_c/D_c)}D_c^{0.201}. \tag{16.26}$$

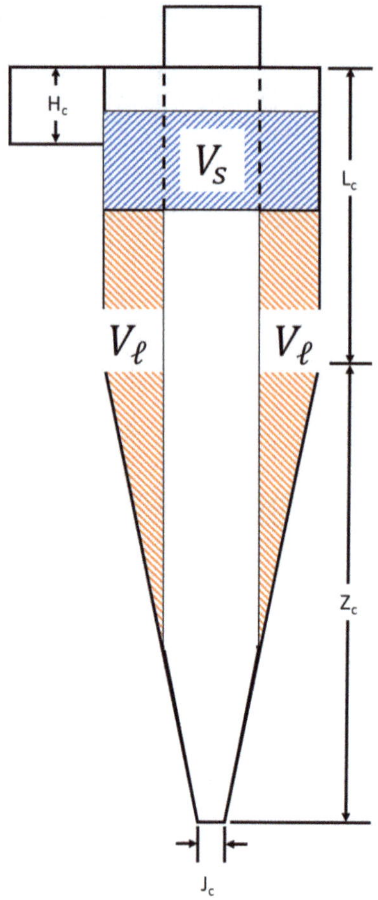

Fig. 16.5 Schematic diagram of V_s and V_ℓ

16.4.2 Step 2. To Determine Dimensions

As body diameter D_c is a fundamental parameter in the cyclone design, other dimension parameters can be readily calculated in accordance with relations listed in Table 16.2. For a given operating condition where gas flow rate Q (acfs), temperature (K), the viscosity of gas stream (μ, lbm/ft s), and density of gas stream and particles (ρ_G and ρ_p, lbm/ft^3) are all predetermined, the diameter can be estimated by Eq. (16.27):

$$D_c = 0.0502\left[\frac{Q\rho_G^2}{\mu\rho_p}\frac{(1 - B_c/D_c)}{(H_c/D_c)(B_c/D_c)^{2.2}}\right]^{0.454}. \tag{16.27}$$

The values of B_c/D_c and H_c/D_c are taken from Table 16.2 for different equipment types and operating conditions.

Table 16.2 Standard cyclone dimensions [9–13]

Dimensions	High efficiency		Conventional		High throughput	
Body diameter D_c/D_c	1.0	1.0	1.0	1.0	1.0	1.0
Inlet height H_c/D_c	0.5	0.44	0.5	0.5	0.75	0.8
Inlet width B_c/D_c	0.2	0.21	0.25	0.25	0.375	0.35
Outlet diameter D_e/D_c	0.5	0.4	0.5	0.5	0.75	0.75
Body length L_c/D_c	1.5	1.4	2.0	1.75	1.5	1.7
Cone length Z_c/D_c	2.5	2.5	2.0	2.0	2.5	2.0
Dust exit diameter J_c/D_c	0.375	0.4	0.25	0.4	0.375	0.4

Once the body is determined, other dimension parameters are readily estimated. According to the hydrodynamic theory, the length of the cyclone should be sufficiently long to provide enough space to generate an inward vortex which is called "natural length" (ℓ) measured from the bottom edge of the gas outlet as estimated by Eq. (16.28):

$$\ell = 2.3D_e\left(D_c^2/H_cB_c\right)^{1/3}. \tag{16.28}$$

This natural length should be shorter than the physical length, i.e., $Z_c + L_c - H_c - S_e$. Otherwise, the vortex will extend into the dust hood, leading to undesired re-entrainment.

Attention should also be paid to the number of gas revolutions. As particles are collected by centrifugal force, the designer should ensure that the cyclone can provide enough gas revolutions to enhance removal efficiency, with a longer cone. In addition, according to the flow regime analysis, both the outer and inner vortexes exist in a cyclone separator. The outer vortex is responsible for particle removal and related to the number of gas revolutions which is a function of cyclone height evaluated as follows:

$$N_e = \frac{1}{H_c}\left(L_c + \frac{Z_c}{2}\right). \tag{16.29}$$

If the dimensions of a cyclone are given, the value of N_e will increase along with inlet velocity. As a rule of thumb, N_e should vary from 0.3 to 10 with an average value of 5.

16.4.3 Step 3. To Validate the Removal Efficiency

Single-trajectory Model

One approach to evaluating the removal efficiency of particulate matter is the single-particle trajectory approach where critical particle diameter is pre-defined and predetermined. There are two categories of critical particles, which are defined as follows:

(1) Static particle: the particle is suspended at the edge of the central core which is in a static state as the centrifugal force is balanced by the inward dragging force. Particles larger than this will be span out and be collected while smaller ones penetrate the gas stream.
(2) Time-of-flight particle: the particle enters the collector at r_{in} then hits the wall at time t. Combining entry position and particle size, we can estimate if the particle is 100% or 50% collected by Eq. (16.6).

To define the relative particle velocity (u_{rel}) of a particle traveling toward the outer wall [18], Eq. (16.30) is used:

$$u_{rel} = \frac{\rho_p d_p^2 u_\theta^2}{9\mu D_c}. \quad (16.30)$$

As a result, the collection efficiency (η) can be estimated as follows:

$$\eta = \frac{1}{B_c} u_{rel} \frac{N_e \pi D_c}{u_\theta} = \frac{\pi N_e \rho_p d_p^2 u_\theta}{9\mu B_c D_c}. \quad (16.31)$$

Radial Back-mix Model

The radial back-mix model was developed by Leith [19] which assumes that uncollected dust at any depth in the cyclone is completely remixed to maintain a uniform concentration in any horizontal cross-section. The resulting collection efficiency (η_i) is a function of the particle residence time in the cyclone, shown in Eq. (16.32):

$$\eta_i = 1 - e^{-2\left[2(n+1)\tau_i\left(\frac{u_\theta}{D_c/2}\right)^2 t\right]^{\frac{1}{2n+2}}}. \quad (16.32)$$

The next step is to determine the residence time t. A popular way is to assume that the residence time equals the sum of the time that the gas needs to descend from the average level of entrance, $H_c/2$, to the bottom of the exit and the time in residence below this point. The mathematical expression is given as Eq. (16.33):

$$t = (V_s + V_\ell/2)/Q, \quad (16.33)$$

where V_s is the annular volume around the exit duct from the bottom duct up to half of the inlet whose calculation is expressed by Eq. (16.34):

$$V_s = \pi(S_e + H_c/2)\left(D_c^2 - D_e^2\right)/4. \quad (16.34)$$

And V_l is the annular volume around the central core zone from the bottom of the exit duct down to the natural length defined by Eq. (16.28):

$$\begin{cases} V_\ell = \frac{\pi D_c^2}{4}(L_c - H_c - S_e) + \frac{\pi}{4}\frac{(\ell + H_c + S_e - L_c)}{3}\left(1 + \frac{d}{D_c} + \frac{d^2}{D_c^2}\right) - \frac{\pi D_e^2 \ell}{4}, \\ d = D_c - \frac{(D_c - J_c)(\ell + H_c + S_e - L_c)}{Z_c}, \end{cases} \quad (16.35)$$

$$\eta_i = 1 - e^{-2\left[\frac{(n+1)\tau_i Q}{D_c^2}\frac{8K_c}{K_a^2 K_b^2}\right]^{\frac{1}{2n+2}}}, \quad (16.36)$$

where k (-) is the design number for cyclone which reflects the shape of the equipment and K_c is a function of the proportions of various dimensions. The estimations of k and K_c are given by Eqs. (16.37) and (16.38):

$$k = 8K_c/K_a^2 K_b^2, \quad (16.37)$$

$$K_c = (V_s + V_{nl}/2)/D_c^3. \quad (16.38)$$

Ultimately, the grade efficiency expressed by Eq. (16.36) can be simplified and given by Eq. (16.39):

$$\eta_i = 1 - e^{-M d_{Pi}^N}, \quad (16.39)$$

where

$$M = 2\left[\frac{KQ}{D_c^3}\frac{\rho_p}{18\mu}(n+1)\right]^{\frac{1}{2n+2}}, \quad (16.40)$$

$$\mathrm{N} = 1/(n+1). \quad (16.41)$$

The diameter of the particles at the cut size, i.e., $\eta_i = 50\%$, can be estimated by Eq. (16.42)

$$d_{pc} = \left(\frac{0.6931}{M}\right)^{n+1}. \quad (16.42)$$

As a result, the grade efficiency can also be estimated by Eq. (16.43):

$$\eta_i = 1 - e^{-0.6931\left(\frac{d_p}{d_{pc}}\right)^{\frac{1}{n+1}}}. \tag{16.43}$$

16.4.4 Step 4. To Determine the Pressure Drop

Licht [20] has suggested that there are four major sources of pressure drop in a cyclone system including friction, the volume expansion, turbulence, and elevation of the gas stream. Frictional loss occurs mainly due to the interaction of the gas stream with shells of the equipment. When the gas stream enters the cyclone and/or the duct, volume expands, and momentum is consumed. Turbulence, which is proportional to the ratio of the area of the inlet duct to that of the outlet duct, is regarded as the major source of pressure drop. Lastly, gas stream elevation should also be taken into consideration. The total pressure drop is given by Eq. (16.44):

$$\Delta P = 0.0030 \rho_G U_\theta^2 N_H. \tag{16.44}$$

Note that British units should be used in this equation and N_H (-) is the inlet velocity which is estimated according to Eq. (16.45):

$$N_H = \frac{16ab}{D_e^2} = \frac{16 K_a K_b}{K_e^2}. \tag{16.45}$$

A more precise approach to estimating N_H was obtained by Casal and Martinez-Benet [21] as follows:

$$N_H = \frac{16ab}{D_e^2} = 11.3\left(K_a K_b / K_e^2\right)^2 + 3.3. \tag{16.46}$$

In addition, dust loading exhibits a profound effect on pressure drop [22]. Masin and Koch [23] concluded that this effect can be described by:

$$\Delta P|_c = \Delta P|_{c=0}\left(1 - 0.013 c^{0.5}\right). \tag{16.47}$$

16.4.5 Step 5. To Verify with Technical Specifications

The first item on the check list should be configuration (shape and dimension) of the proposed cyclone which can be compared with other cyclones. According to existing experience and experiments, Licht [20] concluded that (1) the value of S_e should be positive, meaning that the height of the inlet should be less than that of the outlet.

In this case, short circuiting can be prevented; (2) pressure drop can be effectively controlled if $B_c \leq (D_c - D_e)/2$; (3) the total height $(L_c + Z_c)$, even though it is usually limited by the head-room, should be greater than the natural length $(H_c + S_e + \ell)$ to make sure that the vortex is constrained in the cyclone; (4) the angle of the conical base is approximately 7–8° to provide downward slippage for dust; and (5) the overall mass efficiency, η_M, keeps increasing until $D_c/D_e \geq 2.5$ or $(L_c + Z_c)/D_e \geq 10$.

16.4.6 *Case Study: A Rapid Method for Prediction of Overall Collection Efficiency*

1. Mathematical Simplification

The mass median diameter, the geometric standard deviation of the dust size distribution, the cut diameter (50% efficiency diameter), and geometric standard deviation of the collector fractional efficiency curve are the four parameters which can be utilized to develop the simplified method for predicting the overall collection efficiency [24].

If the particle size distribution of the dust and the fractional efficiency of the collector can be presented by log-normal functions, the overall collection efficiency can be completely determined from the above four parameters expressed by Eq. (16.49):

$$\eta_{OA} = \int_{-\infty}^{+\infty} H_1(X) dH_2(X). \tag{16.49}$$

$H_1(X)$ represents the log-normal function for the fractional efficiency, $H_2(X)$ denotes the log-normal function for the particle size distribution, and X is the particle diameter.

Consequently, the overall collection efficiency can be computed by Eq. (16.50) or obtained from the Error Function Table. A portion of the Error Function Table was presented in Table 16.3.

$$\eta_{OA} = \mathrm{erf}\left[\frac{\mathrm{In}\frac{D_m}{D_c}}{\sqrt{\mathrm{In}^2\sigma_g + \mathrm{In}^2\sigma_{gc}}}\right]. \tag{16.50}$$

2. Sample Calculations

Figure 16.6 presents cumulative distribution plots of two classifications of ground silica and a "fine dust" from Stairmand. Figure 16.7 presents a fractional efficiency plot of a 16 in. diameter high-efficiency cyclone. The significant parameters needed

Table 16.3 Error function

Areas under the Standard Normal Curve form—∞ to X

$\mathrm{erf}(X) = \frac{1}{2\pi}\int_{-\infty}^{x} e^{-\frac{1}{2}f^2 df}$

X	0	2	4	6	8
0.0	0.5000	0.5080	0.5160	0.5239	0.5319
0.1	0.5398	0.5478	0.5557	0.5636	0.5714
0.2	0.5793	0.5871	0.5948	0.6026	0.6103
0.3	0.6179	0.6255	0.6331	0.6406	0.6480
0.4	0.6554	0.6628	0.6700	0.6772	0.6844
0.5	0.6915	0.6985	0.7054	0.7123	0.7190
0.6	0.7258	0.7324	0.7389	0.7454	0.7518
0.7	0.7580	0.7642	0.7704	0.7764	0.7823
0.8	0.7881	0.7939	0.7996	0.8051	0.8106
0.9	0.8159	0.8212	0.8264	0.8315	0.8365
1.0	0.8413	0.8461	0.8508	0.8554	0.8599
1.1	0.8643	0.8686	0.8729	0.8770	0.8810
1.2	0.8849	0.8888	0.8925	0.8962	0.8997
1.3	0.9032	0.9066	0.9099	0.9131	0.9162
1.4	0.9192	0.9222	0.9251	0.9279	0.9306
1.5	0.9332	0.9357	0.9382	0.9406	0.9429
1.6	0.9452	0.9474	0.9495	0.9515	0.9535
1.7	0.9554	0.9573	0.9591	0.9608	0.9625
1.8	0.9641	0.9656	0.9671	0.9686	0.9699
1.9	0.9713	0.9726	0.9738	0.9750	0.9761
2.0	0.9772	0.9783	0.9793	0.9803	0.9312
2.1	0.9821	0.9830	0.9838	0.9846	0.9854
2.2	0.9861	0.9868	0.9875	0.9881	0.9887
2.3	0.9893	0.9898	0.9904	0.9909	0.9913
2.4	0.9918	0.9922	0.9927	0.9931	0.9934
2.5	0.9938	0.9941	0.9945	0.9948	0.9951
2.6	0.9953	0.9956	0.9959	0.9961	0.9963
2.7	0.9965	0.9967	0.9969	0.9971	0.9973
2.8	0.9974	0.9976	0.9977	0.9979	0.9980
2.9	0.9981	0.9982	0.9984	0.9985	0.9986

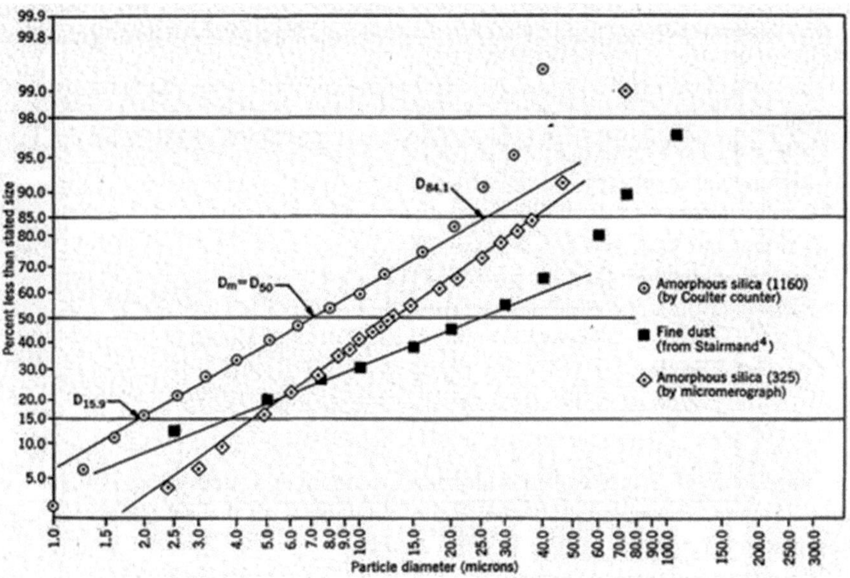

Fig. 16.6 Particle size distributions

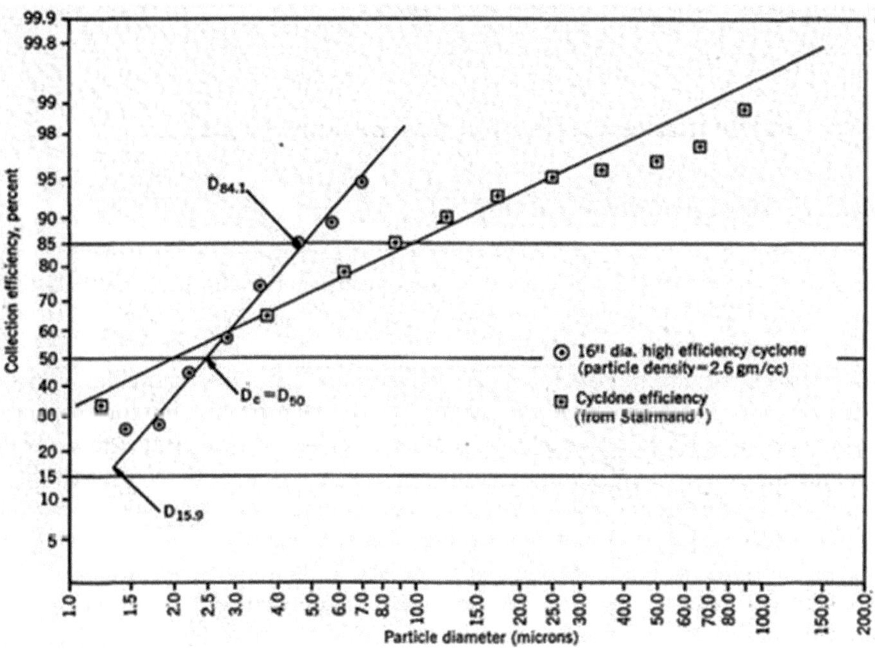

Fig. 16.7 Cyclone fractional efficiency curves

to specify the curve are at the 50% point, in this case termed the cut diameter (*Dc*) and the geometric standard deviation (σ_{gc}).

The collection efficiency of the 1160 silica on the 16 in. diameter cyclone is computed as follows.

From Figs. 16.8 and 16.9:

Mass median diameter $D_m = 7.2\mu$.

Geometric standard deviation $\sigma_g = \frac{25}{7.2} = 3.47$.

Cyclone cut diameter $D_c = 2.5\mu$.

Geometric standard deviation $\sigma_{gc} = \frac{4.6}{2.5} = 1.84$.

Fractional efficiency can be determined by Eq. (16.50):

$$\eta_{OA} = \text{erf}\left[\frac{\text{In}\frac{7.2}{2.5}}{\sqrt{\text{In}^2 1.84 + \text{In}^2 3.472}}\right] = \text{erf}(0.7631).$$

From Table 16.3 the overall collection efficiency is found to be 77.73% which can be compared with the value of 77.5% obtained from the histogram calculation presented in Fig. 16.8.

The above calculation is repeated by using the graphical solution of Eq. 16.50 presented in Fig. 16.8. The denominator of Eq. 16.50 is first obtained from the lower graph of Fig. 16.9 as $D = 1.4$ at $\sigma_g = 3.47$ and $\sigma_{gc} = 1.84$. $D = 1.4$ is used in the upper graph to read the overall efficiency of 77.5% at $D_m/D_c = 7.2/2.5 = 2.88$.

16.5 Performance Evaluation and Enhancement

16.5.1 Performance Evaluation

In general, the cyclone can be classified as conventional, high-efficiency, and high-throughput types. In general, conventional cyclones are able to achieve 70–90% PM removal which is mainly contributed by the high removal efficiency of large particles. High-efficiency cyclones can collect more than 90% of smaller particles with an aerodynamic diameter of less than 5 μm attributing to the high-pressure drop throughout the collector. For high-throughput cyclone, it is noted that this type of cyclone can only efficiently collect particles larger than 20 μm. To meet the stringent air emission regulations in the future, the conventional cyclone cannot remove fine particles successfully and should incorporate with the wet scrubber, bag-house filter, and electrostatic precipitator which will be stated as the following charters.

The power requirement for a cyclone collector mainly comes from the energy that is used to drive the gas flowing through the collector. Based on the pressure drop across the collector, the energy consumption by a cyclone can be calculated as:

$$\text{Electricity}_{\text{fan}} = \frac{Q \times \Delta P}{\varepsilon}. \tag{16.48}$$

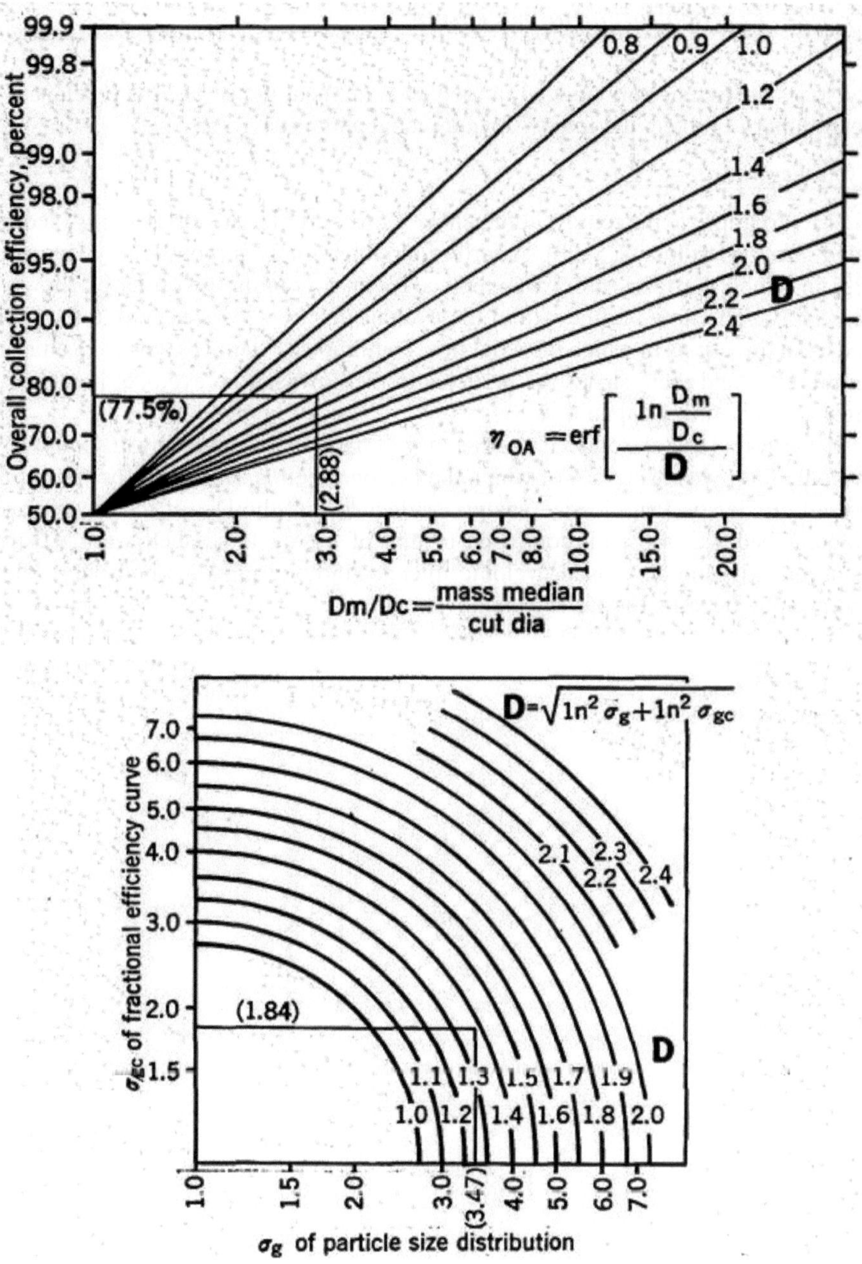

Fig. 16.8 Overall efficiency calculator, graphical solution of Eq. 16.50

16.5.2 *Performance Improvement*

Cyclones are extensively utilized in food and chemical industries as a precleaner or conditioner to protect the follow-up equipment and/or devices.

Cyclonic Filter

In order to improve the removal efficiency of fine particles, a fabric filter is incorporated into the cyclone [25–28]. A further modification was proposed by Gao et al. [29] in which the fabric filter was replaced with a granular bed filter as shown in Fig. 16.10. The flow regime within the cyclone can be separated into to the radial and axial directions in which the axial flow continuously removes dust cake attached on the filter and then carries the particles to the downstream cyclone; it exhibits a great self-sustaining capability.

It was found that the collection efficiency of particulate matters by this innovative cyclonic filter was over 95% of which the cyclone shell is accounted for 80% of the overall efficiency. Particle size analysis indicated that the D_{50} of the incorporated granular bed was 4.6 μm, compared to that of the cyclonic shell was 18.0 μm. Consequently, the developed cyclonic filter technology provides both fabric filter and cyclone benefits for particles removal.

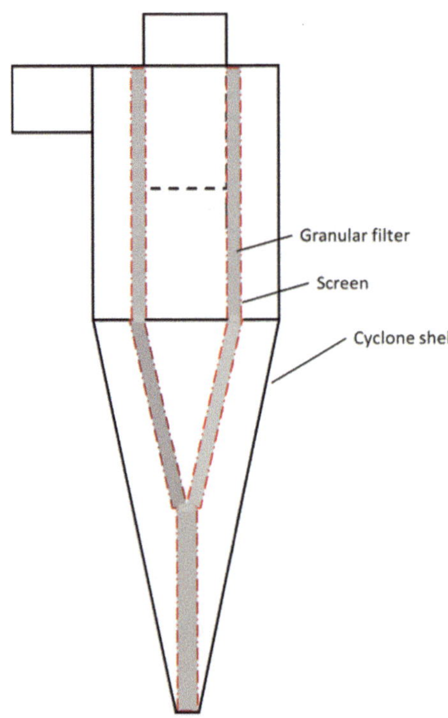

Fig. 16.10 Schematic diagram of the cyclone with circulating granular bed filter

A traditional gas cyclone in cooperated with an electroacoustic ultrasonic nebulizer can also help fine particles grow and agglomerate into bigger particles for separating fine particles more effectively. In addition, particles can also be wetted by the vapor, which increases their changes to agglomerate due to the condensational growth and collision coalescence.

For instance, wetting agents such as Sodium dodecyl benzene sulfonate (SDBS), Polyethylene glycol-4000 (PEG), and Sodium dodecyl sulfate (SDS) can increase the wettability of dust particles especially for hydrophobic dust thereby enhancing particle agglomeration and improving the dust removal performance [27, 28].

Consequently, fine particles can be effectively enlarged and separated by the centrifugal force field in the cyclone. More specifically, the removal efficiency of the molecular sieve dust increased from 66.2 to 92.9% by adding SDS to the vapor as shown in Fig. 16.11.

Electrocyclone

The electrocyclone is a combination of the cyclone and electrostatic precipitator where the efficiency can be enhanced by generating the electrostatic force which is incorporated into the centrifugal force. As seen in Fig. 16.12, the discharge electrode or vortex finder installed in the central axis of cyclone serves as the corona generator to produce high-voltage electricity [30, 31].

Since it is the combination of the cyclone and electrostatic precipitator, the particle collection of the electrocyclone increased along the residence time [32], the applied voltage [30], dust loading [30], and the effective length [33]. Lim, Lee [34] suggested that the collection efficiency for small-size particles is drastically enhanced which can reach up to 99% for 0.5–10 μm particles and the cut diameter is typically below 2.0 μm. Chen and Wang [35] performed a cost-benefit analysis and concluded that

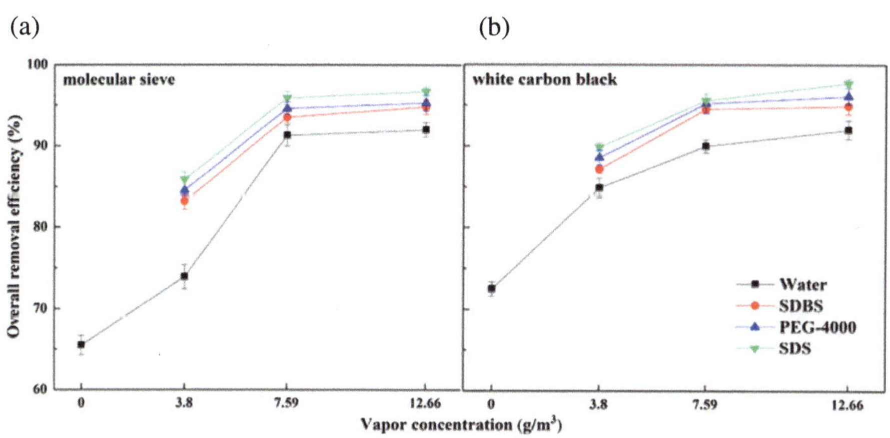

Fig. 16.11 The overall removal efficiencies for different concentrations of ultrasonic vapors of molecular sieve dust (**a**) and white carbon black (**b**)

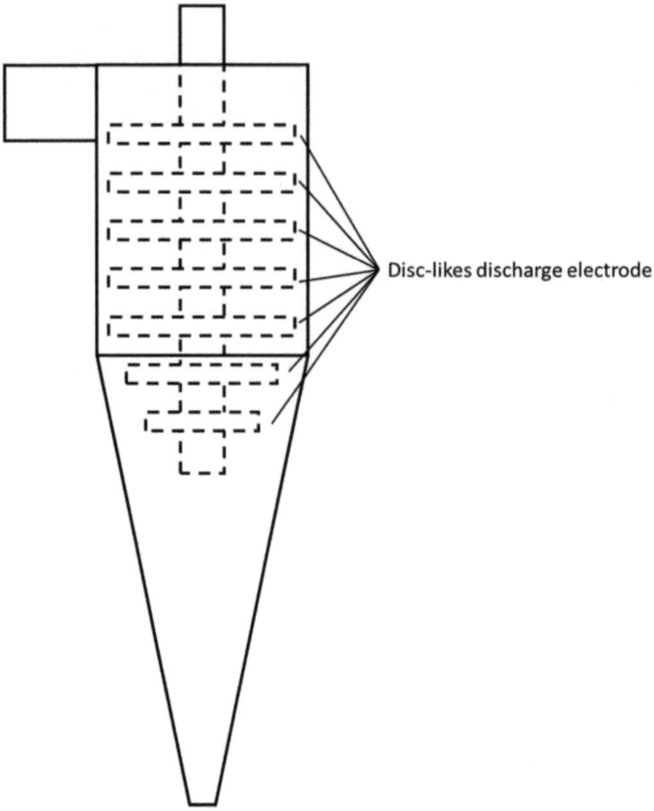

Fig. 16.12 Schematic diagram of electro cyclone

under the same dust loadings the operating cost of electrocyclone was only one-third of the conventional cyclone.

16.5.3 Innovative Technologies

A cyclone is a gas cleaning device that employs centrifugal force, which spins a gas stream into a swirling action to remove particles throughout it. Modifications of the cyclone separator dimensions with higher efficient and lower cast were investigated. For instance, the dimensions of a 1D3D cyclone have a vertical height equal to one-cylinder diameter, and a cone height equivalent to three-cylinder diameter. In was evident that inlet velocity decreased with increasing inlet dimensions, with an optimum ratio of inlet width to inlet height in the range of 0.5–0.7 [36]. Incorporating filters such as Al_2O_3 foams and zeolites with microscopic pores and channels to

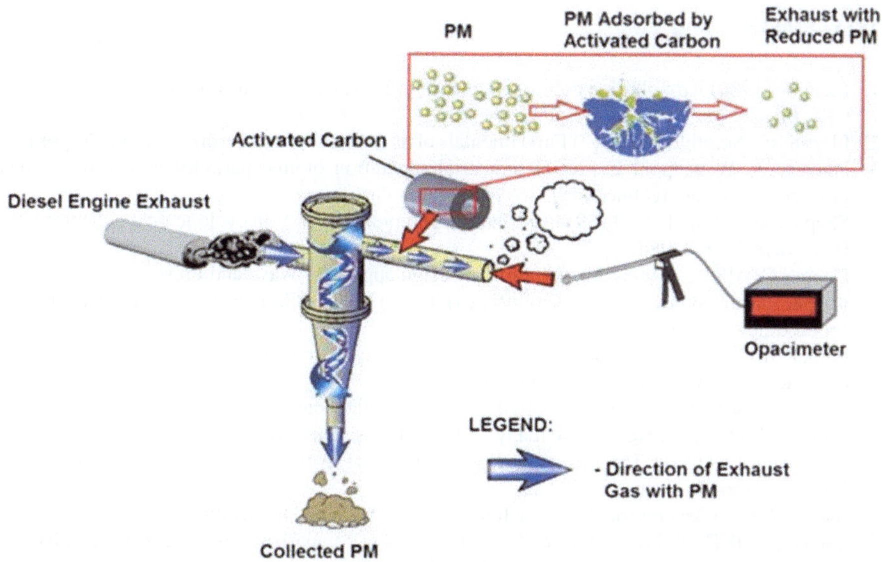

Fig. 16.13 Schematic diagram of cyclone equipped with an activated carbon for PM reduction

the outlet of the cyclone may help collect the remaining particulates with smaller aerodynamic diameters.

Another example reported by Duran and Caldona [37] which indicates that the design, fabrication, and operation of a cyclone equipped with an activated carbon having the co-benefit of reducing the pressure drop four times of the original value.

Exercises

1. Define the following terms and explain why they should be considered for design use: natural length, number of gas revolutions, grade efficiency, and cyclone fractional efficiency curve.
2. Please describe the general procedures for designing a cyclone.
3. List the major factors affecting the design of a cyclone.
4. Briefly describe a rapid method for prediction of overall collection efficiency suggested by Ronald E. Sundberg.
5. A cyclone is designed to run at 60% efficiency. A gas stream consists of monodispersed particles with a diameter of 10 mm. The viscosity of the gas, the specific gravity of the particle, the effective number of turns, and width of cyclone inlet are 1.17 × 1025 lb/ft s, 2.0, 5.0 and 3 ft, respectively. When is the inlet gas velocity, assuming that the cyclone must operate at its design efficiency?

References

1. Licht W (1988) Air pollution control engineering: Basic calculations for particulate collection. CRC Press
2. Flagan RC, Seinfeld JH (2012) Fundamentals of air pollution engineering. Courier Corporation
3. Peukert W, Wadenpohl C (2001) Industrial separation of fine particles with difficult dust properties. Powder Technol 118:136–148
4. Shephered C, Lapple C (1939) Flow pattern and pressure drop in cyclone dust collectors. Ind Eng Chem 31:972–984
5. Cooper C, Alley F (1994) Air pollution: a design approach. Waveland Press
6. Kaspar P, Mihalski K, Parnell Jr C (1993) Evaluation and development of cyclone design theory. Proceedings (USA)
7. Parnell Jr C (1996) Cyclone design for air pollution abatement associated with agricultural operations. In: Beltwide cotton conferences 1996
8. Wark K, Warner CF (1981) Air pollution: its origin and control
9. Lapple CE (1951) Processes use many collector types. Chem Eng 58:144–151
10. Stairmand CJ (1951) The design and performance of cyclone separators. Trans Inst Chem Engrs. 29:356–362
11. Swift P (1986) An empirical approach to cyclone design and application. Filtr Sep 23:24–27
12. Dirgo J, Leith D (1985) Cyclone collection efficiency: comparison of experimental results with theoretical predictions. Aerosol Sci Technol 4:401–415
13. Whitby KT, Lundgren D, Peterson C (1965) Homogeneous aerosol generators. Int J Air Water Poll 9:263–277
14. Mihalski K, Kaspar P, Parnell Jr C (1993) Design of pre-separators for cyclone collectors. In: Proceedings (USA)
15. Simpson S (1995) New low-pressure cyclone design for cotton gins. In: Proceedings of the belt wide cotton production conferences, Memphis, Tenn National Cotton Council, pp 680–7
16. Wang L, Parnell Jr CB, Shaw BW (1999) Performance characteristics for the 1D2D, 2D2D, 1D3D, and barrel cyclones. ASAE Paper
17. Kalen B, Zenz F (1974) Theoretical-empirical approach to saltation velocity in cyclone design. In: AIChE symposium Series, p 388
18. Ranz WE (1985) Wall flows in a cyclone separator: a description of internal phenomena. Aerosol Sci Technol 4:417–432
19. Leith D (1972) The collection efficiency of cyclone type particle collectors-a new theoretical approach. In: AIChE symposium Series, pp 196–206
20. Licht W (1988) Air pollution control engineering basic calculations for particulate collection, 2nd edn
21. Casal J, Martinez-Benet JM (1983) Better way to calculate cyclone pressure drop. Chem Eng 90:99–100
22. Yuu S, Jotaki T, Tomita Y, Yoshida K (1978) The reduction of pressure drop due to dust loading in a conventional cyclone. Chem Eng Sci 33:1573–1580
23. Masin J, Koch W (1986) Cyclone efficiency and pressure drop correlations in oil shale retorts. Environ Prog Sustain Energy 5:116–122
24. Sundberg RE (1974) The prediction of overall collection efficiency of air pollution control devices from fractional efficiency curves. J Air Pollut Control Assoc 24:758–764
25. Sibanda V, Greenwood R, Seville J (2001) Particle separation from gases using cross-flow filtration. Powder Technol 118:193–202
26. Sibanda V, Greenwood R, Seville J, Ding Y, Iyuke S (2010) Predicting particle segregation in cross-flow gas filtration. Powder Technol 203:419–427
27. Sharma SD, Dolan M, Ilyushechkin A, McLennan K, Nguyen T, Chase D (2010) Recent developments in dry hot syngas cleaning processes. Fuel 89:817–826
28. Sharma SD, McLennan K, Dolan M, Nguyen T, Chase D (2013) Design and performance evaluation of dry cleaning process for syngas. Fuel 108:42–53

29. Gao S, Zhang D, Fan Y, Lu C (2018) A novel gas-solids separator scheme of coupling cyclone with circulating granular bed filter (C-CGBF). J Hazard Mater
30. Shrimpton JS, Crane R (2001) Small electrocyclone performance. Chem Eng Technol 24:951–955
31. Gradoń L, Luckner H, Podgorski A, Wertejuk Z (1998) Separation of neutral and charged aerosol particles in cyclones with external electric field. J Aerosol Sci, S927-S8.
32. Chen C-J (2001) Enhanced collection efficiency for cyclone by applying an external electric field. Sep Sci Technol 36:499–511
33. Lim K, Kim H, Lee K (2004) Comparative performances of conventional cyclones and a double cyclone with and without an electric field. J Aerosol Sci 35:103–116
34. Lim K, Lee K, Kuhlman M (2001) An experimental study of the performance factors affecting particle collection efficiency of the electrocyclone. Aerosol Sci Technol 35:969–977
35. Chen C-J, Wang LF (2001) Cost-benefit analysis of electrocyclone and cyclone. Resour Conserv Recycl 31:285–292
36. Wang C, Zhang Y, Dong K, Wang B, Li S, Xin R, Jiang Y (2020) Enhanced collection of fine particles in a cyclone using ultrasonic vapor with surfactants. Adv Powder Technol (Article in press)
37. Duran JZ, Caldona EB (2019) Design of an activated carbon equipped-cyclone separator and its performance on particulate matter removal. Part Sci Technol. https://doi.org/10.1080/02726351.2019.1607637

Chapter 17
Wet Scrubber

Nomenclature

a	Radius of the collecting sphere
A_c	The cross-sectional area of the unit
A_t	The cross-section area of the throat
b	Radius of the outer boundary
b'	The dimensionless radius of the boundary layer
C_D	The drag coefficient
d_i	The particle diameter
dN_{p}	The number of particles collected by the collection unit
D_{e}	Equivalent diameter
$\overline{D_D}$	The Sauter mean diameter
f_{t}	Turbulent friction factor
G	The volumetric gas flow rate
K_P	The characteristic inertia parameter in the throat
ℓ_t	The Venturi throat length
n_{p}	The particle concentration in the laden gas
p_L	The inlet liquid pressure
Pe	Peclet number
P_n	Fractional penetration ratios
P_T	The total contacting power of the scrubber
ΔP	The pressure drop
Q_L	Liquid and gas feed rates
Q_G	Liquid and gas feed rates
r	Position coordinates in the radial directions
u	The flow velocity at the concentric boundary
u_{De}	The droplet velocity at the exit of the throat
u_{Gt}	The gas velocity in the throat
v_{D}	The velocity of the droplet
v_{g}	The gas velocity

P.-C. Chiang and X. Gao, *Air Pollution Control and Design*,
https://doi.org/10.1007/978-981-13-7488-3_17

x	The liquid injection
X	The characteristic throat length
z	The longitude position of the gas inlet
η_{fan}	The fan efficiency
η_i	The fractional collection efficiency
η_{int}	Single-sphere interception efficiency
η_{overall}	The overall removal efficiency
θ	Position coordinates in the circumferential directions
$\mu_D$nn	The fluid viscosity
ρ_D	The density of the particles
ρ_{sat}n	The density of gas at saturation
σnn	The surface tension
ψ_b	Stream function at r = b.

17.1 Introduction

17.1.1 Features of Wet Scrubber

Wet scrubbers refer to air pollution control equipment which can be divided into low energy scrubber (L1, L2, L3), medium (M1, M2, M3), and high energy scrubber (H1, H2, H3), 1 presented in Table 17.1. The scrubber selection should be based on economic, environment, and engineering considerations.

According to this, technical guidelines for selecting the wet scrubber should be considered as follows:

(1) The introduction of liquid to the gas is permissible to the process.

Table 17.1 Summary of scrubber performance

	Low energy scrubber			Medium energy scrubber			High energy scrubber
Types of wet scrubber	L1 gravity spray	L2 packed bed	L3 mechanically induced spray	M1 centrifugal or wet cyclone	M2 self-induced spray	M3 impingement plate	H1 venture
Efficiency	80	99	99	87	93	97	99
Pressures drop (in water gauge)	0.5–2.0	0.5–1.0 (in.wg/ft bed)	1.5–4.0	1.0–15.0	2.0–15.0	1.0–8.0	5.0–70.0
Liquid consumption range (Gal/1000 ft^3)	5–20	8–20	4–5	5–15	1–5	3–5	2–40

(2) The liquid can be purged from the process without causing water pollution. A satisfactory effluent treatment system must be provided.
(3) The gas must be cooled in any event.
(4) Combustible particles or gasses must be treated with minimum risk.
(5) Vapors or gaseous matter and particulates must be removed from the gas.

In wet scrubbers, chemical compounds can be mixed with liquid, and the scrubbers can be used to absorb gaseous pollutants rather than just be limited to particle collection. Since the gaseous pollutants by absorption processes have been introduced in Chap. 13, this chapter, wet scrubber, was only focused on PM removal. The structure of a wet scrubber system is presented in Fig. 17.1 which indicates that the pretreated or conditioned flue gas flows into the scrubber. Meanwhile, water or solvent (for removal of gaseous pollutant) is injected at the top of the scrubber and flows downward. The effluent is collected and partially treated by the wastewater treatment process and recycled. In order to reduce the possibilities of the white plume, the scrubber is usually equipped with a mist eliminator and users can also install a reheater that is used to raise the temperature of the gas above the dew point. The operation of the system including gas and the liquid flow rate is controlled based on collected data.

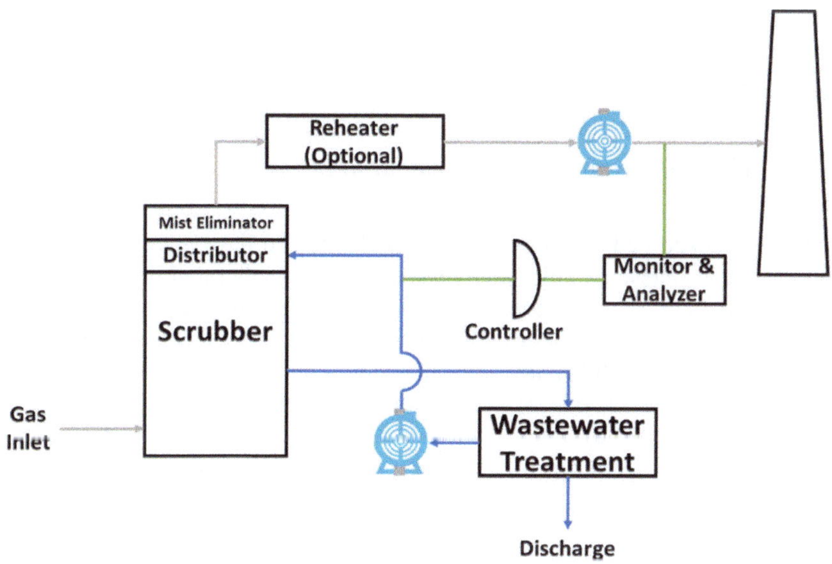

Fig. 17.1 Schematic diagram of wet scrubber system

17.1.2 Applications of Wet Scrubber

A well designed wet scrubber can be used to remove several gaseous pollutants simultaneously so that the flue gas treatment process is simplified. In addition, when a liquid, usual water, is used as the agent, wet scrubbers can reliably collect flammable and explosive dusts. Re-entrainment of particles is reduced to a large extent once the particles are trapped in the liquid. If the treatment process has a specific temperature requirement, then the temperature of the wet scrubber can be adjusted in advance. Other advantages of wet scrubbers include the ability to handle high-temperature and high-humidity gas streams, moderate capital cost, relatively low-pressure drop, and small space requirement.

The most significant disadvantage of the wet scrubber is that it requires wastewater treatment, otherwise secondary pollution may occur. In addition, particles are collected in the form of sludge which makes it difficult and expensive to convey and dispose of. Corrosion problems are more severe compared to dry FGD systems. Wet scrubbers may cause additional energy loss, for instance, momentum loss is inevitable during the cooling process prior to the wet scrubber. On the other hand, reheating may be required to enhance the momentum and heat flu of flue gas for plume rise and eliminate the mist plume for visibility.

17.2 Principles

17.2.1 Hydrodynamics in Wet Scrubbers

In general, packed beds are filled with granules which have a spherical or nearly spherical shape. These granules are closely packed together and mutually interfere with the flow field around the neighboring spheres. For this reason, the flow field model developed by Kuwabara [1] should be introduced by considering the flow field around a single solid sphere with boundary conditions that the velocity at the collecting sphere surface (r = a) and the vorticity at the outer boundary surface (r = b) are zero (see Fig. 17.2).

As a result, the stream function around the solid is given below [2]:

$$\psi = \frac{ua^2}{K}\left(\frac{1}{4}\left(1-\frac{2}{5}\alpha\right)\frac{a}{r}-\frac{3r}{4a}+\frac{1}{2}\left(1+\frac{1}{2}\alpha\right)\frac{r^2}{a^2}-\frac{3}{20}\alpha\frac{r^4}{a^4}\right)(\sin\theta)^2, \quad (17.1)$$

where a and b are radius of the collecting sphere and the outer boundary, respectively;α is the ratio of the volumes of collecting media to the space defined by the outer boundary, which expressed as $\alpha = a^3/b^3$; K is estimated by $K = 1-(9/5)\alpha^{1/3}+\alpha-(1/5)\alpha^2$; u is the flow velocity at the concentric boundary and at $\theta = 0$; r and θ are position coordinates in the radial and circumferential directions.

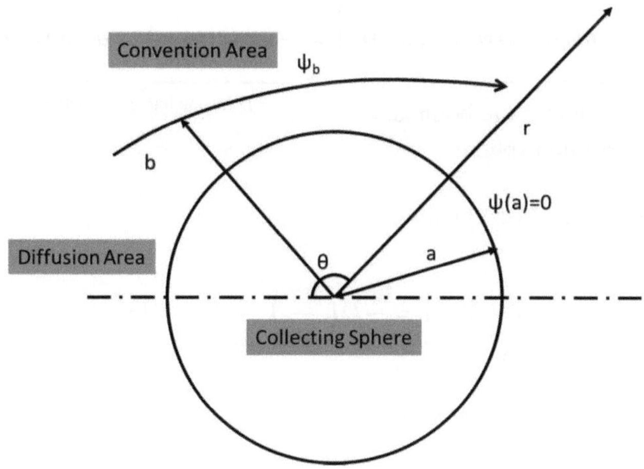

Fig. 17.2 Schematic diagram of stream around a sphere

This model predicts that with an increase in the packing density of droplets, i.e., α, the radius of the outer boundary is reduced, thereby making the individual collection efficiency of a single droplet higher.

Since Eq. (17.1) is too complicated to be performed in the calculation, the formula is simplified. First, by assigning $\psi' = \frac{\psi}{a^2 u}$, and $r' = r/a$, we have the dimensionless stream function which is

$$\psi' = \frac{1}{K}\left(\frac{1}{4}\left(1 - \frac{2}{5}\alpha\right)\frac{1}{r'} - \frac{3r'}{4} + \frac{1}{2}\left(1 + \frac{1}{2}\alpha\right)r'^2 - \frac{3}{20}\alpha r'^4\right)(\sin\theta)^2. \quad (17.2)$$

Equation (17.2) can be further reduced into

$$\psi' = \frac{3(1-\alpha)}{4K}\frac{(r'-1)^2}{r'^{[(1+2\alpha)/3-3\alpha]}}(\sin\theta)^2. \quad (17.3)$$

Experimental results have revealed that the value of α falls in the range of around 0.6. Equation (17.3) thus is approximately:

$$\psi' = \frac{3(1-\alpha)}{4K}\left(\frac{r'-1}{r'}\right)^2(\sin\theta)^2. \quad (17.4)$$

Introducing the stream function, we find the particle concentration balance around the collecting sphere:

$$\underbrace{d\left[-\int_0^{2\pi\psi_b} n(\psi,\theta)d\psi + n_0 2\pi\psi_b\right]}_{\text{Convection into the boundary around sphere}} = \underbrace{-2\pi D\left(\frac{\partial n}{\partial r}\right)_{r=a} a^2 \sin\theta d\theta}_{\text{Diffusion into inner surface}}. \quad (17.5)$$

Integration of Eq. (17.5) yields

$$d\int_0^{n_0} \psi dn = -D\left(\frac{\partial n}{\partial r}\right)_{r=a} a^2 \sin\theta d\theta. \quad (17.6)$$

In dimensionless form, it becomes

$$d\int_0^1 \psi' dn' = \frac{-2}{Pe}\left(\frac{\partial n'}{\partial r'}\right)_{r=a} \sin\theta d\theta = \left(\frac{3\pi}{4Pe}\right)^{2/3}\left(\frac{1-\alpha}{K}\right)^{1/3}, \quad (17.7)$$

where n' represents an arbitrary particle concentration profile which can be estimated by Friedlander [3] suggested as follows:

$$n' = \frac{(r'-1)/r'}{(b'-1)/b'}, \quad (17.8)$$

where b' is the dimensionless radius of the boundary layer which is calculated by $b' = b/\alpha$. According to the above analysis, the collection efficiency can be solved.

17.2.2 Collection Efficiency of Wet Scrubber

Wet scrubbers can be categorized into two types where (1) liquid droplets directly form a collecting medium and (2) liquid and packing materials produce a wetted collecting medium. As indicated in Fig. 17.3, no matter which type is, three mechanisms dominate particle removal including inertia impaction, interception, and Brownian diffusion.

Removal efficiency by diffusion is a function of particle size, droplet diameter, and intensity of diffusion which is a function of temperature. Diffusion is more obvious for particles smaller than 0.1 μm due to their higher diffusion coefficient. The efficiency of particles collected by an individual sphere is theoretically expressed as follows [2]:

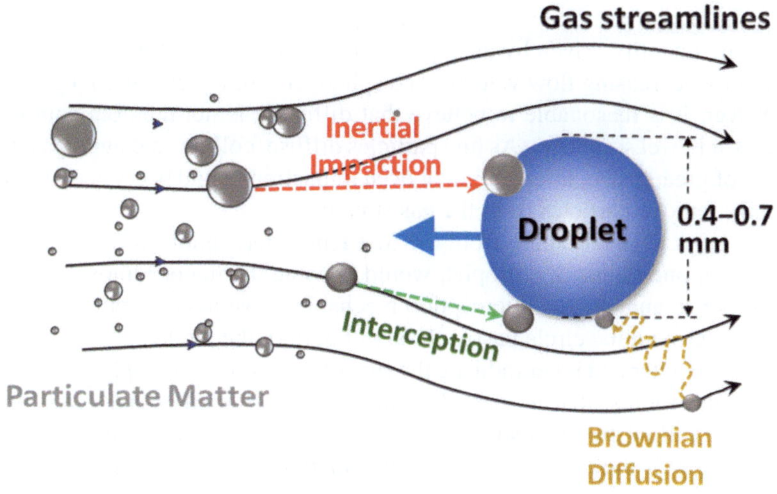

Fig. 17.3 Schematic diagram of removal mechanisms of particulate matters by wet scrubber [4]

$$\eta_{\mathrm{diff}} = \frac{2\pi Da^2 \int_0^{\pi} \left(\frac{\partial n}{\partial r}\right)_{r=a} \sin\theta d\theta}{\pi u a^2 n_0} = \frac{4}{Pe} \int_0^{\pi} \left(\frac{\partial n'}{\partial r'}\right)_{r'=1} \sin\theta d\theta. \tag{17.9}$$

Taking the stream function into consideration, Eq. (17.9) becomes

$$\eta_{\mathrm{diff}} = 2\left(\frac{3\pi}{4Pe}\right)^{2/3} \left(\frac{1-\alpha}{K}\right)^{1/3}. \tag{17.10}$$

Equation (17.10) is graphically presented in Fig. 17.4. The diffusion efficiency is dependent on the two-thirds power of the Peclet number (*Pe*) which represents the

Fig. 17.4 The variation of single-sphere diffusion efficiency with different *Pe* values and solidities

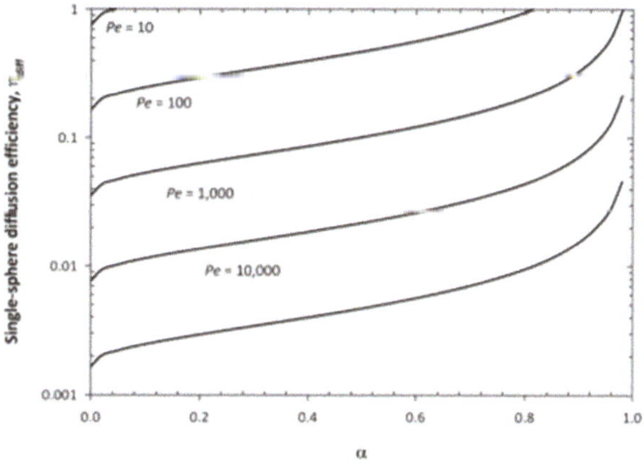

ratio of the rate of advection of a physical quantity by flow to the rate of diffusion of the same quantity driven by an appropriate gradient. Therefore, we can enhance diffusion by decreasing flow velocity and sphere size (characteristic length).

However, it is reasonable to believe that diffusion is not the dominant removal mechanism in wet scrubbers. As fine particles diffuse, collide, and aggregate to form particles of greater size, interception or even impaction would become dominant and efficiently remove particles from the gas stream.

When particle size ranging from 0.1 to 1 μm, interception, which occurs in the space within one radius of a droplet, would become the major removal mechanism [5]. In order to analyze the interception mechanism, we assume an imaginary flow field that is around the circle passes through not only the circle but also the annular region whose outer and inner radii are the sum of the radius of the sphere and particle, and the sphere radius, all at $\theta = \pi/2$. In addition, it is assumed that particles subject to interception follow the exact streamline of the flue gas so single-sphere interception efficiency (η_{int}) is the ratio of the above imaginary circle to the projected area of the sphere

$$\eta_{\mathrm{int}} = \frac{2\pi\psi_b}{\pi a^2 u}, \tag{17.11}$$

where ψ_b is the stream function at $r = b$. For the purpose of enhancing interception, we increase the spraying intensity in the system [6].

Impaction is the dominant collection mechanism in wet scrubbers for particles with large masses, especially for particles whose diameters are larger than 5.0 mm. In general, particles tend to follow the streamline of the exhaust gas stream. However, when liquid droplets exist in the space, the streamline of the exhaust gas is distorted so large particles are unable to maintain their original trajectory due to the inertia momentum. As a result, the particles diverge around the droplet and do not always follow the original stream and hit onto droplets and other obstacles. It has been found that inertial impaction is a function of particle mass and velocity, predominantly with high gas velocities and large particles [4].

17.3 Design Methods

17.3.1 Theoretical Approach

A theoretical approach for the scrubber design can be performed by the Calvert Cut Diameter Method which was developed by assuming that the fractional collection efficiency is a function of particle diameters. Figure 17.5 presents an example of the particle size distribution profile which exhibits a Gaussian distribution transforming into a probability-log form [7] for particle size estimation. In this case, the geometric mean diameter is indicated by the 50% in size distribution on mass base while the standard deviation is quantified as follows:

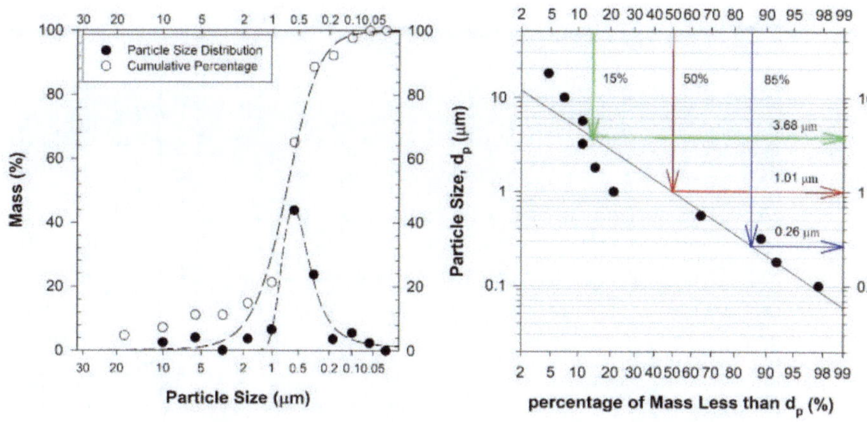

Fig. 17.5 Examples of **a** Frequency for analyzing particle size distribution of PM in the flue gas. **b** Probability-log plot for analyzing particle size distribution of PM [8]

$$\sigma = \frac{d_{85}}{d_{50}}, \tag{17.12}$$

where d_{85} and d_{50} are 85th and 50th percentile particle sizes on a mass basis.

In designing a particle controlling device, one important parameter is the cut diameter denoted as d_{p50} which is defined as the size of particles whose collection efficiency is 50%. In general, the collection efficiency is higher for particles greater than d_{p50} and lower if the particles are small. As a result, the relationship between collection efficiency and particle size can be expressed as follows:

$$\eta_{\mathrm{i}} = 1 - e^{-A \cdot d_{\mathrm{i}}^{B}}, \tag{17.13}$$

where η_i (-) is the fractional collection efficiency, d_i (μm) represents the particle diameter, and A and B are constants dominated by the scrubber design. For venturi, packed bed, and tray wet scrubbers B is equal to 2.0, while for cyclonic wet scrubber B is 0.67.

The cut/power relationship was developed to clarify the power demand for the desired cut diameter, as shown in Eq. (17.14).

$$\eta = 1 - e^{-A \cdot d_p^{2}}. \tag{17.14}$$

Figure 17.6 presents the relationship between pressure drop and the fractional efficiency of the wet scrubber produced by MikroPul company. The higher the desired efficiency, the more energy is consumed. In addition, for a given pressure drop or power supply, the removal efficiency of large particles would be higher than that of a small one.

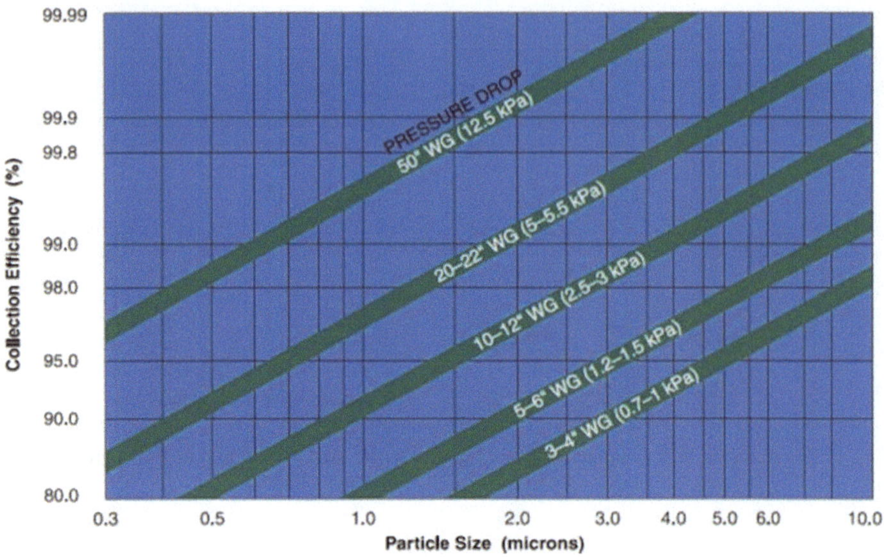

Fig. 17.6 Performance plot of pressure drop against fractional removal efficiency [9]

By observation of Fig. 17.6, the appropriate wet scrubber can be selected by the performance curves providing by the vendor. It is clear that the pressure drop in the scrubber could not be acceptable, if a high collection efficiency for fine particles is required.

17.3.2 Contact Power Theory

Like designing other kinds of particle collection equipment, dust load, particle size distribution, and the physicochemical properties of laden gas should be determined in advance. In addition, legislation requirements will influence the selection of equipment to a very large extent. Due to more stringent regulations and standards for $PM_{2.5}$, the performance of both spray towers and tray towers cannot meet the requirements.

Geometric dimensions, i.e., length and cross-sections, determine the performance of the Venturi scrubber. Figure 17.7 presents the schematic structure of a Venturi scrubber composing of (1) liquid injection valve, at 1; (2) converging section, from 0 to 2; and (3) throat, from 2 to 3. In most cases the liquid injection is positioned at 2, otherwise acceleration of drops should be carefully considered in the converging section and throat. For simplicity, the longitude position of the gas inlet is denoted as z while the liquid injection is x. Gas flows into the scrubber at $z = z_0$ and gets accelerated as the cross-section of the scrubber narrows along the longitude. The liquid is then injected into the gas stream at $z = z_1$, or $x = 0$ and gains velocity due to the fast gas stream. Subsequently, the gas–liquid mixture enters the scrubber though the throat in which particles collide with droplets and are removed.

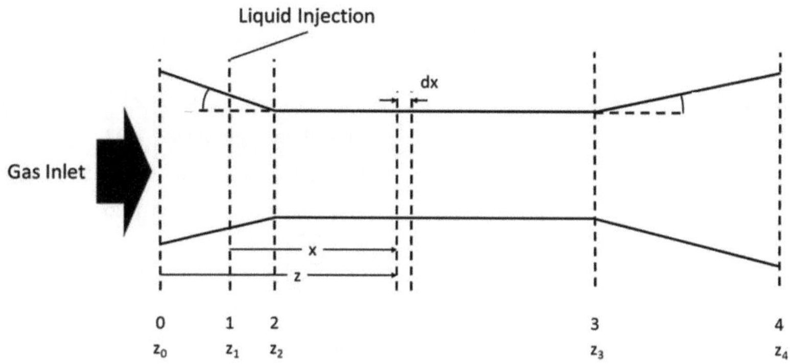

Fig. 17.7 Schematic diagram of a Venturi scrubber

Lapple and Kamack [10] summarized that the collection efficiency of a wet scrubber is dominated by provided power rather than the profile of the equipment. Further, a mathematical expression of this theory has been developed by Semrau [11] and is given as Eq. (17.15):

$$\eta = 1 - e^{-\alpha {P_T}^{\beta}}, \tag{17.15}$$

where α (-) and β (-) are constants related to the characteristics of particles, P_T (kWh/1,000 m^3) represents the total contacting power of the scrubber which equals the sum of power from a gas stream P_G (kWh/1,000 m^3) and the one from liquid P_L (kWh/1,000 m^3). As a result, P_T can be expressed by Eq. (17.16) [12]:

$$P_T = 2.724 \times 10^{-4} \Delta P + 0.28 p_L (Q_L / Q_G), \tag{17.16}$$

where ΔP (in H_2O) is the pressure drop, p_L (100 kPa) is the inlet liquid pressure, and Q_L and Q_G (m^3/h) are liquid and gas feed rates. Semrau [13] has obtained values of α and β regarding a Venturi scrubber, which are available in either metric or English units as shown in Table 17.2.

17.4 Design Procedures

In the context of escalating concerns about $PM_{2.5}$, venturi scrubbers have drawn much attention since it provides a great balance between performance and pressure drop [14]. For this reason, a brief introduction regarding the design of this innovative wet scrubber is included in this section. Figure 17.8 illustrates the procedures for designing a venturi scrubber starting with determining the operating conditions or

Table 17.2 Empirical parameters for different kinds of wet scrubbers in contact power theory [13]

Scrubber		Dust	α	β
Venturi scrubber		Talc dust	2.97	0.362
		Phosphoric acid mist	1.33	0.647
		Foundry cupola dust	1.35	0.621
		Open-hearth steel furnace fume	1.26	0.569
		Odorous mist	0.363	1.41
Venturi evaporator		Hot black liquor gas	0.522	0.861
Venturi and cyclonic spray		Lime kiln dust	1.47	1.05
		Black liquor furnace fume	1.75	0.620
		Ferrosilicon furnace fume	0.870	0.459
		Lime kiln dust	0.915	1.05
		Black liquor fume	0.740	0.861
Venturi condensation scrubber	Mechanical spray generation	Copper sulfate	0.390	1.14
	Hydraulic nozzles		0.562	1.06
Orifice		Talc dust	2.70	0.362
Cyclone		Talc dust	1.16	0.655

the L/G (liquid/gas) ratio. A proper L/G ratio ensures a satisfactory collection efficiency while reducing unnecessary water and energy consumption. Thereafter, the dimensions of the scrubber are estimated by assigning the gas velocity within the throat. Since geometric parameters and operating conditions are known, collection efficiency and pressure drop should be examined. Collection efficiency is validated by the infinite-throat model which is technical feasible when the characteristic throat length is greater than 2.0. Pressure drop analysis can be based on momentum balance or an empirical model developed by Yung et al. [15]. Several auxiliaries are needed to support the operation of the venturi scrubber including the induced draft fan system, mist eliminator, and wastewater collection and treatment system. In some cases, where the white plume is prohibited, a heat exchanger is necessary to raise the temperature above the dew point.

17.4.1 Step 1: To Determine the Appropriate Liquid to Gas Ratio

Table 17.3 summarizes the performances and operational conditions of different types of wet scrubbers, indicating that nearly 99% of particles can be collected in

Determination of the Requirement for Venturi Scrubber Operation

- Investigate characteristics of the pollutant that need to be removed.
- Probe the physico-chemical properties of exhausted gas stream.
- Study the emission regulation on the target substance.
- Important Design Factor: gas velocity, L/G ratio.

Step 1: To Determine the Appropriate Liquid-to-Gas Ratio

- Dust loading, pressure drop, collection efficiency, allowable air flow
- L/G Ratio: $\overline{D_D} = [283{,}000+793(L/G)^{1.932}]/v_g^{1.602}$........(17-17)

Step 2: To Determine the Geometric Dimensions

- Assumption: Particles are collected by infinite units in series
- Collection in a finite unit: $P_t = e^{-\eta\frac{A_c Z}{V}}$ ....(17-22)
- For Venturi scrubber:
  - Throat Length: $\ell_t = 3d_t$..(17-25)
  - Diverging Section Length: $\ell_{\text{div}} = 4d_t$....................(17-25)

Step 3: To Validate the Collection Efficiency

- Examine the collection efficiency: $\eta_{\text{overall}} = 1 - \sum_{d_p=1}^{n} P_n \times d_{p_n}$...(17-27)
- For venturi scrubber:
  - Infinite-throat Modelln $P = -B\frac{4K_P+4.2-5K_P^{0.5}\left(1+\frac{0.7}{K_P}\right)\tan^{-1}\sqrt{\frac{K_P}{0.7}}}{K_P+0.7}$...(17-28)

Step 4: To Estimate the Pressure Drop

- For gravitational scrubber, the estimation used for absorption is applicable.
- For venturi scrubber:
  - $\Delta P = P_{z3} - P_{z2} = -\rho_D u_{Gt}\frac{Q_L}{Q_G}u_{De} = -2\rho_D u_{Gt}^2\frac{Q_L}{Q_G}\left(1 - X^2 + X\sqrt{X^2-1}\right)$...(17-33)

Step 5. To Install Auxiliary Equipment

- Induced draft fan system
- Mist eliminator
- Wastewater collection and treatment system
- Heat exchanger

Fig. 17.8 The framework of designing a venturi scrubber

proper operating conditions. It is also noted that orifice and condensation scrubbers exhibit higher efficiency for fine particle collection. Meanwhile, spray tower, tray tower, and Venturi scrubbers can deal with massive gas loads and can be used as a pretreatment equipment to removal large particles. Furthermore, temperature of the flue gas should be regulated to prevent the liquid from freezing and/or boiling, which compromises the collection efficiency.

Table 17.3 Performance comparison among different types of wet scrubbers

Equipment	Removal Performance	Operating Conditions	Remarks	Ref
Spray Tower	• Particulate Matters: 70–99% • Inorganic Gases: 95–99% • Soluble VOC >90%	• Air Flow: 1,500–100,000 scfm • L/G ratio: 5.0–20.0 gal/1000 ft^3 • Temperature: 4–370 °C for PM control and 4–38 °C for gas absorption • Pollutant Loading: PM loading is dependent on liquid feeding and gaseous pollutant loading ranges from 250 to 10,000 ppmv	• Large amount of liquid is acquired for fine particle collection • Cheap	[16]
Tray Tower	• Particulate Matter: 50%–99% • Inorganic Gases: 95–99%	• Air Flow: 1,000–75,000 scfm • Temperature: 4–370 °C for PM control and 4–38 °C for gas absorption • Pollutant Loading: No practical limits to inlet PM concentrations	• Low removal efficiency for particles smaller than 1 μm • Relatively cheap	[17]
Venturi Scrubber	• Particulate Matter: 70–99%	• Air Flow: 500–100,000 scfm • L/G ratio: 3.0–40.0 gal/1000 ft^3 • Temperature: 4–400 °C • Pollutant Loading: Venturi scrubber is operational under 1–115 g/m^3	• Effective to control particles ranging from 0.5 to 5 mm • Expensive	[18]
Orifice Scrubber	• Particulate Matter: 80–99%	• Air Flow: 10,000–50,000 scfm • 0.5–5.0 gal/1000 ft^3 • Temperature: 4–150 °C • Pollutant Loading: Orifice scrubber is operational under 1–23 g/m^3	• Efficient to control $PM_{2.5}$ • Relatively expensive	[19]
Condensation Scrubber	• Particulate Matter: >99%	• Air Flow: 21,000 scfm • Temperature: 20–26 °C	• Specific to fine particles • Cheapest	[20]

The collection performance of a wet scrubber is affected not only by particle size distribution and dust loading but also by the liquid–gas ratio. Regarding liquid injection, the droplet diameter is an important parameter used to determine the rate of collision. For this reason, droplet diameters should be carefully estimated. One approach is shown as Eq. (17.17):

$$\overline{D_D} = \left[283,000 + 793(L/G)^{1.932}\right]/v_{\mathrm{g}}^{1.602}, \tag{17.17}$$

where $\overline{D_D}$ (μm) is the Sauter mean diameter, v_{g} (ft/s) is the gas velocity, and the L/G ratio is often expressed in gal/Mft^3. This equation is feasible for gas atomized sprays where droplets are atomized by the shattering impact of the high-speed gas stream.

In some cases, when a pneumatic nozzle is used and the relative velocity between gas and droplets need to be considered, the following equation is used for estimation of droplet diameter:

$$\overline{D_D} = \frac{585}{v_{\mathrm{rel}}}\sqrt{\frac{\sigma}{\rho_D}} + 597\left(\frac{\mu_D}{\sqrt{\sigma\rho_D}}\right)^{0.45}\left(\frac{1000Q_L}{Q_G}\right)^{1.5}, \tag{17.18}$$

where $v_{\mathrm{rel}} = v_{\mathrm{g}} - v_0$ in (m/s), σ (dynes/cm) is the surface tension, ρ_D (g/cm^3) is the density of the particles, μ_D (poises) is the fluid viscosity, and Q_L and Q_G are expressed in m^3/min which makes the L/G ratio a dimensionless factor.

For a chosen gas velocity (v_{g}) and droplet diameter, the gas flow surrounding the droplet can be described by Eq. (17.19):

$$Re_{\overline{D}} = \frac{\overline{D_D}\left(v_{\mathrm{g}} - v_{\mathrm{D}}\right)\rho_{\mathrm{g}}}{\mu_{\mathrm{g}}}, \tag{17.19}$$

where v_{D} is the velocity of the droplet, and ρ_{g} and μ_{g} are density and viscosity of flue gas, respectively. In summary, as long as we have the L/G ratio, the geometric dimensions of the scrubber can be readily determined (see Table 17.4).

Table 17.4 Calculation of drag coefficient [21]

Value of $Re_{\overline{D}}$	$\mathbf{C_D}$
$Re_{\overline{D}} < 0.1$	$C_D = \frac{24}{Re_{\overline{D}}}$
$0.1 \leq Re_{\overline{D}} < 0.5$	$C_D = \frac{24}{Re_{\overline{D}}} + 4.5$
$0.5 \leq Re_{\overline{D}} < 3.0$	$C_D = \frac{24}{Re_{\overline{D}}} + \frac{3.6}{Re_{\overline{D}}^{0.313}}$
$3.0 \leq Re_{\overline{D}} < 400$	$C_D = \frac{24}{Re_{\overline{D}}} + \frac{4}{Re_{\overline{D}}^{1/3}}$
$400 \leq Re_{\overline{D}} < 800$	0.44

17.4.2 Step 2: To Determine the Geometric Dimensions

Assuming that the ideal L/G ratio has been attained, the number of particles collected in a finite collection unit (Fig. 17.9), $dV = A_c dz$, is [22]

$$dN_p = -\eta n_p A_c v_g dt = -\eta n_p A_c dz, \tag{17.20}$$

where dN_p is the number of particles collected by the collection unit; n_p is the particle concentration in the laden gas; and A_c is the cross-sectional area of the unit. If the concentrations and velocities of collectors and particles are uniform within the unit, then

$$\frac{dn_p}{n_p} = -\eta \frac{A_c dz}{V}. \tag{17.21}$$

Integration of Eq. (17.21) leads to

$$P_t = e^{-\eta \frac{A_c Z}{V}}. \tag{17.22}$$

As a result, the dimensions of the wet scrubber could be calculated by the performance curve which shows the relationship of the scrubber efficiency to the contacting power as shown in Eq. (17.23):

$$NTU = \alpha \cdot {P_T}^{\beta}, \tag{17.23}$$

$$\eta = 1 - \exp(-N_t), \tag{17.24}$$

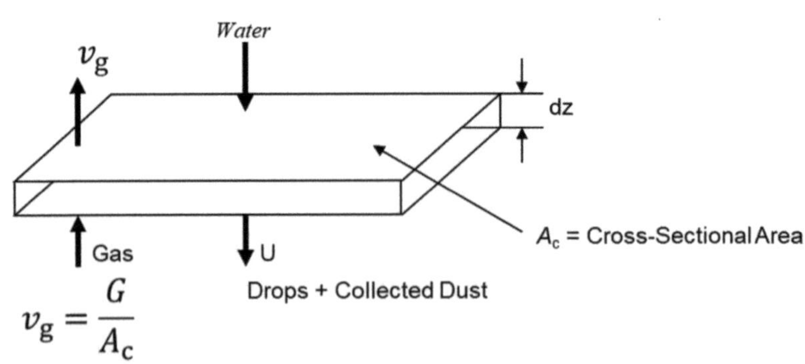

Fig. 17.9 Graphic presentation of a function unit within a wet scrubber

Table 17.5 Estimation of penetration number for various kinds of wet scrubbers

Wet scrubber	Penetration number, P_t	References
Spray tower	$P_t = e^{-\eta_1 3Lh/2GD}$	[23]
Packed column	$P_t = e^{-3.5\psi L/eD_c}$	[23]
Tray tower	$P_t = e^{-80F_L{}^2\psi}$	[24]
Impingement plate	$P_t = e^{-0.693d_{pa}{}^2/d_{pc}{}^2}$	[23]
Venturi scrubber	$P_t = e^{-B\frac{4K_P+4.2-5K_P^{0.5}\left(1+\frac{0.7}{K_P}\right)\tan^{-1}\sqrt{\frac{K_P}{0.7}}}{K_P+0.7}}$	[25]

where α and β are empirical constants that depend primarily on the aerosol (dust or mist) collected. Table 17.5 summarizes the mathematic presentations of penetration numbers for different scrubber types which can be used for estimating the number of transfer units.

The emphasis is placed on venturi scrubber throat which plays a central role in the collision among particles and droplets. In practice, designing a throat with proper cross-sectional area and length for venturi scrubber is somewhat empirical. The cross-sectional area should be sized to ensure that the gas stream can flow through at high speeds, v_t, which is estimated by the Bernoulli equation:

$$v_t = \frac{G}{A_t} = C\sqrt{\frac{\Delta P}{\rho_{sat}}}, \tag{17.25}$$

where G (m³/s) is the volumetric gas flow rate, A_t (m²) is the cross-section area of the throat, ρ_{sat} (kg/m³) is the density of gas at saturation, and C is the constant that correlates with L/G ratio which is defined as $C = 1060e^{-0.279L/G}$.

In general, increasing velocity not only increases the momentum of the particle and enhances the removal of fine particles by impaction but also produces a higher pressure drop. As a rule of thumb, if gas velocity at the throat ranges from 45 to 150 m/s, the corresponding pressure drop is 10–80 in w.c.

Based on the cross-sectional area of the throat, the length of that, ℓ_t, can be empirically estimated by Eq. (17.26)

$$\begin{cases} \ell_t = 3d_t, \\ \ell_{div} = 4d_t, \end{cases} \tag{17.26}$$

where d_t is the diameter or width of the throat, and ℓ_{div} represents the length of diverging section. It is noteworthy that a proper throat length and diverging section can reduce the pressure drop, thus reducing economic cost.

17.4.3 Step 3: To Validate the Collection Efficiency

An approximation of overall removal efficiency can be obtained by integrating the penetration along with the increase in particle size which is

$$\eta_{\text{overall}} = 1 - \int P_n d\left(d_{p_n}\right), \tag{17.27}$$

where η_{overall} is the overall removal efficiency, and P_n and d_{p_n} are fractional penetration ratios and weight fraction of the particle with aerodynamic diameter n, respectively. For discrete particle size distribution, Eq. (17.27) translates into

$$\eta_{\text{overall}} = 1 - \sum_{d_p=1}^{n} P_n \times d_{p_n}. \tag{17.28}$$

As a collision among the particulates and droplets is enhanced, the collection efficiency of a Venturi scrubber is improved for the fine particles. Allen [26] reported that the cut size of a Venturi scrubber can be lower than 0.5 μm. The collection efficiency of a Venturi scrubber can be estimated by assuming that the throat is the major component where collection occurs. This method adopts the Infinite-throat Model which was developed by Calvert [27] and refined by Yung et al. [25] as shown in Eq. (17.29):

$$\ln P = -B \frac{4K_P + 4.2 - 5K_P^{0.5}\left(1 + \frac{0.7}{K_P}\right)\tan^{-1}\sqrt{\frac{K_P}{0.7}}}{K_P + 0.7}, \tag{17.29}$$

where B is the characteristic L/G parameter calculated by $B = (L/G)\frac{\rho_L}{\rho_G C_D}$ and K_P is the characteristic inertia parameter in the throat which can be estimated by $K_P = \frac{d_p^2 v_G}{9\mu_G \overline{D_D}}$. Note that this equation is valid only when the characteristic throat length is greater than 2.0 which is defined by Eq. (17.30):

$$X = \frac{3\ell_t C_D \rho_G}{2\overline{D_D}\rho_L}, \tag{17.30}$$

where X is the characteristic throat length, ℓ_t (cm) is the Venturi throat length, C_D is the drag coefficient, ρ_G and ρ_L (g/cm^3) are the gas and liquid densities, and $\overline{D_D}$ (cm) is the droplet diameter defined by Eq. (17.17) or (17.18).

The relationship between B, K_{p}, and P_{t} can be visualized in Fig. 17.10 which provides a simple estimation of collection efficiency under specific operating conditions. When the amount of droplet (indicated by B) as well as particle size (indicated by K_{p}) increase, chances of collision between particles and droplets are enhanced. It

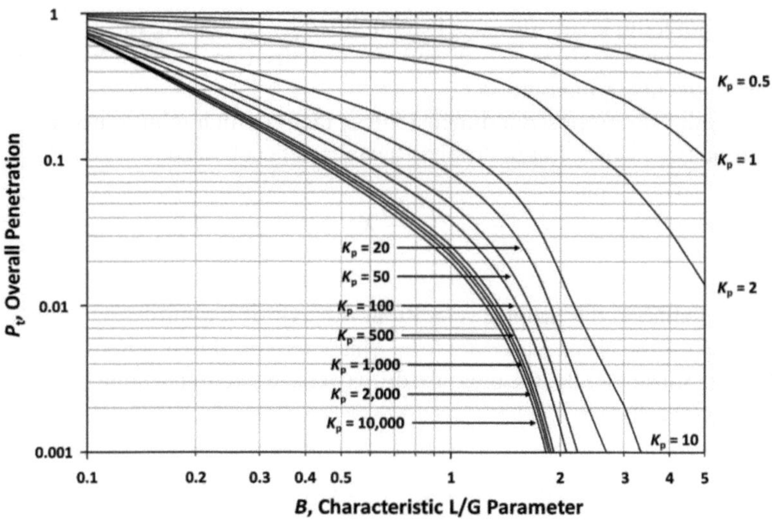

Fig. 17.10 Relationships among P_t, B, and K_p [25]

was found that increasing the L/G ratio would significantly influence the penetration and that it is necessary to maintain the value of B higher than 1.

17.4.4 Step 4: To Estimate the Pressure Drop

Pressure drop estimation for absorption reactors, given in Chap. 13, is applicable to gravitational scrubbers. As for venturi scrubbers, that information can be obtained by analyzing momentum balance within the control volume of length dx, which is

$$\underbrace{A_t dP}_{\text{Dragging Force}} + \underbrace{M_g dv_g + M_L dv_L}_{\text{Momentum}} + \underbrace{(M_g + M_L)\frac{f_t v_g^2}{2d_t}dx}_{\text{Wall Friction}} = 0, \tag{17.31}$$

where M_g and M_L are mass fluxes of gas and liquid streams, and f_t is the turbulent friction factor. This equation describes the velocity gain of gas streams and droplets under the pressure gradient. It should be noted that the last term represents wall friction based on the equivalent diameter D_e. The total pressure drop in the Venturi scrubber is thus the result of integrating Eq. (17.31) from $x = z_1$ to $x = z_4$:

$$\Delta P = -\int_{z_1}^{z_4} \frac{M_g dv_g + M_L dv_L + (M_g + M_L)\frac{f_t v_g^2}{2d_t}}{A_t} dx. \tag{17.32}$$

Furthermore, the integration can be simplified by neglecting the momentum loss of the gas stream or $M_g dv_g = 0$, and the wall friction is offset by the pressure recovery in the diverging section.

Yung et al. [15] proposed a dimensionless factor which represented the fraction of gas velocity to which liquid droplets have been accelerated

$$\mathrm{u}_c^* = \frac{u_{De}}{u_{Gt}} = 2\left(1 - X^2 + X\sqrt{X^2 - 1}\right), \tag{17.33}$$

where u_{De} (cm/s) is the droplet velocity at the exit of the throat, u_{Gt} (cm/s) is the gas velocity in the throat, and X is the characteristic throat length derived from Eq. (17.30). Subsequently, the pressure drop over the throat that contributes the most to total pressure drop is

$$\Delta P = P_{z3} - P_{z2} = -\rho_{\mathrm{D}} u_{Gt} \frac{Q_L}{Q_G} u_{De} = -2\rho_{\mathrm{D}} u_{Gt}^2 \frac{Q_L}{Q_G}\left(1 - X^2 + X\sqrt{X^2 - 1}\right). \tag{17.34}$$

Equation (17.34) can be readily expressed in a simple form by introducing the number of inlet velocity heads lost, N_{H}:

$$N_{\mathrm{H}} = \frac{2|\Delta P|}{\rho_G u_{Gt}^2} = 4\frac{\rho_{\mathrm{D}} Q_L}{\rho_G Q_G}\left(1 - X^2 + X\sqrt{X^2 - 1}\right) = 2\frac{\rho_{\mathrm{D}} Q_L}{\rho_G Q_G}\mathrm{u}_c^*. \tag{17.35}$$

17.4.5 Step 5: To Install the Auxiliary Equipment

An induced draft fan (ID fan) is an important component of a wet scrubber and is used to overcome pressure drop. Apart from the power of the equipment mentioned above, the location of the fan placement is dependent on the characteristics of flue gas and should be carefully considered. For instance, if a high dust load is expected, fans should be placed after the wet scrubber. It is noted that the selection of an ID fan should also consider corrosion-resistance to prolong the service time.

Mist eliminators are a necessary component for any kind of wet scrubber; otherwise, the remaining droplets in the gas stream will erode the equipment or even cause a significant pressure drop. In general, mist eliminators are chevron baffles or meshes which can prevent more than 90% of droplets from escaping the scrubber. As the gas stream passes through the mist eliminator, the droplets coalesce on it and grow larger. Finally, the droplet will fall by gravity and be collected by the wastewater collection system. It should be noted that mist eliminators require periodic washing to prevent clogging. In some cases, cyclone separators are used as entrainment controlling devices.

To prevent secondary pollution due to dissolved hazardous substances, a wastewater treatment system is needed. The collected wastewater after treatment can be partially recycled to reduce fresh water demand. Another portion of the water should be drained out of the system to maintain a proper solid concentration and prevent clogging. Attention should be paid to the final disposal of the drained solid or sludge: (1) if non-hazardous or inert, it can be put in landfill and/or recycled with or without treatment; (2) sophisticated treatments, i.e., solidification and stabilization, are usually required if the solid content is hazardous or toxic.

Heat exchangers including coolers and heaters are required if the temperature of the flue gas needs to be optimized; otherwise, the liquid would freeze or evaporate. In addition, the visible white plume would rise if flue gas is emitted directly, which may not be favorable for regulatory compliance or public acceptance. As a result, a re-heater is needed to heat the temperature above dew point thereby eliminating white plume.

17.4.6 A Rapid Method for Sizing Basic Components of Venturi Scrubber

A rapid method for sizing basic components of venture scrubber suggested by USEPA was presented in Fig. 17.11. Figure 17.12 shows the relationship of scrubber dimensions to the amount of contaminated gas treated.

d is the diameter of inlet waste process gas nozzle; D is diameter of separator; – indicates configuration for erosive material; and xxx indicates surfaces to be protected against erosion.

With the data observed in Fig. 17.12, it would be much easier to determine the area of separator (As), venture threat (Av), and elbow cross over (Ae) based on this correlation:

$$\mathrm{Av} : \mathrm{Ae} : \mathrm{As} = 0.12 : 0.18 : 0.78.$$

On the other hand, the height of separator (H), the diameter of separator (D), the diameter of inlet gas nozzle (d), and total length (A) and width (B) of venture scrubber could be estimated by observation of Fig. 17.12, once the waste inlet gas in thousands of ACFM is known.

For instance, if the amount of waste inlet gas is measured as 120,000 ACFM, then the required internal surface area is observed to be 1,800 ft^2 as indicated in Fig. 17.12. With this available information, the separator (As), venture throat (Av), and elbow cross over (Ae) will be determined as 196, 180, and 1424 ft^2, respectively, based on the ratio of Av: Ae: As = 0.12: 0.18: 0.78. The height of separator (H), the diameter of separator (D), the diameter of inlet gas nozzle (d), and total length (A) and width (B) of venture scrubber is also found to be 30, 14, 6, 16, and 28 ft, respectively, from Fig. 17.12.

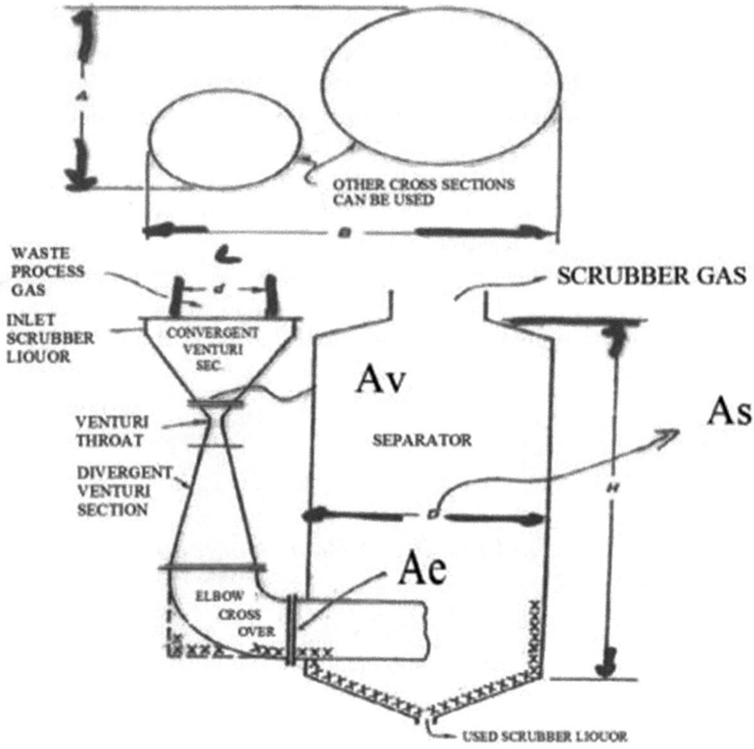

Fig. 17.11 Basic components of high energy scrubber

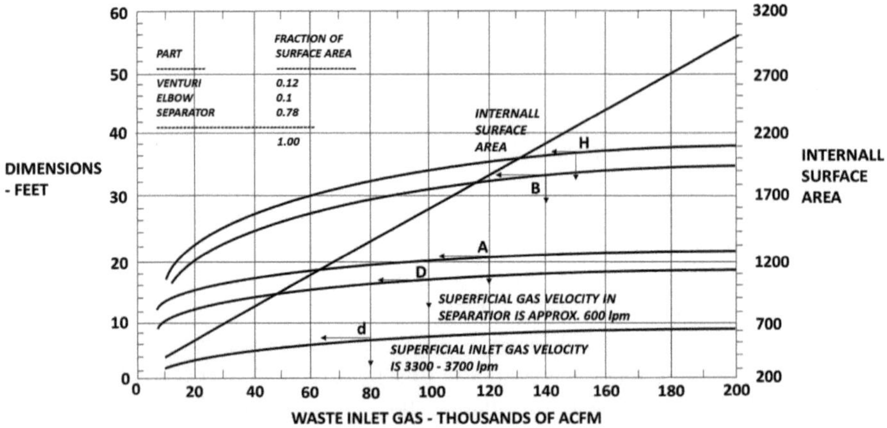

Fig. 17.12 Relationship of scrubber dimensions to the amount of contaminated gas treated

17.5 Performance Evaluation and Enhancement

17.5.1 Performance Evaluation

Wet scrubbers provide a simple but effective method for simultaneously removing particles and gaseous pollutants. The major disadvantage of the wet scrubber is the poor removal efficiency of 0.1–2 μm sized particles, since neither inertia impaction nor diffusion is effective in removing particles in this range [28]. In addition, it should be noted that the efficiency might not be satisfactory if the wet scrubber is meant to be used for removing gaseous pollutants and the particles larger than 1.0 μm. This is because the removal mechanism for large particles is inertia impaction which requires high gas velocities, but gaseous pollutant removal is enhanced under lower velocities due to the improved mass transfer efficiency and the extended residence time. For this reason, wet scrubbers are usually equipped with a supplementary collection technology or other equipment such as ESP and baghouse.

According to the contact power theory, higher removal efficiency is achieved through higher energy utilization. Specifically, electricity consumption by the fan system, i.e., HP_{fan} (hp) should be estimated by Eq. (17.36):

$$HP_{\text{fan}} = \frac{\Delta P Q_{\text{in}}}{6356\eta_{\text{fan}}}, \tag{17.36}$$

where ΔP (in w.c.) is the pressure in the fan; Q_{in} (acfm) represents the gas flow rate; and η_{fan} (-) indicates the fan efficiency. Electricity needed for the water pump is given as follows:

$$HP_{\text{pump}} = \frac{\Delta P_{\text{pump}} \cdot L/G \cdot \frac{Q_{\text{in}}}{1000} \cdot \gamma}{3952.6\eta_{\text{pump}}}, \tag{17.37}$$

where ΔP_{pump} (in w.c.) is the pressure in the pump; η_{pump} (-) is the pump efficiency; L/G is expressed in gal/1000 feet^3; Q_{in} (acfm) is the inlet flow rate; and γ (-) is the specific gravity of the liquid.

17.5.2 Performance Improvement

In general, wet scrubber systems take advantages over fabric filters and ESPs, because they are smaller and simpler, with lower capital and maintenance costs. Collection efficiency of wet scrubbers varies with the particulate size distribution and scrubber type and can be higher than 99% for sub-micron particles with optimum design. The main operating parameters are particle size distribution, gas velocity or gas flow rate, liquid to gas ratio, droplet size distribution, temperature, and pressure drop [35].

Venturi scrubbers are therefore more expensive than washing towers for operational costs (while capital and maintenance costs are comparable), but removal efficiency for fine PM is generally greater. Tray tower scrubbers contain several perforated plates with different geometrical shapes in providing more gas–liquid mixing and contact with a high efficiency level (>97%) for particles larger than 5 μm in diameter. Capital and operational costs are moderately higher than simple spray towers, but maintenance costs can be greater due to clogging of perforations by large PM. Packed scrubbers contain packing material, having specific characteristics, such as specific surface area, pressure drop, weight, corrosion resistance and cost, structured or randomly arranged, which provides a large wetted surface for gas–liquid contact. Bubble-column wet scrubbers represent a promising alternative for nanoparticle collection due to its unique mechanism of PM removal, i.e., diffusion control. With this evidence, in principle, a bubble-column scrubber can be competitive in terms of nanoparticle removal compared with fabric filters and ESPs.

The efficiency of a wet scrubber, especially for removing gaseous pollutants, is dependent on the concentration gradient and proper addition of fresh solution. In general, there are two techniques for solution replenishment, which are overflow and blowdown. The overflow method is more common and simple to operate since only a flow meter is required. However, in either method the replenishment rate must be high enough to compensate for evaporation losses, otherwise the concentration gradient might be insufficient.

Plugging occurs in the mist eliminators and nozzles. For this reason, solids should not be injected into the system, and filters should be routinely washed and kept clean. In addition, care should be taken to control biological growth in the scrubber and the following mist eliminators which will adversely affect performance. This is especially problematic in acid scrubbers, where pH is typically in the 8–9 range, which is favorable for biological growth. Therefore, it is common to use algae inhibitors and pH buffers to prepare the solution. With the adverse effects caused by the maldistribution of gas and liquid flows, components within the scrubber unit including auxiliaries must be checked and maintained routinely after installation and start-up.

Maldistribution is usually due to improper pump size. Adequate pump size is determined based on performing hydraulic calculations for the recirculation system. The three parameters that affect the design of a pump head are friction loss through piping and fittings, pumping height, and pressure loss of nozzles. It is recommended that an oversize ratio of 125% should be considered to ensure adequate capacity and operation [29]. Maldistribution can also be caused by channel, which could be the result of plugged spray nozzles, poor air distribution, or rectangular housings.

17.5.3 Innovative Technologies

HiGee

As mentioned previously, conventional wet scrubbers are not effective in simultaneously controlling gaseous pollutants and particles. HiGee, which was originally invented to enhance the mass transfer between two flow phases, has proved to be a powerful wet scrubber for simultaneously removing contaminant gas and particles. Pan et al. [4] reviewed particle control technologies and concluded that HiGee provides an effective control method for reducing particle emissions in a large size range (see Fig. 17.13). The most significant advantage over conventional wet scrubbers is that cutting sizes for the HiGee system are typically small, ranging from 0.02 to 0.3 μm.

In addition, it was found that collection efficiency could be presented as the following [30]:

$$\eta_i = 1 - e^{-a\times\left(\frac{d_p}{D}\right)^{b}}, \tag{17.38}$$

where a and b are constants correlated with operating conditions. The contact power theory suggests that interactions between particles and the droplets should be enhanced by imposing centrifugal force on the droplet. According to this, similar collection efficiencies can be achieved with less water consumption (usually the L/G ratio is less than 1.0 L/m^3) by using the HiGee system. Under these operating conditions, the pressure drop in the RPB system is less than 0.8 kPa [31].

For overall efficiency for particle removal, some literatures mentioned about RPB reactors, including their types, reactor size, PM source, operating conditions, and the highest efficiency, as shown in Table 17.6. In general, the particle removal efficiency of counter-current flow RPB (i.e., 98.5%) was higher than that of cross-flow RPB

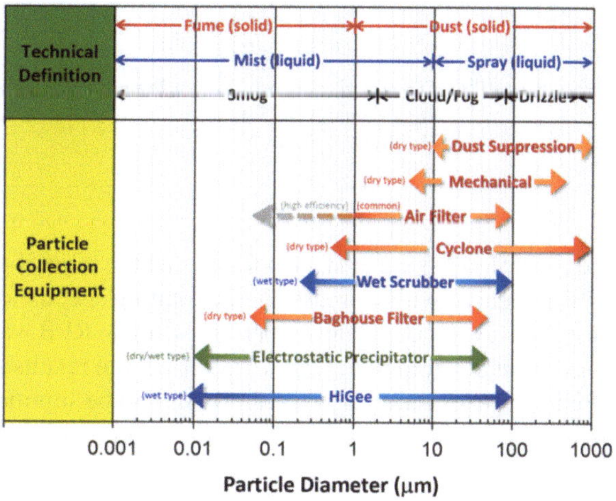

Fig. 17.13 Relationship between various types of particulate control equipment and their applicable particle size of the collection

Table 17.6 Performance evaluation of PM removal using HiGee RPB

RPB type	Size	Particulate matter (PM) source	Liquid phase	Operating conditions	Highest efficiency	References
Tapered rotating fluidized bed	D: 25 cm; H: 2 cm;	Fine cohesive corn starch Powder (d: 16 μm; density: 1630 kg/m^3)	5 wt% hydroxypropyl cellulose	224 rpm (7 g); Q_G:10 m^3/h; Q_L: 6.67 × 10^{-5} kg/s	96	[33]
Counter-current flow RPB	D_o: 6 cm; D_i: 19 cm; H: 20	Talcum powder; PM: 4 g/m^3; d_{max}: 4.73 μm	Water	1400 rpm; Q_G/Q_L: 1000; Q_G: 600 m^3/h; D_c: 0.02–0.03 mm	97.4	[34–36]
	D_o: 37.5 cm; D_i: 19 cm; H: 17.8 cm	Fly ash; PM: 5 g/m^3; d_{max}: 2.25 μm	Water	700 rpm; Q_G/Q_L: 1430; Q_G: 900 m^3/h	98.5	[28]
	D_o: 32 cm; D_i: 16 cm; H: 5 cm	Coal combustion fly ash	Water	1000 rpm; Q_G/Q_L: 1600; Q_G: 240 m^3/h	99.9	[32]
	D_o: 30 cm; D_i: 10 cm; H: 5 cm	Coal combustion fly ash; PM: 3–50 g/m^3; d_{50}: 2.55 μm	Water	900-1500 rpm; Q_G/Q_L:100–400; Q_G: 200 m^3/h; Dc: 0.02–0.3 mm	>99	[30]
	D: 30 cm; H: 40 cm	Thermal power plant flue gas; PM: 20 g/m^3	Water	1400 rpm; Q_G/Q_L: 167; Q_G: 200 m^3/h	99.9	
	D_o:10 cm; D_i: 2.4 cm; H: 7 cm	E-waste combustion process; PM: 38–105 g/m^3; d_{50}: 1.2 μm	Water	1740 rpm; Q_G: 6 L/min Q_L: 1448 ml/min	99.5	
Cross-flow RPB	D_o: 40 cm; D_i: 19 cm; H: 21 cm	Fly ash; PM: 5 g/m^3; d_{max}: 2.25 mm	Water	800–1200 rpm; Q_G: 300 -900 m^3/h; Q_L: 0.5–2 m^3/h	96	[29]

(i.e., 97.5%) [28]. Fu et al. also found that the ratio of particle removal in a cross-flow RPB increased with increases in the rotating speed, liquid-to-solid ratio and gas flow rate [29]. It was noted that the pressure drop in RPB was comparatively lower than the conventional equipment [30]. In addition, the results indicated that the performance of fine particles removal in the RPBs should be significantly higher than that by other dedusting equipment, such as cyclone, wet scrubber, and electrostatic precipitator [29, 31, 32].

As a result, performance of particular matter reduction and energy consumption was seriously considered by the recent research work [36].

Figure 17.14a shows the effect of the high-gravity factor (β) on PM removal ratio and energy consumption under the L/G ratios of 0.01, 0.02, and 0.03, respectively. Obviously, the PM removal ratio increased along with the increase of the β and L/G ratio. The PM collection performance was mainly related to the droplets formed in the RPB. At a higher gravity field, higher droplet numbers and smaller droplet size were thrown off from the reactor hub to the outer area, and thus the possibility of collision between droplets and particles increased. Meanwhile, the higher value of L/G ratio represents greater liquid flow rate into the RPB. This contributed to increased droplet formation. The highest PM removal ratio of 95.9% was obtained at the L/G ratio of 0.03 and β of 235.4

Figure 17.14b shows the effect of L/G ratio on PM removal ratio and energy consumption under the β of 235.4. The profile between removal ratio and energy consumption reveals a symmetrical trend, showing that improved PM removal ratio led to lower energy consumption. However, the removal ratio and energy consumption savings could not be enhanced when the L/G ratio reached 0.03. This suggests that the suitable high gravity level and L/G ratio are 235.4 and 0.03, respectively.

Dyna Wave

Dyna Wave scrubber is a reverse jet scrubber where laden gas enters into the scrubber from the top and collides with the scrubbing slurry, which is injected upwardly through abrasion-resistant injectors [32]. The result of this interaction between the gas and the liquid is the generation of a fourth zone characterized by a standing wave of highly turbulent flow thereby resulting in a very high rate of liquid surface renewal. Subsequently, gas is efficiently cooled to its adiabatic saturation temperature while pollutants are removed. After scrubbing, gas enters the disengagement vessel where impingement is dominated to separate entrained liquids from the gas. In the end, the remaining droplets are removed by a mist eliminator and the cleaned gas is released (see Fig. 17.15).

Reverse jet scrubber provides a high collection efficiency by generating a large gas–liquid interfacial area, long residence times, and high differential velocities. Gas-phase pressure drop through the reverse jet scrubber was only half that of conventional wet scrubbers while achieving satisfactory collection efficiency [33].

Electrostatic spray wet scrubber (ESWS)

Almuhanna et al. [27, 28] evaluated an electrostatically assisted particulate wet scrubber (EPWS) was effective in reducing airborne PM concentrations, with PM removal efficiencies of about 92% for PM10 and 33–60% for $PM_{2.5}$ at swine facilities. Anatol et al. [20] confirmed electrostatically enhanced industrial WS technology is effective in fine PM control, with less water consumption, and generates a lower pressure drop.

Recently, an electrostatic spray wet scrubber (ESWS) technology designed for removal of PM (PM_{10} and $PM_{2.5}$) emissions from poultry facilities was evaluated by Ru et al. [34] as shown in Fig. 17.16, which consists of a spray scrubber body,

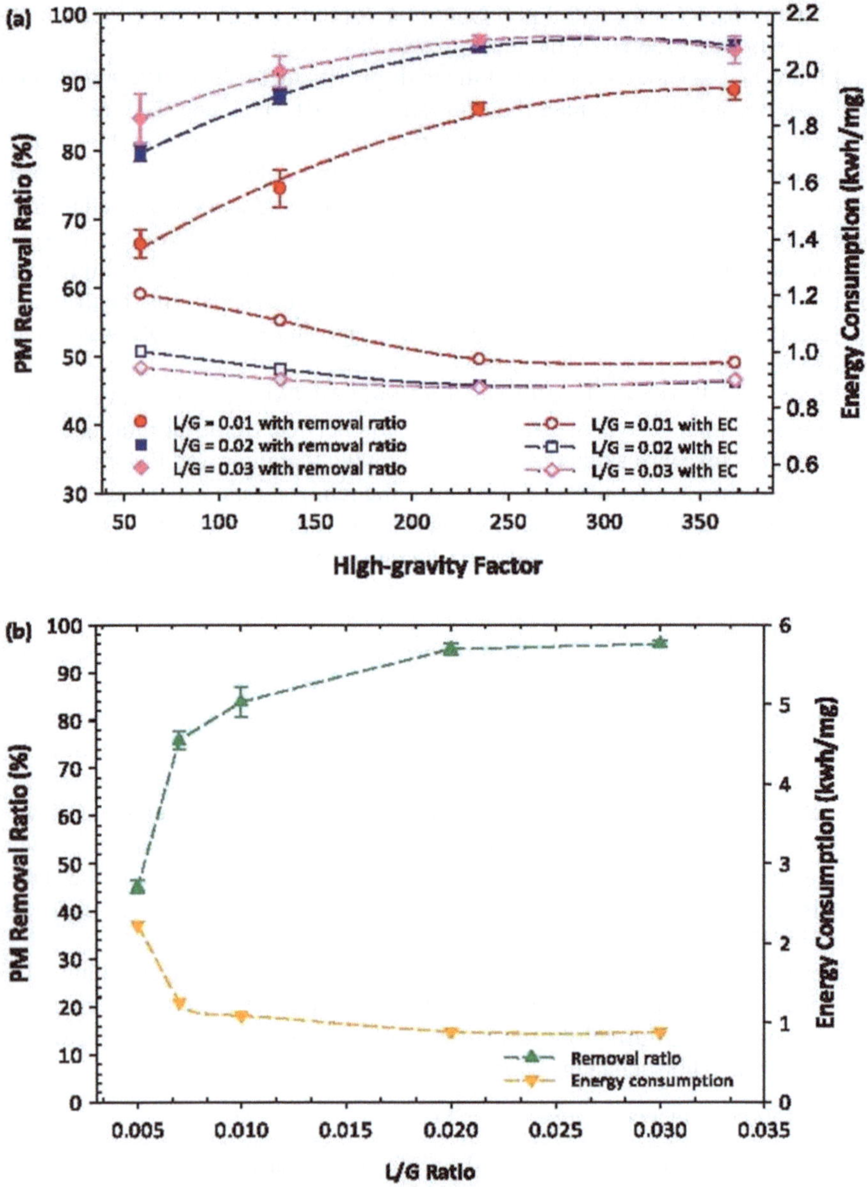

Fig. 17.14 Effect of **a** high-gravity factor and **b** L/G ratio (operation condition: temperature of 40 °C, and high-gravity factor of 235.4) on PM removal and energy consumption

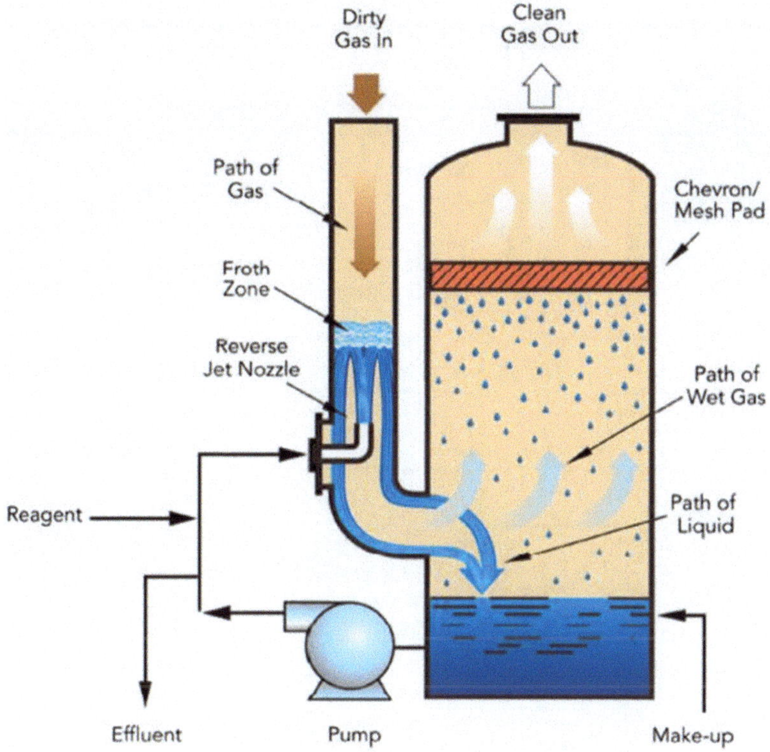

Fig. 17.15 Schematic diagram of Dyna Wave scrubber [32]

a spray nozzle, a ring shape electrode for charging droplets, and a liquid collector at the bottom of the spray column. This study demonstrated that the ESWS technology reduced emissions of PM10 by 85–94% and PM2.5 by 85–88% with a charging voltage of 7 kV, air speed of 0.5–0.75 m s^{-1}, and droplet diameter of 100–150 μm. The water consumption rate was approximately 2.35 L min^{-1}, while the total power consumed was 270 W. It was thus deduced that the ESWS could be a potentially effective and feasible tool in controlling PM emissions for commercial poultry facilities.

Exercises

1. Define the following terms and explain why they should be considered for design use: Brownian diffusion, particle seize distribution, liquid to gas ratio, drag coefficient.
2. Please describe the general procedures for designing a wet scrubber.
3. Briefly describe the "contact power theory" and explain how it can be used for design of wet scrubber.
4. Compare the performance of conventional wet scrubber and HiGee for PM reduction.

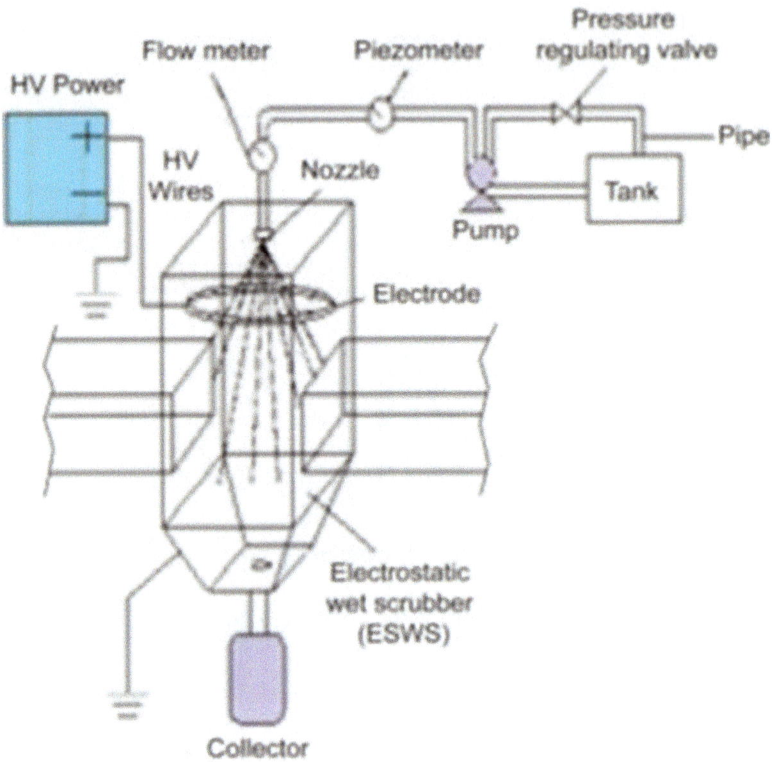

Fig. 17.16 Schematic of the ESWS unit

5. Calculate the number of transfer units Nt at given conditions:

   $\alpha = 1.47$;
   $\beta = 1/05$;
   $Pt = 2.20$.

References

1. Kuwabara S (1959) The forces experienced by randomly distributed parallel circular cylinders or spheres in a viscous flow at small Reynolds numbers. J Phys Soc Jpn 14:527–532
2. Gal-Or B (1970) On motion of bubbles and drops. Can J Chem Eng 48:526–531
3. Friedlander S (1957) Mass and heat transfer to single spheres and cylinders at low Reynolds numbers. AIChE J 3:43–48
4. Pan S-Y, Wang P, Chen Q, Jiang W, Chu Y-H, Chiang P-C (2017) Development of high-gravity technology for removing particulate and gaseous pollutant emissions: principles and applications. J Clean Prod 149:540–556

5. Kreith F, Schnelle K, Brown C (2002) Air pollution control technology handbook. Laple, USEPA AP, vol 40, pp 94–99
6. Schifftner KC, Hesketh HE (1983) Wet scrubbers
7. Sundberg RE (1974) The prediction of overall collection efficiency of air pollution control devices from fractional efficiency curves. J Air Pollut Control Assoc 24:758–764
8. Pei S-L, Pan S-Y, Li Y-M, Gao X, Chiang P-C (2018) Performance evaluation of integrated air pollution control with alkaline waste valorization via high-gravity technology. J Taiwan Inst Chem Eng 87:165–173
9. MikroPul (2009) Wet scrubbers
10. Lapple C, Kamack H (1955) Performance of wet dust scrubbers. Chem Eng Progr 51
11. Semrau KT (1963) Dust scrubber design—a critique on the state of the art. J Air Pollut Control Assoc 13:587–594
12. Richards JR, Segall RR (1983) Wet scrubber inspection and evaluation manual. Wet scrubber inspection and evaluation manual: EPA. Office of Air Quality Planning and Standards
13. Semrau KT (1960) Correlation of dust scrubber efficiency. J Air Pollut Control Assoc 10:200–207
14. Ali M, Yan C, Sun Z, Gu H, Mehboob K (2013) Dust particle removal efficiency of a venturi scrubber. Ann Nucl Energy 54:178–183
15. Yung S-C, Barbarika HF, Calvert S (1977) Pressure loss in venturi scrubbers. J Air Pollut Control Assoc 27:348–351
16. USEPA (2010) Air pollution control technology fact sheet: spray-chamber/spray-tower wet scrubber
17. USEPA (2010) Air pollution control technology fact sheet: impingement-plate/tray-tower scrubber
18. USEPA (2011) Air pollution control technology fact sheet: venturi scrubber
19. USEPA (2003) Air pollution control technology fact sheet: orifice scrubber
20. USEPA (2010) Air pollution control technology fact sheet: condensation scrubber
21. Theodore L (2008) Air pollution control equipment calculations. Wiley
22. Cooper DW (1997) Wet scrubber particulate collection. In: Fayed ME, Otten L (eds) Handbook of powder science & technology. Springer, US, Boston, MA, pp 803–844
23. Calvert SJAP (1977) Scrubbing 4:257–291
24. Taheri M, Calvert S (1968) Removal of small particles from air by foam in a sieve-plate column 18:240–245
25. Yung S-C, Calvert S, Barbarika HF (1978) Venturi scrubber performance model. Environ Sci Technol 12:456–459
26. Allen R (1996) Prediction of Venturi scrubber grade efficiency curves using the contacting power law. Powder Technol 86:137–144
27. Calvert S (1972) Wet scrubber system study. Volume I. Scrubber handbook
28. Calvert S, Goldschmid J, Leith D, Mehta D (1972) Wet scrubber system study. Scrubber Handbook, vol 1, pp 5–81
29. Today M. Factors Affecting Wet Scrubber Performance. 2018.
30. Pei SL, Pan SY, Li YM, Chiang PC (2017) Environmental benefit assessment for the carbonation process of petroleum coke fly ash in a rotating packed bed. Environ Sci Technol 51:10674–10681
31. Fu J, Qi G-S, Liu Y-Z, Tian J-X, Guo Q, Dong M-Y (2015) Removal of fine particles by high gravity wet cleaning (in Chinese). Chem Eng 43:6–10
32. Safariyeganeh S, Lovett CJ (2014) Cement plant environmental technology for achieving high SO_2 removal. In: Cement industry technical conference (CIC), 2014 IEEE-IAS/PCA. IEEE, pp 1–8
33. Holmes TL, Meyer CF, DeGarmo JL (1983) Reversejet scrubber for control of fine particulates. Chem Eng Prog 79:60–66
34. Ru Y, Zhao L, Hadlocon LJ, Zhu H, Ramdon SK (2016) Laboratory evaluation of electrostatic spray wet scrubber to control particulate matter emissions scrubber to control particulate matter emissions from poultry facilities. Environ Technol. https://doi.org/10.1080/09593330.2016.1184319

35. Bianchini A, Pellegrini M, Rossi J, Saccani C (2018) Theoretical model and preliminary design of an innovative wet scrubber for the separation of fine particulate matter produced by biomass combustion in small size boilers. Biomass Bioenerg 116:60–71
36. Chen TL, Xiong YX, Chen YH, Chiang PC, Chen YH (2020) Performance evaluation and process simulation for synergetic removal of NO_x, CO_2 and PM using green alkaline solution in a high-gravity rotating packed bed. Fuel 280:118643

Chapter 18
Baghouse System

Nomenclature

A_H	Hamaker constant of interacting media
A_{tc}	Gross cloth area
D	Mass mean diameter of the particles
D_f	Uniform diameter of the cylinder
d_g	Diameter of single cylinder fiber
D_i	Particle diffusion coefficient
D_p	Depths of the particulate layer
E	Strength of the electric field
L	Thickness of single cylinder fiber
L_f	Total length of cylinder per unit bed volume
ΔP	Total pressure drop
ΔP_f	Part pressure contributed by fabric
ΔP_p	Part pressure contributed by particulate layer
Q	Volumetric gas flow rate
$Q_{e,a}$	Stream emission rate
q_p	Charge state of the particle
S	Total filter drag
S_E	Effective residual drag
T	Temperature
t_c	Cleaning time
t_f	Filtration time
t_r	Running time
u_∞	Flow velocity within the bed
v	Relative velocity between fiber and particles
v_t	Terminal velocity defined by Stock's law
W	Fabric load which is a function of the operating time and the dust load
ε	Solid fraction of the filter

P.-C. Chiang and X. Gao, *Air Pollution Control and Design*,
https://doi.org/10.1007/978-981-13-7488-3_18

η_f Fractional collection efficiency
λ Mean free path of gas molecules
μ Viscosity of gas stream
ρ_g Density of gas stream

18.1 Introduction

18.1.1 Features of Baghouse Filter

Baghouse filters are high-efficiency dust collection equipment where laden gas is forced through and filtered by several layers of cloth-like material. Baghouse offers incredible versatile solutions for particle collection by varying the structures and construction materials of the filters. As a result, they have been widely used for air pollution control and material reclamation. The filter also serves as an excellent carrier for catalysts and/or other chemical agents to enhance the removal of gaseous pollutants [1].

While efficient, capital and O&M costs are relatively higher than other dust collection equipment due to the higher pressure drop and the cost of filter replacements. In addition, baghouse filters are vulnerable to heat, gas flow, and corrosive chemicals in the flue gas which require maintenance and result in a higher economic cost.

18.1.2 Application of Baghouse Filter

As commonly used dust collection equipment, the most significant advantage of the baghouse is the high collection efficiency, which makes the equipment more useful in the context of increased haze situations. For instance, Yi et al. [2] reported that the collection efficiency of $PM_{2.5}$ by baghouse was higher than 99.7%. Meanwhile, baghouse filters are more resistant to the fluctuations of gas flow and dust loading. In addition, the collected dust can be recovered while the clean gas can be recirculated for heat conservation which is beneficial toward achieving a circular economy and sustainable development. Since no liquid is needed to feed the filter, the secondary pollution due to wastewater could be avoided. Regarding maintenance, the deteriorated unit can be easily replaced due to the accessible modular design and construction, resulting in shorter startup time. In addition, other advantages include flexibility and reliability under various operating conditions.

However, high removal efficiency is achieved at the expense of more energy overcoming the higher pressure drop. Meanwhile, operation and maintenance costs of baghouse filter are relatively higher than that of other dust collection equipment due to its higher pressure drop and the deteriorated filters replacement. In addition, the temperature of the gas flow should be monitored frequently, especially when it exceeds 550 °F (~290 °C) under which refractory mineral or metallic fabrics might be required. Furthermore, some materials in the fabric filter are flammable, which requires additional attention to the concentration of flammable dust and more frequent clean-ups.

18.2 Principles

18.2.1 Fundamentals of Fabric Filtration

The collection mechanisms of a fabric filter categorized into cake and depth filtrations are largely dependent on the material of the fabric bag. In general, materials can be categorized as woven and felted fabrics. Among them, woven fabrics are the most popular fabrics used for particle collection since their collection efficiency (permeability) can be easily adjusted by changing the weaving methods and patterns. However, the collection efficiency of particles by the original woven filter is usually limited. As a result, some modifications to the woven fabric are made. Felted or nonwoven fabric is built by assembling the fiber through chemical, mechanical, heat, or solvent treatment thereby generating many needle-like fibers resulting in enhancement of the collection efficiency. In practice, woven fabric provides a base which is called scrim, while the nonwoven fibers are attached to the woven fibers.

Non-cake filtration

A fabric filter is a packed bed consisting of cylinders which collect particles independently as many individual collectors. By assuming the flue gas flows through the filter normally, the collection of an individual cylinder can be mathematically estimated. The first step is to determine the solid fraction of the filter ε which is as follows:

$$\varepsilon = \frac{\pi D_f^2 L_f}{4}, \tag{18.1}$$

where D_f is the uniform diameter of the cylinder and L_f represents the total length of the cylinder per unit bed volume. Assuming that the gas stream is incompressible and it approaches the bed filter with a velocity of $\overline{u}$, the flow velocity within the bed u_∞ is thus calculated by Eq. (18.2):

$$u_\infty = \frac{Q}{A(1-\varepsilon)} = \frac{\overline{u}}{1-\varepsilon}, \tag{18.2}$$

where Q is the volumetric gas flow rate and A represents the cross-sectional area of the control unit. Considering the control unit dx, the concentration of particles flowing into and out are denoted as $N|_x$ and $N|_{x+dx}$, respectively. As a result, the mass balance of particles in the control unit is expressed as follows:

$$\underbrace{A \cdot \bar{u} \cdot N|_x}_{\text{Flow in}} = \underbrace{A \cdot \bar{u} \cdot N|_{x+dx}}_{\text{Flow out}} + \underbrace{\left(D_f L_f \eta_f\right) u_\infty \cdot N|_x (A dx)}_{\text{Collected by the filter}} \tag{18.3}$$

where η_f is the fractional collection efficiency. Integrating Eq. (18.3) over the entire length, L, and introducing the initial condition $N|_{L=0} = N_{(0)}$, we have the following:

$$\frac{N_{(L)}}{N_{(0)}} = e^{-\frac{4\varepsilon\eta_f L}{\pi(1-\varepsilon)D_f}}. \tag{18.4}$$

Finally, the overall collection efficiency is as follows:

$$\eta = 1 - \frac{N_{(L)}}{N_{(0)}} = 1 - e^{-\frac{4\varepsilon\eta_f L}{\pi(1-\varepsilon)D_f}}. \tag{18.5}$$

Note that $\frac{\pi(1-\varepsilon)D_f}{4\varepsilon\eta_f}$ can be regarded as the characteristic penetration depth of the suspended particles in the bed (Fig. 18.1).

Cake Filtration

Removal efficiency also depends on the development of a dust cake that enhances fabric interstices. It has been reported that particles with similar shape and size will

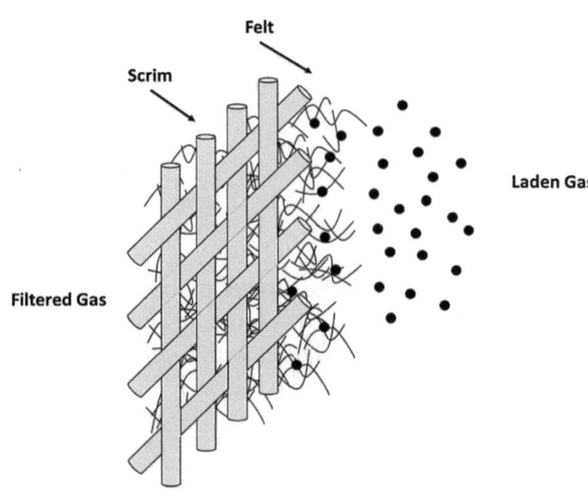

Fig. 18.1 Schematic diagram of non-cake filtration on a felted fabric

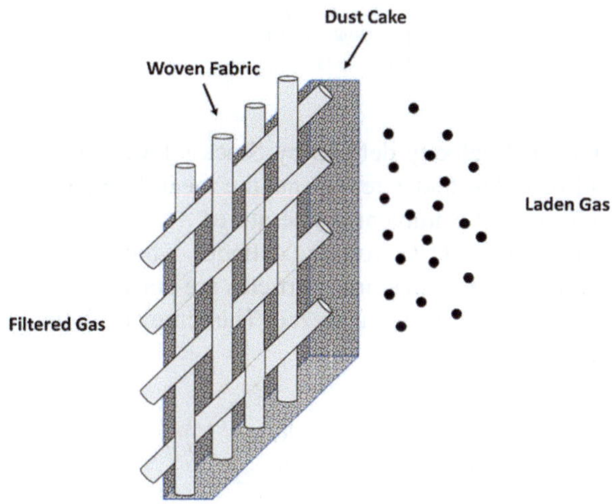

Fig. 18.2 Schematic diagram of cake filtration on a woven fabric

produce a dense dust cake that restricts airflow while the subsequent particulate matter escapes [3]. It is widely acknowledged that the dust cake growing on the filter plays a central role in PM collection, which is also known as cake filtration. As shown in Fig. 18.2, the collected particles aggregate on the surface of the fabric filter and build up a dust cake during the operation cycle, which effectively reduces the mesh size of the filter. As a result, the removal efficiency, along with pressure drop, increases. It is thus concluded that the cleaning process should not be conducted too frequently as the collection efficiency would be compromised due to the destruction of the dust cake. Mechanical shaking, reverse air cleaning, and other low-energy cleaning methods are suitable for protecting the suitable dust cake.

18.2.2 Collection Efficiency of Baghouse Filter

Conventional fabric filter consists of cylinders whose major mechanisms of PM collection include impaction, interception, diffusion, sieving, and electrostatic attraction.

Impaction occurs when large particles (>1 μm) are unable to follow through the stream at high speeds, thus hitting the fabric filter directly. It is assumed that particles are collected by impact with mass, but the size is neglected. A rough estimation of the removal efficiency by impaction is presented in Eq. (18.6):

$$\begin{cases} \eta_{\text{Imp}} = \frac{C_c \cdot v_t \cdot v}{D_f}, \\ C_c = 1 + 1.7\frac{\lambda}{d_p}, \end{cases} \quad (18.6)$$

where v_t is the terminal velocity defined by Stock's law, v is the relative velocity between fiber and particles, and λ represents the mean free path of gas molecules, which is about 0.6 μm at the ambient temperatures.

An interception occurs at $D_p/2$ from the cylinder, which is located in the path of the particles. Contrary to the impaction, particles are considered as weightless balls when the interception mechanism is analyzed. The removal efficiency resulting from interception is given by Eq. (18.7):

$$\begin{cases} \eta_{\text{Int}} = (2 - \ln Re_f)^{-1}\left(\frac{d_p}{d_f}\right)^2, \\ Re_f = \frac{D_f \cdot v \cdot \rho_g}{\mu}, \end{cases} \quad (18.7)$$

where D_f is the fiber diameter and ρ_g and μ are the density and viscosity of the gas stream, respectively.

Diffusion dominates the removal of fine particles with sizes roughly smaller than 0.5 μm. As the particles in this size range are able to stick closely to the gas stream, interception and impaction cannot effectively remove them. In the analysis, it is reasonably assumed that these particles are massless so that momentum is negligible. Quantitatively, removal efficiency due to diffusion is a function of particle size as well as velocity profile which are shown as equations developed by Tufenkji and Elimelech [4]:

$$\begin{cases} \eta_{\text{Diff}} = 2.4A_s{}^{\frac{1}{3}}\left(\frac{D_p}{d_g}\right)^{-0.081} Pe^{-0.715} N_{\text{vdw}}{}^{0.052}, \\ A_s = \frac{2-2(1-\varepsilon)^{\frac{5}{3}}}{\left[2-3(1-\varepsilon)^{\frac{1}{3}}+3(1-\varepsilon)^{\frac{5}{3}}-2(1-\varepsilon)^2\right]^{\frac{1}{3}}}, \\ Pe = \frac{d_g U}{D_i}, \\ N_{\text{vdw}} = \frac{A_H}{KT}, \end{cases} \quad (18.8)$$

where D_i represents the particle diffusion coefficient; A_H is the Hamaker constant of interacting media which is 0.85×10^{-19} for glass beads and 0.79×10^{-19} for NaCl particles.

Sieving occurs when the voids among the fibers are smaller than the particle so the particle is stopped and trapped.

Electrostatic attraction leads to an enhanced collection efficiency and usually results from abrasion among cylinders when laden gas passes through the fabric filter. Guillory et al. [5] suggested that the enhanced collection efficiency (η_E) could be expressed as Eq. (18.9):

$$\eta_E = 1 - \exp\left[\frac{-2(1-\varepsilon)Lq_pC_cE}{\pi\mu Ud_gd_p}\right], \tag{18.9}$$

where ε, L, and d_g represent the porosity, the thickness, and the diameter of a single-cylinder fiber; q_p is the charge state of the particle, while E is the strength of the electric field. According to Eq. (18.9), single collector efficiency due to electrostatic attraction is a function of the dimensions of the cylinder fiber as well as the characteristics of the electric field. It was also found that electrostatic attraction is responsible for collecting particles of about 100 nm [6].

As a result, an estimation of the overall removal efficiency is conducted by considering the following four mechanisms:

$$\eta_{\text{overall}} = 1 - (1-\eta_{\text{Int}})\big(1-\eta_{\text{Imp}}\big)(1-\eta_{\text{Diff}})(1-\eta_{\text{Sie}})(1-\eta_{\text{E}}). \tag{18.10}$$

18.3 Design Methods

18.3.1 Air-Cloth Ratio Approach

A proper filtration rate is essential to balance the collection efficiency and the service time of the filter. In the baghouse design, the proper filtration rate is expressed as the air-to-cloth ratio or A/C ratio, as shown in Eq. (18.11):

$$\text{A/C ratio} = \frac{\text{gas volume rate}}{\text{cloth area}} \tag{18.11}$$

In practice, either the gross ratio or the net ratio is used as follows:

$$\text{grossA/Cratio} = \frac{\text{total inlet gas volume rate}}{\text{total cloth area in baghouse}}, \tag{18.12}$$

$$\text{net A/C ratio} = \frac{\text{total inlet gas volume rate + cleaning volume rate}}{\text{online cloth area}}. \tag{18.13}$$

Gross A/C ratio refers to the total cloth area used to filter the entire flue gas stream, while the net ratio is used to describe the amount of cloth area available for filtering. For instance, some bag filters could be offline for cleaning or maintenance. Thus, the estimation based on the net A/C ratio is more precise and reliable as it subtracts the number of unavailable filters.

In practice, A/C ratio is dictated by the structure of the baghouse and material of the filter thereby estimating the filtration velocity if the filter is predetermined. For instance, a combination of reverse-air cleaning and woven fiberglass bag filters could be successfully implemented to collect particles from a small combustion plant which suggests that the filtration velocity should be higher than 1 cm/s. This is because a filtration velocity that is too high may generate significant drag force on the filter, leading to premature deterioration. Correction plans can either be performed by increasing the number of filters to reduce filtration velocity or by using stronger filters. Fortunately, A/C is usually available from the vendor and typical values are presented in the next section.

18.3.2 Design Criteria

Using a well-designed baghouse system could easily meet the emission requirement. The key to the success of baghouse design underlines not only the appropriate equipment but also a good estimation of the air-to-gas ratio and the cleaning frequency. Specifically, air-to-cloth (A/G) ratios recommended as a rule of thumb for mechanical shaking, reverse air, and pulse-jet baghouses are between 2.0:1 and 2.5:1, 1.75:1 and 2.5:1, and 3.25:1 and 4.0:1, respectively. The detailed A/G is dependent on the type of fabric material used and usually can be obtained by the equipment vendor.

18.4 Design Procedures

Like other particle collectors, the particle size distribution, particle load, and the physicochemical properties of the particles are required to gather for design use. In addition, other technical information regarding fabric type, the cleaning method, air-to-cloth ratio, and equipment configuration are also needed for design purpose.

In general, the filter material is largely dependent on the properties of the pollutant and operating conditions. For instance, the porosity of filter materials dictates the collection efficiency and the pressure drop; meanwhile, operating conditions including temperature, filtration velocity, pressure, and cleaning techniques affect the filter material lifespan.

Equipment configuration of baghouse filters refers to the number of compartments and their arrangement. In general, the more compartments in the system, the less flow fluctuation there is during cleaning. However, too many compartments would occupy huge space and require sophisticated flow distribution techniques to allocate the load to each compartment evenly (Fig. 18.3).

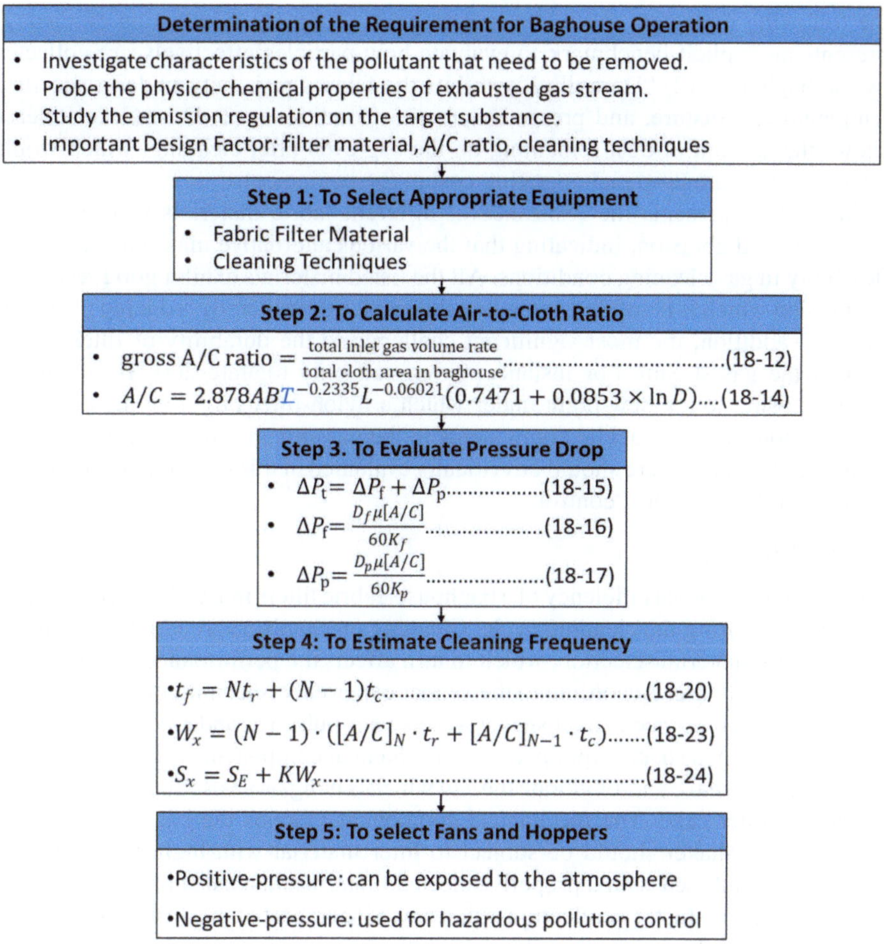

Fig. 18.3 Design framework of baghouse system

18.4.1 Step 1: To Select the Appropriate Equipment

The design of a baghouse begins from the selection of proper equipment including filter material and cleaning methods. Filter material selection is subject to the laden gas selection, operating conditions, and economic considerations. However, the cleaning process also affects the selection of the filter material and operation mode.

Fabric Filter Material

Selecting the filter material is the most important procedure thereby balancing the removal efficiency, lifespan, and economic cost. Regarding engineering performance, the following indicators should be incorporated [7, 8]: (1) "Collection efficiency" presents an explicit benchmark to evaluate how particles effectively are collected by the baghouse; (2) "Durability" exhibits the fabrics resistivity under a specified temperature, moisture, and pressure operating conditions; (3) Dust Cake Release shows the amount of dust remaining on the bag even after cleaning, which would affect the pressure drop and operating cost of the equipment.

Table 18.1 presents the resistance of different fabric materials to heat, acidic corrosion, and abrasion, indicating that the various alternative materials have great flexibility in gas-cleaning conditions. All the listed materials exhibit good resistivity to abrasion which is helpful in extending the service time thereby reducing the O&M costs. In addition, the most significant challenge to the durability of filters is the heat of the gas stream. For instance, nylon can only tolerate 200 °F or roughly 90 °C, closing to the dew point under which a nylon-filter might be damaged due to the condensation of acidic components in the gas stream. Therefore, the fabric's resistance to temperature should be critically evaluated in this situation, Ryton would be more suitable for dust control.

Cleaning Techniques

To sustain the removal efficiency of a baghouse, fabric filters must release particulates periodically during the cleaning cycle. Cleaning mechanisms are largely dependent on the fabric material selection, which in turn affects the performance and the structure of the baghouse. According to the vender experience, cleaning mechanisms can be characterized as mechanical shaker, reverse air, pulse jet, and sonic cleaning.

Among them, a shaker equipped with mechanical components is commonly used to periodically shake and/or vibrate the bag which is hung on an oscillating framework to remove filter cake. The mechanical component vibrates the bag, so the use of a mechanical shaker should be subject to filter material with higher strength and durability. Combined with a properly selected filter, mechanical shaking techniques can achieve a satisfactory collection efficiency. Since a part of the dust should remain on the bag thereby generating the cake filtration reducing the re-entrainment of collected dust and maintaining the constant emission quality. The recommended air-to-cloth ratio for baghouses equipped with mechanical shaking is between 2.0:1 and 2.5:1 (Fig. 18.4).

Reverse air cleaning removes dust cake by using airflow that is opposite to the influent gas stream, which is like the backwash process. As presented in Fig. 18.5, the laden gas flows through the filter while the particles are intercepted. When the compartment needs to be cleaned, cleaning gas flows in the direction opposite of the laden gas, resulting in the collapse of the dust cake. As low-pressure air is usually introduced in the cleaning cycle, the required force is much lower than the mechanical shaking, which makes it more suitable for fragile bags possessing a longer lifespan. The recommended air to cloth ratio for these baghouses is between 1.75:1 and 2.5:1.

Table 18.1 Summary of characteristics for different fabric materials [9–13]

Material	Temperature (°F)	Resistance to acidic corrosion	Resistance to abrasion	Characteristics
Acrylic	260		Excellent	• Acrylic offers good hydrolytic resistance over a limited temperature range, i.e., 127 °C for continuous and 135 °C for surge application • Acrylic can also be integrated into wet filtration applications
Fiberglass	500	Fair to good	Fair	**Advantages**: • Exhibiting excellent resistance to high-temperature while noncombustible, fiberglass fabrics are typically used in coal-fired boilers and metals refinery industries • Fiberglass fabrics are highly flexible which can be produced either in woven or felted forms. In addition, fiberglass can combine with many variations of yarn construction, fabric weaves, and fabric finishes to provide versatile functions • Fiberglass fabrics are not only chemically inert but also dimensional stable **Disadvantages**: • Woven glass fabrics exhibit low flex strength which results in low abrasion resistance • The finish may decompose at higher temperatures, while the mechanical wear of the glass fibers may get accelerated • Corrosive compounds in the flue gas may get fiberglass fabrics damaged as the impurities in the glass fibers are attacked. Thus the fiberglass fabrics should be operated above the dew point of acids

(continued)

Table 18.1 (continued)

Material	Temperature (°F)	Resistance to acidic corrosion	Resistance to abrasion	Characteristics
Nomex	375	Fair	Excellent	**Advantages**: • Nomex exhibits great thermal stability that less than 1% of shrinkage could be acquired at 177 °C • Nomex fabrics are inflammable which would be helpful to ensure the safety of the equipment **Disadvantages**: • The life cycle of Nomex fabrics is usually short due to the corrosion of acidic pollutants in the flue gas, which limits the application in coal-fired boilers • Combustible dust may accumulate resulting in potential concern on fire hazard
P84	475	Good	Good	• P84 is a kind of felted fabric made up of aromatic polymer fiber which is helpful to improve the capture efficiency • P84 exhibits great flexibility by which P84 can combine with other felted material to produce fabric composite and to improve the performance while saving the economic cost
Polypropylene	200	Excellent	Excellent	**Advantages**: • Widely used in food, detergent, chemical processing, pharmaceutical, and tobacco industries, polypropylene exhibits great ability to resist the damage resulting from static buildup, humidity and abrasion • Polypropylene also shows good dust release property during the cleaning process as the inherent smooth surface **Disadvantages**: • Polypropylene fabrics are vulnerable to oxidizing agents, copper, and related salts

(continued)

Table 18.1 (continued)

Material	Temperature (°F)	Resistance to acidic corrosion	Resistance to abrasion	Characteristics
Polyester	275 [11]		Excellent	**Advantages**: • Polyester fabric provides impressive strength at the cost of the very lightweight. As a result, polyester can retain its shape very well • Polyester exhibits great durability since it resists the deterioration from many kinds of chemicals and mechanical abrasion **Disadvantages**: • Polyester may hydrolyze when subject to humid gas streams • The structure of polyester may deteriorate by concentrated sulfuric, nitric, and carbolic acids
Teflon				Teflon® is unique among synthetics in its ability to resist chemical attack across the entire pH range throughout its operating temperature range of 232 °C continuous. This fluorocarbon fiber is non-adhesive, has zero moisture absorption, and is unaffected by mildew or ultraviolet light. The primary shortcomings of Teflon® are its high cost and relatively poor abrasion resistance. Applications of Teflon® include coal-fired boilers, waste-to-energy incinerators, carbon black, titanium dioxide, primary and secondary smelting operations, and chemical processing
Ryton	375	Excellent	Good	**Advantages**: • Ryton fabrics are much more chemically stable to acids and alkalis, which makes it competent in combustion-control applications **Disadvantages**: • Subject to high-temperatures (>190 °C) and high humidity environment, Ryton fabrics will hydrolyze

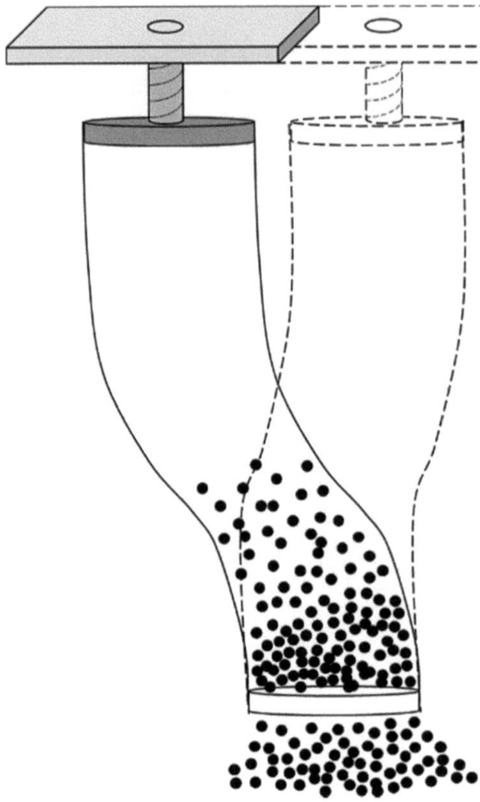

Fig. 18.4 Schematic structure of mechanical shaking

Pulse-jet cleaning uses compressed air pulses instead of a low-pressure air stream to dislodge the dust cake. The unique benefit of using this technique is that it can be performed while the bag is still online, so no extra bag is needed for backup. However, the cleaning frequency of this technique is much higher, which results in a higher air-to-cloth ratio during the collection process. It is noted that the pulse jet cleaning is not suitable for the hazardous pollutant because (1) the cleaning may cause severe pollutant emissions; (2) no dust cake remains on the bag which reduces the removal efficiency contributed by cake filtration. Furthermore, a pulse-jet cleaned filter collects particles from the outside which flows into the filter. The recommended air to cloth ratio for these filters is between 3.25:1 and 4.0:1. In order to maintain the geometric dimensions of the filter under this high filtration velocity, the embedded framework is usually implemented (Fig. 18.6).

Table 18.2 summarizes the features of the cleaning techniques which indicate that selection of the appropriate cleaning technique determines the performance and operation mode to a large extent, especially the operation mode (intermittent or continuous) and the collation orientation (inside or outside). Intermittent systems

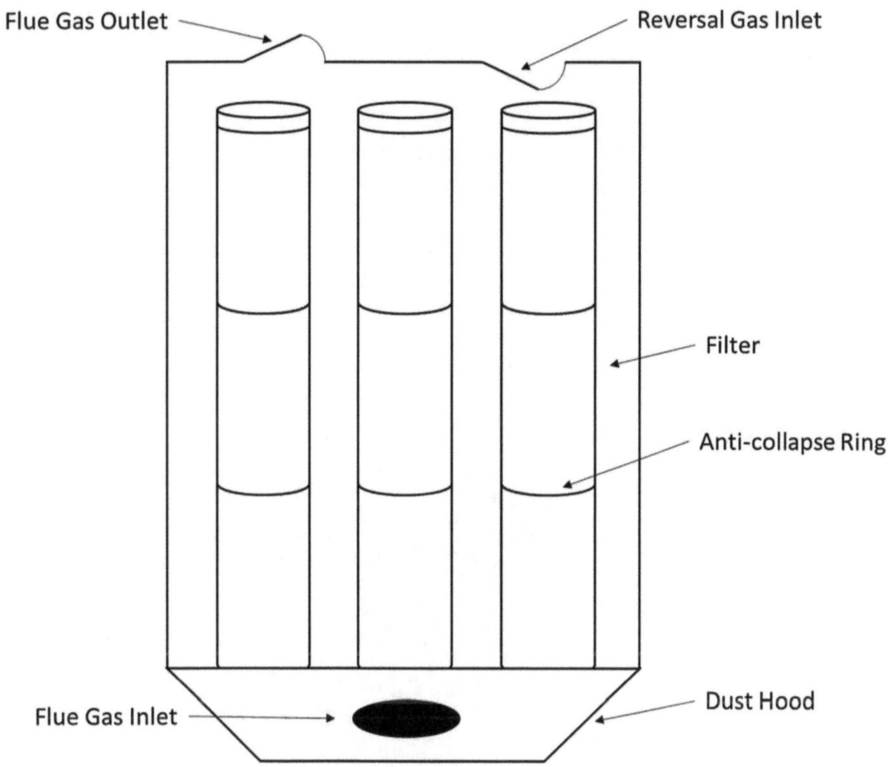

Fig. 18.5 Schematic structure of reverse-air cleaning

are usually used for single compartment baghouses or shaker type baghouses where the process should be stopped once bags are cleaned. A continuous system consists of several compartments and cleaning involves taking individual compartments offline to clean, which suspends the normal operation of the system during cleaning. Pulse-jet cleaning allows the process to continue during cleaning and is termed as a "Continuous Online System".

18.4.2 Step 2: To Determine the Air-to-Cloth Ratio

After the filter and cleaning technique is selected, the required dimensions of the bag should be determined to provide a sufficient cloth area for particle removal. If the gas flow rate exceeds the treating capacity, both pressure drop and penetration would rise significantly, resulted in reducing the lifespan of the filter. On the other hand, if the gas flow is lower than the treatment capacity, the baghouse would be underutilized costly.

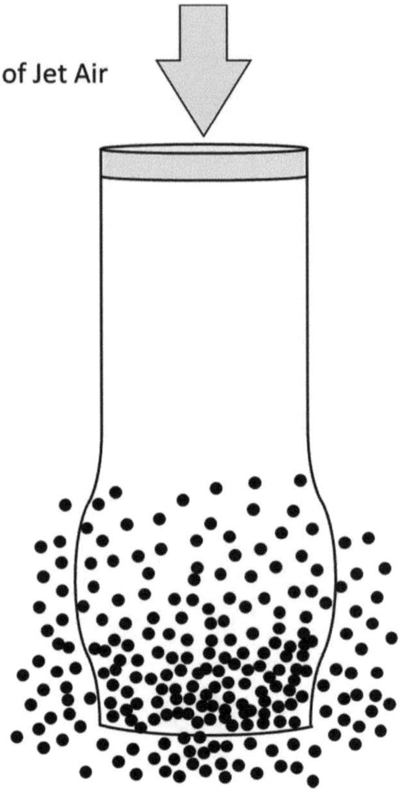

Fig. 18.6 Schematic diagram of pulse-jet cleaning

It is difficult to determine an appropriate A/C ratio through theoretical approaches, as the A/C ratio is a function of multiple variables including filter material, physio-chemical properties of the gas stream and dust, and operation conditions. Specifically, for pulse jet cleaned baghouses, the A/C ratio can be empirically estimated based on particle size, dust loading, and operating conditions [15]:

$$A/C = 2.878ABT^{-0.2335}L^{-0.06021}(0.7471 + 0.0853 \times \ln D), \quad (18.14)$$

where A and B are material and application factors used to characterize the filter material and the operating scenarios, T (°F) is the temperature, L (gr/ft^3) is the dust load, and D (μm) represents the mass mean diameter of the particles. However, acquiring data and information from similar applications or tests is more reliable. Finally, we introduce a safety factor (typically 10%) to ensure the safety and reliability of the equipment [16].

Table 18.3 lists the various A/C ratio in ft^3/(ft^2/min) for particle control which clearly shows that A/C ratios are highly dependent on the characteristics of the particles and cleaning methods. It is noted that the A/C ratios for pulse jet fabrics

Table 18.2 Summary of different cleaning mechanisms [9, 12, 14]

Cleaning mechanism	Operation mode	Construction structure	Advantages & disadvantages
Mechanical shaker	• Inside Collection • Manual or automatic cleaning • Cleaning lasts about 30 s • Intermittent or continuous Operation	• Woven filter • Small to large volume • Often compartment	**Advantages** • Easiness in Design and operation • Can be compartmentalized to allow one compartment to be taken offline for cleaning **Disadvantages** • Not applicable for sticky dust • Cannot operate at high-temperature • Requires more energy and longer cleaning time • Premature bag failure may occur when subject to improper cleaning frequency
Reverse air	• Inside Collection • Automatic cleaning • Cleaning lasts 10 to 30 s • Continuous Offline Operation	• Woven filter • Medium to large volume • Often compartment	**Advantages** • Longer bag life due to gentle cleaning action • Applicable for high-temperature flue gas **Disadvantages** • Cleaning air must be filtered to prevent the re-entrainment • Limited effectiveness for removing residual dust buildup
Pulse jet	• Outside Collection • Automatic cleaning • Cleaning lasts about 0.3 ~ 0.5 s • Continuous Online Operation	• Felted filter • Small to medium volume • Often compartment	**Advantages** • Thorough cleaning, lengthening bag life • Lowering system pressure drops and space requirements **Disadvantages** • Lower collection efficiency due to the compromised dust cake • Vulnerable to humidity in the exhaust gases

Table 18.3 Summary of the A/C ratios of different filter materials with various operation conditions [17]

Dust	Mechanical shaking/reverse air	Pulse jet/reverse air
	Woven	Felt
Alumina	2.5	8
Asbestos	3.0	10
Bauxite	2.5	8
Carbon Black	1.5	5
Coal	2.5	8
Cocoa, Chocolate	2.5	12
Clay	2.5	9
Cement	2.0	8
Cosmetics	1.5	10
Enamel Frit	2.5	9
Feeds, Grain	3.5	14
Feldspar	2.2	9
Fertilizer	3.0	8
Flour	3.0	12
Fly Ash	2.5	5
Graphite	2.0	5
Gypsum	2.0	10
Iron Ore	3.0	11
Iron Oxide	2.5	7
Iron Sulfate	2.0	6
Lead Oxide	2.0	6
Leather Dust	3.5	12
Lime	2.5	10
Limestone	2.7	8
Mica	2.7	9
Paint Pigment	2.5	7
Paper	3.5	10
Plastics	2.5	7
Quartz	2.8	9
Rock Dust	3.0	9
Sand	2.5	10
Sawdust (Wood)	3.5	12
Silica	2.5	7
Slate	3.5	12
Soap Detergent	2.0	5

(continued)

Table 18.3 (continued)

Dust	Mechanical shaking/reverse air	Pulse jet/reverse air
	Woven	Felt
Spices	2.7	10
Starch	3.0	8
Sugar	2.0	7
Talc	2.5	10
Tobacco	3.5	13
Zinc Oxide	2.0	5

are typically 2–3 times of those for mechanical shaking. Bear in mind that the fabric for the pulse jet cleaning method requires higher strength to resist the abrasion.

18.4.3 Step 3: To Estimate the Pressure Drop

Pressure drop can be separated into three parts including fabric, particulate layer, and baghouse structure during the entire service cycle. In general, the pressure drop due to the structure is too low to be considered, while the pressure drop due to fabric and particulate layer can be estimated by Darcy's law [18], which are expressed from Eqs. (18.15) to (18.17):

$$\Delta P_{\mathrm{t}} = \Delta P_{\mathrm{f}} + \Delta P_{\mathrm{p}}, \tag{18.15}$$

$$\Delta P_{\mathrm{f}} = \frac{D_f \mu [A/C]}{60 K_f}, \tag{18.16}$$

$$\Delta P_{\mathrm{p}} = \frac{D_p \mu [A/C]}{60 K_p}, \tag{18.17}$$

where ΔPt, ΔP_{f}, and ΔP_{p} (N/m^2) are the total pressure drop and the part contributed by fabric and particulate layer, respectively; D_f (m) and D_p (m) are depths of the filter and the particulate layer; μ (kg/m-s) is gas viscosity; and v (m/min) is the A/C ratio. The pressure drop is a function of gas velocity, thus the above equations can be integrated and simplified by dividing the term v, which is as follows:

$$\begin{cases} S = S_E + KW, \\ W = C_{\mathrm{i}}[A/C]\Delta t, \end{cases} \tag{18.18}$$

where S is the total filter drag, S_E is called "effective residual drag" which describes the drag for the clean fabric surface, K is a constant, and W represents the fabric load which is a function of the operating time and the dust load. As a result, the relationship between the fabric load (operating time) and the filter drag (pressure drop) is hypothesized to be linear as shown in Fig. 18.7.

For example, assuming a fabric filter is used to treat flue gas with a particle concentration of 5 g/m^3 at an A/C ratio of 0.9 m/min. According to the experimental test, K and S_E are 1.85 N-min/g-m and 115 Pa-min/m, respectively. If system pressure is not to exceed 1 kPa, it is estimated that the operating cycle should be less than 120 min.

Figure 18.7 presents the relationship between filter drag and fabric loading which reveals that the linear relationship only holds in higher W range. At lower fabric load, the curve shifts downward. This is because in real situations the dust layer is initially developed through filling the void between the fibers, which leads to a fast buildup and pressure drop. In some cases, the upward curvature of fabric drag, S, along with the fabric load, W, would occur, which could be due to the compression of the dust layer. Consequently, we can modify K by correlating the A/C ratio and fabric loading so that Eq. (18.18) can fit the pressure drop much better:

$$\begin{cases} K \propto [A/C]^m \\ K \propto [A/C]^n \end{cases}. \tag{18.19}$$

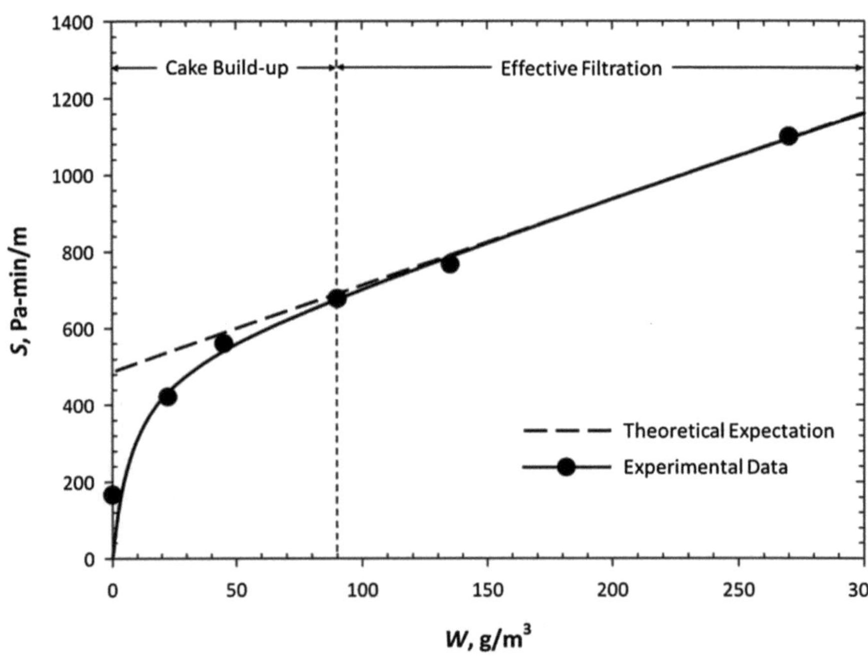

Fig .18.7 The theoretical and experimental results of the filter drag versus fabric loading

18.4.4 Step 4: To Propose the Cleaning Frequency

Just like other filtration techniques, baghouse systems require periodic "back-washes", otherwise the pressure drop rises and the safety of the system is much concerned. In practice, a baghouse is comprised of several compartments arranged in parallel to ensure reliability during fluctuating gas flow while reducing the impact of off-line compartments. In other words, when one or several compartments are off-line for cleaning, the remaining compartments can still function well and filter the dust load.

Let us use a baghouse filter with N compartments to treat a flue gas flow at the rate of Q. If the maximum pressure that the selected fabric material can withstand is $\Delta P_{\max}$, which occurs when there is a compartment undergoing cleaning. The pressure profile against time is presented in Fig. 18.8 which indicates that pressure drop increases suddenly when one compartment is removed from the system for cleaning.

To describe the pressure drop during the operation, we define (1) filtration time (t_f): the duration that the filter can operate before needing to be cleaned; (2) running time (t_r): the duration that all compartments are online for normal operation; and (3) cleaning time (t_c): the time required for a cleaning cycle. Therefore, Eq. (18.20) reveals that the filtration time equals the sum of the total running time and the cleaning time of $(N-1)$ compartments:

$$t_f = Nt_r + (N-1)t_c. \tag{18.20}$$

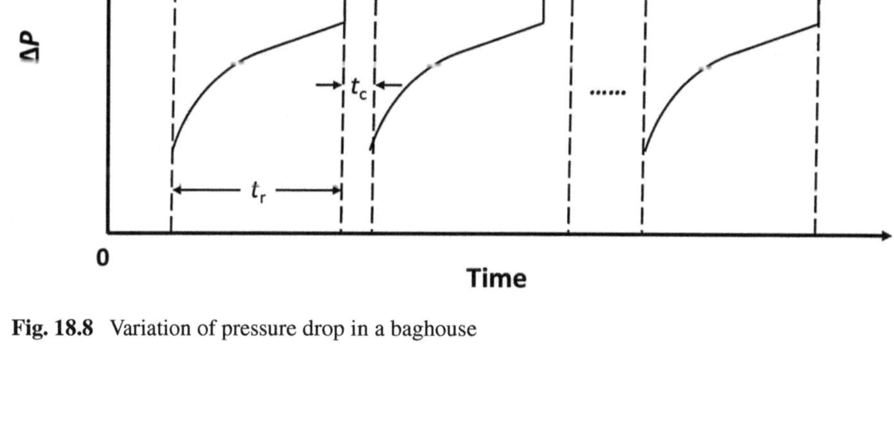

Fig. 18.8 Variation of pressure drop in a baghouse

Next, let us consider a specific compartment x, the service time represents the time between two cleaning cycles of compartment x, denoted as t_x which is equal to

$$t_x = t_f - t_r. \tag{18.21}$$

Combining Eqs. (18.20) and (18.21) we have

$$t_x = (N - 1)(t_r + t_c). \tag{18.22}$$

As a result, the fabric load of compartment x in an operation-cleaning cycle is given by Eq. (18.23):

$$W_x = (N - 1) \cdot \left(\left[\frac{A}{C} \right]_N \cdot t_r + \left[\frac{A}{C} \right]_{N-1} \cdot t_c \right), \tag{18.23}$$

while the corresponding filter drag is expressed as Eq. 18.24:

$$S_x = S_E + K W_x. \tag{18.24}$$

18.4.5 Step 5: To Select Fans and Hoppers

Fans and hoppers are responsible for the stable operation of baghouse filters. However, excessive fan vibration, worn bearing, and malfunctions of the belt, drive, gear, and electrical control may occur if the fan system is not properly designed and maintained.

According to the way the gas stream is oriented, the baghouse can be categorized as a positive-pressure baghouse or a negative-pressure (or suction) baghouse. A positive-pressure baghouse pushes the gas stream through the system by a forced draft fan which is located on the upstream. On the other hand, a negative-pressure baghouse pulls the gas stream using an induced draft fan located on the downstream. Due to the distinct mechanical structure, positive-pressure baghouses can be exposed to the atmosphere without a casing, while negative-pressure baghouses can only be installed in closed environments. As a result, negative-pressure baghouses can be used for hazardous air pollutant control to prevent accidental leakage of the captured pollutant. Special care must be taken to prevent the leakage of closed systems (negative-pressure baghouse), otherwise ambient air could infiltrate and disturb the system and create a dew point issue resulting in the system corrosion and deterioration.

18.4.6 A Rapid Method for Sizing Baghouse Filter

18.4.6.1 Essential Information Required (Customer-Inquiry Data Sheet)

(1) Description of application ______________________________

(2) Gas inlet volume ______________ ACFM; Humidity __________

(3) Gas inlet temp __________ ℉; Pressure __________

(4) Description of dust ______________________________

(5) Space available for equipment (sketch or dwgs) ______________________________

(6) Availability of adequate utilities (i.e., power, water, etc.) and ancillary system facilities (i.e., waste treatment and disposal, etc.) ______________________________

(7) Maximum allowable omissions (air pollution regulations) ______________________________

(8) Aesthetic considerations (i.e., visible steam or water vapor plume, impact on scenic vistas, etc.) ______________________________

(9) Contribution of air pollution control system to plant noise levels ______________

(10) Particulate pollutants: Type of dust: ________ Particle size _______Particle shape_

(a) Abrasive?______________ Explosive? __________ Inflammable? _____

(b) Sticky?______ Corrosive? _____ Tixuc? _______ Moisture content (wt%) __

(c) Bulk density ______________________ Specific gravity _______

(d) Inlet loading ______________________________ grains/SCF of lb/min

(e) Allowable outlet loading ______________________________ grains/SCF or lb/min

(11) Gaseous pollutants: Type(s) and Concentration(s) ____________ Composition Viscosity ________ Reactivity ______ Density ________ Combustibility ______

(12) Additional comments (i.e., location, etc) ______________________________

(13) Type of duty: Intermittent ______________ Continuous

(14) Operating Pressure ______________ in water gauge

(15) Existing electrical and compressed air capabilities ______________________________

18.4.6.2 Determination of Air-to-Cloth-Ratio

1. Equation of air-to-cloth-ratio

$$V_a = V_m{\cdot}C_1 \cdot C_2{\cdot}C_3{\cdot}C_4{\cdot}C_5{\cdot}C_6{\cdot}C_7,$$

where
V_a : Design rating
V_m : Normal rating by considering material characteristics
C_1 : Coefficient of the type of bag filter
C_2 : Coefficient of the dust loading
C_3 : Coefficient of the particle size distribution
C_4 : Coefficient of the gas temperature
C_5 : Coefficient of the required efficiency of separation
C_6 : Coefficient of the abrasive properties of the dust
C_7 : Coefficient of the type of process

2. Normal rating V_m (m^3/min.m^2)

V_m	Material
4.6	Cake mix, Cardboard dust, Cocoa, Feeds, Flour, Grain, Leather dust, Tobacco
3.7	Asbestos, Buffing dust, Fibrous and Cellulosic materials, Soda ash, Hydrated lime, Perlite, Salt, Sand, Sand blast dust, Talc
3.0	Alumina, Fine carbon black, Cement, Ceramic pigments, Clay and brick dusts, Coal, Fluorspar, Kaolin, Limestone, Rock dust, Ores and Minerals, Silica, Sugar, Cement raw meal, Clinker dust
2.7	Ammonium phosphate fertilizer, Coke, Diatomaceous earth, Dyes, Fly ash, metal powder, Metal oxides, Pigments, Plastics, Resins, Silicates, Starch, Stearates, Clay dust, Ground sand, Gypsum
1.8	Activated carbon, Molecular carbon black, Detergents, Fumes and Other dispersed products directed from reactions, Powder milk, Soaps

3. Coefficient of the type of bag filter (C1)

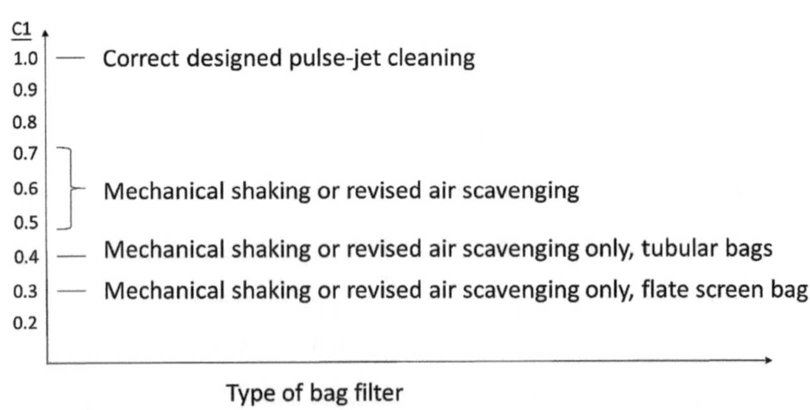

4. Coefficient of the dust loading (C2)

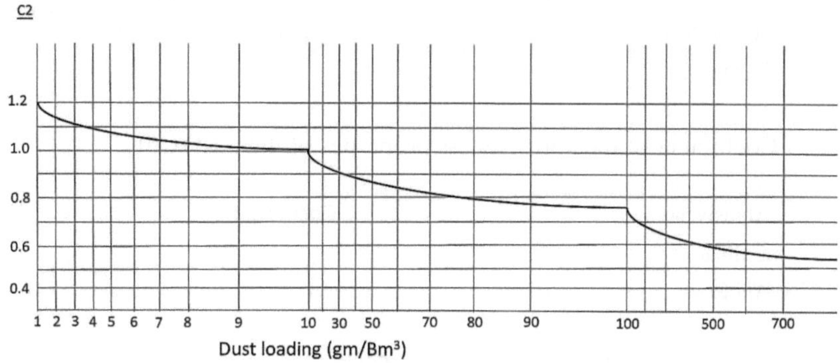

5. Coefficient of the particle size distribution (C3)

Particle size distrioution (μ)	Coefficient C3
Over 100	1.2
50–100	1.1
10–50	1.0
3–10	0.9
Under 3	0.8

6. Coefficient of the gas temperature (C4)

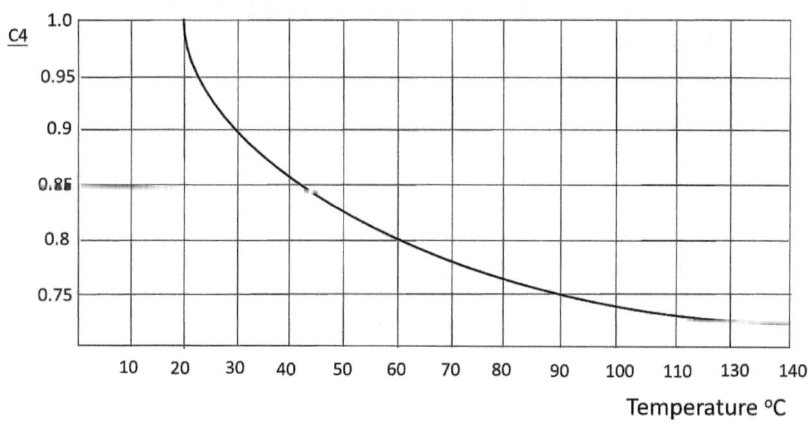

7. Coefficient of the required efficiency of separation (C5)

Dust loading of exit gas	Coefficient C5
50 mg /Nm^3	1
20 mg/Nm^3	<1

8. Coefficient of the abrasiveness properties of the dust (C6)

Abrasiveness	Coefficient C6
Ordinary	1
Abrasive	<1

9. Coefficient of the type of process (C7)

Process	Coefficient C7
Nuisance venting Relief of transfer points, Conveyors, packing stations, etc	1.0
Product collection Air-conveying venting mill, flash dryers classifiers, etc	0.9
Process gas filtration Spray dryers, kilns, reactors, etc	0.8

18.5 Performance Evaluation and Improvement Enhancement

18.5.1 Performance Evaluation

In general, the performance metric of a baghouse system is the PM concentration at the outlet rather than the "collection efficiency" since the outlet concentration is constant over a large inlet loading range. In normal operations, outlet concentrations typically reported varying from 0.001 to 0.01 g/scf [19] under which the dust cake builds up resulting in enhanced collection performance. Consequently, by applying appropriate configurations and cleaning techniques, a good engineering practice having higher collection efficiency and lower pressure drop should be achievable.

The power required to support the operation of the baghouse systems largely depends on cleaning methods and the filtration velocity. Assuming the fan efficiency is 60% while the specific gravity of the fluid is 1.00, the general fan power requirement can be estimated by Eq. (18.25):

$$Power_{\text{Fan}} = 1.81 \times 10^{-4}\left(Q_{e,a}\right)(\Delta P)(T), \quad (18.25)$$

where $Q_{e,a}$ (acfm) is the stream emission rate, T (hr/year) is the operation time per year, and ΔP (in. H_2O) represents the total pressure drop of the system. In addition, we would also consider the electricity consumption in the cleaning process. For mechanically cleaned baghouses, the additional power requirement is estimated by Eq. (18.26):

$$\Delta Power_{\text{m}} = 6.05 \times 10^{-6}(T)(A_{tc}), \quad (18.26)$$

where A_{tc} (ft^2) is the gross cloth area which can be calculated by Eq. (18.27):

$$A_{tc} = \left(Q_{e,a}\right)/(A/C). \quad (18.27)$$

The total power consumption for mechanical baghouse is given by Eq. (18.28):

$$Power_{total} = Power_{Fan} + \Delta Power_{m}. \quad (18.28)$$

In the case of a pulse-jet cleaning method, the additional electricity requirement of compressed air can also be estimated by the equation. In general, a pulse-jet system that treats 1000 scfm of emission stream would consume about 2 scfm of compressed air. It is noted that pulse-jet systems cannot continuously operate because the designer should consider the operating time and cleaning processes separately.

Compared with other particle collectors, the baghouse system is relatively expensive with respect to the capital cost. For instance, 1 ft^2 of floor space is acquired for each 5 ft^3/min treatment capacity. Large capital investment usually goes into the fabric filter which accounts for about 20–40% of the total expenses. The costs of fabric may get higher if special treatment to the filter for performance enhancement is required.

18.5.2 Performance Enhancement

Since the lifespans of the baghouse filters are highly dependent on the operating conditions and maintenance, the routine inspection and replacement of filters should be carried out according to the required schedule to prevent malfunction like filter failure.

Consequently, a holistic monitoring system to ensure the temperature, pressure drop, gas flow rate, and chemical compositions within threshold values should be built. Inspection and maintenance items usually vary with the construction material, structure, and operating conditions. Otherwise, Table 18.4 provides universal maintenance instructions.

Table 18.4 Summary of inspection and maintenance items [20]

Frequency	Items
Daily	• Check pressure drop • Monitor gas flow rate • Observe stack outlet visually or with a continuous monitor • Monitor cleaning cycle, pilot lights, or meters on control panel • Check compressed air on pulse-jet baghouses • Monitor discharge system; make sure dust is removed as needed • Walkthrough baghouse to check for normal or abnormal visual and audible conditions
Weekly	• Check all moving parts on the discharge system including screw-conveyor bearings • Check damper operation; bypass, isolation, etc • Spot check bag tensioning for reverse-air and shaker bags • Check compressed air lines including line oilers and filters • Blow out any dust from manometer lines • Verify temperature-indicating equipment • Check bag-cleaning sequence to see that all valves are seating properly • Check drive components on fan
Monthly	• Spot check bag-seating condition • Check all moving parts on shaker baghouses • Check fan for corrosion and blade wear • Check all hoses and clamps • Spot check for bag leaks and holes • Inspect baghouse housing for corrosion
Quarterly	• Thoroughly inspect bags • Check duct for dust buildup • Observe damper valves for proper seating • Check gaskets on all doors • Inspect paint on baghouse • Calibrate opacity monitor • Inspect baffle plate for wear

To improve the cleaning performance of a filter bag, filters can be subjected to several types of surface treatment by changing the properties of the filter medium and modifying the cleaning mechanism of filter bags during pulse-jet cleaning. In general, the initial medium movement leads to maximum acceleration values resulted in dislodgement of dust on the surface of the filter bag; meanwhile a high-velocity airflow jets into the bag exerting the detachment of dust cake. Thus, the cleaning mechanism is a combined effect of sudden expansion of the filter medium and reverse airflow permeation from the inside to the outside of the filter bag during pulse-jet cleaning [5, 7, 14–16]. Kanaoka et al. [17, 18] found that the structure or quality of filter material will affect the cleaning performance of a filter in a series study of the rigid ceramic filter. Both the structure and quality of filter material have an important influence on the cleaning mechanism of the filter bag.

The vibration frequency of acceleration can be used as an index to evaluate the cleaning effect of filter bags. The higher the vibration frequency of the acceleration is, the better the cleaning effect will be.

18.5.3 Innovative Technologies

Catalytic Filter

As emission regulations get more and more strict, baghouse systems need to be more versatile in reducing air pollution, especially gaseous pollutant removal. Ultra-Cat™ Catalyst Filter, developed by the Tri-Mer corporation, is a new generation of lightweight, ductile ceramic filters for efficient control of not only particles, but also NOx and dioxins [21]. The filter is comprised of fibrous ceramic materials mixed with nano-bits of the proprietary catalyst which is presented in Fig. 18.9. Ultracat filters can be operated together with a dry injection system. Test results revealed that PM concentration at the outlet was less than 2.0 mg/Nm^3, while the removal efficiency of NO_x and SO_x were 97–99% and 90–98%, respectively.

Such performance was achieved under moderate operating conditions of 350 °F, indicating the reactivity of the catalyst. In addition, the pressure drop was reported to be 6–8 in. w. c. which is very energy-efficient.

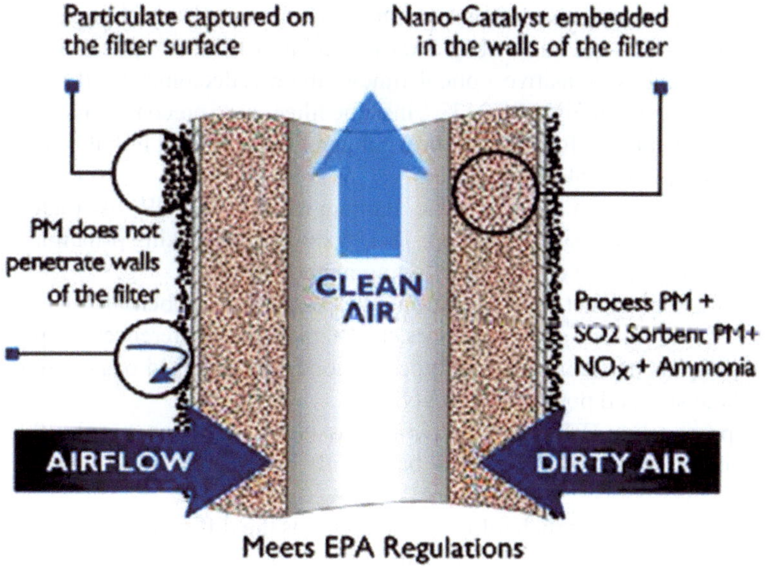

Fig. 18.9 Ceramic filter coated with nano-catalyst materials

Table 18.5 Characteristics of the precoat materials

Material		Hydrated limestone	Limestone powder	iFIL coating N
Median diameter/μm	By number	~0.78	~0.47	~0.5
	By mass	~3.1	~10.2	~36.0
Density/kg/m^3		2274	2768	1968
Bulk density/kg/m^3		345	787	117
Porosity/%		85	72	94

Filter using Precoat Materials

The precipitation of sticky and ultrafine particles has become increasingly important. To prevent clogging of the filter cloth by sticky and ultrafine particles, it is necessary to use a precoat layer, i.e., limestone powder, hydrated limestone, and iFIL coating N® (mainly consists of silicon dioxide) as shown in Table 18.5 [22].

The measured total collection efficiencies were > 99% for all precoat materials, and the grade efficiencies were equally matched for the highly efficient separation of ultrafine dust.

Intelligent detection technology for leakage bag

Due to the complicated structure and unpleasant environment of bag filter, the traditional method of bag damage detection and location having long working time which high cost and cannot identify and locate the damaged filter bag on-line in real time.

In order to overcome the deficiency of traditional methods an intelligent detection technology for leak age bag of bag filter based on distributed optical fiber sensing technology of phase-sensitive optical time-domain reflectometer (Φ-OTDR) was developed as shown in Fig. 18.10 [3]. Since the filter using precoat materials exhibits the long-term stable performance to precipitate ultrafine dust, it might be considered as an economic and ecological process.

Figure 18.10a shows the schematic diagram of the Φ-OTDR distributed optical fiber sensing detection system. Figure 18.10b shows the working principle diagram of the baghouse and the optical fiber laying method inside it.

Using wavelet packet decomposition (WPD) method to obtain the energy spectrum and energy entropy as eigenvectors, the characteristic differences of optical fiber vibration signals between different types of good and damaged bags with different leakage hole size and position are analyzed.

Back propagation (BP) neural network classifier is used to recognize the optical fiber vibration signals in damaged bags with different leakage hole sizes and positions.

With these developed technologies and methods the difference between the characteristics of the non-damaged and damaged events is very obvious by comparing the characteristic parameters between the optical fiber vibration signals in the non-damaged filter bags and in various types of damaged filter bags.

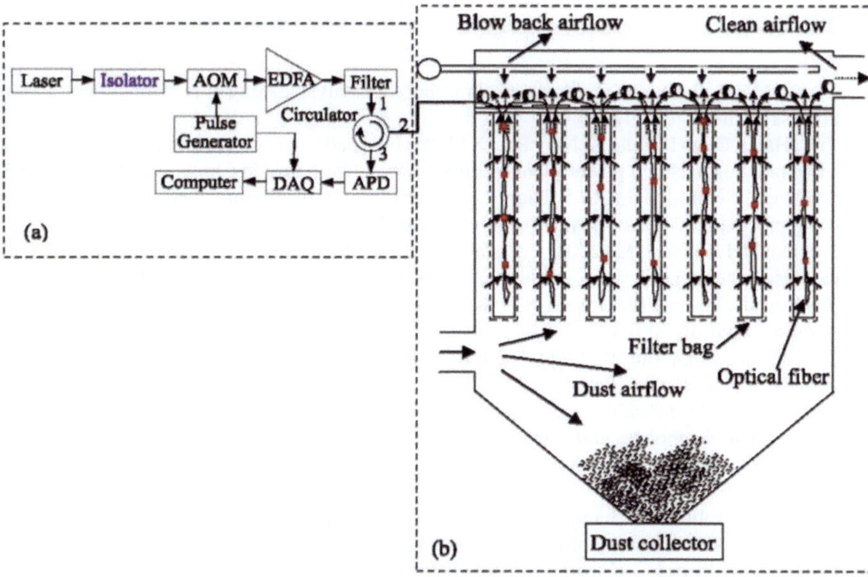

Fig .18.10 Filter bag leakage monitoring and positioning system based on Φ-OTDR distributed optical fiber technology. **a** Φ-OTDR distributed optical fiber sensing system. **b** Working principle diagram of the baghouse and the optical fiber laying method inside it

It is evident that the proposed technology can maintain higher recognition stability and rate (>90%) for different types of damaged bags.

Exercises

1. Define the following terms and explain why they should be considered for design use: air to cloth ratio, pulse-jet cleaning, effective residual drag, coefficient of the dust loading.
2. Please describe the general procedures for designing a baghouse.
3. List the major factors affecting the design of a baghouse.
4. Describe and compare the performance of conventional and catalytic filters.
5. Tests showed that filtration of a dusty air stream containing 2 grains of particulate matter per cubic foot of air gave a maximum pressure drop of 10 inches of water at a flow rate of 3 ft^3/min per square foot of filtering surface. If the exhaust volume rate is 10,000 acfm, what is the number of 1 ft (diameter) by 20 ft (length) filtering bags required?

References

1. Cavalli MN, Lilke N (2009) Mechanical milling for catalyst loading of polymer filters. J Mater Process Technol 209:5423–5428
2. Yi H, Hao J, Duan L, Tang X, Ning P, Li X (2008) Fine particle and trace element emissions from an anthracite coal-fired power plant equipped with a bag-house in China. Fuel 87:2050–2057
3. Schiller S, Schmid H-J (2015) Highly efficient filtration of ultrafine dust in baghouse filters using precoat materials. Powder Technol 279:96–105
4. Tufenkji N, Elimelech M (2004) Correlation equation for predicting single-collector efficiency in physicochemical filtration in saturated porous media. Environ Sci Technol 38:529–536
5. Guillory JL, Placer FM, Grace DS (1981) Electrostatic enhancement of moving-bed granular filtration. Environ Int 6:387–395
6. Givehchi R, Li Q, Tan Z (2015) The effect of electrostatic forces on filtration efficiency of granular filters. Powder Technol 277:135–140
7. Mycock JC (1999) Pick the right baghouse material. Power Engineering 103:43–46
8. 360 E (2018) Baghouses and Baghouse Filters Information
9. McKenna J, Furlong D (1992) Fabric filters. Van Nostrand Reinhold, Air Pollution Engineering Manual New York
10. Belba VH, Grubb WT, Chang R (1992) The potential of pulse-jet baghouses for utility boilers. Part 1: a worldwide survey of users. J Air Waste Manage Assoc 42:209–17
11. Wang LK, Williford C, Chen WY (2004) Fabric filtration. Humana Press
12. McKenna JD, Turner JH (1989) Fabric Filter--Baghouses I: theory, design, and selection:(a Reference Text): ETS
13. Greiner GP (1993) Fabric filter--baghouses II: Operation, maintenance, and troubleshooting (a Users Manual): ETS
14. Theodore L, Buonicore AJ (1982) Air pollution control equipment: selection, design, operation and maintenance
15. Frey R, Reinauer T (1964) New filter rate guide. Air Eng 30
16. Industrial air filtration: Solving problems in baghouse filtration. Filtration + Separation. 2012; 49: 28–31
17. Mussatti DC, Srivastava R, Hemmer PM, Strait R (2002) EPA air pollution control cost manual. Air quality strategies and standards division of the office of air quality planning and standards, US Environmental Protection Agency, Research Triangle Park, NC. 27711
18. Ensor D, VanOsdell D, Viner A, Donovan R, Hovis L (1984) Modeling baghouse performance. Fifth Symposium on the Transfer and Utilization of Particulate Control Technology
19. Wang LK, Pereira NC, Hung Y-T (2004) Air pollution control engineering: Springer
20. Young RA, Cross Jr F (1980) Operation and maintenance for air particulate control equipment
21. Tri-Mer (2018) UltraCat Catalyst Filters Control PM,SO2 ,HCl, NOx & Dioxins
22. Qian Y, Bi Y, Zhang M, Chen H, Guanjie Xu (2015) Effect of filtration operation and surface treatment on pulse-jet cleaning performance of filter bags. Powder Technol 277:82–88
23. Liu X, Li J, Shi B, Ding G, Dong F, Zhang Z (2019) Intelligent detection technology for leakage bag of baghouse based on distributed optical fiber sensor. Optical Fiber Technology 52:101947

Chapter 19
Electrostatic Precipitator

Nomenclature

A	Particle radius
A	Area of collecting electrode
D	Constant of ion diffusivity
D	Outer cylinder radius
D_p	Diameter of individual particle
E	Applied electric field
E_0	External electric field strength
E_c	Corona onset field at the wire surface
E_{spark}	Electric strength at which spark would occur
$j_r^{\pm}(a, \theta)$	Radial ion current density at the particle surface
J	Current density
K	Boltzmann's constant
K	Particle dielectric constant
MMD	Corresponding mass mean diameter
m_p	Weight of individual particle
N	Number of charges on the particle
N_d	Specific number of ducts to regulate gas flow and prevent maldistribution
P	Penetration ratio of particles with various diameters
Q	Unit charge
q_e	Elementary ion charge
Q_L	Water flow rate
q_s	Saturation charge
R	Particle radius
R	Resistivity of the dust layer on the collecting electrode
RR	Reentrainment
r_w	Wire radius
S_g	Specific gravity of water being pumped
S_N	Gas sneakage

P.-C. Chiang and X. Gao, *Air Pollution Control and Design*,
https://doi.org/10.1007/978-981-13-7488-3_19

T	Absolute temperature
U	Velocities of particles and fluid
V	Velocities of particles and fluid
Z	Water head
Δ	Motor efficiency of the pump
$\in$	Permittivity of space
θ	Angle in interest
M	Viscosity of fluid
ρ_r	Relative air density
T	Characteristic time
ϕ	Electric potential which is related to the electrical strength

19.1 Introduction

19.1.1 Features of Electrostatic Precipitator

Electrostatic precipitators, also known as ESP, have been widely used for material recovery and particulate removal for over a century since it was perfected by Fredrick Cottrell [1]. ESP can efficiently capture up to 99.9% of particles, while an older existing equipment has a range of actual operating efficiencies from 90 to 99.9%.

A precipitator consists of discharge electrodes, collection electrodes, high voltage electrical systems, rapping systems, hoppers, and a shell. Discharging electrodes ionize gas molecules and create a strong electrical field that charges the particles. Collecting electrodes are usually flat plates with charges opposite to the charging electrode which are used to collect the charged particulate matter. As the collection process proceeds, particles accumulated on the collecting plate thereby reducing the electric strength and then particle charging capacity. The accumulated particles are removed by rappers which vibrate the electrodes resulted in collapsing the dust layer. The alternative of the rapper cleaning technique is the water spray, which is implemented in wet precipitators for droplets and sticky particles. The removed particles fall into the hopper located at the bottom of the precipitator and are temporarily stored. The above-mentioned components are enclosed in a shell which provides a base support and protection.

19.1.2 Applications of Electrostatic Precipitator (ESP)

ESP has its unique advantages compared to other aerosol collection devices [2]. For instance, it exhibits extremely high collection efficiency if the equipment is well designed and properly operated and maintained. It is also capable of handling high-temperature gases exhausted from the boilers and steel furnaces. In addition, the direct impact of particles onto the ESP can be prevented and result in a lower pressure drop and reduced fan cost. Furthermore, it is able to withstand a variation of particulate sizes and a fluctuation of dust load.

However, there are three disadvantages regarding ESPs: (1) the capital cost is the highest of any particulate collection system and electrodes require sophisticated maintenance; (2) the footprint is significant due to the space requirement of auxiliaries; and (3) implementation is limited if particles are flammable, combustible, or explosive.

19.2 Principles

Figure 19.1 presents the movement of particles within the cell of an electrostatic precipitator, which is comprised of a charging electrode and a collecting electrode. In an electrostatic precipitator, the discharging electrode generates a powerful electric field thereby ionizing gas molecules and producing free ions or electrons. Thereafter, particles from the flue gas are charged by colliding with ions or electrons. Due to the electrostatic force, the charged particles deviate away from the stream trajectory and move toward the collecting electrode.

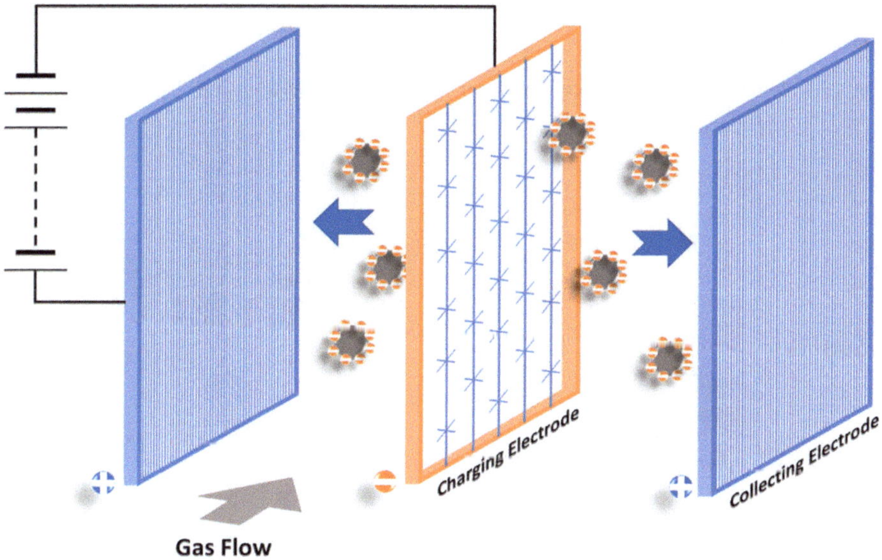

Fig. 19.1 Scheme of particle removal by electrostatic precipitator

19.2.1 Generation of Corona

The generation of a corona requires applying a high voltage on a wire and producing a highly nonuniform electric field between the discharging and collecting electrode. As shown in Fig. 19.2, gas molecules have a few free electrons and an equal number of positive ions, which would move toward opposite directions. If the electric field keeps increasing, the coulombic forces imposed on ions and electrons would increase and eventually the gas molecule would collapse.

Electrons and ions would accelerate the particle separation because of the coulombic forces. When moving to an electrode with a different charge, the electrons and ions are accelerated due to the electric force, producing more electrons and ions which is the phenomenon so-called "Avalanche multiplication" (see Fig. 19.3). On the other hand, many electrons or ions collide directly onto the surface of the metal,

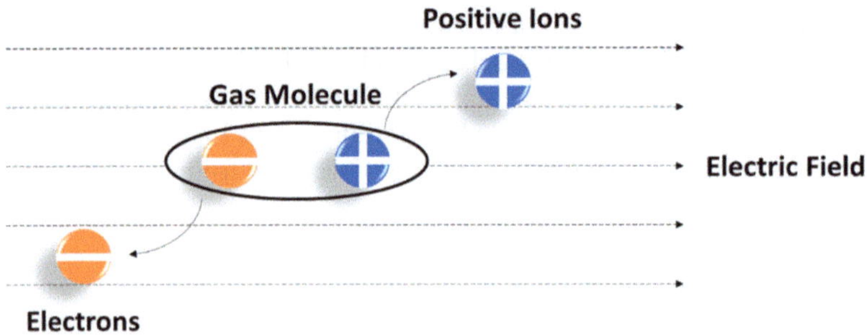

Fig. 19.2 Ionization of gas molecules under intensive electric field

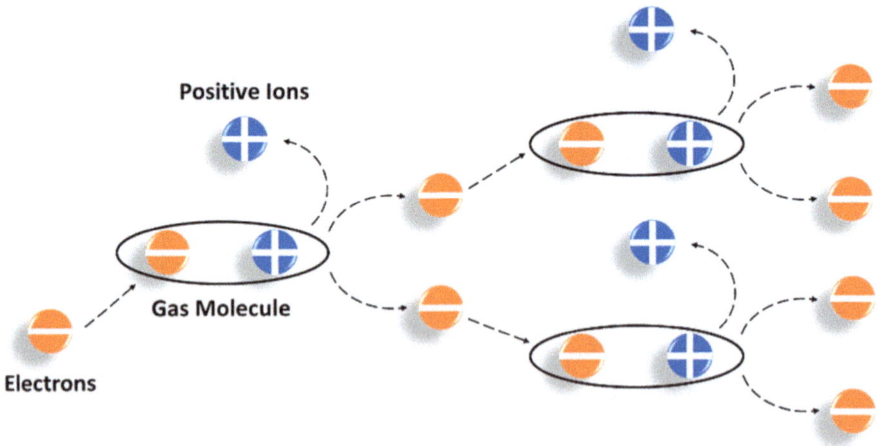

Fig. 19.3 Scheme of Avalanche multiplication initiated by electrons

discharging electrodes or flow into space around the wire resulting in additional electrons knocking off. This phenomenon is called secondary emission which produces positive ions and a large number of free electrons.

19.2.2 Charging of Particles

After gas molecules are ionized, particles get charged through the collision with the ions and electrons. During the charging process, two major forces act on a gas ion in the vicinity of a particle which are attractive forces due to the field strength and the ionic image and repulsive forces due to the Coulomb effect. As a result, charging will continue until the other two forces reach a balanced state [3]. Here we assume ions that are generated by the corona possess positive charges, thus there would be three distinct scenarios: (1) particles have a positive charge; (2) particles have zero charges; and (3) particles have a negative charge, which is depicted in Fig. 19.4.

With the assumption that the charge on particles is a continuous variable, we can express the charging rate as Eq. (19.1):

$$\frac{dn}{dt} = -\oint j_r^{\pm}(a, \theta)dA, \tag{19.1}$$

where $j_r^{\pm}(a, \theta)$ (ions/cm^2/s) is the radial ion current density on the particle surface, n is the number of charges on the particle, a (cm) is the particle radius, and θ (rad) is the angle in interest. In addition, we assume that the ion density on the particle surface is zero, thus $j_r^{\pm}(a, \theta)$ is given by Eq. (19.2):

$$j_r^{\pm}(a, \theta) = -D \cdot \left. \frac{\partial N\left(r, \overline{\theta}\right)}{\partial r} \right|_{r=a}, \tag{19.2}$$

where D (–) is the constant of ion diffusivity and $N\left(r, \overline{\theta}\right)$ represents the ion density which can be solved by the following equation:

$$\nabla^2 N(\overline{r}, \theta) \pm \frac{q_e}{kT} \nabla \phi(\overline{r}, \theta) \cdot \nabla N(\overline{r}, \theta) = 0, \tag{19.3}$$

where q_e is the unit charge, k is Boltzmann's constant, T is the absolute temperature, and $\phi(\overline{r}, \theta)$ is the electric potential which is related to the electrical strength and can be expressed as follows [4]:

Fig. 19.4 Schemes of electric fields near charged particles

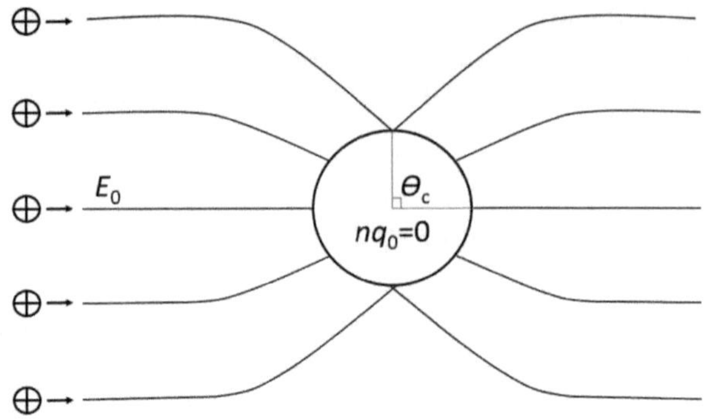

(a) Zero charged particles

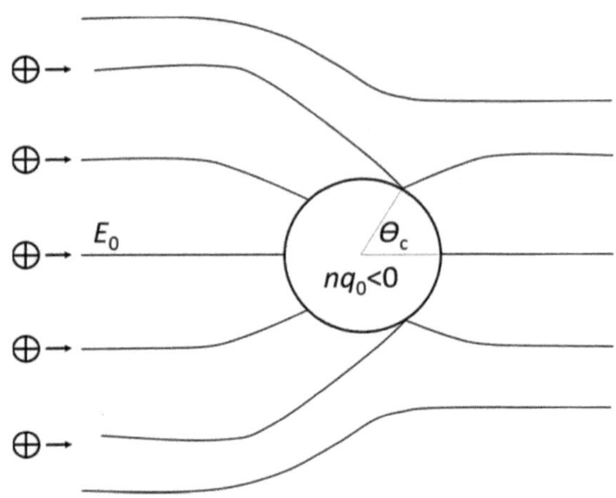

(b) Negatively charged particles

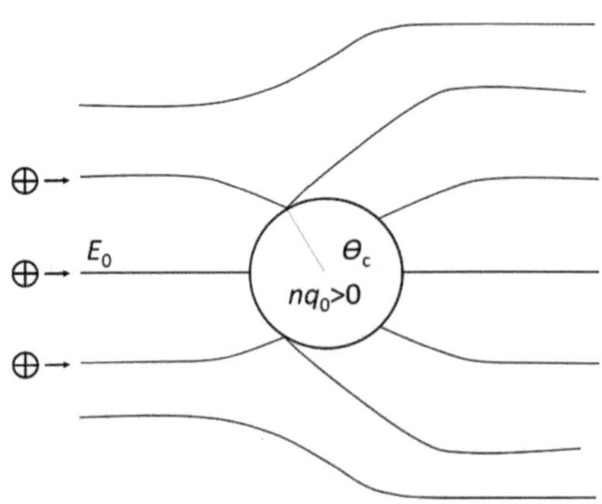

(c) Positively charged particles

$$\phi(r,\theta) = \underbrace{-E_0 r\cos\theta + \left(\frac{K-1}{K+2}\right)\frac{a^3}{r^2}E_0\cos\theta}_{\text{Field Charge}} + \underbrace{\frac{nq_e}{4\pi \in r}}_{\substack{\text{Diffusion}\\ \text{Charge}}}, \tag{19.4}$$

where E_0 is the external electric field strength, K is the particle dielectric constant, and ϵ is the permittivity. To obtain the charging rate, Eq. (19.4) can be substituted for Eq. (19.3) to determine the ion density. Subsequently, the radial ion density is calculated by Eq. (19.2). Finally, the charging rate, which is a function of time, is estimated by Eq. (19.1).

For simplicity, a dimensionless-notation system is established to describe the electric field within the electrostatic precipitator [5]:

$$\begin{cases} \upsilon = \frac{n\cdot q_e{}^2}{4\pi\cdot\epsilon\cdot r\cdot k\cdot T} \text{ dimensionless charge,} \\ w = \frac{r\cdot E}{(k\cdot T/q_e)} \text{ dimensionless electric field,} \\ \tau = \frac{q_e\cdot\mu\cdot N\cdot t}{\epsilon} \text{ dimensionless exposure time,} \end{cases} \tag{19.5}$$

where n is the number of ions or electrons on the particle, q_e is the elementary ion charge which is 1.6×10^{-19}C, ϵ (8.845×10^{-12} F/m) is the permittivity of space, r is the particle radius, k ($1.38 \times 10^{-23}\text{J} \cdot \text{K}^{-1}$) is Boltzmann's constant, T is the absolute temperature, and E is the applied electric field. As a result, $k \cdot T/q_e$ is about 26 mV at room temperature.

Particles are charged through field charging and/or diffusion charging. In field charging, the particles are directly charged through collisions with ions in the gas stream which is very effective for particles larger than 3 μm in diameter. Once they collide, the ions stick to the particle as it travels through the gas stream driven by the electrical field. Field charging is achieved by neglecting diffusion effects and calculating the charging rate from the transport of ions along the electric field lines. In other words, the charging process is restricted to angles within $\theta_c < \theta < \pi$ as shown in Fig. 19.4. Pauthenier, Cochet [6] provided the solution in Eq. (19.6):

$$\left(\frac{d\upsilon}{d\tau}\right)_{\text{field}} = \begin{cases} \frac{3w}{4}\left(1 - \frac{\upsilon}{3w}\right)^2, v < 3w, \\ 0, v \geq 3w, \end{cases} \tag{19.6}$$

where $3w$ represents the saturation charge which is determined by assuming that conductive particles are smaller for the dielectric particles.

Diffusion charging is effective for small particles since the particles are charged by collision with ions through Brownian motion. For larger particles, the Brownian motion is overwhelmed as the particles can hold more charge. The combination of field and diffusion charging is at a minimum near 0.5 μm. A mathematical expression for diffusion charge is calculated by neglecting the effects of the external electric field which are the first two terms in Eq. (19.7) [7–10] and is shown as follows:

$$\left(\frac{d\upsilon}{d\tau}\right)_{\text{diff}} = \frac{\upsilon}{e^{\upsilon} - 1}. \tag{19.7}$$

Note that $\frac{d\upsilon}{d\tau}$ is 1 when υ approaches to zero.

As a result, the total charging rate for the combined field and diffusion charging is shown in Eq. (19.8):

$$\left(\frac{d\upsilon}{d\tau}\right)_{\text{total}} = \begin{cases} \frac{3w}{4}\left(1 - \frac{\upsilon}{3w}\right)^2 + 1, v < 3w, \\ \frac{\upsilon - 3w}{e^{\upsilon - 3w} - 1}, v \geq 3w. \end{cases} \tag{19.8}$$

It is noted that neither a field nor diffusion charge can effectively impose a charge on particles ranging from 0.5 to 3 μm, making it a defect in the fractional collection curve.

19.2.3 Migration

The migration of particles is dominated by the forces imposed on them, which is difficult to describe since the electric force is not uniform or constant. For instance, particles around the discharging electrode exhibit the greatest acceleration rate when flowing away from the electrode. As they go further and further, the electric force gets weaker, resulted in a decreasing acceleration rate. Meanwhile, electric force on a particle is also influenced by the electron, ions, and other charged particles in the space, which makes it more complex.

For this reason, the terminal velocity of particles in a steady-state should be introduced. According to Newton's law, the movement of particles subject to an external force is expressed as follows:

$$m_p \frac{dv}{dt} = \frac{3\pi \mu D_p}{C_c}(u - v) + \sum_i F_i, \tag{19.9}$$

where C_c is the Cunningham slip correction factor, m_p and D_p are the weight and diameter of an individual particle, respectively, v and u are the velocities of particles and fluid, respectively, μ represents the viscosity of the fluid. As a result, the first item on the right side is the Stokes drag force and the second item represents external forces, which are primarily electric forces in this chapter. It is noted that the gas flow vertical to the collecting electrode in the electrostatic precipitator can be neglected so that the fluid velocity u is zero.

In electrostatic precipitators, the external force imposed by the electric field is a product of the charge and the potential gradient of the electric field, then Eq. (19.9) can be rearranged as follows:

$$m_p \frac{dv}{dt} = \frac{3\pi \mu D_p}{C_c}(-v) + qE. \tag{19.10}$$

As the particles are accelerated due to the electric force, the friction force from the fluid is also increased which is opposite to the direction of motion. Ultimately, the particle velocity could be balanced by the drag force and expressed as follows:

$$v_e = \frac{qEC_c}{3\pi \mu D_p}. \tag{19.11}$$

19.3 Design Method

19.3.1 Theoretical Approach

The theoretical approach to design an electrostatic precipitator is based on the electric field generating charge particles. Specifically, the electric strength is a function of the distance between charging and collecting electrodes and the resistivity of the particles. White [11] developed a simplified Eq. (19.12) for the estimation of diffusion and field charging:

$$\begin{cases} q_{\text{diff}} = \left(\frac{rkT}{e}\right)\ln(1+\tau), \\ q_{\text{field}} = \frac{q_s\theta}{\theta+\tau'}, \end{cases} \tag{19.12}$$

where τ and τ' are characteristic terms which are expressed as $\frac{\pi r \upsilon N (q_e)^2 \theta}{kT}$ and $\frac{4\epsilon}{N\mu q_e}$, respectively; θ is the real-time during which the particles are exposed in the electric field; q_s is the saturation charge defined as $12\pi\epsilon r^2 E$.

The scales of ESP are dependent on electrodes whose sizes are based on four assumptions: (1) no longitudinal gas stream mixing; (2) uniform electric field and particle distribution; (3) instant particle terminal velocity; and (4) negligible re-entrainment of collected particles. When charged particles are transported to a collecting electrode, where they are held by electrostatic attraction. Assuming flue gas enters an ESP at a velocity of U while the distance between charging and collecting electrodes is $d/2$, and the length and height of collecting electrodes are L and W, respectively (see Fig. 19.5).

If the ESP operation is performed under a steady state, which means that the particle loading remains constant during the operation period, a finite section dx in the direction of gas flow would lead to

$$\frac{dC}{dt} = \underbrace{(UWd)C(x)}_{\text{Mass in}} - \underbrace{(UWd)C(x+dx)}_{\text{Mass out}} - \underbrace{(v_e W dx) C\left(x + \frac{dx}{2}\right)}_{\text{Mass Removed}} = 0. \tag{19.13}$$

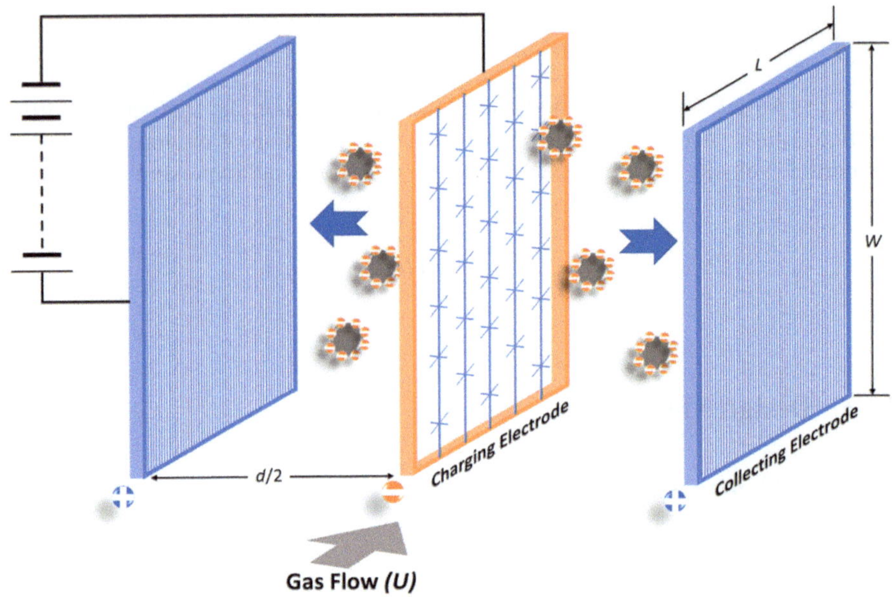

Fig. 19.5 Graphic illustration of the design parameters in ESP system

After rearranging the terms, Eq. (19.13) turns into

$$(Ud)\frac{C(x+dx)-C(x)}{dx} = -(v_e)C\left(x+\frac{dx}{2}\right). \tag{19.14}$$

When the limit of $dx \to 0$, Eq. (19.14) transforms into

$$\frac{dC}{dx} = -\frac{v_e}{Ud}C. \tag{19.15}$$

Note that the initial boundary of this equation is that

$$C|_{x=0} = C_{\text{in}}. \tag{19.16}$$

Finally, the solution to this equation is readily obtained

$$C_{\text{out}} = C_{\text{in}}e^{-v_e L/Ud}, \tag{19.17}$$

or

$$\eta = 1 - e^{-v_e L/Ud} = 1 - e^{-v_e WL/Q}. \tag{19.18}$$

With the developed Eq. (19.18), it suggests that higher removal efficiencies for the electrostatic precipitators can be achieved by extending the length while narrowing the space for collecting electrodes.

19.3.2 Good Engineering Practice

The design of an ESP depends on the dust load (in g/m^3, lb/ft^3, or gr/ft^3) and the gas flow rate (in m^3/min, ft^3/min or acfm), like other particle collection technologies. In addition, the applied voltage and current densities (the combined term as the operation point), as well as the specific collection area is the principal design parameters for ESP. However, the application of these parameters is dependent on the type of ESP that is used. For instance, multiple sections are implemented in ESP design for more effective and flexible operations.

In general, the applied voltage ranging from 20,000 to 100,000 V is needed to generate a sufficient electrical field for transporting particles to collecting plates. If the applied voltage is lower, collection efficiencies may be unsatisfactory. Meanwhile, the increase in collection efficiency by increasing the corona power under higher applied voltages would not always have a positive response. Therefore, practical current densities should be maintained from 50 to 750 μA/m^2 under which a reasonable corona power is 1.75 to 17.5 W/m^3 min.

Specific collecting area (SCA), expressed as the area of collection plate per 1000 m^3/hr of gas flow, is dominated by the required collecting efficiency and is usually 11 to 45 h/m with a typical value of 22 h/m due to the resistivity of particles. In addition, the profile of the collecting plate is constrained by the aspect ratio from 1.3 to 2.0 thereby ensuring sufficient retention time for the gas stream in the collecting plate. Attention should also be paid to the power supply system of the collecting plates. In order to achieve a reliable performance of ESP, the proper electrical distribution is needed to produce a uniform electric field. As a rule-of-thumb, approximately one T-R set for every 930 to 2970 m^2 of the collection-plate area is recommended.

Since a particle's trajectory can be predicted based on its size, position, and charge, the collection efficiency can be estimated by the empirical formula including, effective length, plate spaces, and field strength [12, 13] in the form of Deutsch-Anderson equations as shown in Eq. (19.19):

$$\begin{cases} \eta = 1 - a_1 e^{-a_2 L}, \text{ effective length,} \\ \eta = 1 - b_1 e^{-b_2/d}, \text{ plate distance,} \\ \eta = 1 - c_1 e^{-c_2 E^2}, \text{ field strength.} \end{cases} \tag{19.19}$$

Reynolds [14] thus identified the coefficients of three pairs of parameters listed in Table 19.1 which indicates that the greater the effective length, the narrower the plate space, and the stronger the electric field required to remove the fly ash.

Table 19.1 Summary of empirical parameters for ESP design

Applications	Empirical design factors					
	a_1	a_2	b_1	b_2	c_1	e_2
Fly ash (EP)	0.2590	0.1596	0.1594	15.58	0.7052	0.2047
Municipal incinerator	0.1342	0.1121	0.1556	14.82	0.1407	0.0942
Cement dust	0.4191	0.1010	0.4362	13.40	0.4743	0.0918
Blast furnace	0.2050	0.2166	0.4996	35.59	0.4133	0.2330
Coke oven deterring	1.0542	0.1150	1.1722	14.95	1.0637	0.1015
Petroleum cracking	0.3320	0.0856	0.3366	10.88	0.3891	0.0633

19.4 Design Procedures

Figure 19.6 presents the design procedures for an ESP system in which the characteristics of the dust and flue gas should be investigated. According to the characteristics of the particles and flue gas as well as required emission standards, the proper device type, configuration, operation mode and installation position should be evaluated. If the optimum voltage level of an ESP is determined, the collecting area should become a function of the desired collection efficiency, gas stream flow rate, and particle drift velocity [15]. In order to ensure reliability, the estimated collection area should be examined while sectionalizing the ESP. Finally, cleaning techniques including the method, intensity and frequency should also be assessed.

19.4.1 Step 1: To Select the Appropriate Equipment

Electrostatic precipitators provide highly flexible, reliable, and effective techniques for particle collection. Based on different designs of ESP such as tubular or plate, ESP can operate in wet or dry and cold or hot conditions.

Configuration

Typical configurations of ESP include plate and tubular precipitators. The original ESP was tubular, with the high-voltage wire acting as the collecting electrode. As the technology improves, the performance of the plate ESP exhibits much better than that of the tubular ESP, which is typically operated in a wet mode thereby avoiding leakage and re-entrainment phenomenon. The diameter of a tubular ESP varied from 0.15 to 0.31 m (0.5–1 ft), and length ranged from 1.85 to 4.0 m (6–15 ft).

Plate ESPs refer to precipitators with collecting electrodes in the plate form; while charging electrodes can be wire, rigid-frame or plate. In particular, plate-wire is used in a wide variety of industrial applications with large gas flow rates (typically 100–500 m^3/s in standard conditions) [16] and temperatures as high as 700 °C. The treatment capacity of a plate-wire ESP is typically 2–110 g/m^3. Discharging wires

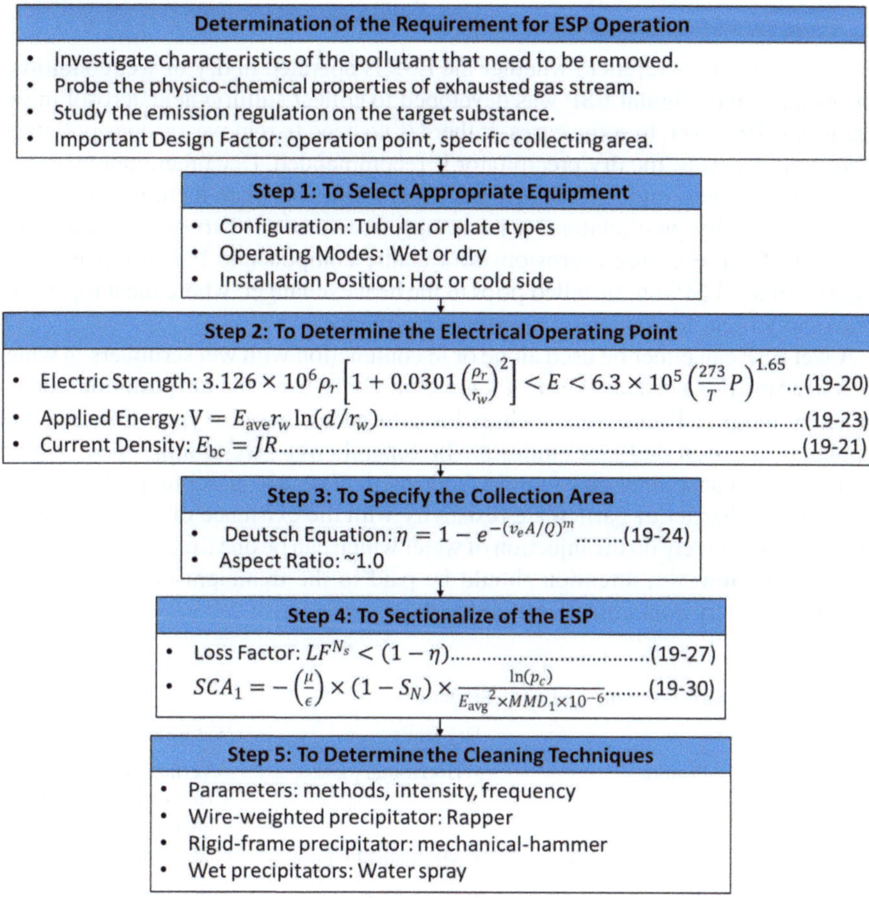

Fig. 19.6 Design framework of electrostatic precipitator

and collecting plates in a plate-wire ESP are approximately 0.13–0.38 cm thick and 6 and 12 m high. In addition, the space between the plates is usually between 15 and 30 cm.

For flue gas with flow rates ranges from 50 m^3/s to 100 m^3/s, the flat plate ESP is recommended rather than the plate-wire ESP (AWMA, 1992). In general, the plate ESP is more effective for collecting particles with high resistivity and small (1–2 μm) mass median diameters (MMDs). However, a higher collection efficiency of fly ash can be achieved if the low gas velocity is maintained to prevent penetration due to the rapping losses. The plates in this type of ESP are typically spaced 30–38 cm (12–15 in.) apart and 8–12 m (30–40 ft) in height.

Operating Modes

The operating mode refers to whether the ESP is operated in dry or wet conditions. For instance, the original ESP was developed to collect sulfuric acid aerosol in wet conditions. However, in some cases, if there is no need to remove gaseous pollutants or aqueous aerosols, the dry precipitator is recommended. Dry precipitators usually have vertical plates, under which the gas stream flows between them in a horizontal direction. For dry precipitators, it is strongly recommended to operate above the dew point for preventing corrosion from acidic components. For this reason, the majority of dry ESPs are installed prior to the heat exchanger, where the temperature is favorably high, for achieving a better performance.

A wet ESP can either be used alone or in conjunction with wet scrubbers in which no water spray is needed. Wet ESPs are commonly used for air pollution control for many industrial processes such as blast furnaces, basic oxygen furnaces, and cupolas in the iron and steel industry. In general, wet precipitators have distinct advantages because they can remove both particulate and gaseous pollutants and overcome the barrier of particulate resistivity with the existence of the water agent. Wet precipitators rely on the injection of water which can be one time flowing through or recycled. However, attention should be paid to the treatment of wastewater to prevent secondary contamination (Table 19.2).

Table 19.2 Comparisons of dry and wet ESP devices

		Dry ESPs	Wet ESPs
Configurations	Profiles	• Horizontal plate	• Vertical tube • Horizontal/vertical plate
	Cleaning techniques	• Mechanical rapping	• Water spray
	Position	• Forepart of APC sequence	• Rear end of APC sequence
Operating conditions	Temperature	• Above dew point	• Below dew point (~60 °C)
	Humidity	• 5–20%	• ~100%
	Gas flow	• 0.9–1.5 m/s	• 1.8–3.0 m/s
Engineering performance	Collecting efficiency	• > 99% for coarse particles • ~ 90% for $PM_{2.5}$	• > 99% for $PM_{2.5}$
	Water usage	• No	• Yes
	Pressure drop	• < 0.5 kPa	• <0.5 kPa
	Others		• Effective to soluble gaseous pollutant
Costs		• Lower	• Higher

Installation Positions

According to the installation location, ESP can be categorized as hot and cold sides. Hot-side precipitators are usually used in high-temperature applications including cement kilns, steel refining furnaces, and other industrial processes where dry precipitators dominate. The advantage of a hot-side precipitator is the suppressed potential of corrosion and hopper plugging. On the other hand, devices used on the hot side should be designed to be capable of dealing with large amounts of gas volume.

In practice, both dry and wet devices can be used on cold-side applications. However, with the promotion of using low-sulfur coals, the dry precipitator has a relatively lower collection efficiency due to the high resistivity of particulate. As a result, the wet precipitator becomes prevalent in these applications. However, frequently monitoring the potency of corrosion is necessary to avoid the condensation of acidic components.

19.4.2 Step 2: To Determine the Electrical Operating Point

The operating point refers to the applied voltage and corresponding current under which the optimum removal efficiency could be achieved when the applied electric field and voltage is strong enough to generate a corona. In practice, the operating point is correlated to the resistivity of the collected material. The higher the resistivity, the higher the voltage applied. As indicated in Eqs. (19.9–19.20) the electric strength should be powerful enough to produce a glowing corona, but it will be strained by the sparking point [16]:

$$\underbrace{3.126 \times 10^{6}\rho_r\left[1 + 0.0301\left(\frac{\rho_r}{r_w}\right)^{2}\right]}_{\text{Minimum electric strength for generating glow corona, } E_c} < E < \underbrace{6.3 \times 10^{5}\left(\frac{273}{T}P\right)^{1.65}}_{\text{Maximum electric strength at sparking point, } E_{spark}} \tag{19.20}$$

where E_c (V/m) is the corona onset field at the wire surface, ρ_r (−) is the relative air density at 1 atm 20 °C, r_w (m) is the wire radius, E_{spark} (V/m) is the electric strength at which spark would occur, and T (K) and P (atm) are temperature and pressure of the gas stream. Attention should also be paid on the back corona which impairs the electric strength as the dust is accumulated. The occurrence of the back corona is dependent on the resistivity of the dust layer, which is given by

$$E_{\text{bc}} = JR, \tag{19.21}$$

where J (A/m^2) is the current density and R (Ω-m) represents the resistivity of the dust layer on the collecting electrode.

In practice, the average electric strength is used to estimate the required voltage to generate a sufficient electric field:

$$\begin{cases} E_{\text{ave,plate-wire}} = \frac{E_{\text{spark}}}{1.75}, \\ E_{\text{ave,flatplate}} = \frac{5\times E_{\text{spark}}}{6.3}. \end{cases} \tag{19.22}$$

Note that the above formula is only good for cases where the back corona is negligible. If there is a severe back corona effect, the calculated value should multiply by a safety factor of 0.7 for practical consideration. Such an electric field should be maintained by the corona onset voltage (V) which is calculated by integrating the electric field from the wire to the plate:

$$\text{V} = E_{\text{ave}} r_w \ln(d/r_w), \tag{19.23}$$

where d is the outer cylinder radius for the tubular ESP (m) or $4/\pi$ wire-plate separation for the plate-wire ESP (m).

The electric field that is generated by charging electrodes can be produced from a metal wire that hangs vertically in the precipitator, several wires attached together in rigid frames, or a rigid electrode made from a single piece of metal.

19.4.3 Step 3: To Determine the Collection Area

As mentioned earlier, the removal efficiency is a function of the collection area, gas flow rate, and drift velocity of the particles which can be more easily illustrated by introducing the specific collection area (SCA), i.e., the ratio of the collecting electrodes to the gas flow rate shown in Fig. 19.7. It is apparent that the greater the SCA is, the higher the collection efficiency becomes. On the other hand, for particles with a larger drift velocity, less SCA is required to achieve the same collection efficiency. Furthermore, if the SCA is fixed, the higher voltage should be applied so that particles can get larger drift velocity for higher collection efficiency.

The drift velocity used for ESP design can refer to the available data obtained from the laboratory tests as shown in Table 19.3 which lists the drift velocities of various particles in different types of ESPs. Drift velocities are impaired by back coronas so that the specific collection area should be greater than the estimated value by providing longer retention time. For this reason, the flat plate ESP generates the higher particle drift velocity which would reduce the requirement for specific collection areas since the operation is less susceptible to the back corona.

However, if the relationship between the drift velocity and collection efficiency could not be expressed by Eq. (19.18) in practice, a modified Matts-Öhnfeldt et al. [17] equation should be introduced as shown in Eq. (19.18):

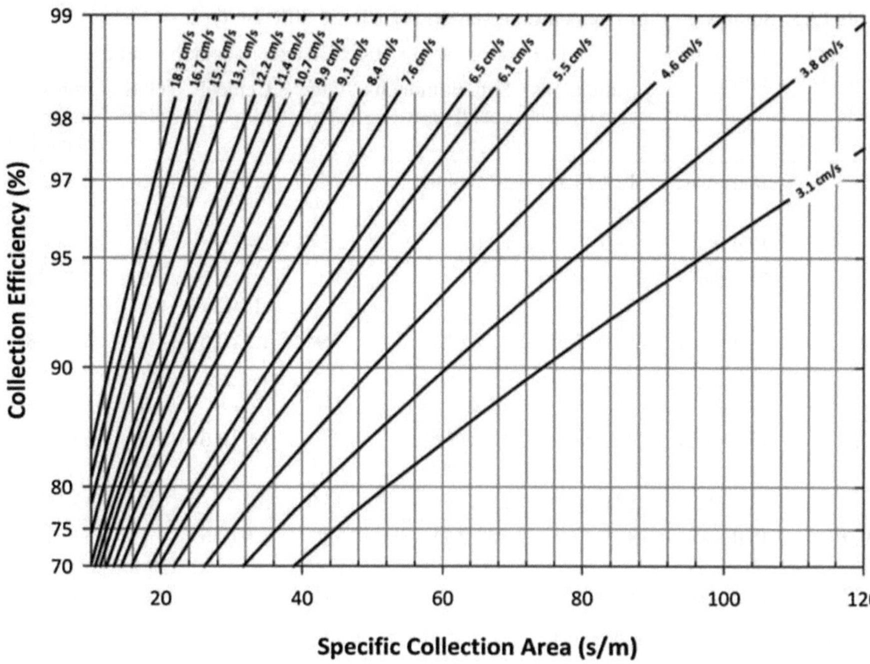

Fig. 19.7 Relationship among the collection efficiency, drift velocity, and the specific collection area

$$\eta = 1 - e^{-(v_e A/Q)^m}, \tag{19.24}$$

where $A = WL$ is the area of the collecting electrode; m ranges from 0.4 to 0.7 and has a typical value of 0.5.

The dimensions of the collection electrode are mostly determined by the aspect ratio (AR) which is defined as the ratio of effective length to height. It is recommended that the aspect ratio for ESPs should range from 0.5 to 2.0. In particular, if a collection efficiency higher than 99.5% is desired, the precipitator design should have an aspect ratio of greater than 1.0.

19.4.4 Step 4: To Sectionalize the ESP

As a rule-of-thumb, the optimum gas velocity for the best ESP performance should be maintained in the range of 3.5–5.5 ft/s. Therefore, a specific number of ducts to regulate gas flow and prevent maldistribution can be determined as follows:

Table 19.3 Summary of drift velocities of various particles in different types of ESPs [18]

Collection efficiency (%)	Precipitator types	Drift velocities for various collected materials in the unit of cm/s					
		Bituminous coal fly ash	Sub-bituminous coal fly ash	Other coal	Cement kiln	Glass plant	Iron/steel sinter plant dust
95	Plate-wire	12.6 (3.1)	17.0 (4.9)	9.7 (2.9)	1.5 (0.6)	1.6 (0.5)	6.8 (2.2)
	Flat plate	13.2	28.6	15.5	2.4	1.8	13.4
	Tubular				2.1–5.1 (1.0–2.6)	1.3 (0.7)	
99	Plate-wire	10.1 (2.5)	11.8 (3.1)	7.9 (2.2)	1.5 (0.6)	1.6 (0.5)	6.2 (1.8)
	Flat plate	15.1	18.2	11.2	2.3	1.9	12.1
99.5	Plate-wire	9.3 (2.4)	10.3 (2.6)	7.9 (2.1)	1.8 (0.6)	1.5 (0.5)	6.6 (1.8)
	Flat plate	18.6	21.2	15.1	3.2	2.6	13.1
99.9	Plate-wire	8.2 (2.1)	8.8 (2.2)	7.2 (1.9)	1.8 (0.6)	1.5 (0.5)	6.3 (1.7)
	Flat plate	16.0	17.7	13.5	3.1	2.6	12.4

* The numbers in braces are drift velocities subject to back corona

$$N_d = \frac{Q}{uDH}. \tag{19.25}$$

In practice, a single ESP unit cannot meet air emission requirements due to the gas leakage (S_N) and re-entrainment (RR), and the ESP and the total specific collection area must be sectionalized and verified, respectively. In particular, the leakage is a function of precipitator types which values 0.07 and 0.10 for plate-wire and flat plate precipitators, respectively. Meanwhile, re-entrainment is correlated with characterizations of the particles and precipitators and the typical value is 0.14. As a result, the loss factor (LF) can be calculated as Eq. (19.26):

$$LF = S_N + RR(1 - S_N). \tag{19.26}$$

The number of sections is estimated by comparing the loss factor with the desired collection efficiency which is

$$LF^{N_s} < (1 - \eta) \tag{19.27}$$

p is the penetration ratio of particles with various diameters. The SCA in each section is varied with the corresponding mass mean diameter (MMD) as shown in Eq. (19.28):

$$SCA_n \times MMD_n = SCA_{n-1} \times MMD_{n-1} = \text{Constant}. \quad (19.28)$$

Equation (19.28) indicates that if the SCA and MMD for the first section are known, then the following can be readily estimated by iteration which implies that the MMD_1 is equal to the MMD measured at the inlet. Meanwhile, the subsequent MMDs can be determined by Eq. (19.29):

$$MMD_n = \{MMD_{n-1} \times S_N + [(1 - p_c)MMD_p + p_c \times MMD_{n-1}] \times p_c/D\} + MMD_{rp}. \quad (19.29)$$

The definitions of the parameters are listed in Table 19.4.

On the other hand, the SCA in the first section is given as follows:

$$SCA_1 = -\left(\frac{\mu}{\epsilon}\right) \times (1 - S_N) \times \frac{\ln(p_c)}{{E_{\text{avg}}}^2 \times MMD_1 \times 10^{-6}}, \quad (19.30)$$

where μ (kg/m·s) is the gas viscosity expressed as $\mu = 1.72 \times 10^{-5}\left(\frac{T}{273}\right)^{0.71}$; ϵ (F/m) is the free space permittivity, i.e., 8.845×10^{-12} F/m denotes a typical value. Ultimately, the total specific collection area is a sum of the individual SCA in each section which can be expressed as follows:

$$SCA_{\text{total}} = \sum_{i=1}^{n} SCA_i. \quad (19.31)$$

Table 19.4 Definitions of the parameters used in Eq. (19.29)

Parameter	Definition	Remarks
p_c	Collection penetration	$p_c = \frac{p_s - LF}{1 - LF}$
p_s	Average section penetration	$p_s = p^{1/n} = (1 - \eta)^{1/n}$
MMD_p	Most peneuation *MMD*	Assumed to be 2 μm in general
D	Particle size change factors	$D = S_N + p_c(1 - S_N) + RR(1 - S_N)(1 - p_c)$
MMD_{rp}	*MMD* changes among sections	$MMD_{rp} = RR(1 - S_N)(1 - p_c)MMD_r/D$
MMD_r	Rapping puff *MMD*	Take the value of 5 μm for the particle whose inlet *MMD* is greater than 5 μm, and 3 μm for that less than 5 μm

19.4.5 Step 5: To Determine the Cleaning Techniques

As the collection process proceeds, particles keep accumulating on collection plates and must be removed periodically. The cleaning process is essential to the overall performance because it prevents the re-entrainment of collected particles and sparking due to excessive particulate buildup on the collection plates or discharge electrodes. For dry precipitators, the dust layer is removed by a mechanical vibration component. Meanwhile, electrically or pneumatically motivated rappers are used for wire-weighted precipitators. During cleaning operations, a pulse through the rapper supplies the energy to move the steel plunger. Thereafter, the shock wave transmits to the electrode thereby dislodging the dust layer. For rigid-frame precipitators, mechanical-hammer rappers are commonly used. When the cleaning cycle is engaged, each frame is rapped by one hammer hung on a shaft whose intensity and frequency are governed by hammer weight and shaft rotation, respectively.

The dust layer is removed by automatic spraying in wet precipitators [19], thus the re-entrainment is effectively controlled. However, to prevent secondary pollution, a simple system with a solid-water separator was introduced to treat wastewater containing innocuous dusts. However, a complex system including skimming, sludge removal, clarification equipment, pH adjustment, and dissolved-solids separation should be employed for treating hazardous air pollutants.

19.4.6 A Rapid Method for Calculating Collection Efficiencies of Electrostatic

A model for calculating collection efficiencies was developed based on the electrostatic precipitator (EP) performance curves for twelve industries such as electric power, cement, pulp and paper, steel, chemical, and petroleum industries. Twelve particular industries were chosen for two reasons: (1) they represent the electrostatic precipitator operated in the United States, and (2) the required data for the efficiency calculations via the Theodore model are available in the literature [14].

The EP performance curves describe the predicted collection efficiency as a function of precipitator length, plate-to-plate spacing and average field strength which could be presented in graphical, tabular, and equation forms. For instance, to convert the data generated by the model into a more usable and simplified form, a regression analysis method is used to produce two-constant equations representing the results listed in Tables 19.5, 19.6, and 19.7. These 'best-fitting' equations take the following forms:

Efficiency versus precipitator length:

$$\eta = 1 - a_1 e^{-a_2 Z}. \tag{19.32}$$

Efficiency versus plate spacing (or tube diameter):

Table 19.5 Collection efficiency as a function of precipitator length

Base case: plate-to plate distance, P = 9 in. precipitator length, Z = 14 ft
Field strength, E = 4 kV/cm

		Precipitator length, ft					$\eta = 1 - a_1 e^{-a_2 Z}$	
Industry	Type	10	14	17	20	24	a_1	a_2
Fly ash (electric power)	Plate	94.60	97.23	98.40	98.90	99.43	0.2590	0.1596
Municipal incinerator	Plate	95.93	97.00	97.83	98.67	99.10	0.1342	0.1121
Gypsum dust	Plate	95.43	97.33	98.20	98.77	99.30	0.1735	0.1332
Cement dust	Plate	83.83	90.27	92.53	94.80	96.03	0.4191	0.1010
Pulp salt cake	Plate	54.53	66.53	73.97	79.97	87.10	1.1517	0.0893
Sintering plant	Plate	95.63	98.70	99.63	100.00	100.00	7.9914	0.4928
Open hearth furnace	Plate	83.10	87.60	89.83	91.87	94.03	0.3525	0.0737
Blast furnace	Tubular	97.40	98.80	99.70	99.70	99.87	0.2050	0.2166
Coke over deterring	Tubular	66.60	78.80	85.23	89.87	93.90	1.0542	0.1150
Sulfuric acid mist	Tubular	79.10	89.93	94.20	96.63	98.33	1.2353	0.1808
Phosphorous mist	Tubular	99.27	99.83	99.97	100.00	100.00	0.1468	0.3321
Petroleum cracking	Plate	85.63	90.03	92.77	93.57	95.80	0.3320	0.0856

Table 19.6 Collection efficiency as a function plate spacing

Base case: plate-to plate distance, P = 9 in. precipitator length, Z = 14 ft
Field strength, E = 4 kV/cm

		Plate-to plate distance, in. (or tube diameter, in.)					$\eta = 1 - b_1 e^{-\frac{b_2}{P}}$	
Industry	Type	6	8	9	11	13	b_1	b_2
Fly ash (electric power)	Plate	98.70	98.07	97.23	96.03	94.87	0.1594	15.58
Municipal incinerator	Plate	98.70	97.53	97.00	95.77	95.23	0.1556	14.82
Gypsum dust	Plate	99.23	98.17	97.33	96.33	95.30	0.2348	20.34
Cement dust	Plate	95.20	92.17	90.27	87.13	84.00	0.4362	13.40
Pulp salt cake	Plate	82.67	71.37	66.53	59.37	53.40	1.1130	11.04
Sintering plant	Plate	100.00	99.33	98.70	97.57	95.27	14.2475	67.66
Open hearth furnace	Plate	92.27	89.27	87.60	84.63	81.80	0.3649	9.49
Blast furnace	Tubular	99.87	99.47	98.80	98.20	96.83	0.4996	35.59
Coke over deterring	Tubular	91.10	81.73	78.80	71.67	65.63	1.1122	14.95
Sulfuric acid mist	Tubular	96.87	92.70	89.93	84.10	78.00	1.1441	21.74
Phosphorous mist	Tubular	100.00	100.00	99.83	99.60	99.23	0.3788	53.16
Petroleum cracking	Plate	94.30	91.97	90.03	87.17	85.17	0.3366	10.88

Table 19.7 Collection efficiency as a function of field strength

Base case: plate-to plate distance, P = 9 in. precipitator length, Z = 14 ft Field strength, E = 4 kV/cm

		Field strength, kv/cm					$\eta = 1 - c_1 e^{-c_2 E^2}$	
Industry	Type	3	4	4.5	5	6	c_1	c_2
Fly ash (electric power)	Plate	92.40	97.23	98.47	99.33	99.97	0.7052	0.2044
Municipal incinerator	Plate	93.60	97.00	98.20	98.47	99.53	0.1407	0.0942
Gypsum dust	Plate	93.73	97.33	98.57	99.07	99.87	0.2543	0.1423
Cement dust	Plate	76.80	90.27	92.90	95.23	98.17	0.4743	0.0918
Pulp salt cake	Plate	43.53	66.53	75.27	83.40	92.53	1.1136	0.0752
Sintering plant	Plate	92.73	98.70	99.73	100.00	100.00	0.5572	0.2671
Open hearth furnace	Plate	77.63	87.60	90.67	92.70	96.00	0.3564	0.0625
Blast furnace	Tubular	95.93	98.80	99.40	99.80	100.00	0.4133	0.2233
Coke over deterring	Tubular	56.87	78.80	86.30	92.03	97.17	1.0637	0.1015
Sulfuric acid mist	Tubular	70.50	89.93	94.53	97.33	99.57	1.2444	0.1561
Phosphorous mist	Tubular	98.70	99.83	100.00	100.00	100.00	0.0221	0.1816
Petroleum cracking	Plate	82.00	90.03	92.87	93.87	96.90	0.2891	0.0633

$$\eta = 1 - b_1 e^{-\frac{b_2}{P}}. \tag{19.33}$$

Efficiency versus field strength:

$$\eta = 1 - c_1 e^{-c_2 E^2}. \tag{19.34}$$

EP performance curves presented in Tables 19.5, 19.6 and 19.7 demonstrate the effects on the calculated efficiency of perturbing, in turn, each of the three operating conditions, namely, precipitator length (Z), field strength (E), and plate-to-plate spacing or, in the case of the tubular precipitator, diameter (P). The above-developed models can be regarded as more simple but reliable approaches to the prediction of collection efficiencies in an electrostatic precipitator design.

An electrostatic precipitator was designed for the flue gas flowing at 3.5×10^6 Nm^3/h from a coal-fired power plant equipped with two 600 MW generators. Composition analysis revealed that the particle concentration at the inlet was 45.49 g/Nm^3 while the regulated emit concentration should be lower than 50 mg/Nm^3, thus a collection efficiency at 99.9% was desired.

19.5 Performance Evaluation and Enhancement

19.5.1 Performance Evaluation

The wet ESP has been regarded as the maximum available control technology (MACT) for PM_{10} and $PM_{2.5}$ controls to improve ambient air quality [18]. In addition, the ESP has the lower pressure drop about 1/5 to 1/10 that of the baghouse [20]. However, sophisticated cleaning techniques are required to install for effectively controlling re-entrainment.

The annually required electricity (kWh/yr) for the TR-set and pump, fan systems, and particle charging during the ESP operation can be estimated as follows:

$$E_{\text{TR-set}} = 1.94 \times 10^{-3} A\theta, \tag{19.35}$$

where A (ft^2) is the plate area in the ESP, and θ (hr/yr) represents the operating hours in one year. Additional electricity consumed by the ESP includes the fan or blower systems which are determined by the gas flow rate, pressure drop, and annual operating time as shown in Eq. (19.33):

$$E_{\text{Fan}} = 0.000181 \times Q\Delta P\theta. \tag{19.36}$$

Furthermore, the water to be pumped into wet precipitators is driven by a pump system whose energy consumption is given as follows:

$$E_{\text{Pump}} = \frac{0.746 Q_L Z S_g \theta}{3960\delta}, \tag{19.37}$$

where Q_L (gal/min) is the water flow rate, Z (ft) is the water head, S_g (−) is the specific gravity of water being pumped compared to water at 70oF and 29.92 in. Hg, and δ (−) is the motor efficiency of the pump. In practice, a typical power consumption of ESP is estimated ranging from 25 to 100 W/kacfm.

Compared to other kinds of particle collection equipment, the ESP requires relatively higher capital costs of construction materials and configurations and operation and maintenance costs of electricity, chemical conditioning agents, and water use and treatment in wet precipitators.

In general, the collection efficiency of particles by ESP is the result of the competition of the pressure and the temperature. In detail, higher pressure would lead to higher precipitation efficiency as the result of the reduced ion mobility which increases the potential required for corona and sparking. On the other hand, the increased temperature reduces collection efficiency because of the reduced ion mobility, the lowered critical potentials, as well as the increased gas viscosity, which together reduce migration velocities. As a result, the effects of temperature and pressure on the collection efficiency are connected through the relative gas density and were empirically developed by Robinson as [21]:

$$\frac{E_c}{\rho'} = A\frac{B}{\sqrt{d_w\rho'/2}}, \quad (19.38)$$

where E_c (kW/cm) is the corona-starting field, ρ' is the relative gas density, d_w (cm) is the diameter of the charging wire, and A and B are constants characterizing the gas, respectively.

19.5.2 Performance Improvement

Just like other particle collection equipment, ESP is effective in capturing coarse particles. In particular, the defect particle size range for ESP is 0.2–0.4 μm, where neither field nor diffusion charging mechanisms are effective in charging the particles. In addition, particles smaller than 1 μm can reduce the potential of corona generation which would impair the capability of particle collection. It is thus suggested that a complementary collection unit be implemented. Many designs incorporate baffles to provide quiescent zones where the gas stream is stagnant and particle collection is enhanced.

The resistivity is an important factor affected the performance of the dry ESP operation and should be in the range of 10^8–10^{10} Ω/m. Severe back corona occurs if the resistivity reaches to 10^{12} Ω/m. In cases where the resistivity is outside of the above range, chemical agents are used to altering the characteristics of the particles. For instance, the most popular agent is sulfur trioxide which is widely used in the coal-fired process. The condensed sulfur is adsorbed on fly ash surfaces, which increases the cohesiveness of the fly ash while reducing resistivity. Other agents commonly in use include ammonia, ammonium sulfate, triethylamine, sodium compounds, potassium sulfate, and sodium chloride. A typical dose rate for any of the gaseous agents should range from 10 to 30 ppmv.

Temperature and moisture are the other two important factors affecting the resistivity that resulted in influencing the collection efficiency of particles [22]. In general, the resistivity rises with the temperature until a peak is reached, thereafter the resistivity falls if the temperature keeps elevating. For this reason, ESPs should be operated under the optimum temperature thereby providing the low resistivity of particulate. In addition, the performance would be enhanced by introducing appropriate conditioning agents to generate higher moisture during the ESP operation. The higher the moisture, the lower the resistivity.

It is thus concluded that chemical agglomeration technology can improve the removal of fine particles throughout the coagulation and flocculation processes thereby changing the physio-chemical properties of fine particles. Typical agglomerates commonly used in the ESP for removal of fine particles are shown in Table 19.8 [23].

The concentration and pH value of the agglomeration solution, flue gas temperature, and volumetric rate, and diameter of the spray droplets can influence the

Table 19.8 Characteristics of Agglomerants

Agglomerant	Abbreviation	Molecular weight	Characteristic
Pectin gum	PG	50,000–150,000	Natural organic polysaccharide
Sodium carboxymethyl cellulose	CMC	2,420,000	Natural organic polysaccharide
Xanthan gum	XTG	1,000,000	Natural organic polysaccharide
Polyaluminum chloride	PAC	1500–3000	Inorganic polymer flocculant
Polyferric sulfate	PFS	2000–5000	Inorganic polymer flocculant
Sodium alginate	SA	100,000–400,000	Natural organic polysaccharide

agglomeration effects of the fine particles. In general, the average diameter of the particles could grow more than four times, meanwhile the PM2.5 concentration at the ESP outlet could decrease by 40% because of the effects of wetting, the liquid bridge force, and adsorption bridging. The fine particle concentration can be further reduced from 5 to 10% a by adding the phosphoric acid solution to lower pH value.

Proper voltage control is critical to charging and collecting particles, thus improving the collection efficiency. It is suggested that the average sparkover rate for the optimum precipitation operation should be ranged from 50 to 100 sparks per minute. As a result, the collection efficiency can be improved by increasing the voltage-operated under the appropriate spark rate.

Furthermore, the performance and reliability of the ESP are also influenced by the cleaning intensity and frequency. In general, a well-designed cleaning system can operate with dust buildup ranging from 6 to 25 mm [24] under which the re-entrainment is effectively controlled. In practice, one rapper is responsible for cleaning 1200 to 1600 ft^2 of collecting area and 1000 to 7000 ft of discharging wire, respectively. Regarding cleaning parameters, intensity, and rapping intervals range from 5 to 50 ft/lb and 30 to 600 s, respectively.

19.5.3 Innovative Technologies

WESP (Wet ESP)

In the cases where the material to be collected is wet, sticky, flammable, explosive, or has a high resistivity, wet type ESPs (WESP) are used instead of the dry ones. WESP require water, which replaces the rapping mechanism usually used by dry ESPs, to be injected or sprayed near the top of the collector pipes either continuously or at timed intervals. The water flows with the collected particles into a sump from which the fluid is pumped or drained. As a result, the use of WESP would lead to higher collection efficiencies.

In addition, WESP exhibit great flexibility since they are capable of handling relatively large gas flow rates under either high pressure or vacuum conditions. The WESP with the atomization of charged water drop significantly reduced the water consumption because the high density, highly charged droplets, and the large relative velocity between droplets and particles resulted from the electrodynamic breakup of water drop, thereby capturing more particles with a smaller amount of water supply in the WESP [25].

For instance, the removal efficiency of the WESP was elevated to 91.69%–95.88% from 91.52 to 93.89% of the dry ESP, as the electric field intensity increased from 2 kV/cm to 5 kV/cm. The synergistic effect of the longer residence time and higher applied voltage significantly improved the grade removal efficiency in the WESP, which increased from 91.86–95.36% to 97.86–98.89% as they were enhanced from 2.14 s and 2 kV/cm to 4.04 s and 5 kV/cm, respectively. It was evident that the atomization of charged water drop in the WESP significantly reduced water consumption, meanwhile, maintaining the removal efficiency as the level of traditional water treated processes.

Low-Low Temperature Electrostatic Precipitator (LLT-ESP)

A traditional low temperature ESP was operated around 150 °C to avoid corrosion; however, the electric resistivities of coal dust were rather high in this temperature range. To enhance the performance of the traditional ESP, a low-low temperature electrostatic precipitator (LLT-ESP) was introduced by using a heat exchanger prior to a traditional electrostatic precipitator in which the temperature of the flue gas could be reduced to below the dew point, resulted in reducing electric resistivity of the dust and increasing removal efficiency of the dust. It was evident that the LLT-ESP showed excellent removal efficiency for filterable particulate matter (FPM), and condensable particulate matter (CPM) was more than 99.9 and 60.9%, respectively [26].

Exercises

1. Define the following terms and explain why they should be considered for design use: particle charging, migration velocity, specific collecting area, and Deutch Equation.
2. Please describe the general procedures for designing an electrostatic precipitator.
3. List the major factors affecting the design of an electrostatic precipitator.
4. Describe and compare the performance of wet and dry electrostatic precipitator.
5. A small coal-fired power plant sends 2400 acfm through its electrostatic precipitation. The particle migration velocity is known to be 0.35 ft/s. What is the collection area if the overall ESP efficiency is 99.78%?

References

1. Katz J (1979) The art of electrostatic precipitation. Precipitator Technology. Inc Munhall, Pennsylvania
2. Parker KR (2012) Applied electrostatic precipitation. Springer Science & Business Media
3. Deutsch W (1922) Bewegung und Ladung der Elektrizitätsträger im Zylinderkondensator 373:335–344
4. Reitz JR, Milford FJ, Christy RW (2008) Foundations of electromagnetic theory. Addison-Wesley Publishing Company
5. Fjeld RA, McFarland AR (1989) Evaluation of select approximations for calculating particle charging rates in the continuum regime. Aerosol Sci Technol 10:535–549
6. Pauthenier M, Cochet R, Dupuy J (1956) Probleme general de la charge acquise par une particule spherique dans un champ electrique bi-ionise. Comptes Rendus Hebdomadaires Des Seances De l Academie Des Sciences 243:1606–1608
7. Fuchs N (1963) On the stationary charge distribution on aerosol particles in a bipolar ionic atmosphere. Geofis Pura Appl 56:185–193
8. Fuchs N (1947) The charges on the particles of aerocolloids. Izv Akad Nauk SSSR. Ser Geogr Geofiz 11:341–348
9. Bricard PJ (1949) L'Equlibre ionique De La basse atmosphere. J Geophys Res 54:39–52
10. Gunn R (1954) Diffusion charging of atmospheric droplets by ions, and the resulting combination coefficients. J Meteorol 11:339–347
11. White HJ (1963) Industrial electrostatic precipitation. Industrial electrostatic precipitation. Addison-Wesley
12. Theodore L, Pardini J (1971) Application of modeling and simulation techniques to the design of electrostatic precipitators. Presented as Paper, pp 71–124
13. Theodore L, Eastment T (1972) Simulation of an electrostatic precipitator—effect of velocity, particle size, particle mass flow rate, and electrostatic force distribution on collection efficiency. In: 65th Annual Meeting of the Air Pollution Control Association
14. Reynolds JP, Theodore L, Marino J (1975) Calculating collection efficiencies for electrostatic precipitators. J Air Pollut Control Assoc 25:610–616
15. Danielson JA (1973) Air pollution engineering manual. Environmental Protection Agency
16. Turner JH, Lawless PA, Yamamoto T, Coy DW, Greiner GP, McKenna JD et al (1988) Sizing and costing of electrostatic precipitators: Part I. Sizing considerations. Japca 38:458–471
17. Matts S, Öhnfeldt P (1963) Efficient gas cleaning with SF electrostatic precipitators. Flakten Rev 6:93–110
18. Mussatti DC, Srivastava R, Hemmer PM, Strait R. EPA air pollution control cost manual. Air Quality Strategies and Standards Division of the Office of Air Quality Planning and Standards, US Environmental Protection Agency, Research Triangle Park, NC. 2002;27711
19. Perry RH, Green DW, Maloney JO (1984) Perry's chemical engineers 'handbook. Perry's chemical engineer's handbook" McGraw-Hill Book
20. Billings CE, Wilder J (1970) Handbook of fabric filter technology. Volume I. Fabric filter systems study
21. Theodore L (2008) Air pollution control equipment calculations. Wiley
22. Liu DH, Liptak BG (1997) Environmental engineers' handbook. CRC Press
23. Liu Y, Bin H, Lei Z, Yezheng J, Yang L (2016) Improving the removal of fine particles with an electrostatic precipitator by chemical agglomeration. Energy Fuels 30:8441−8447
24. Robinson MJ (1969) Critical pressures of the positive corona discharge between concentric cylinders in air. JoAP 40:5107–12
25. Teng C, Fan X, Li J (2020) Effect of charged water drop atomization on particle removal performance in plate type wet electrostatic precipitator. J Electrostat 104:103426
26. Qi Z, Li J, Wu D, Xie W, Li X, Liu C (2017) Particulate matter emission characteristics and removal efficiencies of a low-low temperature electrostatic precipitator. Energy Fuels 31:1741−1746

References

Chapter 20
Prospective and Perspective

Portfolio solutions can be provided along with the multiple air pollutant control. This Chapter provides the perspective and prospective of key strategies for effective air pollution control, i.e., innovative technology and integrated management for multiple air pollutants control as green solutions. Cleaner production technology should be developed for pollution and accident prevention and energy and resource sustainability. Green energy practice should be implemented by using the green energy utilization and best available green energy. ICA (instrumentation, control, automation), IOT (internet of things), and MIS (management information system) should be introduced to optimize the air pollution control system. Sustainable attainment strategy should be proposed throughout the ABaCAS (air benefit and cost and attainment assessment system) study.

The effective control strategies related to ambient air quality should be a set of specific available control techniques and measures identified and implemented to achieve reductions in air pollution to attain the air quality standards. Along the way, life cycle assessment (LCA) and 4E (environmental, engineering, economic, and energy) evaluation should be applied as a structured basis for evaluating the performance of environmental impacts and benefits of the integrated air pollution control. The win–win philosophy indicates that a prosper economy and healthy ecosystem/environment should be co-existed. More detailed interpretation regarding the development of novel technologies for air pollution control, deployment of multiple pollutant control technology, integration of green energy and emergy materials for air pollution control, development of IOT technologies for ambient air quality management, and mitigation and adaptation to climate change for carbon neutrality are desaried as follows.

P.-C. Chiang and X. Gao, *Air Pollution Control and Design*,
https://doi.org/10.1007/978-981-13-7488-3_20

20.1 Development of Novel Technologies for Air Pollution Control

Great concerns about serious damages caused by air pollutant emissions and rapid GHG accumulation in the atmosphere are raised. In many cases, the costs of separated air pollution control are too high for large-scale industrial deployment. As a result, effective technologies and tools to controlling air pollution emissions are required to achieve the goal of clean air act.

Another challenge on the ambient air pollution control may be the issue of cost-allocation in the case of GHG emission reductions. For instance, the management of a variety of CO_2 emission sources within a single industrial plant and the selection of appropriate process and technology for CO_2 sequestration should be the major concern to lowering the industrial capture costs. Moreover, simultaneous removal of criterial pollutants and GHG in one unit process also needs to be investigated.

SO_x control

It is expected to further increase the SO_2 removal efficiency to achieve the ultra-low emission of SO_2 by introducing the FGD technologies from coal-fired power plants to other industries. In consideration of SO_3 emission, it is essential to investigate the formation mechanism of SO_3, as well as its control principles. Monitoring devices for accurate measurement of SO_3 emission should be developed. Also a low-cost novel technology for efficient SO_3 removal is urgently needed to solve the problem of SO_3 pollution.

NO_x control

Due to the low-temperature of the flue gas from other stationary industrial sources, the novel catalyst for the low-temperature SCR should be developed for efficient NO_x removal. This effort can boost more installation of SCR unit for treating low-temperature gases. Besides, it is also beneficiary for NO_x control from mobile sources, especially at the cold start. Moreover, catalyst modification to achieve the simultaneous oxidation of Hg^0 and dioxin is also essential to control the toxics.

VOCs control

Due to the wide ranges of types of and complicated physicochemical characterizations VOCs, more targeted and efficient monitoring methods and novel technologies of VOCs emissions control should be promoted. For the occasion of multi-point source, low VOCs emission rate and intermittent emission, a new VOCs governance model of "decentralized collection and centralized treatment" is proposed in order to reduce the cost of pollution control for such enterprises or single enterprises in industrial parks. The control technology should be more energy conservation and high efficiency, such as developing VOCs catalyst at low-temperature, encourage the recovery and utilization of VOCs.

PM control

PM control technology should be developed on the basis of comprehensive multidisciplinary collaboration thereby identifying the physicochemical characterizations of PM, understanding the mechanisms of PM removed by baghouse filter and electrostatic precipitator and enhancing the energy efficiency for PM control device. In order to cope with practical industrial applications, it is necessary to study the carbon particle removal performance for each PM control device. For instance, the mechanism of the effect of resistivity on the electrostatic removal based on the force analysis of the dust layer and the interaction between the particles and the plates should be explored.

CO_2 control

Apart from the technological selection for adsorption (or mineralization), absorption, membrane, and chemical looping processes, engineering prospective regarding thermodynamic properties of CO_2, essential requirements for long-term storage site integrity, and technical specifications are created by regulatory board. Among them, the key of success for chemical looping technology are their materials which are adequate to react at temperature below melting point ensures the energy conversion efficiency. The only real challenge is a political one after the technical one. Finding and deploying politically acceptable policy instruments, which motivate industry and enable commercially viable business models to execute, are very essential rather on paper plan. The Paris Agreement and its consequential outcomes will help drive that evolution.

20.2 Deployment of Multiple Pollutant Control Technology

The hazardous air pollutants, including nitric oxides (NO_x), sulfur dioxide (SO_2), particulate matters (PM), and heavy metals from coal-fired plants are believed to be one of the main reasons for the complex air pollution worldwide. To eliminate the pollutant emission level, various air pollutant control technologies have been continuously developed to meet strict emission standards in recent years. In power plants, the implementation of stricter legislation and regulations to limit the emission of pollutants also promote the development of multiple pollutant control technology. However, there are still challenges ahead: (i) higher-efficiency and low-cost technologies to further reduce emissions from coal-fired power plants; (ii) higher-reliability technologies to handle complex flue gases generated from high-ash, high-sulfur, and high-alkali coals; and (iii) higher-applicability technologies to cover more energy-consuming industries, including power plants, sintering machines, cement kiln, coking plants, industrial boilers, iron smelting, etc.

In addition, the change of coal quality (such as sulfur content, ash content, and chemical components), the load fluctuations of power plants, and the rigorous air quality requirement all contribute to higher requirements on flue gas pollutant control technology. Consequently, single-pollutant removal technology is no longer the solution. As a promising approach, the multi-pollutants synergistic removal technology can fulfill the requirement because multi-pollutants (NOx, PM, SO_2, SO_3, $PM_{2.5}$, heavy metals, VOCs, etc.) can be simultaneously removed through each individual device.

Typically, a multi-pollutant simultaneous removal technology is defined as one that removes two or more pollutants in a single reactor or a single system, which aims to carry out the goals of high efficiency, high reliability, low cost, etc. This issue is currently the focus of attention among all researchers and engineers worldwide. Currently, the simultaneous removal of multi-pollutants is mainly realized by taking advantage of the synergies of the existing air pollution control devices. Selective catalytic reduction (SCR), flue gas desulfurization (FGD), and electrostatic precipitator (ESP) are originally developed to control the emission of NO_x, SO_2, and PM.

Recent researches indicate that one or more other pollutants could be removed at the same time as the efficient removal of a certain kind of pollutant, so as the significant improvement of the removal performance in the system. For instance, as a widely commercialized technology for the abatement of NO_x emissions from coal-fired power plants, SCR could also be beneficial to the control of Hg and VOCs. Meanwhile, wet flue gas desulfurization is developed to remove SO_2, Hg^{2+}, and particulate matter from flue gas. Dry and wet ESPs can remove not only particulate matters but also other pollutants, such as SO_3 and Hg. A typical ultra-low emission system for coal-fired power plants is equipped with sub-systems of SCR, dry ESP, WFGD, and wet ESP. During operation, each system casts significant synergetic effects on the removal of pollutants, including NO_x, SO_2, PM, SO_3, Hg, etc.

In recent years, numerous research institutes and enterprises have carried out in-depth researches to improve the simultaneous removal performance of the existing air pollution control devices. For instance, efforts have been invested to develop novel catalysts for synergistic catalytic conversion of NOx and mercury; novel pulsed power supply have been designed for conversion/capture of multi-pollutants (SOx, NOx, trace elements) in electrostatic field; "environmentally friendly" oxidation method have been applied to oxidize NO and synergistically absorbs and remove NO_x and SO_2 in the WFGD system. With these synergistic pollutant control technologies for coal-fired power plants, the pollutant emissions can be close to that of natural gas combustion.

In the future, the development trends of multi-pollutants synergistic removal technology will follow the qualifications of higher efficiency, stronger adaptability, higher reliability, and lower cost. Four possible trends can be forecasted. Specific research work should be focused on understanding the synergistic effect of the existing pollutant control equipment and then developing the innovative technology for synergistic removal of unconventional pollutants such as SO_3, $PM_{2.5}$, heavy metals, and VOCs via an integrated system.

20.3 Integration of Green Energy and Emergy Materials for Air Pollution Control

Energy and the environment are closely interlinked and their relationships are popular subjects to study. As an irreplaceable part of human life, most of the world's energy consumption comes from fossil fuel combustion despite the recent rapid expansion of renewable sources of electricity. The exploitation and consumption of energy have altered the original environment/ecosystem. Globally, the burning of fossil fuels at a massive scale such as coal, oil, and natural gas makes power generation one of the most prominent sources of air pollutants.

With further economic developments, the demand is unleashed for more electricity for appliances and oil for transport, potentially resulting in higher emissions of SO_2, NO_x, PM, and other pollutants. Meanwhile, a series of severe consequences such as acid rains was caused by sulfur oxides. More prominently, global warming brought by the highly concentrated CO_2 in the atmosphere is seriously threatening the ecosystem due to high dependency on fossil fuel. In general, an increase in urban population often correlates to economic development, which can boost and concentrate energy-related air pollution as a result. Among all pollutants, the emission of SO_2, NO_x, and PM is almost entirely attributed to energy production and utilization.

Coping with the global challenge in climate change and air pollution, it is necessary not only to develop novel strategies and techniques to increase the pollutant reduction efficiency but also to modify the energy structure, i.e., encouraging the utilization of renewable and clean energy. The same as other large-scale power plants, renewable energy power plants are also constructed in remote areas thus requiring the construction of a new transmission infrastructure. Considering the instability and intermittence of renewable energy, for example, the solar energy sensitive to weather condition, additional support equipment is needed to reduce the impacts on electric grids. Challenge of constructing a renewable energy generation plant is the regulatory process. An abundant amount of capital is needed by related companies to support the projects and get them through the multi-year interagency process ranging from local, provinces, and the nation level. It is expected that corporations sourcing of renewable energy would continuously increase and they would play a significant role on spreading renewable energy to new regions in the future.

Batteries and their related technologies have become more and more important for daily life. To support the application of the advanced cathode materials, developing novel electrolyte systems with high anodic stability is very important. Considering bulk production and cost reduction, the fuel cells for vehicles or for stationary power are needed to achieve mass production, for which the advanced technology for automated mass production is of great significance. In mid-2012, the first efficient solid-state perovskite cells were reported. The confirmed efficiency over the perovskite cells is rapidly increasing with the effort worldwide. In addition, a variety of different fabrication techniques has been developed to further increase the efficiency, which also indicates that performance is still far from being fully optimized.

20.4 Development of IOT Technologies for Ambient Air Quality Management

The Internet of Things (IoT) generally refers to the goals achieved by following established network communication protocols, connecting items through various networks and using radio frequency identification (RFID) technology, various sensors, location recognition systems, and information processing mechanisms. From the technical level analysis, IoT can be divided into perception, network, and application layers, as shown in Fig. 20.1 and stated as follows:

(1) Perception layer includes diverse sensor nodes and gateways, such as various sensors, electronic tags, read/write devices, cameras, and GPS, whose main function is to identify, collect, and transmit information of objects.
(2) Network layer is composed of various internet, wireless communication networks, mobile air networks (3G/4G/5G), private networks, network management systems, cloud computing platforms, etc. Its main role is to carry out real-time transmission and fast various data information obtained by the perception layer.
(3) Application layer provides a convenient information interaction system for IoT and users according to the actual needs of users.

Ambient air quality management refers to the activities a regulatory authority undertakes to protect wellbeing health and the ecosystems/environmental quality from the adverse impacts of air pollution.

The Air Benefit and Cost and Attainment Assessment System (ABaCAS; www.abacas-dss.com) is a new policy-oriented integrated scientific assessment system and

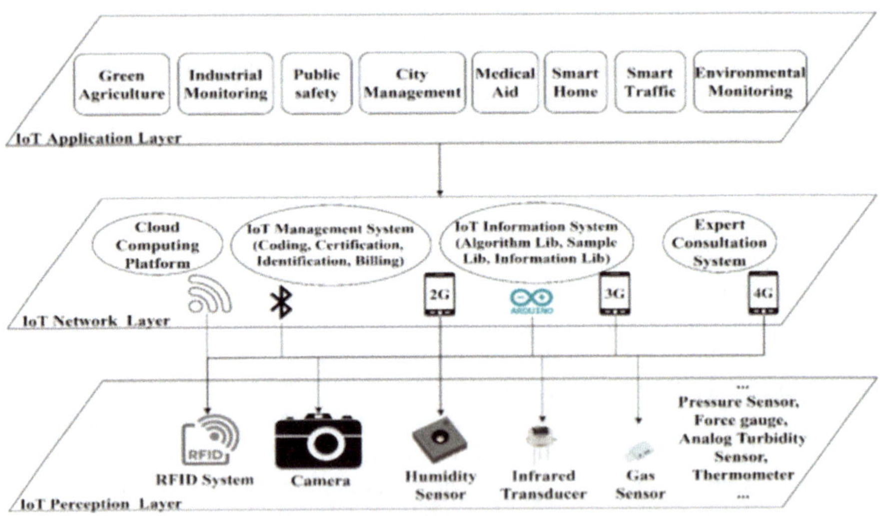

Fig. 20.1 Three layers of IoT

was first developed by the U.S. Environmental Protection Agency in 2012, which aims to address the optimum control strategy for ambient air quality management there by maximizing the socio-economic benefits and minimizing the environmental impacts resulting air quality benefit will be cost-efficient.

In general, the ABaCAS has a set of decision-support tools including International Cost Estimate Tool (ICET), Response Surface Model (RSM), Software of Model Attainment Test (SMAT), and Environmental Benefits Mapping and Analysis Program (BenMAP), which are briefly stated as follows:

- The International Cost Estimate Tool (ICET) shows control strategies based on cost information of control technologies applied in specific emission sectors.
- The Response Surface Model (RSM) provides a real-time estimated response of pollution concentrations to emissions changes.
- The Software of Model Attainment Test (SMAT) performs attainment tests to examine whether an emission reduction strategy will lower future ambient air pollution concentrations to a certain level.
- The Environmental Benefits Mapping and Analysis Program (BenMAP) estimates monetized human health effects resulting from the change in ambient air pollution.

20.5 Mitigation and Adaptation to Climate Change for Carbon Neutrality

In order to maintain the balance of the natural environment system and promote sustainable development, climate change measures are divided into Mitigation and Adaptation.

1. Mitigation: Reduce GHG emissions to lessen global warming
2. Adaptation: Adjust and adapt to a changing climate in the natural or human environment and to reduce the impact of climate change

To transit carbon neutrality, there are several key elements, including enhance of energy efficiency, optimization of renewable energy, deployment of carbon capture storage and utilization, and development of green fuel. Among them, CCS technologies commercialized development, facing barriers include energy-intensive technology, lack of CO_2 transport infrastructure, long-term CO_2 storage safety, public acceptance, high operation and investment costs, regulatory framework, and CO_2 pricing which should be overcome. Therefore, it is necessary to establish regulatory framework of CO_2 market, set price of CO_2, and then develop high-performance and low-cost CCS technologies. Once CCS technologies achieve deployment successfully, carbon neutrality can be expected to reach.

Mitigation efforts can be quantified by tons of greenhouse gas emissions, but adaptation contributions cannot be easily measured because it is influenced by different factors such as knowledge, economic development, technology, human capital, or governance structures. Adaptation can be viewed as a process, adjustment, or outcome and is a continuum from coping with short term and urgent needs through disaster risk reduction, to longer-term national planning and efforts to deal with damages.

In general, the research work should be focused on the following:

1. Investigating state of art of international and national climate adaptation plans and environmental policies.
2. Understanding the relationships among international and national climate change adaptation plans, environmental policies, and sustainable development goals (SDGs) in the viewpoint of Environmental Protection Administration.
3. Proposing a National Environmental Action Plan with key performance indicators (KPIs) to evaluate the performance of the proposed Environmental Action Plan.

Especially, environmental actions related to air and thermal stress should.

1. Collect scientific research on the combined effects of climate change and water and air pollution, as well as the impact on vulnerable groups;
2. Review infrastructure operation and management plans, existing technical regulations and standards related to the effects of climate change, with a series of advanced legislative, policy, and regulatory frameworks and tools to help minimize climate and disasters Risk or potential climate and disaster risk;
3. Promote heat island effect countermeasures, create shaded green spaces and parks, use vegetation and water bodies, urban structures, urban farmland and lifestyles to improve urban paving and public spaces, and reduce the impact on human health;
4. Develop the passive micro-climate control to adapt system to climate change;
5. Improve equipment performance and introduce robot technology and information communication technology (ICT) thereby providing information and early warning systems for different groups and conduct expert training.

MIX
Papier aus verantwortungsvollen Quellen
Paper from responsible sources
FSC® C105338

If you have any concerns about our products,
you can contact us on
ProductSafety@springernature.com

In case Publisher is established outside the EU,
the EU authorized representative is:
Springer Nature Customer Service Center GmbH
Europaplatz 3, 69115 Heidelberg, Germany

Printed by Libri Plureos GmbH
in Hamburg, Germany